ABOUT THE AUTHOR

Mike Seeds is Professor of Astronomy at Franklin and Marshall College, where he has taught astronomy since 1970. His research interests have focused on peculiar variable stars and the automation of astronomical telescopes. He is continuing his research by serving as Principal Astronomer in charge of the Phoenix 10, the first fully robotic telescope, located in southern Arizona. In 1989, he received the Christian R. and Mary F. Lindback Award for Distinguished Teaching. In addition to teaching, writing, and research, Mike has published educational tools for use in computer-smart classrooms. His interest in the history of astronomy led him to offer upper-level courses "Archaeoastronomy" and "Changing Concepts of the Universe," which he continues to develop. He has also published educational software for preliterate toddlers. Mike was Senior Consultant in the creation of the 26-episode telecourse *Universe: The Infinite Frontier.* He is the author of *Astronomy: The Solar System and Beyond,* Fourth Edition (2005), *Foundations of Astronomy,* Eighth Edition (2005), and *Horizons: Exploring the Universe,* Eighth Edition (2004), published by Brooks/Cole.

ABOUT THE COVER

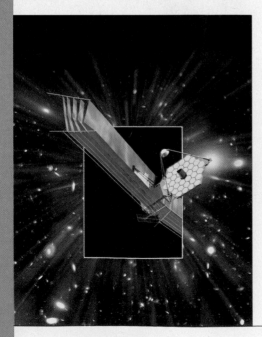

Modern observations are revealing the secrets of the expanding universe. We are learning how stars are born and how they die; we can understand how galaxies collide and interact; and we are even discovering how the universe began in a big bang. The James Webb Space Telescope, the successor of the Hubble Space Telescope, will someday tell us more.

STARS AND GALAXIES

FOURTH EDITION

Michael A. Seeds
Joseph R. Grundy Observatory, Franklin and Marshall College

THOMSON
BROOKS/COLE

Australia • Canada • Mexico • Singapore • Spain
United Kingdom • United States

THOMSON

BROOKS/COLE

Astronomy Editor: *Keith Dodson*
Assistant Editor: *Carol Benedict*
Editorial Assistant: *Melissa Newt*
Publisher/Executive Editor: *David Harris*
Development Editor: *Marie Carigma-Sambilay*
Technology Project Manager: *Sam Subity*
Marketing Managers: *Erik Evans and Kelley McAllister*
Marketing Assistant: *Leyla Jowza*
Advertising Project Manager: *Nathaniel Bergson-Michelson*
Project Manager, Editorial Production: *Lisa Weber*
Executive Art Director: *Vernon Boes*
Print Buyer: *Karen Hunt*

Permissions Editor: *Kiely Sexton*
Production Service: *Heckman & Pinette*
Text Designer: *Jeanne Calabrese*
Copy Editor: *Margaret Pinette*
Illustrator: *Precision Graphics*
Cover Designer: *Denise Davidson*
Cover Images: *Deep space image courtesy of NASA and the Hubble Space Telescope; digital illustration and image design by Bob Western*
Compositor: *Thompson Type*
Text and Cover Printer: *Transcontinental Printing/Interglobe*

Printed in Canada
1 2 3 4 5 6 7 08 07 06 05 04

For more information about our products, contact us at:
Thomson Learning Academic Resource Center
1-800-423-0563
For permission to use material from this text, contact us by:
Phone: 1-800-730-2214
Fax: 1-800-730-2215
Web: http://www.thomsonrights.com

Library of Congress Control Number: 2003114973

ISBN 0-534-42093-1

Brooks/Cole–Thomson Learning
10 Davis Drive
Belmont, CA 94002
USA

Asia
Thomson Learning
5 Shenton Way #01-01
UIC Building
Singapore 068808

Australia/New Zealand
Thomson Learning
102 Dodds Street
Southbank, Victoria 3006
Australia

Canada
Nelson
1120 Birchmount Road
Toronto, Ontario M1K 5G4
Canada

Europe/Middle East/Africa
Thomson Learning
High Holborn House
50/51 Bedford Row
London WC1R 4LR
United Kingdom

Latin America
Thomson Learning
Seneca, 53
Colonia Polanco
11560 Mexico D.F.
Mexico

Spain/Portugal
Paraninfo
Calle/Magallanes, 25
28015 Madrid, Spain

For Loretta Michael

BRIEF CONTENTS

CONTENTS

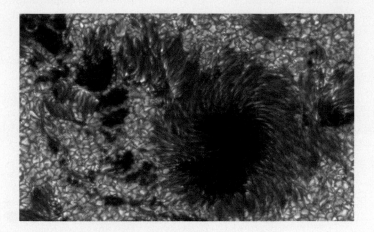

A NOTE TO THE STUDENT

FROM MIKE SEEDS

Astronomy is about us. Although an astronomy course covers planets and stars and galaxies, it is really about you and me. Astronomy helps us answer the ultimate question of human existence: What are we? Great minds like Plato, Beethoven, and Hemingway have tried to give their personal answers, but we must each find our own answer. Astronomy helps us understand the meaning of our own existence.

Of course, part of our personal answer must include philosophical and cultural issues. We define ourselves by what we create, what we worship, what we admire, and what we expect of each other. But a major part of our answer is embedded in the physical universe. We cannot hope to find a complete answer until we understand how the birth of the universe, the origin of galaxies, the evolution of stars, and the formation of planets created and brought together the atoms of which we are made. A basic understanding of astronomy is fundamental to knowing what we are. Astronomy is exciting because it is about us.

Part of the excitement of astronomy is the discovery of new things. Astronomers expect to be astonished. That excitement is reflected in this book in the form of new images, new discoveries, and new understandings that take us, in an introductory course, to the frontier of human knowledge. Here we explore the new evidence that galaxies collide and merge. We will study the newest images of dying stars, and we will struggle to understand new evidence that the expansion of the universe is speeding up. The Hubble Space Telescope, the Chandra X-Ray Observatory, and giant Earth-based telescopes on remote mountaintops provide us with a daily dose of excitement that goes far beyond sensationalism. These new discoveries in astronomy are exciting because they are about us. They tell us more and more about what we are.

I hope you will share in the excitement and satisfaction that scientists feel as they puzzle out nature's secrets. That is why I have created 22 special art spreads located throughout this book. Each presents a carefully selected subject visually, not only to make it exciting but to challenge you to synthesize your own undersanding. Rather than memorize facts and predigested descriptions of nature, you can create your own understanding through the guided-discovery design of these special features. The understanding we create ourselves is the most exciting and the most durable learning that we will ever experience.

In addition, I have used the principles of guided discovery to create over 16 new figures that will guide you to discover for yourself some of nature's most interesting secrets.

As a teacher, my quest is simple. I want you to understand your place in the universe—not just your location in space, but your location in the unfolding history of the physical universe. Not only do I want you to know where you are and what you are in the universe, but I want you to understand how we know. By the end of this book, I want you to know that the universe is very big, but that it is described by a small set of rules and that we have found a way to figure out the rules—a method called science.

The most important concept in introductory astronomy is the process of science, the process by which scientists ask questions of nature and gradually puzzle out the beautiful secrets of the physical world. You should not memorize facts or believe principles because some astronomer says it is so. Of course not! To be a participant in the adventure and not a mere observer, you need to understand how evidence and hypotheses interact to create new understanding. Those who understand how science works are part of the adventure.

Another reason you need to understand how science works is that you live in a technological world that is changing rapidly through the practical application of new scientific discoveries. To control your life in such a world, to guide and care for others, you need to understand the power and the limitation of scientific study.

Science is based on the interplay of evidence and hypothesis, and that interplay is the principal organizing theme for this book at every level—from the individual sentences to the order of the chapters. We will deal with each topic by surveying basic observational facts, synthesizing hypotheses, and then discussing further observations as evidence that supports or contradicts those hypotheses.

When we compare evidence and hypotheses, we make science logical and easy to understand, and that approach is reflected at every level of this book. At the highest level, the order of the chapters leads us from our home on Earth out among the galaxies. Then we can understand our place in the universe as we discuss life on other worlds in the final chapter. At a lower level, individual chapters, such as Chapter 14, Neutron Stars and Black Holes, are organized to describe basic observations and compare those with theories to build understanding. Look at the section on the search for black holes and notice how it, like most of the sections in this book, discusses observations compared with theories. Even at the level of single paragraphs, this book avoids a dry recitation of facts and instead compares observations with hypotheses.

Part of the logic of science is understanding how we know what we know. Rather than asking you to remember that our galaxy's disk contains clouds, this book explains how we know that fact. We can't use facts if we can't depend on them, and we can't depend on facts if we don't know how they were understood. Notice how this book helps you understand how we know.

I want you to understand that science is nothing more than a logical attempt to understand nature. Scientists think in logical arguments, and this book presents the usual facts in carefully organized logical arguments, not because the facts are easier to remember that way, but because the real goal is understanding nature, not remembering facts.

To help you grasp the logic of astronomy, I have created a number of features within the book.

- *Special art spreads,* mentioned earlier, provide an opportunity for you to create your own understanding and share in the satisfaction that scientists feel as they uncover the secrets of natural processes.

- *Guided discovery figures* illustrate important ideas visually and guide students to understand relationships and contrasts interactively.

- *Windows on Science* provide a parallel commentary on how all of science works. For example, the Windows point out where we are using statistical evidence, where we are reasoning by analogy, and where we are building a scientific model rather than a scientific hypothesis.

- *Guideposts* on the opening page of each chapter help you see the organization of the book. The Guidepost connects the chapter with preceding and following chapters to provide an overall organizational guide.

- *Review/Critical Inquiry* at the end of each text section is a carefully designed question to help you review and synthesize concepts and understand the scientific procedures used in the section. A short answer follows to show how scientists construct logical arguments (from observations, evidence, theories, and natural laws) that lead to a conclusion. A further question then gives you a chance to construct your own argument on a related issue.

- *End-of-Chapter Review Questions* are designed to help you review and test your understanding of the material.

- *End-of-Chapter Discussion Questions* go beyond the text and invite you to think critically and creatively about scientific questions. You can think about these questions yourself or discuss them in class.

- *Critical Inquiries for the Web* conclude each chapter by challenging you to use the World Wide Web to explore further and to think creatively and analytically about astronomy.

- *Exploring* **TheSky** are experiments you can perform with the software *TheSky,* experiments that will allow you to see the sky and celestial bodies in new ways on your own computer screen.

As additional aids, be aware that this book also includes, free of charge, the following electronic enhancements:

- ***Virtual Astronomy Laboratories.*** This set of 20 online labs is free with a passcode included with every new copy of this textbook. The labs cover topics from helioseismology to dark matter and allow you to submit your results electronically to your instructor or print them out to hand in. The first page of each chapter in this textbook notes which labs correlate to that chapter.

- ***AceAstronomy.*** Take charge of your learning with the first assessment-centered student learning tool for astronomy. Access AceAstronomy free via the Web with the passcode card bound into this book and begin to maximize your study time with a host of interactive tutorials and quizzes that help you focus on what you need to learn to master astronomy.

- **TheSky *Student Edition CD-ROM.*** With this CD-ROM, a personal computer becomes a powerful personal planetarium. Loaded with data on 118,000 stars and 13,000 deep-sky objects with images, it allows you to view the universe at any point in time from 4000 years ago to 8000 years in the future, to see the sky in motion, to view constellations, to print star charts, and much more.

- ***InfoTrac® College Edition.*** You receive free unlimited access for one academic term to this online database of complete articles and images from more than 700 popular and scholarly periodicals of the past four years, updated daily. An online student guide gives you tips for using InfoTrac and correlates each chapter in this book to InfoTrac articles. (Check with your instructor for availability outside North America.)

- ***The Brooks/Cole Astronomy Resource Center (www.brookscole.com/astronomy).*** Thousands of students and instructors visit this free Web site every week. The Astronomy Resource Center offers—among other things—online study aids, virtual field trips, and links to other sites.

Do not be humble. Astronomy tells us that the universe is vast and powerful, but it also tells us that we are astonishing creatures. We humans are the parts of the universe that think. The human brain is the most complex piece of matter known, so as you explore the universe, remember that it is your human brain that is capable of understanding the depth and beauty of the cosmos.

To appreciate your role in this beautiful universe, you must learn more than just the facts of astronomy. You must understand what we are and how we know. Every page of this book reflects that ideal.

Mike Seeds
mike.seeds@fandm.edu

ACKNOWLEDGMENTS

I started writing astronomy textbooks in 1973, and over the years I have had the guidance of a great many people who care about astronomy and teaching. I would like to thank all of the students and teachers who have responded so enthusiastically to *Foundations of Astronomy*. Their comments and suggestions have been very helpful in shaping this book. I would especially like to thank the reviewers whose careful analysis and thoughtful suggestions have been invaluable in completing this new edition: Dr. Martha S. Gilmore, Wesleyan University; Dr. Charles Lindsey, Solar Physics Research Group; Dr. Rebecca Eby Williams, Malin Space Science Systems; Dr. David Wilner, Harvard University.

The following institutions have provided the many photos and diagrams in this book: 2dF Galaxy Redshift Survey Team, 2Mass Sky Survey, Anglo-Australian Telescope Board, Arizona State Computer Science and Geography Departments, Associated Universities for Research in Astronomy, AT&T Archives, BOOMERANG Collaboration, California Association for Research in Astronomy, Caltech, Celestron International, Chandra X-Ray Observatory Center, European Southern Observatory, Gemini Observatory, Goddard Space Flight Center, High Altitude Observatory, Hubble Heritage Team, Institute for Astronomy, IPAC, Johns Hopkins Applied Physics Laboratory, JPL, Keck Observatories, The Large Binocular Telescope Project, Malin Space Science Systems, Max-Planck Institute for Extraterrestrial Physics, National Aeronautics and Space Administration, National Geophysical Data Center, National Museum of Natural History, National Oceanic and Atmospheric Administration, National Optical Astronomy Observatories, National Radio Astronomy Observatory, National Science Foundation, New England Meteoritical, The Observatories of the Carnegie Institution of Washington, Royal Swedish Academy of Sciences, Sacramento Peak Observatory, SETI Institute, Sloan Digital Sky Survey, Solar and Heliospheric Observatory, Space Telescope Science Institute, Steward Observatory, Sudbury Neutrino Observatory, Transition Region and Coronal Explorer, United States Geological Survey, University of Hawaii, and Yerkes Observatory.

I would especially like to thank the following individuals who went out of their way to help me locate materials for *Foundations of Astronomy:* H. A. Abt, David P. Anderson, Andy Aperala, Anthony Aveni, Dana Backman, Mike Bailey, Sam Barden, Allen Beechel, Charles Blue, David Bradstreet, Jack Burns, Linda Bustos, Bruce Campbell, Duncan Chesley, Chip Clark, Jan Curtis, David Hathaway, Cheryl and Don Davis, Frank Drake, Reginald Dufour, John A. Eddy, Michelangelo Fazio, Robert Fosbury, Sidney Fox, Henry Freudenreich, Bruce Fryxell, Dan Gezari, Daniel Good, Bradford Greeley, Randall Grubbs, John Grula, Calvin J. Hamilton, Kevin Hand, F. R. Harnden, William Hartmann, John Harvey, Phil Harvey, John Hayes, Floyd Herbert, Jeff Hester, John Hill, John P. Hughes, Shawn Hughes, Vincent Icke, George Jacoby, Paul Kalas, Victoria Kaspi, William Keel, Russell Kempton, Dean Ketelsen, Stephen M. Larson, Tod Lauer, Deborah Levine, Elsa Luna, Kwok-Yung Lo, John W. Mackenty, David Malin, Jack B. Marling, Ursula B. Marvin, Alfred S. McEwen, Stanley Miller, Reinhard Mundt, Harold Nations, David Parker, Richard Perley, Carolyn C. Porco, Ron Reese, Vera Rubin, Laurence Rudnick, Rudolph E. Schild, Maarten Schmidt, J. William Schopf, François Schweizer, Michael Seldner, Nigel Sharp, Seth Shostak, L. D. Simon, Damon Simonelli, Mike Skrutskie, T. Scott Smith, Steve Snowden, Hyron Spinrad, Alan Toomre, Joachim Trümper, Tony Tyson, Joseph Ververka, Mike Werner, Michael J. West, Gart Westerhout, Jimmy Westlake, Richard E. White, Simon White, Reed Wicander, and Robin Witmore.

Certain unique diagrams in Chapters 11, 12, 13, and 14 are based on figures I designed for my article "Stellar Evolution," which appeared in *Astronomy*, February 1979.

I am happy to acknowledge the use of images and data from a number of important programs. In preparing materials for this book I used NASA's Sky View facility located at NASA Goddard Space Flight Center. I have used atlas images and mosaics obtained as part of the Two Micron All Sky Survey (2MASS), a joint project of the University of Massachusetts and the Infrared Processing and Analysis Center/California Institute of Technology, funded by the National Aeronautics and Space Administration and the National Science Foundation. A number of solar images are used by the courtesy of the SOHO consortium, a project of international cooperation between ESA and NASA.

It is always a pleasure to work with the Thomson Learning team at Brooks/Cole and Wadsworth. Special thanks go to all of the people who have contributed to this project including Kelley McAllister, Leyla Jowza, David Harris, Carol Benedict, Melissa Newt, Nathaniel Bergson-Michelson, Sam Subity, Lisa Torri, and Lisa Weber. I have enjoyed working on production with Margaret Pinette and Bill Heckman of Heckman & Pinette, and I appreciate their understanding and goodwill. I would especially like to thank my editor Keith Dodson for his help and guidance on this project.

Most of all, I would like to thank my wife, Janet, and my daughter, Kate, for putting up with "the books." They know all too well that textbooks are made of time.

Mike Seeds

NEW MEDIA

We encourage you to use the media resources that accompany this textbook. It is far easier to understand astronomy if you see it in action and these new materials will enable you to become a part of that action. The media listed below can be found at **http://val.brookscole.com** (Virtual Astronomy Labs) and **http://astronomy.brookscole.com/seeds8e** (AceAstronomy).

ANIMATIONS

Chapter 2
Daily Motion in the Sky
Constellations from Different Latitudes
Constellations in Different Seasons
The Celestial Sphere
Seasons
Path of the Sun

Chapter 3
Small-Angle Formula
Lunar Phases

Chapter 4
Epicycles
Eratosthenes

Chapter 5
Gravity and Orbits
Orbital Motion
Center of Mass
Satellites

Chapter 6
Different Telescopes
Resolution and Telescopes

Chapter 7
Kirchhoff's Laws

Chapter 9
Apparent Brightness

Chapter 10
Scattering

Chapter 11
Conduction, Convection, and Radiation

Chapter 12
Cutaway of the Sun
The H–R Diagram
Cluster Turnoff
Future of the Sun

Chapter 13
Cutaway of High/Low Mass Stars

Chapter 14
Schwarzschild Radius
Neutron Star

Chapter 15
Explorable Milky Way

Chapter 16
Galaxy Types
Hubble

Chapter 17
Jet Deflection

Chapter 18
Hubble
The Raisin Bread Model
Cosmic Microwave Background

Chapter 19
The Drake Equation
Earth Calendar
Future of the Sun
Interstellar Communications

VIRTUAL ASTRONOMY LABS

Chapter 1
Lab 1: Units, Measure, and Conversion

Chapter 3
Lab 6: Tides and Tidal Forces

Chapter 7
Lab 2: Properties of Light and Its Interaction with Matter
Lab 3: The Doppler Effect
Lab 11: The Spectral Sequence and the H–R Diagram

Chapter 8
Lab 4: Solar Wind and Cosmic Rays
Lab 10: Helioseismology

Chapter 9
Lab 11: The Spectral Sequence and the H–R Diagram
Lab 12: Binary Stars
Lab 16: Astronomical Distance Scales

Chapter 13
Lab 13: Stellar Explosions, Novae, and Supernovae

Chapter 14
Lab 14: Neutron Stars and Pulsars
Lab 15: General Relativity and Black Holes

Chapter 16
Lab 16: Astronomical Distance Scales
Lab 19: The Hubble Law

Chapter 17
Lab 18: Active Galactic Nuclei

Chapter 18
Lab 17: Evidence for Dark Matter
Lab 20: Fate of the Universe

Chapter 19
Lab 8: Extrasolar Planets

THE SCALE OF THE COSMOS

The longest journey begins with a single step.

Confucius

GUIDEPOST

How can we study something so big it includes every-thing, even us? The cosmos, or the universe as it is more commonly called, is our subject in astronomy. Perhaps the best way to begin our study is to grab a quick impression as we zoom from things our own size up to the largest things in the universe.

That cosmic zoom, the subject of this chapter, gives us our first glimpse of the objects we will study in the rest of this book. In the next chapter, we will return to Earth to think about the appearance of the sky, and subsequent chapters will discuss stars, galaxies, and worlds that fill our universe.

Our quick survey of the universe in this chapter is more than just a listing of objects. It will illustrate the relations between objects—which are big, which are small, and which are contained inside others. In other words, this chapter will give us perspective for all of our exploration to follow. It is easy to learn a few facts, but it is the relationships between facts that are interesting. The relationships illustrated in this chapter will give us a perspective on our place in the cosmos.

While we study the cosmos, we will observe the process by which we learn. That process, science, gives us a powerful way to understand not only the universe but also ourselves.

VIRTUAL LABORATORY

UNITS, MEASURE, AND UNIT CONVERSION

We are going on a voyage out to the end of the universe. Marco Polo journeyed east, Columbus west, but we will travel outward away from our home on Earth, out past the moon, sun, and other planets, past the stars we see in the sky, and past billions more that we cannot see without the aid of the largest telescopes. We will journey through great whirlpools of stars to the most distant galaxies visible from Earth—and then we will continue on, carried only by experience and imagination—looking for the structure of the universe itself.

Besides journeying through space, we will also travel in time. We will explore the past to see the sun and planets form and search for the formation of the first stars and the origin of the universe. We will also explore forward in time to watch the sun die and Earth wither. Our imagination will become a scientific time machine searching for the ultimate end of the universe.

Although we will discover a beginning for our universe, a time when it came into existence, we will not find an edge in space. However far we voyage, we will not find a limit, and we will see evidence that our universe may be infinite and extends in all directions without limit. Such vastness dwarfs our human dimensions but not our human intelligence or imagination.

Astronomy—this imagined voyage—is more than the study of stars and planets. It is the study of the universe in which we exist. Our personal lives are confined to a small part of a small planet circling a small sun drifting through the universe, but astronomy can take us out of ourselves and thus help us understand what we are.

We live small lives on our planet, but our study of astronomy introduces us to sizes, distances, and times far beyond our usual experience. Our task in this chapter is to grasp the meaning of these unfamiliar sizes, distances, and times. The solution lies in a single word, *scale.* We will compare objects of different sizes to grasp the scale of the universe.

We begin with a human figure whose size we recognize. ▌Figure 1-1 shows a region about 52 feet across occupied by a human being, a sidewalk, and a few trees—all objects whose size we can understand. Each successive picture in this chapter will show a region of the universe that is 100 times wider than the preceding picture. That is, each step will widen our view by a factor of 100.

In ▌Figure 1-2 we widen our field of view by a factor of 100 and see an area 1 mile in diameter. The arrow points to the scene shown in the preceding photograph. People, trees, and sidewalks have vanished, but now we can see a college campus and the surrounding streets and houses. The dimensions of houses and streets are familiar to us. We have been in houses, crossed streets, and walked or run a mile. This is the familiar world around us, and we can relate such objects to the scale of our bodies.

FIGURE 1-2
(USGS)

FIGURE 1-1
(Michael A. Seeds)

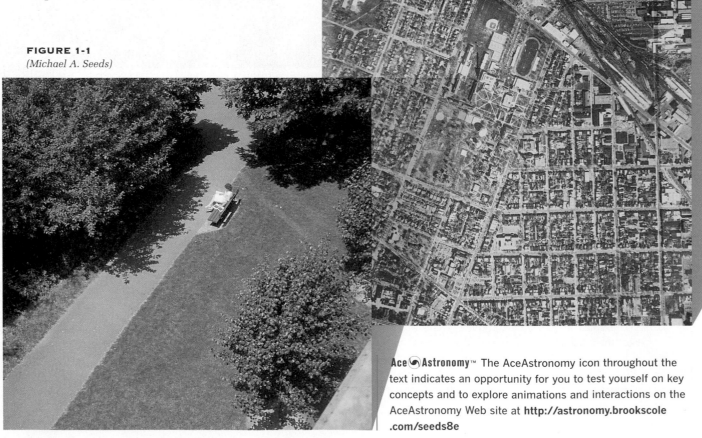

Ace✿Astronomy™ The AceAstronomy icon throughout the text indicates an opportunity for you to test yourself on key concepts and to explore animations and interactions on the AceAstronomy Web site at **http://astronomy.brookscole .com/seeds8e**

Infrared-wavelength image

FIGURE 1-3
(NASA infrared photograph)

sensitive only to a narrow range of colors. As we explore the universe, we must learn to use a wide range of "colors," which astronomers refer to as wavelengths. We will discuss wavelengths in detail in a later chapter, but for the moment we should note that Figure 1-3 was made at infrared wavelengths. Photographs in this book are carefully labeled to show the approximate wavelength at which the images were made. Astronomical images range from X-ray to radio wavelengths and reveal sights invisible to our eyes.

At the next step in our journey, we see our entire planet (Figure 1-4), which is 12,756 km in diameter. Earth rotates on its axis once a day, exposing half of

Visual-wavelength image

FIGURE 1-4
(NASA)

We have begun our adventure using feet and miles, but we should use the metric system of units. Not only is it used by all scientists around the world, but it makes calculations much easier. If you are not already familiar with the metric system, or need a review, study Appendix A before reading on.

The photo in Figure 1-2 is 1 mile in diameter. A mile equals 1.609 kilometers, so we can see in the photo that a kilometer is a bit over two thirds of a mile—a short walk across a neighborhood.

Our view in Figure 1-3 spans 160 kilometers. Green foliage of different kinds appears as shades of red in this infrared photo. The college campus is now invisible. The gray blotches are small cities, and the suburbs of Philadelphia are visible at the lower right. At this scale, we see the natural features of Earth's surface. The Allegheny Mountains of southern Pennsylvania cross the image in the upper left, and the Susquehanna River flows southeast into Chesapeake Bay. What appear as white bumps are a few puffs of clouds.

These features remind us that we live on the surface of a changing planet. Forces in Earth's crust pushed the mountain ranges up into parallel folds, like a rug wrinkled on a polished floor. The clouds remind us that Earth's atmosphere is rich in water, which falls as rain and erodes the mountains, washing material down the rivers and into the sea. The mountains and valleys that we know are only temporary features; they are constantly changing. As we explore the universe, we will see that it, like Earth's surface, is evolving. Change is the norm in the universe.

Look again at Figure 1-3 and note the red color. This photo is an infrared photograph in which healthy green leaves and crops show up as red. Our eyes are

its surface to daylight at any particular moment. The photo shows most of the daylight side of the planet. The blurriness at the extreme right is the sunset line. The rotation of Earth carries us eastward, and as we cross the sunset line into darkness, we say the sun has set. Thus, the rotation of our planet causes the cycle of day and night.

We know that Earth's interior is made mostly of iron and nickel and that its crust is mostly silicate rocks. Only a thin layer of water makes up the oceans, and the atmosphere is only a few hundred kilometers deep. On the scale of this photograph, the depth of the atmosphere on which our lives depend is less than the thickness of a piece of thread.

Again we enlarge our field of view by a factor of 100, and we see a region 1,600,000 km wide (Figure 1-5). Earth is the small white dot in the center, and the moon, whose diameter is only one-fourth that of Earth, is an even smaller dot along its orbit 380,000 km from Earth.

These numbers are so large that it is inconvenient to write them out. Astronomy is the science of big numbers, and we will use numbers much larger than these to discuss the universe. Rather than writing out these numbers as in the previous paragraph, it is convenient to write them in **scientific notation.** This is nothing more than a simple way to write numbers without writing a great many zeros. In scientific notation, we write 380,000 as 3.8×10^5. If you are not familiar with scientific notation, read the section on Powers of 10 Notation in the Appendix. The universe is too big to discuss without using scientific notation.

When we once again enlarge our field of view by a factor of 100 (❙ Figure 1-6), Earth, the moon, and the moon's orbit all lie in the small red box at lower left. But now we see the sun and two other planets that are part of our solar system. Our **solar system** consists of the sun, its family of planets, and some smaller bodies such as moons and comets.

Like the Earth, Venus and Mercury are **planets,** small, nonluminous bodies that shine by reflected light. Venus is about the size of Earth, and Mercury is a bit larger than our moon. On this diagram, they are both too small to be seen as anything but tiny dots. The sun is a **star,** a self-luminous ball of hot gas that generates

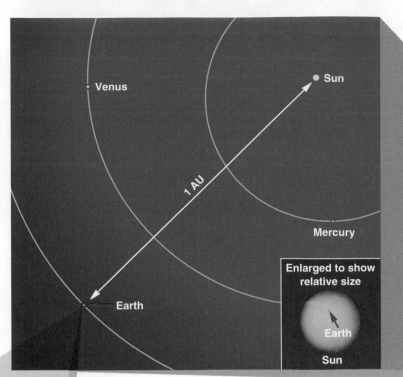

FIGURE 1-6
(NOAO)

its own energy. The sun is 109 times larger in diameter than Earth (inset), but it too is nothing more than a dot in this diagram.

This diagram has a diameter of 1.6×10^8 km. One way astronomers deal with large numbers is to define new units. The average distance from Earth to the sun is a unit of distance called the **astronomical unit (AU),** a distance of 1.5×10^{11} m. Using this unit we can say that the average distance from Venus to the sun is about 0.7 AU. The average distance from Mercury to the sun is about 0.39 AU.

The orbits of the planets are not perfect circles, and this is particularly apparent for Mercury. Its orbit carries it as close to the sun as 0.307 AU and as far away as 0.467 AU. Earth's orbit is more circular, and its distance from the sun varies by only a few percent.

Our first field of view was only 52 feet (about 16 m) in width. After only six steps of enlarging by a factor of 100, we now see the entire solar system (❙ Figure 1-7). Our field of view is 1 trillion (10^{12}) times wider than in our first view. The details of the preceding figure are now lost in the red square at the center of this diagram. We see only the brighter, more widely separated objects as we enlarge our view. The sun, Mercury, Venus, and Earth lie so close together that we cannot separate them at this scale. Mars, the next outward planet, lies only 1.5 AU from the sun. In contrast, Jupiter, Saturn, Uranus, Neptune, and Pluto are so far from the sun that they are easy to place in this diagram. These are cold worlds far from the sun's warmth. Light from the sun reaches Earth in only 8 minutes, but it takes over 4 hours to reach Neptune. Notice that Pluto's orbit is so elliptical

FIGURE 1-5
(NASA)

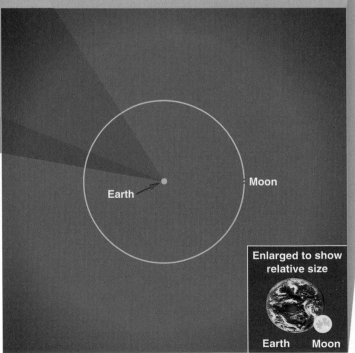

that Pluto can come closer to the sun than Neptune does, as Pluto did between 1979 and 1999.

When we again enlarge our field of view by a factor of 100, our solar system vanishes (▐ Figure 1-8). The sun is only a point of light, and all the planets and their orbits are now crowded into the small, red square at the center. The planets are too small and reflect too little light to be visible so near the brilliance of the sun.

Nor are any stars visible except for the sun. The sun is a fairly typical star, and it seems to be located in a fairly average neighborhood in the universe. Although

nearest stars. These stars are so distant that it is not reasonable to give their distances in astronomical units. We must define a new unit of distance, the light-year. One **light-year (ly)** is the distance that light travels in one year, roughly 10^{13} km or 63,000 AU. The diameter of our field of view is, in our new unit, 17 ly. The nearest star to the sun, Proxima Centauri, is 4.2 ly from Earth. In other words, light from Proxima Centauri takes 4.2 years to reach Earth.

Although these stars are roughly the same size as the sun, they are so far away that we cannot see them as anything but points of light. Even with the largest telescopes on Earth, we still see only points of light when we look at stars. Many of these stars probably have planets, but the planets reflect too little light and are too close to their bright stars to be visible from Earth. Astronomers have found over 100 such planets by indirect means, but planets would not be visible in Figure 1-9.

In our diagram, the sizes of the dots represent not the size of the stars but their brightness. This is the custom in astronomical diagrams, and it is also how star images are recorded on photographs. Bright stars make larger spots on a photograph than faint stars. Thus, the size of a star image in a photograph tells us not how big the star is but only how bright it looks.

In ▐ Figure 1-10, we expand our field of view by another factor of 100, and the sun and its neighboring stars vanish into the background of

FIGURE 1-7

FIGURE 1-8

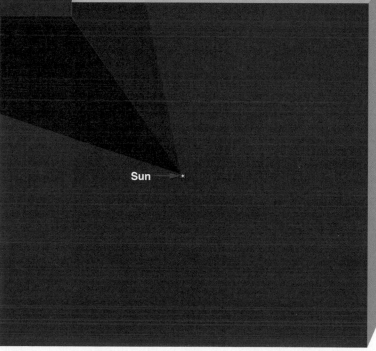

there are many billions of stars like the sun, none is close enough to be visible in this diagram, which shows an area only 11,000 AU in diameter. The stars are typically separated by distances about 10 times larger than the distance represented in this diagram. We will see stars in our next field of view, but, except for the sun at the center, this diagram is empty.

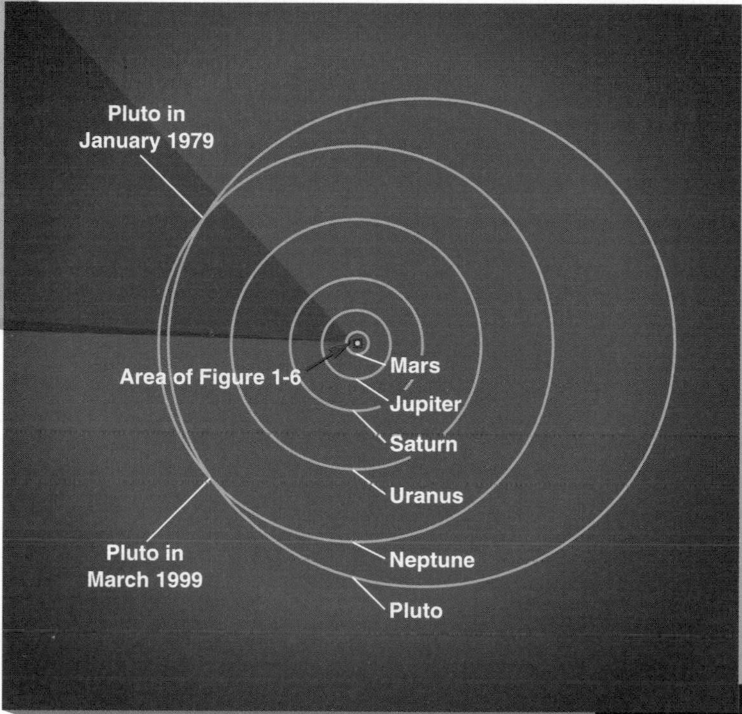

It is difficult to grasp the isolation of the stars. If the sun were represented by a golf ball in New York City, the nearest star would be another golf ball in Chicago. Except for the widely scattered stars and a few atoms of gas drifting between the stars, the universe is nearly empty.

In ▐ Figure 1-9, our field of view has expanded to a diameter a bit over 1 million AU. The sun is at the center, and we see a few of the

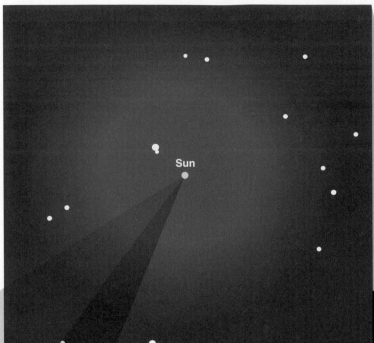

FIGURE 1-9

photo, but if we could see it, we would find it in the outer parts of the disk of the galaxy.

Our galaxy, like many others, has graceful **spiral arms** winding outward through the disk; we will discover that stars are born in great clouds of gas and dust as they pass through the spiral arms.

Ours is a fairly large galaxy, roughly 75,000 ly in diameter. Only a century ago astronomers thought it was the entire universe—an island universe of stars in an otherwise empty vastness. Now we know that our galaxy is not unique. Indeed ours is only one of many billions of galaxies scattered throughout the universe.

As we expand our field of view by another factor of 100, our galaxy appears as a tiny lumi-

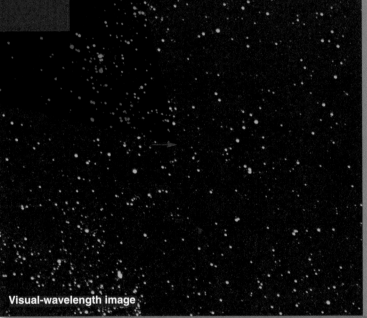

Visual-wavelength image

FIGURE 1-10
This box ■ represents the relative size of the previous frame. (NOAO)

thousands of stars. The field of view is now 1700 ly in diameter. Of course, no one has ever journeyed thousands of light-years from Earth to look back and photograph the solar neighborhood, so we use a representative photo of the sky. The sun is a relatively faint star that would not be easily located in a photo at this scale. Looking at this photograph, we notice a loose cluster of stars in the lower left quarter of the image. We will discover that many stars are born in clusters and that both old and young clusters exist in the sky. Star clusters are forming right now.

What we do not see is critically important. We do not see the thin gas that fills the spaces between the stars. Although those clouds of gas are thinner than the best vacuum on Earth, it is those clouds that give birth to new stars. Our sun formed from such a cloud about 5 billion years ago. We will see more star formation in our next view.

If we expand our field of view by a factor of 100, we see our galaxy (∥ Figure 1-11). A **galaxy** is a great cloud of stars, gas, and dust bound together by the combined gravity of all the matter. Galaxies range from 1500 to over 300,000 ly in diameter and can contain over 100 billion stars. In the night sky, we see our galaxy as a great, cloudy wheel of stars ringing the sky as the **Milky Way,** and we refer to our galaxy as the **Milky Way Galaxy.** Of course, no one can journey far enough into space to look back and photograph our galaxy, so the photo in Figure 1-11 shows a galaxy similar to our own. Our sun would be invisible in such a

nous speck surrounded by other specks (∥ Figure 1-12). This diagram includes a region 17 million ly in diameter, and each of the dots represents a galaxy. Notice that our galaxy is part of a cluster of a few dozen galaxies. We will find that galaxies are commonly grouped together in clusters. Some of these galaxies have beautiful spiral patterns like our own galaxy, but others do not. Some are strangely distorted. One of the mysteries we will try to solve is what produces these differences among the galaxies.

If we again expand our field of view, we see that the clusters of galaxies are connected in a vast network (∥ Figure 1-13). Clusters are grouped into superclus-

Our problem in studying astronomy is to keep a proper sense of scale. Remember that each of the billions of galaxies contains billions of stars. Most of those stars probably have families of planets like our solar system, and on some of those billions of planets liquid-water oceans and protective atmospheres may have sheltered the spark of life. It is possible that some other planets are inhabited by intelligent creatures who share our curiosity and our wonder at the scale of the cosmos.

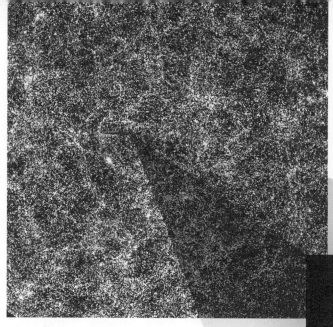

FIGURE 1-13

This box ■ represents the relative size of the previous frame. *(Detail from galaxy map from M. Seldner, B. L. Siebers, E. J. Groth, and P. J. E. Peebles,* Astronomical Journal 82 *[1977])*

ters—clusters of clusters—and the superclusters are linked to form long filaments and walls outlining voids that seem nearly empty of galaxies. These appear to be the largest structures in the universe. Were we to expand our field of view another time, we would probably see a uniform fog of filaments and voids. When we puzzle over the origin of these structures, we are at the frontier of human knowledge.

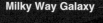

Milky Way Galaxy ➞

FIGURE 1-12

Visual-wavelength image

FIGURE 1-11

(© Anglo-Australian Telescope Board)

SUMMARY

Our goal in this chapter is to preview the scale of astronomical objects. To do so, we journey outward from a familiar campus scene by expanding our field of view by factors of 100. Only 12 such steps take us to the largest structures in the universe.

The numbers in astronomy are so large it is not convenient to express them in the usual way. Instead, we use the metric system to simplify our calculations and scientific notation to simplify the writing of big numbers. The metric system and scientific notation are discussed in Appendix A.

We live on the rotating planet Earth, which orbits a rather typical star we call the sun. A unit of distance convenient for discussing our solar system, the astronomical unit (AU), equals the average distance from Earth to the sun. Of the other planets in our solar system,

Mercury is closest to the sun, only 0.39 AU away, and Neptune is about 30 AU.

The sun, like most stars, is very far from its neighboring stars, and this leads us to use another unit of distance, the light-year. A light-year is the distance light travels in one year. The star nearest to the sun is Proxima Centauri, at a distance of 4.2 ly.

As we enlarge our field of view, we discover that the sun is only one of about 100 billion stars in our galaxy and that our galaxy is only one of many billions of galaxies in the universe. Galaxies appear to be grouped together in clusters, superclusters, and filaments, the largest structures known.

As we explore, we note that the universe is evolving. Earth's surface is evolving, and so are stars. Stars form from the gas in space, grow old, and eventually die.

Among the billions of stars in each of the billions of galaxies, many probably have planets, but detecting planets orbiting other stars is very difficult, and only a few more than a hundred have been found. We suppose that among the many planets in the universe, there must be some that are like Earth. We wonder if a few are inhabited by intelligent beings like ourselves.

NEW TERMS

scientific notation	light-year (ly)
solar system	galaxy
planet	Milky Way
star	Milky Way Galaxy
astronomical unit (AU)	spiral arm

REVIEW QUESTIONS

Ace⊛Astronomy™ Assess your understanding of this chapter's topics with additional quizzing and animations at **http:// astronomy.brookscole.com/seeds8e**

1. What is the largest dimension you have personal knowledge of? Have you run a mile? Hiked 10 miles? Run a marathon?

2. In Figure 1-4, the division between daylight and darkness is at the right on the globe of Earth. How do we know this is the sunset line and not the sunrise line?

3. What is the difference between our solar system, our galaxy, and the universe?

4. Look at Figure 1-6. How can you tell that Mercury follows an elliptical orbit? Can you detect the elliptical shape of any other orbits in this figure or the next?

5. Which is the outermost planet in our solar system? Why does that change?

6. Why are light-years more convenient than miles, kilometers, or astronomical units for measuring certain distances?

7. Why is it difficult to detect planets orbiting other stars?

8. What does the size of the star image in a photograph tell us?

9. What is the difference between the Milky Way and the Milky Way Galaxy?

10. What are the largest known structures in the universe?

PROBLEMS

1. The diameter of Earth is 7928 miles. What is its diameter in inches? In yards?

2. If a mile equals 1.609 km and the moon is 2160 miles in diameter, what is its diameter in kilometers?

3. One astronomical unit is about 1.5×10^8 km. Explain why this is the same as 150×10^6 km.

4. Venus orbits 0.7 AU from the sun. What is that distance in kilometers?

5. Light from the sun takes 8 minutes to reach Earth. How long does it take to reach Mars?

6. The sun is almost 400 times farther from Earth than is the moon. How long does light from the moon take to reach Earth?

7. If the speed of light is 3×10^5 km/s, how many kilometers are in a light-year? How many meters?

8. How long does it take light to cross the diameter of our Milky Way Galaxy?

9. The nearest galaxy to our own is about 2 million light-years away. How many meters is that?

10. How many galaxies like our own would it take laid edge to edge to reach the nearest galaxy? (*Hint:* See Problem 9.)

CRITICAL INQUIRIES FOR THE WEB

1. Locate photographs of Earth taken from space. What do cities look like? Can you see highways? Is the presence of our civilization detectable from space?

2. Locate photographs of nearby galaxies and compare them with photos of very distant galaxies. What kind of detail is invisible for distant galaxies?

3. One of the biggest clusters of galaxies is the Virgo cluster. Find out how many and what kind of galaxies are in the cluster. Is it nearby or far away?

EXPLORING *THESKY*

1. Locate and center one example of each of three different types of objects:

a. A planet, such as Saturn. Find its rising and setting time. Such objects have distances measured in astronomical units (AU).

How to proceed: Decide on the object you want to locate. Then find and center the object by pressing the **Find** button on the **Object Toolbar**. The second method is to press the **F** key. The third is to click **Edit,** then

Find. Once you have the **Object Information** window, press the **center** button.

b. A star. All stars in *TheSky* belong to our Milky Way Galaxy. Give the star's name, its magnitude, and its distance in light-years.

How to proceed: Click on any star, which brings up an **Object Information** window containing a variety of information about the star.

c. A galaxy; give its name and/or its designation.

How to proceed: To show galaxies, click on the **Galaxies** button in the **Object Toolbar,** then click on any galaxy. Distances to galaxies are on the order of millions and billions of light years.

2. Look at the solar system from beyond Pluto by clicking on **View** and then on **3D Solar System Mode.** Tip the solar system edge-on and then face-on. Zoom in to see the inner planets. Under **Tools,** set the **Time Skip Increment** to 1 day and then go **forward in time** to watch the planets move.

3. Identify some of the brightest constellations located along the Milky Way. (*Hint:* See **View, Reference Lines.**)

 Visit the Seeds *Foundations of Astronomy* companion Web site for critical thinking exercises, articles, and additional readings from InfoTrac College Edition, Brooks/Cole's online student library.

> The Southern Cross I saw every night abeam. The sun every morning came up astern; every evening it went down ahead. I wished for no other compass to guide me, for these were true.
>
> *Captain Joshua Slocum,* Sailing Alone Around the World

Chapter 2

William Hartmann

GUIDEPOST

Astronomy is about us. As we learn about astronomy, we learn about ourselves. We search for an answer to the question "What are we?" The quick answer is that we are thinking creatures living on a planet that circles a star we call the sun. In this chapter, we begin trying to understand that answer. What does it mean to live on a planet?

The preceding chapter gave us a quick overview of the universe, and chapters later in the book will discuss the details. This chapter and the next help us understand what the universe looks like seen from the surface of our spinning planet.

But appearances are deceiving. We will see in Chapter 4 how difficult it has been for humanity to understand what we see in the sky every day. In fact, we will discover that modern science was born when people tried to understand the appearance of the sky.

The night sky is the rest of the universe as seen from our planet. When we look up at the stars, we look out through a layer of air only a few hundred kilometers deep. Beyond that, space is nearly empty, with the stars scattered light-years apart. We begin our search for the natural laws that govern the universe by trying to understand what that universe looks like.

What we see in the sky is influenced dramatically by a simple fact. We live on a planet. The stars are scattered into the void all around us, most very distant and some closer. Our planet rotates on its axis once a day, so from our viewpoint the sky appears to rotate around us each day. Not only does the sun rise in the east and set in the west, but so also do the stars. That apparent daily motion of the sky is caused by the rotation of our planet. As you read this chapter and learn about the sky, keep in mind that you are also learning what it means to live on a planet.

Ace✸Astronomy™ Go to AceAstronomy and click Active Figures to see "Rotation of the Sky." Notice the angles at which stars rise in the east and set in the west.

2-1 THE STARS

On a dark night far from city lights, we can see a few thousand stars in the sky. Like ancient astronomers, we will try to organize what we see by naming groups of stars and individual stars and by specifying the brightness of individual stars.

CONSTELLATIONS

All around the world, ancient cultures celebrated heroes, gods, and mythical beasts by naming groups of stars—**constellations** (▌Figure 2-1). We should not be surprised that the star patterns do not look like the creatures they represent any more than Columbus, Ohio, looks like Christopher Columbus. The constellations "celebrate" the most important mythical figures in each culture. The constellations we know in Western culture originated in Mesopotamia over 5000 years ago, with other constellations added by Babylonian, Egyptian, and Greek astronomers during the classical age. Of these ancient constellations, 48 are still in use.

Ace✸Astronomy™ The AceAstronomy icon throughout the text indicates an opportunity for you to test yourself on key concepts and to explore animations and interactions on the AceAstronomy Web site at **http://astronomy.brookscole .com/seeds8e**

FIGURE 2-1

The constellations are an ancient heritage handed down for thousands of years as celebrations of great heroes and mythical creatures. Here Sagittarius and Scorpius hang above the southern horizon.

Different cultures grouped stars and named constellations differently. The constellation we know as Orion was known as Al Jabbar, the giant, to the ancient Syrians, as the White Tiger to the Chinese, and as Prajapati in the form of a stag in ancient India. The Pawnee Indians knew the constellation Scorpius as two groupings. The long tail of the scorpion was the Snake, and the two bright stars at the tip of the scorpion's tail were the Two Swimming Ducks.

Many ancient cultures around the world, including the Greeks, northern Asians, and Native Americans, associated the stars of the Big Dipper with a bear. The concept of the celestial bear may have crossed the land bridge into North America with the first Americans over 10,000 years ago. Some of the constellations we know today may be among the oldest surviving traces of human culture.

To the ancients, a constellation was a loose grouping of stars. Many of the fainter stars were not included in any constellation, and regions of the southern sky not visible to the ancient astronomers of northern latitudes were not identified with constellations. Constellation boundaries, when they were defined at all, were only approximate (▌Figure 2-2a), so a star like Alpheratz could be thought of as part of Pegasus or part of Andromeda. In recent centuries astronomers have added 40 modern constellations to fill gaps, and in 1928 the International Astronomical Union established 88 official constellations with clearly defined boundaries (Figure 2-2b). Thus a constellation now represents not a group of stars but an area of the sky, and any star within the region belongs to one and only one constellation.

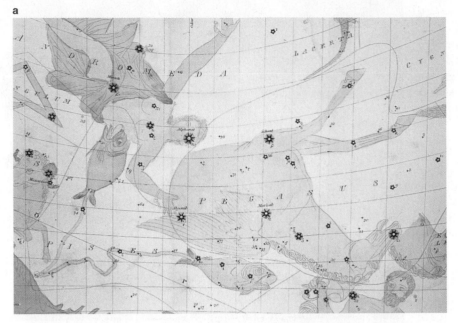

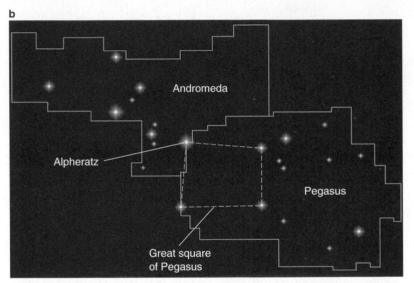

FIGURE 2-2

(a) In antiquity, constellation boundaries were poorly defined, as shown on this map by the curving dotted lines that separate Pegasus from Andromeda. *(From Duncan Bradford,* Wonders of the Heavens, *Boston: John B. Russell, 1837)* (b) Modern constellation boundaries are precisely defined by international agreement.

In addition to the 88 official constellations, the sky contains a number of less formally defined groupings called **asterisms.** The Big Dipper, for example, is a well-known asterism that is part of the constellation Ursa Major (the Great Bear). Another asterism is the Great Square of Pegasus (Figure 2-2b), which includes three stars from Pegasus and one from Andromeda. The star charts at the end of this book will introduce you to the brighter constellations and asterisms.

Although we name constellations and asterisms, most are made up of stars that are not physically associated with one another. Some stars may be many times

farther away than others and moving through space in different directions. The only thing they have in common is that they lie in approximately the same direction from Earth (▮ Figure 2-3).

THE NAMES OF THE STARS

The brightest stars were named thousands of years ago, and these names, along with the names of the constellations, found their way into the first catalogs of stars. We take our constellation names from Greek versions translated into Latin—the language of science from the fall of Rome to the 19th century—but most star names come from ancient Arabic. Names such as Sirius (the Scorched One), Capella (the Little She Goat), and Aldebaran (the Follower of the Pleiades) are beautiful additions to the mythology of the sky.

Giving the stars individual names is not very helpful, because we see thousands of stars, and these names do not help us locate the star in the sky. In which constellation is Antares, for example? In 1603, Bavarian lawyer Johann Bayer published an atlas of the sky called *Uranometria* in which he assigned lowercase Greek letters to the brighter stars of each constellation in approximate order of brightness. Astronomers have used those Greek letters ever since. (See the Appendix table with the Greek alphabet.) Thus, the brightest star is usually designated α (alpha), the second-brightest β (beta), and so on (▮ Figure 2-4). To identify a star in this way, we give the Greek letter followed by the possessive form of the constellation name, such as α Scorpii (for Antares). Now we know that Antares is in the constellation Scorpius and that it is probably the brightest star in that constellation.

This method of identifying a star's brightness, however, is only approximate. To be precise, we must have an accurate way of referring to the brightness of stars, and for that we must consult one of the first great astronomers.

THE BRIGHTNESS OF STARS

Astronomers measure the brightness of stars using the **magnitude scale,** a system that first appeared in the writings of the ancient astronomer Claudius Ptolemy about 140 AD. The system may have originated earlier than Ptolemy, and most astronomers attribute it to the Greek astronomer Hipparchus (160–127 BC) (▮ Figure 2-5).

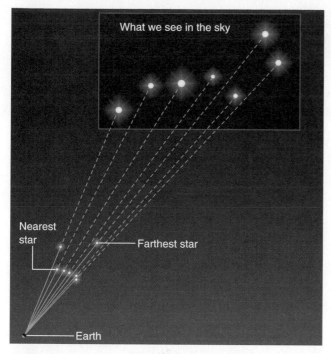

FIGURE 2-3

The stars we see in the Big Dipper—the brighter stars of the constellation Ursa Major, the Great Bear—are not at the same distance from Earth. We see the stars in a group in the sky because they lie in the same general direction as seen from Earth. The size of the star dots in the star chart represents the apparent brightness of the stars.

FIGURE 2-5

Hipparchus (2nd century BC) was the first great observational astronomer. He is believed to have compiled the first star catalog and may have invented the magnitude scale. He is honored here on a Greek stamp, which also shows one of his observing instruments.

The ancient astronomers divided the stars into six classes. The brightest were called first-class stars and those that were fainter, second-class. The scale continued downward to sixth-class stars, the faintest visible to the human eye. Thus, the larger the magnitude number, the fainter the star. This makes sense if we think of the bright stars as first-class stars and the faintest stars visible as sixth-class stars.

Hipparchus is believed to have compiled the first star catalog, and he may have used the magnitude system in that catalog. Almost 300 years later Ptolemy used the magnitude system in his own catalog, and successive generations of astronomers have continued to use the system. Later astronomers had to revise the ancient magnitude system to include very faint and very bright stars. Telescopes reveal many stars fainter than those the eye can detect, so the magnitude scale was extended to include these faint stars. The Hubble Space Telescope, currently the most sensitive astronomical telescope, can detect stars

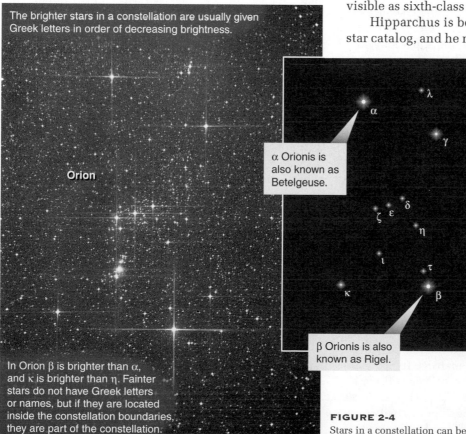

The brighter stars in a constellation are usually given Greek letters in order of decreasing brightness.

Orion

α Orionis is also known as Betelgeuse.

β Orionis is also known as Rigel.

In Orion β is brighter than α, and κ is brighter than η. Fainter stars do not have Greek letters or names, but if they are located inside the constellation boundaries, they are part of the constellation.

FIGURE 2-4

Stars in a constellation can be identified by Greek letters and by names derived from Arabic. *(William Hartmann)*

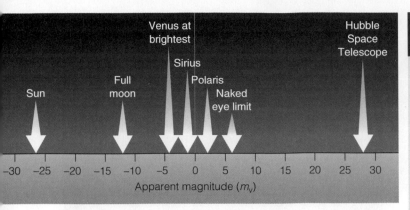

FIGURE 2-6
The scale of apparent visual magnitudes extends into negative numbers to represent the brightest objects and to positive numbers larger than 6 to represent objects fainter than the human eye can see.

TABLE 2-1
Magnitude and Intensity

Magnitude Difference	Intensity Ratio
0	1
1	2.5
2	6.3
3	16
4	40
5	100
6	250
7	630
8	1600
9	4000
10	10,000
⋮	⋮
15	1,000,000
20	100,000,000
25	10,000,000,000
⋮	⋮

fainter than +28 magnitude. In contrast, the brightest stars are actually brighter than the first brightness class in the ancient magnitude system. Thus, modern astronomers extended the magnitude system into negative numbers to account for brighter objects. For instance, Vega (alpha Lyrae) is almost zero magnitude at 0.04, and Sirius, the brightest star in the sky, has a magnitude of −1.42. We can even place the sun on this scale at −26.5 and the moon at −12.5 (Figure 2-6).

These numbers are known as **apparent visual magnitudes (m_v),** and they describe how the stars look to human eyes observing from Earth. Although some stars emit large amounts of infrared or ultraviolet light, we can't see it, and it is not included in the apparent visual magnitude. The subscript "v" reminds us that we are only including light we can see. Another problem is the distance to the stars. Very distant stars look fainter, and nearby stars look brighter. Apparent visual magnitude ignores the effect of distance and tells us only how bright the star looks as seen from Earth.

MAGNITUDE AND INTENSITY

Nearly every star catalog from today back to the time of the ancients uses the magnitude scale of stellar brightness. However, brightness is subjective, depending on such things as the physiology of the eye and the psychology of perception. To be accurate in measurements and calculations astronomers use the more precise term *intensity*—a measure of the light energy from a star that hits 1 square meter in 1 second. Thus astronomers need a way to convert between magnitudes and intensities.

The human eye senses the brightness of objects by comparing the ratios of their intensities. If two stars have intensities I_A and I_B, we can compare them by writing the ratio of their intensities, I_A/I_B. Hipparchus's magnitude system is based on a constant intensity ratio of about 2.5 for each magnitude. That is, if two stars differ by 1 magnitude, then they have an intensity ratio

of about 2.5. If they differ by 2 magnitudes, their ratio is 2.5 × 2.5, and so on.

When 19th-century astronomers began measuring starlight, they realized they needed to define a mathematical magnitude system that was precise, but for convenience they wanted a system that agreed at least roughly with that of Hipparchus. The stars that ancient astronomers classified as first and sixth differ by 5 magnitudes and have an intensity ratio of almost exactly 100, so the modern system of magnitudes specifies that a magnitude difference of 5 magnitudes corresponds to an intensity ratio of 100. That means that 1 magnitude corresponds to an intensity ratio of precisely 2.512, the fifth root of 100. That is, $100 = (2.512)^5$.

Now we can compare the light we receive from two stars. The light from a first-magnitude star is 2.512 times more intense than that from a second-magnitude star. The light from a third-magnitude star is $(2.512)^2$ times more intense than the light from a fifth-magnitude star. In general, the intensity ratio equals 2.512 raised to the power of the magnitude difference. That is:

$$I_A/I_B = (2.512)^{(m_B - m_A)}$$

For instance, if two stars differ by 6.32 magnitudes, we can calculate the ratio of their intensities as $2.512^{6.32}$. A pocket calculator tells us that the ratio is 337.

This magnitude system has some advantages. It compresses a tremendous range of intensity into a small range of magnitudes (Table 2-1) and makes it possible for modern astronomers to compare their measurements with all the recorded measurements of the past, right back to the first star catalogs over 2000 years ago.

Nonastronomers sometimes complain that the magnitude scale is awkward. Why would they think it is awkward, and how did it get that way?

Two things might make the magnitude scale seem awkward. First, it is backwards; the bigger the magnitude number the fainter the star. Of course, that arose because ancient astronomers were not measuring the brightness of stars but rather classifying them, and first-class stars would be brighter than second-class stars. The second awkward feature of the magnitude scale is its mathematical relation to intensity. If two stars differ by one magnitude, one is about 2.5 times brighter than the other. But if they differ by two magnitudes, one is 2.5×2.5 times brighter. This mathematical relationship arises because of the way we perceive brightness as ratios of intensity.

If the magnitude scale is so awkward, why do you suppose astronomers have used it for over two millennia?

In this section, we have found a way to identify stars by name and by brightness. Next we must look at the sky as a whole and notice its motion.

2-2 THE SKY AND ITS MOTION

The sky above us seems to be a great blue dome in the daytime and a sparkling ceiling at night. Learning to look at the sky requires that we begin thousands of years ago.

THE CELESTIAL SPHERE

Ancient astronomers believed the sky was a great sphere surrounding Earth with the stars stuck on the inside like thumbtacks in the ceiling. Modern astronomers know that the stars are scattered through space at different distances, but it is still convenient to think of the sky as a great celestial sphere.

As you study ▮ "The Sky Around Us" on pages 16 and 17, notice two important points. First, the sky appears to rotate westward around Earth each day, but that is a consequence of the eastward rotation of Earth. Second, what we can see of the sky depends on where we are on Earth. If we lived in Australia, we would see many constellations and asterisms invisible from North America, but we would never see the Big Dipper.

The celestial sphere is an example of a scientific model, a common feature of scientific thought (Window on Science 2-1). Notice that a scientific model does not have to be true to be useful. We will discuss many scientific models in the chapters that follow.

In addition to the daily motion of the sky, Earth's rotation adds a second motion to the sky that can be detected only over centuries.

PRECESSION

Over 2000 years ago, Hipparchus compared a few of his star positions with those made nearly two centuries before and realized that the celestial poles and equator were slowly moving across the sky. Later astronomers understood that this motion is caused by Earth's top-like motion.

WINDOW ON SCIENCE 2-1

Frameworks for Thinking About Nature: Scientific Models

In everyday language, we use the word *model* in various ways—fashion model, model airplane, model student—but scientists use the word in a very specific way. A **scientific model** is a carefully devised mental conception of how something works, a framework that helps scientists think about some aspect of nature. For example, astronomers use the celestial sphere as a way to think about the rotations of the sky, sun, moon, and stars.

Although we will think of a scientific model as a mental conception, it can take many forms. Some models are quite abstract—the psychologist's model of how the human mind processes visual information into images, for instance. But other models are so specific that they can be expressed as a set of mathematical equations. For example, an astronomer might use a set of equations to describe in detail how gas falls into a black hole. We could refer to such a calculation as a model. Of course, we could use metal and plastic to build a celestial globe, but the thing we build wouldn't really be the model any more than the equations are a model. A scientific model is a mental conception, an idea in our heads, that helps us think about nature.

On the other hand, a model is not meant to be a statement of truth. The celestial sphere is not real; we know the stars are scattered through space at various distances, but we can imagine a celestial sphere and use it to help us think about the sky. A scientific model does not have to be true to be useful.

Chemists, for example, think about the atoms in molecules by visualizing them as balls joined together by bonds. This model of a molecule is not fully correct, but it is a helpful way to think about molecules; it gives chemists a framework within which to organize their ideas.

Because scientific models are not meant to be totally correct, we must always remember the assumptions on which they are based. If we begin to think a model is true, it can mislead us instead of helping us. The celestial sphere, for instance, can help us think about the sky, but we must remember that it is only a mental crutch. The universe is much larger and much more interesting than this ancient scientific model of the heavens.

The Sky Around Us

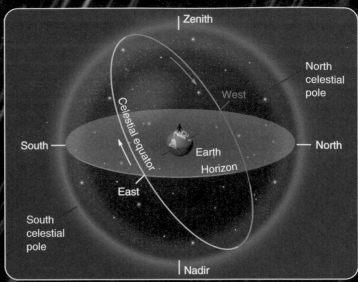

Zenith

West

North celestial pole

South

North

Celestial equator

Earth

Horizon

East

South celestial pole

Nadir

The sky above appears to be part of a sphere, the celestial sphere, with Earth at its center. From any location on Earth we see only half of the celestial sphere, the half above our horizon. The zenith marks the top of the sky above our head, and the nadir marks the bottom of the sky directly under our feet. What we see in the sky depends on where we are on Earth. The drawing at right shows the view for an observer in North America. An observer in South America would have a dramatically different horizon, zenith, and nadir.

The eastward rotation of Earth causes the sun, moon, and stars to move westward in our sky as if the celestial sphere were rotating westward around Earth. The apparent pivot points are the **north celestial pole** and the **south celestial pole** located directly above Earth's north and south poles. Halfway between the celestial poles lies the **celestial equator**. Earth's rotation defines the directions we use every day. The **north point** and **south point** are the points on the horizon closest to the celestial poles. The **east point** and the **west point** lie halfway between the north and south points. The celestial equator always touches the horizon at the east and west points.

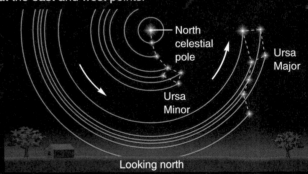

North celestial pole

Ursa Major

Ursa Minor

Looking north

AURA/NOAO/NSF

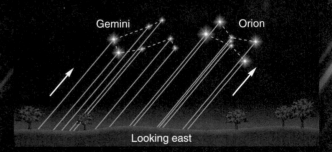

Gemini

Orion

Looking east

This time exposure of about 30 minutes shows stars as streaks, called star trails, rising behind an observatory dome. The camera was facing northeast to take this photo. The motion we see in the sky depends on which direction we look, as shown at right. Looking north, we see the star Polaris, the North Star, located near the north celestial pole. As the sky appears to rotate westward, Polaris hardly moves, but other stars circle the celestial pole. Looking south from a location in North America, we can see stars circling the south celestial pole, which is invisible below our southern horizon.

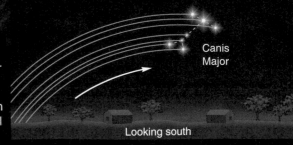

Canis Major

Looking south

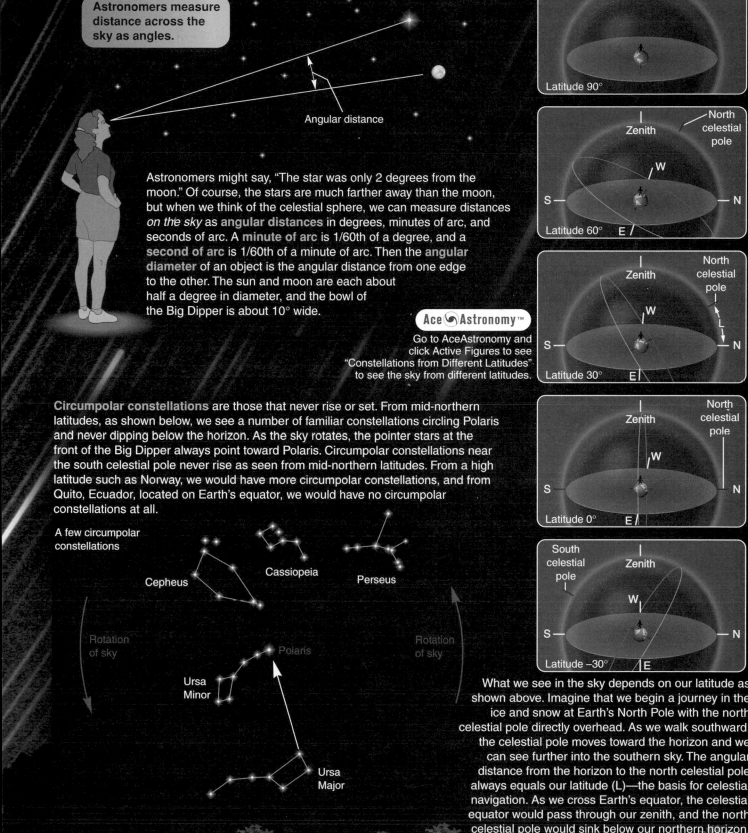

Astronomers measure distance across the sky as angles.

Angular distance

Astronomers might say, "The star was only 2 degrees from the moon." Of course, the stars are much farther away than the moon, but when we think of the celestial sphere, we can measure distances *on the sky* as **angular distances** in degrees, minutes of arc, and seconds of arc. A **minute of arc** is 1/60th of a degree, and a **second of arc** is 1/60th of a minute of arc. Then the **angular diameter** of an object is the angular distance from one edge to the other. The sun and moon are each about half a degree in diameter, and the bowl of the Big Dipper is about 10° wide.

Ace✪Astronomy™

Go to AceAstronomy and click Active Figures to see "Constellations from Different Latitudes" to see the sky from different latitudes.

Circumpolar constellations are those that never rise or set. From mid-northern latitudes, as shown below, we see a number of familiar constellations circling Polaris and never dipping below the horizon. As the sky rotates, the pointer stars at the front of the Big Dipper always point toward Polaris. Circumpolar constellations near the south celestial pole never rise as seen from mid-northern latitudes. From a high latitude such as Norway, we would have more circumpolar constellations, and from Quito, Ecuador, located on Earth's equator, we would have no circumpolar constellations at all.

A few circumpolar constellations

Cepheus

Cassiopeia

Perseus

Rotation of sky

Polaris

Rotation of sky

Ursa Minor

Ursa Major

North celestial pole
Zenith
Latitude 90°

North celestial pole
Zenith
W
S — — N
Latitude 60° E

North celestial pole
Zenith
W
L
S — — N
Latitude 30° E

North celestial pole
Zenith
W
S — — N
Latitude 0° E

South celestial pole
Zenith
W
S — — N
Latitude −30° E

What we see in the sky depends on our latitude as shown above. Imagine that we begin a journey in the ice and snow at Earth's North Pole with the north celestial pole directly overhead. As we walk southward the celestial pole moves toward the horizon and we can see further into the southern sky. The angular distance from the horizon to the north celestial pole always equals our latitude (L)—the basis for celestial navigation. As we cross Earth's equator, the celestial equator would pass through our zenith, and the north celestial pole would sink below our northern horizon

If you have ever played with a gyroscope or top, you have seen how the spinning mass resists any change in the direction of its axis of rotation. The more massive the top and the more rapidly it spins, the more difficult it is to change the direction of its axis of rotation. But you probably recall that the axis of even the most rapidly spinning top sweeps around in a conical motion. The weight of the top tends to make it tip over, and this combines with its rapid rotation to make its axis sweep around in a conical motion called **precession** (▌Figure 2-7a).

Earth spins like a giant top, but it does not spin upright in its orbit; it is inclined 23.5° from vertical. Earth's large mass and rapid rotation keep its axis of rotation pointed toward a spot near the star Polaris, and the axis would not wander if Earth were a perfect sphere. However, Earth, because of its rotation, has a slight bulge around its middle, and the gravity of the sun and moon pull on this bulge, tending to twist Earth upright in its orbit. The combination of these forces and Earth's rotation causes Earth's axis to precess in a conical motion, taking about 26,000 years for one cycle (Figure 2-7b).

Because the celestial poles and equator are defined by Earth's rotational axis, precession moves these reference marks. We notice no change at all from night to night or year to year, but precise measurements reveal the precessional motion of the celestial poles and equator.

Over centuries, precession has dramatic effects. Egyptian records show that 4800 years ago the north celestial pole was near the star Thuban (α Draconis). The pole is now approaching Polaris and will be closest to it about 2100. In about 12,000 years, the pole will have moved to within 5° of Vega (α Lyrae). Figure 2-7c shows the path followed by the north celestial pole. We will discover in later chapters that precession is common among rotating celestial bodies.

As you study astronomy, notice the special terms used to describe such things as precession and the celestial sphere. We need to know those terms, but science is about understanding nature and not about naming its parts (Window on Science 2-2). Science is more than just vocabulary.

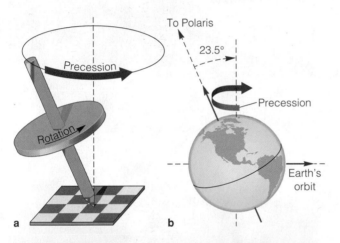

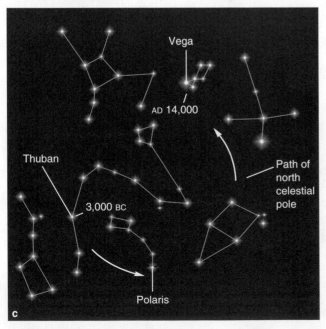

REVIEW CRITICAL INQUIRY

Does everyone see the same circumpolar constellations?
Here we can use the celestial sphere as a convenient model of the sky. A circumpolar constellation is one that does not set or rise. Which constellations are circumpolar depends on our latitude. If we live on Earth's equator, we see all the constellations rising and setting, and there are no circumpolar constellations at all. If we live at Earth's North Pole, all the constellations north of the celestial equator never set, and all the constellations south of the celestial equator never rise. In that case, every constellation is circumpolar. At intermediate latitudes, the circumpolar regions are caps whose angular radius equals the latitude of the observer. If we live in Iceland, the caps are very large, and if we live in Egypt, near the equator, the caps are much smaller.

   Locate Ursa Major and Orion on the star charts at the end of this book. For people in Canada, Ursa Major is circumpolar, but people in Mexico see most of this constellation slip below the horizon. From much of the United States, some of the stars of Ursa Major set, and some do not. In contrast, Orion rises and sets as seen from nearly everywhere on Earth. Explorers at Earth's poles, however, never see Orion rise or set. How would you improve the definition of a circumpolar constellation to clarify the status of Ursa Major? Would your definition help in the case of Orion?

FIGURE 2-7
Precession. (a) A spinning top precesses in a conical motion around the perpendicular to the floor because its weight tends to make it fall over. (b) Earth precesses around the perpendicular to its orbit because the gravity of the sun and moon tend to twist it upright. (c) Precession causes the north celestial pole to drift among the stars, completing a circle in 26,000 years.

The motions of the celestial sphere may seem to have little to do with our lives, but at the end of this chapter we will see how precession may be one of the causes of the ice ages. Before we can think about ice ages, however, we must consider Earth's orbital motion around the sun and the resulting apparent motion of the sun.

2-3 THE CYCLES OF THE SUN

Perhaps the most obvious cycles in the sky are those of the sun, but those apparent motions are produced by Earth's rotation and revolution. **Rotation** is the turning of a body about an axis through its center. Thus we say that Earth rotates on its axis once a day. **Revolution** is the motion of a body around a point located outside the body, and we say that Earth revolves around the sun once a year.

The cycle of day and night is caused by the rotation of Earth on its axis. In Figure 2-8 we see that four people in different places on Earth have different times of day, and that Earth's rotation carries us across the daylight side and then across the night side of Earth, producing the cycle of day and night.

Earth's annual revolution around the sun produces a cycle in the sky that we experience as the seasons. To understand that motion, we must imagine that we can make the sun fainter.

THE ANNUAL MOTION OF THE SUN

Even in the daytime, the sky is filled with stars, but the glare of sunlight fills our atmosphere with scattered light, and we can only see the brilliant sun. If the sun were fainter and we could see the stars, we would no-tice that day by day the sun was moving slowly eastward against the background of stars. This motion is caused by Earth's motion as it revolves around the sun. In Figure 2-9, we can see how the sun would appear in front of the stars of the constellation Sagittarius on January 1 and how Earth's motion along its orbit would cause the sun to appear to move eastward against the constellations. By March 1 the sun appears to be in Aquarius.

Of course, it is not quite correct to say that the sun is "in Aquarius." The sun is only 93 million miles away, and the stars of Aquarius are at least a million times farther away. But in March of each year, we would see the sun against the background of the stars in Aquarius, and thus we could say, "The sun is in Aquarius."

If we continue watching the sun against the background of stars throughout the year, we could plot its path on a star chart. After one full year, we would see the sun begin to retrace this line as it continued its

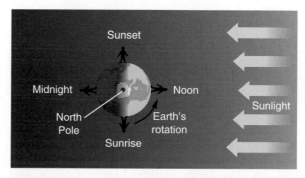

FIGURE 2-8
Looking down on Earth from above the North Pole shows how the time of day or night depends on our location on Earth.

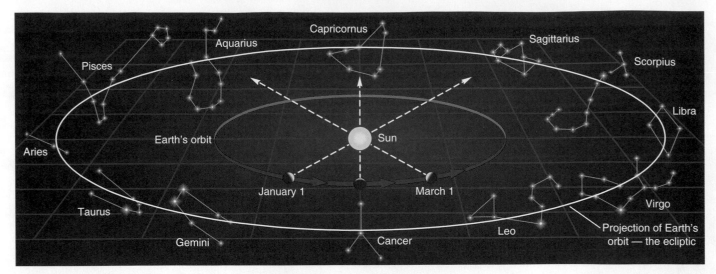

ACTIVE FIGURE 2-9

Earth's motion around the sun makes the sun appear to move against the background of the stars. Earth's circular orbit is thus projected on the sky as the circular path of the sun, the ecliptic.

Ace✪Astronomy™ Go to AceAstronomy and click Active Figures to see "Constellations in Different Seasons." Notice how the visible constellations change as the sun moves eastward along the ecliptic.

annual cycle of motion around the sky. This line, the apparent path of the sun in its yearly motion around the sky, is called the **ecliptic**. Another way to define the ecliptic is to say it is the projection of Earth's orbit on the sky. If the sky were a great screen illuminated by the sun at the center, then the shadow cast by Earth's orbit would be the ecliptic. Yet a third way to define the ecliptic is to refer to it as the plane of Earth's orbit. These three definitions of the ecliptic are equivalent, and it is worth considering them all because the ecliptic is one of the most important reference lines on the sky.

Earth circles the sun in 365.25 days, and consequently the sun appears to circle the sky in the same period. That means the sun, traveling 360° around the ecliptic in 365.25 days, travels about 1° eastward each day, about twice its angular diameter. We don't notice this motion because we can't see the stars in the daytime, but the motion of the sun has an important consequence that we do notice—the seasons.

THE SEASONS

The seasons are caused by a simple fact: Earth's axis of rotation is tipped 23.5° from the perpendicular to its orbit (▮ Figure 2-10). As you study "▮ The Cycle of the Seasons" on pages 22 and 23, notice two important principles. First, the seasons are not caused by any variation in the distance from Earth to the sun. Earth's orbit is nearly circular, so it is always about the same distance from the sun. Second, the seasons are caused by the changes in solar energy that Earth's northern and southern hemispheres receive at different times of the year. Because of circulation patterns in Earth's atmosphere, the northern and southern hemispheres are

mostly isolated from each other and exchange little heat. When one hemisphere receives more solar energy than the other, it grows rapidly warmer.

In ancient times, the cycle of the seasons and the solstices and equinoxes were celebrated with rituals and festivals. Shakespeare's play *A Midsummer Night's Dream* describes the enchantment of the summer solstice night. (In Shakespeare's time, the equinoxes and solstices were taken to mark the midpoint of the seasons.) Many North American Indians marked the summer solstice with ceremonies and dances. Early church officials placed Christmas day in late December to coincide with an earlier pagan celebration of the winter solstice.

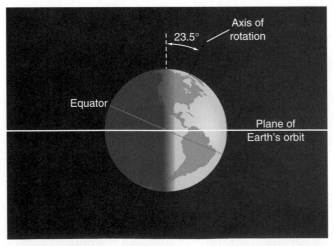

FIGURE 2-10

Earth's axis is inclined 23.5° from the perpendicular to the plane of its orbit. As it moves along its orbit, its axis of rotation remains fixed, pointing at the North Star.

If Earth had a significantly elliptical orbit, how would its seasons be different?

Suppose Earth had an elliptical orbit so that perihelion occurred in July and aphelion in January. At perihelion, Earth would be closer to the sun, and the entire surface of Earth would be a bit warmer. If that happened in July, it would be summer in the northern hemisphere and winter in the southern hemisphere, and both would be warmer than they now are. It could be a dreadfully hot summer in Canada, and it might not snow at all in southern Argentina. Six months later, aphelion would occur in January, which means Earth would be farther from the sun. Winter in northern latitudes would be frigid, and summers in Argentina would be cool.

Of course, this doesn't happen. Earth's orbit is nearly circular, and the seasons are caused not by a variation in the distance of Earth from the sun but by the inclination of Earth in its orbit.

Nevertheless, Earth's orbit is slightly elliptical. Earth passes perihelion about January 4 and aphelion about July 4. Although Earth's oceans tend to store heat and reduce the importance of this effect, this very slight variation in distance does affect the seasons. Does it make your winters warmer or cooler?

The sun is not the only body that moves along the ecliptic. The moon and planets maintain constant traffic along the sun's highway.

2-4 THE MOTION OF THE PLANETS

The ecliptic is important to us because it is the path of the sun around the sky, and that motion gives rise to the seasons. In a later chapter, we will consider the moon's cycle around the ecliptic, but here it is worthwhile to mention the motion of the planets. Not only are most of the planets visible to the naked eye, but their motion is the basis for one of the longest-living superstitions in human history—that our fate is written in the stars.

THE MOVING PLANETS

Most of the planets of our solar system are visible to the unaided eye, though they produce no light of their own. We see them by reflected sunlight. Mercury, Venus, Mars, Jupiter, and Saturn are all visible to the naked eye, but Uranus is usually too faint to be seen (magnitude 5.6 at its brightest), and Neptune is never bright enough. Pluto is even fainter, and we need a large telescope to find it. Although Uranus, Neptune, and Pluto are usually too faint to see, their motions are the same as those of the other planets.

All the planets of our solar system move in nearly circular orbits around the sun. If we were looking down on the solar system from the north celestial pole, we would see the planets moving in the same counterclockwise direction around their orbits (Chapter 1). The farther they are from the sun, the more slowly the planets move.

When we look for planets in the sky, we always find them near the ecliptic, because their orbits lie in nearly the same plane as the orbit of Earth. As they orbit the sun, they appear to move generally eastward along the ecliptic.* In fact, the word *planet* comes from a Greek word meaning "wanderer." Mars moves completely around the ecliptic in slightly less than two years, but Saturn, being farther from the sun, takes nearly 30 years.

As seen from Earth, Venus and Mercury can never move far from the sun because their orbits are inside Earth's orbit. They sometimes appear near the western horizon just after sunset or near the eastern horizon just before sunrise. Venus is easier to locate because its larger orbit carries it higher above the horizon than Mercury (▌Figure 2-11). Mercury's orbit is so small that it can never get farther than about 28° from the sun. Consequently, it is usually hard to see against the sun's glare and is often hidden in the clouds and haze near the horizon. At certain times when it is farthest from the sun, however, Mercury shines brightly and can be located near the horizon in the evening or morning sky. (See the Appendix for the best times to observe Venus and Mercury.)

By tradition, any planet visible in the evening sky is called an **evening star,** although planets are not stars. Any planet visible in the sky shortly before sunrise is called a **morning star.** Perhaps the most beautiful is Venus, which can become as bright as minus fourth magnitude. As Venus moves around its orbit, it can dominate the western sky each evening for about half the year, but eventually its orbit carries it back toward the sun, and it is lost in the haze near the horizon. In a few weeks it reappears in the dawn sky as a brilliant morning star.

ASTROLOGY

Seen from Earth, the planets move gradually eastward along the ecliptic, but they don't follow the ecliptic exactly. Also, each travels at its own pace and seems to speed up and slow down at various times. To the ancients, this complex motion reflected the moods of the sky gods, and astrology was born.

Ancient astrologers defined a **zodiac,** a band 18° wide centered on the ecliptic, as the highway the planets follow. They divided this band into 12 segments named for the constellations along the ecliptic—the

---

*We will discuss occasional exceptions to this eastward motion in Chapter 4.

The Cycle of the Seasons

We can use the celestial sphere to help us think about the seasons. The celestial equator is the projection of Earth's equator on the sky, and the ecliptic is the projection of Earth's orbit on the sky. Because Earth is tipped in its orbit, the ecliptic and equator are inclined to each other by 23.5°. As the sun moves eastward around the sky, it spends half the year in the southern half of the sky and half of the year in the northern half. That causes the seasons.

The sun crosses the celestial equator going northward at the point called the **vernal equinox**. The sun is at its farthest north at the point called the **summer solstice**. It crosses the celestial equator going southward at the **autumnal equinox**, and reaches its most southern point at the **winter solstice**.

The seasons are defined by the dates when the sun crosses these four points, as shown in the table at the right. *Equinox* comes from the word for "equal"; on the day of an equinox, we have equal amounts of daylight and darkness. *Solstice* comes from the words meaning "sun" and "stationary." *Vernal* comes from the word for "green." The "green" equinox marks the beginning of spring.

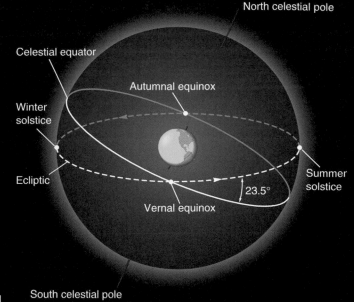

Event	Date*	Season
Vernal equinox	March 20	Spring begins
Summer solstice	June 22	Summer begins
Autumnal equinox	September 22	Autumn begins
Winter solstice	December 22	Winter begins

* Give or take a day due to leap year and other factors.

Ace Astronomy™

Go to AceAstronomy and click Active Figures to see "Seasons" and watch Earth orbiting the sun.

On the day of the summer solstice in late June, Earth's northern hemisphere is inclined toward the sun, and sunlight shines almost straight down at northern latitudes. At southern latitudes, sunlight strikes the ground at an angle and spreads out. North America has warm weather, and South America has cool weather.

Earth's axis of rotation points toward Polaris, and, like a top, the spinning Earth holds its axis fixed as it orbits the sun. On one side of the sun, Earth's northern hemisphere leans toward the sun; on the other side of its orbit, it leans away. However, the direction of the axis of rotation does not change.

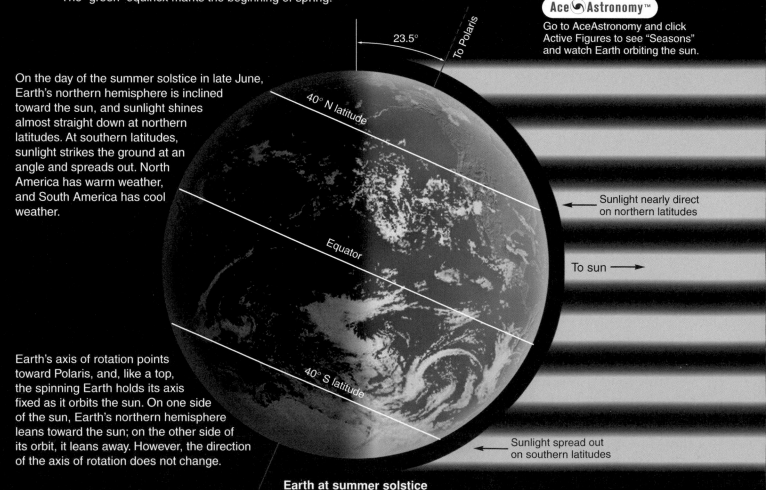

Earth at summer solstice

Summary solstice light · **Winter solstice light**

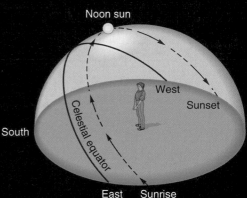

At summer solstice

Light striking the ground at a steep angle spreads out less than light striking the ground at a shallow angle. Light from the summer-solstice sun strikes northern latitudes from nearly overhead and is concentrated. Light from the winter-solstice sun strikes northern latitudes at much steeper angle and spreads out. The same amount of energy is spread over a larger area, so the ground receives less energy from the winter sun.

The two causes of the seasons are shown at right for someone in the northern hemisphere. First, the noon summer sun is higher in the sky and the winter sun is lower, as shown by our longer winter shadows. Thus winter sunlight is more spread out. Second, the summer sun rises in the northeast and sets in the northwest, spending more than 12 hours in the sky. The winter sun rises in the southeast and sets in the southwest, spending less than 12 hours in the sky. Both of these effects mean that northern latitudes receive more energy from the summer sun, and summer days are warmer than winter days.

At winter solstice

Ace Astronomy™

Go to AceAstronomy and click Active Figures to see "Path of the Sun" and see this figure from the inside.

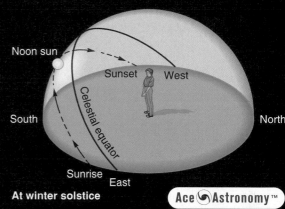

On the day of the winter solstice in late December, Earth's northern hemisphere is inclined away from the sun, and sunlight strikes the ground at an angle and spreads out. At southern latitudes, sunlight shines almost straight down and does not spread out. North America has cool weather and South America has warm weather.

Earth's orbit is only very slightly elliptical. About January 4th, Earth is at **perihelion**, its closest point to the sun, when it is only 1.7 percent closer than average. About July 4th, Earth is at **aphelion**, its most distant point from the sun, when it is only 1.7 percent farther than average. This small variation does not significantly affect the seasons.

Earth at winter solstice

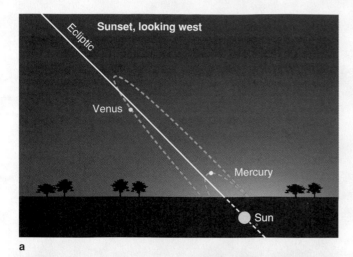

Sunset, looking west

Ecliptic

Venus

Mercury

Sun

a

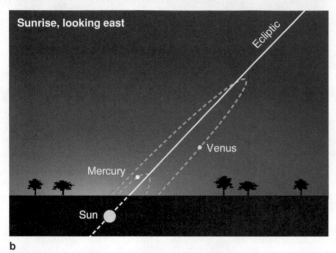

Sunrise, looking east

Ecliptic

Venus

Mercury

Sun

b

FIGURE 2-11
Mercury and Venus are sometimes visible in the western sky just after sunset (a) or in the eastern sky just before sunrise (b).

or between November 30 and December 17, the sun was passing through a corner of the nonzodiacal constellation Ophiuchus, and you have no official zodiacal sign.* Furthermore, astronomers like to point out, there is no mechanism by which the planets could influence us. The gravitational influence of a doctor who is delivering a baby is many times more powerful than the gravitational influence of the planets.

The arguments in the preceding paragraph actually miss the point. Astrology is not related to the physical world at all. It does not matter what constellation the sun occupies, because astrology divides the zodiac into equal mathematical sections, sometimes called houses, and it does not matter to the believer in astrology that the constellations don't match. Furthermore, the physical mechanism is beside the point for the true believer. Astrology is not so much an astronomical superstition as it is a mathematical superstition.

Astrology makes sense only when we think of the world as the ancients did. They believed in multiple sky gods whose moods altered events on Earth. They believed in mystical influences between natural events and human events. The ancients did not understand natural forces such as gravity as the mechanisms that cause things to happen, and thus they did not believe in cause and effect as we do. If a house burned, they might conclude that it caught fire because it was cursed and not because someone was careless with an oil lamp.

Modern science left astrology behind centuries ago, but it survives as a fascinating part of human history—an early attempt to understand the meaning of the sky.

*The author of this book was born on December 14 and thus has no astrological sign. An astronomer friend claims that the author must therefore have no personality.

signs of the zodiac. A **horoscope** shows the location of the sun, moon, and planets among the zodiacal signs with respect to the horizon at the moment of a person's birth as seen from that longitude and latitude. Even if astrology worked, the generalized horoscopes published in newspapers and tabloids can't have been calculated accurately for individual readers.

Astrology buffs argue that a person's personality, life history, and fate are revealed in his or her horoscope, but the evidence contradicts this belief. Astrology has been tested many times over the centuries, and it just doesn't work. Believers, however, don't give up on it. Thus, astrology is a superstition that depends on blind belief and not a science that depends on evidence (Window on Science 2-3).

One reason astronomers find astrology irritating is that it has no link to the physical world. For example, precession has moved the constellations so that they no longer match the zodiacal signs. Whatever sign you were "born under," the sun was probably in the previous zodiacal constellation. In fact, if you were born on

Astrology and Pseudoscience

Astronomers have a low opinion of astrology not so much because it is groundless but because it pretends to be a science. It is a pseudoscience, from the prefix *pseudo,* meaning false. There are many examples of pseudoscience, and it is illuminating to consider the difference between pseudoscience and science.

A pseudoscience is a set of beliefs that appear to be based on scientific ideas but that fail to obey the most basic rules of science. For example, some years ago a fad arose in which people placed objects under pyramids made of paper, plastic, wire, and so on. The claim was that the pyramidal shape would focus cosmic forces on anything inside and so preserve fruit, sharpen razor blades, and do other miraculous things. Many books promoted this idea, but simple experiments showed that any shape would protect a piece of fruit from airborne spores and allow it to dry without rotting. Likewise, any shape would allow oxidation to improve the cutting

edge of a razor blade. In short, experimental evidence contradicted the claim. Nevertheless, supporters of the theory declined to abandon or revise their claims. Thus, the fad of pyramid power was a pseudoscience.

One characteristic of a pseudoscience is that it appeals to our needs and desires. Thus, some pseudoscientific claims are self-fulfilling. For example, some people bought pyramidal tents to put over their beds and thus improve their rest. While there is no logical mechanism by which such a tent could affect a sleeper, because people wanted and expected the claim to be true they slept more soundly. Many pseudoscientific claims involve medical cures, ranging from magnetic bracelets and crystals to focus spiritual power to astonishingly expensive and illegal treatments for cancer. Logic is a stranger to pseudoscience, but human fears and needs are not.

Astrology is a pseudoscience. Over the centuries, astrology has been tested

repeatedly, and no correlation has been found. But it survives, and its supporters disregard any evidence that it doesn't work. Like all pseudosciences, astrology is not open to revision in the face of contradictory evidence. Furthermore, astrology fulfills our human need to believe that there is order and meaning to our lives. It may comfort us to believe that our sweetheart has rejected us because of the motions of the planets rather than to admit that we behaved badly on our last date. Comfort aside, astrology is a poor basis for life decisions.

Human nature and human needs probably ensure that pseudoscientific beliefs will continue to plague us like emotional viruses propagating from person to person. But if we recognize them for what they are, we can more easily guide our lives by rational principles and not by giving credit for our successes and blame for our failings to the stars.

Modern astronomers have a low tolerance for astrological superstition, yet the motions of the heavenly bodies do affect our lives. The motion of the sun produces the seasons, and, as we will see in Chapter 3, the moon governs the tides. In addition, small changes in Earth's orbit may partially control global climate.

2-5 ASTRONOMICAL INFLUENCES ON EARTH'S CLIMATE

Weather is what happens today; climate is the average of what happens over decades and centuries. We know that Earth has gone through past episodes, called ice ages, when the worldwide climate was cooler and dryer and thick layers of ice covered northern latitudes. The earliest known ice age occurred about 570 million years ago and the next about 280 million years ago. The most recent ice age began only about 3 million years ago and is still going on. We are living in one of the periodic episodes when the glaciers melt and Earth grows slightly warmer. The current warm period began about 20,000 years ago.

Ice ages seem to occur with a period of roughly 250 million years, and cycles of glaciation within ice ages

occur with a period of about 40,000 years. These cyclic changes have an astronomical origin.

THE HYPOTHESIS

Sometimes a theory or hypothesis is proposed long before scientists can find the critical evidence to test it. That situation happened in 1920 when Yugoslavian meteorologist Milutin Milankovitch proposed what became known as the **Milankovitch hypothesis**—that changes in the shape of Earth's orbit, in precession, and in inclination affect Earth's climate and trigger ice ages. We will examine each of these three motions in turn.

First, astronomers know that the elliptical shape of Earth's orbit varies slightly over a period of about 100,000 years. At present, Earth's orbit carries it 1.7 percent closer than average to the sun during northern-hemisphere winters and 1.7 percent farther away in northern-hemisphere summers. This makes the northern winters warmer, and that is critical—most of the land mass where ice can accumulate is in the northern hemisphere. If Earth's orbit became more elliptical, Milankovitch suggested, northern winters might be warm enough to prevent the accumulation of snow and ice from forming glaciers, and Earth's climate would warm.

A second factor is also at work. Precession causes Earth's axis to sweep around a cone with a period of about 26,000 years, and that changes the location of the

The Foundation of Science: Evidence

Science is based on evidence. Every theory and conclusion must be supported by evidence obtained from experiments or from observation. If a theory is supported by many pieces of evidence but is clearly contradicted by a single experiment or observation, scientists quickly abandon it. For a theory to be true, there can be no contradictory evidence.

Of course, scientists argue about the significance of particular evidence and often disagree on the interpretation of evidence. Some observations may seem significant at first glance, but upon closer examination we may find that the procedure was flawed and so the piece of evidence is not important. Or we might conclude that the observational fact is correct but is being misinterpreted. The observation may not mean what it seems. Some of the most famous disagreements in science, such as those surrounding Galileo and Darwin, have arisen over the interpretation of well-established factual evidence.

Furthermore, scientists are not allowed to be selective in considering evidence. A lawyer in court can call a certain witness and intentionally fail to ask a critical question that would reveal evidence harmful to the lawyer's case. But a scientist may not ignore any known evidence. The difference in their methods is revealing. The lawyer is attempting to prove only one side of the case and rightly may ignore contradictory evidence. The scientist, however, is searching for the truth and so must test any theory against all available evidence. In a sense, the scientist, in dealing with evidence, must act as both the prosecution and the defense.

As you read about any science, look for the evidence in the form of measurements or observations. Every theory or conclusion should have supporting evidence. If you can find and understand the evidence, the science will make sense. All scientists, from astronomers to zoologists, demand evidence. You should, too.

---

seasons around Earth's orbit. Northern winters now occur when Earth is 1.7 percent closer to the sun, but in 13,000 years northern winters will occur on the other side of Earth's orbit where Earth is farther from the sun. Northern winters will be colder, and glaciers may grow.

The third factor is the inclination of Earth's equator to its orbit. Currently at 23.5°, this angle varies from 22° to 24° with a period of roughly 41,000 years. When the inclination is greater, seasons are more severe.

In 1920, Milankovitch proposed that these three factors cycled against each other to produce complex periodic variations in Earth's climate and the advance and retreat of glaciers (❚ Figure 2-12a). But no evidence was available to test the theory in 1920, and scientists treated it with skepticism. Many thought it was laughable.

THE EVIDENCE

By the middle 1970s, Earth scientists could collect the data that Milankovitch needed. Oceanographers could drill deep into the seafloor and collect samples, and geologists could determine the age of the samples from the natural radioactive atoms they contained. From all this, scientists constructed a history of ocean temperatures that convincingly matched the predictions of the Milankovitch hypothesis (Figure 2-12b).

The evidence seemed very strong, and by the 1980s, the Milankovitch hypothesis was widely discussed as the leading hypothesis. But science follows a mostly unstated set of rules that holds that a hypothesis must be tested over and over against all available evidence (Window on Science 2-4). In 1988, scientists discovered contradictory evidence.

While scuba diving in a water-filled crack in Nevada called Devil's Hole, scientists drilled out samples of calcite, a mineral that contains oxygen atoms. For 500,000 years, layers of calcite have built up in Devil's Hole, recording in their oxygen atoms the temperature of the atmosphere when rain fell there. Finding the ages of the mineral samples was difficult, but the results seemed to show that the previous ice age ended thousands of years too early to have been caused by Earth's motions.

These contradictory findings are irritating because we naturally prefer certainty, but such circumstances are common in science. The disagreement between ocean floor samples and Devil's Hole samples triggered a scramble to understand the problem. Were the ages of one or the other set of samples wrong? Were the ancient temperatures wrong? Or were scientists misunderstanding the significance of the evidence?

In 1997, a new study of the ages of the samples confirmed that those from the ocean floor are correctly dated. This seems to give scientists renewed confidence in the Milankovitch hypothesis. But the same study found that the ages of the Devil's Hole samples are also correct. Evidently the temperatures at Devil's Hole tell us about local climate changes in the region that became the southwestern United States. The ocean floor samples tell us about global climate changes, and they fit well with the Milankovitch hypothesis. In this way, the Milankovitch hypothesis, though widely accepted today, is still being tested as we try to understand the world we live on.

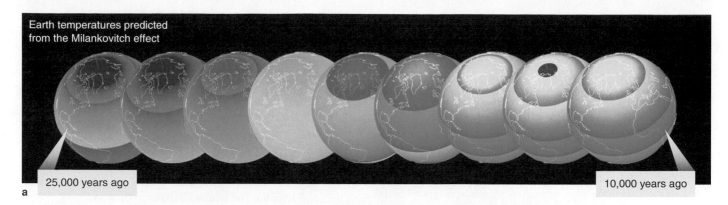

Earth temperatures predicted
from the Milankovitch effect

25,000 years ago

10,000 years ago

a

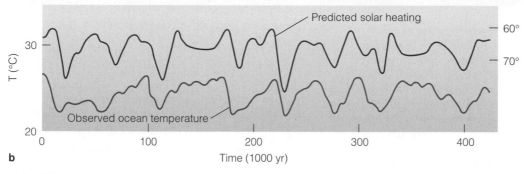

b

FIGURE 2-12

(a) Mathematical models of the Milankovich effect can be used to predict temperatures on Earth over time. In these Earth globes, cool temperatures are represented by violet and blue and warm temperatures by yellow and red. *(Courtesy Arizona State University, Computer Science and Geography Departments)*
(b) Over the last 400,000 years, changes in ocean temperatures measured from fossils found in sediment layers from the seabed match calculated changes in solar heating. *(Adapted from Cesare Emiliani)*

REVIEW CRITICAL INQUIRY

How do precession and the shape of Earth's orbit interact to affect Earth's climate?

One technique that is useful in the critical analysis of an idea is exaggeration. If we exaggerate the variation in the shape of Earth's orbit, we can see dramatically the influence of precession. At present, Earth reaches perihelion during winter in the northern hemisphere and aphelion during summer. The variation in distance is only about 1.7 percent, and that difference doesn't cause much change in the severity of the seasons. But if Earth's orbit were much more elliptical, then winter in the northern hemisphere would be much warmer, and summer would be much cooler.

Now we can see the importance of precession. As Earth's axis precesses, it points gradually in different directions, and the seasons occur at different places in Earth's orbit. In 13,000 years, northern winter will occur at aphelion, and if Earth's orbit were highly elliptical, northern winter would be terrible. Similarly, summer would occur at perihelion, and the heat would be awful. Such extremes might deposit large amounts of ice in the winter but then melt it away in the hot summer, thus preventing the accumulation of glaciers.

Continue this analysis by exaggeration. What effect would precession have if Earth's orbit were more circular?

This chapter is about the sky, but it has told us a great deal about Earth. We have discovered that Earth's orbital motion produces the apparent motion of the sun along the ecliptic, causes the seasons, and may even cause the cycle of the ice ages. As is often the case, astronomy helps us understand what we are and where we are. In this case, our study of the sky has helped us understand what it means to live on a planet. However, we have omitted one of the most dramatic objects in the sky—the moon. We will repair that omission in the next chapter.

SUMMARY

Astronomers divide the sky into 88 areas called constellations. Although the ancient constellations originated in Greek mythology, the names are Latin. Even the modern constellations, added to fill in the areas between the ancient figures, have Latin names. The names of stars usually come from ancient Arabic, though modern astronomers often refer to a star by constellation and Greek letters assigned according to brightness within each constellation.

The magnitude system is the astronomers' brightness scale. First-magnitude stars are brighter than second-magnitude stars, which are brighter than third-magnitude stars, and so on. The magnitude we see when we look at a star in the sky is its apparent visual magnitude.

The celestial sphere is a model of the sky, carrying the celestial objects around Earth. Because Earth rotates eastward, the celestial sphere appears to rotate westward on its axis. The northern and southern celestial poles are the pivots on which the sky appears to rotate. The celestial equator, an imaginary line around the sky above Earth's equator, divides the sky in half.

Because Earth orbits the sun, the sun appears to move eastward around the sky following the ecliptic. Because the ecliptic is tipped 23.5° to the celestial equator, the sun spends half the year in the northern celestial hemisphere and half the year in the southern celestial hemisphere, producing the seasons. The seasons are reversed south of Earth's equator. That is, while the northern hemisphere is experiencing a warm season, the southern hemisphere is experiencing a cold season.

Of the nine planets in our solar system, Mercury, Venus, Mars, Jupiter, and Saturn are visible to the naked eye. Their orbital motion around the sun carries them along the zodiac, a band 18° wide centered on the ecliptic. The positions of the sun, the moon, and these five planets form the basis of astrology, an ancient superstition that originated in Babylonia about 1000 BC.

Earth's motion may change in ways that can affect the climate. Changes in orbital shape, in precession, and in axial tilt can alter the planet's heat balance and may be responsible for the ice ages and glacial periods.

NEW TERMS

constellation	angular diameter
asterism	circumpolar constellation
magnitude scale	scientific model
apparent visual magnitude (m_v)	precession
	rotation
celestial sphere	revolution
horizon	ecliptic
zenith	vernal equinox
nadir	summer solstice
north celestial pole	autumnal equinox
south celestial pole	winter solstice
celestial equator	perihelion
north point	aphelion
south point	evening star
east point	morning star
west point	zodiac
angular distance	horoscope
minute of arc	Milankovitch hypothesis
second of arc	

REVIEW QUESTIONS

Ace ✪ Astronomy™ Assess your understanding of this chapter's topics with additional quizzing and animations at **http://astronomy.brookscole.com/seeds8e**

1. Why are most modern constellations composed of faint stars or located in the southern sky?

2. What does a star's Greek-letter designation tell us that its ancient Arabic name does not?

3. From your knowledge of star names and constellations, which of the following stars in each group is the brighter and which is the fainter? Explain your answers.

   a. α Ursae Majoris; θ Ursae Majoris

   b. λ Scorpii; β Pegasus

   c. β Telescopium; β Orionis

4. Give two reasons why the magnitude scale might be confusing.

5. Why do modern astronomers continue to use the celestial sphere when they know that stars are not all at the same distance?

6. How do we define the celestial poles and the celestial equator?

7. From what locations on Earth is the north celestial pole not visible? the south celestial pole? the celestial equator?

8. If Earth did not turn on its axis, could we still define an ecliptic? Why or why not?

9. Give two reasons why winter days are colder than summer days.

10. How do the seasons in Earth's southern hemisphere differ from those in the northern hemisphere?

11. Why should the eccentricity of Earth's orbit make winter in Earth's northern hemisphere different from winter in the southern hemisphere?

DISCUSSION QUESTIONS

1. Have you thought of the sky as a ceiling? as a dome overhead? as a sphere around Earth? as a limitless void?

2. How would the seasons be different if Earth were inclined 90° instead of 23.5°? 0° instead of 23.5°?

PROBLEMS

1. If one star is 6.3 times brighter than another star, how many magnitudes brighter is it?

2. If one star is 40 times brighter than another star, how many magnitudes brighter is it?

3. If two stars differ by 7 magnitudes, what is their intensity ratio?

4. If two stars differ by 8.6 magnitudes, what is their intensity ratio?

5. If star A is third magnitude and star B is fifth magnitude, which is brighter and by what factor?

6. If star A is magnitude 4 and star B is magnitude 9.6, which is brighter and by what factor?

7. By what factor is the sun brighter than the full moon? (*Hint:* See Figure 2-6.)

8. What is the angular distance from the north celestial pole to the summer solstice? to the winter solstice?

9. As seen from your latitude, what is the angle between the north celestial pole and the northern horizon? between the southern horizon and the noon sun at the summer solstice?

CRITICAL INQUIRIES FOR THE WEB

1. Nearly all cultures have populated the sky with gods, heroes, animals, and objects. What can you learn on the Web about non-Western constellations?

2. Who was Orion? How is he related to the scorpion in the sky?

3. What holidays, rituals, special foods, and beliefs are associated with the winter solstice?

4. What can you find out about Milutin Milankovitch? What is the latest news about the Milankovitch hypothesis?

EXPLORING *THESKY*

1. As discussed in this chapter, Earth's rotation about its own axis gives us the impression that the whole sky rotates around the north celestial pole in a period of one day. This apparent motion of the celestial sphere is difficult to notice because it happens so slowly. However, *TheSky* makes it possible to simulate this motion at a pace that is easy to observe by using a feature called **Time Skip**. Observe and describe the apparent motion of the sky as you see it looking north, east, south, and west.
   How to proceed: Set the **Time Skip Increment** (a dropdown menu on the **Time Skip Toolbar**) to 1 minute, and click on the **Go Forward** button to begin the simulation. View the sky from the four cardinal directions, due north, south, east, and west. (You'll find **Time Skip** under the **Tools** menu as well.)

2. In which constellation was the sun located on the date of your birth? (*Hint:* Click **Data** and then **Site Information** to set the location, date, and time of your birth. Turn on constellation figures, constellation boundaries, and labels. Then find the sun.)

3. Set your location to Earth's North Pole and the date to the summer solstice. Turn on the ecliptic and then step forward in time through a day to see what happens to the sun. Repeat for the autumnal equinox and the winter solstice.

4. Repeat Problem 3 above for a location on Earth's equator.

 Visit the Seeds *Foundations of Astronomy* companion Web site for critical thinking exercises, articles, and additional readings from InfoTrac College Edition, Brooks/Cole's online student library.

THE CYCLES OF THE MOON

Even a man who is pure in heart
And says his prayers by night
May become a wolf when the wolfbane blooms
And the moon shines full and bright.

Proverb from old Wolfman movies

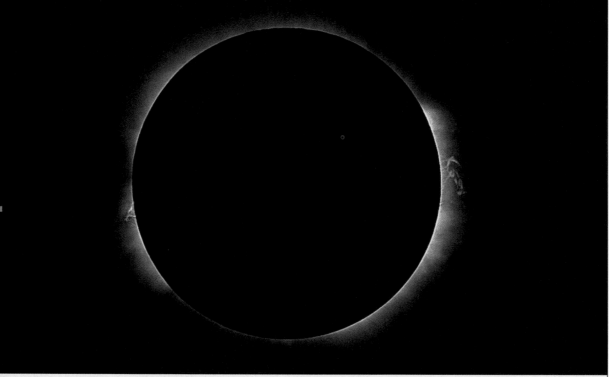

Daniel Good

Chapter 3

GUIDEPOST

In the preceding chapter, we saw how the sun dominates our sky and determines the seasons. The moon is not as bright as the sun, but the moon passes through dramatic phases and occasionally participates in eclipses. The sun dominates the daytime sky, but the moon rules the night.

As we try to understand the appearance and motions of the moon in the sky, we discover that what we see is a product of light and shadow. To understand the appearance of the universe, we must understand light. Later chapters will show that much of astronomy hinges on the behavior of light.

In the next chapter, we will see how Renaissance astronomers found a new way to describe the appearance of the sky and the motions of the sun, moon, and planets.

VIRTUAL LABORATORY

TIDES AND TIDAL FORCES

"Don't stare at the moon—you'll go crazy," more than one child has been warned.* The word *lunatic* comes from a time when even doctors thought that the insane were "moonstruck." A "mooncalf" is someone who has been crazy since birth, and the word is probably related to the belief that moonlight can harm unborn children. Everyone "knows" that people act less rationally when the moon is bright, but everyone is wrong. Careful studies of hospital and police records show that there is no real correlation between the moon and erratic behavior. The moon is so bright and its cycles through the sky are so dramatic we *expect* it to influence us, and many people are disappointed to learn the moon does not influence us. Some simply refuse to believe the evidence.

The cycles of the moon are indeed dramatic. Many different cultures around the world explain eclipses of

*When I was very small, my grandmother told me if I gazed at the moon, I might go crazy. But it was too beautiful, and I ignored her warning. I secretly watched the moon from my window, became fascinated by the sky, and became an astronomer.

the sun as invisible monsters devouring the sun (■ Figure 3-1). A Chinese story tells of two astronomers, Hsi and Ho, who were too drunk to predict the solar eclipse of October 22, 2137 BC, or perhaps failed to conduct the proper ceremonies to scare away the dragon snacking on the sun. When the emperor recovered from the terror of the eclipse, he had the two astronomers beheaded. The cycles of the moon provide a wealth of beauty in the sky and an important part of the cultural beliefs of Earth's peoples.

This chapter discusses the lunar cycles of phases, tides, and eclipses. Studying these events will help us understand what we see in the night sky. They will also

Ace◐Astronomy™ The AceAstronomy icon throughout the text indicates an opportunity for you to test yourself on key concepts and to explore animations and interactions on the AceAstronomy Web site at **http://astronomy.brookscole .com/seeds8e**

FIGURE 3-1
(a) A 12th-century Mayan symbol believed to represent a solar eclipse. The black-and-white sun symbol hangs from a rectangular sky symbol, and a voracious serpent approaches from below. (b) Chinese representation of a solar eclipse as a dragon flying in front of the sun. *(From the collection of Yerkes Observatory)* (c) A wall carving from the ruins of a temple at Vijayanagara in southern India. It symbolizes a solar eclipse as two snakes approaching the disk of the sun. *(T. Scott Smith)*

WINDOW ON SCIENCE 3-1

Gravity: The Universal Force

Isn't it weird that Isaac Newton is said to have "discovered" gravity in the late 17th century—as if people didn't have gravity before that, as if people in the 16th century floated around holding onto tree branches? Newton's accomplishment was that he realized that gravity was universal. The same force that makes an apple fall holds the moon in its orbit, guides the planets around the sun, and shapes the motion of the universe. We will discuss Newton's accomplishment in its proper historical place in Chapter 5, but here we need to recognize gravity for what it is, the master of the universe.

Newton realized that gravity was a property of matter, and consequently every object made of matter has to exert a gravitational attraction on every other object. Earth attracts the moon and the moon attracts Earth, but they also attract you and your book. Further, you and your book exert a gravitational force on each other and on every other object in the universe.

The size of an object's gravitational effect depends on the amount of matter it contains. A small amount of matter has a small gravitational effect, which is why you don't have satellites orbiting around you. Most objects in astronomy contain more matter, and their gravitational effects are larger. Moons, planets, stars, and galaxies each exert their gravitational attraction in proportion to the amount of matter they contain.

Of course, objects that are far away have less of an effect than objects that are nearby. That is another important point that Newton recognized. Earth's gravitational attraction presses us to our chairs, and the gravitational attraction of a distant galaxy containing much more matter than Earth is undetectably small because the galaxy is so far away.

The universe is filled with these gravitational influences, and every object in the universe moves under their guidance. The moon orbits Earth, and Earth orbits the sun. The sun orbits our galaxy, which moves under the influence of the gravitation of all the other galaxies in the universe. The universe is a swirling waltz of matter dancing to the music of gravity.

---

introduce us to some of the most basic concepts of astronomy. We will discover how gravity produces tides, how light and shadow move through space, how the size and distance of an object affect what we see, and how cyclic events can be analyzed by searching for their patterns. Most of all, this chapter forces us to begin thinking of our home as a world in space.

3-1 THE CHANGEABLE MOON

Starting this evening, look for the moon in the sky. If it is a cloudy night or if the moon is in the wrong part of its orbit, you may not see it, but keep trying on successive evenings, and within a week or two you will see the moon. Then watch for the moon on following evenings, and you will see it following its orbit around Earth and cycling through its phases as it has done for billions of years (Window on Science 3-1).

THE MOTION OF THE MOON

When we watch the moon night after night, we notice two things about its motion. First, we see it moving eastward against the background of stars; second, we notice that the markings on its face don't change. These two observations help us understand the motion of the moon and the origin of the moon's phases.

The moon moves rapidly among the constellations. If you watch the moon for just an hour, you can see it move eastward against the background of stars by slightly more than its angular diameter. In the previous chapter, we discovered that the moon is about 0.5° in angular diameter, so it moves eastward a bit more than

0.5° per hour. In 24 hours, it moves 13°. Each night when we look at the moon, we see it about 13° eastward of its location the night before. This eastward movement is the result of the motion of the moon along its orbit around Earth.

THE CYCLE OF PHASES

The changing shape of the moon as it orbits Earth is one of the most easily observed phenomena in astronomy. We have all noticed the full moon rising dramatically or a thin crescent moon hanging in the evening sky.

Study "▌ The Phases of the Moon" on pages 34 and 35 and notice three important points. First, the moon always keeps the same side facing Earth, and we never see the far side of the moon. "The Man in the Moon" is produced by the familiar features on the moon's near side. Second, the changing shape of the moon as it passes through its cycle of phases is produced by sunlight illuminating different parts of the side of the moon we can see. The third thing to notice is the difference between the orbital period of the moon around Earth and the length of the lunar phase cycle. That difference is a good illustration of how our view from Earth is produced by the combined motions of Earth and other heavenly bodies such as the sun and moon.

You can make a moon-phase dial from the middle diagram on page 34 by covering the lower half of the moon's orbit with a sheet of paper, aligning the edge of the paper to pass through the word "Full" at the left and the word "New" at the right. Push a pin through the edge of the paper at Earth's North Pole to make a pivot, and under the word "Full" write on the paper

"Eastern Horizon." Under the word "New" write "Western Horizon." The paper now represents the horizon we see facing south.

You can set your moon-phase dial for a given time by rotating the diagram behind the horizon paper. Set the dial to sunset by turning the diagram until the human figure labeled "sunset" is standing at the top of the Earth globe; the dial shows, for example, that the full moon at sunset would be at the eastern horizon.

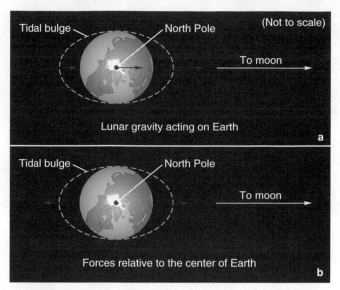

FIGURE 3-2
(a) Because one side of Earth is closer to the moon than the other side, the moon's gravity does not attract all parts of Earth with the same force. (b) Relative to the center of Earth, small differences in the forces cause tidal bulges (much exaggerated here) on the sides of Earth facing toward and away from the moon.

> | **REVIEW** CRITICAL INQUIRY |
>
> **Why do we sometimes see the moon in the daytime?**
> The full moon rises at sunset and sets at sunrise, so it is visible in the night sky but never in the daytime sky. But at other phases, it is possible to see the moon in the daytime. For example, when the moon is a waxing gibbous moon, it rises a few hours before sunset. If you look in the right spot in the sky, you can see it in the late afternoon in the southeast sky. It looks pale and washed out because of sunlight in our atmosphere, but it is quite visible once you notice it. You can also locate the waning gibbous moon in the morning sky. It sets a few hours after sunrise, so you would look for it in the southwestern sky in the morning.
>
> If you look in the right place, you might even see the first- or third-quarter moon in the daytime sky, but you have probably never seen a crescent moon in the daytime. Why not? Where would you have to look?

The phases of the moon don't affect us directly—we don't act crazier than usual at full moon. But the moon does have an important influence on Earth—the periodic advance and retreat of the ocean tides.

3-2 THE TIDES

Anyone who lives near the sea is familiar with a dramatic lunar effect—the ebb and flow of the tides. These periodic changes in the ocean are caused by the moon's gravity.

THE CAUSE OF THE TIDES

We feel Earth's gravity drawing us downward with a force we refer to as our weight, but the moon also exerts a gravitational force on us. Because the moon is less massive and more distant, its gravitational force acting on an object on Earth's surface is only about 0.0003 percent of Earth's gravitational force. That is a tiny force, but it is enough to cause tides in Earth's oceans.

Tides are produced by a *difference* between the gravitational force acting on different parts of an object.

The side of Earth facing the moon is about 6400 km (4000 miles) closer to the moon than is Earth's center, and the moon's gravity pulls more strongly on the oceans on the near side than on Earth's center. The difference is small, only about 3 percent of the moon's total gravitational force on Earth, but it is enough to make the ocean waters flow into a bulge on the side of Earth facing the moon.

A bulge also forms on the side of Earth facing away from the moon. Earth's far side is about 6400 km farther from the moon than is Earth's center, and the moon's gravity pulls on it less strongly than it does on Earth's center. Thus, relative to Earth's center, a small force makes the ocean waters on Earth's far side flow away from the moon (Figure 3-2).

We can see dramatic evidence of tidal forces if we watch the ocean shore for a few hours. Although Earth rotates on its axis, the tidal bulges remain fixed along the Earth–moon line. As the turning Earth carries us into a tidal bulge, the ocean water deepens, and the tide crawls up the beach. To be precise, the tide does not "come in." Rather, we are "carried into" the tidal bulge. Because there are two bulges on opposite sides of Earth, the tides rise and fall twice a day, and the times of high and low tide depend on the phase of the moon.

In reality, the tide cycle at any given location can be quite complex, depending on the latitude of the site, the shape of the shore, the north–south location of the moon, and so on. Tides in the Bay of Fundy occur twice a day and can exceed 12 m. The northern coast of the Gulf of Mexico has only one tidal cycle of roughly 30 cm each day.

The Phases of the Moon

We see the moon go through a month-long cycle of phases. This cycle is caused by the play of light and shadow on the moon as it follows its orbit around Earth.

As the moon orbits Earth, it rotates to keep the same side facing Earth. Consequently we always see the same features on the moon, and we never see the far side of the moon. A mountain on the moon that points at Earth will always point at Earth as the moon revolves and rotates.

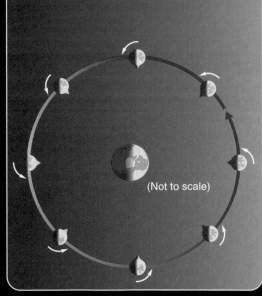

(Not to scale)

First quarter

Waxing gibbous

Waxing crescent

Sunset

North Pole

Midnight — Noon

Earth's rotation

Sunrise

Full

New

SUNLIGHT

Waning gibbous

Waning crescent

Third quarter

Ace Astronomy™

Go to AceAstronomy and click Active Figures to see "Lunar Phases" and watch the moon orbit Earth.

Sunlight always illuminates half of the moon. Because we see different amounts of this sunlit side, we see the moon cycle through phases. At the phase called "new moon," sunlight illuminates the far side of the moon, and the side we see is in darkness. At new moon we see no moon at all. At full moon, the side we see is fully lit, and the far side is in darkness. How much we see depends on where the moon is in its orbit.

Notice that there is no such thing as the "dark side of the moon." All parts of the moon experience day and night in a month-long cycle.

In the diagram at the left, we see that the new moon is close to the sun in the sky, and the full moon is opposite the sun. The time of day depends on the observer's location on Earth. (Compare with Figure 2-8.)

Gibbous comes from the Latin word for humpbacked.

The first quarter moon is one week through its 4-week cycle.

The full moon is two weeks through its 4-week cycle.

The first two weeks of the cycle of the moon is shown below by its position at sunset on 14 successive evenings. As the moon grows fatter from new to full, it is said to wax.

Waxing gibbous

Waxing crescent

THE SKY AT SUNSET

Full moon

New moon is invisible

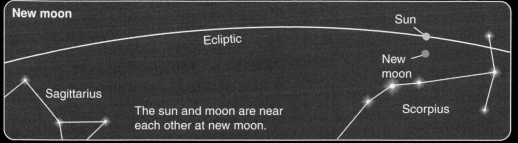

New moon

Ecliptic

Sun

New moon

Sagittarius

Scorpius

The sun and moon are near each other at new moon.

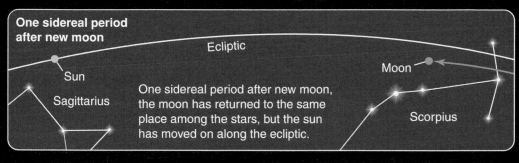

One sidereal period after new moon

Ecliptic

Sun

Sagittarius

Moon

Scorpius

One sidereal period after new moon, the moon has returned to the same place among the stars, but the sun has moved on along the ecliptic.

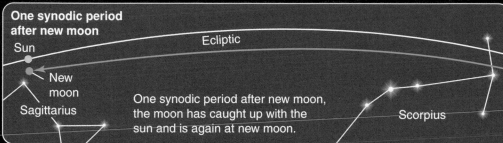

One synodic period after new moon

Ecliptic

Sun

New moon

Sagittarius

Scorpius

One synodic period after new moon, the moon has caught up with the sun and is again at new moon.

The moon orbits eastward around Earth in 27.32 days, its **sidereal period.** This is how long the moon takes to circle the sky once and return to the same position among the stars.

A complete cycle of lunar phases takes 29.53 days, the moon's **synodic period.** (Synodic comes from the Greek words for "together" and "path.")

To see why the synodic period is longer than the sidereal period, study the star charts at the left.

Although we think of the lunar cycle as being about 4 weeks long, it is actually 1.53 days longer than 4 weeks. Our calendar divides the year into 30-day periods called months (literally "moonths") in recognition of the 29.53 day synodic cycle of the moon.

You can use the diagram on the opposite page to determine when the moon rises and sets at different phases.

Earth   Moon

This diagram is to scale. The moon's orbit is about 30 Earth diameters in radius. Because the orbit is slightly elliptical, the moon's angular diameter in the sky can vary by plus or minus 6 percent. Because the orbit is tipped a bit over 5 degrees, the moon does not follow the ecliptic exactly.

TIMES OF MOONRISE AND MOONSET

Phase	Moonrise	Moonset
New	Dawn	Sunset
First quarter	Noon	Midnight
Full	Sunset	Dawn
Third quarter	Midnight	Noon

The third quarter moon is 3 weeks through its 4-week cycle.

The last two weeks of the cycle of the moon is shown below by its position at sunrise on 14 successive mornings. As the moon shrinks from full to new, it is said to wane.

Waning crescent

Waning gibbous

New moon is invisible near the sun

Full moon sets at sunrise

THE SKY AT SUNRISE

East   **South**   **West**

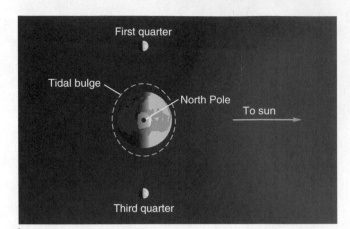

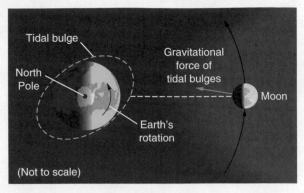

FIGURE 3-4

Earth's rotation drags the tidal bulges ahead of the Earth–moon line (exaggerated here). The gravitational attraction of these masses of water does not pull directly toward Earth's center but rather pulls the moon slightly forward in its orbit (green arrow), forcing its orbit to grow larger.

FIGURE 3-3

Looking down on Earth from above the North Pole, we can draw the shape of the tidal bulges (exaggerated here for clarity). (a) When the moon and sun pull along the same line, their tidal forces combine, and the tidal bulges are larger. (b) When the moon and sun pull at right angles, their tidal bulges do not combine, and the tidal bulges are smaller.

The sun, too, produces tides on Earth. The sun is roughly 27 million times more massive than the moon, but it lies almost 400 times farther from Earth. Consequently, tides on Earth caused by the sun are less than half those caused by the moon. At new moon and full moon, the moon and sun produce tidal bulges that join together to cause extreme tidal changes (▌Figure 3-3a); high tide is very high, and low tide is very low. Known as **spring tides,** these occur at every new and full moon. **Neap tides** occur at first- and third-quarter moon, when the moon and sun pull at right angles to each other. Then the smaller tide caused by the sun slightly reduces the tide caused by the moon, and the resulting high and low tides are less extreme (Figure 3-3b).

Though the ocean has been used as an example, you should note that tides would occur even if Earth had no oceans. The difference in the gravitational forces acting on different parts of Earth causes slight bulges in the rocky shape of Earth itself. We do not feel the mountains and plains rising and falling by a few centimeters; the changes in the fluid oceans are much more obvious.

TIDAL EFFECTS

Tidal forces can have surprising effects on both rotation and orbital motion. The friction of Earth's ocean waters against the seabeds slows Earth's rotation, and our days are getting longer by 0.0015 second per century. Some marine animals deposit layers in their shells in phase with the tides and the day–night cycle, and fossils of these shells confirm that only 900 million years ago Earth's day was 18 hours long.

In addition, Earth's gravitational field exerts tidal forces on the moon; and, although there are no oceans on the moon, tides do flex its rocky bulk. The resulting friction in the rock has slowed the moon's rotation. It once rotated much faster, but the tides caused by Earth have slowed the moon until it now rotates to keep one side permanently facing Earth. It is thus tidally locked to Earth. We will discover that most of the moons in the solar system are tidally locked to their planets.

Tidal forces can also affect orbital motion. For example, friction with the rotating Earth drags the tidal bulges eastward out of a direct Earth–moon line (▌Figure 3-4). These tidal bulges contain a large amount of mass, and the gravitation of the bulge near the moon pulls the moon slightly forward in its orbit. The bulge on Earth's far side pulls the moon slightly backward in its orbit, but because it is farther from the moon, that bulge is less effective. The net effect is to drag the moon forward in its orbit. As a result, the moon's orbit is growing larger, and the moon is receding from Earth at about 3.8 cm per year, an effect that astronomers can measure by bouncing laser beams off reflectors left on the lunar surface by the Apollo astronauts. Thus, we have clear evidence that tides are changing the moon's orbit.

REVIEW CRITICAL INQUIRY

Would Earth have tides if it had no moon?

Yes, there would be tides, but they would be much smaller. Don't forget that the sun also produces tides on Earth. The solar tides are less extreme than the lunar tides, but we would still see the advance and retreat of the oceans if the moon did not exist.

There would be an interesting difference, however, if the only tides were solar tides. The sun rises every 24 hours, but the moon moves rapidly eastward in the sky and rises, on average, every 24.8 hours. How many hours would separate high tides caused by the sun compared with high tides caused by the moon?

Tides and tidal forces are important in many areas of astronomy. In later chapters, we will see how tidal forces can pull gas away from stars, rip galaxies apart, and melt the interiors of moons orbiting near massive planets. Now, however, we must consider yet another kind of lunar phenomenon—eclipses.

3-3 LUNAR ECLIPSES

A lunar eclipse occurs at full moon when the moon moves through Earth's shadow. Because the moon shines only by reflected sunlight, it gradually darkens as it enters the shadow.

EARTH'S SHADOW

Earth's shadow consists of two parts. The **umbra** is the region of total shadow. If we were floating in space in the umbra of Earth's shadow, we would see no portion of the sun. However, if we moved into the **penumbra**, we would be in partial shadow and would see part of the sun peeking around Earth's edge. In the penumbra, the sunlight is dimmed but not extinguished.

We can construct a model of this by pressing a map tack into the eraser of a pencil and holding the tack between a lightbulb a few feet away and a white cardboard screen (Figure 3-5). The lightbulb represents the sun, and the map tack represents Earth. When we hold the screen close to the tack, we see that the umbra is nearly as large as the tack and that the penumbra is only slightly larger. However, as we move the screen away from the tack, the umbra shrinks and the penumbra expands. Beyond a certain point, the shadow has no dark core at all, indicating that the screen is beyond the end of the umbra.

The umbra of Earth's shadow is about 1.4 million km (860,000 miles) long and points directly away from the sun. A giant screen placed in the shadow at the average distance of the moon would reveal a dark umbra about 9000 km (5700 miles) in diameter, and the faint outer edges of the penumbra would mark a circle about 16,000 km (10,000 miles) in diameter. For comparison, the moon's diameter is only 3476 km (2160 miles). Thus, when the moon's orbit carries it through the umbra, it has plenty of room to become completely immersed in shadow.

TOTAL LUNAR ECLIPSES

A lunar eclipse occurs when the moon passes through Earth's shadow and grows dark. If the moon passes through the umbra and no part of the moon remains outside the umbra in the partial sunlight of the penumbra, we say the eclipse is a **total lunar eclipse.**

 Figure 3-6 illustrates the stages of a total lunar eclipse as a three-dimensional diagram (Window on Science 3-2) showing Earth, its shadows, and the path of the moon. As the moon begins to enter the penumbra it is only slightly dimmed, and a casual observer may not notice anything odd. After an hour, the moon

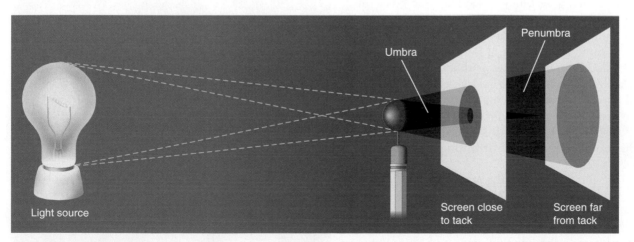

FIGURE 3-5
The shadows cast by a map tack resemble those of Earth and the moon. The umbra is the region of total shadow; the penumbra is the region of partial shadow.

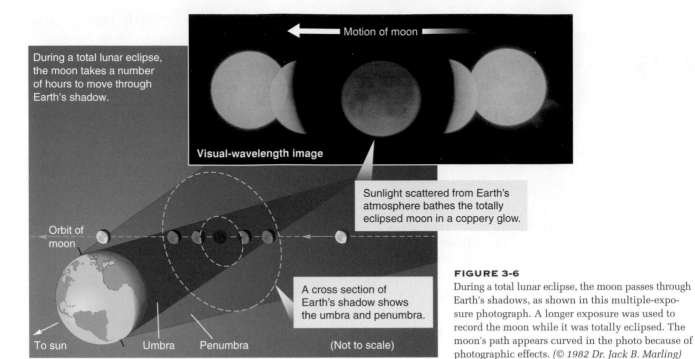

During a total lunar eclipse, the moon takes a number of hours to move through Earth's shadow.

Motion of moon

Visual-wavelength image

Sunlight scattered from Earth's atmosphere bathes the totally eclipsed moon in a coppery glow.

Orbit of moon

A cross section of Earth's shadow shows the umbra and penumbra.

To sun   Umbra   Penumbra   (Not to scale)

FIGURE 3-6

During a total lunar eclipse, the moon passes through Earth's shadows, as shown in this multiple-exposure photograph. A longer exposure was used to record the moon while it was totally eclipsed. The moon's path appears curved in the photo because of photographic effects. (© 1982 Dr. Jack B. Marling)

is deeper in the penumbra and dimmer; and, once it begins to enter the umbra, we see a dark bite on the edge of the lunar disk. The moon travels its own diameter in an hour, so it takes about an hour to enter the umbra completely and become totally eclipsed.

Even when the moon is totally eclipsed, it does not disappear completely. Sunlight, bent by Earth's atmosphere, leaks into the umbra and bathes the moon in a faint glow. Because blue light is scattered by Earth's atmosphere more easily than red light, it is red light that penetrates to illuminate the moon in a coppery glow. If we were on the moon during totality and looked back at Earth, we would not see any part of the sun because it would be entirely hidden behind Earth. However, we would see Earth's atmosphere illuminated from behind by the sun in a spectacular sunset completely ringing Earth. It is the red glow from this sunset that gives the totally eclipsed moon its reddish color.

How dim the totally eclipsed moon becomes depends on a number of things. If Earth's atmosphere is especially cloudy in those regions that must bend light into the umbra, the moon will be darker than usual. An unusual amount of dust in Earth's atmosphere (from volcanic eruptions, for instance) also causes a dark eclipse. Also, total lunar eclipses tend to be darkest when the moon's orbit carries it through the center of the umbra (▌ Figure 3.7).

As the moon moves through Earth's umbral shadow, we can see that the shadow is circular. From this the Greek philosopher Aristotle (384–322 BC) concluded that Earth had to be a sphere, because only a sphere could cast a shadow that was always circular.

Depending on the geometry of the eclipse, the moon can take as long as 1 hour 40 minutes to cross the umbra

and another hour to emerge into the penumbra. Still another hour passes as it emerges into full sunlight. A total eclipse of the moon, including the penumbral stage, can take almost six hours from start to finish.

PARTIAL AND PENUMBRAL LUNAR ECLIPSES

Not all eclipses of the moon are total. Because the moon's orbit is inclined a bit over 5° to the plane of Earth's or-

FIGURE 3-7

During a total lunar eclipse, the moon turns coppery red. In this photo, the moon is darkest toward the lower right, the direction toward the center of the umbra. The edge of the moon at upper left is brighter because it is near the edge of the umbra. (Celestron International)

3-D Relationships

Much of science is explained in diagrams; one particular type of diagram can be confusing. When artists draw three-dimensional diagrams on flat sheets of paper, they use lots of artistic clues that have been discovered over the centuries. Perspective, shading, color, and shadows help us see the three-dimensional figure if the drawing is familiar—a house, for example. But when a drawing shows something that is unfamiliar, as is often the case in sci-

ence, the artist's clues don't work as well. When we look at drawings of a molecule, a nerve cell, layers of rock under the ocean, or Earth and its shadows, we must pay special attention to the three-dimensional nature of the figure.

When you see a three-dimensional diagram in any science book, it helps to decide what the point of view is. For example, if the diagram were a photograph, where was the camera? In the

drawing in Figure 3-6, the camera would have to have been in space looking back at Earth and the moon. Of course, astronauts have never been that far from Earth, but we can make such a voyage in our imagination, and that helps us understand the geometry of eclipses. Other three-dimensional diagrams have other points of view, so always be sure to first determine the point of view when you look at any three-dimensional diagram.

bit, the moon might not pass through the center of the umbra.

If the moon's orbit carries the full moon too far north or south of the umbra, the moon may only partially enter the umbra. The resulting **partial lunar eclipse** is usually not as dramatic as a total lunar eclipse. Because part of the moon remains outside the umbra, it receives some sunlight and looks bright in contrast with the dark part of the moon inside the umbra. Unless the moon almost completely enters the umbra, the glare

from the illuminated part of the moon drowns out the fainter red glow inside the umbra. Partial lunar eclipses are interesting because part of the full moon is darkened, but they are not as beautiful as a total lunar eclipse.

If the orbit of the moon carries the moon far enough north or south of the umbra, the moon may only pass through the penumbra and never reach the umbra. Such **penumbral eclipses** are not dramatic at all. In the partial shadow of the penumbra, the moon is only partially dimmed. Most people glancing at a penumbral eclipse would not notice any difference from a full moon.

Total, partial, or penumbral, lunar eclipses are interesting events in the night sky and are not difficult to observe. When the moon passes through Earth's shadow, the eclipse is visible from anywhere on Earth's dark side. One or two lunar eclipses occur in most years. Consult ▌ Table 3-1 to find the next lunar eclipse visible in your part of the world.

TABLE 3-1
Total and Partial Eclipses of the Moon, 2004 to 2013

Year	Date	Time* of Mideclipse (GMT)	Length of Totality (Hours:Min)	Length of Eclipse† (Hours:Min)
2004	May 4	20:32	1:16	3:22
2004	Oct. 28	3:05	1:20	3:38
2005	Oct. 17	12:04	Partial	0:56
2006	Sept. 7	18:52	Partial	1:30
2007	Mar. 3	23:22	1:14	3:40
2007	Aug. 28	10:38	1:30	3:32
2008	Feb. 21	3:27	0:50	3:24
2008	Aug. 16	21:11	Partial	3:08
2009	Dec. 31	19:24	Partial	1:00
2010	June 26	11:40	Partial	2:42
2010	Dec. 21	8:18	1:12	3:28
2011	June 15	20:13	1:40	3:38
2011	Dec. 10	14:33	0:50	3:32
2012	June 4	11:03	Partial	2:08
2013	April 25	20:10	Partial	0:28

*Times are Greenwich Mean Time. Subtract 5 hours for Eastern Standard Time, 6 hours for Central Standard Time, 7 hours for Mountain Standard Time, and 8 hours for Pacific Standard Time. From your time zone, lunar eclipses that occur between sunset and sunrise will be visible, and those at midnight will be best placed.

†Does not include penumbral phase.

REVIEW CRITICAL INQUIRY

Why doesn't Earth's shadow on the moon look red during a partial lunar eclipse?

During a partial lunar eclipse, part of the moon protrudes from Earth's umbral shadow into sunlight. This part of the moon is very bright compared to the fainter red light inside Earth's shadow, and the glare of the reflected sunlight makes it difficult to see the red glow. If a partial eclipse is almost total, so that only a small sliver of moon extends out of the shadow into sunlight, we can sometimes detect the red glow in the shadow.

Of course, this red glow does not happen for every planet–moon combination in the universe. Suppose a planet had a moon but no atmosphere. Would the moon glow red during a total eclipse? Why not?

Lunar eclipses are slow and stately. For drama and excitement, there is nothing like a solar eclipse.

3-4 SOLAR ECLIPSES

A solar eclipse occurs when the moon moves between Earth and the sun. If the moon covers the disk of the sun completely, the eclipse is a **total solar eclipse.** If the moon covers only part of the sun, the eclipse is a **partial solar eclipse.** During a particular solar eclipse, people in one place on Earth may see a total eclipse, while people only a few hundred kilometers away see a partial eclipse.

These spectacular sights are possible because we on Earth are very lucky. Our moon has the same angular diameter as our sun, so it can cover the sun almost exactly. That lucky coincidence allows us to see total solar eclipses.

THE ANGULAR DIAMETER OF THE SUN AND MOON

We discussed the angular diameter of an object in Chapter 2; now we need to think carefully about how the size and distance of an object like the moon determine its angular diameter. This is the key to understanding solar eclipses.

Linear diameter is simply the distance between an object's opposite sides. We use linear diameter when we order a 16-inch pizza—the pizza is 16 inches in diameter. The linear diameter of the moon is 3476 km. The angular diameter of an object is the angle formed by lines extending toward us from opposite sides of the object and meeting at our eye (Figure 3-8). Clearly, the farther away an object is, the smaller its angular diameter.

The **small-angle formula** gives us a way to figure out the angular diameter of any object, whether it is a pizza, the moon, or a galaxy. In the small-angle formula, we always express angular diameter in seconds of arc,* and we always use the same units for distance and linear diameter:

$$\frac{\text{angular diameter}}{206{,}265''} = \frac{\text{linear diameter}}{\text{distance}}$$

Of course, we can use this formula to find any one of these three quantities if we know the other two; here we are interested in finding the angular diameter of the moon.

The moon has a linear diameter of 3476 km and a distance from Earth of about 384,000 km. What is its angular diameter? The moon's linear diameter and distance are both given in the same units, so we can put them directly into the small-angle formula:

$$\frac{\text{angular diameter}}{206{,}265''} = \frac{3476\ \text{km}}{384{,}000\ \text{km}}$$

*The number 206,265″ is the number of seconds of arc in a radian. When we divide by 206,265″, we convert the angle from seconds of arc to radians.

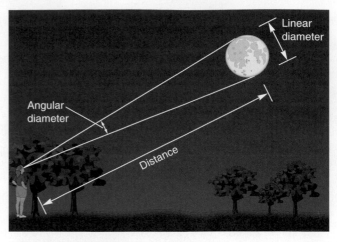

ACTIVE FIGURE 3-8
Angular diameter is the angle formed by lines extending from our eye to opposite sides of the object—in this figure, the moon. Linear diameter and distance are typically measured in kilometers or meters.

Ace Astronomy™ Go to AceAstronomy and click Active Figures to see "Small-Angle Formula." You can take control of this diagram.

To solve for angular diameter, we multiply both sides by 206,265 and find that the angular diameter is 1870 seconds of arc. If we divide by 60, we get 31 minutes of arc or, dividing by 60 again, about 0.5°. The moon's orbit is slightly elliptical, so it can sometimes look a bit larger or smaller, but its angular diameter is always close to 0.5°.

We can repeat this calculation for the angular diameter of the sun. The sun is 1.39×10^6 km in linear diameter and 1.50×10^8 km from Earth. Using the small-angle formula, we discover that the sun has an angular diameter of 1900 seconds of arc, which is 32 minutes of arc or about 0.5°. Earth's orbit is slightly elliptical, and thus the sun can sometimes look slightly larger or smaller, but it, like the moon, is always close to 0.5° in angular diameter.

By fantastic good luck, we live on a planet with a moon that is almost exactly the same angular diameter as our sun. When the moon passes in front of the sun, it is almost exactly the right size to block the brilliant surface of the sun. Then we see the most exciting sight in astronomy—a total solar eclipse. There are few other worlds where this can happen, because the angular diameters of the sun and a moon rarely match so closely. To see this beautiful sight, all we have to do is arrange to be in the moon's shadow when the moon crosses in front of the sun.

THE MOON'S SHADOW

Like Earth's shadow, the moon's shadow consists of a central umbra of total shadow and a penumbra of par-

TABLE 3-2
Total and Annular Eclipses of the Sun, 2004** to 2014

Date	Total/Annular (T/A)	Time of Mideclipse* (GMT)	Maximum Length of Total or Annular Phase (Min:Sec)	Area of Visibility
2005 Apr. 8	AT	21^h	0:42	Pacific, N. of S. America
2005 Oct. 3	A	11^h	4:32	Atlantic, Spain, Africa
2006 Mar. 29	T	10^h	4:07	Atlantic, Africa, Turkey
2006 Sept. 22	A	12^h	7:09	N.E. of S. America, Atlantic
2008 Feb. 7	A	4^h	2:14	S. Pacific, Antarctica
2008 Aug. 1	T	10^h	2:28	Canada, Arctic, Siberia
2009 Jan. 26	A	8^h	7:56	S. Atlantic, Indian Ocean
2009 July 22	T	3^h	6:40	Asia, Pacific
2010 Jan. 15	A	7^h	11:10	Africa, Indian Ocean
2010 July 11	T	20^h	5:20	Pacific, S. America
2012 May 20	A	23^h	5:46	Japan, N. Pacific, W. US
2012 Nov. 13	T	22^h	4:02	N. Australia, S. Pacific
2013 May 10	A	0^h	6:04	Australia, Pacific
2013 Nov. 3	AT	13^h	1:40	Atlantic, Africa

The next major total solar eclipse visible from the United States will occur on August 21, 2017.

*Times are Greenwich Mean Time. Subtract 5 hours for Eastern Standard Time, 6 hours for Central Standard Time, 7 hours for Mountain Standard Time, and 8 hours for Pacific Standard Time.

^{h}hours.

**There are no total or annular eclipses of the sun during 2004 or 2014.

tial shadow. What we see when the moon crosses in front of the sun depends on where we are in the moon's shadow. The moon's umbral shadow produces a spot of darkness roughly 269 km (167 miles) in diameter on Earth's surface (▌ Figure 3-9). (The exact size of the umbral shadow depends on the location of the moon in its elliptical orbit and the angle at which the shadow strikes Earth.) If we are in this spot of total shadow, we see a total solar eclipse. If we are just outside the umbral shadow but in the penumbra, we see part of the sun

peeking around the moon, and the eclipse is partial. Of course, if we are outside the penumbra, we see no eclipse at all. Because of the orbital motion of the moon, its shadow sweeps across Earth at speeds of at least 1700 km/h (1060 mph). To be sure of seeing a total solar eclipse, we must select an appropriate eclipse (▌ Table 3-2), plan far in advance, and place ourselves in the **path of totality,** the path swept out by the umbral spot.

TOTAL SOLAR ECLIPSES

A total solar eclipse begins when we first see the edge of the moon encroaching on the sun. This is the moment when the edge of the penumbra sweeps over our location.

During the partial phase, part of the sun remains visible, and it is hazardous to look at the eclipse without protection. Dense filters and exposed film do not necessarily provide protection, because some do not block the invisible heat radiation (infrared) that can burn the retina of our eyes. This has led officials to warn the public not to look at solar eclipses and has even frightened some people into locking themselves and their children into windowless rooms during eclipses. In fact, the sun is a bit less dangerous than usual during an eclipse because part of the bright surface is covered by the moon. But an eclipse is dangerous

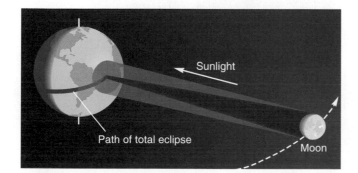

FIGURE 3-9
Observers in the path of totality see a total solar eclipse when the umbral shadow sweeps over them. Those in the penumbra see a partial eclipse.

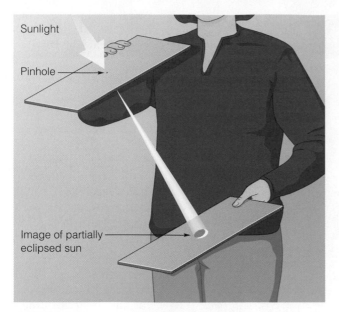

FIGURE 3-10

A safe way to view the partial phases of a solar eclipse. Use a pinhole in a card to project an image of the sun on a second card. The greater the distance between the cards, the larger (and fainter) the image will be.

in that it can tempt us to look at the sun directly and burn our eyes.

The safest and simplest way to observe the partial phases of a solar eclipse is to use pinhole projection. Poke a small pinhole in a sheet of cardboard. Hold the sheet with the hole in the sunlight, and allow light to pass through the hole to a second sheet of cardboard (▌Figure 3-10). On a day when there is no eclipse, the result is a small, round spot of light that is an image of the sun. During the partial phases of a solar eclipse, the image shows the dark silhouette of the moon obscuring part of the sun. These pinhole images of the partially eclipsed sun can also be seen in the shadows of trees as the sunlight peeks through the tiny openings between the leaves and branches. This can produce an eerie effect just before totality as the remaining sliver of sun produces thin crescents of light on the ground under trees.

Throughout the partial phases of a solar eclipse, the moon gradually covers the bright disk of the sun (▌Figure 3-11). Totality begins as the last sliver of the sun's bright surface disappears behind the moon. This is the moment when the edge of the umbra sweeps over our location. So long as any of the sun is visible, the countryside is bright; but, as the last of the sun disappears, dark falls in a few seconds. Automatic streetlights come on, car drivers switch on their headlights, and birds go to roost. The darkness of totality depends on a number of factors, including the weather at the observing site, but it is usually dark enough to make it difficult to read the settings on cameras.

The totally eclipsed sun is a spectacular sight. With the moon covering the bright disk of the sun, called the

A Total Solar Eclipse

The moon moving from the right just begins to cross in front of the sun.

The disk of the moon gradually covers the disk of the sun.

Sunlight begins to dim as more of the sun's disk is covered.

During totality, pink prominences are often visible.

A longer-exposure photograph during totality shows the fainter corona.

FIGURE 3-11

This sequence of photos shows the first half of a total solar eclipse. *(Daniel Good)*

photosphere, * we can see the sun's faint outer atmosphere, the **corona,** glowing with a pale, white light so faint we can safely look at it directly. This corona is

*The photosphere, corona, chromosphere, and prominences will be discussed in detail in Chapter 8. Here the terms are used as the names of features we see during a total solar eclipse.

a

b

FIGURE 3-12

(a) During a total solar eclipse, the moon covers the photosphere, and the white corona and pink prominences are visible. Note the streamers in the corona caused by the sun's magnetic field. *(Daniel Good)*
(b) The diamond ring effect can sometimes occur momentarily at the beginning or end of totality if a small segment of the photosphere peeks out through a valley at the edge of the lunar disk. *(National Optical Astronomy Observatory)*

made of low-density, hot gas, which is given a wispy appearance by the solar magnetic field as shown in the last frame of Figure 3-11. Also visible just above the photosphere is a thin layer of bright gas called the **chromosphere.** The chromosphere is often marked by eruptions on the solar surface called **prominences** (▌ Figure 3-12a), which glow with a clear, pink color due to the high temperature of the gases involved. The small-angle formula tells us that a large prominence is about 3.5 times the diameter of Earth.

Totality cannot last longer than 7.5 minutes under any circumstances, and the average is only 2 to 3 minutes. Totality ends when the sun's bright surface reappears at the trailing edge of the moon. This corresponds to the moment when the trailing edge of the moon's umbra sweeps over the observer.

Just as totality begins or ends, a small part of the photosphere can peek out from behind the moon through a valley at the edge of the lunar disk. Although it is intensely bright, such a small part of the photosphere does not completely drown out the fainter corona, which forms a silvery ring of light with the brilliant spot of photosphere gleaming like a diamond (Figure 3-12b). This **diamond ring effect** is one of the most spectacular of astronomical sights, but it is not visible during every solar eclipse. Its occurrence depends on the exact orientation and motion of the moon.

Once totality is over, daylight returns quickly, and the corona and chromosphere vanish. Astronomers travel great distances to place their instruments in the path of totality to study the faint outer corona and make other measurements possible only during the few minutes of a total solar eclipse.

Sometimes when the moon crosses in front of the sun, it is too small to fully cover the sun, and we see an **annular eclipse,** a solar eclipse in which a ring (or annulus) of the photosphere is visible around the disk of the moon (▌ Figure 3-13). With a portion of the photosphere visible, the eclipse never becomes total; it never quite gets dark; and we can't see the prominences, chromosphere, and corona. Annular eclipses occur because the moon follows a slightly elliptical orbit around Earth, and thus its angular diameter can vary (▌ Figure 3-14). When it is at **perigee,** its point of closest approach to Earth, it looks significantly larger than when it is at **apogee,** the most distant point in its orbit. Furthermore, Earth's orbit is slightly elliptical, so the Earth–sun distance varies slightly, and thus the diameter of the solar disk varies slightly. If the moon is in the farther part of its orbit during totality, its angular diameter will be less than the angular diameter of the sun, and thus we see the annular eclipse. Such an annular eclipse of the sun swept across the United States on May 10, 1994.

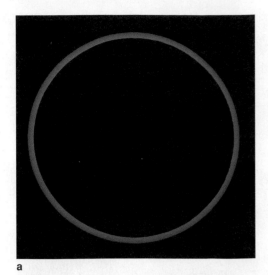

a

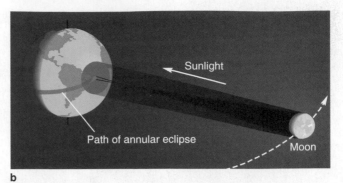

b

FIGURE 3-13

(a) The annular eclipse of 1994. A bright ring of photosphere remains visible around the moon, and the corona and prominences are not visible. *(Daniel Good)* (b) An annular eclipse occurs when the moon is in the farther part of its orbit and its umbral shadow does not reach Earth. From Earth, we see an annular eclipse because the moon's angular diameter is smaller than the angular diameter of the sun.

⎯| **REVIEW** CRITICAL INQUIRY |⎯

If people on Earth were seeing a total solar eclipse, what would astronauts on the moon see when they looked at Earth?

Astronauts on the moon could see Earth only if they were on the side that faces Earth. Because solar eclipses always happen at new moon, the near side of the moon would be in darkness, and the far side of the moon would be in full sunlight. The astronauts would be standing in darkness, and they would be looking at the fully illuminated side of Earth. They would see a "full Earth." The moon's shadow would be crossing Earth; and, if the astronauts looked closely, they might be able to see the spot of darkness where the moon's umbral shadow touched Earth. It would take hours for the shadow to cross Earth.

Standing on the moon and watching the moon's umbral shadow sweep across Earth would be a cold, tedious assignment. Perhaps it would be more interesting for astronauts on the moon to watch Earth while people on Earth were seeing a total lunar eclipse. What would the astronauts see then?

Eclipses of the sun and moon are often dramatic and mysterious. Ancient astronomers studied them and found ways to predict the coming of an eclipse. In the section that follows, we will see how eclipses can be predicted from a basic understanding of the motion of the sun and the moon.

3-5 PREDICTING ECLIPSES

To make exact eclipse predictions, you would need to calculate the precise motions of the sun and moon, and that requires a computer and proper software. Such software is available for desktop computers, but it isn't necessary if you are satisfied with making less exact predictions. In fact, many primitive peoples, such as the builders of Stonehenge and the ancient Maya, are believed to have made eclipse predictions.

We examine eclipse prediction for three reasons. First, it is an important part of the history of science. Second, it illustrates how apparently complex phenomena can be analyzed in terms of cycles. Third, eclipse prediction exercises our mental muscles and

---

FIGURE 3-14

The angular diameter of the moon (left) varies by a total of almost 12 percent, because its orbit is elliptical, and its distance from Earth varies from perigee (closest) to apogee (farthest). The angular diameter of the sun (right) varies by a total of only 3.4 percent, because Earth's orbit is more nearly circular, and its distance from the sun varies only slightly from perihelion (closest) to aphelion (farthest).

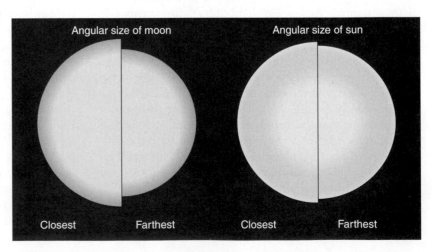

Angular size of moon   Angular size of sun

Closest   Farthest   Closest   Farthest

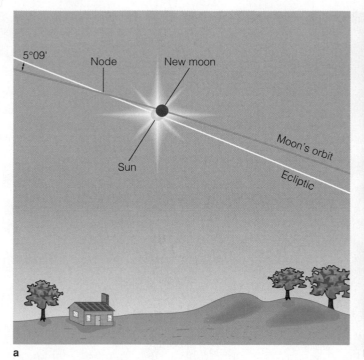

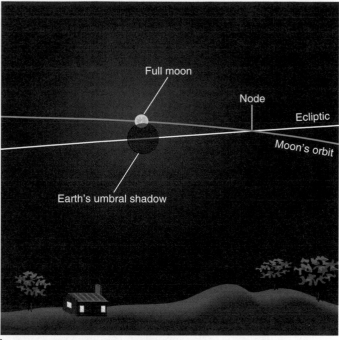

FIGURE 3-15

Eclipses can occur only near the nodes of the moon's orbit. (a) A solar eclipse occurs when the moon meets the sun near a node. (b) A lunar eclipse occurs when the sun and moon are near opposite nodes. Partial eclipses are shown here for clarity.

forces us to see Earth, the moon, and the sun as objects moving through space.

CONDITIONS FOR AN ECLIPSE

We can predict eclipses by understanding the conditions that make them possible. As we begin to think about these conditions, we must be sure we understand our point of view. (See Window on Science 3-2.) Later we will change our point of view, but to begin we will imagine that we can look up into the sky from our home on Earth and see the sun moving along the ecliptic and the moon moving along its orbit.

The orbit of the moon is tipped 5°8′43″ to the plane of Earth's orbit, so we see the moon follow a path tipped by that angle to the ecliptic. Each month, the moon crosses the ecliptic at two points called **nodes.** At one node it crosses going southward, and two weeks later it crosses at the other node going northward.

Eclipses can only occur when the sun is near one of the nodes of the moon's orbit. A solar eclipse is caused by the moon passing in front of the sun. Most new moons pass too far north or too far south of the sun to cause an eclipse. Only when the sun is near a node in the moon's orbit can the moon cross in front of the sun, as shown in ▌Figure 3-15a. A lunar eclipse doesn't happen at every full moon because most full moons pass too far north or too far south of the ecliptic and miss Earth's shadow. The moon can enter Earth's shadow only when the shadow is near a node in the moon's orbit, and that means the sun must be near the other node. This is shown in Figure 3-15b.

Thus, there are two conditions for an eclipse: The sun must be crossing a node, and the moon must be crossing either the same node (solar eclipse) or the other node (lunar eclipse). Clearly, solar eclipses can occur only when the moon is new, and lunar eclipses can occur only when the moon is full.

An **eclipse season** is the period during which the sun is close enough to a node for an eclipse to occur. For solar eclipses, an eclipse season is about 32 days. Any new moon during this period will produce a solar eclipse. For lunar eclipses, the eclipse season is a bit shorter, about 22 days. Any full moon in this period will be eclipsed.

This makes eclipse prediction easy. We simply keep track of where the moon crosses the ecliptic, and when the sun is near one of these nodes we predict that the nearest new moon will cause a solar eclipse and the nearest full moon will cause a lunar eclipse. This system works fairly well, and ancient astronomers such as the Maya may have used such a system. But we can do better if we change our point of view.

THE VIEW FROM SPACE

Let us change our point of view and imagine that we are looking at the orbits of Earth and the moon from a

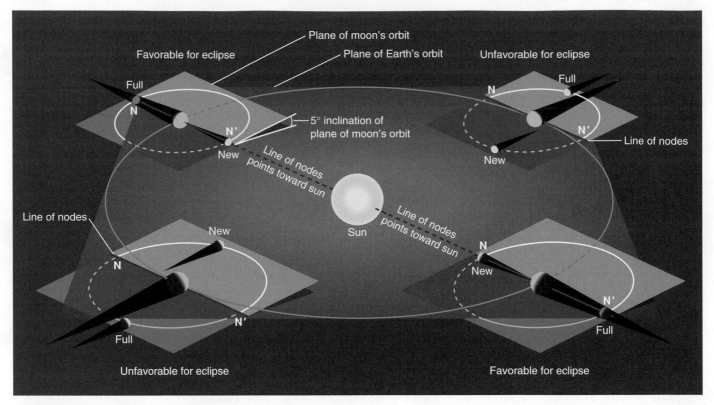

FIGURE 3-16

The moon's orbit is tipped about 5° to Earth's orbit. The nodes N and N' are the points where the moon passes through the plane of Earth's orbit. If the line of nodes does not point at the sun, the shadows miss, and there are no eclipses at new moon and full moon. At those parts of Earth's orbit where the line of nodes points toward the sun, eclipses are possible at new moon and full moon.

point far away in space. We see the moon's orbit as a smaller disk tipped at an angle to the larger disk of Earth's orbit. As Earth orbits the sun, the moon's orbit remains fixed in direction. The nodes of the moon's orbit are the points where it passes through the plane of Earth's orbit; an eclipse season occurs each time the line connecting these nodes, the **line of nodes,** points toward the sun (▌Figure 3-16).

The shadows of Earth and moon, seen from space, are very long and thin (▌Figure 3-17). Only at the time of an eclipse season, when the line of nodes points toward the sun, do the shadows produce eclipses.

From our point of view in space, we would see the orbit of the moon precess like a hubcap spinning

on the ground. This precession is caused mostly by the gravitational influence of the sun, and it makes the line of nodes rotate once every 18.6 years. People back on Earth see the nodes slipping westward along the ecliptic 19.4° per year, and the sun takes only 346.62 days (an **eclipse year**) to return to a node. This means that, according to our calendar, the eclipse seasons begin about 19 days earlier every year (▌Figure 3-18).

The cyclic pattern of eclipses shown in Figure 3-18 makes eclipse prediction simple. We know that the eclipse seasons occur 19 days earlier each year because the moon's orbit precesses. Any new moon during an eclipse season will cross in front of the sun and cause

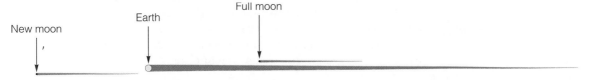

FIGURE 3-17

Umbral shadows of Earth and the moon. Notice how easy it is for the shadows to miss their mark at full moon and at new moon and fail to produce eclipses. That is, it is easy for the moon to reach full phase and not enter Earth's shadow. It is also easy for the moon to reach new phase and not cast its shadow on Earth. (The diameters of Earth and the moon are exaggerated by a factor of 2 for clarity.)

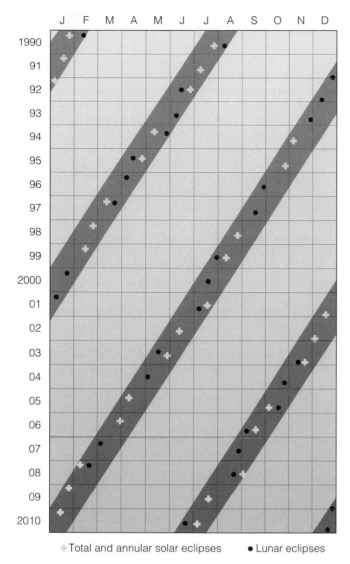

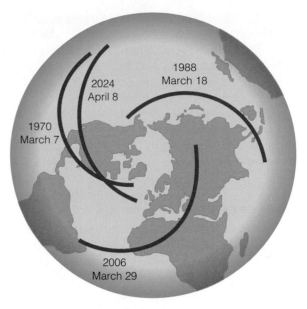

FIGURE 3-19

The saros cycle at work. The total solar eclipse of March 7, 1970, recurred after 18 years $11\frac{1}{3}$ days over the Pacific Ocean. After another interval of 18 years $11\frac{1}{3}$ days, the same eclipse will be visible from Asia and Africa. After a similar interval, the eclipse will again be visible from the United States.

+ Total and annular solar eclipses   • Lunar eclipses

FIGURE 3-18

A calendar of eclipse seasons. Each year the eclipse seasons begin about 19 days earlier. Any new moon or full moon that occurs during an eclipse season results in an eclipse. Not all eclipses are shown here.

a solar eclipse, and any full moon will enter Earth's shadow and cause a lunar eclipse.

THE SAROS CYCLE

Ancient astronomers could predict eclipses in a crude way using the eclipse seasons, but they could have been much more accurate if they recognized that eclipses occur following certain patterns. The most important of these is the **saros cycle** (sometimes referred to simply as the saros). After one saros cycle of 18 years $11\frac{1}{3}$ days, the pattern of eclipses repeats. In fact, *saros* comes from a Greek word that means "repetition."

The eclipses repeat because, after one saros cycle, the moon and the nodes of its orbit return to the same place with respect to the sun. One saros contains 6585.321 days, which is equal to 223 lunar months.

Therefore, after one saros cycle the moon is back to the same phase it had when the cycle began. But one saros is nearly equal to 19 eclipse years. After one saros cycle, the sun has returned to the same place it occupied with respect to the nodes of the moon's orbit when the cycle began. If an eclipse occurs on a given day, then 18 years $11\frac{1}{3}$ days later the sun, the moon, and the nodes of the moon's orbit return to nearly the same relationship, and the eclipse occurs all over again.

Although the eclipse repeats almost exactly, it is not visible from the same place on Earth. The saros cycle is one-third of a day longer than 18 years 11 days. When the eclipse recurs, Earth will have rotated one-third of a turn farther east, and the eclipse will occur eight hours of longitude west of its earlier location (❙ Figure 3-19). Thus, after three saros cycles—a period of 54 years 1 month—the same eclipse occurs in the same part of Earth.

One of the most famous predictors of eclipses was Thales of Miletus (about 640–546 BC), who supposedly learned of the saros cycle from the Chaldeans, who had discovered it. No one knows which eclipse Thales predicted, but some scholars suspect the eclipse of May 28, 585 BC. In any case, the eclipse occurred at the height of a battle between the Lydians and the Medes, and the mysterious darkness in midafternoon so startled the two factions that they concluded a truce.

In fact, many historians doubt that Thales actually predicted the eclipse. It would have been very difficult to gather enough information about past eclipses, because total solar eclipses are rare. If you stay in one city, you will see a total solar eclipse about once in 360 years. Also, 585 BC is very early for the Greeks to have known of the saros cycle. The important point is not that Thales did it, but that he could have done it. If he had had records of past eclipses of the sun visible from the area, he could have discovered that they tended to recur with a period of 54 years 1 month (three saros cycles). Indeed, he could have predicted the eclipse without ever understanding what the sun and moon were or how they moved.

REVIEW CRITICAL INQUIRY

Why can't two successive full moons be totally eclipsed?
A total lunar eclipse occurs when the moon passes through Earth's shadow, and that can happen only when the sun is near one node and the moon crosses the other node. Earth's shadow always points toward the ecliptic exactly opposite the sun, so most of the time the moon will pass north or south of Earth's shadow, and there will be no eclipse. Only when the shadow is near one of the nodes of the moon's orbit can the moon enter the shadow and be eclipsed.

An eclipse season for a total lunar eclipse is only 22 days long. If the moon crosses a node more than 11 days before or after the sun crosses the other node, there will be no eclipse. The moon takes 29.5 days to go from one full moon to the next, so if one full moon is totally eclipsed, the next full moon 29.5 days later will occur too late for an eclipse.

Use your knowledge of the cycles of the sun and moon to explain why the sun can be eclipsed by two successive new moons.

Predicting eclipses isn't very hard, and many ancient peoples were probably familiar enough with the cycles of the sun and moon to know when eclipses were likely. We might call those first predictions early science, but science involves understanding nature, not just predicting events. In the next chapter, we will see how modern science was born out of astronomers' attempts to understand the cycles they saw in the sky.

SUMMARY

Because we see the moon by reflected sunlight, its shape appears to change as it orbits Earth. The lunar phases wax from new moon to first quarter to full moon and wane from full moon to third quarter to new moon. A complete cycle of lunar phases takes 29.53 days.

The moon's gravitational field exerts tidal forces on Earth that pull the ocean waters up into two bulges, one on the side of Earth facing the moon and the other on the side away from the moon. As the rotating Earth carries the continents through these bulges of deeper water, the tides rise and fall. Friction with the seabeds slows Earth's rotation, and the gravitational force the bulges exert on the moon forces its orbit to grow larger.

A lunar eclipse occurs when the moon enters Earth's shadow. If the moon becomes completely immersed in the umbra, the central shadow, the eclipse is termed total. The moon glows coppery red due to sunlight bent as it passes through Earth's atmosphere. If the moon enters the umbral shadow only partly, the eclipse is termed partial, and the reddish glow is not visible. A penumbral eclipse occurs when the moon passes through the penumbra, the region of partial shadow. Penumbral eclipses are not very noticeable.

A solar eclipse occurs when Earth passes through the moon's shadow. To see a total solar eclipse, observers must place themselves in the path of totality, the path swept by the umbra of the moon. As the umbra sweeps over the observers, they see the bright surface of the sun, the photosphere, blotted out by the moon. Then the fainter chromosphere and corona, higher layers of the sun's atmosphere, become visible. Eruptions on the solar surface, called prominences, may be visible peeking around the edge of the moon.

The corona, chromosphere, and prominences are not visible to observers outside the path of totality. Observers in the path of the penumbra of the moon will see a partial solar eclipse, but the photosphere will never be completely hidden.

If an eclipse of the sun occurs when the moon is not in the nearer part of its orbit, the moon's umbra does not reach Earth's surface. In these cases, the eclipse is not total. The moon's angular diameter is too small to cover the sun completely. Even at maximum eclipse, therefore, a bright ring, or annulus, of the photosphere is visible around the edge of the moon. This annular eclipse is not as dramatic as a total eclipse.

Ancient astronomers could predict eclipses because they occur not randomly but in a pattern. Two conditions must be met if an eclipse is to occur. First, the moon must be on or near the ecliptic. The two points where the moon crosses the ecliptic are called the nodes of the moon's orbit. The second condition is that the sun must be at or near one of the nodes. This means that eclipses can occur only at new moon or full moon during two eclipse seasons that are about a month long and that occur almost 6 months apart.

Because the moon's orbit precesses, the nodes slip westward along the ecliptic, and the eclipse seasons begin 19 days earlier each year. The moon, sun, and nodes return to the same relative positions every 18 years $11\frac{1}{3}$ days in what is called the saros cycle. After the passage of a saros cycle, the pattern of eclipses begins to repeat. This means that ancient astronomers could predict eclipses just by examining the dates of previous eclipses. The Chaldeans discovered the saros cycle, and the ancient Greeks used it. Many other primitive cultures also used this method.

NEW TERMS

sidereal period	penumbra
synodic period	total eclipse (lunar or solar)
spring tides	partial eclipse (lunar or solar)
neap tides	
umbra	penumbral eclipse

small-angle formula

path of totality

photosphere

corona

chromosphere

prominence

diamond ring effect

annular eclipse

perigee

apogee

node

eclipse season

line of nodes

eclipse year

saros cycle

REVIEW QUESTIONS

Ace ◑Astronomy™ Assess your understanding of this chapter's topics with additional quizzing and animations at **http:// astronomy.brookscole.com/seeds8e**

1. Which lunar phases would be visible in the sky at dawn? at midnight?

2. If you looked back at Earth from the moon, what phase would you see when the moon was full? new? a first-quarter moon? a waxing crescent?

3. Give examples to show how tides can alter the rotation and revolution of celestial bodies. (*Hint:* Recall from Chapter 2 the difference between rotation and revolution.)

4. Could a solar-powered spacecraft generate any electricity while passing through Earth's umbral shadow? the penumbral shadow?

5. Draw the umbral and penumbral shadows onto the diagram in the middle of page 34. Explain why lunar eclipses can occur only at full moon and solar eclipses can occur only at new moon.

6. How did lunar eclipses lead Aristotle to conclude that Earth was a sphere?

7. Why isn't the corona visible during partial or annular solar eclipses?

8. Why can't the moon be eclipsed when it is halfway between the nodes of its orbit?

9. Why aren't solar eclipses separated by one saros cycle visible from the same location on Earth?

10. How could Thales of Miletus have predicted the date of a solar eclipse without observing the location of the moon in the sky?

DISCUSSION QUESTIONS

1. If the moon were closer to Earth such that it had an orbital period of 24 hours, what would the tides be like?

2. How would eclipses be different if the moon's orbit were not tipped with respect to the plane of Earth's orbit?

3. Are there other planets in our solar system from whose surface we could see a lunar eclipse? a total solar eclipse?

4. Can you detect the Saros cycle in Figure 3-18?

PROBLEMS

1. Identify the phases of the moon if on March 21 the moon is located at the point on the ecliptic called (a) the vernal equinox, (b) the autumnal equinox, (c) the summer solstice, (d) the winter solstice.

2. Identify the phases of the moon if at sunset the moon is (a) near the eastern horizon, (b) high in the southern sky, (c) in the southeastern sky, (d) in the southwestern sky.

3. About how many days must elapse between first-quarter moon and third-quarter moon?

4. How many hours would elapse between successive high tides if tides at a given location were caused only by the sun's gravity? only by the moon's gravity? Why is there a difference?

5. How many times larger than the moon is the diameter of Earth's umbral shadow at the moon's distance? (*Hint:* See Figure 3-6.)

6. Use the small-angle formula to calculate the angular diameter of Earth as seen from the moon.

7. During solar eclipses, large solar prominences are often seen extending 5 minutes of arc from the edge of the sun's disk. How far is this in kilometers? in Earth diameters?

8. If a solar eclipse occurs on October 3: (a) Why can't there be a lunar eclipse on October 13? (b) Why can't there be a solar eclipse on December 28?

9. A total eclipse of the sun was visible from Canada on July 10, 1972. When did this eclipse occur next? From what part of Earth was it total?

10. When will the eclipse described in Problem 9 next be total as seen from Canada?

11. When will the eclipse seasons occur during the current year? What eclipse(s) will occur?

CRITICAL INQUIRIES FOR THE WEB

1. Search the Web for myths about the moon. What common themes do different cultures ascribe to the moon and its cycle of phases?

2. A total solar eclipse swept across Europe and into Asia in the summer of 1999. Search for Web pages showing photos and observations of the eclipse. Can you find Web pages that show cultural responses to the eclipse such as celebrations or religious ceremonies?

3. Most people see more total lunar eclipses in a lifetime than total solar eclipses. Why is this so? Compare the regions of visibility for a number of past and upcoming eclipses and determine which future events will be visible from your area.

4. How do the tides on Earth vary with the phases of the moon? Use the Internet to explore the range of tides for a particular location through the next few weeks. Also, look up moon phase information for that same interval. How do the tidal amplitudes correlate with moon phase during that period?

EXPLORING *THESKY*

1. How many days must elapse between new moons? Use the **Moon Phase Calendar** under the **Tools** menu to locate new moons.

2. Where would you have to go to see a total solar eclipse in the next year or two? Use the **Eclipse Finder** under the **Tools** menu to select a total solar eclipse. Check **Show Path of Totality** to see a map of Earth.

3. View the total solar eclipse you located in activity 2. In the **Eclipse Finder,** click **View** to see the sun and moon from your present location. Then use **Site Information** under the **Data** menu to change your location to a spot in the path of totality. Set the **Time Step** to 5 minutes and click the **arrows** to move the moon across the sun.

4. Locate and observe an annular eclipse.

5. Locate and observe a total lunar eclipse.

 Visit the Seeds *Foundations of Astronomy* companion Web site for critical thinking exercises, articles, and additional readings from InfoTrac College Edition, Brooks/Cole's online student library.

THE ORIGIN OF MODERN ASTRONOMY

How you would burst out laughing, my dear Kepler, if you would hear
what the greatest philosopher of the Gymnasium told the Grand Duke about me . . .

From a letter by Galileo Galilei

Chapter 4

GUIDEPOST

The sun, moon, and planets sweep out a beautiful and complex dance across the heavens. Previous chapters have described that dance; this chapter describes how astronomers learned to understand what they saw in the sky and how that changed humanity's understanding of what we are.

In learning to interpret what they saw, Renaissance astronomers invented a new way of knowing about nature, a way of knowing that we recognize today as modern science.

This chapter tells the story of heavenly motion from a cultural perspective. In the next chapter, we will give meaning to the motions in the sky by adding the ingredient Renaissance astronomers were missing—gravity.

The history of astronomy is more like a soap opera than a situation comedy. In a simple half-hour sitcom, only a few characters carry the story, but the plot of a soap opera is so complex, so filled with characters and subplots only tenuously related to one another, that the action jumps from character to character, from subplot to subplot, with almost no connection. The history of astronomy is similarly complex, filled with brilliant people who lived at different times and worked in different parts of the world.

Two subplots twine through our story. One is the human quest to understand the place of the Earth. That plot will lead us from Aristotle to Copernicus and finally to the trial of Galileo before the Inquisition. But the second subplot, the puzzle of planetary motion, is equally important to our story. That plot will involve an astronomer with a false nose and another whose mother was tried for witchcraft. We will see that the true place of the Earth could be understood only when the puzzle of planetary motion was solved.

Our story goes far beyond the birth of astronomy. The quest for the place of the Earth and the puzzle of planetary motion became, in the 16th and 17th centuries, the center of a storm of controversy over the best way to understand our world and ourselves. That conflict is now known as the scientific revolution. Thus, our story describes not only the birth of astronomy but also the birth of modern science.

Of course, astronomy had its beginnings long before the 16th century. It began when the first semihuman creature looked up at the moon and stars and wondered what they were. Since then, hundreds of generations of talented astronomers have lived and worked on this planet and left almost no record of their accomplishments. Only through archaeology can we get a hint of their insights and triumphs. Other cultures, such as ancient Greece, have left written records from which we can reconstruct the sophisticated astronomy of the ancient world.

4-1 THE ROOTS OF ASTRONOMY

Astronomy has its origin in that most noble of all human traits, curiosity. Just as modern children ask their parents what the stars are and why the moon changes, so did ancient humans ask themselves these questions. The answers, often couched in mythical or religious terms, reveal a great reverence for the order of the heavens.

Ace✺Astronomy™ The AceAstronomy icon throughout the text indicates an opportunity for you to test yourself on key concepts and to explore animations and interactions on the AceAstronomy Web site at **http://astronomy.brookscole .com/seeds8e**

ARCHAEOASTRONOMY

The study of the astronomy of ancient peoples, **archaeoastronomy,** came to public attention in 1965 when Gerald Hawkins published a book called *Stonehenge Decoded.* He reported that Stonehenge, the prehistoric ring of stones, was a sophisticated astronomical observatory.

Stonehenge, standing on Salisbury Plain in southern England, was built in stages from about 3000 BC to about 1800 BC, a period extending from the late Stone Age into the Bronze Age. Though the public is most familiar with the massive stones of Stonehenge (❚ Figure 4-1), those were added late in its history. In its first stages, Stonehenge consisted of a circular ditch slightly larger in diameter than the length of a football field, with a concentric bank just inside the ditch and a long avenue leading away toward the northeast. A massive stone, the Heelstone, stood then, as it does now, outside the ditch in the opening of the avenue.

As early as AD 1740, the English scholar W. Stukely suggested that the avenue pointed toward the rising sun at the summer solstice, but few accepted the idea. More recently, astronomers have recognized significant astronomical alignments at Stonehenge. Seen from the center of the monument, the summer-solstice sun rises behind the Heelstone. Other sight lines point toward the most northerly and most southerly risings of the moon.

The significance of these alignments has been debated. Hawkins claimed that the Stone Age people who built Stonehenge were using it as a device to predict lunar eclipses. Others have had more conservative notions, but the truth may never be known. The builders of Stonehenge had no written language and left no records of their intentions. Nevertheless, the presence of solar and lunar alignments at Stonehenge and at many other Stone Age monuments dotting England and continental Europe shows that so-called primitive peoples were paying detailed attention to the sky. The roots of astronomy lie not in sophisticated science and logic but in human curiosity and wonder.

The early inhabitants of North America were also interested in astronomy. The Big Horn Medicine Wheel,* located on a 3000-m shoulder of the Big Horn Mountains of Wyoming, was used by the Plains Indians about AD 1500–1750. This arrangement of rocks marks a 28-spoke wheel about 27 m in diameter. Piles of rocks, called cairns, mark the center and six locations on the circumference. Astronomer John Eddy has discovered that these cairns mark sight lines toward a number of important points on the eastern horizon and could have been used as a calendar to help schedule hunting, planting, harvesting, and celebrations (❚ Figure 4-2).

Many other American Indian sites have astronomical alignments. More than three dozen medicine wheels are known, although most do not have the sophistication of the Big Horn Medicine Wheel. The Moose Mountain Wheel in Saskatchewan, Canada, for in-

*Medicine is used here to mean magical power.

a

b

FIGURE 4-1

(a) A stamp issued to mark the return of Comet Halley shows the central horseshoe of upright stones at Stonehenge. (b) The best-known astronomical alignment at Stonehenge is the summer solstice sun rising over the Heelstone. (c) Although a number of astronomical alignments have been found at Stonehenge, experts debate their significance.

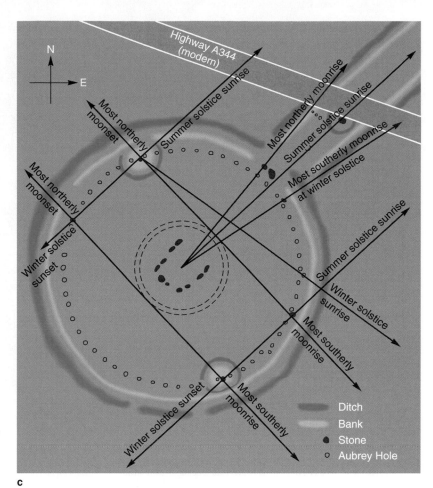

c

stance, seems to have been in use as early as AD 100. Alignments at some Mound Builder sites of the midwest show that these peoples were familiar with the sky. In the southwest, the Anasazi Indian ruins in Chaco Canyon, New Mexico, among others, have alignments that point toward the rising and setting of the sun at the summer and winter solstices.

Primitive astronomy also flourished in Central and South America. The Mayan and Aztec empires built many temples aligned with the solstice rising of the sun and with the extreme points where Venus rose and set. The Caracol temple in the Yucatán is a good example (❙ Figure 4-3). It is a circular tower containing complicated passageways and a spiral staircase (thus the name *Caracol*—"the snail's shell"). The tubelike windows at the top of the tower point toward the equinox sunset point and the most northerly and most southerly setting points of Venus. Unfortunately, only about one-third of the tower top survives, so the directions of any other windows are forever lost.

Archaeoastronomers are uncovering the remains of ancient astronomical observatories around the world. Some temples in the jungles of southeast Asia, for instance, have astronomical alignments.

Other scholars are looking not at temples but at small artifacts from thousands of years ago. Scratches on certain bone and stone implements seem to follow a pattern and may be an attempt to keep a record of the phases of the moon (❙ Figure 4-4). Some scientists contend that humanity's first attempts at writing were stimulated by a desire to record and predict lunar phases.

Archaeoastronomy is uncovering the earliest roots of astronomy and simultaneously revealing some of the first human efforts at systematic inquiry. The most important lesson of archaeoastronomy is that humans don't have to be technologically sophisticated to admire and study the universe.

One thing about archaeoastronomy is especially sad. Although we are learning how ancient people observed the sky, we may never know what they thought about their universe. Many had no written language. In other cases, the written record has been lost. Dozens, perhaps hundreds, of beautiful Mayan manuscripts, for instance, were burned by Spanish missionaries who believed that the books were the work of Satan. Only four of these books have survived, and all four contain astronomical references. One contains sophisticated tables that allowed the Maya to predict the motion of

FIGURE 4-2

The Big Horn Medicine Wheel in Wyoming was built by Native Americans before the westward spread of European settlers. Its rock cairns appear to be aligned with the rising and setting of the summer solstice sun (red) and with the rising of three bright stars (blue). Such connections with the sky are found in many Native American structures and ceremonies. *(John A. Eddy)*

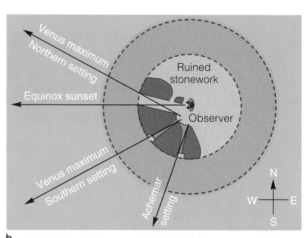

FIGURE 4-3

The Caracol at the Mayan city of Chichén Itzá (a) is a 1000-year-old observatory. The remaining windows in the partially ruined tower contain sight lines that point toward the equinox sunset, the setting places of Venus at its most northerly and most southerly positions, and the setting place of the star Achernar (b). Various evidence suggests that the Caracol was directly associated with the worship of the Venus god, Quetzalcoatl. *(Anthony Aveni)*

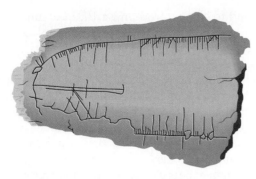

FIGURE 4-4
A fragment of a 27,000-year-old mammoth tusk found at Gontzi in Ukraine contains scribe marks on its edge, simplified in this drawing. These markings have been interpreted as a record of four cycles of lunar phases. Although controversial, such finds suggest that some of the first human attempts at recording events in written form were stimulated by astronomical phenomena

Venus and eclipses of the moon. We will never know what was burnt.

The fate of the Mayan books illustrates one reason why histories of astronomy usually begin with the Greeks. Some of their writing has survived, and we can discover what they thought about the shape and motion of the heavens.

THE ASTRONOMY OF GREECE

Greek astronomy was based on the astronomy of Babylonia and Egypt, but these astronomies were heavily influenced by religion and astrology. Those ancient astronomers studied the motions of the heavens as a way of worship and divination. The Greek astronomers studied astronomy in an entirely new way—they tried to understand the universe.

This new attitude toward the heavens, a truly scientific attitude, was made possible by two early Greek philosophers. Thales of Miletus (c. 624–547 BC) lived and worked in what is now Turkey. He taught that the universe is rational and that the human mind can understand why the universe works the way it does. This view contrasts sharply with those of earlier cultures, which believed that the ultimate causes of things are mysteries beyond human understanding. To Thales and his followers, the mysteries of the universe are mysteries because they are unknown, not because they are unknowable.

The second philosopher who made the new scientific attitude possible was Pythagoras (c. 570–500 BC). He and his students noticed that many things in nature seem to be governed by geometrical or mathematical relations. Musical notes, for example, are related in a regular way to the lengths of plucked strings. This led Pythagoras to propose that all nature was underlain by musical principles, by which he meant mathematics. One result of this philosophy was the later belief that the harmony of the celestial movements produced actual music, the music of the spheres. But at a deeper level, the teachings of Pythagoras made Greek astronomers look at the universe in a new way. Thales said that the universe could be understood, and Pythagoras said that the underlying rules were mathematical.

In trying to understand the universe, Greek astronomers did something that Babylonian astronomers had never done—they tried to construct descriptions based on geometrical forms. Anaximander (c. 611–546 BC) described a universe made up of wheels filled with fire: The sun and moon are holes in the wheels through which we see the flames. Philolaus (5th century BC) argued that Earth moves in a circular path around a central fire (not the sun), which is always hidden behind a counterearth located between the fire and Earth. This was the first theory to suppose that Earth is in motion.

Plato (428–347 BC) was not an astronomer, but his teachings influenced astronomy for 2000 years. Plato argued that the reality we see is only a distorted shadow of a perfect, ideal form. If our observations are distorted, then observation can be misleading, and the best path to truth is through pure thought on the ideal forms that underlie nature.

Plato also argued that the most perfect form was the circle and that therefore all motions in the heavens should be made up of combinations of circular motion. Because the most perfect motion is uniform motion, later astronomers tried to describe the motions of the heavens using the principle of uniform circular motion.

Pythagoras had taught that Earth is a sphere and that the other heavenly bodies were divine, perfect spheres moving in perfect circles. Eudoxus of Cnidus (409–356 BC), a student of Plato, combined a system of 27 nested spheres rotating at different rates about different axes with the concept of uniform circular motion to produce a mathematical description of the motions of the universe (❙ Figure 4-5).

At the time of the Greek philosophers, it was common to refer to systems such as that of Eudoxus as descriptions of the world, where the word *world* included not only Earth but all of the heavenly spheres. The reality of these spheres was open to debate. Some thought of the spheres as nothing more than mathematical ideas that described motion in the world model, while others began to think of the spheres as real objects made of perfect celestial material. Aristotle, for example, seems to have thought of the spheres as real.

Aristotle (384–322 BC) taught and wrote on philosophy, history, politics, ethics, poetry, drama, and so on (❙ Figure 4-6). Because of his sensitivity and insight, he became the great authority of antiquity, and astronomers for almost 2000 years cited him as their authority in adopting the Greek model of the universe.

Much of what Aristotle wrote about scientific subjects was wrong, but he was not a scientist. He was trying to understand the universe by creating a system of

FIGURE 4-5

The spheres of Eudoxus explain the motions in the heavens by means of nested spheres rotating about various axes at different rates. Earth is located at the center.

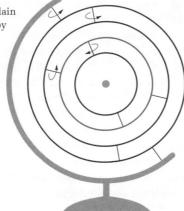

spheres had to be centered on Earth. Like most other Greek philosophers, Aristotle viewed the universe as a perfect heavenly machine that was not many times larger than Earth itself.

About a century after Aristotle, the Alexandrian philosopher Aristarchus proposed a theory that Earth rotated on its axis and revolved around the sun. This theory is, of course, correct, but most of the writings of Aristarchus were lost, and his theory was not well known. Later astronomers rejected any suggestion that Earth could move, because it conflicted with the teachings of Aristotle.

Aristotle had taught that Earth had to be a sphere because it always casts a round shadow during lunar eclipses, but he could only estimate its size. About 200 BC, Eratosthenes, working in the great library in Alexandria, found a way to calculate Earth's radius. He learned from travelers that the city of Syene (Aswan) in southern Egypt contained a well into which sunlight shone vertically on the day of the summer solstice. Thus, the sun was at the zenith at Syene, but on that same day in Alexandria, he noted that the sun was one-fiftieth of the circumference of the sky (about 7°) south of the zenith. Because sunlight comes from such a great distance, its rays arrive at Earth traveling almost parallel. Thus, Eratosthenes could conclude from simple geometry (▌Figure 4-7) that the distance from Alexandria to Syene was one-fiftieth Earth's circumference.

To find Earth's circumference, Eratosthenes had to know the distance from Alexandria to Syene. Travelers told him it took 50 days to cover the distance, and he knew that a camel can travel about 100 stadia per day. Thus, the total distance was about 5000 stadia. If 5000 stadia is one-fiftieth Earth's circumference, then Earth must be 250,000 stadia around, and, dividing by 2π, Eratosthenes found Earth's radius to be 40,000 stadia.

We don't know how accurate Eratosthenes was. The stadium (singular of *stadia*) had different lengths in ancient times. If we assume 6 stadia to the kilometer, then Eratosthenes's result was too big by only 4 percent. If he used the Olympic stadium, his result was 14 percent too big. In any case, this was a much better measurement of Earth's radius than Aristotle's estimate, which was only about 40 percent of the true radius.

The greatest of the ancient observers was Hipparchus (see Figure 2-5), who lived during the 2nd century BC, about two centuries after Aristotle. He is usually credited with the invention of trigonometry, he compiled the first star catalog, and he discovered precession (Chapter 2). Instead of describing the motion of the sun and moon using nested spheres, as most Greek philosophers did, Hipparchus proposed that the sun and moon traveled around circles with Earth near, but not at, their centers. These off-center circles are now known as **eccentrics.** Hipparchus recognized that he could produce the same motion by having the sun, for instance, travel around a small circle that followed a larger circle around Earth. The compounded circular

formal philosophy. Modern science depends on evidence and hypothesis, but scientific methods had not been invented at the time of Aristotle. Rather, he attempted to use the most basic observations combined with first principles (ideas that he believed were obviously true) to understand the world. The perfection of the heavens was, for Aristotle, a first principle.

Aristotle believed that the universe was divided into two parts—Earth, corrupt and changeable; and the heavens, perfect and immutable. Like most of his predecessors, he believed that Earth was the center of the universe, so his model is called a geocentric (Earth-centered) universe. The heavens surrounded Earth, and he added more crystalline spheres to bring the total to 55. The lowest sphere, that of the moon, marked the boundary between the imperfect region of Earth and the perfection of the celestial realm above the moon.

Because he believed Earth to be immobile, he had to make these spheres whirl westward around Earth each day and move with respect to one another to produce the motions of the sun, moon, and planets. Because his model was geocentric, he taught that Earth could be the only center of motion. All of his whirling

FIGURE 4-6

Aristotle, honored on this Greek stamp, wrote on such a wide variety of subjects and with such deep insight that he became the great authority on all matters of learning. His opinions on the nature of Earth and the sky were widely accepted for almost two millennia.

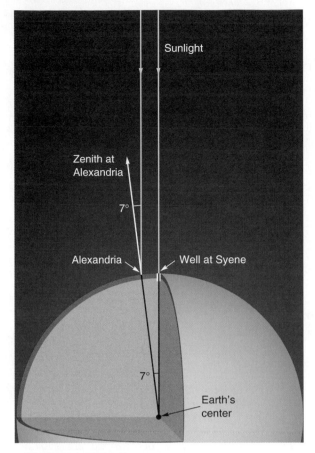

Sunlight

Zenith at
Alexandria

7°

Alexandria

Well at Syene

7°

Earth's
center

ACTIVE FIGURE 4-7

On the day of the summer solstice, sunlight fell to the bottom
of a well at Syene, but the sun was about one-fiftieth of a circle
(7°) south of the zenith at Alexandria. From this Eratosthenes
was able to calculate Earth's radius.

Ace✸Astronomy™ Go to AceAstronomy and click Active Figures to see
"Eratosthenes' Experiment."

motion that he devised became the key element in the
masterpiece of the last great astronomer of classical
times, Ptolemy.

THE PTOLEMAIC UNIVERSE

Claudius Ptolemaeus was one of the great astronomer-
mathematicians of antiquity. His nationality and birth
date are unknown, but he lived and worked in the
Greek settlement at Alexandria about AD 140. He en-
sured the survival of Aristotle's universe by fitting to it
a sophisticated mathematical model.

Study "▮ The Ancient Universe" on pages 58 and
59 and notice three important ideas. First, ancient
philosophers and astronomers accepted as a first principle that Earth was located at the center of the universe. Although a few writers mentioned the possibility that Earth might move, they did so mostly to point
out the obvious objections.

Second, notice that the ancients accepted without
question that the heavens were perfect and that Earth
was imperfect. This was a first principle of ancient as-

tronomy, and it led to the conclusion that the motions
of the heavens must be made up of combinations of circles rotating uniformly.

The third thing to notice is how Ptolemy weakened
the first principles of Plato and Aristotle by moving
Earth slightly off-center and adjusting the speed of the
planets as they circled Earth. Ptolemy lived roughly five
centuries after Aristotle, and although Ptolemy believed
in the Aristotelian universe, he was interested in a different problem. He was a brilliant mathematician, and
he was mainly interested in creating a mathematical description of the motion of the planets. For him, first
principles took second place to mathematical precision.

Aristotle's universe, as embodied in the mathematics of Ptolemy, dominated ancient astronomy, but it
was wrong. The planets don't follow circles at uniform
speeds. At first the Ptolemaic system predicted the positions of the planets well; but, as centuries passed, errors accumulated. A watch that gains only one second
a year will keep time well for many years, but the error
gradually accumulates. After a century the watch will
be 100 seconds fast. So, too, did the errors in the Ptolemaic system gradually accumulate as the centuries
passed. Islamic and later European astronomers tried
to update the system, computing new constants and adjusting epicycles. In the middle of the 13th century, a
team of astronomers supported by King Alfonso X of
Castile studied the *Almagest* for 10 years. Although
they did not revise the theory very much, they simplified the calculation of the positions of the planets using
the Ptolemaic system and published the result as the
Alfonsine Tables, the last great attempt to make the
Ptolemaic system of practical use.

REVIEW CRITICAL INQUIRY

**How did the astronomy of Hipparchus and Ptolemy violate
the principles of the early Greek philosophers Plato and
Aristotle?**

Hipparchus and Ptolemy lived very late in the history of
classical astronomy, and they concentrated more on the
mathematical problems and less on philosophical principles. They replaced the perfect spheres of Plato with nested
circles in the form of epicycles and deferents. Earth was
moved slightly away from the center of the deferent, so their
models of the universe were not exactly geocentric, and the
epicycles moved uniformly only as seen from the equant.
The celestial motions were no longer precisely uniform, and
the principles of geocentrism and uniform circular motion
were weakened.

The work of Hipparchus and Ptolemy led eventually to a
new understanding of the heavens, but first astronomers
had to abandon the principle of uniform circular motion.
The great philosopher Plato had argued for uniform circular
motion from what seemed the most basic of first principles.
What were those first principles?

The Ancient Universe

For 2000 years, the minds of astronomers were shackled by a pair of ideas. The Greek philosopher Plato argued that the heavens were perfect. Because the only perfect geometrical shape is a circle and the only perfect motion is uniform motion, Plato concluded that all motion in the heavens must be made up of combinations of circles turning at uniform rates. This was called uniform circular motion.

Plato's student Aristotle argued that Earth was imperfect and lay at the center of the universe. That is, he argued for a geocentric universe. He devised a model universe with 55 spheres turning at different rates and at different angles to carry the seven known planets (the moon, Mercury, Venus, the sun, Mars, Jupiter, and Saturn) across the sky.

Aristotle was known as the greatest philosopher in the ancient world, and his authority, lasting for 2000 years, chained the minds of astronomers and forced them to expect the universe to be geocentric and to move in uniform circular motion.

From *Cosmographica* by Peter Apian (1539).

Seen by left eye

Seen by right eye

Ancient astronomers believed that Earth did not move because they saw no **parallax,** the apparent motion of an object because of the motion of the observer. To demonstrate parallax, close one eye and cover a distant object with your thumb held at arm's length. Switch eyes, and your thumb appears to shift position. If Earth moves, ancient astronomers reasoned, we should see the sky from different locations at different times of the year, and we should see parallax distorting the shapes of the constellations. They saw no parallax, so they concluded Earth could not move. Actually, the parallax of the stars is too small to see with the unaided eye.

Planetary motion was a big problem for ancient astronomers. In fact, the word *planet* comes from the Greek word for "wanderer," referring to the eastward motion of the planets against the background of the fixed stars. The planets did not, however, move at a constant rate, and they could occasionally stop and move westward for a few months before resuming their eastward motion. This backward motion is called **retrograde motion.**

East

Ecliptic

Oct. 3, 2005

Dec. 12, 2005

Taurus

West

Position of Mars at 5-day intervals

Orion

Every 2.14 years, Mars passes through a retrograde loop. Shown here is its path through Orion and Taurus during the fall of 2005.

Simple uniform circular motion centered on Earth could not explain retrograde motion, so ancient astronomers combined uniformly rotating circles much like gears in a machine to try to reproduce the motion of the planets.

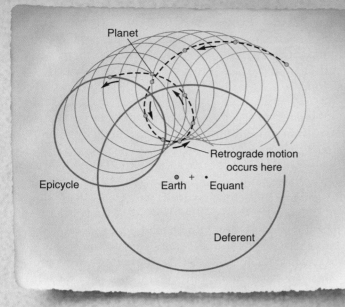

Uniformly rotating circles were key elements of ancient astronomy. Claudius Ptolemy created a mathematical model of the Aristotelian universe in which the planet followed a small circle called the **epicycle** which slid around a larger circle called the **deferent.** By adjusting the size and rate of rotation of the circles, he could approximate the retrograde motion of a planet.

To adjust the speed of the planet, Ptolemy supposed that Earth was slightly off center and that the center of the epicycle moved such that it appeared to move at a constant rate as seen from the point called the **equant.**

To further adjust his model, Ptolemy added small epicycles riding on top of larger epicycles (not shown here), producing a highly complex model.

Ace ⊗ Astronomy™

Go to AceAstronomy and click Active Figures to see "Epicycles" and watch how the epicycle moved around the deferent.

Ptolemy's great book *Mathematical Syntaxis* (c. AD 140) contained the details of his model. Islamic astronomers preserved and studied the book through the Middle Ages, and they called it *Al Magisti* (The Greatest). When the book was found and translated from Arabic to Latin in the 12th century, it became known as *Almagest.*

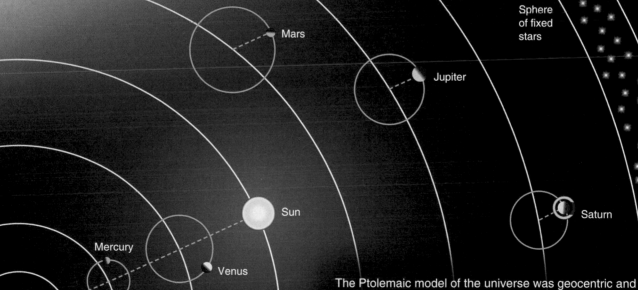

The Ptolemaic model of the universe was geocentric and based on uniform circular motion. Note that Mercury and Venus were treated differently from the rest of the planets. The centers of the epicycles of Mercury and Venus had to remain on the Earth–Sun line as the sun circled Earth through the year.

Equants and smaller epicycles are not shown here. Some versions contained nearly 100 epicycles as generations of astronomers tried to fine-tune the model to better reproduce the motion of the planets.

Notice that this modern illustration shows rings around Saturn and sunlight illuminating the globes of the planets, features that could not be known before the invention of the telescope.

Ptolemy was the last of the great classical astronomers, and his work dominated astronomical thought for almost 1500 years. The collapse of the Ptolemaic model and the rise of a new model of the universe make an exciting story, but it is not just the story of an astronomical idea. It includes the invention of science as a new way of knowing and understanding what we are and where we are.

4-2 THE COPERNICAN REVOLUTION

Nicolaus Copernicus (Figure 4-8) triggered an Earth-shaking revision in human thought by proposing that the universe is not centered on Earth, but rather is centered on the sun. Such a model is called a **heliocentric universe.** That idea eventually brought Galileo before the Inquisition and changed forever how we think of our world and ourselves. This Copernican Revolution was much more than an upheaval in astronomy; it marks the birth of modern science.

COPERNICUS THE REVOLUTIONARY

When Copernicus proposed that the universe was heliocentric, he risked controversy. According to Aristotle, the most perfect region was the starry sphere, and the most imperfect was Earth's center. Thus, the classical geocentric universe matched the commonly held Christian geometry of heaven and hell, and anyone who criticized the geometry of the Aristotelian universe challenged Christian belief and thus risked a charge of heresy.

Throughout his life, Copernicus was associated with the church. His uncle, by whom he was raised and educated, was an important bishop in Poland; and, after studying canon law and medicine in some of the finest universities in Europe, Copernicus became a canon of the church at the unusually young age of 24. He was secretary and personal physician to his powerful uncle for 15 years. When his uncle died, Copernicus moved to quarters adjoining the cathedral in Frauenburg. No doubt his long association with the church added to his reluctance to publish controversial ideas.

He first wrote about a heliocentric universe sometime before 1514 in a short pamphlet that he distributed in handwritten form and in some cases anonymously. Over the years, Copernicus worked on his book *De Revolutionibus Orbium Coelestium* (Figure 4-9a), but he hesitated to publish it even though other astronomers knew of his theories.

This was a time of rebellion—Martin Luther was speaking harshly about fundamental church teachings, and others, scholars and scoundrels, questioned the authority of the church. Even astronomy could stir argument; moving Earth from its central place was a controversial and perhaps heretical idea. Copernicus was concerned about how his ideas would be received, but in 1540 he allowed the visiting astronomer Joachim Rheticus (1514–1576) to publish an account of the Copernican universe in Rheticus's book *Prima Narratio* (*First Narrative*). In 1542, Copernicus sent the manuscript for *De Revolutionibus* off to be printed. He died in the spring of 1543 before the printing was completed.

The most important idea in the book was the placement of the sun at the center of the universe. That single innovation had an astonishing consequence—the retrograde motion of the planets was immediately explained in a straightforward way without the large epicycles that Ptolemy used. In the Copernican system, Earth moves faster along its orbit than the planets that lie further from the sun. Consequently, Earth periodically overtakes and passes these planets, and they appear to slow and fall behind (Figure 4-10). Because the planetary orbits do not lie in precisely the same plane, a planet does not resume its eastward motion in precisely the same path it followed earlier. Consequently, it describes a loop whose shape depends on the angle between the orbital planes.

Copernicus could explain retrograde motion without epicycles, and that was impressive. But he could not do away with epicycles completely. Copernicus, a classical astronomer, had tremendous respect for the old concept of uniform circular motion. In fact, he objected strongly to Ptolemy's use of the equant. It seemed arbitrary to Copernicus, a direct violation of the elegance of Aristotle's philosophy of the heavens. Copernicus called equants "monstrous" in that they violated both geocentrism and uniform circular motion. In his model, Copernicus returned to a strong belief in uniform circular motion. Although he did not need epicycles to explain retrograde motion, Copernicus discovered that the sun, moon, and planets suffered small variations in their motions—variations that he could not explain

FIGURE 4-8
Copernicus proposed that the sun and not Earth was the center of the universe. Notice the heliocentric model on this stamp issued in 1973 to commemorate the 500th anniversary of his birth.

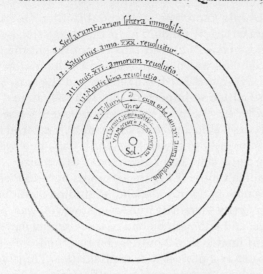

NICOLAI COPERNICI

net, in quo terram cum orbe lunari tanquam epicyclo contineri diximus. Quinto loco Venus nono mense reducitur. Sextum denisq locum Mercurius tenet, octuaginta dierum spacio circu currens. In medio uero omnium residet Sol. Quis enim in hoc

I. Stellarum Fixarum sphæra immobilis.

pulcherrimo templo lampadem hanc in alio uel meliori loco po neret, quàm unde totum simul possit illuminare? Siquidem non inepte quidam lucernam mundi, alij mentem, alij rectorem uo cant. Trimegistus uisibilem Deum, Sophoclis Electra intuente omnia. Ita profecto tanquam in solio regali Sol residens circum agentem gubernat Astrorum familiam. Tellus quoque minime fraudatur lunari ministerio, sed ut Aristoteles de animalibus ait, maximam Luna cum terra cognatione habet. Concipit interea à Sole terra, & impregnatur annuo partu. Inuenimus igitur sub hac

a

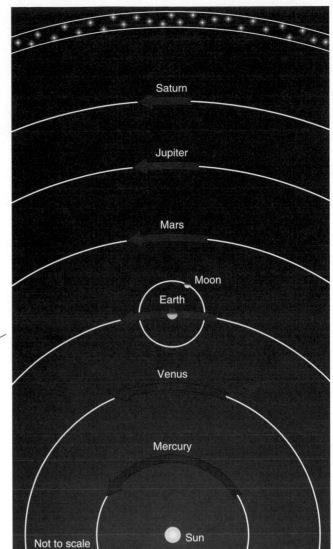

b

FIGURE 4-9

(a) The Copernican universe as reproduced in *De Revolutionibus*. Earth and all the known planets revolve in separate circular orbits about the sun (Sol) at the center. The outermost sphere carries the immobile stars of the celestial sphere. Notice the orbit of the moon around Earth. *(Yerkes Observatory)* (b) The model was elegant not only in its arrangement of the planets but in their motions. Orbital velocities (red arrows) decreased from Mercury, the fastest, to Saturn, the slowest. Compare the elegance of this model with the complexity of the Ptolemaic model on page 59.

with the concept of uniform circular motion centered on the sun. Today, we recognize those variations as typical of objects following elliptical orbits, but Copernicus held firmly to uniform circular motion, so he had to use small epicycles to produce these minor variations in the motions of the sun, moon, and planets.

Because Copernicus imposed uniform circular motion on his model, it could not accurately predict the motions of the planets. The *Prutenic Tables* (1551) were based on the Copernican model, and they were not significantly more accurate than the *Alfonsine Tables* (1251), which were based on Ptolemy's model. Both could be in error by as much as 2°, four times the angular diameter of the full moon.

The Copernican *model* was inaccurate, but the Copernican *hypothesis* that the universe is heliocentric

was correct. There are probably a number of reasons why the hypothesis gradually won acceptance in spite of the inaccuracy of the epicycles and deferents. The most important factor may be the elegance of the idea. Placing the sun at the center of the universe produced a symmetry among the motions of the planets that was pleasing to the eye and to the intellect (Figure 4-9b). All of the planets moved in the same way at speeds that were simply related to their distance from the sun. In the Ptolemaic model, Venus and Mercury were treated differently from the other planets. In the Copernican model, all planets were treated the same. Thus, the model may have won support not for its accuracy, but for its elegance.

The most astonishing consequence of the Copernican hypothesis was not what it said about the sun, but

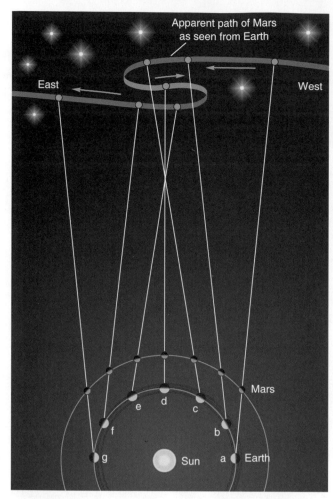

FIGURE 4-10

The Copernican explanation of retrograde motion. As Earth overtakes Mars (a–c), Mars appears to slow its eastward motion. As Earth passes Mars (d), Mars appears to move westward. As Earth draws ahead of Mars (e–g), Mars resumes its eastward motion against the background stars. Compare with the illustration of retrograde motion on page 58. The positions of Earth and Mars are shown at equal intervals of one month.

what it said about Earth. By placing the sun at the center, Copernicus made Earth move along an orbit just as the other planets did. By making Earth a planet, Copernicus revolutionized humanity's view of our place in the universe and triggered a controversy that would eventually bring Galileo before the Inquisition.

Although astronomers throughout Europe read and admired *De Revolutionibus,* they did not usually accept the Copernican hypothesis. The mathematics was elegant, and the astronomical observations and calculations were of tremendous value. Yet few astronomers believed, at first, that the sun actually was the center of the planetary system and that Earth moved. How the Copernican hypothesis became gradually recognized as correct has been named the Copernican Revolution because it involved not just the adoption of a new idea,

but a total revolution in the way astronomers thought about the place of the Earth (Window on Science 4-1).

GALILEO THE DEFENDER

Most people know two facts about Galileo, and both facts are wrong. We should begin our discussion by getting those facts right: Galileo did not invent the telescope, and he was not condemned by the Inquisition for believing that Earth moved around the sun. Then why is Galileo so important that in 1979, almost 400 years after his trial, the Vatican reopened his case? As we discuss Galileo, we will discover that his trial concerned not just the place of the Earth but a new way of finding truth, a method known today as science.

Galileo Galilei (❙ Figure 4-11) was born in Pisa (in what is now Italy) in 1564, and he studied medicine at the university there. His true love, however, was mathematics, and although he had to leave school early because of financial difficulties, he returned only four years later as a professor of mathematics. Three years later, he became professor of mathematics at the university at Padua. He remained there for 18 years.

During this time, Galileo seems to have adopted the Copernican model, although he admitted in a 1597 letter to Kepler, a German astronomer, that he did not support Copernicanism publicly because of the criticism such a declaration would bring. It was the telescope that drove Galileo to publicly defend the heliocentric model.

Galileo did not invent the telescope. It was apparently invented around 1608 by lens makers in Holland. Galileo, hearing descriptions in the fall of 1609, was able to build telescopes in his workshop. Galileo was not the first person to look at the sky through a telescope, but he was the first person to observe the sky systematically and apply his observations to the theoretical problem of the day—the place of the Earth.

What Galileo saw through his telescopes was so amazing he rushed a small book into print. *Sidereus Nuncius* (*The Sidereal Messenger*) reported three major discoveries. First, the moon was not perfect. It had mountains and valleys on its surface, and Galileo used the shadows to calculate the height of the mountains. The Ptolemaic model held that the moon was perfect, but Galileo showed that it was not only imperfect, it was a world like Earth.

The second discovery reported in the book was that the Milky Way was made up of a myriad of stars too faint to see with the unaided eye. While intriguing, this discovery could not match the third discovery. Galileo's telescope revealed four new "planets" circling Jupiter, planets known today as the Galilean moons of Jupiter.

The moons of Jupiter supported the Copernican model over the Ptolemaic model. Critics of Copernicus

Creating New Ways to See Nature: Scientific Revolutions

The Copernican Revolution is often cited as the perfect example of a scientific revolution. Over a few decades, astronomers abandoned a way of thinking about the universe that was almost 2000 years old and adopted a new set of ideas and assumptions. The American philosopher of science Thomas Kuhn has referred to a commonly accepted set of scientific ideas and assumptions as a scientific **paradigm.** Thus, the pre-Copernican astronomers had a geocentric paradigm that included uniform circular motion and the perfection of the heavens. That paradigm survived for many centuries until a new generation of astronomers was able to overthrow the old paradigm and establish a new paradigm that included heliocentrism and Earth's motion.

A scientific paradigm is powerful because it shapes our perceptions. It determines what we judge to be important questions and what we judge to be significant evidence. Thus, it is often difficult for us to recognize how our paradigms limit what we can understand. For example, the geocentric paradigm contained problems that seem obvious to us, but because astronomers before Copernicus lived and worked inside that paradigm, they had difficulty seeing the problems. Overthrowing an outdated paradigm is not easy, because we must learn to see nature in an entirely new way. Galileo, Kepler, and Newton saw nature from a new paradigm that would have been almost incomprehensible to astronomers of earlier centuries.

We can find examples of scientific revolutions in many fields, including biology, geology, genetics, and psychology. These scientific revolutions have been difficult and controversial because they have involved the overthrow of accepted paradigms. But that is why scientific revolutions are exciting. They give us not just a new idea or a new theory, but an entirely new insight into how nature works—a new way of seeing the world.

---

had said Earth could not move because the moon would be left behind. But Jupiter moved yet kept its satellites, so Galileo's discovery proved that Earth, too, could move and keep its moon. Also, the Ptolemaic model included the Aristotelian belief that all heavenly motion was centered on Earth at the center of the universe. Galileo showed that Jupiter's moons revolve around Jupiter, so there could be other centers of motion.

Sometime after *Sidereus Nuncius* was published, Galileo made another discovery that made Jupiter's moons even stronger evidence supporting the Copernican universe. Eventually, he was able to measure the

FIGURE 4-11
Galileo, remembered as the great defender of Copernicanism, also made important discoveries in the physics of motion. He is honored here on an Italian 2000-lira note. The reverse side shows one of his telescopes at lower right and a modern observatory above it.

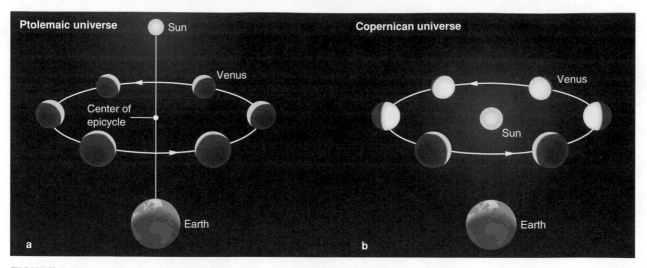

FIGURE 4-12

(a) If Venus moved in an epicycle centered on the Earth–sun line (see page 59), it would always appear as a crescent. (b) Galileo's telescope showed that Venus goes through a full set of phases, proving that it must orbit the sun.

orbital periods of the four moons, and he found that the innermost moon had the shortest period and that the moons further from Jupiter had proportionally longer periods. In this way, Jupiter's moons made up a harmonious system ruled by Jupiter, just as the planets in the Copernican universe were a harmonious system ruled by the sun (Figure 4-9b). The similarity isn't proof, but Galileo saw it as an argument that the solar system was sun centered and not Earth centered.

Soon after *Sidereus Nuncius* was published, Galileo made two additional discoveries. When he observed the sun, he discovered sunspots, raising the suspicion that the sun was less than perfect. Further, by noting the movement of the spots, he concluded that the sun was a sphere and that it rotated on its axis. When he observed Venus, Galileo saw that it was going through phases like those of the moon. In the Ptolemaic model, Venus moves around an epicycle centered on a line between Earth and the sun. Thus, it would always be seen as a crescent (▌Figure 4-12a). But Galileo saw Venus go through a complete set of phases, proving that it did indeed revolve around the sun (Figure 4-12b).

Sidereus Nuncius was very popular and made Galileo famous. He became chief mathematician and philosopher to the Grand Duke of Tuscany in Florence. In 1611, Galileo visited Rome and was treated with great respect. He had long, friendly discussions with the powerful Cardinal Barberini, but he also made enemies. Personally, Galileo was outspoken, forceful, and sometimes tactless. He enjoyed debate, but most of all he enjoyed being right. Thus, in lectures, debates, and letters he offended important people who questioned his telescopic discoveries.

By 1616, Galileo was the center of a storm of controversy. Some critics said he was wrong, and others said he was lying. Some refused to look through a tele-scope lest it mislead them, and others looked and claimed to see nothing (hardly surprising given the awkwardness of those first telescopes) (▌Figure 4-13). Pope Paul V decided to end the disruption, so when Galileo visited Rome in 1616 Cardinal Bellarmine interviewed him privately and ordered him to cease debate. There is some controversy today about the nature of Galileo's instructions, but he did not pursue astronomy for some years after the interview. Books relevant to Copernicanism, including *De Revolutionibus,* were banned.

In 1621 Pope Paul V died, and his successor, Pope Gregory XV, died in 1623. The next pope was Galileo's friend Cardinal Barberini, who took the name Urban VIII. Galileo rushed to Rome hoping to have the prohibition of 1616 lifted; and, although the new pope did not revoke the orders, he did encourage Galileo. Thus, Galileo began to write his great defense of the Copernican model, finally completing it on December 24, 1629. After some delay, the book was approved by both the local censor in Florence and the head censor of the Vatican in Rome. It was printed in February 1632.

Called *Dialogo Dei Due Massimi Sistemi* (*Dialogue Concerning the Two Chief World Systems*), it confronts the ancient astronomy of Aristotle and Ptolemy with the Copernican model and with telescopic observations as evidence. Galileo wrote the book as a debate among three friends. Salviati, a swift-tongued defender of Copernicus, dominates the book; Sagredo is intelligent but largely uninformed. Simplicio is the dismal defender of Ptolemy. In fact, he does not seem very bright.

The publication of *Dialogo* created a storm of controversy, and it was sold out by August 1632, when the Inquisition ordered sales stopped. The book was a clear defense of Copernicus, and, either intentionally or unintentionally, Galileo exposed the pope's authority to

FIGURE 4-13

Galileo's telescope made him famous, and he demonstrated his telescope and discussed his observations with powerful people. Some thought the telescope was the work of the devil and would deceive anyone who looked. In any case, Galileo's discoveries produced intense, and in some cases angry, debate. *(Yerkes Observatory)*

ridicule. Urban VIII was fond of arguing that, as God was omnipotent, He could construct the universe in any form while making it appear to us to have a different form, and thus we could not deduce its true nature by mere observation. Galileo placed the pope's argument in the mouth of Simplicio, and Galileo's enemies showed the passage to the pope as an example of Galileo's disrespect. The pope thereupon ordered Galileo to face the Inquisition.

Galileo was interrogated by the Inquisition four times and was threatened with torture. He must have thought often of Giordano Bruno, tried, condemned, and burned at the stake in Rome in 1600. One of Bruno's offenses had been Copernicanism. But the trial did not center on Galileo's belief in Copernicanism. *Dialogo* had been approved by two censors. Rather, the trial centered on the instructions given Galileo in 1616. From his file in the Vatican, his accusers produced a record of the meeting between Galileo and Cardinal Bellarmine that included the statement that Galileo was "not to hold, teach, or defend in any way" the principles of Copernicus. Some historians believe that this document, which was signed neither by Galileo nor by Bellarmine nor by a legal secretary, was a forgery. Others suspect it may be a draft that was never used. In any case, it is possible that Galileo's true instructions were much less restrictive. But Bellarmine was dead and could not testify at Galileo's trial.

The Inquisition condemned him not for heresy but for disobeying the orders given him in 1616. On June 22, 1633, at the age of 70, kneeling before the Inquisition, Galileo read a recantation admitting his errors. Tradition has it that as he rose he whispered *"E pur si muove"* ("Still it moves"), referring to Earth. Although he was sentenced to life imprisonment, he was actually confined at his villa for the next 10 years, perhaps through the intervention of the pope. He died there on January 8, 1642, 99 years after the death of Copernicus.

Galileo was not condemned for heresy, nor was the Inquisition interested when he tried to defend Copernicanism. He was tried and condemned on a charge we might call a technicality. Then why is his trial so important that historians have studied it for almost four centuries? Why have some of the world's greatest authors, including Bertolt Brecht, written about Galileo's trial? Why in 1979 did Pope John Paul II create a commission to reexamine the case against Galileo?

To understand the trial, we must recognize that it was the result of a conflict between two ways of understanding our universe. Since the Middle Ages, scholars had taught that the only path to true understanding was through religious faith. St. Augustine (AD 354–430) wrote *"Credo ut intelligame,"* which can be translated as "Believe in order to understand." But Galileo and other scientists of the Renaissance used their own observations to try to understand the universe; and, when their observations contradicted Scripture, they assumed their observations of reality were correct (❚ Figure 4-14). Galileo paraphrased Cardinal Baronius in saying, "The Bible tells us how to go to heaven, not how the heavens go."

Various passages of Scripture seemed to contradict observation. For example, Joshua is said to have commanded the sun to stand still, not Earth to stop rotating (Joshua 10:12–13). In response to such passages, Galileo argued that we should "read the book

FIGURE 4-14
Although he did not invent it, Galileo will always be associated with the telescope because it was the source of the observational evidence he used to try to understand the universe. By depending on evidence instead of first principles, Galileo led the way to the invention of modern science as a way to know about the natural world.

of nature"—that is, we should observe the universe with our own eyes.

The trial of Galileo was not about the place of the Earth. It was not about Copernicanism. It wasn't really about the instructions Galileo received in 1616. It was about the birth of modern science as a rational way to understand our universe. The commission appointed by John Paul II in 1979, reporting its conclusions in October 1992, said of Galileo's inquisitors, "This subjective error of judgment, so clear to us today, led them to a disciplinary measure from which Galileo 'had much to suffer.'" Galileo was not found innocent in 1992 so much as the Inquisition was forgiven for having charged him in the first place.

The gradual change from reliance on personal faith to scientific observation came to a climax with the trial of Galileo. Since that time, scientists have increasingly reserved Scripture and faith for ethical guidance and personal comfort and have depended on systematic observation to describe the physical world. The final verdict of 1992 was an attempt to bring some balance to this conflict. In his remarks on the decision, Pope John

Paul II said, "A tragic mutual incomprehension has been interpreted as the reflection of a fundamental opposition between science and faith. . . . this sad misunderstanding now belongs in the past." Galileo's trial is over, but it continues to echo through history as part of our struggle to understand the place of the Earth in our universe.

REVIEW CRITICAL INQUIRY

How were Galileo's observations of the moons of Jupiter evidence against the Ptolemaic model?

Galileo presented his arguments in the form of evidence and conclusions, and the moons of Jupiter were key evidence. Ptolemaic astronomers argued that Earth could not move or it would lose its moon, but even in the Ptolemaic universe Jupiter moved and kept its moons. Evidently, Earth could move and not leave its moon behind. Furthermore, moons circling Jupiter did not fit the classical belief that all motion was centered on Earth. Obviously there could be other centers of motion. Finally, the orbital periods of the moons were related to their distance from Jupiter, just as the orbital periods of the planets were, in the Copernican system, related to their distance from the sun. This similarity suggested that the sun rules its harmonious family of planets just as Jupiter rules its harmonious family of moons.

Of all of Galileo's telescopic observations, the moons of Jupiter caused the most debate. But the craters on the moon and the phases of Venus were also critical evidence. How did they argue against the Ptolemaic model?

Galileo faced the Inquisition because of a conflict between two ways of knowing and understanding the world. The Church taught faith and understanding through revelation, but the scientists of the age were inventing a new way to understand nature that relied on evidence. In astronomy, evidence means observations, so our story turns now from the philosophical problem of the place of the Earth to the observational problem of the motion of the planets.

4-3 THE PUZZLE OF PLANETARY MOTION

While Galileo was demonstrating his telescope in Florence and Rome, two other astronomers were working in northern Europe, beyond the sway of the Inquisition. Although they, too, struggled to understand the place of the Earth, they approached the problem differently. They used the most accurate observations of the positions of the planets to try to discover the rules that govern planetary motion. When that puzzle was finally solved, astronomers understood why the Ptolemaic model of the universe does not work well and how to

make the Copernican universe a precise predictor of planetary motion.

TYCHO THE OBSERVER

The great observational astronomer of our story is a Danish nobleman, Tycho Brahe, born December 14, 1546, only three years after the publication of *De Revolutionibus* (▌Figure 4-15). As a son of a noble family, he was well known for his vanity and lordly manners.

Tycho's college days were eventful. He was officially studying law with the expectation that he would enter Danish politics, but he made it clear to his family that his real interest was astronomy and mathematics. It was also during his college days that he fought a duel and received a wound that disfigured his nose. For the rest of his life, he wore false noses made of gold and silver and stuck on with wax. The disfigurement probably did little to improve his disposition.

Tycho's first astronomical observations were made while he was a student. In 1563, Jupiter and Saturn passed very near each other in the sky, nearly merging into a single point on the night of August 24. Tycho found that the *Alfonsine Tables* were a full month in error and that the *Prutenic Tables* were in error by a number of days. These discrepancies dismayed Tycho and sparked his interest in the motions of the planets.

In 1572, a brilliant "new star" (now called Tycho's supernova) appeared in the sky. Such changes in the stars puzzled classically trained astronomers. Aristotle had argued that the starry sphere was perfect and unchanging, and therefore such new stars had to lie closer to Earth than the moon. Layers below the moon were thought to be less perfect and thus more changeable.

Tycho carefully measured the position of the star time after time through the night and found that it displayed no parallax. To understand the significance of this observation, we must note that Tycho believed the heavens rotated westward around Earth. The new star, according to classical astronomy, represented a change

in the heavens and therefore had to lie below the sphere of the moon. In that case, Tycho reasoned, the new star would appear slightly too far east as it rose and slightly too far west as it set (▌Figure 4-16). That is parallax. Tycho saw no parallax in the position of the new star, so he concluded that it must lie above the sphere of the moon and was probably on the starry sphere itself. This contradicted Aristotle's belief that the starry sphere was perfect and unchanging.

No one before Tycho could have made this discovery because no one had ever measured the positions of stars so accurately. Tycho had great confidence in the precision of his measurements, so when he failed to detect parallax for the new star, he knew it was important evidence against the Ptolemaic theory. He announced his discovery in a small book, *De Stella Nova* (*The New Star*), published in 1573.

The book attracted the attention of astronomers throughout Europe, and soon Tycho was summoned to the court of the Danish King Frederik II and offered funds to build an observatory on the island of Hveen just off the Danish coast. Tycho also received a steady source of income as landlord of a coastal district from which he collected rents. (He was not a popular landlord.) On Hveen, Tycho constructed a luxurious home with six towers specially equipped for astronomy and populated it with servants, assistants, and a dwarf to act as jester. Soon Hveen was an international center of astronomical study.

TYCHO BRAHE'S LEGACY

Tycho Brahe made no direct contribution to astronomical theory. Because he could measure no parallax for the stars, he concluded that Earth had to be stationary, thus rejecting the Copernican hypothesis. However, he

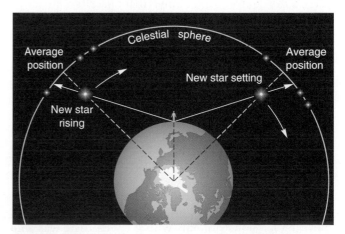

FIGURE 4-16
If an object lay lower than the celestial sphere, reasoned Tycho Brahe, then it should be seen east of its average position as it was rising and west of its average position when it was setting. Because he did not detect this daily parallax, he concluded the new star of 1572 had to lie on the celestial sphere.

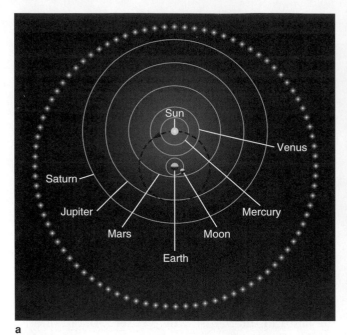

a

b

FIGURE 4-17

(a) Tycho Brahe's model of the universe held that Earth was fixed at the center of the starry sphere. The moon and sun circled Earth while the planets circled the sun. (b) Much of Tycho's success was due to his skill in designing large, accurate instruments. In this engraving of his mural quadrant, the figure of Tycho, his dog, and the background scene are painted on the wall within the arc of the quadrant. The observer (Tycho himself) at the extreme right peers through a sight out the loophole in the wall at the upper left to measure an object's angular distance above the horizon. *(Granger Collection, New York)*

also rejected the Ptolemaic model because of its inaccurate predictions. Instead, he devised a complex model in which Earth was the immobile center of the universe around which the sun and moon moved. The other planets circled the sun. This model thus incorporated part of the Copernican model, but Earth—rather than the sun—was stationary. In this way, Tycho preserved the central, immobile Earth described in scripture (❚ Figure 4-17a). Although Tycho's model, the Tychonic Universe, was very popular at first, the Copernican model replaced it within a century.

The true value of Tycho's work was observational. Because he was able to devise new and better instruments, he was able to make highly accurate observations of the positions of the stars, sun, moon, and planets. Tycho had no telescopes—they were not invented until the next century—so his observations were made by the naked eye peering along sights. All of his instruments were designed to measure angles in the sky. For example, his quadrant could measure angles up to 90° above the horizon (Figure 4-17b). By designing and building large instruments with great care, he was able to measure angles to high precision. He measured the positions of 777 stars to better than 4 minutes of arc and regularly measured the positions of the sun, moon, and planets for the 20 years he stayed on Hveen.

Unhappily for Tycho, King Frederik II died in 1588, and his young son took the throne. Suddenly Tycho's temper, vanity, and noble presumptions threw him out of favor. In 1596, taking most of his instruments and books of observations, he went to Prague, the capital of Bohemia, and became imperial mathematician to the Holy Roman Emperor Rudolph II. His goal was to revise the *Alfonsine Tables* and publish the revision as a monument to his new patron. It would be called the *Rudolphine Tables*.

Tycho did not intend to base the *Rudolphine Tables* on the Ptolemaic system but rather on his own Tychonic system, proving once and for all the validity of his hypothesis. To assist him, he hired a few mathematicians and astronomers, including one Johannes Kepler. Then in November 1601, Tycho collapsed at a nobleman's home. Before he died, 11 days later, he asked Rudolph II to make Kepler imperial mathematician. Thus the newcomer, Kepler, became Tycho's replacement (though at one-sixth Tycho's salary).

KEPLER THE ANALYST

No one could have been more different from Tycho Brahe than Johannes Kepler (❚ Figure 4-18). He was born December 27, 1571, to a poor family in a region

FIGURE 4-18

Johannes Kepler (1571–1630) derived three laws of planetary motion from Tycho Brahe's observations of the positions of the planets. This Romanian stamp commemorates the 400th anniversary of Kepler's birth. Ironically, it contains an error—the horns of the crescent moon should be inclined upward from the horizon.

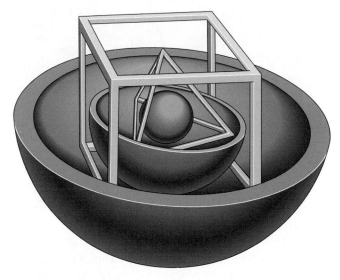

FIGURE 4-19

In this diagram, based on one drawn by Kepler, the sizes of the celestial spheres carrying the outer three planets, Saturn, Jupiter, and Mars, are determined by spacers (blue) consisting of a cube and a tetrahedron. Inside the sphere of Mars, the remaining regular solids (not shown here) separate the spheres of the Earth, Venus, and Mercury. The sun lay at the very center of this Copernican universe based on geometrical spacers.

now included in southwestern Germany. His father was unreliable and shiftless, principally employed as a mercenary soldier fighting for whoever paid enough. He finally failed to return from a military expedition, either because he was killed or because he found circumstances more to his liking. Kepler's mother was apparently an unpleasant and unpopular woman. She was accused of witchcraft in her later years, and Kepler had to defend her in a trial that dragged on for three years. She was finally acquitted but died the following year.

Kepler was the oldest of six children, and his childhood was no doubt unhappy. The family was not only poor, but suffered from an absentee father whom Kepler described as "vicious, inflexible, quarrelsome, and doomed to a bad end." In addition, Kepler was never healthy, even as a child, so it is surprising that he did well in school, eventually winning promotion to a Latin school and finally a scholarship to the university at Tübingen, where he studied to become a Lutheran pastor.

During his last year of study, Kepler accepted a job in Graz teaching mathematics and astronomy. Evidently, he was not a good teacher. He had few students his first year and none at all his second. His superiors put him to work teaching a few introductory courses and preparing an annual almanac that contained astronomical, astrological, and weather predictions. Through good luck, in 1595 some of his weather predictions were fulfilled, and he gained a reputation as an astrologer and seer; even in later life he earned money from his almanacs.

While still a college student, Kepler had become a believer in the Copernican theory, and at Graz he used his extensive spare time to study astronomy. By 1596, the same year Tycho left Hveen, Kepler was ready to solve the mystery of the universe. That year he pub-

lished a book called *The Forerunner of Dissertations on the Universe, Containing the Mystery of the Universe.* Like nearly all scientific works, the book was in Latin, and it is now known as *Mysterium Cosmographicum.*

The book begins with a long appreciation of Copernicanism and then goes on to speculate on the spacing of the planetary orbits. Kepler, as a Copernican, knew of six planets circling the sun—Mercury, Venus, Earth, Mars, Jupiter, and Saturn. According to his model, the spheres containing the six planets were separated from one another, and their relative sizes were fixed by the five regular solids*—the cube, tetrahedron, dodecahedron, icosahedron, and octahedron (▌Figure 4-19). Kepler gave astrological, numerological, and musical arguments for his theory and so followed the tradition of Pythagoras in believing that the order of the universe is underlain by musical (meaning mathematical) principles.

In the second half of the book, Kepler tried to fit the five solids to the planetary orbits and so demonstrated that he was a talented mathematician well versed in astronomy. He sent copies to Tycho and to Galileo, and both recognized his talents.

Life was unsettled for Kepler because of the persecution of Protestants in the region, so when Tycho invited him to Prague in 1600, he went readily, anxious to work with the famous astronomer. Tycho's sudden death in 1601 left Kepler in a position to use the obser-

---

*A regular solid is a three-dimensional body each of whose faces is the same. For example, a cube is a regular solid each of whose faces is a square.

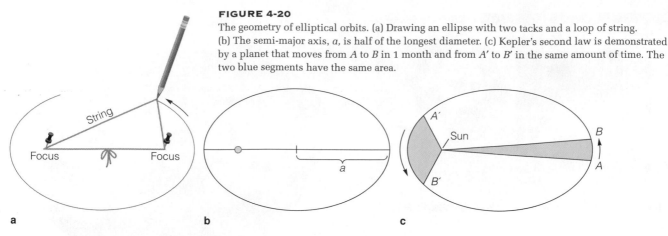

FIGURE 4-20

The geometry of elliptical orbits. (a) Drawing an ellipse with two tacks and a loop of string. (b) The semi-major axis, *a*, is half of the longest diameter. (c) Kepler's second law is demonstrated by a planet that moves from *A* to *B* in 1 month and from *A'* to *B'* in the same amount of time. The two blue segments have the same area.

vations from Hveen to analyze the motions of the planets and complete the *Rudolphine Tables.* Tycho's family, recognizing that Kepler was a Copernican and guessing that he would not follow the Tychonic system in completing the *Rudolphine Tables,* sued to recover the instruments and books of observations. The legal wrangle went on for years. The family did recover the instruments Tycho had brought to Prague, but Kepler had the books, and he kept them.

Whether Kepler had any legal right to Tycho's records is debatable, but he put them to good use. He began by studying the motion of Mars, trying to deduce from the observations how the planet moves. By 1606, he had solved the puzzle of planetary motion. The orbit of Mars is an ellipse and not a circle. Thus he abandoned the 2000-year-old belief in the circular motion of the planets. But the mystery was even more complex. The planets do not move at a uniform speed along their elliptical orbits. Kepler recognized that they move faster when close to the sun and slower when farther away. Kepler therefore abandoned both circular motion and uniform motion.

Kepler published his results in 1609 in a book called *Astronomia Nova* (*New Astronomy*). Like Copernicus's book, *Astronomia Nova* did not become an instant success. It is written in Latin for other scientists and is highly mathematical. In some ways, the book is surprisingly advanced. For instance, Kepler discusses the force that holds the planets in their orbits, a question that Isaac Newton considered later in his recognition of gravity as a natural force.

Despite the abdication of Rudolph II in 1611, Kepler continued his astronomical studies. He wrote about a supernova that had appeared in 1604 (now known as Kepler's supernova) and about comets, and he wrote a textbook about Copernican astronomy. In 1619, he published *Harmonice Mundi* (*The Harmony of the World*), in which he returned to the cosmic mysteries of *Mysterium Cosmographicum.* The only thing of note in *Harmonice Mundi* is his discovery that the radii of the planetary orbits are related to the planets' orbital periods. That and his two previous discoveries

are now recognized as Kepler's three laws of planetary motion.

KEPLER'S THREE LAWS OF PLANETARY MOTION

Although Kepler dabbled in the philosophical arguments of the day, he was a mathematician, and his triumph was the solution of the problem of the motion of the planets. The key to his solution was the ellipse.

An **ellipse** is a figure drawn around two points called the *foci* in such a way that the distance from one focus to any point on the ellipse and back to the other focus equals a constant. This makes it easy to draw ellipses with two thumbtacks and a loop of string. Press the thumbtacks into a board, loop the string about the tacks, and place a pencil in the loop. If you keep the string taut as you move the pencil, it traces out an ellipse (❙ Figure 4-20a).

The geometry of an ellipse is described by two simple numbers. The **semimajor axis, *a*,** is half of the longest diameter. The **eccentricity** of an ellipse, *e,* is the distance from either focus to the center of the ellipse divided by the semimajor axis. To draw a circle with the string and tacks as shown in Figure 4-20a, we would move the two thumbtacks together, which shows that a circle is really just an ellipse with eccentricity equal to zero. As we move the thumbtacks farther apart, the ellipse becomes flatter, and the eccentricity moves closer to 1.

Kepler used ellipses to describe the motion of the planets in three fundamental rules that have been tested and confirmed so many times that astronomers now refer to them as natural laws (Window on Science 4-2). They are commonly called Kepler's laws of planetary motion (❙ Table 4-1).

Kepler's first law states that the orbits of the planets around the sun are ellipses with the sun at one focus. Thanks to the precision of Tycho's observations and the sophistication of Kepler's mathematics, Kepler was able to recognize the elliptical shape of the orbits, even though they are nearly circular. Of the planets known to

WINDOW ON SCIENCE 4-2

Hypothesis, Theory, and Law: Levels of Confidence

Even scientists misuse the words *hypothesis*, *theory*, and *law*. We must try to distinguish these terms from one another because they are key elements in science.

A **hypothesis** is a single assertion or conjecture that must be tested. It could be true or false. "All Texans love chili" is a hypothesis. To know whether it is true or false, we need to test it against reality by making observations or performing experiments. Copernicus asserted that the universe was heliocentric; his assertion was a hypothesis subject to testing.

In Chapter 2, we saw that a model (Window on Science 2-1) is a description of some natural phenomenon; it can't be right or wrong. A model is not a conjecture of truth but merely a convenient way to think about a natural phenomenon. Consequently, a model such as the celestial sphere cannot be a hypothesis. Copernicus used his hypothesis to build a model, but they are not the same thing.

A **theory** is a system of rules and principles that can be applied to a wide variety of circumstances. A theory may have begun as one or more hypotheses, but it has been tested, expanded, and generalized. Many textbooks refer to the "Copernican theory," but some historians argue that it was not complete and had not been tested enough to be a theory. It is probably better to call it the Copernican hypothesis.

A **natural law** is a theory that has been refined, tested, and confirmed so often scientists have great confidence in it. Natural laws are the most fundamental principles of scientific knowledge. Kepler's laws are good examples.

Confidence is the key to understanding these terms. Scientists have more confidence in a theory than in a hypothesis and great confidence in a natural law. Nevertheless, scientists are not always consistent about these words. For example, Einstein's theory of relativity is much more accurate than Newton's laws of gravity and motion, but tradition dictates that we do not refer to Einstein's laws and Newton's theories. Darwin's theory of evolution has been tested many times, and scientists have great confidence in it, but no one refers to Darwin's law. These distinctions are subtle and sometimes depend more on custom than on levels of confidence.

Kepler, Mercury has the most elliptical orbit, but even it deviates only slightly from a circle (Figure 4-21).

Kepler's second law states that a line from the planet to the sun sweeps over equal areas in equal intervals of time. This means that when the planet is closer to the sun and the line connecting it to the sun is shorter, the planet moves more rapidly to sweep over the same area that is swept over when the planet is farther from the sun. Thus the planet in Figure 4-20c would move from point A' to point B' in one month, sweeping over the area shown. But when the planet is farther from the sun, one month's motion would be shorter, from A to B.

The time that a planet takes to travel around the sun once is its orbital period, P, and its average distance from the sun equals the semimajor axis of its orbit, a. Kepler's third law tells us that these two quantities are related; the orbital period squared is proportional to the semimajor axis cubed. If we measure P in years and a in astronomical units, we can summarize the third law as:

$$P_y^2 = a_{AU}^3$$

The subscripts remind us that we must measure the period in years and the semimajor axis in astronomical units (AU).

We can use Kepler's third law to make simple calculations. For example, Jupiter's average distance from the sun is 5.20 AU. What is its orbital period? If a equals 5.20, then a^3 equals 140.6. The orbital period

---

TABLE 4-1
Kepler's Laws of Planetary Motion

I. The orbits of the planets are ellipses with the sun at one focus.

II. A line from a planet to the sun sweeps over equal areas in equal intervals of time.

III. A planet's orbital period squared is proportional to its average distance from the sun cubed:

$$P_y^2 = a_{AU}^3$$

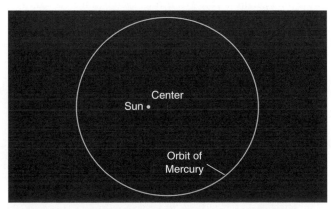

FIGURE 4-21

The orbits of the planets are nearly circular. In this scale drawing of the orbit of Mercury, the horizontal diameter and the vertical diameter of the orbit differ by less than 3 percent.

must be the square root of 140.6, which equals about 11.8 years.

It is important to notice that Kepler's three laws are empirical. That is, they describe a phenomenon without explaining why it occurs. Kepler derived them from Tycho's extensive observations, not from any fundamental assumption or theory. In fact, Kepler never knew what held the planets in their orbits or why they continued to move around the sun. His books are a fascinating blend of careful observation, mathematical analysis, and mystical theory.

THE *RUDOLPHINE TABLES*

In spite of Kepler's recurrent involvement with astrology and numerology, he continued to work on the *Rudolphine Tables*. At last, in 1627, they were ready, and he financed their printing himself, dedicating them to the memory of Tycho Brahe. In fact, Tycho's name appears in larger type on the title page than Kepler's own. This is especially surprising when we recall that the tables were based on the heliocentric model of Copernicus and the elliptical orbits of Kepler and not on the Tychonic system. The reason for Kepler's evident deference was Tycho's family, still powerful and still intent on protecting the memory of Tycho. They even demanded a share of the profits and the right to censor the book before publication, though they changed nothing but a few words on the title page and added an elaborate dedication to the emperor.

The *Rudolphine Tables* were Kepler's masterpiece. They could predict the positions of the planets 10 to 100 times more accurately than previous tables. Kepler's tables were the precise model of planetary motion that Copernicus had sought but failed to find. The accuracy of the *Rudolphine Tables* was strong evidence that both Kepler's model for planetary motion and the Copernican hypothesis for the place of Earth were correct. Copernicus would have been pleased.

Kepler solved the problem of planetary motion, and his *Rudolphine Tables* demonstrated his solution. Although he did not understand why the planets moved or why they followed ellipses, insights that had to wait half a century for Isaac Newton, Kepler's three rules worked. In science the only test of a theory is, "Does it describe reality?" Kepler's laws have been used for almost four centuries as a true description of orbital motion.

Kepler died November 15, 1630. During his life he had been an astrologer, a mystic, a numerologist, and a seer, but he became one of the world's great astronomers. His work not only overthrew the concept of uniform circular motion that had shackled the minds of astronomers for over 2000 years but also opened the way to the empirical study of planetary motion, a study that lies at the heart of modern astronomy.

4-4 MODERN ASTRONOMY

We date the origin of modern astronomy from the 99 years between the deaths of Copernicus and Galileo (1543 to 1642) because it was an age of transition. That period marked the transition between the Ptolemaic model of the universe and the Copernican with the attendant controversy over the place of the Earth. But that same period also marked a transition in the nature of astronomy in particular and science in general, a transition illustrated in the resolution of the puzzle of planetary motion. The puzzle was solved by precise observation and careful computation, techniques that are the foundation of modern science.

The discoveries of Kepler and Galileo found acceptance in the 1600s because the world was in transition. Astronomy was not the only thing changing during this period. The Renaissance is commonly taken to be the period between 1300 and 1600, and thus these 99 years of astronomical history lie at the culmination of the reawakening of learning in all fields (▌ Figure 4-22). The world was open to new ideas and new observations. Martin Luther remade religion, and other philosophers and scholars re-formed their areas of human knowledge. Had Copernicus not published his hypothesis, someone else would have suggested that the uni-

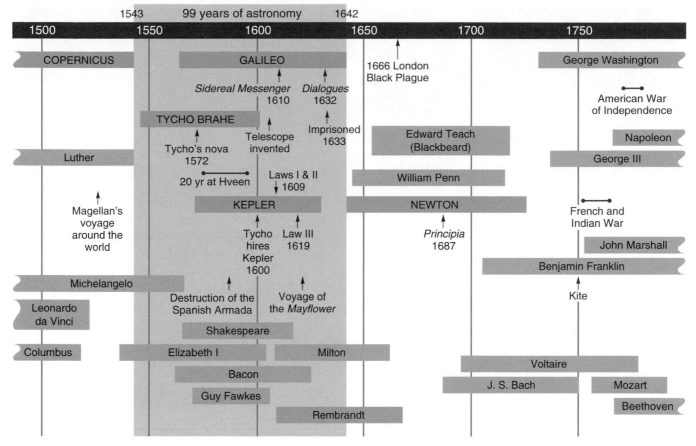

FIGURE 4-22

The 99 years between the death of Copernicus in 1543 and the birth of Newton in 1642 marked the transition from the ancient astronomy of Ptolemy and Aristotle to the revolutionary theory of Copernicus. This period saw the birth of modern scientific astronomy.

verse is heliocentric. History was ready to shed the Ptolemaic system.

In addition, this period marks the beginning of the modern scientific method. Beginning with Copernicus, Tycho, Kepler, and Galileo, scientists depended more and more on evidence, observation, and measurement. This, too, is coupled to the Renaissance and its advances in metalworking and lens making. Before our story began, no astronomer had looked through a telescope, because one could not be made. By 1642, not only telescopes but also other sensitive measuring instruments had transformed science into something new and precise. Also, the growing number of scientific societies increased the exchange of observations and hypotheses among scientists and stimulated more and better work. However, the most important advance was the application of mathematics to scientific questions. Kepler's work demonstrated the power of mathematical analysis, and as the quality of these numerical tools improved, the progress of science accelerated. Our story, therefore, is the story of the birth not just of modern astronomy but of modern science as well.

We will continue our story in the next chapter with the life of Isaac Newton. We will see how his accomplishments had their origins in the work of Galileo and Kepler.

SUMMARY

Archaeoastronomy, the study of the astronomy of ancient cultures, is revealing that primitive astronomers could make sophisticated astronomical observations. The Stone Age people of Britain who built Stonehenge aligned its long avenue and Heelstone to point toward the rising sun at the summer solstice. Other stones mark sight lines toward the rising moon. It seems clear that Stonehenge was a calendrical device, and some believe that its builders used it to predict eclipses.

Other archaeoastronomy sites span the world. A great many stone circles dot England and northern Europe, and related patterns, called medicine wheels, show that North American Indians also practiced astronomy. The Aztecs, Maya, and Incas of Central and South America also studied astronomy.

Traditional histories of astronomy usually begin with the Greeks. The Greek philosopher Aristotle held that Earth is fixed at the center of the universe, and Ptolemy based a mathematical model of the moving planets on this geocentric universe. To preserve the concept of uniform circular motion, he used epicycles, deferents, and equants for each of the planets, the sun, and the moon.

Near the end of his life, in 1543, Nicolaus Copernicus published his hypothesis that the sun is the center of the universe. Because the teachings of Aristotle had been adopted as official church doctrine, the heliocentric universe of Copernicus was considered radical.

Galileo Galilei was the first astronomer to use the newly invented telescope for a systematic study of the heavens, and he found that the appearance of the sun, moon, and planets supported the heliocentric model. After he described his findings in a book, the church, in 1616, ordered him to stop teaching Copernicanism. After he published another book on the subject in 1632, he was tried by the Inquisition and ordered to recant. He was imprisoned in his home until his death in 1642.

In northern Europe, beyond the power of the Church, Tycho Brahe rejected the Ptolemaic and Copernican systems. In Tycho's own system, Earth is fixed at the center of the universe, and the sun and moon revolve around Earth. The planets orbit the moving sun. Although his hypothesis is wrong, he was a brilliant observer and accumulated 20 years of precise measurements of the positions of the sun, moon, and planets.

At Tycho's death, one of his assistants, Johannes Kepler, became his successor. Analyzing Tycho's observations, Kepler finally discovered how the planets move. His three laws of planetary motion overthrew uniform circular motion. The first law says the orbits of the planets are elliptical rather than circular, and the second law says the planets do not move at a uniform rate but move faster when near the sun and slower when more distant. The third law relates orbital period to orbital size. These studies of the place of the Earth and the nature of planetary motion led to the birth of modern astronomy.

NEW TERMS

archaeoastronomy

eccentric

uniform circular motion

geocentric universe

parallax

retrograde motion

epicycle

deferent

equant

heliocentric universe

paradigm

ellipse

semimajor axis, a

eccentricity, e

hypothesis

theory

natural law

REVIEW QUESTIONS

Ace Astronomy™ Assess your understanding of this chapter's topics with additional quizzing and animations at **http://astronomy.brookscole.com/seeds8e**

1. What evidence do we have that early human cultures observed astronomical phenomena?

2. Why did Plato propose that all heavenly motion was uniform and circular?

3. How do the epicycles of Mercury and Venus differ from those of Mars, Jupiter, and Saturn?

4. Why did Copernicus have to keep small epicycles in his system?

5. Explain how each of Galileo's telescopic discoveries contradicted the Ptolemaic theory.

6. Galileo was condemned, but Kepler, also a Copernican, was not. Why not?

7. When Tycho observed the new star of 1572, he could detect no parallax. Why did that result undermine belief in the Ptolemaic system?

8. Does Tycho's model of the universe explain the phases of Venus that Galileo observed? Why or why not?

9. How do the first two of Kepler's three laws overthrow one of the basic beliefs of classical astronomy?

10. What is the difference between a hypothesis, a theory, and a law?

11. How did the *Alfonsine Tables,* the *Prutenic Tables,* and the *Rudolphine Tables* differ?

DISCUSSION QUESTIONS

1. Historian of science Thomas Kuhn has said that *De Revolutionibus* was a revolution-making book but not a revolutionary book. How was it an old-fashioned, classical book?

2. Why might Tycho Brahe have hesitated to hire Kepler? Why do you suppose he appointed Kepler his scientific heir?

3. How does the modern controversy over creationism and evolution reflect two ways of knowing about the physical world?

PROBLEMS

1. Draw and label a diagram of the eastern horizon from northeast to southeast and label the rising point of the sun at the solstices and equinoxes. (See page 23 and Figure 4-1).

2. If you lived on Mars, which planets would describe retrograde motion? Which would never be visible as crescent phases?

3. Galileo's telescope showed him that Venus has a large angular diameter (61 seconds of arc) when it is a crescent and a small angular diameter (10 seconds of arc) when it is nearly full. Use the small-angle formula to find the ratio of its maximum distance to its minimum distance. Is this ratio compatible with the Ptolemaic universe shown in on page 59?

4. Galileo's telescopes were not of high quality by modern standards. He was able to see the moons of Jupiter, but he never reported seeing features on Mars. Use the small-

angle formula to find the angular diameter of Mars when it is closest to Earth. How does that compare with the maximum angular diameter of Jupiter?

5. If a planet has an average distance from the sun of 4 AU, what is its orbital period?

6. If a space probe is sent into an orbit around the sun that brings it as close as 0.5 AU and as far away as 5.5 AU, what will be its orbital period?

7. Pluto orbits the sun with a period of 247.7 years. What is its average distance from the sun?

CRITICAL INQUIRIES FOR THE WEB

1. The trial of Galileo is an important event in the history of science. We now know, and the Roman Catholic Church now recognizes, that Galileo's view was correct, but what were the arguments on both sides of the issue as it was unfolding? Research the Internet for documents that chronicle the trial, Galileo's observations and publications, and the position of the Church. Use this information to outline the case for and against Galileo in the context of the times in which the trial occurred.

2. Take a "virtual tour" of Stonehenge by browsing the Web for information related to this megalithic site and the possibility that it was used for astronomical purposes.

(Be careful to use legitimate scientific and historical Web sites in your survey.) Summarize the various astronomical alignments evident in the layout of the site.

EXPLORING *THESKY*

1. Go to Stonehenge in southern England and watch the sun rise on the morning of the summer solstice. Where does it rise on the morning of the winter solstice?

2. Observe Mars going through its retrograde motion. (*Hint:* Use **Reference Lines** under the **View** menu to turn on the ecliptic. Be sure you are in **Free Rotation** under the **Orientation** menu. Locate Mars and use the time skip arrows to watch it move.)

3. Compare the size of the retrograde loops made by Mars, Jupiter, and Saturn.

4. Can you recognize the effects of Kepler's second law in the orbital motion of any of the planets? (*Hint:* Use **3D Solar System Mode** under the **View** menu.)

5. Can you recognize the effects of Kepler's third law in the orbital motion of the planets?

 Visit the Seeds *Foundations of Astronomy* companion Web site for critical thinking exercises, articles, and additional readings from InfoTrac College Edition, Brooks/Cole's online student library.

NEWTON, EINSTEIN, AND GRAVITY

Nature and Nature's laws lay hid in night:
God said, "Let Newton be!" and all was light.

Alexander Pope

Chapter 5

GUIDEPOST

Astronomers are gravity experts. All of the heavenly motions described in the preceding chapters are dominated by gravitation. Isaac Newton gets the credit for discovering gravity, but even Newton couldn't explain what gravity was. Einstein proposed that gravity is a curvature of space, but that only pushes the mystery further away. "What is curvature?" we might ask.

This chapter shows how scientists build theories to explain and unify observations. Theories can give us entirely new ways to understand nature, but no theory is an end in itself. Astronomers continue to study Einstein's theory, and they wonder if there is an even better way to understand the motions of the heavens.

The principles we discuss in this chapter will be companions through the remaining chapters. Gravity is universal.

Isaac Newton was born in Woolsthorpe, England, on December 25, 1642, and on January 4, 1643. This was not a biological anomaly but a calendrical quirk. Most of Europe, following the lead of the Catholic countries, had adopted the Gregorian calendar, but Protestant England continued to use the Julian calendar. Thus, December 25 in England was January 4 in Europe. If we take the English date, then Newton was born in the same year that Galileo Galilei died.

Newton went on to become one of the greatest scientists who ever lived (■ Figure 5-1), but even he admitted the debt he owed to those who had studied nature before him. He said, "If I have seen farther than other men, it is because I stood on the shoulders of giants."

One of those giants was Galileo (■ Figure 5-2). Although we remember Galileo as the defender of Copernicanism, he was also a talented scientist who studied the motions of falling bodies.

Newton based his own work on the discoveries of Galileo and others. During his life he studied optics, invented calculus, developed three laws of motion, and discovered the principle of mutual gravitation. Of these, the last two are the most important to us in this chapter because they make it possible to understand the orbital motion of the moon and planets.

Newton's laws of motion and gravity were astonishingly successful in describing the heavens. A friend of Newton, Edmond Halley, used Newton's laws to calculate the orbits of comets and discovered that certain comets that had been seen throughout recorded history were actually a single object returning every 77 years, now known as Comet Halley.

For over two centuries following the publication of Newton's works, astronomers used his laws to describe the universe. Then, early in the 20th century, Albert Einstein proposed a new way to describe gravity. The new theory did not replace Newton's laws but rather showed that they were only approximately correct and could be seriously in error under special circumstances. We will see how Einstein's theories further extend our understanding of the nature of gravity. Just as Newton had stood on the shoulders of Galileo, Einstein stood on the shoulders of Newton.

5-1 GALILEO AND NEWTON

Johannes Kepler discovered three laws of planetary motion, but he never understood why the planets move along their orbits. At one place in his writings, he wonders if they are pulled along by magnetic forces

a

b

FIGURE 5-1
Isaac Newton (1642–1727) was the founder of modern physics. He made important discoveries in optics, developed three laws of motion, invented differential calculus, and discovered the law of mutual gravitation. These stamps were issued in 1987 to mark the 300th anniversary of the publication of his book *Principia*.

emanating from the sun. At another place, he considers and dismisses the idea that the planets are pushed along their orbits by angels.

Newton refined Kepler's model of planetary motion but did not perfect it. In science, a model is an intellectual conception of how nature works (Window on Science 2-1). No model is perfect. Kepler's model was better than Aristotle's, but Newton improved Kepler's model

FIGURE 5-2
Although Galileo is often associated with the telescope, as on this Italian stamp, he also made systematic studies of the motion of falling bodies and made discoveries that led to the law of inertia.

Ace✦Astronomy™ The AceAstronomy icon throughout the text indicates an opportunity for you to test yourself on key concepts and to explore animations and interactions on the AceAstronomy Web site at **http://astronomy.brookscole .com/seeds8e**

by expanding it into a general theory of motion and gravity. In fact, most scientists now refer to Newton's *law* of gravity. Whether we call it a model, a theory, or a law, we must recognize that it was not perfect. Newton never understood what gravity was. It was as mysterious as an angel pushing the moon inward toward Earth instead of forward along the moon's orbit.

To understand science, we must understand the nature of scientific descriptions. The scientist studies nature by either creating new theories or refining old theories. Yet a theory can never be perfect, because it can never represent the universe in all its intricacies. Instead, a theory must be a limited approximation of a single phenomenon, such as orbital motion. It is fitting that Newton's discoveries all began with Kepler's fellow Copernican, Galileo.

GALILEO AND MOTION

Even before Galileo built his first telescope, he had begun studying the motion of freely moving bodies. After the Inquisition condemned and imprisoned him in 1633, he continued his study of motion. He seems to have realized that he would have to understand motion before he could truly understand the Copernican system. That he was eventually able to formulate principles that later led Newton to the laws of motion and the theory of gravity is a tribute to Galileo's ability to set aside authority and think for himself.

The authority of the age was Aristotle, whose ideas on motion were hopelessly confused. Aristotle said that the world is made up of four classical elements: earth, water, air, and fire. According to this idea, all earthly things—wood, rock, flesh, bone, metals, and so on—are made up of different mixtures of the four classical elements, with earth and water dominating. The motions of bodies are determined by the natural tendencies of the four elements to move toward their proper place in the cosmos. All objects dominated by earth and water, for example, move toward the center of the cosmos, which, in Aristotle's geocentric universe, is Earth's center. Thus, objects fall downward because they are moving toward their proper place.* On the other hand, fire and air move upward toward the heavens, their proper place in the cosmos.

Aristotle called these motions **natural motions** to distinguish them from **violent motions** produced, for instance, when we push on an object and make it move other than toward its proper place. According to Aristotle, such motions stop as soon as the force is removed. To explain how an arrow could continue to move upward even after it had left the bowstring, he said currents in the air around the arrow carried it forward even though the bowstring was no longer pushing it.

*This is one reason why Aristotle had to have a geocentric universe. If Earth's center had not also been the center of the cosmos, his explanation of gravity would not have worked.

These ideas about the proper places of objects, natural and violent motion, and the necessity of a force to preserve motion were still accepted theory in Galileo's time. In fact, in 1590, when Galileo was 26, he wrote a short work called *De Motu* (*On Motion*) that deals with the proper places of objects and their natural motions.

In Galileo's time and for the two preceding millennia, scholars had commonly tried to resolve problems of science by referring to authority. To analyze the flight of a cannonball, for instance, they would turn to the writings of Aristotle and other classical philosophers and try to deduce what those philosophers would have said on the subject. This generated a great deal of discussion but little real progress. Galileo broke with this tradition and conducted his own experiments.

He began by studying the motions of falling bodies, but he quickly discovered that the velocities were so great and the times so short that he could not measure them accurately. Consequently, he began using polished bronze balls rolling down gently sloping inclines. In that instance, the velocity is lower and the time longer. Using an ingenious water clock, he was able to measure the time the balls took to roll given distances down the incline, and he correctly recognized that these times are proportional to the times taken by falling bodies.

He found that falling bodies do not fall at constant rates, as Aristotle had said, but are accelerated. That is, they move faster with each passing second. Near Earth's surface, a falling object will have a velocity of 9.8 m/s (32 ft/s) at the end of 1 second, 19.6 m/s (64 ft/s) after 2 seconds, 29.4 m/s (96 ft/s) after 3 seconds, and so on. Each passing second adds 9.8 m/s (32 ft/s) to the object's velocity (❚ Figure 5-3). In modern terms, this is called the **acceleration of gravity** at Earth's surface.

Galileo also discovered that the acceleration does not depend on the weight of the object. This, too, is contrary to the teachings of Aristotle, who believed that heavy objects, containing more earth and water, fall with higher velocity. Galileo found that the acceleration of a falling body is the same whether it is heavy or light. According to some accounts, he demonstrated this by dropping balls of iron and wood from the top of the Leaning Tower of Pisa to show that they would fall together and hit the ground at the same time (❚ Figure 5-4a). In fact, he probably didn't perform this experiment. It would not have been conclusive anyway because of air resistance. More than 300 years later, Apollo 15 astronaut David Scott, standing on the airless moon, demonstrated Galileo's discovery by dropping a feather and a steel geologist's hammer. They hit the lunar surface at the same time (Figure 5-4b).

Having described natural motion, Galileo turned his attention to violent motion—that is, motion directed other than toward an object's proper place in the cosmos. He pointed out that an object rolling down an incline is accelerated toward Earth and that an object rolling up the same incline is decelerated. If the surface

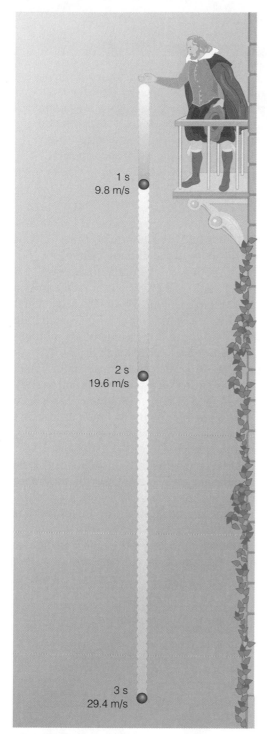

FIGURE 5-3
Galileo found that a falling object is accelerated downward.
Each second, its velocity increases by 9.8 m/s (32 ft/s).

were perfectly horizontal and frictionless, he reasoned, there could be no acceleration or deceleration to change the object's velocity, and, in the absence of friction, the object would continue to move forever. In his own words, "any velocity once imparted to a moving body will be rigidly maintained as long as the external causes of acceleration or retardation are removed."

a

b

FIGURE 5-4
(a) According to tradition, Galileo demonstrated that the acceleration of a falling body is independent of its weight by dropping balls of iron and wood from the Leaning Tower of Pisa. In fact, air resistance would have confused the result. (b) In a historic television broadcast from the moon on August 2, 1971, David Scott dropped a hammer and a feather at the same instant. In the vacuum of the lunar surface, they fell together. *(NASA)*

This is contrary to Aristotle's belief that motion can continue only if a force is present to maintain it. In fact, Galileo's statement is a perfectly valid summary of the law of inertia, which became Newton's first law of motion.

Galileo published his work on motion in 1638, two years after he had become entirely blind and only four years before his death. The book was called *Mathematical Discourses and Demonstrations Concerning Two New Sciences, Relating to Mechanics and to Local Motion*. It is known today as *Two New Sciences*.

The book is a brilliant achievement for a number of reasons. To understand motion, Galileo had to abandon the authority of the ancients, devise his own experiments, and draw his own conclusions. In a sense, this was the first example of experimental science. But Galileo also had to generalize his experiments to discover how nature worked. Though his apparatus was finite and plagued by friction, he was able to imagine an infinite, totally frictionless plane on which a body moves at constant velocity. In his workshop, the law of inertia was obscure; but, in his imagination, it was clear and precise.

NEWTON AND THE LAWS OF MOTION

Newton's three laws of motion (▮ Table 5-1) are critical to our understanding of orbital motion. They apply to any moving object, from an automobile driving along a highway to galaxies colliding with each other.

The first law is really a restatement of Galileo's law of inertia. An object continues at rest or in uniform motion in a straight line unless acted upon by some force. Astronauts drifting in space will travel at constant rates in straight lines forever if no forces act on them (▮ Figure 5-5a).

Newton's first law also explains why a projectile continues to move after all forces have been removed—for instance, why an arrow continues to move after leaving the bowstring. The object continues to move

| **TABLE 5-1** |
| Newton's Three Laws of Motion |

I. A body continues at rest or in uniform motion in a straight line unless acted upon by some net force.

II. The acceleration of a body is inversely proportional to its mass, directly proportional to the net force, and in the same direction as the net force.

III. To every action, there is an equal and opposite reaction.

because it has momentum. We can think of an object's **momentum** as a measure of the amount of motion.

Momentum depends on velocity and mass.* A low-velocity object such as a paper clip tossed across a room has little momentum, and we could easily catch it in our hand. But the same paper clip fired at the speed of a rifle bullet would have tremendous momentum, and we would not dare try to catch it.

Momentum also depends on the mass of an object (Parameters of Science 5-1). To see how, imagine that, instead of tossing us a paper clip, someone tosses us a bowling ball. A bowling ball contains much more mass than a paper clip and therefore has much greater momentum at the same velocity.

Newton's first law explains the consequences of the conservation of momentum. When we say that momentum is conserved, we mean that it remains constant until something acts to change it. A moving object has a given amount of momentum. To change that momentum, we must exert some force on the object to change either the speed or the direction. Newton's first law and the concept of momentum came from the work of Galileo.

*Mathematically, momentum is the product of mass and velocity.

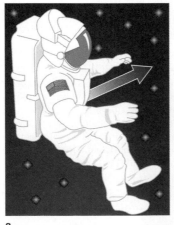
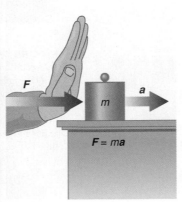
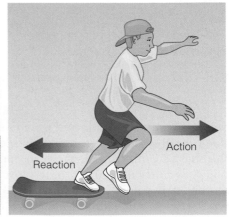

FIGURE 5-5
Newton's three laws of motion.   a   b   c

Mass

One of the most fundamental parameters in science is **mass,** a measure of the amount of matter in an object. A bowling ball, for example, contains a large amount of mass; but a child's rubber ball contains less matter than the bowling ball, and we say it is less massive.

Mass is not the same as weight. Our weight is the force that Earth's gravity exerts on the mass of our bodies. Because gravity pulls us downward, we press against the bathroom scale and we can measure our weight. Floating in space, we would have no weight at all; a bathroom scale would be useless. But our bodies would still contain the same amount of matter, so we would still have mass.

Sports analogies illustrate the importance of mass in dramatic ways. A bowling ball, for example, must be massive in order to have a large effect on the pins it strikes. Imagine trying to knock down all the pins with a bowling ball no more massive than a balloon. Even in space, where the bowling ball would be weightless, a low-mass bowling ball would have little effect on the pins. On the other hand, runners want track shoes that have low mass and thus are easy to move. Imagine trying to run a 100-meter dash wearing track shoes that were as massive as bowling balls. They would be very hard to move, and it would be difficult to accelerate away from the starting block. The shot put takes muscle because the shot is massive, not because it is heavy. Imagine throwing the shot in space where it would have no weight. It would still be massive, and it would take great effort to start it moving.

Mass is a unique measure of the amount of material in an object. Using the metric system (Appendix), we will measure mass in kilograms.

Newton's second law of motion discusses forces, and Galileo did not talk about forces. He spoke instead of accelerations. Newton saw that an acceleration is the result of a force acting on a mass (Figure 5-5b). Newton's second law is commonly written as:

$$F = ma$$

Once again, we must carefully define terms. An **acceleration** is a change in velocity, and a **velocity** is a directed rate of motion. By rate of motion, we mean, of course, a speed, but the word *directed* has a special meaning. Speed itself does not have any direction associated with it, but velocity does. If you drive a car in a circle at 55 mph, your speed is constant, but your velocity is changing. Thus, an object experiences an acceleration if its speed changes or if its direction of motion changes. Every automobile has three accelerators—the gas pedal, the brake pedal, and the steering wheel. All three change the auto's velocity.

The acceleration of a body is proportional to the force applied. This is reasonable. If we push gently against a grocery cart, we do not expect a large acceleration. The second law of motion also says that the acceleration is inversely proportional to the mass of the body. This, too, is reasonable. If the cart were filled with bricks and we pushed it gently, we would expect very little result. If it were full of inflated balloons, however, it would move easily in response to a gentle push. Finally, the second law says that the resulting acceleration is in the direction of the force. This is also what we would expect. If we push on a cart that is not moving, we expect it to begin moving in the direction we push.

The second law of motion is important because it establishes a precise relationship between cause and effect (Window on Science 5-1). Objects do not just move. They accelerate due to the action of a force. Moving objects do not just stop. They decelerate due to a force. Also, moving objects don't just change direction for no reason. Any change in direction is a change in velocity and requires the presence of a force. Aristotle said that objects move because they have a tendency to move. Newton said that objects move due to a specific cause, a force.

Newton's third law of motion specifies that for every action there is an equal and opposite reaction. In other words, forces must occur in pairs directed in opposite directions. For example, if you stand on a skateboard and jump forward, the skateboard will shoot away backward. As you jump, your feet exert a force against the skateboard, which accelerates it toward the rear. But forces must occur in pairs, so the skateboard must exert an equal but opposite force on your feet that accelerates your body forward (Figure 5-5c).

MUTUAL GRAVITATION

The three laws of motion led Newton to consider the force that causes objects to fall. The first and second laws tell us that falling bodies accelerate downward because some force must be pulling downward on them. Newton wondered what that force could be.

Newton was also aware that some force has to act on the moon. The moon follows a curved path around Earth, and motion along a curved path is accelerated motion. The second law says that a force is required in order to make the moon follow that curved path.

Newton wondered if the force that holds the moon in its orbit could be the same force that causes rocks to roll downhill—gravity. He was aware that gravity extends at least as high as the tops of mountains, but he

The Unstated Assumption of Science: Cause and Effect

One of the most often used and least often stated principles of science is cause and effect, and we could argue that Newton's second law of motion was the first clear statement of the principle. Ancient philosophers such as Aristotle argued that objects moved because of tendencies. They said that earth and water, and objects made of earth and water, had a natural tendency to move toward the center of the universe. This natural motion had no cause but was inherent in the nature of the objects. But Newton's second law says $F = ma$. If an object (of mass m) changes its motion (a in the equation), then it must be acted on by a force (F in the equation). Any effect (a) must be the result of a cause (F).

The principle of cause and effect goes far beyond motion. The principle of cause and effect gives scientists confidence that every effect has a cause. Hearing loss in certain laboratory rats, color changes in certain chemical dyes, and explosions on certain stars are all effects that must have causes. All of science is focused on understanding the causes of the effects we see. If the universe were not rational, then we could never expect to discover causes. Newton's second law of motion was arguably the first statement that the behavior of the universe depends rationally on causes.

did not know if it could extend all the way to the moon. He believed that it could, but he thought it would be weaker at greater distances. He also guessed that its strength would decrease as the square of the distance increased.

This relationship, the **inverse square law,** was familiar to Newton from his work on optics, where it applied to the intensity of light. A screen set up 1 unit from a candle flame receives a certain amount of energy on each square meter. However, if that screen is moved to a distance of 2 units, the light that originally illuminated 1 m^2 must cover 4 m^2 (❙ Figure 5-6). Thus, the intensity of the light is inversely proportional to the square of the distance to the screen.

Newton made two assumptions that enabled him to predict the strength of Earth's gravity at the distance of the moon. He assumed that the strength of gravity follows the inverse square law and that the critical distance is not the distance from Earth's surface but the distance from Earth's center. Because the moon is about 60 Earth radii away, Earth's gravity at the distance of the moon should be about 60^2 times less than at Earth's surface. Instead of being 9.8 m/s^2 at Earth's surface, it should be about 0.0027 m/s^2 at the distance of the moon.

Now, Newton wondered, could this acceleration keep the moon in orbit? He knew the moon's distance and its orbital period, so he could calculate the actual acceleration needed to keep it in its curved path. The answer is 0.0027 m/s^2. To the accuracy of Newton's data for Earth's radius, it was exactly what his assumptions predicted. The moon is held in its orbit by gravity, and gravity obeys the inverse square law.

Newton's third law says that forces always occur in pairs, and this leads to the conclusion that gravity is mutual. If Earth pulls on the moon, then the moon must pull on Earth. Gravitation is a general property of the universe. The sun, the planets, and all their moons must also attract each other by mutual gravitation. In fact, every particle in the universe must attract every other particle, and Newtonian gravity is often called universal mutual gravitation.

Clearly the force of gravity depends on mass. Although our bodies are made of mass, our personal gravitational fields do not attract personal satellites. Larger masses have stronger gravity. From an analysis of the third law of motion, Newton realized that the mass that resists acceleration in the first law must be the same as the mass associated with gravity. Then when two bodies attract one another and accelerate toward each other, the two masses must be equally involved. Newton performed precise experiments with pendulums and confirmed this equivalence between the mass that resists acceleration and the mass that causes gravity. From this, combined with the inverse square law, he

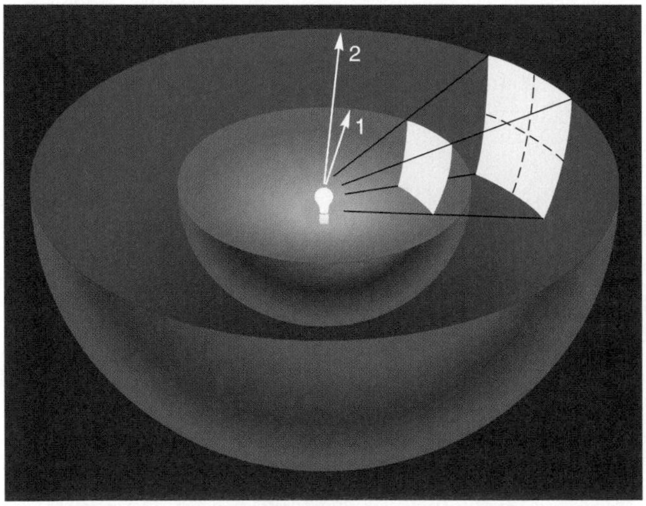

FIGURE 5-6

The inverse square law. A light source is surrounded by spheres with radii of 1 unit and 2 units. The light falling on an area of 1 m^2 on the inner sphere spreads to illuminate an area of 4 m^2 on the outer sphere. Thus, the brightness of the light source is inversely proportional to the square of the distance.

was able to write the famous formula for the gravitational force between two masses, M and m.

$$F = -G\frac{Mm}{r^2}$$

The constant G is the gravitational constant, and r is the distance between the masses. The negative sign tells us the force is attractive, pulling the masses together and making r decrease. In plain English, Newton's law of gravitation states: The force of gravity between two masses M and m is proportional to the product of the masses and inversely proportional to the square of the distance between them.

Newton's description of gravity was a difficult idea for physicists of his time to accept because it is an example of action at a distance. Earth and moon exert forces on each other although there is no physical connection between them. Modern scientists resolve this problem by referring to gravity as a **field.** Earth's presence produces a gravitational field directed toward Earth's center. The strength of the field decreases according to the inverse square law. Any particle of mass in that field experiences a force that depends on the mass of the particle and the strength of the field at the particle's location. The resulting force is directed toward the center of the field.

The field is an elegant way to describe gravity, but it does not tell us what gravity is. For that we must wait until later in this chapter, when we discuss Einstein's theory of curved space-time.

Newton first came to understand gravity by thinking about the orbital motion of the moon. Gravitation is tremendously important in astronomy, and its most important manifestation is orbital motion.

5-2 ORBITAL MOTION

Newton's laws of motion and gravitation make it possible to understand why the moon orbits Earth and how the planets move along their orbits. We can even discover why Kepler's laws work.

ORBITS

To understand how an object can orbit another object, we must see orbital motion as Newton did. Objects in orbit are falling. We will explore Newton's insight by analyzing the motion of objects orbiting Earth.

Study "❚ Orbiting Earth" on pages 84 and 85, and notice three important ideas. First, an object orbiting Earth is actually falling (being accelerated) toward Earth's center. The object continuously misses Earth because of its orbital velocity. Second, notice that objects orbiting each other actually revolve around their center of mass. Finally, notice the difference between closed orbits and open orbits. If you want to leave Earth never to return, you must give your spaceship a high enough velocity so it will follow an open orbit.

ORBITAL VELOCITY

If we were about to ride a rocket into orbit, we would have a critical question. How fast do we have to travel to stay in orbit? An object's circular velocity is the lateral velocity the object must have to remain in a circular orbit. If we assume the mass of our spaceship is small compared with the mass of the object we expect to orbit, Earth in this case, then the circular velocity is:

$$V_c = \sqrt{\frac{GM}{r}}$$

In this formula, M is the mass of the central body in kilograms, r is the radius of the orbit in meters, and G is the gravitational constant, 6.67×10^{-11} m^3/s^{2}kg. This formula is all we need to calculate how fast an object must travel to stay in a circular orbit.

For example, how fast does the moon travel in its orbit? Earth's mass is 5.98×10^{24} kg, and the radius of the moon's orbit is 3.84×10^8 m. Then the moon's velocity is:

$$V_c = \sqrt{\frac{6.67 \times 10^{-11} \times 5.98 \times 10^{24}}{3.84 \times 10^8}} = \sqrt{\frac{39.9 \times 10^{13}}{3.84 \times 10^8}}$$

$$= \sqrt{1.04 \times 10^6} = 1020 \text{ m/s} = 1.02 \text{ km/s}$$

This calculation shows that the moon travels 1.02 km along its orbit each second. That is the circular velocity at the distance of the moon.

Orbiting Earth

We can understand orbital motion by thinking of a cannon ball falling around Earth in a circular path.

Imagine a cannon on a high mountain aimed horizontally. A little gunpowder gives the cannonball a low velocity, and it doesn't travel very far before falling to Earth.

More gunpowder gives the cannonball a higher velocity, and it travels farther.

With enough gunpowder, the cannonball travels so fast it never strikes the ground. Earth's gravity pulls it toward Earth's center, but Earth's surface curves away from it at the same rate it falls. It is in orbit.

North Pole

A satellite above Earth's atmosphere feels no friction and will fall around Earth indefinitely.

The velocity needed to stay in a circular orbit is called the **circular velocity**. Just above Earth's atmosphere, circular velocity is 7790 m/s or about 17,400 miles per hour, and the orbital period is about 90 minutes.

Earth satellites eventually fall back to Earth if they orbit too low and experience friction with the upper atmosphere.

Ace Astronomy™

Go to AceAstronomy and click Active Figures to see "Newton's Cannon" and fire your own version of Newton's cannon.

A Geosynchronous Satellite

At a distance of 42,250 km (26,260 miles) from Earth's center, a satellite orbits with a period of 24 hours.

The satellite orbits eastward, and Earth rotates eastward under the moving satellite.

The satellite remains fixed above a spot on Earth's equator.

A **geosynchronous satellite** orbits eastward with the rotation of Earth and remains above a fixed spot — ideal for communications and weather satellites.

Ace Astronomy™

Go to AceAstronomy and click Active Figures to see "Geosynchronous Orbit" and place your own satellite in geosynchronous orbit.

According to Newton's first law of motion, the moon should follow a straight line and leave Earth forever. Because it follows a curve, Newton knew that some force must continuously accelerate it toward Earth — gravity. Each second the moon moves 1020 m (3350 ft) eastward and falls about 1.6 mm (about 1/16 inch) toward Earth. The combination of these motions produces the moon's curved orbit. The moon is falling.

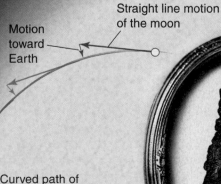

Straight line motion of the moon

Motion toward Earth

Curved path of moon's orbit

Earth

Astronauts in orbit around Earth feel weightless, but they are not "beyond Earth's gravity," to use a term from old science fiction movies. Like the moon, the astronauts are accelerated toward Earth by Earth's gravity, but they travel fast enough along their orbits that they continually "miss the Earth." They are literally falling around Earth. Inside or outside a spacecraft, astronauts feel weightless because they and their spacecraft are falling at the same rate. Rather than saying they are weightless, we should more accurately say they are in free fall.

To be precise we should not say that an object orbits Earth. Rather the two objects orbit each other. Gravitation is mutual, and if Earth pulls on the moon, the moon pulls on Earth. The two bodies revolve around their common **center of mass,** the balance point of the system.

Two bodies of different mass balance at the center of mass, which is located closer to the more massive object. As the two objects orbit each other, they revolve around their common center of mass.

Center of mass

The center of mass of the Earth–moon system lies only 4708 km (2926 miles) from the center of Earth — inside the Earth. As the moon orbits the center of mass on one side, the Earth swings around the center of mass on the opposite side.

Closed orbits return the orbiting object to its starting point. The moon and artificial satellites orbit Earth in closed orbits. Below, the cannonball could follow an elliptical or a circular closed orbit.

Ace **Astronomy**™ Go to AceAstronomy and click Active Figures to see "Center of Mass" and change the mass ratio in this diagram.

If the cannonball travels as fast as **escape velocity,** the velocity needed to leave a body, it will enter an open orbit. An **open orbit** does not return the cannonball to Earth. It will escape.

Hyberbola

A cannonball with a velocity greater than escape velocity will follow a hyperbola and escape from Earth.

Parabola

North Pole

A cannonball with escape velocity will follow a parabola and escape.

As described by Kepler's Second Law, an object in an elliptical orbit has its lowest velocity when it is farthest from Earth (apogee), and its highest velocity when it is closest to Earth (perigee). Perigee must be above Earth's atmosphere, or friction will rob the satellite of energy and it will eventually fall back to Earth.

Ellipse

Circle

Ellipse

A satellite just above Earth's atmosphere is only about 200 km above Earth's surface, or 6578 km from Earth's center, so Earth's gravity is much stronger and the satellite must travel much faster to stay in a circular orbit. We can use the formula above to find that the circular velocity just above Earth's atmosphere is about 7790 m/s, or 7.79 km/s. This is about 17,400 miles per hour and shows why putting satellites into Earth orbit takes such large rockets. Not only must the rocket lift the satellite above Earth's atmosphere, but the rocket must tip over and accelerate the satellite to circular velocity.

CALCULATING ESCAPE VELOCITY

If we launch a rocket upward, it will consume its fuel in a few moments and reach its maximum speed. From that point on, it will coast upward. How fast must a rocket travel to coast away from Earth and escape? We know, of course, that no matter how far it travels, it can never escape from Earth's gravity. The effects of Earth's gravity extend to infinity. It is possible, however, for a rocket to travel so fast initially that gravity can never slow it to a stop. Thus, it could leave Earth.

The escape velocity is the velocity required to escape from the surface of an astronomical body. Here we are interested in escaping from Earth or a planet; in later chapters, we will consider the escape velocity from stars, galaxies, and even a black hole.

The escape velocity, V_e, is given by a simple formula:

$$V_e = \sqrt{\frac{2GM}{r}}$$

Here G is the gravitational constant 6.67×10^{-11} m^3/s^{2}kg, M is the mass of the astronomical body in kilograms, and r is its radius in meters. (This formula is very similar to the formula for circular velocity.)

We can find the escape velocity from Earth by looking up its mass, 5.98×10^{24} kg, and its radius, 6.38×10^6 m. Then the escape velocity is:

$$V_e = \sqrt{\frac{2 \times 6.67 \times 10^{-11} \times 5.98 \times 10^{24}}{6.38 \times 10^6}} = \sqrt{\frac{7.98 \times 10^{14}}{6.38 \times 10^6}}$$

$$= \sqrt{1.25 \times 10^8} = 11{,}200 \text{ m/s} = 11.2 \text{ km/s}$$

This is equal to about 25,000 miles per hour.

Notice that the formula tells us that escape velocity depends on both mass and radius. A massive body might have a low escape velocity if it has a very large radius. We will meet such objects when we discuss giant stars. On the other hand, a rather low-mass body could have a very large escape velocity if it had a very small radius, a condition we will discuss when we meet black holes.

Circular velocity and escape velocity are two aspects of Newton's laws of gravity and motion. Once Newton understood gravity and motion, he could do what Kepler had failed to do—he could explain why the planets obey Kepler's laws of planetary motion.

KEPLER'S LAWS REEXAMINED

Now that we understand Newton's laws, gravity, and orbital motion, we can understand Kepler's laws of planetary motion in a new way.

Kepler's first law says that the orbits of the planets are ellipses with the sun at one focus. Kepler wondered why the planets keep moving along these orbits, and now we know the answer. They move because there is nothing to slow them down. Newton's first law says that a body in motion stays in motion unless acted on by some force. The gravity of the sun accelerates the planets inward toward the sun and holds them in their orbits, but it doesn't pull backward on the planets, so they don't slow to a stop. With no friction, they must continue to move.

The orbits of the planets are ellipses because gravity follows the inverse square law. In one of his most famous problems, Newton proved that if a planet moves in a closed orbit under the influence of an attractive force that follows the inverse square law, then the planet must follow an elliptical path.

Kepler's second law says that a planet moves faster when it is near the sun and slower when it is farther away. Once again, Newton's discoveries explain why. Earlier we saw that a body moving on a frictionless surface will continue to move in a straight line until it is acted on by some force; that is, the object has momentum. In a similar way, an object set rotating on a frictionless surface will continue rotating until something acts to speed it up or slow it down. Such an object has **angular momentum,** a measure of the rotation of the body about some point. A planet circling the sun has a given amount of angular momentum; and, with no outside influences to speed it up or slow it down, it must conserve its angular momentum. That is, its angular momentum must remain constant.

Mathematically, a planet's angular momentum is the product of its mass, velocity, and distance from the sun. This explains why a planet must speed up as it comes closer to the sun along an elliptical orbit. Its angular momentum is conserved, so as its distance from the sun decreases, its velocity must increase. In the same way, the planet's velocity must decrease as its distance from the sun increases.

This conservation of angular momentum is actually a common human experience. Skaters spinning slowly can draw their arms and legs closer to their axis of rotation and, through conservation of angular momentum, spin faster (▌Figure 5-7). To slow their rotation, they again extend their arms. Similarly, divers can spin rapidly in the tuck position and slow their rotation by stretching into the extended position.

Energy

One of the most fundamental parameters in science is **energy.** Physicists define energy as the ability to do work, but we might paraphrase that definition as the ability to produce a change. Certainly a moving body has energy. A planet moving along its orbit, a cement truck rolling down the highway, and a golf ball sailing down the fairway all have the ability to produce a change. Imagine colliding with any of these objects! But energy need not be represented by motion. Sunlight falling on a green plant, on photographic film, or on unprotected skin can produce chemical changes, and thus light is a form of energy.

Energy can take many forms. Batteries and gasoline are examples of chemical energy, and uranium fuel rods contain nuclear energy. A tank of hot water contains thermal energy. Even a weight on a high shelf can represent stored energy. Imagine a bowling ball falling off a high shelf onto your desk. That would produce significant change.

Much of science is the study of how energy flows from one place to another place, producing changes. A biologist might study the way a nerve cell transmits energy along its length to a muscle, while a geologist might study how energy flows as heat from Earth's interior and deforms Earth's surface. In such processes, we see energy being transformed from one state to another. Sunlight (energy) is absorbed by ocean plants and stored as sugars and starches (energy). When the plant dies, it and other ocean life are buried and become oil (energy), which we pump to the surface and burn in automobile engines to produce motion (energy).

Aristotle believed that all change originated in the motion of the starry sphere and flowed down to Earth. Modern science has found a more sophisticated description of the continual change we see around us, but we are still interested in the way energy flows through the world and produces change. Energy is the pulse of the natural world.

Using the metric system (Appendix), we will measure energy in **joules** (abbreviated **J**). One joule is about as much energy as that given up when an apple falls from a table to the floor.

Kepler's third law is also explained by a conservation law, but in this case it is the law of conservation of energy (Parameters of Science 5-2). A planet orbiting the sun has a specific amount of energy that depends only on its average distance from the sun. That energy can be divided between energy of motion and energy stored in the gravitational attraction between the planet and the sun. The energy of motion depends on how fast the planet moves, and the stored energy depends on the size of its orbit. The relation between these two kinds of energy is fixed by Newton's laws. That means there has to be a fixed relationship between the rate at which a planet moves around its orbit and the size of the orbit—between its orbital period P and the orbit's semimajor axis a. This is just Kepler's third law.

NEWTON'S VERSION OF KEPLER'S THIRD LAW

The equation for circular velocity is actually a version of Kepler's third law, as we can prove with three lines of simple algebra. The result is one of the most useful formulas in astronomy.

The equation for circular velocity, as we have seen, is:

$$V_c = \sqrt{\frac{GM}{r}}$$

The orbital velocity of a planet is simply the circumference of its orbit divided by the orbital period:

$$V = \frac{2\pi r}{P}$$

If we substitute this for V in the equation for circular velocity and solve for P^2, we get:

$$P^2 = \frac{4\pi^2}{GM} r^3$$

Here M is just the total mass of the system in kilograms. For a planet orbiting the sun, we can use the mass of the sun for M, because the mass of the planet is negligible compared to the mass of the sun. In a later chapter,

FIGURE 5-7
Skaters demonstrate conservation of angular momentum when they spin faster by drawing their arms and legs closer to their axis of rotation.

we will apply this formula to two stars orbiting each other, and then the mass M will be the sum of the two masses. For a circular orbit, r equals the semimajor axis a, so this formula is a general version of Kepler's third law, $P^2 = a^3$. In Kepler's version, we use the units AU and years, but in this formula, known as Newton's version of Kepler's third law, we use units of meters, seconds, and kilograms. G, of course, is the gravitational constant.

This is a very powerful formula. Astronomers use it to find the masses of bodies by observing orbital motion. If, for example, we observe a moon orbiting a planet and we can determine the size of its orbit, r, and the orbital period, P, we can use this formula to solve for M, the total mass of the planet plus the moon. There is no other way to find masses in astronomy, and we will use this formula in later chapters to find the masses of stars, galaxies, and planets.

This discussion is a good illustration of the power of Newton's work. By carefully defining motion and gravity and by giving them mathematical expression, Newton was able to derive new truths, among them Newton's version of Kepler's third law. His work changed science and made it into something new, precise, and exciting.

ASTRONOMY AFTER NEWTON

Newton published his work in July 1687 in a book called *Philosophiae Naturalis Principia Mathematica* (*Mathematical Principles of Natural Philosophy*), now known simply as *Principia* (▮ Figure 5-8). It is one of the most important books ever written. The principles changed astronomy, changed science, and changed the way we think about nature.

Principia changed astronomy and ushered in the age of gravitational astronomy. No longer did astronomers appeal to the whim of the gods to explain things in the heavens. No longer did they speculate on why the planets move. They now knew that the motions of the heavenly bodies are governed by simple, universal rules that describe the motions of everything from planets to falling apples. Suddenly the universe was understandable in simple terms.

Newton's laws of motion and gravity made it possible for astronomers to calculate the orbits of planets and moons. Not only could they explain how the heavenly bodies move, they could predict future motions (Window on Science 5-2). This subject, known as gravitational astronomy, dominated astronomy for almost 200 years and is still important. It included the calculation of the orbits of comets and asteroids and the theoretical prediction of the existence of two planets, Neptune and Pluto.

Principia also changed science in general. The works of Copernicus and Kepler had been mathematical, but no book before had so clearly demonstrated the power of mathematics as a language of precision. Newton's arguments in *Principia* were so powerful an illustration of the quantitative study of nature that scientists around the world adopted mathematics as their most powerful tool.

Also, *Principia* changed the way we think about nature. Newton showed that the rules that govern the universe are simple. Particles move according to three rules of motion and attract each other with a force called gravity. These motions are predictable, and that makes the universe a vast machine based on a few simple rules. It is complex only in that it contains a vast number of particles. In Newton's view, if he knew the location and motion of every particle in the universe, he could, in principle, derive the past and future of the universe in every detail. This mechanical determinism has been undermined by modern quantum mechanics, but it dominated science for more than two centuries during which scientists thought of nature as a beautiful clockwork that would be perfectly predictable if we knew how all the gears meshed.

FIGURE 5-8
Newton, working from the discoveries of Galileo and Kepler, derived three laws of motion and the principle of mutual gravitation. He and some of his discoveries are honored on this English pound note. Notice the diagram of orbital motion in the background and the open copy of *Principia* in Newton's hands.

Testing a Theory by Prediction

When you read about any science, you should notice that scientific theories face in two directions. They look back into the past and explain phenomena we have previously observed. For example, Newton's laws of motion and gravity explained how the planets moved. But theories also face forward in that they enable us to make predictions about what we should find as we explore further. Thus, Newton's laws allowed astronomers to calculate the orbits of comets, predict their return, and eventually understand their origin.

Scientific predictions are important in two ways. First, if a theory leads to a prediction and scientists later discover the prediction was true, the theory is confirmed, and scientists gain confidence that it is a true description of nature. But predictions are important in science in a second way. Using an existing theory to make a prediction may lead us into an unexplored avenue of knowledge. Thus, the first theories of genetics made predictions that confirmed the genetic theory of inheritance, but those predictions also created a new understanding of how living creatures evolve.

As you read about any scientific theory, think about both what it can explain and what it can predict.

Most of all, Newton's work broke the last bonds between science and formal philosophy. Newton did not speculate on the good or evil of gravity. He did not debate its meaning. Not more than a hundred years before, scientists would have argued over the "reality" of gravity. Newton didn't care for these debates. He wrote, "It is enough that gravity exists and suffices to explain the phenomena of the heavens."

REVIEW CRITICAL INQUIRY

How do Newton's laws of motion explain the orbital motion of the moon?

If Earth and the moon did not attract each other, the moon would move in a straight line in accord with Newton's first law of motion and vanish into space. Instead, gravity pulls the moon toward Earth's center, and the moon accelerates toward Earth. This acceleration is just enough to pull the moon away from its straight-line motion and cause it to follow a curve around Earth. In fact, it is correct to say that the moon is falling, but because of its lateral motion it continuously misses Earth.

Every orbiting object is falling toward the center of its orbit but is moving laterally fast enough to compensate for the inward motion, and it follows a curved orbit. If this is true, then how can astronauts float inside spacecraft in a "weightless" state? Why might "free fall" be a more accurate term?

Newton's laws of motion and gravitation are critical in astronomy not only because they describe orbital motion but also because they describe the interaction of astronomical bodies. Newton made other discoveries, but we reserve these for later chapters. Now we move forward two centuries to see how Einstein described gravity in a new and powerful way.

5-3 EINSTEIN AND RELATIVITY

In the early years of the last century, Albert Einstein (1879–1955) (❚ Figure 5-9) began thinking about how motion and gravity interact. He soon gained international fame by showing that Newton's laws of motion and gravity were only partially correct. The revised theory became known as the theory of relativity. As we will see, there are really two theories of relativity.

SPECIAL RELATIVITY

Einstein began by thinking about how moving observers see events around them. His analysis led him to the first postulate of relativity, also known as the principle of relativity:

> **First postulate** (the principle of relativity): Observers can never detect their *uniform* motion except relative to other objects.

FIGURE 5-9

Einstein's greatest accomplishments, the special theory of relativity and the general theory of relativity, were developed as he analyzed the nature of motion and time.

You may have experienced the first postulate while sitting on a train in a station. You suddenly notice that the train on the next track has begun to creep out of the station. However, after several moments you realize that it is your own train that is moving and that the other train is still motionless on its track.

Consider another example. Suppose you are floating in a spaceship in interstellar space and another spaceship comes coasting by (Figure 5-10a). You might conclude that it is moving and you are not, but someone in the other ship might be equally sure that you are moving and it is not. The principle of relativity says that there is no experiment you can perform to decide which ship is moving and which is not. This means that there is no such thing as absolute rest—all motion is relative.

Because neither you nor the people in the other spaceship could perform any experiment to detect your absolute motion through space, the laws of physics must have the same form in both spaceships. Otherwise, experiments would give different results in the two ships, and you could decide who was moving. Thus, a more general way of stating the first postulate refers to these laws of physics:

> **First postulate** (alternate version): The laws of physics are the same for all observers, no matter what their motion, so long as they are not *accelerated*.

The words *uniform* and *accelerated* are important. If either spaceship were to fire its rockets, then its velocity would change. The crew of that ship would know it because they would feel the acceleration pressing them into their couches. Accelerated motion, therefore, is different—we can always tell which ship is accelerating and which is not. The postulates of relativity discussed here apply only to observers in uniform motion. That is why the theory is called **special relativity.**

The first postulate fit with Einstein's conclusion that the speed of light must be constant for all observers. No matter how you move, your measurement of the speed of light has to give the same result (Figure 5-10b). This became the second postulate of special relativity:

> **Second postulate:** The velocity of light is constant and will be the same for all observers independent of their motion relative to the light source.

Once Einstein had accepted the basic postulates of relativity, he was led to some startling discoveries. Newton's laws of motion and gravity

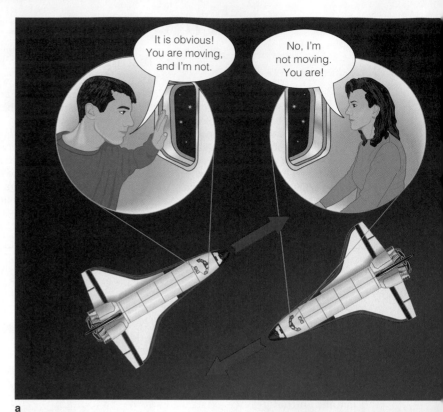

a

b

FIGURE 5-10

(a) The principle of relativity says that observers can never detect their uniform motion, except relative to other observers. Thus, neither of these travelers can decide who is moving and who is not. (b) If the principle of relativity is correct, the velocity of light must be a constant for all observers. If the velocity of light depended on the motion of the observer through space, then these travelers could decide who was moving and who was not.

worked well as long as distances were small and velocities were low. But when we begin to think of very large distances or very high velocities, Newton's laws are no longer adequate to describe what happens. Instead, we must use relativistic physics. For example, special relativity shows that the observed mass of a moving particle depends on its velocity. The higher the velocity, the greater the mass of the particle. This is not significant at low velocities, but it becomes very important as the velocity approaches the velocity of light. Such increases in mass are observed whenever physicists accelerate atomic particles to high velocities (▌ Figure 5-11).

This discovery led to yet another insight. The relativistic equations that describe the energy of a moving particle predict that the energy of a motionless particle is not zero. Rather, its energy at rest is $m_0 c^2$. This is of course the famous equation:

$$E = m_0 c^2$$

The c is the speed of light and the m_0 is the mass of the particle when it is at rest. (We must specify mass this way because one of the consequences of relativity is that a particle's mass depends on its velocity.) This simple formula suggests that mass and energy are related, and we will see in later chapters how nature can convert one into the other inside stars.

For example, suppose that we convert 1 kg of matter into energy. We must express the velocity of light as 3×10^8 m/s, and our result is 9×10^{16} joules (J) (approximately equal to a 20-megaton nuclear bomb). (Recall that a joule is a unit of energy roughly equivalent to the energy given up when an apple falls from a table to the floor.) Our simple calculation shows that the energy equivalent of even a small mass is very large.

Other relativistic effects include the slowing of moving clocks and the shrinkage of lengths measured in the direction of motion. A detailed discussion of the major consequences of the special theory of relativity is beyond the scope of this book. Instead, we must consider Einstein's second advance, the general theory.

THE GENERAL THEORY OF RELATIVITY

In 1916, Einstein published a more general version of the theory of relativity that dealt with accelerated as well as uniform motion. This **general theory of relativity** contained a new description of gravity.

Einstein began by thinking about observers in accelerated motion. Imagine an observer sitting in a windowless spaceship. Such an observer cannot distinguish between the force of gravity and the inertial forces produced by the acceleration of the spaceship (▌ Figure 5-12). This led Einstein to conclude that gravity and acceleration through space-time are related, a conclusion now known as the equivalence principle:

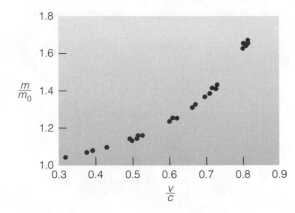

FIGURE 5-11

The observed mass of moving electrons depends on their velocity. As the ratio of their velocity to the velocity of light, v/c, gets larger, the mass of the electrons in terms of their mass at rest, m/m_0, increases. Such relativistic effects are quite evident in particle accelerators, which accelerate atomic particles to very high velocities.

> **Equivalence principle:** Observers cannot distinguish locally between inertial forces due to acceleration and uniform gravitational forces due to the presence of a massive body.

The importance of the general theory of relativity lies in its description of gravity. Einstein concluded that gravity, inertia, and acceleration are all associated with the way space and time are related. This relation is often referred to as curvature, and a one-line description of general relativity explains a gravitational field as a curved region of space-time:

> **Gravity according to general relativity:** Mass tells space-time how to curve, and the curvature of space-time (gravity) tells mass how to accelerate.

Thus, we feel gravity because Earth's mass causes a curvature of space-time. The mass of our bodies responds to that curvature by accelerating toward Earth's center. According to general relativity, all masses cause curvature, and the larger the mass, the more severe the curvature.

CONFIRMATION OF THE CURVATURE OF SPACE-TIME

Einstein's general theory of relativity has been confirmed by a number of experiments, but two are worth mentioning here because they were among the first tests of the theory. One involves Mercury's orbit, and the other involves eclipses of the sun.

Johannes Kepler understood that the orbit of Mercury is elliptical, but only since 1859 have astronomers known that the long axis of the orbit sweeps around the sun in a motion called precession (▌ Figure 5-13). The total observed precession is 5600.73 seconds of arc per

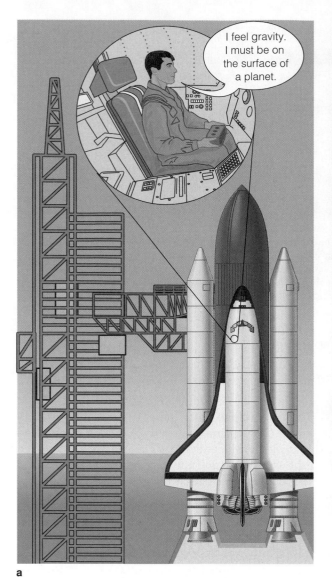

FIGURE 5-12

(a) An observer in a closed spaceship on the surface of a planet feels gravity. (b) In space, with the rockets smoothly firing and accelerating the spaceship, the observer feels inertial forces that are equivalent to gravitational forces.

century (as seen from Earth), or about 1.5° per century. This precession is produced by the gravitation of Venus, Earth, and the other planets. However, when astronomers used Newton's description of gravity, they calculated that the precession should amount to only 5557.62 seconds of arc per century. Thus, Mercury's orbit is advancing 43.11 seconds of arc per century faster than Newton's law predicted.

This is a tiny effect. Each time Mercury returns to perihelion, its closest point to the sun, it is about 29 km (18 miles) past the position predicted by Newton's laws. This is such a small distance compared with the planet's diameter of 4850 km that it could never have been detected had it not been cumulative. Each orbit, Mercury gains 29 km, and in a century it gains over 12,000 km—more than twice its own diameter. Thus,

this tiny effect, called the advance of perihelion of Mercury's orbit, accumulated into a serious discrepancy in the Newtonian description of the universe.

The advance of perihelion of Mercury's orbit was one of the first problems to which Einstein applied the principles of general relativity. First he calculated how much the sun's mass curves space-time in the region of Mercury's orbit, and then he calculated how Mercury moves through the space-time. The theory predicted that the curved space-time should cause Mercury's orbit to advance by 43.03 seconds of arc per century, well within the observational accuracy of the excess (Figure 5-13b).

Einstein was elated with this result, and he would be even happier with modern studies that have shown that Mercury, Venus, Earth, and even Icarus, an asteroid that comes close to the sun, have orbits observed to

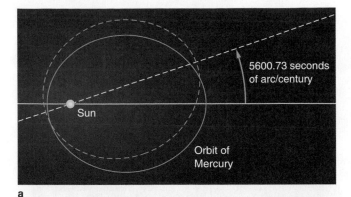

a

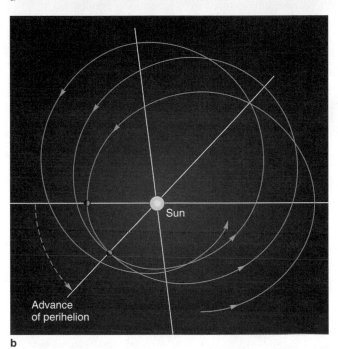

b

FIGURE 5-13

(a) Mercury's orbit precesses 5600.73 seconds of arc per century—43.11 seconds of arc per century faster than predicted by Newton's laws. (b) Even when we ignore the influences of the other planets, Mercury's orbit is not a perfect ellipse. Curved space-time near the sun distorts the orbit from an ellipse into a rosette. The advance of Mercury's perihelion is exaggerated about a million times in this figure.

TABLE 5-2
Precession in Excess of Newtonian Physics

Planet	Observed Excess Precession (seconds of arc per century)	Relativistic Prediction (seconds of arc per century)
Mercury	43.11 ± 0.45	43.03
Venus	8.4 ± 0.48	8.6
Earth	5.0 ± 1.2	3.8
Icarus	9.8 ± 0.8	10.3

by curved space-time just as a rolling golf ball is deflected by undulations in a putting green. Einstein predicted that starlight grazing the sun's surface would be deflected by 1.75 seconds of arc (▌ Figure 5-14). Starlight passing near the sun is normally lost in the sun's glare, but during a total solar eclipse stars beyond the sun could be seen. As soon as Einstein published his theory, astronomers rushed to observe such stars and thus test the curvature of space-time.

The first solar eclipse following Einstein's announcement in 1916 was June 8, 1918. It was cloudy at some observing sites, and results from other sites were inconclusive. The next occurred on May 29, 1919, only months after the end of World War I, and was visible from Africa and South America. British teams went to both Brazil and Príncipe, an island off the coast of Africa. First, they photographed that part of the sky where the sun would be located during the eclipse and measured the positions of the stars on the plates. Then, during the eclipse, they photographed the same star field with the eclipsed sun located in the middle. After measuring the plates, they found slight changes in the

be slipping forward due to the curvature of space-time near the sun (▌ Table 5-2).

This same effect has been detected in pairs of stars that orbit each other. In some cases, the advance of perihelion agrees with general relativity; in many cases, the sizes and masses of the stars are not well enough known for us to be certain of the theoretical rate of advance we should expect. But in a few cases, the stars' orbits appear to be changing faster than predicted. This may be a critical test for Einstein's theory. This shows how science continues to test theories over and over even after they are widely accepted.

A second test of the curvature of space-time was directly related to the motion of light through the curved space-time near the sun. The equations of general relativity predicted that light would be deflected

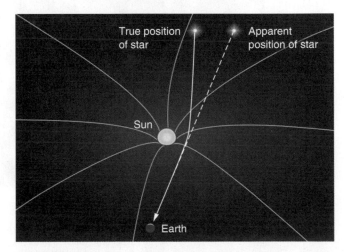

FIGURE 5-14

Like a depression in a putting green, the curved space-time near the sun deflects light from distant stars and makes them appear to lie slightly farther from the sun than their true positions.

positions of the stars. During the eclipse, the positions of the stars on the plates were shifted outward, away from the sun (▌Figure 5-15). If a star had been located at the edge of the solar disk, it would have been shifted outward by about 1.8 seconds of arc. This represents good agreement with the theory's prediction.

This test has been repeated at many total solar eclipses since 1919, with similar results. The most accurate results were obtained in 1973 when a Texas–Princeton team measured a deflection of 1.66 ± 0.18 seconds of arc—good agreement with Einstein's theory.

The general theory of relativity is critically important in modern astronomy. We will discuss it again when we meet black holes, distant galaxies, and the big bang universe. The theory revolutionized modern physics by providing a theory of gravity based on the geometry of curved space-time. Thus, Galileo's inertia and Newton's mutual gravitation are shown to be fundamental properties of space and time.

no longer be in a closed spaceship. As long as we make no outside observations, we can't tell whether our spaceship is firing its rockets and accelerating through space or resting on the surface of a planet where gravity gives us weight.

Einstein took the equivalence principle to mean that gravity and acceleration through space-time are somehow related. The general theory of relativity gives that relationship mathematical form and shows that gravity is really a disturbance in space-time that physicists refer to as curvature. Thus, we say "mass tells space-time how to curve, and space-time tells mass how to move." The equivalence principle led Einstein to an explanation for gravity.

But what about the second postulate of special relativity? Why does it have to be true if the first postulate is true? And what does the second postulate tell us about the nature of uniform motion?

REVIEW CRITICAL INQUIRY

What does the equivalence principle tell us?

The equivalence principle says that there is no observation we can make inside a closed spaceship to distinguish between uniform acceleration and gravitation. Of course, we could open a window and look outside, but then we would

Our discussion of the origin of astronomy began with the builders of Stonehenge and reaches the modern day with Einstein's general theory of relativity. Now that we have seen where astronomy came from, we are ready to see how it helps us understand the nature of the universe. Our first question should be "How do astronomers get information?" The answer involves the astronomer's most basic tool, the telescope; that is the subject of the next chapter.

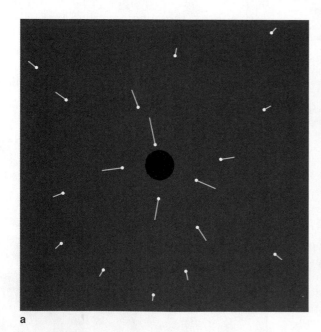
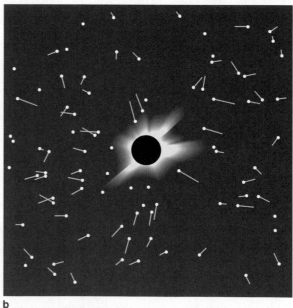

FIGURE 5-15

(a) Schematic drawing of the deflection of starlight by the sun's gravity. Dots show the true positions of the stars as photographed months before. Lines point toward the positions of the stars during the eclipse. (b) Actual data from the eclipse of 1922. Random uncertainties of observation cause some scatter in the data, but in general the stars appear to move away from the sun by 1.77 seconds of arc at the edge of the sun's disk. The deflection of stars is magnified by a factor of 2300 in both (a) and (b).

SUMMARY

Galileo took the first step toward understanding motion and gravity when he began to study falling bodies. He found that a falling object is accelerated; that is, it falls faster and faster with each passing second. The rate at which it accelerates, termed the acceleration of gravity, is 9.8 m/s^2 (32 ft/s^2) at Earth's surface and does not depend on the weight of the object. According to tradition, Galileo demonstrated this by dropping balls of iron and wood from the Leaning Tower of Pisa to show that they would fall together. Finally, Galileo stated the law of inertia. In the absence of friction, a moving body on a horizontal plane will continue moving forever.

Newton adopted Galileo's law of inertia as his first law of motion. The second law of motion establishes the relationship between the force acting on a body, its mass, and the resulting acceleration. The third law says that forces occur in pairs acting in opposite directions.

Newton also developed an explanation for the accelerations that Galileo had discovered—gravity. By considering the motion of the moon, Newton was able to show that objects attract each other with a gravitational force that is proportional to the product of their masses and inversely proportional to the square of the distance between them.

If we understand Newton's laws of motion and gravity, we can better understand orbital motion. An object in space near Earth would move along a straight line and quickly leave Earth were it not for Earth's gravity accelerating the object toward Earth's center and forcing it to follow a curved path, an orbit. If there is no friction, the object will fall around its orbit forever.

Newton's laws also illuminate the meaning of Kepler's three laws of planetary motion. The planets follow elliptical orbits because gravity follows the inverse square law. The planets move faster when closer to the sun and slower when farther away because they conserve angular momentum. The same law makes ice skaters spin faster when they draw their arms and legs nearer their bodies. A planet's orbital period squared is proportional to its orbital radius cubed because the moving planet conserves energy.

In fact, Newtonian gravity and motion show that an ellipse is only one of a number of orbits that a body can follow. The circle and ellipse are closed orbits that return to their starting points. If an object moves at circular velocity, V_c, it will follow a circular orbit. If its velocity equals or exceeds the escape velocity, V_e, it will follow a parabola or hyperbola. These orbits are termed "open" because the object never returns to its starting place.

Newton's laws changed astronomy and our view of nature. They made it possible for astronomers to predict the motions of the heavenly bodies using the analytic power of mathematics. Newton's laws also show that the apparent complexity of the universe is based on a few simple principles, or natural laws.

Einstein published two theories that extended Newton's laws of motion and gravity. The special theory of relativity, published in 1905, applies to observers in uniform motion. The theory holds that the speed of light is a constant for all observers and that mass and energy are related by the expression $E = m_0c^2$.

The general theory of relativity, published in 1916, holds that a gravitational field is a curvature of space-time caused by the presence of a mass. Thus, Earth's mass curves space-time, and the mass of our bodies responds to that curvature by accelerating toward Earth's center. This curvature of space-time was confirmed by the slow advance in perihelion of the orbit of Mercury and by the deflection of starlight observed during a 1919 total solar eclipse.

NEW TERMS

natural motion	geosynchronous satellite
violent motion	center of mass
acceleration of gravity	closed orbit
momentum	escape velocity
mass	open orbit
acceleration	angular momentum
velocity	energy
inverse square law	joule (J)
field	special relativity
circular velocity	general theory of relativity

REVIEW QUESTIONS

Ace⟲Astronomy™ Assess your understanding of this chapter's topics with additional quizzing and animations at **http:// astronomy.brookscole.com/seeds8e**

1. Why wouldn't Aristotle's explanation of gravity work if Earth was not the center of the universe?

2. According to the principles of Aristotle, what part of the motion of a baseball pitched across the home plate is natural motion? What part is violent motion?

3. If we drop a feather and a steel hammer at the same moment, they should hit the ground at the same instant. Why doesn't this work on Earth, and why does it work on the moon?

4. What is the difference between mass and weight? between speed and velocity?

5. Why did Newton conclude that some force had to pull the moon toward Earth?

6. Why do we conclude that gravity has to be mutual and universal?

7. How does the concept of a field explain action at a distance? Name another kind of field also associated with action at a distance.

8. Why can't a spacecraft go "beyond Earth's gravity"?

9. What is the center of mass of the Earth–moon system? Where is it?

10. How do planets orbiting the sun and skaters conserve angular momentum?

11. Why is the period of an open orbit undefined?

12. How does the first postulate of special relativity imply the second?

13. When we ride a fast elevator upward, we feel slightly heavier as the trip begins and slightly lighter as the trip ends. How is this phenomenon related to the equivalence principle?

14. From your knowledge of general relativity, would you expect radio waves from distant galaxies to be deflected as they pass near the sun? Why or why not?

DISCUSSION QUESTIONS

1. How did Galileo idealize his inclines to conclude that an object in motion stays in motion until it is acted on by some force?

2. Give an example from everyday life to illustrate each of Newton's laws.

PROBLEMS

1. Compared with the strength of Earth's gravity at its surface, how much weaker is gravity at a distance of 10 Earth radii from Earth's center? at 20 Earth radii?

2. Compare the force of gravity on the surface of the moon with the force of gravity at Earth's surface.

3. If a small lead ball falls from a high tower on Earth, what will be its velocity after 2 seconds? after 4 seconds?

4. What is the circular velocity of an Earth satellite 1000 km above Earth's surface? (*Hint:* Earth's radius is 6380 km.)

5. What is the circular velocity of an Earth satellite 36,000 km above Earth's surface? What is its orbital period? (*Hint:* Earth's radius is 6380 km.)

6. What is the orbital period of an imaginary satellite orbiting just above Earth's surface? Ignore friction with the atmosphere.

7. Repeat the previous problem for Mercury, Venus, the moon, and Mars.

8. Describe the orbit followed by the slowest cannonball on page 84 on the assumption that the cannonball could pass freely through Earth. (Newton got this problem wrong the first time he tried to solve it.)

9. If you visited an asteroid 30 km in radius with a mass of 4×10^{17} kg, what would be the circular velocity at its surface? A major league fastball travels 90 mph. Could a good pitcher throw a baseball into orbit around the asteroid?

10. What is the orbital period of a satellite orbiting just above the surface of the asteroid in Problem 9?

11. What would be the escape velocity at the surface of the asteroid in Problem 9? Could a major league pitcher throw a baseball off of the asteroid?

CRITICAL INQUIRIES FOR THE WEB

1. Einstein's general theory of relativity predicts the curvature of space-time, but here on Earth we have little opportunity to observe such effects. Find an astronomical situation in which space-time curvature is evident from our observations, and describe the effect of the curvature on what we see when we view these objects.

2. Communications satellites are obvious uses of the geosynchronous orbit, but can you think of other uses for such orbits? Find an Internet site that uses or displays information gleaned from geosynchronous orbit that provides a useful service.

 Visit the Seeds *Foundations of Astronomy* companion Web site for critical thinking exercises, articles, and additional readings from InfoTrac College Edition, Brooks/Cole's online student library.

LIGHT AND TELESCOPES

He burned his house down
for the fire insurance
And spent the proceeds
on a telescope.

Robert Frost, The Star-Splitter

Courtesy William Keel

Chapter 6

GUIDEPOST

Previous chapters have described the sky as it appears to our unaided eyes, but modern astronomers turn powerful telescopes on the sky. Chapter 6 introduces us to the modern astronomical telescope and its delicate instruments.

The study of the universe is so challenging, astronomers cannot ignore any source of information; that is why they use the entire spectrum, from gamma rays to radio waves. This chapter shows how critical it is for astronomers to understand the nature of light.

In each of the chapters that follow, we will study the universe using information gathered by the telescopes and instruments described in this chapter.

What do fleas living on rats have to do with modern astronomy? That may sound like the beginning of a bad joke, but it is actually related to the subject of this chapter. We will examine the tools that modern astronomers use, and those tools are connected by an interesting sequence of events to rats and their fleas.

The most horrible disease in history—the black plague—was spread by fleabites, and the fleas lived on the rats that infested the cities. Plague broke out in London in 1665, and, although no one knew how the disease spread, people who lived in the country were less likely to get the plague than city dwellers. All who could left the cities. When the plague reached Cambridge, the colleges were closed, and both students and faculty fled to the English countryside. One who fled was the young Isaac Newton. From the summer of 1665 to 1667, he spent most of his time in his mother's cottage in the small village of Woolsthorpe. While he was there, he conducted an experiment that changed the history of science.

Boring a hole in a shutter, he admitted a thin beam of sunlight into his darkened room. A glass prism placed in the beam threw a rainbow of color—a spectrum—across the wall. When he used a second prism to recombine the colors, they produced white light. From this and other experiments conducted in his bedroom, he concluded that white light was made up of a mixture of all the colors of the rainbow.

When the plague abated, Newton returned to the university, where he began experimenting with telescopes. He discovered that telescopes made of lenses produced colored fringes around bright objects in the field of view because the glass lenses broke the light into colors, just as his prism broke up the sunlight. To solve the problem, Newton designed and built a telescope containing a mirror instead of a lens. Although his first model hardly exceeded 1 inch in diameter, when Newton presented it to the Royal Society in 1671, it established his reputation as a scientist.

For a century after Newton's first telescope, astronomers did little with such devices, but as instrument makers grew more skilled, large telescopes became the principal tool of the astronomer. Telescopes are important in astronomy because they gather light and concentrate it for study. The larger the telescope, the more light it gathers. Thus astronomers are still striving to build bigger telescopes to gather more light from the objects in the sky (∎ Figure 6-1). Like Newton's original telescope, almost all modern telescopes use mirrors rather than lenses to avoid spreading the light into its component colors.

6-1 RADIATION: INFORMATION FROM SPACE

Just as a book on bread baking might begin with a discussion of flour, our chapter on telescopes begins with a discussion of light—not just visible light, but the entire range of radiation from the sky.

LIGHT AS A WAVE AND A PARTICLE

Light is merely one form of radiation, called **electromagnetic radiation** because it is associated with changing electric and magnetic fields that travel through

FIGURE 6-1
Astronomers build giant telescopes on high mountains where the air is clear and steady. (a) In this time exposure, star trails are visible behind the dome of the northern Gemini Telescope. The tail lights of a vehicle illuminate the road and a shooting star (a meteor) is visible near the horizon. (b) Inside the dome, the telescope holds a mirror 8.2 meters in diameter to focus light. Note the figure at right for scale. *(NOAO/AURA/NSF)*

a   b

space and transfer energy from one place to another. When light enters our eye, the fluctuating electric and magnetic fields carry energy that stimulates nerve endings, and we see what we call light.

The oscillating electric and magnetic fields that constitute electromagnetic radiation move through space at about 300,000 km/s (186,000 mi/s). This speed is commonly referred to as the speed of light, *c*, but it is in fact the speed of all such radiation in a vacuum.

It may seem odd to use the word "radiation" when we speak of light. The word can be used to refer to high-energy particles emitted from radioactive atoms, and we have all leaned to be a little bit concerned when we see the word "radiation." But it really refers to anything that spreads outward from a source. Light radiates from a source, so we can correctly refer to light as a form of radiation.

Electromagnetic radiation is a wave phenomenon; that is, it is associated with a periodically repeating disturbance, or wave. We are familiar with waves in water. If we disturb a quiet pool of water, waves spread across the surface. Imagine that we use a meter stick to measure the distance between the successive peaks of a wave. This distance is the **wavelength,** usually represented by the Greek letter lambda (λ). If we were measuring ripples in a pond, we might find that the wavelength is a few centimeters, whereas the wavelength of ocean waves might be a hundred meters or more. There is no restriction on the wavelength of electromagnetic radiation. Wavelengths can range from smaller than the diameter of an atom to larger than that of Earth.

Because all electromagnetic radiation travels at the speed of light, wavelength is related to **frequency,** the number of cycles that pass in one second. Short-wavelength radiation has a high frequency; long-wavelength radiation has a low frequency. To understand this, imagine watching an electromagnetic wave race past us while we count its peaks (❙ Figure 6-2). If the wavelength is short, we will count many peaks in one second; if the wavelength is long, we will count few peaks per second. The dials on radios are marked in frequency, but they could just as easily be marked in wavelength. The relation between wavelength and frequency is a simple one:

$$\lambda = \frac{c}{f}$$

That is, the wavelength equals the speed of light *c* divided by the frequency *f*. Notice that the larger (higher) the frequency, the smaller (shorter) the wavelength. In most cases, astronomers use wavelength rather than frequency.

Radio waves can have wavelengths from a few millimeters for microwaves to kilometers. In contrast, the wavelength of light is so short that we must use more convenient units. In this book, we will use **nanometers (nm)** because this unit is consistent with the International System of units. One nanometer is 10^{-9} meter,

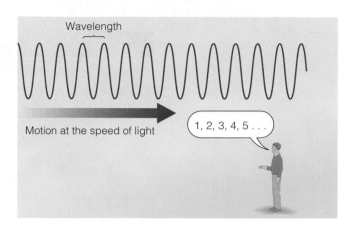

FIGURE 6-2

All electromagnetic waves travel at the speed of light. The wavelength is the distance between successive peaks. The frequency of the wave is the number of peaks that pass us in one second.

and visible light has wavelengths that range from about 400 nm to about 700 nm. Another unit that astronomers commonly use, and a unit that you will see in many references on astronomy, is the **Angstrom (Å).** One Angstrom is 10^{-10} meter, and visible light has wavelengths between 4000 Å and 7000 Å.

You may find radio astronomers describing wavelengths in centimeters or millimeters, and infrared astronomers often refer to wavelengths in micrometers (or microns). One micrometer (μm) is 10^{-6} meter. Whatever unit is used to describe the wavelength, we must keep in mind that all electromagnetic radiation is the same phenomenon.

What exactly is electromagnetic radiation? Although we have been discussing its wavelength, it is incorrect, or at least incomplete, to say that electromagnetic radiation is a wave. It sometimes has the properties of a wave and sometimes has the properties of a particle. For instance, the beautiful colors in a soap bubble arise from the wave nature of light. On the other hand, when light strikes the photoelectric cell in a camera's light meter, it behaves like a stream of particles carrying specific amounts of energy. Throughout his life, Newton believed that light was made up of particles, but we now recognize that light can behave as both particle and wave. Our model of light is thus more complete than Newton's. We will refer to "a particle of light" as a **photon,** and we can recognize its dual nature by thinking of it as a bundle of waves.

The amount of energy a photon carries depends on its wavelength. The shorter the wavelength, the more energy the photon carries; the longer the wavelength, the less energy it contains. This is easy to remember because short wavelengths have high frequencies, and we expect rapid fluctuations to be more energetic. We can express this relationship in a simple formula:

$$E = \frac{hc}{\lambda}$$

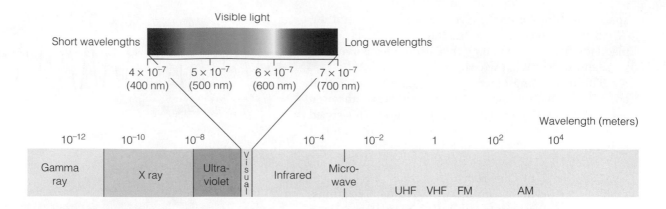

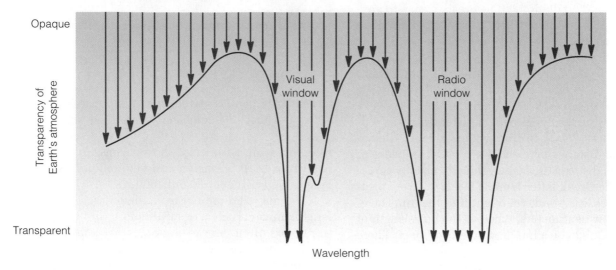

FIGURE 6-3

The spectrum of visible light, extending from red to violet, is only part of the electromagnetic spectrum. Most radiation is absorbed in Earth's atmosphere, and only radiation in the visual window and the radio window can reach Earth's surface

Here h is Planck's constant (6.6262×10^{-34} joule s), c is the speed of light (3×10^8 m/s), and λ is the wavelength in meters. A photon of visible light carries a very small amount of energy, but a photon with a very short wavelength can carry much more.

THE ELECTROMAGNETIC SPECTRUM

A spectrum is an array of electromagnetic radiation in order of wavelength. We are most familiar with the spectrum of visible light, which we see in rainbows. The colors of the spectrum differ in wavelength, with red having the longest wavelength and violet the shortest. The visible spectrum is shown at the top of ▌Figure 6-3.

The average wavelength of visible light is about 0.00005 cm. We could put 50 light waves end to end across the thickness of a sheet of household plastic wrap. Measured in nanometers, the wavelength of visible light ranges from about 400 to 700 nm. Just as we sense the wavelength of sound as pitch, we sense the wavelength of light as color. Light near the short-wavelength end of the visible spectrum (400 nm) looks violet to our

eyes, and light near the long-wavelength end (700 nm) looks red.

Figure 6-3 shows how the visible spectrum makes up only a small part of the entire electromagnetic spectrum. Beyond the red end of the visible spectrum lies **infrared radiation,** where wavelengths range from 700 nm to about 0.1 cm. Our eyes are not sensitive to this radiation, but our skin senses it as heat. A "heat lamp" is just a bulb that gives off principally infrared radiation.

Beyond the infrared part of the electromagnetic spectrum lie radio waves. Microwaves have wavelengths of a millimeter to a few centimeters and are used for radar and long-distance telephone communication. Longer wavelengths are used for UHF and VHF television transmissions. FM, military, governmental, and ham radio signals have wavelengths up to a few meters, and AM radio waves can have wavelengths of kilometers.

The distinction between the wavelength ranges is not sharp. Long-wavelength infrared radiation and the shortest microwave radio waves are the same. Similarly, there is no clear division between the short-wavelength infrared and the long-wavelength part of the visible spectrum. It is all electromagnetic radiation.

Look once again at the electromagnetic spectrum in Figure 6-3, and notice that electromagnetic waves shorter than violet are called **ultraviolet.** Electromagnetic waves even shorter are called X rays, and the shortest are gamma rays. Again, the boundaries between these wavelength ranges are not clearly defined.

X rays and gamma rays can be dangerous, and even ultraviolet photons have enough energy to do us harm. Small doses produce a suntan and larger doses sunburn and skin cancers. Contrast this to the lower-energy infrared photons. Individually they have too little energy to affect skin pigment, a fact that explains why you can't get a tan from a heat lamp. Only by concentrating many low-energy photons in a small area, as in a microwave oven, can we transfer significant amounts of energy.

We are interested in electromagnetic radiation because it brings us clues to the nature of stars, planets, and other celestial objects. Earth's atmosphere is opaque to most electromagnetic radiation, as shown by the graph at the bottom of Figure 6-3. Gamma rays, X rays, and some radio waves are absorbed high in Earth's atmosphere, and a layer of ozone (O_3) at an altitude of about 30 km absorbs ultraviolet radiation. Water vapor in the lower atmosphere absorbs the longer wavelength infrared radiation. Only visible light, some shorter wavelength infrared, and some radio waves reach Earth's surface through two wavelength regions called **atmospheric windows.** Obviously, if we wish to study the sky from Earth's surface, we must look out through one of these windows.

REVIEW CRITICAL INQUIRY

What could we see if our eyes were sensitive only to X rays?
Sometimes the critical analysis of an idea is easier if we try to imagine a totally new situation. In this case, we might at first expect to be able to see through walls, but remember that our eyes detect only light that already exists. There are almost no X rays bouncing around at Earth's surface, so if we had X-ray eyes, we would be in the dark and would be unable to see anything. Even when we look up at the sky, we would see nothing, because Earth's atmosphere is not transparent to X rays. If Superman can see through walls, it is not because his eyes can detect X rays.

But suppose our eyes were sensitive only to infrared waves or to radio waves. Would we be in the dark?

Now that we know something about electromagnetic radiation, we can study the tools astronomers use to gather and analyze that radiation.

6-2 OPTICAL TELESCOPES

Astronomers build optical telescopes to gather light and focus it into sharp images. This requires sophisticated optical and mechanical designs, and it leads astronomers to build gigantic telescopes on the tops of high mountains. To begin, we need to understand the terminology of telescopes, but it is more important to understand how different kinds of telescopes work and why some are better than others.

TWO KINDS OF TELESCOPES

Astronomical telescopes focus light into an image in one of two ways, as shown in ▌Figure 6-4. A lens bends (refracts) the light as it passes through the glass and brings it to a focus to form a small inverted image. A mirror—a concave piece of glass with a reflective surface—forms an image by reflecting the light. In either case, the **focal length** is the distance from the lens or mirror to the image formed of a distant light source, such as a star (▌Figure 6-5). Short-focal-length lenses and mirrors must be strongly curved, and long-focal-length lenses and mirrors are less strongly curved. Grinding the proper shape on a lens or mirror is a delicate, time-consuming, and expensive process.

Because there are two ways to focus light, there are two kinds of astronomical telescopes. **Refracting telescopes** use a large lens to gather and focus the light, whereas **reflecting telescopes** use a concave mirror. The advantages of the reflecting telescope have made it the preferred design for modern observatories.

The main lens in a refracting telescope is called the **primary lens,** and the main mirror in a reflecting telescope is called the **primary mirror.** These are also called the **objective lens** and **mirror.** Both kinds of telescopes form a very small, inverted image that is difficult to observe directly, so astronomers use a small lens called the **eyepiece** to magnify the image and make it convenient to view (▌Figure 6-6).

Refracting telescopes suffer from a serious optical distortion that limits their usefulness. When light is refracted through glass, shorter wavelengths bend more than longer wavelengths, and blue light comes to a focus closer to the lens than does red light (▌Figure 6-7a). If we focus the eyepiece on the blue image, the red light is out of focus, and we see a red blur around the image. If we focus on the red image, the blue light blurs. The color separation is called **chromatic aberration.** Telescope designers can grind a telescope lens of two components made of different kinds of glass and so bring two different wavelengths to the same focus (Figure 6-7b). This does improve the image, but these **achromatic lenses** are not totally free of chromatic aberration, because other wavelengths still blur. Telescopes made with such lenses were popular until the end of the 19th century.

FIGURE 6-4
(a) To see how a lens can focus light, we trace four light rays from the flame and base of a candle through a lens, where they are refracted to form an inverted image.
(b) A mirror forms an image by reflection from a concave surface. Notice that the light is reflected from the aluminized front surface of the mirror and does not enter the glass.

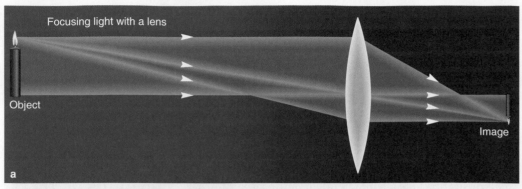

Focusing light with a lens

Object

Image

a

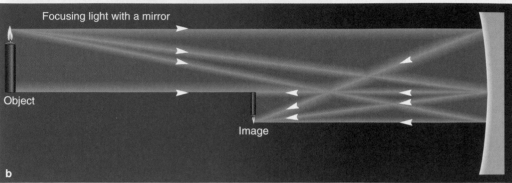

Focusing light with a mirror

Object

Image

b

The primary lens of a refracting telescope is very expensive to make because it must be achromatic, and the glass must be pure and flawless because the light passes through the lens. The four surfaces must be ground precisely, and the lens can be supported only along its edge. The largest refracting telescope in the world was completed in 1897 at Yerkes Observatory in Wisconsin. Its lens is 1 m (40 in.) in diameter and weighs half a ton. Larger refracting telescopes are prohibitively expensive.

Reflecting telescopes are much less expensive because the light reflects from the front surface of the mirror. Consequently only the front surface need be ground to precise shape. Also, the glass of the mirror need not be perfectly transparent, and the mirror can be supported over its back surface to reduce sagging.

Most important, reflecting telescopes do not suffer from chromatic aberration because the light is reflected toward the focus before it can enter the glass. Thus, every large astronomical telescope built since the beginning of the 20th century has been a reflecting telescope.

THE POWERS OF A TELESCOPE

A telescope can aid our eyes in only three ways—the three powers of a telescope. They make images brighter, more detailed, and larger.

Most interesting celestial objects are faint sources of light, so we need a telescope that can gather large amounts of light to produce a bright image. **Light-gathering power** refers to the ability of a telescope to collect light. Catching light in a telescope is like catch-

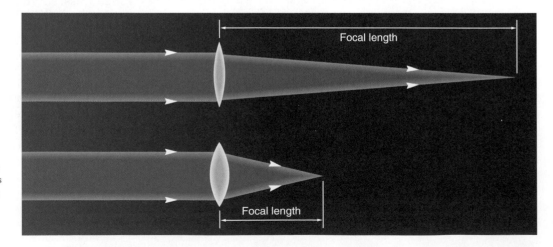

FIGURE 6-5
The focal length of a lens is the distance from the lens to the point where parallel rays of light come to a focus. The lens at the top has a longer focal length than does the lens at the bottom.

Focal length

Focal length

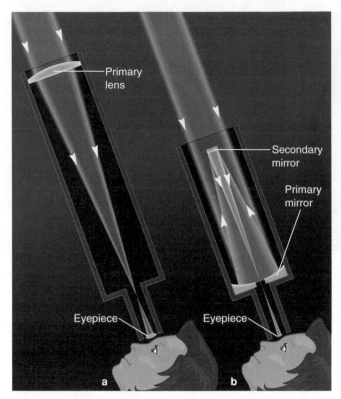

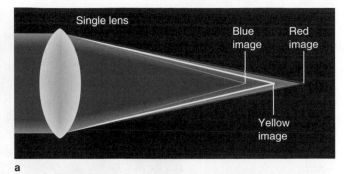

a

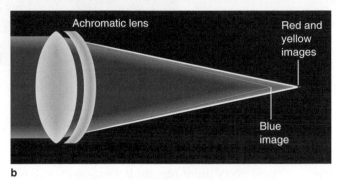

b

ACTIVE FIGURE 6-6

(a) A refracting telescope uses a primary lens to focus starlight into an image that is magnified by a lens called an eyepiece. The primary lens has a long focal length, and the eyepiece has a short focal length. (b) A reflecting telescope uses a primary mirror to focus the light by reflection. A small secondary mirror reflects the starlight back down through a hole in the middle of the primary mirror to the eyepiece.

Ace ◐ Astronomy™ Go to AceAstronomy and click Active Figures to see "Refractors and Reflectors." You can watch light flowing through the telescopes.

FIGURE 6-7

(a) A normal lens suffers from chromatic aberration because short wavelengths bend more than long wavelengths. (b) An achromatic lens, made in two parts, can bring any two colors to the same focus, but other colors remain slightly out of focus.

ing rain in a bucket—the bigger the bucket, the more rain it catches (▌ Figure 6-8). Light-gathering power is proportional to the area of the telescope objective. A lens or mirror with a large area gathers a large amount of light. The area of a circular lens or mirror of diameter D is $\pi\left(\frac{D}{2}\right)^2$. To compare the relative light-gathering powers (*LGP*) of two telescopes A and B, we can calculate the ratio of the areas of their objectives, which reduces to the ratio of their diameters (*D*) squared.

$$\frac{LGP_A}{LGP_B} = \left(\frac{D_A}{D_B}\right)^2$$

For example, suppose we compare a telescope 24 cm in diameter with a telescope 4 cm in diameter. The ratio of the diameters is 24/4, or 6, but the larger telescope does not gather 6 times as much light. Light-gathering power increases as the ratio of diameters squared, so it gathers 36 times more light than the smaller telescope. This example shows the importance of diameter in astronomical telescopes. Even a small increase in diameter produces a large increase in light-gathering power and allows astronomers to study much fainter objects.

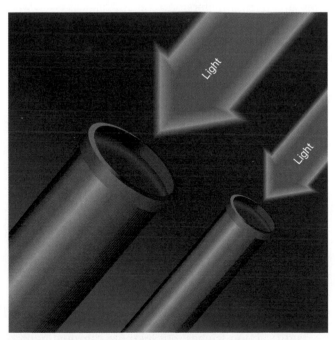

FIGURE 6-8

Gathering light is like catching rain in a bucket. A large-diameter telescope gathers more light and has a brighter image than a smaller telescope of the same focal length.

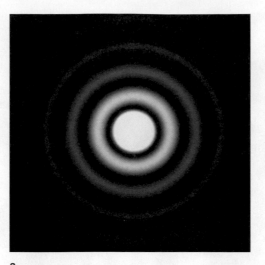
a

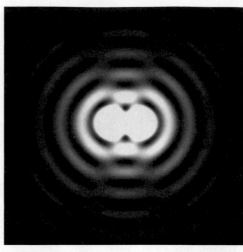
b

The second power, **resolving power,** refers to the ability of the telescope to reveal fine detail. Because light acts as a wave, it produces a small **diffraction fringe** around every point of light in the image, and we cannot see any detail smaller than the fringe (▌ Figure 6-9). Astronomers can't eliminate diffraction fringes, but the larger a telescope is in diameter, the smaller the diffraction fringes are. Thus the larger the telescope, the better its resolving power.

For optical telescopes, we estimate the resolving power by calculating the angular distance between two stars that are just barely visible through the telescope as two separate images. The resolving power, α, in seconds of arc, equals 11.6 divided by the diameter of the telescope in centimeters:

$$\alpha = \frac{11.6}{D}$$

For example, the resolving power of a 25-cm telescope is 11.6 divided by 25, or 0.46 second of arc. No matter how perfect the telescope optics, this is the smallest detail we can see through that telescope.

In addition to resolving power, two other factors—lens quality and atmospheric conditions—limit the detail we can see through a telescope. A telescope must contain high-quality optics to achieve its full potential resolving power. Even a large telescope shows us little detail if its optics are marred with imperfections. Also, when we look through a telescope, we are looking through miles of turbulent air in Earth's atmosphere, which makes the image dance and blur, a condition called **seeing.** On a night when the atmosphere is unsteady and the images are blurred, the seeing is bad (▌ Figure 6-10). Even under good seeing conditions, the detail visible through a large telescope is limited, not by its diffraction fringes, but by the air through which the telescope must look. A telescope performs better on a high mountaintop where the air is thin and steady,

but even there Earth's atmosphere limits the detail the best telescopes can reveal to about 0.5 second of arc.

This limitation on the amount of information in an image is related to the limitation on the accuracy of a measurement. All measurements have some built-in uncertainty (Window on Science 6-1), and scientists must learn to work within those limitations.

The third and least important power of a telescope is **magnifying power,** the ability to make the image bigger. Because the amount of detail we can see is limited

Visual-wavelength image

FIGURE 6-10

The left half of this photograph of a galaxy is from an image recorded on a night of poor seeing. Small details are blurred. The right half of the figure is from an image recorded on a night when Earth's atmosphere above the telescope was steady and the seeing was better. Much more detail is visible under good seeing conditions. *(Courtesy William Keel)*

Resolving Power and the Resolution of a Measurement

Have you ever seen a movie in which the hero magnifies a newspaper photo and reads some tiny detail? It isn't really possible, because newspaper photos are made up of tiny dots of ink, and no detail smaller than a single dot will be visible no matter how much you magnify the photo. In fact, all images are made up of elements of some sort, and that means there is a limit to the amount of detail you can see in an image. In an astronomical image, the resolution is often set by seeing. It is foolish to attempt to see a detail in the image that is smaller than the resolution.

This limitation is true of all measurements in science. A zoologist might be trying to measure the length of a live snake, or a sociologist might be trying to measure the attitudes of people toward drunk driving, but both face limits to the resolution of their measurements. The zoologist might specify that the snake was 43.28932 cm long, and the sociologist might say that 98.2491 percent of people oppose drunk driving, but a critic might point out that it isn't possible to make these measurements that accurately. The resolution of the techniques does not justify the accuracy implied.

Science is based on measurement, and whenever we make a measurement we should ask ourselves how accurate that measurement can be. The accuracy of the measurement is limited by the resolution of the measurement technique, just as the amount of detail in a photograph is limited by the resolution of the photo

---

by the seeing conditions and the resolving power, very high magnification does not necessarily show us more detail. Also, we can change the magnification by changing the eyepiece, but we cannot alter the telescope's light-gathering power or resolving power.

We calculate the magnification of a telescope by dividing the focal length of the objective by the focal length of the eyepiece:

$$M = \frac{F_o}{F_e}$$

For example, if a telescope has an objective with a focal length of 80 cm and we use an eyepiece whose focal length is 0.5 cm, the magnification is 80/0.5, or 160 times.

The search for light-gathering power and high resolution explains why nearly all major observatories are located far from major cities and usually on high mountains. Astronomers avoid cities because **light pollution,** the brightening of the night sky by light scattered from artificial outdoor lighting, can make it impossible to see faint objects (❚ Figure 6-11). In fact, many residents of cities are unfamiliar with the beauty of the night sky because they can see only the brightest stars. Nevertheless, nature's own light pollution, the moon, is so bright it drowns out fainter objects, and astronomers are often unable to observe on the nights near full moon when faint objects cannot be observed even with the largest telescopes on high mountains.

Astronomers prefer to place their telescopes on carefully selected high mountains. The air there is thin and more transparent, but, most important, astronomers select mountains where the air flows smoothly and is not turbulent. This produces the best seeing. Building an observatory on top of a high mountain far from civilization is difficult and expensive, but the dark sky and steady seeing make it worth the effort (❚ Figure 6-12).

BUYING A TELESCOPE

Thinking about how we should shop for a new telescope will not only help us if we decide to buy one but will also illustrate some important points about astronomical telescopes.

Assuming we have a fixed budget, we should buy the highest-quality optics and the largest-diameter telescope we can afford. Of the two things that limit what we see, optical quality is under our control. We can't make the atmosphere less turbulent, but we should buy good optics. If we buy a telescope from a toy store and it has plastic lenses, we shouldn't expect to see very much. Also, we want to maximize the light-gathering power of our telescope, so we want to purchase the largest-diameter telescope we can afford. Given a fixed budget, that means we should buy a reflecting telescope rather than a refracting telescope. Not only will we get more diameter per dollar, but our telescope will not suffer from chromatic aberration.

We can safely ignore magnification. Department stores and camera stores may advertise telescopes by quoting their magnification, but it is not an important number. What we can see is fixed by light-gathering power, optical quality, and Earth's atmosphere. Besides, we can change the magnification by changing eyepieces.

Other things being equal, we should choose a telescope with a solid mounting that will hold the telescope steady and allow us to point at objects easily. Computer-controlled pointing systems are available for a price on many small telescopes. A good telescope on a poor mounting is almost useless.

We might be buying a telescope to put in our backyard, but we must think about the same issues astronomers consider when they design giant telescopes to go on mountaintops. In fact, some of the newest telescopes solve these traditional problems in new ways.

Visual-wavelength image

FIGURE 6-11

This satellite view of the continental United States at night shows the light pollution produced by outdoor lighting. Not only does the glare drown out the fainter stars and interfere with astronomy, but it wastes electrical power. Many astronomers work with city governments to enact laws that improve lighting on the ground and reduce light scattered into the night sky. *(NOAA)*

FIGURE 6-13

Astronomers have begun building multiple telescopes to solve different problems. Two Gemini
telescopes have been built, one in Hawaii and one
in Chile, to observe the entire sky. The Large Binocular Telescope (LBT) carries two 8.4-m mirrors that
combine their light. The four telescopes of the VLT
are housed in separate domes, but they can combine
to work as a very large telescope. *(Gemini: NOAO/
AURA/NSF; LBT: Large Binocular Telescope Project
and European Industrial Engineer; VLT: ESO)*

The LBT has the light-
gathering power of a 11.8-m
telescope and the resolving
power of a 22.8-m telescope.

The mirrors in the VLT
telescopes are 8.2 m in diameter.

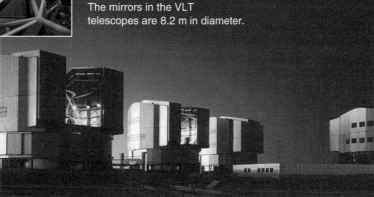

The two Gemini telescopes
have 8.1-m mirrors with
active optics.

NEW-GENERATION TELESCOPES

For most of the 20th century, astronomers faced a serious limitation on the size of astronomical telescopes.
Traditional telescope mirrors were made thick to avoid
sagging that would distort the reflecting surface, but
those thick mirrors were heavy. The 5-m (200-in.) mirror on Mount Palomar weighs 14.5 tons. These traditional telescopes were big, heavy, and expensive.

Modern astronomers have solved these problems
in a number of ways. Study "▐ Modern Astronomical
Telescopes" (pages 108 and 109) and notice three important advances in telescope design made possible by
high-speed computers. First, astronomers can now
build simpler, stronger telescope mountings and depend on computers to move the telescope and follow
the westward motion of the stars as Earth rotates.

Second, notice that computer control of the shape
of telescope mirrors allows the use of thin, lightweight

FIGURE 6-12
The domes of four giant telescopes are visible at upper left at Paranal
Observatory, built by the European Southern Observatory at an
altitude of 2635 m (8660 ft) atop the mountain Cerro Paranal in the
Atacama desert of northern Chile. Located 120 km (75 mi) from the
nearest city, the mountaintop was chosen for its steady air and isolation from lights. It is believed to be the driest pace on Earth. *(ESO)*

mirrors—either "floppy" mirrors or segmented mirrors.
Lowering the weight of the mirror lowers the weight of
the rest of the telescope and makes it stronger and less
expensive. Also, thin mirrors cool faster at nightfall
and produce better images.

The third advance to notice is the way astronomers
use high-speed computers to reduce seeing distortion
caused by Earth's atmosphere. Only a few decades ago,
many astronomers argued that it wasn't worth building
more large telescopes on Earth's surface because of the
limitations set by seeing. Now a number of new giant
telescopes have been built, and more are in development that can partially overcome the seeing problem.

An international collaboration of astronomers have
built the Gemini telescopes with 8.1-m thin mirrors
(▐ Figure 6-13). One is located in the northern hemisphere and one in the southern hemisphere to cover the
entire sky. The European Southern Observatory has
built the Very Large Telescope (VLT) high in the remote Andes mountains of northern Chile (Figure 6-13).
The VLT consists of four telescopes with computer-
controlled mirrors 8.2 m in diameter and only 17.5 cm
(6.9 in.) thick. The four telescopes can work singly
or can combine their light to work as one large telescope. Italian and American astronomers are building
the Large Binocular Telescope, which carries a pair of

Modern Astronomical Telescopes

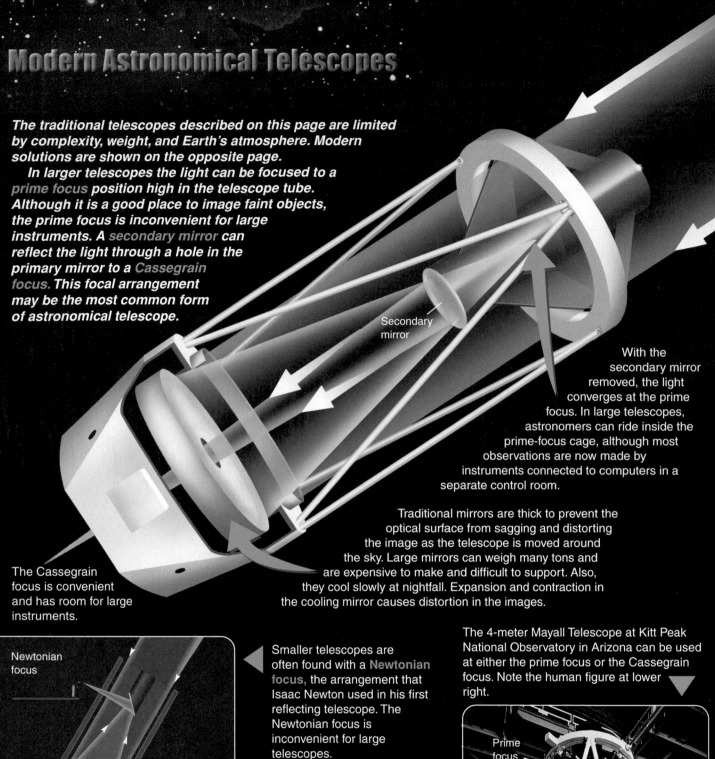

The traditional telescopes described on this page are limited by complexity, weight, and Earth's atmosphere. Modern solutions are shown on the opposite page.

In larger telescopes the light can be focused to a *prime focus* position high in the telescope tube. Although it is a good place to image faint objects, the prime focus is inconvenient for large instruments. A *secondary mirror* can reflect the light through a hole in the primary mirror to a *Cassegrain focus*. This focal arrangement may be the most common form of astronomical telescope.

Secondary mirror

With the secondary mirror removed, the light converges at the prime focus. In large telescopes, astronomers can ride inside the prime-focus cage, although most observations are now made by instruments connected to computers in a separate control room.

The Cassegrain focus is convenient and has room for large instruments.

Traditional mirrors are thick to prevent the optical surface from sagging and distorting the image as the telescope is moved around the sky. Large mirrors can weigh many tons and are expensive to make and difficult to support. Also, they cool slowly at nightfall. Expansion and contraction in the cooling mirror causes distortion in the images.

Newtonian focus

Smaller telescopes are often found with a **Newtonian focus,** the arrangement that Isaac Newton used in his first reflecting telescope. The Newtonian focus is inconvenient for large telescopes.

The 4-meter Mayall Telescope at Kitt Peak National Observatory in Arizona can be used at either the prime focus or the Cassegrain focus. Note the human figure at lower right.

Many small telescopes use a **Schmidt-Cassegrain focus.** A thin correcting plate improves the image but is too slightly curved to introduce serious chromatic aberration.

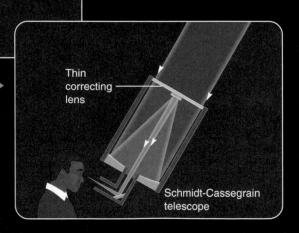

Thin correcting lens

Schmidt-Cassegrain telescope

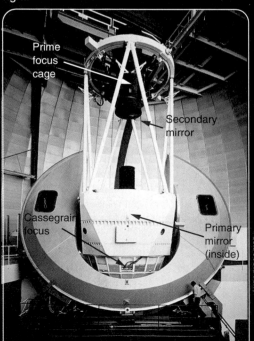

Prime focus cage

Secondary mirror

Cassegrain focus

Primary mirror (inside)

Equatorial mounting

Westward rotation about polar axis follows stars.

Polar axis

North Pole

Eastward rotation of Earth

Alt-azimuth mounting

Computer control of motion about both axes follows stars.

To north celestial pole

North Pole

Eastward rotation of Earth

Telescope mountings must contain a **sidereal drive** to move smoothly westward and counter the eastward rotation of Earth. The traditional **equatorial mounting** has a **polar axis** parallel to Earth's axis, but the modern **alt-azimuth mounting** moves like a cannon — up and down and left to right. Such mountings are simpler to build but need computer control to follow the stars.

Unlike traditional thick mirrors, thin mirrors, sometimes called floppy mirrors, weigh less and require less massive support structures. Also, they cool rapidly at nightfall and there is less distortion from uneven expansion and contraction.

Floppy mirror

Computer-controlled thrusters   Support structure

Grinding a large mirror may remove tons of glass and take months, but new techniques speed the process. Some large mirrors are cast in a rotating oven that causes the molten glass to flow to form a concave upper surface. Grinding and polishing such a preformed mirror is much less time consuming.

Segmented mirror

Computer-controlled thrusters   Support structure

The two largest telescopes in the world, the Keck I and Keck II telescopes in Hawaii, contain segmented mirrors 10 m in diameter.

Mirrors made of segments are economical because the segments can be made separately. The resulting mirror weighs less and cools rapidly.

The thrusters are located behind the mirror segments in this face-on photo of the Keck I mirror. The technician kneels in the light baffle in front of the Cassegrain hole in the center of the mirror.

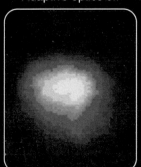

Both floppy mirrors and segmented mirrors sag under their own weight. Their optical shape must be controlled by computer-driven thrusters under the mirror in what is called **active optics.**

Adaptive optics off   Adaptive optics on

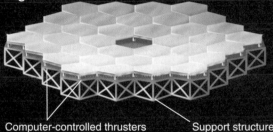

Object revealed as a pair of stars.

1 second of arc

Adaptive optics uses high-speed computers to monitor the image distortion caused by Earth's atmosphere and adjust the optics many times a second to compensate. This can reduce the blurring due to seeing and dramatically improve image quality in Earth-based telescopes.

8.4-m mirrors on a single mounting (Figure 6-13). Around the world astronomers are drawing plans for large telescopes, including truly gigantic instruments with segmented mirrors 30 m and even 100 m in diameter. Computer control of the optics makes such huge telescopes worth considering.

Computer control has made massive surveys of the sky practical. We have mentioned earlier the Hipparchos satellite that surveyed the entire sky, measuring the parallax of over a million years. Other surveys are being made. The Sloan Digital Sky Survey is mapping the sky, measuring the position and brightness of 100 million stars and galaxies at a number of wavelengths. The Two-Micron All Sky Survey (2MASS) has mapped the entire sky at three wavelengths in the infrared. Other surveys are being made at many other wavelengths. Every night large telescopes scan the sky, and billions of bytes of data are compiled automatically in immense sky atlases. Astronomers will study those data banks for decades to come.

The days when astronomers worked beside their telescopes through long, dark, cold nights are nearly gone. The complexity and sophistication of telescopes require a battery of computers, and almost all research telescopes are run from control rooms that astronomers call warm rooms. Astronomers don't need to be kept warm, but computers demand comfortable working conditions (▌ Figure 6-14).

INTERFEROMETRY

One of the reasons astronomers build big telescopes is to increase resolving power, and astronomers have been able to achieve very high resolution by connecting multiple telescopes together to work as if they were

FIGURE 6-14

In a telescope control room, computers monitor everything from the shape of the optics to the weather forecast. Furthermore, most instruments such as cameras and spectrographs are controlled by computers and store their data in computer memory. Astronomers work through the night controlling the computers that control the telescope and its instruments. *(AAO)*

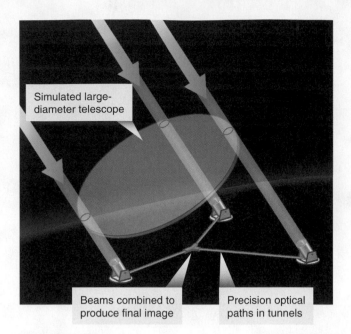

Simulated large-diameter telescope

Beams combined to produce final image

Precision optical paths in tunnels

FIGURE 6-15

In an astronomical interferometer, smaller telescopes can combine their light through specially designed optical tunnels to simulate a larger telescope with a resolution set by the separation of the smaller telescopes

a single telescope. This method of synthesizing a larger telescope is known as **interferometry** (▌ Figure 6-15).

To work as an interferometer, the separate telescopes must combine their light through a network of mirrors, and the path that each light beam travels must be controlled so that it does not vary more than some small fraction of the wavelength. Turbulence in Earth's atmosphere constantly distorts the light, and the high-speed computers must continuously adjust the light paths. Recall that the wavelength of light is very short, roughly 0.0005 mm, so building optical interferometers is one of the most difficult technical problems that astronomers face. Infrared- and radio-wavelength interferometers are slightly easier to build because the wavelengths are longer. In fact, as we will discover later in this chapter, the first astronomical interferometers worked at radio wavelengths.

The VLT shown in Figure 6-13 consists of four 8.2-m telescopes that can operate separately, but they can be linked together through underground tunnels with three 1.8-m telescopes on the same mountaintop. The resulting optical interferometer provides the resolution of a telescope 200 meters in diameter.

Other telescopes can work as interferometers. The two Keck 10-m telescopes can be used as an interferometer. The Navy Prototype Optical Interferometer built near Flagstaff, Arizona has small telescopes located along three arms up to 250 m in length. It is being used to study technical aspects of optical interferometry and to make high-precision measures of star positions. The CHARA array on Mt. Wilson combines six 1-meter telescopes to create the equivalent of a telescope one-fifth of a mile in diameter. The Large Binocular Telescope shown in Figure 6-13 can be used as an interferometer.

Although turbulence in Earth's atmosphere can be partially averaged out in an interferometer, plans are being made to put interferometers in space. The Space Interferometry Mission, for example, will work at visual wavelengths and study everything from the cores of erupting galaxies to planets orbiting nearby stars.

REVIEW CRITICAL INQUIRY

Why do astronomers build observatories at the tops of mountains?

Astronomers have joked that the hardest part of building a new observatory is constructing the road to the top of the mountain. It certainly isn't easy to build a large, delicate telescope at the top of a high mountain, but it is worth the effort. A telescope on top of a high mountain is above the thickest part of Earth's atmosphere. There is less air to dim the light, and there is less water vapor to absorb infrared radiation. Even more important, the thin air on a mountaintop causes less disturbance to the image, and thus the seeing is better. A large telescope on Earth's surface has a resolving power much better than the distortion caused by Earth's atmosphere. Thus, it is limited by seeing, not by its own diffraction. It really is worth the trouble to build telescopes atop high mountains.

Astronomers not only build telescopes on mountaintops; they also build gigantic telescopes many meters in diameter. What are the problems and advantages in building such giant telescopes?

Astronomers sometimes refer to a telescope that produces distorted images as a "light bucket." In a sense, all astronomical telescopes are light buckets, because the light they focus into images tells us very little until it is recorded and analyzed by special instruments attached to the telescopes.

6-3 SPECIAL INSTRUMENTS

Looking through a telescope doesn't tell us much. To use an astronomical telescope to learn about stars, we must be able to analyze the light the telescope gathers. Special instruments attached to the telescope make that possible.

IMAGING SYSTEMS

The original imaging device in astronomy was the photographic plate. It could record faint objects in long exposures and could be stored for later analysis. But photographic plates have been almost entirely replaced in astronomy by electronic imaging systems.

Most modern astronomers use **charge-coupled devices (CCDs)** to record images. A CCD is a specialized computer chip containing roughly a million microscopic light detectors arranged in an array about the size of a postage stamp. These devices can be used like a small photographic plate, but they have dramatic advantages. They can detect both bright and faint objects in a single exposure, are much more sensitive than a photographic plate, and can be read directly into computer memory for later analysis. Although CCDs for astronomy are extremely sensitive and therefore expensive, less sophisticated CCDs are used in most video cameras and digital cameras.

The image from a CCD is stored as numbers in computer memory, so it is easy to manipulate the image to bring out details that would not otherwise be visible. For example, astronomical images are often reproduced as negatives with the sky white and the stars dark. This makes the faint parts of the image easier to see (Figure 6-16a). Astronomers also manipulate images to produce **false-color images** in which the colors represent different levels of intensity and are not related to the true colors of the object (Figure 6-16b).

Measurements of intensity and color were made in the past using a photometer, a highly sensitive light meter attached to a telescope. Today, however, most

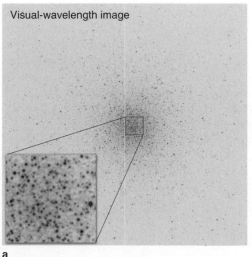

Visual-wavelength image

a

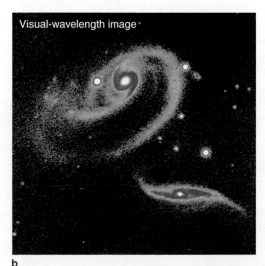

Visual-wavelength image

b

FIGURE 6-16
(a) This CCD image of the star cluster known as M3 has been reproduced as a negative to reveal detail such as faint stars in the cluster visible in the inset. *(A. Saha, NOAO/WIYN/NOAO/NSF)* (b) This CCD image of two galaxies has been given false color according to brightness. The faintest regions are blue and green, and the brighter regions are red and white. The colors in such false-color images are not usually related to the actual colors of the object. *(WIYN/NOAO/NSF)*

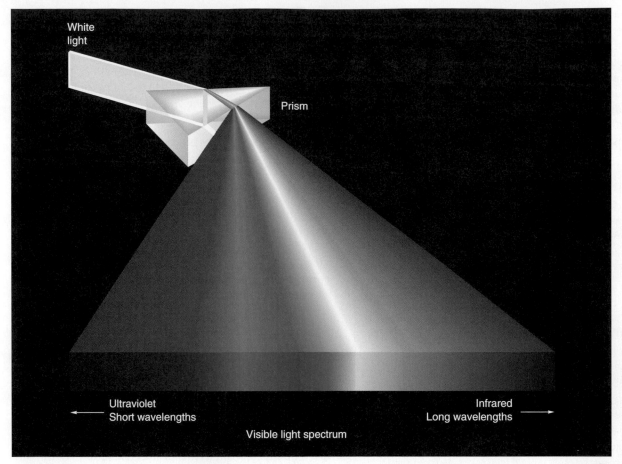

White
light

Prism

Ultraviolet
Short wavelengths

Infrared
Long wavelengths

Visible light spectrum

FIGURE 6-17

A prism bends light by an angle that depends on the wavelength of the light. Short wavelengths bend most, and long wavelengths least. Thus, white light passing through a prism is spread into a spectrum.

such measurements are made on CCD images. Because the CCD image is easily digitized, brightness and color can be measured to high precision.

THE SPECTROGRAPH

To analyze light in detail, we need to spread the light out according to wavelength into a spectrum, a task performed by a **spectrograph.** We can understand how this works if we reproduce the experiment performed by Isaac Newton in 1666. Placing a prism in a beam of sunlight spreads it into a beautiful spectrum. From this Newton concluded that white light was made of a mixture of all the colors.

Newton didn't think in terms of wavelength, but we can use that modern concept to see that the light passing through the prism is bent at an angle that depends on the wavelength. Violet (short wavelength) bends most, and red (long wavelength) least. Thus, the white light entering the prism is spread into a spectrum (▌ Figure 6-17). A typical prism spectrograph contains more than one prism to spread the light farther and lenses to guide the light into the prism and to focus the light into a camera.

Nearly all modern spectrographs use a grating in place of a prism. A **grating** is a piece of glass with thousands of microscopic parallel lines scribed onto its surface. Different wavelengths of light reflect from the grating at slightly different angles, so white light is spread into a spectrum and can be recorded, often by a CCD camera.

Because astronomers must measure the wavelengths of a spectrum, they use a **comparison spectrum** as a calibration of their spectrograph. Special bulbs built into the spectrograph produce emission spectra of such atoms as thorium and argon, or neon. The wavelengths of these spectral lines have been measured to high precision in the laboratory, so astronomers can use the spectra of these light sources like a road map to measure the wavelengths of spectral lines in the spectrum of a planet, star, or galaxy to high accuracy.

One of the newest advances in astronomical instrumentation is called the S-CAM. Developed by the European Space Agency, the S-CAM is a CCD camera in which each pixel records the energy of every photon. That is enough information to create a spectrum for any pixel in the image. In that way the S-CAM combines the advantages of the CCD camera and the spectrograph.

Because astronomers understand how light interacts with matter, a spectrum carries a tremendous amount of information (as we will see in the next chapter), and that makes a spectrograph the astronomer's most powerful instrument. An astronomer recently remarked, "We don't know anything about an object till we get a spectrum," and that is only a slight exaggeration.

REVIEW CRITICAL INQUIRY

What is the difference between light going through a lens and light passing through a prism?

A refracting telescope producing chromatic aberration and a prism dispersing light into a spectrum are two examples of the same thing, but one is bad and one is good. When light passes through the curved surfaces of a lens, different wavelengths are bent by slightly different amounts, and the different colors of light come to focus at different focal lengths. This produces the color fringes in an image called chromatic aberration, and that's bad. But the surfaces of a prism are made to be precisely flat, so all of the light enters the prism at the same angle, and any given wavelength is bent by the same amount wherever it meets the prism. Thus, white light is dispersed into a spectrum. We could call the dispersion of light by a prism "controlled chromatic aberration," and that's good.

CCDs have been very good for astronomy, and they are now widely used. Explain why they are more useful than photographic plates.

So far, our discussion has been limited to visual wavelengths. Now it is time to consider the rest of the electromagnetic spectrum.

6-4 RADIO TELESCOPES

All the telescopes and instruments we have discussed look out through the visible light window in Earth's atmosphere, but there is another window running from a wavelength of 1 cm to about 1 m (see Figure 6-3). By building the proper kinds of instruments, we can study the universe through this radio window.

OPERATION OF A RADIO TELESCOPE

A radio telescope usually consists of four parts: a dish reflector, an antenna, an amplifier, and a recorder (▐ Figure 6-18). The components, working together, make it possible for astronomers to detect radio radiation from celestial objects.

The dish reflector of a radio telescope, like the mirror of a reflecting telescope, collects and focuses radiation. Because radio waves are much longer than light waves, the dish need not be as smooth as a mirror. In some radio telescopes, the reflector may not even be dish-shaped, or the telescope may contain no reflector at all.

FIGURE 6-18
In most radio telescopes, a dish reflector concentrates the radio signal on the antenna. The signal is then amplified and recorded. For all but the shortest radio waves, wire mesh is an adequate reflector (photo). *(Courtesy Seth Shostak/SETI Institute)*

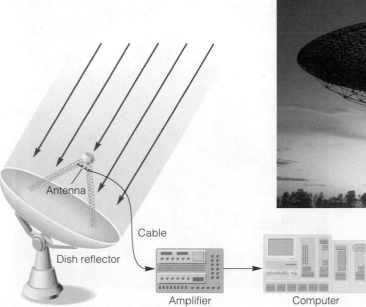

Antenna

Cable

Dish reflector

Amplifier

Computer

Though a radio telescope's dish may be many meters in diameter, the antenna may be as small as your hand. Like the antenna on a TV set, its only function is to absorb the radio energy and direct it along a cable to an amplifier. After amplification, the signal goes to some kind of recording instrument. Most radio observatories record data directly into computer memory. However it is recorded, an observation with a radio telescope measures the amount of radio energy coming from a specific point on the sky.

Because humans can't see radio waves, astronomers must convert them into something perceptible. One way is to measure the strength of the radio signal at various places in the sky and draw a map in which contours mark areas of uniform radio intensity. We might compare such a map to a seating diagram for a baseball stadium in which the contours mark areas in which the seats have the same price (▐ Figure 6-19a). Contour maps are very common in radio astronomy and are often reproduced using false colors (Figure 6-19b).

LIMITATIONS OF THE RADIO TELESCOPE

A radio astronomer works under three handicaps: poor resolution, low intensity, and interference. We saw that the resolving power of an optical telescope depends on the diameter of the objective lens or mirror. It also depends on the wavelength of the radiation. At very long wavelengths, like those of radio waves, images become fuzzy because of the large diffraction fringes. As with an optical telescope, the only way to improve the resolving power is to build a bigger telescope. Consequently, radio telescopes must be quite large.

Even so, the resolving power of a radio telescope is not good. A dish 30 m in diameter receiving radiation with a wavelength of 21 cm has a resolving power of about 0.5°. Such a radio telescope would be unable to show us any details in the sky smaller than the moon. Fortunately, radio astronomers can combine two or more radio telescopes to form a **radio interferometer** capable of much higher resolution. For example, the Very Large Array (VLA) consists of 27 dish antennas spread in a Y-shape across the New Mexico desert (▐ Figure 6-20). In combination, they have the resolving power of a radio telescope 36 km (22 mi) in diameter. The VLA can resolve details smaller than 1 second of arc. Eight new dish antennas being added across New Mexico will give the VLA 10 times better resolving power. Another large radio interferometer, the Very Long Baseline Array (VLBA), consists of matched radio dishes spread from Hawaii to the Virgin Islands and has an effective diameter almost as large as Earth.

The second handicap radio astronomers face is the low intensity of the radio signals. We saw earlier that the energy of a photon depends on its wavelength. Photons of radio energy have such long wavelengths that their individual energies are quite low. In order to get strong signals focused on the antenna, the radio astronomer must build large collecting dishes.

The largest fully steerable radio telescope in the world is at the National Radio Astronomy Observatory in Green Bank, West Virginia (▐ Figure 6-21a). The telescope has a reflecting surface 100 meters in diameter, big enough to hold an entire football field, and can be pointed anywhere in the sky. Its surface consists of 2004 computer-controlled panels that adjust to maintain the shape of the reflecting surface.

The largest radio dish in the world is 300 m (1000 ft) in diameter. So large a dish can't be supported in the usual way, so it is built into a mountain valley in Arecibo, Puerto Rico. The reflecting dish is a thin metallic surface supported above the valley floor by ca-

FIGURE 6-19
(a) A contour map of a baseball stadium shows regions of similar admission prices. The most expensive seats are those behind home plate. (b) A false-color-image radio map of Tycho's supernova remnant, the expanding shell of gas produced by the explosion of a star in 1572. The radio contour map has been color-coded to show intensity. Red is the strongest radio intensity, and violet the weakest. *(NRAO)*

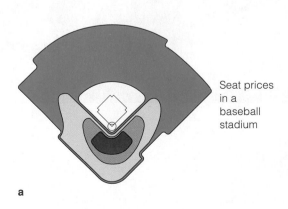

Seat prices in a baseball stadium

a

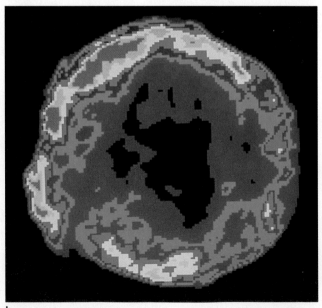

b

FIGURE 6-20
The Very Large Array uses 27 radio dishes, which can be moved to different positions along a Y-shaped set of tracks across the New Mexico desert. They are shown here in the most compact arrangement. Signals from the dishes are combined to create very-high-resolution radio maps of celestial objects. *(NRAO)*

ADVANTAGES OF RADIO TELESCOPES

Building large radio telescopes in isolated locations is expensive, but three factors make it all worthwhile. First, and most important, a radio telescope can show us where clouds of cool hydrogen are located between the stars. Because 90 percent of the atoms in the universe are hydrogen, that is important information. Large clouds of cool hydrogen are completely invisible to normal telescopes, because they produce no visible light of their own and reflect too little to be detected on

a

bles attached near the rim, and the antenna hangs above the dish on cables from three towers built on three mountain peaks that surround the valley (Figure 6-21b). Although this telescope can look only overhead, the operators can change its aim slightly by moving the antenna and by waiting for Earth's rotation to point the telescope in the proper direction. This may sound clumsy, but the telescope's ability to detect weak radio sources, together with its good resolution, makes it one of the most important radio observatories in the world.

The third handicap the radio astronomer faces is interference. A radio telescope is an extremely sensitive radio receiver listening to radio signals thousands of times weaker than artificial radio and TV transmissions. Such weak signals are easily drowned out by interference. Sources of such interference include everything from poorly designed transmitters in Earth satellites to automobiles with faulty ignition systems. To avoid this kind of interference, radio astronomers locate their telescopes as far from civilization as possible. Hidden deep in mountain valleys, they are able to listen to the sky protected from human-made radio noise.

FIGURE 6-21
(a) The largest steerable radio telescope in the world is the GBT located in Green Bank, West Virginia. With a diameter of 100 m, it stands higher than the Statue of Liberty. *(Mike Bailey: NRAO/AUII)* (b) The 300-m (1000-ft) radio telescope in Arecibo, Puerto Rico, hangs from cables over a mountain valley. The Arecibo Observatory is part of the National Astronomy and Ionosphere Foundation. *(David Parker/Science Photo Library)*

b

photographs. However, cool hydrogen emits a radio signal at the specific wavelength of 21 cm. (We will see how the hydrogen produces this radiation when we discuss the gas clouds in space in Chapter 10.) The only way we can detect these clouds of gas is with a radio telescope that receives the 21-cm radiation. These hydrogen clouds are the places where stars are born, and that is one reason that radio telescopes are important.

Nevertheless, there is a second reason. Because radio signals have relatively long wavelengths, they can penetrate the vast clouds of dust that obscure our view at visual wavelengths. Light waves are short, and they interact with tiny dust grains floating in space; thus, the light is scattered and never penetrates the dust to reach optical telescopes on Earth. However, radio signals from far across the galaxy pass unhindered through the dust, giving us an unobscured view.

Finally, a radio telescope can detect objects that are more luminous at radio wavelengths than at visible wavelengths. This includes everything from the coldest clouds of gas to the hottest stars. Some of the most distant objects in the universe, for instance, are detectable only at radio wavelengths.

REVIEW CRITICAL INQUIRY

Why do optical astronomers build big telescopes, while radio astronomers build groups of widely separated smaller telescopes?

Optical astronomers build large telescopes to maximize light-gathering power, but the problem for radio telescopes is resolving power. Because radio waves are so much longer than light waves, a single radio telescope can't see details in the sky much smaller than the moon. By linking radio telescopes miles apart, radio astronomers build a radio interferometer that can simulate a radio telescope miles in diameter and thus increase the resolving power.

The difference between the wavelengths of light and radio waves makes a big difference in building the best telescopes. But why don't radio astronomers want to build their telescopes on mountaintops as optical astronomers do?

Our atmosphere causes trouble for Earth's astronomers in two ways. It distorts images, and it absorbs many wavelengths. The only way to avoid these limitations completely is to send telescopes above the atmosphere, into space.

6-5 SPACE ASTRONOMY

Ground-based telescopes can operate only at wavelengths in the visual and radio windows of the atmosphere. Most of the rest of the electromagnetic radiation—infrared, ultraviolet, X ray, and gamma ray—never reaches Earth's surface. To observe at these wavelengths, telescopes must fly above the atmosphere in high-flying aircraft, rockets, balloons, and satellites. The only exception is some observations of the shorter infrared wavelengths that can be made from high mountains.

INFRARED ASTRONOMY

Some infrared radiation does leak through our atmosphere. This radiation enters narrow, partially open atmospheric windows scattered from 1200 nm to about 40,000 nm. Infrared astronomers usually measure wavelength in micrometers (10^{-6} meters), so they refer to this wavelength range as 1.2 to 40 microns. In this range, called the near infrared, much of the radiation is absorbed by water vapor, carbon dioxide, and oxygen molecules in Earth's atmosphere, so it is an advantage to place telescopes on mountains where the air is thin and dry. A number of important infrared telescopes, for example, observe from the 4150-m (13,600-ft) summit of Mauna Kea in Hawaii (▌Figure 6-22). At this altitude, they are above much of the water vapor, which is the main absorber of infrared.

The far-infrared range, which includes wavelengths longer than 40 microns, can tell us about planets, comets, forming stars, and other cool objects, but these wavelengths are absorbed high in the atmosphere. To observe in the far infrared, telescopes must venture to high altitudes. Remotely operated infrared telescopes suspended under balloons have reached altitudes as high as 41 km (25 mi). For many years, a NASA jet transport carried a 91-cm infrared telescope and a crew of astronomers to altitudes of 12,000 m (40,000 ft) to get above 99 percent of the water vapor in Earth's atmosphere. Now retired from service, that airborne observatory will soon be replaced with the Stratospheric Observatory for Infrared Astronomy (SOFIA), a Boeing 747 that will carry a 2.5-m telescope to the fringes of the atmosphere.

The ultimate solution is to place infrared telescopes in space above the atmosphere. In the early 1980s, the Infrared Astronomy Satellite (IRAS) mapped the sky at infrared wavelengths (▌Figure 6-23). In the middle 1990s, European astronomers launched the Infrared Space Observatory, which carried detectors more sensitive than IRAS.

The newest space telescope observing in the infrared is the Spitzer Space Telescope Facility launched in 2003. It carries a mirror of 85 cm in diameter that is cooled by liquid helium to less than 5.5 K and can observe from 3 to 180 microns. The telescope is so sensitive to heat, it cannot be placed in orbit around Earth (▌Figure 6-24).

We can understand why a telescope observing in the infrared must have its optics cooled if we note that

FIGURE 6-22
Comet Hale-Bopp hangs in the sky over the 3-meter NASA infrared telescope atop Mauna Kea. The air at high altitudes is steady and so dry that it is transparent to shorter infrared photons. Infrared astronomers can observe with the lights on in the telescope dome. Their instruments are usually insensitive to visible light. *(Courtesy William Keel)*

infrared radiation is emitted by warm objects. If the telescope is warm it will emit many times more infrared radiation than that coming from a distant object. Imagine trying to look at a dim, moonlit scene through binoculars that are glowing brightly. In a telescope observing near-infrared wavelengths, only the detector, the element on which the infrared radiation is focused, must be cooled. To observe in the far infrared, however, as IRAS did, the entire telescope must be cooled.

ULTRAVIOLET ASTRONOMY

Ultraviolet radiation with wavelengths shorter than about 290 nm, the far ultraviolet, is completely absorbed by the ozone layer in our atmosphere. The ozone layer extends from 20 km to about 40 km above Earth's surface. Telescopes to observe in the far ultraviolet must get above the ozone layer, and that means they must go into space.

One of the most successful ultraviolet observatories was the International Ultraviolet Explorer (IUE). Launched in 1978, it carried a 45-cm (18-in.) telescope and used TV systems to record spectra from 320 nm to 115 nm. Although IUE was expected to last only a year or two, it survived and was widely used by astronomers from all around the world until, partially shut down because of budget cutbacks, it finally failed in 1996.

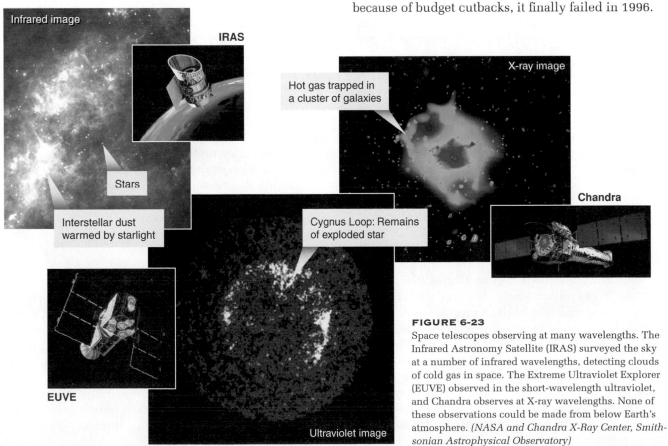

FIGURE 6-23
Space telescopes observing at many wavelengths. The Infrared Astronomy Satellite (IRAS) surveyed the sky at a number of infrared wavelengths, detecting clouds of cold gas in space. The Extreme Ultraviolet Explorer (EUVE) observed in the short-wavelength ultraviolet, and Chandra observes at X-ray wavelengths. None of these observations could be made from below Earth's atmosphere. *(NASA and Chandra X-Ray Center, Smithsonian Astrophysical Observatory)*

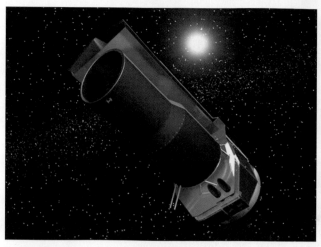

FIGURE 6-24
Because the Spitzer Space Telescope observes at long-infrared wavelengths, it is very sensitive to heat, so the main telescope is shielded from direct sunlight by a sunscreen. Also, the telescope does not orbit Earth, a source of infrared radiation. Rather, the telescope is trailing Earth as it orbits the sun. *(NASA/JPL/Caltech)*

The Extreme Ultraviolet Explorer (EUVE) telescope was launched in 1992 to observe at wavelengths from 100 nm to 10 nm. With four telescopes and more modern detectors than IUE, EUVE surveyed the entire sky and studied specific targets at its shorter wavelength range, as shown in Figure 6-23.

Observations in the infrared reveal cool material, but observations in the ultraviolet tend to show hot, excited regions. Hot stars and hot gas are mapped by such telescopes.

X-RAY ASTRONOMY

Beyond the UV, at wavelengths from 10 nm to 0.01 nm, lie the X rays. These photons can be produced only by high-energy processes, so we see them coming from very hot regions in stars and from violent events such as matter smashing onto a neutron star—a subject we will discuss in Chapter 14. X-ray images can give us information about the heavens that we can get in no other way.

Although early X-ray observations of the sky were made from balloons and small rockets in the 1960s, the age of X-ray astronomy did not really begin until 1970, when an X-ray telescope named Uhuru (Swahili for "freedom") was put into orbit. Uhuru detected nearly 170 separate sources of celestial X rays. In the late 1970s, the three High Energy Astronomy Observatories (HEAO) satellites, carrying more sensitive and more sophisticated equipment, pushed the total to many hundreds. The second HEAO satellite, named the Einstein Observatory, used special optics to produce X-ray images.

For many years, astronomers have worked to create a major X-ray observatory in space. In 1999, the Advanced X-Ray Astrophysics Facility (AXAF) was carried into orbit by the Space Shuttle and renamed Chandra in honor of the late Indian-American Nobel Laureate Subrahmanyan Chandrasekhar, who worked on theoretical astrophysics in many areas of astronomy which the X-ray observatory Chandra studies (Figure 6-23).

Focusing X rays is difficult because the high-energy X-ray photons do not reflect off conventional mirrors but rather penetrate into the mirrors. The optics in Chandra are specially designed mirrors in which the X-ray photons graze the surface of the mirrors at very small angles. Under these circumstances, the photons are reflected to form an image. Just as a CCD chip records visible-light photons, the detector in Chandra has been designed to absorb and record X-ray photons to produce an X-ray image. Chandra can detect X-ray-emitting objects 50 times fainter than any previous X-ray telescope and can resolve details 10 times smaller.

GAMMA-RAY TELESCOPES

Gamma rays have wavelengths even shorter than X rays, and that means they have even higher energies. They are not understood very well, but they appear to be produced by the hottest and most violent objects in the universe: exploding stars, erupting galaxies, neutron stars, and black holes.

We don't understand gamma rays well because they are so difficult to detect. First, they cannot be focused or detected as X rays can. Thus, gamma-ray telescopes are more complex and can see less detail than similar X-ray telescopes. Also, gamma rays are such high-energy photons that natural processes produce only a few. Gamma-ray telescopes must count gamma rays one at a time. For example, a gamma-ray telescope observed an intense source for a month and counted only 3000 gamma-ray photons—one every 15 minutes.

In addition, gamma rays are almost totally absorbed by our atmosphere, so gamma-ray telescopes must observe from orbit. NASA has operated three important gamma-ray telescopes aboard satellites. The 17-ton Gamma Ray Observatory was launched in April 1991. At 10 to 50 times the sensitivity of any previous gamma-ray telescope, it first mapped the entire sky and then observed selected targets.

COSMIC RAYS

All of the radiation we have discussed in this chapter has been electromagnetic radiation. **Cosmic rays,** however, are not really rays; they are subatomic particles traveling at tremendous velocities that strike our atmosphere from space. Almost no cosmic rays reach the ground, but they do smash into gas atoms in the upper atmosphere, and fragments of these collisions shower down on us day and night over our entire lives. These secondary cosmic rays are passing through you as you read this sentence.

Some cosmic-ray research can be done from high mountains or high-flying aircraft; but, to study cosmic rays in detail, detectors must go into space. A number of cosmic-ray detectors have been carried into orbit, but this area of astronomical research is just beginning to bear fruit.

We can't be sure where cosmic rays come from. Because they are atomic particles with electric charges, they are deflected by the magnetic fields spread through our galaxy, and that means we can't tell where they are coming from. The space between the stars is a glowing fog of cosmic rays. Some lower-energy cosmic rays come from the sun, but many cosmic rays are probably produced by the violent explosions of dying stars. At present, cosmic rays largely remain a mystery. We will discuss them again in future chapters.

THE HUBBLE SPACE TELESCOPE

Astronomers dreamed of having a giant telescope in orbit from the 1920s on. Such a telescope would not only allow observation at nonvisible wavelengths but would also avoid the turbulence in Earth's atmosphere. That dream was fulfilled on April 24, 1990, when the Hubble Space Telescope (HST) was released into orbit (❙ Figure 6-25a). Named after the man who discovered the expansion of the universe (Chapter 19), it is the largest orbiting telescope ever built. As big as a large bus, the HST carries a 2.4-m (96-in.) mirror. The light can be directed to any of its spectrographs and cameras.

Although many space telescopes observe at wavelengths that are absorbed in Earth's atmosphere, the Hubble Space Telescope is designed to observe mainly at visual wavelengths. It was placed in orbit to avoid the blurring (seeing) and dimming of starlight caused by Earth's atmosphere. The telescope has far exceeded its original design criteria. It can detect objects so faint it can look 50 times farther into space than can conventional Earth-based telescopes. With its precise optics and its location above Earth's distorting atmosphere, it can resolve details 10 times smaller than can be resolved by any conventional telescope on Earth. Every few years, astronauts visit the telescope in orbit and install new instruments.

HST is controlled from the Space Telescope Science Institute located at Johns Hopkins University in Baltimore, Maryland. Astronomers there plan the telescope's observing schedule to maximize the scientific return while engineers study the operation of the telescope. HST transmits its data to the Institute, where it is analyzed. Because the telescope is a national facility, anyone may propose research projects for the telescope. Competition is fierce, and only the most worthy projects win approval. Nevertheless, even some amateur astronomers have had projects approved for the Space Telescope.

The Hubble Space Telescope has been a phenomenal success, exploring such mysteries as the age of the universe, the birth and death of stars, and weather on nearby planets (Figure 6-25b). Visits by astronauts will

a   b

FIGURE 6-25
The Hubble Space Telescope is the largest orbiting telescope ever launched. It was carried into orbit by the space shuttle in 1990. (b) This image of Mars recorded by the Hubble Space Telescope reveals thin clouds drifting around high volcanoes at the left and details of the polar caps at the top. *(NASA)*

keep it in operation for years to come, but astronomers are already planning an even larger space telescope that will someday orbit above Earth's atmosphere.

| REVIEW CRITICAL INQUIRY |

Why can infrared astronomers observe from high mountaintops, while X-ray astronomers must observe from space? Infrared radiation is absorbed by water vapor in Earth's atmosphere. If we built our infrared telescope on top of a high mountain, we would be above most of the water vapor in the atmosphere, and we could collect some infrared radiation from the stars. The longer-wavelength infrared radiation is absorbed much higher in the atmosphere, so we couldn't observe it from our mountaintop. Similarly, X rays are absorbed in the uppermost layers of the atmosphere, and we would not be able to find any mountain high enough to get an X-ray telescope above those absorbing layers. To observe the stars at X-ray wavelengths, we would need to put our telescope in space, above Earth's atmosphere.

X-ray and far-infrared telescopes must observe from space, but the Hubble Space Telescope observes in the visual wavelength range. Then why must the Hubble Space Telescope observe from orbit?

The tools of the astronomer are designed to gather radiation from the sky and extract information. Perhaps no tool is as important as the spectrograph, because no form of observation is as loaded with information as a spectrum. In the next chapter, we will see how we can harvest the information in a star's spectrum.

SUMMARY

Electromagnetic radiation is an electric and magnetic disturbance that transports energy at the speed of light. The electromagnetic spectrum includes radio waves, infrared radiation, visible light, ultraviolet radiation, X rays, and gamma rays.

We can think of "a particle of light," a photon, as a bundle of waves that sometimes acts as a particle and sometimes as a wave. The energy a photon carries depends on its wavelength. The wavelength of visible light, usually measured in nanometers (10^{-9} m), ranges from 400 nm to 700 nm. Infrared and radio photons have longer wavelengths and carry less energy. Ultraviolet, X-ray, and gamma-ray photons have shorter wavelengths and carry more energy.

Astronomical telescopes are of two types, refractor and reflector. A refractor uses a lens to bend the light and focus it into an image. Because of chromatic aberration, refracting telescopes cannot bring all colors to the same focus, resulting in color fringes around the images. An achromatic lens partially corrects for this, but such lenses are expensive and cannot be made larger than about 1 m in diameter.

Reflecting telescopes use a mirror to focus the light and are less expensive than refracting telescopes of the same diameter. In addition, reflecting telescopes do not suffer from chromatic aberration. Thus, most recently built telescopes are reflectors.

The largest traditional telescopes in the world have thick, solid mirrors of great weight. New telescopes use thin mirrors controlled by computers or mirrors made up of segments. The two Keck Telescopes in Hawaii use segments to make up mirrors 10 m in diameter.

The powers of a telescope are light-gathering power, resolving power, and magnifying power. The first two of these depend on the telescope's diameter; thus, astronomical telescopes often have large diameters.

Special instruments attached to a telescope analyze the light it gathers. The photographic plate records vast amounts of detail for later analysis, but electronic imaging systems such as charge-coupled devices have largely replaced photographic plates. CCDs are more sensitive, record both faint and bright objects, and can be read directly into computer memory. Spectrographs, using prisms or gratings, break the starlight into a spectrum, which then can be photographed or electronically recorded.

To observe radio signals from celestial objects, we need a radio telescope, which usually consists of a dish reflector, an antenna, an amplifier, and a data recorder such as a computer. Such an instrument measures the intensity of radio signals over the sky and constructs radio maps. The poor resolution of the radio telescope can be improved by combining it with another radio telescope to make a radio interferometer. The three principal advantages of radio telescopes are that they can detect the very cold gas clouds in space, they can detect regions of very hot gas produced by exploding stars or erupting galaxies, and they can look through the dust clouds that block our view at optical wavelengths.

Observations at some wavelengths in the near infrared are possible from high mountaintops or from high-flying aircraft. At these altitudes, the air is thin and dry, and the infrared radiation can reach the telescope. At longer infrared wavelengths, the telescope must observe from above Earth's atmosphere—from orbit. To observe at the short ultraviolet wavelengths, a telescope must be above the ozone layers in Earth's atmosphere. That means it must be in orbit. X-ray and gamma-ray telescopes must also be placed in orbit above Earth's absorbing atmosphere.

The largest orbiting telescope is the Hubble Space Telescope, launched in the spring of 1990. The Chandra X-ray telescope was put into orbit in 1999, and the Spitzer Space Telescope, a major infrared observatory in space, began observing in 2003.

NEW TERMS

electromagnetic radiation	photon
wavelength	infrared radiation
frequency	ultraviolet radiation
nanometer (nm)	atmospheric window
Angstrom (Å)	focal length

refracting telescope

reflecting telescope

primary lens, mirror

objective lens, mirror

eyepiece

chromatic aberration

achromatic lens

light-gathering power

resolving power

diffraction fringe

seeing

magnifying power

light pollution

prime focus

secondary mirror

Cassegrain focus

Newtonian focus

Schmidt-Cassegrain focus

sidereal drive

equatorial mounting

polar axis

alt-azimuth mounting

active optics

adaptive optics

interferometry

charge-coupled device (CCD)

false-color image

spectrograph

grating

comparison spectrum

radio interferometer

cosmic ray

REVIEW QUESTIONS

Ace⟲Astronomy™ Assess your understanding of this chapter's topics with additional quizzing and animations at **http://astronomy.brookscole.com/seeds8e**

1. Why do nocturnal animals usually have large pupils in their eyes? How is that related to astronomical telescopes?

2. Astronomers have not constructed a single major refracting telescope since about 1900. Why not?

3. How have computers and new technology changed the design of astronomical telescopes and their mountings?

4. Small telescopes are often advertised as "200 power" or "magnifies 200 times." As someone knowledgeable about astronomical telescopes, how would you improve such advertisements?

5. An astronomer recently said, "Some people think I should give up photographic plates." Why might she change to something else?

6. What purpose do the colors in a false-color image or false-color radio map serve?

7. Why can we observe in the infrared from high mountains and aircraft but must go into space to observe in the far ultraviolet?

8. The moon has no atmosphere at all. What advantages would we have if we had an observatory on the lunar surface?

DISCUSSION QUESTIONS

1. Why does the wavelength response of the human eye match so well the visual window of Earth's atmosphere?

2. Most people like beautiful sunsets with brightly glowing clouds, bright moonlit nights, and twinkling stars. Most astronomers don't. Why?

PROBLEMS

1. The thickness of the plastic in plastic bags is about 0.001 mm. How many wavelengths of red light is this?

2. What is the wavelength of radio waves transmitted by a radio station with a frequency of 100 million cycles per second?

3. Compare the light-gathering powers of one of the Keck telescopes and a 0.5-m telescope.

4. How does the light-gathering power of one of the Gemini telescopes compare with that of the human eye? (*Hint:* Assume that the pupil of your eye can open to about 0.8 cm.)

5. What is the resolving power of a 25-cm telescope? What do two stars 1.5 seconds of arc apart look like through this telescope?

6. Most of Galileo's telescopes were only about 2 cm in diameter. Should he have been able to resolve the two stars mentioned in Problem 5?

7. How does the resolving power of the 5-m telescope compare with that of the Hubble Space Telescope? Why does the HST outperform the 5-m telescope?

8. If we build a telescope with a focal length of 1.3 m, what focal length should the eyepiece have to give a magnification of 100 times?

9. Astronauts observing from a space station need a telescope with a light-gathering power 15,000 times that of the human eye, capable of resolving detail as small as 0.1 second of arc, and having a magnifying power of 250. Design a telescope to meet their needs. Could you test your design by observing stars from Earth?

10. A spy satellite orbiting 400 km above Earth is supposedly capable of counting individual people in a crowd. What minimum-diameter telescope must the satellite carry? (*Hint:* Use the small-angle formula.)

CRITICAL INQUIRIES FOR THE WEB

1. Research in chemistry, physics, and biology is supported in part by industry. Because astronomy has few industrial applications, it is not well supported by industry. Visit Web sites for major observatories and find out who pays their bills.

2. Visit Web sites for the major observatories in space, such as Hubble Space Telescope and Chandra, and check on their latest status. What kinds of observations are they making, and what kinds of discoveries are they making?

3. Astronomers are leaders in efforts to reduce electric power wasted on outdoor lighting because it produces

light pollution. Find Web sites on light pollution and see if your town is making any effort to preserve the beauty of the night sky and simultaneously save electrical power.

EXPLORING *THESKY*

1. Astronomical telescopes using equatorial mountings must be aligned precisely with the north celestial pole. Locate Polaris and determine how far it is from the north celestial pole. (*Hint:* Use **Reference Lines** under the **View** menu and check **Grid** under Equatorial. Be sure the spacing is set to auto/fine. Then locate the Little Dipper and zoom in on Polaris.)

 Visit the Seeds *Foundations of Astronomy* companion Web site for critical thinking exercises, articles, and additional readings from InfoTrac College Edition, Brooks/Cole's online student library.

ATOMS AND STARLIGHT

Awake! for Morning in the Bowl of Night
Has flung the Stone that puts the Stars to Flight:
And Lo! the Hunter of the East has caught
The Sultan's Turret in a Noose of Light.

The Rubáiyát of Omar Khayyám, *Trans. Edward FitzGerald*

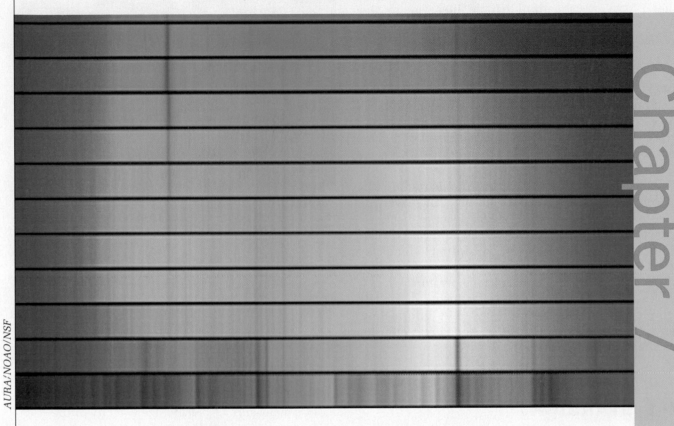

AURA/NOAO/NSF

Chapter 7

GUIDEPOST

Some chapters in textbooks do little more than present facts. The chapters in this book attempt to present astronomy as organized understanding. But this chapter is special. It presents us with a tool. The interaction of light with matter gives astronomers clues about the nature of the heavens, but the clues are meaningless unless astronomers understand how atoms leave their traces on starlight. Thus, we dedicate an entire chapter to understanding how atoms interact with light.

This chapter marks a transition in the way we look at nature. Earlier chapters described what we see with our eyes and explained those observations using models and theories. With this chapter, we turn to modern astrophysics, the application of physics to the study of the sky. Now we can search out secrets of the stars that lie beyond the grasp of our eyes.

If this chapter presents us with a tool, then we should use it immediately. The next chapter will apply our new tool to understanding the sun.

VIRTUAL LABORATORIES

PROPERTIES OF LIGHT AND ITS INTERACTION WITH MATTER

THE DOPPLER EFFECT

THE SPECTRAL SEQUENCE AND THE H–R DIAGRAM

No laboratory jar on Earth holds a sample labeled "star stuff," and no instrument has ever probed inside a star. The stars are far beyond our reach, and the only information we can obtain about them comes to us hidden in light (▌Figure 7-1). Whatever we want to know about the stars we must catch in a noose of light.

We Earthbound humans knew almost nothing about stars until the early 19th century, when the Munich optician Joseph von Fraunhofer studied the solar spectrum and found it interrupted by some 600 dark lines. As scientists realized that the lines were related to the various atoms in the sun and found that stellar spectra had similar patterns of lines, the door to an understanding of stars finally opened.

In this chapter, we will go through that door by considering how atoms interact with light to produce spectral lines. We begin with the hydrogen atom because it is the most common atom in the universe as well as the simplest. Other atoms are larger and more complicated, but in many ways their properties resemble those of hydrogen.

Once we understand how an atom's structure can interact with light to produce spectral lines, we will recognize certain patterns in stellar spectra. By classifying the spectra according to these patterns, we can arrange the stars in a sequence according to temperature. One of the most important pieces of information revealed in a star's spectrum is its temperature.

But, properly analyzed, a stellar spectrum can tell us much more. The spectrum gives us information about the chemical composition of the star and the star's motion relative to Earth.

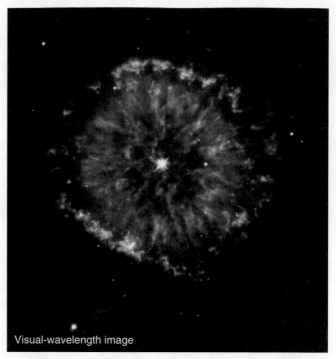

Visual-wavelength image

FIGURE 7-1
What's going on here? The sky is filled with beautiful and mysterious objects that lie far beyond our reach, in the case of the nebula NGC 6751 about 6500 ly beyond our reach. The only way to understand such objects is by analyzing the light we receive from them. Such an analysis reveals that this object is a dying star surrounded by the expanding shell of gas it ejected a few thousand years ago. We will discuss this phenomenon in Chapter 10. *(NASA Hubble Heritage Team/STScI/AURA)*

7-1 STARLIGHT

If you look at the stars in the constellation Orion, you will notice that they are not all the same color (see Figure 2-4). Betelgeuse, in the upper left corner, is quite red; Rigel, in the lower right corner, is blue. These differences in color arise from the way the stars produce light and give us our first clue to the temperatures of stars.

TEMPERATURE AND HEAT

Temperature is one of the defining characteristics of a star. That is, if we know a star's temperature and a few other properties, such as size, we can understand the star. But if we don't know a star's temperature, we can know almost nothing about it.

Ace⟲Astronomy™ The AceAstronomy icon throughout the text indicates an opportunity for you to test yourself on key concepts and to explore animations and interactions on the AceAstronomy Web site at **http://astronomy.brookscole .com/seeds8e**

A gas is made up of particles—atoms and molecules—that are in constant motion, colliding with one another millions of times a second. The **temperature** of a gas is a measure of the average velocity of the particles. If a gas is hot, the particles are moving very rapidly; if it is cool, the particles are moving more slowly. The motion of the particles represents stored energy, and we feel that energy as heat. Read Parameters of Science 7-1 and notice the distinction we make between the temperature, the stored energy, and the heat we feel.

We will measure the temperature of stars on the **Kelvin temperature scale.** The Kelvin scale sets its zero point at **absolute zero,** the temperature at which the particles of a gas have no remaining motion that we can extract as heat. Absolute zero is −273.2°C, which is −459.7°F. (See Appendix A.) The temperatures of the stars range from 2000 K to 40,000 K or more and can be deduced from an analysis of starlight.

THE ORIGIN OF STARLIGHT

The starlight we see comes from gases in the outer surface of the star—the photosphere. The gases deep inside the star also emit light, but it is absorbed before it can escape, and the low-density gas above the photo-

Thermal Energy and Heat

Thermal energy is a measure of the amount of agitation in the particles in an object. Because temperature and thermal energy are often confused, we must be careful to distinguish between them. Temperature is an intensity, and thermal energy is an amount.

Temperature is a measure of the average motion of the atoms and molecules that make up an object. In a hot object, the particles vibrate at higher speeds than in a cool object. If you have your temperature taken, it will probably be 98.6°F, an indication that the atoms and molecules in your body are vibrating at a normal pace. If you measure the temperature of a month-old baby, the thermometer should register the same temperature, showing that the atoms and molecules in the baby's body are moving at the same average velocity as the atoms and molecules in your body. Thus, temperature is a measure of the intensity of the motion.

Thermal energy is a measure of the total energy stored in a body as motion among its particles. Although you and the infant have the same temperature, you contain much more mass than the infant, so you must contain much more thermal energy. Thus, the thermal energy in your body and in the baby's body have the same intensity (temperature) but different amounts.

Many people say "heat" when they should say thermal energy. Heat is the energy that moves from a hot object to a cool object. If two objects have the same temperature, you and the infant for example, there is no transfer of thermal energy and no heat. This is a fine distinction, but when you hear someone say "heat," check to see if he or she doesn't really mean thermal energy.

You may have burned your tongue on cheese pizza, but you probably haven't burnt yourself on green beans. At the same temperature, cheese holds more thermal energy than green beans. It isn't the temperature that burns us, but the thermal energy.

sphere is too thin to emit significant amounts of light. Thus, the photosphere of a star is that layer of gases dense enough to emit significant amounts of light but thin enough to allow the photons to escape. (Recall that we met the photosphere of the sun in Chapter 3.)

To see how the photosphere of a star can produce light, we must consider two things: how photons are produced and how the temperature of a material is related to motion among its atoms.

First, a photon can be produced by a changing electric field. An **electron** is a negatively charged subatomic particle, and if we disturb the motion of an electron, the sudden change in the electric field around it can produce an electromagnetic wave. For example, if you run a comb through your hair while standing near a radio tuned to an AM station, you can produce popping noises on the radio. The moving comb disturbs the electrons in both the comb and your hair, building static electricity. The sudden sparks of static electricity produce electromagnetic waves that the radio picks up as noise. This illustrates an important principle: When we change the motion of an electron, we generate an electromagnetic wave.

Second, recall that temperature is a measure of the average motion among the particles in a material. In the gases of a hot star, the atoms move faster, on average, than the atoms in a cool star.

Now we can understand why a hot object glows. The hotter an object is, the more violent the motion among its particles. The agitated particles collide with electrons, and when electrons are accelerated, part of the energy is carried away as a photon. The radiation emitted by a heated object is called **black body radiation,** a name that refers to the way a perfect emitter of radiation would behave. A perfect emitter would also be a perfect absorber and at room temperature would look black. Thus we often use the term *black body radiation* to refer to objects that glow brightly.

Black body radiation is quite common. In fact, it is responsible for the light emitted by an incandescent lightbulb. Electricity flowing through the filament of the lightbulb heats it to high temperature, and it glows. We also recognize the light emitted by a heated horseshoe in the blacksmith's forge as black body radiation. Many objects in astronomy, including stars, emit radiation approximately as if they were black bodies.

Hot objects emit black body radiation, but so do cold objects. Ice cubes are cold, but their temperature is higher than absolute zero, so they contain some thermal energy and must emit some black body radiation. The coldest gas drifting in space has a temperature only a few degrees above absolute zero, but it too emits black body radiation.

We must discuss two important features of black body radiation. First, the hotter an object is, the more black body radiation it emits. Hot objects emit more radiation because their agitated particles travel faster and collide more often. Thus we expect a glowing coal from a fire to emit more total energy than an ice cube of the same size.

The second feature we must discuss is the relationship between the temperature of the object and the wavelengths of the photons it emits. The wavelength of a photon emitted when a particle collides with an electron depends on the violence of the collision. Only a violent collision can produce a short wavelength (high-energy) photon. Because extremely violent collisions don't occur very often, short-wavelength photons are rare. Similarly, most collisions are not extremely gentle, so long-wavelength (low-energy) photons are also rare. Consequently, black body radiation is made up of photons with a distribution of wavelengths, and very short and very long wavelengths are rare. The **wavelength of maximum intensity (λ_{max})**, the wavelength at which the object emits the most radiation, occurs at some intermediate wavelength.

▌Figure 7-2 shows the intensity of radiation versus wavelength for three objects of different temperatures. The curves are high in the middle and low at either end, telling us that these objects emit most intensely at intermediate wavelengths. The total area under each curve is proportional to the total energy emitted, and we can see that the hotter object emits more total energy than the cooler objects. Closer examination of the curves reveals that the wavelength of maximum intensity depends on temperature. The hotter the object, the shorter the wavelength of maximum intensity. Notice in this figure how temperature determines the color of a glowing black body. The hotter object emits more blue light than red and thus looks blue, and the cooler object emits more red than blue and consequently looks red. We will see this among stars. Hot stars look blue and cool stars red.

Notice that cool objects may emit little visible radiation but are still producing black body radiation. For example, the human body has a temperature of 310 K and emits black body radiation mostly in the infrared part of the spectrum. Infrared security cameras can detect burglars by the radiation they emit, and mosquitoes can track you down in total darkness by homing in on your infrared radiation. Although we humans emit lots of infrared radiation, we rarely emit higher-energy photons; we almost never emit X-ray or gamma-ray photons. Our wavelength of maximum intensity lies in the infrared part of the spectrum.

TWO RADIATION LAWS

Black body radiation can be described by two simple laws that will help us understand the nature of starlight. One law is related to energy and one to color.

As we saw in the previous section, a hot surface emits more black body radiation than a cool surface. That is, it emits more energy. Recall from Chapter 5 that we measure energy in units called joules (J); 1 joule is about the energy of an apple falling from a table to the floor. The total radiation given off by 1 square meter of the object in joules per second equals a constant number, represented by σ, times the temperature raised to the fourth power.* This relationship is often called the Stefan–Boltzmann law:

*For the sake of completeness, we should note that the constant σ equals 5.67×10^{-8} J/m²s degree⁴.

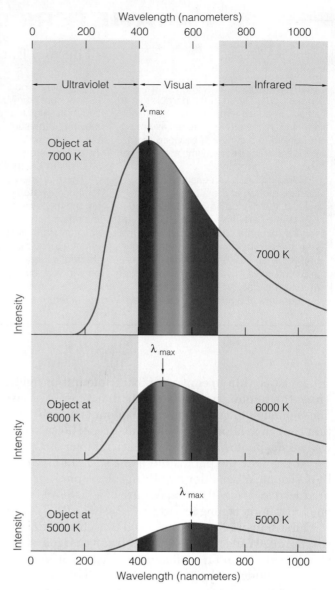

FIGURE 7-2
Black body radiation from three bodies at different temperatures demonstrates that a hot body radiates more total energy and that the wavelength of maximum intensity is shorter for hotter objects. The hotter object here will look blue to our eyes, while the cooler object will look red.

$$E = \sigma\, T^4 \;(\text{J/s/m}^2)$$

How does this help us understand stars? Suppose a star the same size as the sun had a surface temperature that was twice as hot as the sun's surface. Then each square meter of that star would radiate not twice as much energy but 2^4 or 16 times as much energy. From this law we see that a small difference in temperature can produce a very large difference in the amount of energy emitted.

The second radiation law is related to the color of stars. In the previous section, we saw that hot stars look blue and cool stars look red. Wien's law tells us that the wavelength at which a star radiates the most energy, the wavelength of maximum (λ_{max}), depends only on the star's temperature:

$$\lambda_{max} = \frac{3,000,000}{T}$$

That is, the wavelength of maximum radiation in nanometers equals 3 million divided by the temperature on the Kelvin scale.

This is a powerful tool in astronomy, because it means we can relate the temperature of a star and its wavelength of maximum intensity. For example, we might find a star that has a surface temperature of 3000 K. Then its wavelength of maximum intensity would be 3,000,000/3000, or 1000 nm—in the near infrared. Later we will meet objects much hotter than most stars; such objects radiate most of their energy at very short wavelengths. The hottest stars, for instance, radiate most of their energy in the ultraviolet.

THE COLOR INDEX

Hot stars look blue, but if we want to know the temperature of the star's surface, we must have a way of measuring color. In astronomy, the **color index** is a measure of color.

To measure the color index of a star, we need to measure brightness through a set of standard filters. The most commonly used filters are the blue (B) and visual (V). Each filter isolates a specific part of the spectrum. The B filter, for example, will not allow any light through except those photons with wavelengths between about 400 nm and 480 nm. The V filter is transparent to photons with wavelengths between about 500 nm and 600 nm, roughly approximating the sensitivity of the human eye.

If we measure the magnitude of a star through each of these filters, the difference between the two magnitudes is a number related to the color of the star. This is commonly written B-V and is referred to as the B-V color index.

For example, a hot star radiates much more blue light than red and will be brighter through the B filter (Figure 7-2). That means that the B magnitude will be smaller, and the B-V color index will be negative. The bluest stars have a B-V color index of about −0.4 and have surface temperatures of about 50,000 K. Red stars radiate much more red light than blue, so they will be fainter through the blue filter. Then B-V for a red star will be positive. The reddest stars have a B-V color index of about 2 and surface temperatures of about 2000 K.

REVIEW CRITICAL INQUIRY

How could a doctor measure someone's temperature without touching him or her?

Doctors and nurses use a handheld device to measure body temperature by observing the infrared radiation emerging from a patient's ear. We might suspect the device depends on the Stephan–Boltzmann law and measures the intensity of the infrared radiation. A person with a fever will emit more energy than a healthy person. However, a healthy person with a large ear canal would emit more energy than a person with a small ear canal, so measuring intensity won't be accurate. The device actually depends on Wien's law in that it measures the "color index" of the infrared radiation. A patient with a fever will emit at a shorter wavelength of maximum intensity, and the infrared radiation emerging from that person's ear will be a tiny bit "bluer" than normal.

Astronomers can measure the temperatures of stars the same way. Use Figure 7-2 to explain how that works.

The spectrum of a star is bursting with information. To interpret that information, astronomers must think carefully about how atoms and light interact at the surfaces of stars.

7-2 ATOMS

The atoms in the surface layers of stars leave their marks on the light the stars emit. By understanding what atoms are and how they interact with light, we can decode the spectra of the stars. We begin this section by constructing a working model of an atom.

A MODEL ATOM

To think about atoms and how they can interact with light, we must create a working model of an atom. In Chapter 2, we created a working model of the sky, the celestial sphere. We identified and named the important parts and described how they were located and how they interacted. We begin our study of atoms by creating a model of an atom.

In our model atom, we identify a positively charged **nucleus** at the center; this nucleus consists of two kinds of particles. **Protons** carry a positive electrical charge, and **neutrons** have no charge. Thus, the nucleus has a net positive charge.

The nucleus in our model atom is surrounded by a whirling cloud of orbiting electrons, low-mass particles with a negative charge. In a normal atom, the number of electrons equals the number of protons, and the positive and negative charges balance to produce a neutral atom. Because protons and neutrons each have a mass 1836 times greater than that of an electron, most of the mass of the atom lies in the nucleus. The hydrogen atom is the simplest of all atoms. The nucleus is a single proton orbited by a single electron, with a total mass of only 1.67×10^{-27} kg, about a trillionth of a trillionth of a gram.

An atom is mostly empty space. To see this we can construct a simple scale model. The nucleus of a hydrogen atom is a proton with a diameter of about 0.0000016 nm, or 1.6×10^{-15} m. If we multiply this by one trillion (10^{12}) we can represent the nucleus of our model atom with a grape seed, which is about 0.16 cm in diameter. The region of a hydrogen atom containing the whirling electron has a diameter of about 0.4 nm or 4×10^{-10} m. Multiplying by a trillion magnifies the diameter to about 400 m, or about 4.5 football fields laid end to end (▐ Figure 7-3). When you imagine a grape seed in the midst of a sphere 4.5 football fields in diameter, you can see that an atom is mostly empty space.

DIFFERENT KINDS OF ATOMS

There are over a hundred kinds of atoms called chemical elements. Which element an atom is depends only on the number of protons in the nucleus. For example, carbon has six protons in its nucleus. An atom with one more proton than this is nitrogen, and an atom with one fewer proton is boron.

Although the number of protons in an atom of an element is fixed, we can change the number of neutrons in an atom's nucleus without changing the atom significantly. For instance, if we add a neutron to the carbon nucleus, we still have carbon, but it is slightly heavier than normal carbon. Atoms that have the same number of protons but a different number of neutrons are **isotopes.** Carbon has two stable isotopes. One form contains six protons and six neutrons for a total of 12 particles and is thus called carbon-12. Carbon-13 has six protons and seven neutrons in its nucleus (▐ Figure 7-4).

Protons and neutrons are bound tightly into the nucleus, but the electrons are held loosely in the electron cloud. Running a comb through your hair creates a static charge by removing a few electrons from their atoms. This process is called **ionization,** and the atom that has lost one or more electrons is an **ion.** A carbon atom is neutral if it has six electrons to balance the positive charge of the six protons in its nucleus. If we ionize the atom by removing one or more electrons, the atom is left with a net positive charge. Under some circumstances, an atom may capture one or more extra electrons, giving it more negative charges than positive. Such a negatively charged atom is also considered an ion.

Atoms that collide may form bonds with each other by exchanging or sharing electrons. Two or more atoms bonded together form a **molecule.** Atoms do collide in

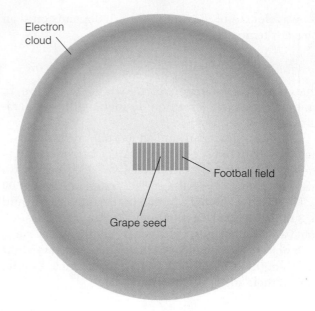

FIGURE 7-3

Magnifying a hydrogen atom by 10^{12} makes the nucleus the size of a grape seed and the diameter of the electron cloud about 4.5 times longer than a football field. The electron itself is still too small to see.

stars, but the high temperatures cause violent collisions that are unfavorable for chemical bonding. Only in the coolest stars are the collisions gentle enough to permit the formation of chemical bonds. We will see later that the presence of molecules such as titanium oxide (TiO) in a star is a clue that the star is cool. In later chapters,

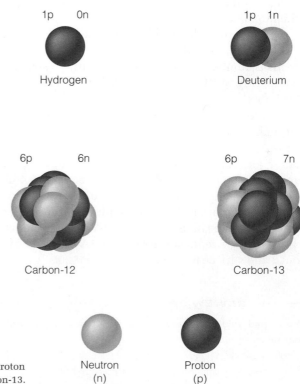

FIGURE 7-4

Some common isotopes. A rare isotope of hydrogen, deuterium, contains a proton and a neutron in its nucleus. Two isotopes of carbon are carbon-12 and carbon-13.

Quantum Mechanics: The World of the Very Small

Quantum mechanics is the set of rules that describe how atoms and subatomic particles behave. When we think about large objects such as stars, planets, aircraft carriers, and hummingbirds, we don't have to think about quantum mechanics, but on the atomic scale, particles behave in ways that seem unfamiliar to us.

One of the principles of quantum mechanics specifies that we can not know simultaneously the exact location and motion of a particle. This is why physicists refer to the electrons in an atom as if they were a cloud of negative charge surrounding the nucleus. Because we can't know the position and motion of the electron, we can't really describe it as a small particle following an orbit. We can use that image as a model to help our imaginations, but the reality is much more interesting, and describing the electrons as a charge cloud gives us a better and more sophisticated model of an atom.

This raises some serious questions about reality. Is an electron really a particle at all? Quantum mechanics describes particles as waves, and waves as particles. If we can't know simultaneously the position and motion of a specific particle, how can we know how it will react to a collision with a photon or another particle? The answer is that we can't know, and that seems to violate the principle of cause and effect (Window on Science 5-1).

Needless to say, we can't explore these discrepancies here. Scientists and philosophers of science continue to struggle with the meaning of reality on the quantum-mechanical level. Here we should note that the reality we see on the scale of stars and hummingbirds is only part of nature. We have constructed some models to help us think about nature on the scale of atoms, but the truth is much more interesting and much more exciting than anything we see on larger scales. Although we use models of atoms to study stars, there is still much to learn about the atoms themselves.

---

we will see that molecules can form in cool gas clouds in space and in the atmospheres of planets.

ELECTRON SHELLS

We mentioned the whirling cloud of orbiting electrons in a general way, but the specific way electrons behave within the cloud is very important in astronomy.

The electrons are bound to the atom by the attraction between their negative charge and the positive charge on the nucleus. This attraction is known as the **Coulomb force,** after the French physicist Charles-Augustin de Coulomb (1736–1806). If we wish to ionize an atom, we need a certain amount of energy to pull an electron away from its nucleus. This energy is the electron's **binding energy,** the energy that holds it to the atom.

An electron may orbit the nucleus at various distances. If the orbit is small, the electron is close to the nucleus, and a large amount of energy is needed to pull it away. Therefore, its binding energy is large. An electron orbiting farther from the nucleus is held more loosely, and less energy will pull it away. It therefore has less binding energy. The size of an electron's orbit is related to the energy that binds it to the atom.

Nature permits atoms only certain amounts (quanta) of binding energy, and the laws that describe how atoms behave are called the laws of **quantum mechanics** (Window on Science 7-1). Much of our discussion of atoms is based on the laws of quantum mechanics.

Because atoms can have only certain amounts of binding energy, our model atoms can have orbits of only certain sizes, called **permitted orbits.** These are like steps in a staircase: You can stand on the number-one step or the number-two step, but not on the number-one-and-one-quarter step. The electron can occupy any permitted orbit but not orbits in between.

The arrangement of permitted orbits depends primarily on the charge of the nucleus, which in turn depends on the number of protons. Thus, each kind of element has its own pattern of permitted orbits (■ Figure 7-5). Isotopes of the same elements have nearly the same pattern because they have the same number of protons. However, ionized atoms have orbital patterns that differ from their un-ionized forms. Thus the arrangement of permitted orbits differs for every kind of atom and ion.

How many hydrogen atoms would it take to cross the head of a pin?

This is not a frivolous question. In answering it, we will discover how small atoms really are, and we will see how powerful physics can be as a way to understand nature. First, we will assume that the head of a pin is about 1 mm in diameter. That is 0.001 m. The size of a hydrogen atom is represented by the diameter of the electron cloud, roughly 0.4 nm. Because 1 nm equals 10^{-9} m, we multiply and discover that 0.4 nm equals 4×10^{-10} m. To find out how many atoms would stretch 0.001 m, we divide the diameter of the pinhead by the diameter of an atom. That is, we divide 0.001 m by 4×10^{-10} m, and we get 2.5×10^6. It would thus take 2.5 million hydrogen atoms lined up side by side to cross the head of a pin.

This shows how tiny an atom is and also how powerful basic physics is. A bit of arithmetic gives us a view of nature beyond the capability of our eyes. Now use another bit of arithmetic to calculate how many hydrogen atoms you would need to add up to the mass of a paper clip (1 g).

FIGURE 7-5

The electron in an atom may occupy only certain, permitted orbits. Because different elements have different charges on their nuclei, the elements have different patterns of permitted orbits.

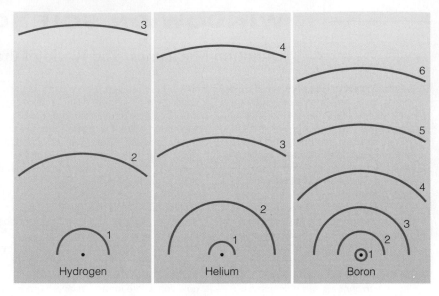

The electron orbits in atoms are familiar territory to astronomers because the electrons in those orbits can interact with light. Such interactions fill starlight with clues to the nature of the stars.

7-3 THE INTERACTION OF LIGHT AND MATTER

If light did not interact with matter, you would not be able to see these words. In fact, you would not exist, because, among other problems, photosynthesis would be impossible and there would be no grass, wheat, bread, beef, cheeseburgers, or any other kind of food. The interaction of light and matter makes our lives possible, and it also makes it possible for us to understand our universe.

We begin our study of light and matter by considering the hydrogen atom. As we noted earlier, hydrogen is both simple and common. Roughly 90 percent of all atoms in the universe are hydrogen.

THE EXCITATION OF ATOMS

Each orbit in an atom represents a specific amount of binding energy, so physicists commonly refer to the orbits as **energy levels.** Using this terminology, we can say that an electron in its smallest and most tightly bound orbit is in its lowest permitted energy level. We can move the electron from one energy level to another by supplying enough energy to make up the difference between the two energy levels. It is like moving a flowerpot from a low shelf to a high shelf; the greater the distance between the shelves, the more energy we need to raise the pot. The amount of energy needed to move the electron is the energy difference between the two energy levels.

If we move the electron from a low energy level to a higher energy level, we say the atom is an **excited atom.** That is, we have added energy to the atom in moving its electron. If the electron falls back to the lower energy level, that energy is released.

An atom can become excited by collision. If two atoms collide, one or both may have electrons knocked into a higher energy level. This happens very commonly in hot gas, where the atoms move rapidly and collide often.

Another way an atom can get the energy that moves an electron to a higher energy level is to absorb a photon. Only a photon with exactly the right amount of energy can move the electron from one level to another. If the photon has too much or too little energy, the atom cannot absorb it. Because the energy of a photon depends on its wavelength, only photons of certain wavelengths can be absorbed by a given kind of atom. ▌Figure 7-6 shows the lowest four energy levels of the hydrogen atom along with three photons the atom could absorb. The longest-wavelength photon only has enough energy to excite the electron to the second energy level, but the shorter-wavelength photons can excite the electron to higher levels. A photon with too much or too little energy cannot be absorbed. Because the hydrogen atom has many more energy levels than shown in Figure 7-6, it can absorb photons of many different wavelengths.

Atoms, like humans, cannot exist in an excited state forever. The excited atom is unstable and must eventually (usually within 10^{-6} to 10^{-9} seconds) give up the energy it has absorbed and return its electron to the lowest energy level. Because the electrons eventually tumble down to this bottom level, physicists call it the **ground state.**

When the electron drops from a higher to a lower energy level, it moves from a loosely bound level to one more tightly bound. The atom then has a surplus of energy—the energy difference between the levels—that it can emit as a photon. Study the sequence of events in ▌Figure 7-7 to see how an atom can absorb and emit photons. Because each type of atom or ion has its

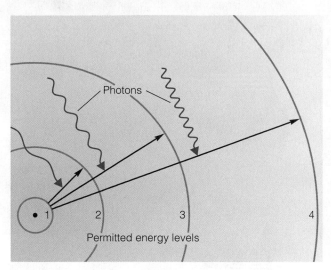

FIGURE 7-6
A hydrogen atom can absorb only those photons that move the atom's electron to one of the higher-energy orbits. Here three different photons are shown along with the change they would produce if they were absorbed.

unique set of energy levels, each type absorbs and emits photons with a unique set of wavelengths. Thus we can identify the elements in a gas by studying the characteristic wavelengths of light absorbed or emitted.

The process of excitation and emission is a common sight in urban areas at night. A neon sign glows when atoms of the neon gas in the tube are excited by electricity flowing through the tube. As the electrons in the electric current flow through the gas, they collide with the neon atoms and excite them. As we have seen, immediately after an atom is excited, its electron drops back to a lower energy level, emitting the surplus energy as a photon of a certain wavelength. The photons emitted by excited neon produce a reddish-orange glow. Signs of other colors, erroneously called "neon," contain other gases or mixtures of gases instead of pure neon.

THE FORMATION OF A SPECTRUM

The spectrum of a star is formed as light passes outward through the gases near its surface. Study ▮ "Atomic Spectra" on pages 132 and 133. Notice three important properties of spectra.

First, there are three kinds of spectra. When we see one of these types of spectra, we can recognize the kind of matter that emitted the light.

Notice also that the wavelengths of the photons that are absorbed or emitted in a spectrum are determined by the atomic energy levels in the atoms. The emitted photons coming from a hot cloud of hydrogen gas have the same wavelengths as the photons absorbed by hydrogen atoms in the gases of a star. Different atoms have different energy levels, so their spectra contain different patterns of spectral lines.

Finally, notice that the hydrogen atom produces many spectral lines from the ultraviolet to the infrared, but only three are visible to our eyes. These hydrogen lines are prominent in the spectra of stars and can tell us the temperature of the stars.

| REVIEW **CRITICAL INQUIRY** |

What spectrum would we see if we observed molten iron?
Molten iron is a dense liquid, and the atoms and molecules collide so often they emit all wavelengths and we would see a continuous spectrum. That is Kirchhoff's first law. But in order to see the molten iron, we would have to look through the hot vapors rising from it. Photons on their way to our spectrograph would pass through these gases, and atoms in the gases would absorb certain wavelengths. So what we would really see would be the continuous spectrum of the molten iron with weak absorption lines caused by the gases above the iron. This is what Kirchhoff's third law describes.

But suppose we had a very sensitive spectrograph that could look at the hot gases above the molten iron from the side so as to avoid looking directly at the molten iron. What kind of spectrum would the hot gases emit?

Whatever kind of spectrum astronomers look at, the most common spectral lines are the Balmer lines of hydrogen, the only hydrogen lines we can study from Earth's surface. In the next section, we will see how Balmer lines can tell us a star's temperature.

7-4 STELLAR SPECTRA

In later chapters, we use spectra to study galaxies and planets, but we begin here by studying the spectra of stars. Such spectra are the easiest to understand, and the nature of stars is central to our study of all celestial objects.

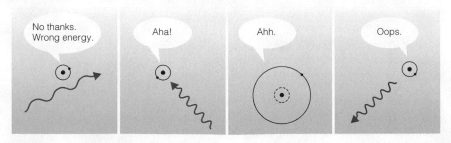

FIGURE 7-7
An atom can absorb a photon only if the photon has the correct amount of energy. The excited atom is unstable and within a fraction of a second returns to a lower energy level, reradiating the photon in a random direction.

Atomic Spectra

To understand how we can analyze a spectrum, we begin with a simple incandescent lightbulb.

The hot filament emits black body radiation, which forms a **continuous spectrum**.

An **absorption spectrum** results when radiation passes through a cool gas. In this case we can imagine that our lightbulb is surrounded by a cool cloud of gas. Atoms in the gas absorb photons of certain wavelengths, which are missing from the spectrum, and we see their positions as dark **absorption lines**. Such spectra are sometimes called **dark-line spectra**.

An **emission spectrum** is produced by photons emitted by an excited gas. We could see **emission lines** by turning our telescope aside so that photons from the bright bulb did not enter our telescope. The photons we would see would be those emitted by the excited atoms near the bulb. Such spectra are also called **bright-line spectra**.

The spectrum of a star is an absorption spectrum. The denser layers of the photosphere emit black body radiation. Gases in the atmosphere of the star absorb their specific wavelengths and form dark absorption lines in the spectrum.

In 1859, long before scientists understood atoms and energy levels, the German scientist Gustav Kirchhoff formulated three rules, now known as **Kirchhoff's laws**, that describe the three types of spectra.

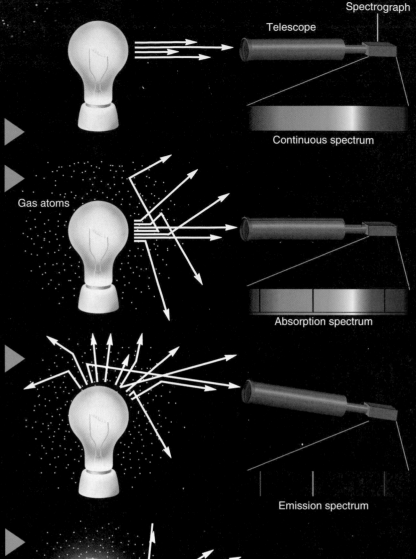

Telescope
Spectrograph

Continuous spectrum

Gas atoms

Absorption spectrum

Emission spectrum

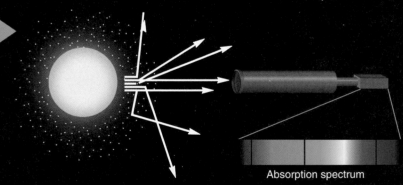

Absorption spectrum

GUSTAV ROBERT KIRCHHOFF
1824
1887
ΣI_s
$\Sigma U_s = 0$
30
DEUTSCHE BUNDESPOST BERLIN

Courtesy Ron Reese and Michelangelo Fazio

KIRCHHOFF'S LAWS

Law I: The Continuous Spectrum

A solid, liquid, or dense gas excited to emit light will radiate at all wavelengths and thus produce a continuous spectrum.

Law II: The Emission Spectrum

A low-density gas excited to emit light will do so at specific wavelengths and thus produce an emission spectrum.

Law III: The Absorption Spectrum

If light comprising a continuous spectrum passes through a cool, low-density gas, the result will be an absorption spectrum.

Ace Astronomy™ Go to AceAstronomy and click Active Figures to see "Kirchhoff's Laws" and the three kinds of spectra formed.

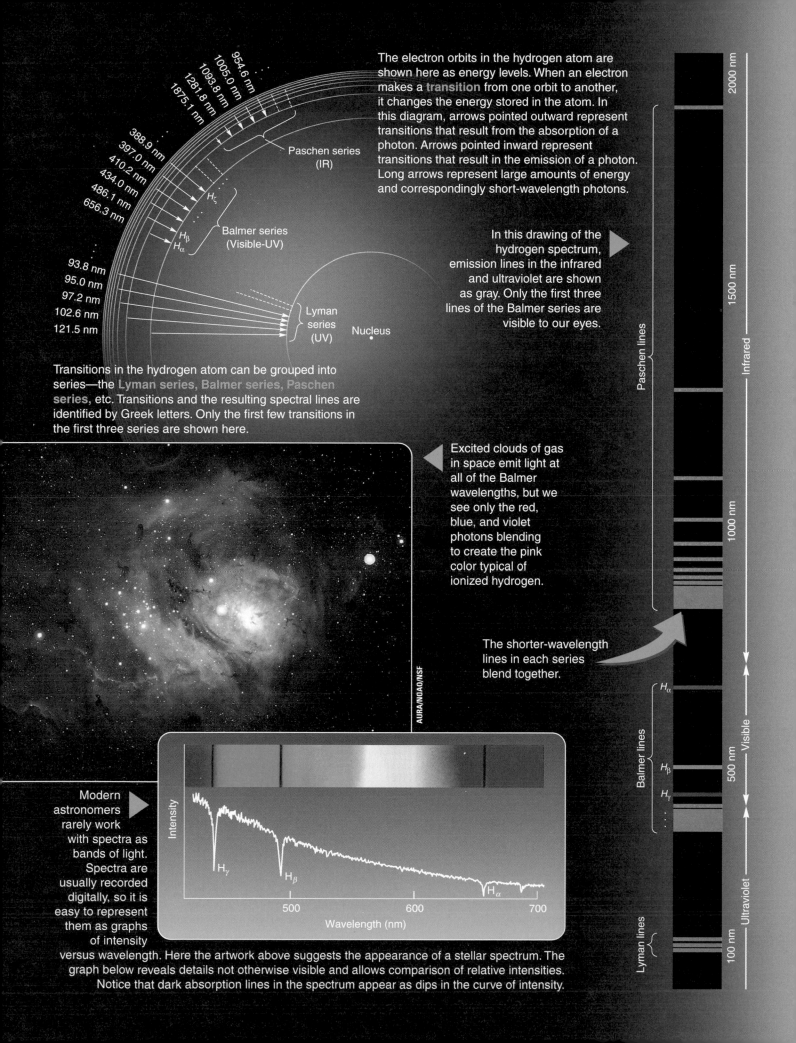

954.6 nm
1005.0 nm
1093.8 nm
1281.8 nm
1875.1 nm

388.9 nm
397.0 nm
410.2 nm
434.0 nm
486.1 nm
656.3 nm

93.8 nm
95.0 nm
97.2 nm
102.6 nm
121.5 nm

Paschen series (IR)

H_ζ

H_β
H_α

Balmer series (Visible-UV)

Lyman series (UV)

Nucleus

The electron orbits in the hydrogen atom are shown here as energy levels. When an electron makes a transition from one orbit to another, it changes the energy stored in the atom. In this diagram, arrows pointed outward represent transitions that result from the absorption of a photon. Arrows pointed inward represent transitions that result in the emission of a photon. Long arrows represent large amounts of energy and correspondingly short-wavelength photons.

In this drawing of the hydrogen spectrum, emission lines in the infrared and ultraviolet are shown as gray. Only the first three lines of the Balmer series are visible to our eyes.

Transitions in the hydrogen atom can be grouped into series—the Lyman series, Balmer series, Paschen series, etc. Transitions and the resulting spectral lines are identified by Greek letters. Only the first few transitions in the first three series are shown here.

Excited clouds of gas in space emit light at all of the Balmer wavelengths, but we see only the red, blue, and violet photons blending to create the pink color typical of ionized hydrogen.

AURA/NOAO/NSF

The shorter-wavelength lines in each series blend together.

Modern astronomers rarely work with spectra as bands of light. Spectra are usually recorded digitally, so it is easy to represent them as graphs of intensity

Intensity

H_γ

H_β

H_α

500   600   700
Wavelength (nm)

versus wavelength. Here the artwork above suggests the appearance of a stellar spectrum. The graph below reveals details not otherwise visible and allows comparison of relative intensities. Notice that dark absorption lines in the spectrum appear as dips in the curve of intensity.

2000 nm

1500 nm

Infrared

Paschen lines

1000 nm

500 nm

Visible

H_α

H_β

H_γ

Balmer lines

100 nm

Ultraviolet

Lyman lines

THE BALMER THERMOMETER

We can use the Balmer absorption lines as a thermometer to find the temperatures of stars. Earlier we saw how to estimate temperature from color, but the strengths of the Balmer lines in a star's spectrum give a much more accurate estimate of the star's temperature.

The Balmer thermometer works because the Balmer absorption lines are produced only by atoms whose electrons are in the second energy level. If the star is cool, there are few violent collisons between atoms to excite the electrons, and most atoms have their electron in the ground state. If most electrons are in the ground state, they can't absorb photons in the Balmer series. As a result, we should expect to find weak Balmer absorption lines in the spectra of cool stars.

In hot stars, on the other hand, there are many violent collisions between atoms, exciting electrons to high energy levels or knocking the electron clear out of some atoms; that is, some atoms are ionized. Thus, few hydrogen atoms have their electron in the second orbit to form Balmer absorption lines, and we should expect hot stars, like cool stars, to have weak Balmer absorption lines.

At some intermediate temperature, the collisions are just right to excite large numbers of electrons into the second energy level. With many atoms excited to the second level, the gas absorbs Balmer-wavelength photons strongly and thus produces strong Balmer lines.

To summarize, the strength of the Balmer lines depends on the temperature of the star's surface layers. Both hot and cool stars have weak Balmer lines, but medium-temperature stars have strong Balmer lines.

Theoretical calculations can predict just how strong the Balmer lines should be for stars of various temperatures. The details of these calculations are not important to us, but the results are. The curve in ▌Figure 7-8a shows the strength of the Balmer lines for various stellar temperatures. We could use this as a temperature indicator except that the curve gives us two answers. A star with Balmer lines of a certain strength might have either of two temperatures, one high and one low. We must examine other spectral lines to choose the correct temperature.

We have seen how the strength of the Balmer lines depends on temperature. Temperature has a similar ef-

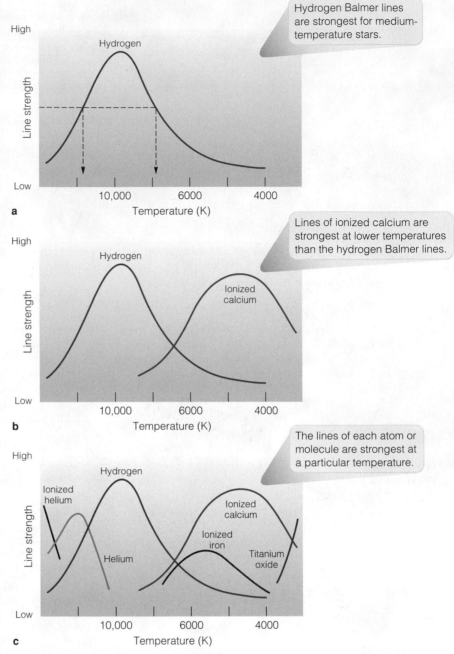

FIGURE 7-8

The strength of spectral lines can tell us the temperature of a star. (a) Balmer hydrogen lines alone are not enough because Balmer lines of a certain strength could be produced by a hotter star or a cooler star. (b) Adding another atom to the diagram helps, and (c) adding many atoms and molecules to the diagram gives us a precise tool to find the temperatures of stars.

fect on the spectral lines of other elements, but the temperature at which they reach maximum strength differs for each element (Figure 7-8b). If we add these elements to our graph, we get a handy tool for finding the stars' temperatures (Figure 7-8c).

How do we use this tool? We determine a star's temperature by comparing the strengths of its spectral lines with our graph. For instance, if we recorded a spectrum of a star and found medium-strength Balmer lines and strong helium lines, we could conclude it had a temperature of about 20,000 K. But if the star had weak hydrogen lines and strong lines of ionized iron, we would assign it a temperature of about 5800 K, similar to that of the sun.

The spectra of stars cooler than about 3000 K contain dark bands produced by molecules such as titanium oxide (TiO). Because of their structure, molecules can absorb photons at many wavelengths, producing numerous, closely spaced spectral lines that blend together to form bands. These molecular bands appear only in the spectra of the coolest stars because, as mentioned before, molecules in cool stars are not subject to the violent collisions that would break up molecules in hotter stars. Thus the presence of dark bands in a star's spectrum indicates that the star is very cool.

From stellar spectra, astronomers have found that the hottest stars have surface temperatures above 40,000 K and the coolest about 2000 K. Compare these with the surface temperature of the sun, which is about 5800 K.

SPECTRAL CLASSIFICATION

We have seen that the strengths of spectral lines depend on the surface temperature of the star. From this we can predict that all stars of a given temperature should have similar spectra. If we learn to recognize the pattern of spectral lines produced by a 6000-K star, for instance, we need not use Figure 7-8c every time we see that kind of spectrum. We can save time by classifying stellar spectra rather than analyzing each one individually.

The first widely used classification system was devised by astronomers at Harvard during the 1890s and 1900s. One of them, Annie J. Cannon, personally inspected and classified the spectra of over 250,000 stars. The spectra were first classified in groups labeled A through Q, but some groups were later dropped, merged with others, or reordered. The final classification includes the seven **spectral classes,** or **types,** still used today: O, B, A, F, G, K, M.*

This sequence of spectral types, called the **spectral sequence,** is important because it is a temperature sequence. The O stars are the hottest, and the temperature continues to decrease down to the M stars, the coolest. For maximum precision, astronomers divide each spectral class into 10 subclasses. For example, spectral class A consists of the subclasses A0, A1, A2, . . . A8, A9. Next comes F0, F1, F2, and so on. This finer division gives us a star's temperature to an accuracy within about 5 percent. The sun, for example, is not just a G star, but a G2 star, with a temperature of 5800 K.

We classify a star by the lines and bands in its spectrum, as shown in ▌Table 7-1. For example, if it has weak Balmer lines and lines of ionized helium, it must be an O star. This table is based on the same information we used in Figure 7-8c.

▌Figure 7-9 shows 13 stellar spectra ranging from the hottest at the top to the coolest at the bottom. Notice how special features change gradually from hot to cool stars. Although these spectra are attractive, astronomers rarely work with spectra as color images. Rather, they display spectra as graphs of intensity versus wavelength with dark absorption lines as dips in the graph (▌Figure 7-10). Such graphs show more detail than photographs. Notice also that the overall curves are similar to black body curves. The wavelength of maximum is in the infrared for the coolest stars and in the ultraviolet for the hottest stars.

Compare Figures 7-9 and 7-10 and notice how the strength of spectral lines depends on temperature. Note that the Balmer lines are strongest in A stars, where the temperature is moderate but still high enough to excite the electrons in hydrogen atoms to the second energy level, where they can absorb Balmer wavelength

*Generations of astronomy students have remembered the spectral sequence using the mnemonic "Oh, Be A Fine Girl (Guy), Kiss Me." More recent suggestions from students include, "Oh Boy, An F Grade Kills Me," and "Only Bad Astronomers Forget Generally Known Mnemonics."

TABLE 7-1
Spectral Classes

Spectral Class	Approximate Temperature (K)	Hydrogen Balmer Lines	Other Spectral Features	Naked-Eye Example
O	40,000	Weak	Ionized helium	Meissa (O8)
B	20,000	Medium	Neutral helium	Achernar (B3)
A	10,000	Strong	Ionized calcium weak	Sirius (A1)
F	7,500	Medium	Ionized calcium weak	Canopus (F0)
G	5,500	Weak	Ionized calcium medium	Sun (G2)
K	4,500	Very weak	Ionized calcium strong	Arcturus (K2)
M	3,000	Very weak	TiO strong	Betelgeuse (M2)

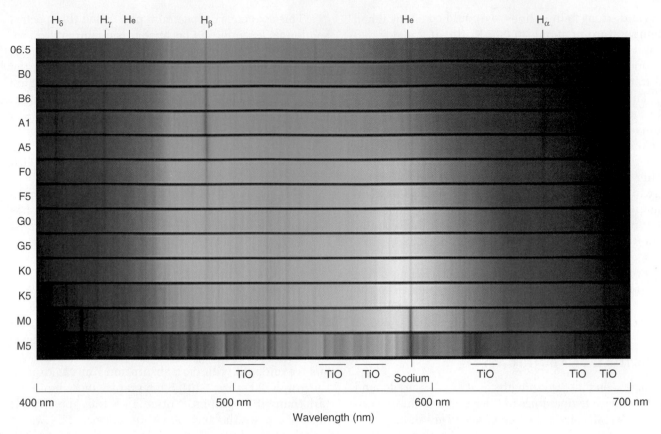

FIGURE 7-9
These spectra show stars from hot O stars at the top to cool M stars at the bottom. The Balmer lines of hydrogen are strongest about A0, but the two closely spaced lines of sodium in the yellow are strongest for very cool stars. Helium lines appear only in the spectra of the hottest stars. Bands produced by the molecule titanium oxide are strong in the spectra of the coolest stars. *(AURA/NOAO/NSF)*

photons. In the hotter stars (O and B), the Balmer lines are weak because the higher temperature excites the electrons to energy levels above the second or ionizes the atoms. The Balmer lines in cooler stars (F through M) are also weak, but for a different reason. The lower temperature cannot excite many electrons to the second energy level, so few hydrogen atoms are capable of absorbing Balmer wavelength photons.

The spectral lines of other atoms also change from class to class. Helium is visible only in the spectra of the hottest classes, and titanium oxide bands only in the coolest. Two lines of ionized calcium increase in strength from A to K and then decrease from K to M. Because the strength of these spectral lines depends on temperature, it requires only a few moments to study a star's spectrum and determine its temperature.

The study of spectral types is a century old, but astronomers continue to discover new types of stars. The **L dwarfs,** found in 1998, are cooler and fainter than M stars. The L dwarfs are clearly a different type of star. The spectra of M stars contain bands produced by metal oxides such as titanium oxide, but L dwarf spectra contain bands produced by chromium hydride and iron hydride (Figure 7-11). The **T dwarfs,** discovered

in 2000, are even cooler and fainter than L dwarfs. The development of giant telescopes and highly sensitive infrared cameras and spectrographs is allowing astronomers to find and study these coolest of stars.

THE COMPOSITION OF THE STARS

It seems as though it should be easy to find the composition of the sun and stars just by looking at their spectra, but it turns out to be a difficult task that wasn't well understood until the 1920s. The story of how astronomers first discovered the composition of the stars is worth telling, not only because it is the story of an important American astronomer who never got proper credit, but also because the story illustrates the temperature dependence of spectral features. The story begins in England.

As a child in England, Cecilia Payne (1900–1979) excelled in classics, languages, mathematics, and literature, but her first love was astronomy. After finishing Newnham College in Cambridge, she left England, sensing that there were no opportunities in England for a woman of science. In 1922, Payne arrived at Harvard, where she eventually earned her Ph.D., although the

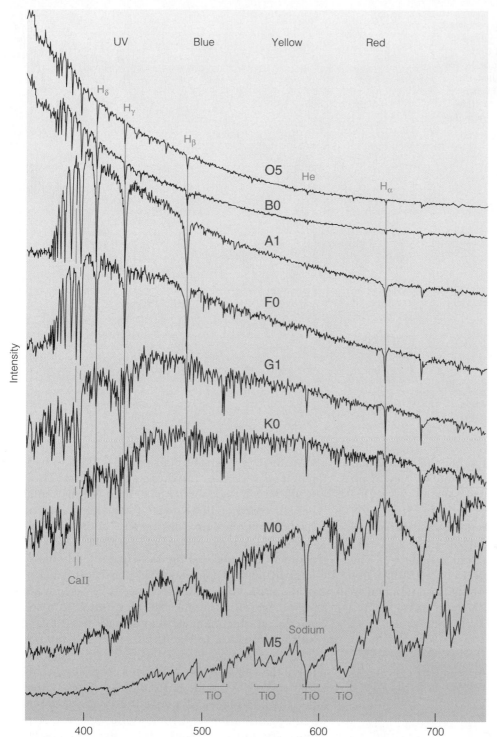

FIGURE 7-10
Modern digital spectra are often represented by graphs of intensity versus wavelength. Dark absorption lines are dips in intensity. The hottest stars are at the top and the coolest at the bottom. Hydrogen Balmer lines are strongest at about A0, while lines of ionized calcium (CaII) are strong in K stars. Titanium oxide (TiO) bands are strongest in the coolest stars. Compare these spectra with Figures 7-8c and 7-9. *(Courtesy NOAO, G. Jacoby, D. Hunter, and C. Christian)*

degree was awarded by Radcliffe because Harvard did not then admit women.

In her thesis, Payne attempted to relate the strength of the absorption lines in stellar spectra to the physical conditions in the atmospheres of the stars. This was not easy because, as we have seen in this chapter, a given spectral line can be weak because the atom is rare or because the temperature is too high or too low for that atom to be able to absorb efficiently. If we see sodium lines in a star's spectrum, we can be sure that the star contains sodium atoms, but if we see no sodium lines, we must consider the possibility that the star is too hot or too cool for sodium to produce spectral lines.

Payne's problem was to untangle these two factors and find the true temperatures of the stars and the true abundance of the atoms in their atmospheres. Recent

FIGURE 7-11

The spectrum of an M star and an L star compared: M-star spectra contain strong bands produced by the titanium oxide molecule (TiO), but L-star spectra contain weak or absent TiO bands. Notice the missing TiO band near 710 nm in the L-star spectrum above. Instead, the spectra of L stars contain bands produced by chromium hydride (CrH) and iron hydride (FeH). *(Adapted from data by J. Davy Kirkpatrick)*

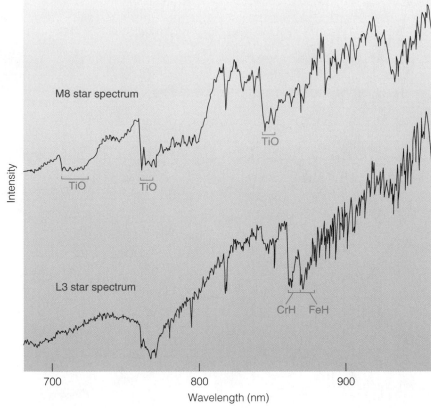

advances in atomic physics gave her the theoretical tools she needed. About the time Payne left Newnham College, Indian physicist Meghnad Saha published his work on the ionization of atoms. Drawing from such theoretical work, Payne was able to show that over 90 percent of the atoms in stars (including the sun) were hydrogen and most of the rest helium (▌ Table 7-2). The heavier atoms seemed more abundant only because they are better at absorbing photons at the temperatures of stars.

TABLE 7-2
The Most Abundant Elements in the Sun

Element	Percentage by Number of Atoms	Percentage by Mass
Hydrogen	91.0	70.9
Helium	8.9	27.4
Carbon	0.03	0.3
Nitrogen	0.008	0.1
Oxygen	0.07	0.8
Neon	0.01	0.2
Magnesium	0.003	0.06
Silicon	0.003	0.07
Sulfur	0.002	0.04
Iron	0.003	0.1

At the time, astronomers found it hard to believe Payne's abundances of hydrogen and helium. They especially found the abundance of helium unacceptable. After all, hydrogen lines are at least visible in most stellar spectra, but helium lines are almost invisible in the spectra of all but the hottest stars. Rather, nearly all astronomers assumed that the stars had roughly the same composition as Earth's surface; that is, they believed that the stars were composed mainly of heavier atoms such as carbon, silicon, iron, aluminum, and so on. Even the most eminent astronomers dismissed Payne's result as illusory. Faced with this pressure and realizing the limited opportunities available to women in science in the 1920s, Payne could not press her discovery.

It was 1929 before astronomers generally understood the importance of temperature on measurements of composition derived from stellar spectra. At that point, astronomers recognized that stars are mostly hydrogen and helium, but Payne received no credit.

Payne worked for many years as a staff astronomer at the Harvard College Observatory with no formal position on the faculty. She married Russian astronomer Sergei Gaposchkin in 1934 and was afterward known as Cecilia Payne-Gaposchkin. In 1956, when Harvard accepted women to its faculty, she was appointed a full professor and chair of the Harvard astronomy department.

Cecilia Payne-Gaposchkin's work on the chemical composition of the stars illustrates the importance of fully understanding the interaction between light and

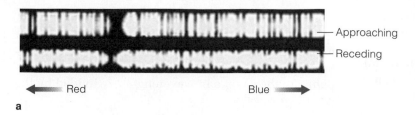

Approaching

Receding

◄— Red   Blue —►

a

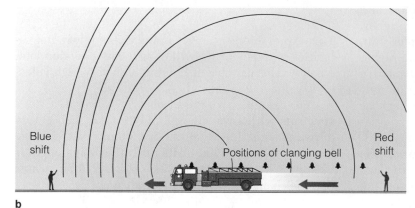

Blue shift

Positions of clanging bell

Red shift

b

c

FIGURE 7-12

The Doppler effect. (a) A blue shift appears in the spectrum of a star approaching Earth (top spectrum). A red shift appears in the spectrum of a star moving away from Earth (bottom spectrum). *(The Observatories of the Carnegie Institution of Washington)* (b) The clanging bell on a moving fire truck produces sounds that move outward (black circles). An observer ahead of the truck hears the clangs closer together, while an observer behind the truck hears them farther apart. (c) A moving source of light emits waves that move outward (black circles). An observer in front of the light source observes a shorter wavelength (a blue shift), and an observer behind the light source observes a longer wavelength (a red shift).

matter. Only a detailed understanding of the physics could lead her to the correct composition. As we turn our attention to other information that can be derived from stellar spectra, we again discover the importance of understanding light.

THE DOPPLER EFFECT

Surprisingly, one of the pieces of information hidden in a spectrum is the velocity of the light source. Astronomers can measure the wavelengths of lines in a star's spectrum and find the velocity of the star. The **Doppler effect** is the apparent change in the wavelength of radiation caused by the motion of the source.

If a star is moving toward Earth, the lines in its spectrum will be shifted slightly toward shorter wavelengths. That is, they are shifted toward the blue end of the spectrum—a **blue shift.** If a star is moving away from Earth, the lines are shifted slightly toward the red end of the spectrum—a **red shift.** These Doppler shifts are small and don't change the color of the star. But it is easy to detect these changes in wavelength in a star's spectrum.

▌Figure 7-12a shows the Doppler effect in two spectra of the star Arcturus. The lines in the top spectrum are slightly blue shifted because the spectrum was recorded when Earth, following its orbit, was moving toward Arcturus. The lines in the bottom spectrum are red shifted because it was recorded six months later, when Earth was moving away from Arcturus.

The Doppler effect tells us how rapidly the distance between us and the source of light is increasing or decreasing. It does not matter whether we are moving or the star is moving. Only the relative velocity is important. Also, the Doppler shift is only sensitive to the part of the velocity directed away from us or toward us. This part of the velocity is called the **radial velocity (V_r).** The **transverse velocity** is the part of the star's velocity perpendicular to our line of sight (▌Figure 7-13). The Doppler effect cannot be used to measure the transverse velocity.

The Doppler effect occurs for any kind of radiation, not just light. Radio astronomers, for instance, observe the Doppler effect when they measure the wavelength of radio waves coming from objects moving toward or away from Earth. Whatever wavelength range we consider—radio, X rays, gamma rays—we will still refer to

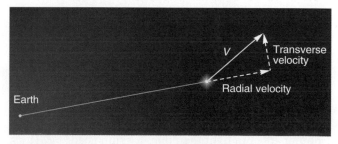

FIGURE 7-13

The radial velocity of a star is the part of the star's velocity *V* that is directed away from or toward Earth. The Doppler effect can tell us a star's radial velocity but cannot tell us the transverse velocity—the part of the velocity perpendicular to the radial direction.

a lengthening of the observed wavelength as a red shift and a shortening of the observed wavelength as a blue shift.

We can even detect the Doppler shift in sound. Sound is not electromagnetic radiation, of course; it is a mechanical wave transmitted through air, but because it is a wave, it is subject to the Doppler effect. Sounds with long wavelengths have low pitches, and sounds with short wavelengths have higher pitches. When we hear a truck pass by, we hear a Doppler shift. Its sound is shifted to shorter wavelengths and higher pitches (a "blue" shift) while it is approaching and is shifted to longer wavelengths and lower pitches (a "red" shift) after it passes. As the truck passes, we hear the pitch drop.

We have described how the Doppler shift works, but we haven't explained why it works. To see why the Doppler shift occurs, we can think about a fire truck approaching us with a bell clanging once a second. When the bell clangs, the sound travels ahead of the truck to reach our ears. One second later, the bell clangs again, but not at the same place. During that one second, the fire truck moved closer to us, so the bell is closer at its second clang. Now the sound has a shorter distance to travel and reaches our ears a little sooner than it would have if the fire truck were not approaching. The third time the bell clangs, it is even closer. By timing the bell, we could observe that the clangs are slightly less than 1 second apart, all because the fire truck is approaching. If the fire truck were moving away from us, we would hear the clangs sounding more than one second apart, because each successive clang of the bell occurs farther from us.

Figure 7-12b shows a fire truck moving toward one observer and away from another observer. The position of the bell at each clang is shown by the small black bells with the sound spreading outward as black circles. We can see that the clangs are squeezed together ahead of the fire truck and stretched apart behind.

Now we can substitute a source of light waves for the clanging of the bell. If the source is approaching, then each time the source emits the peak of a wave it will be slightly closer to us, and we will observe a shorter wavelength. If it is moving away, we will observe a longer wavelength. This is shown in Figure 7-12c, where the peaks of the waves appear compressed in front of the moving light source and stretched out behind the source.

Of course, how great the change in wavelength is depends on the velocity. Just as a slowly moving car has a smaller Doppler effect in sound than does a high-speed airplane, a slowly moving star has a smaller Doppler effect in light than does a high-speed star. Thus, we can measure the velocity by measuring the amount by which the spectral lines are shifted in wavelength. A large Doppler shift means a high radial velocity.

CALCULATING THE DOPPLER VELOCITY

It is easy to calculate the radial velocity of an object from its Doppler shift. The formula is a simple ratio relating the radial velocity V_r divided by the speed of light c to the change in wavelength $\Delta\lambda$ divided by the unshifted wavelength λ_0:

$$\frac{V_r}{c} = \frac{\Delta\lambda}{\lambda_0}$$

This expression is quite accurate for the low radial velocities of stars, but we will need a better version later when we discuss objects moving with very high velocities.

For example, we might observe a line in a star's spectrum with a wavelength of 600.1 nm. Laboratory measurements show that the line should have a wavelength of 600 nm. That is, its unshifted wavelength is 600 nm. What is the star's radial velocity? First we note that the change in wavelength is 0.1 nm:

$$\frac{V_r}{c} = \frac{0.1}{600} = 0.000167$$

Multiplying by the speed of light, 3×10^5 km/s, gives the radial velocity, 50 km/s. Because the wavelength is shifted to the red (lengthened), the star must be receding from us.

You may be quite familiar with this method of speed measurement. Police radar uses the Doppler shift in reflected radio waves to determine the velocity of approaching cars.

Understanding the Doppler shift leads us to a final illustration of the information hidden in stellar spectra. Even the shapes of the spectral lines can tell us secrets about the stars.

THE SHAPES OF SPECTRAL LINES

When astronomers refer to the shape of a spectral line, they mean the variation of intensity across the line. An absorption line, for instance, is darkest in the center and grows brighter to each side. This shape is represented by a **line profile,** a graph of brightness as a function of wavelength across a spectral line (❚ Figure 7-14).

The exact shape of a line profile can tell us a great deal about a star, but the most important characteristic is the width of the line. Spectral lines are not perfectly narrow; if they were, we could not see them. They have a natural width because nature allows an atom some leeway in the energy it may absorb or emit. The quantum mechanics behind this effect is beyond the scope of our discussion, but the result is very simple. In the absence of all other effects, spectral lines have a natural width of about 0.001 to 0.00001 nm—very narrow indeed.

The natural widths of spectral lines are not important in most branches of astronomy because other effects smear out the lines and make them much broader. For example, if a star spins rapidly, the Doppler effect

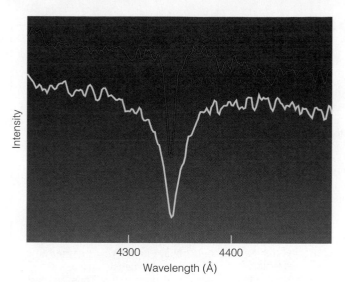

FIGURE 7-14

A line profile is a graph that shows the intensity of starlight as a function of wavelength across a spectral line. In these two examples, the spectra show the same spectral line from the spectra of two A1 stars. Thus, both stars have the same temperature. Differences in the density of the gas in the atmospheres of the stars cause the differences in the widths of the spectral lines. Such differences in the width of the spectral lines are easily measured from such line profiles. *(Courtesy NOAO, G. Jacoby, D. Hunter, and C. Christian)*

will broaden the spectral lines. As the star rotates, one side will recede from us, and the other side will approach. Light from the receding side will be red shifted, and light from the approaching side will be blue shifted, so any spectral lines will be broadened.

Another important process is called **Doppler broadening.** To consider this process, let us imagine that we photograph the spectrum of a jar full of hydrogen atoms (▌ Figure 7-15). Because the gas has some thermal energy (it is not at absolute zero), the gas atoms are in motion. Some will be coming toward our spectrograph, and some will be receding. Most, of course, will not be traveling very fast, but some will be moving very quickly. The photons emitted by the atoms approaching us will have slightly shorter wavelengths because of the Doppler effect, and photons emitted by atoms receding from us will have slightly longer wavelengths. Thus, the Doppler shifts due to the motions of the individual atoms will smear the spectral line out and make it broader. We have described the Doppler broadening of an emission line, but the effect is the same for absorption lines.

The extent of Doppler broadening depends on the temperature of the gas. If the gas is cold, the atoms travel at low velocities, and the Doppler shifts are small (Figure 7-15a). If the gas is hot, however, the atoms travel faster, Doppler shifts are larger, and the lines will be wider (Figure 7-15b).

Another form of broadening, **collisional broadening,** is caused by collisions between atoms, and thus it depends on the density of the gas (Parameters of Science 7-2). Densities in astronomy cover an enormous range, from one atom per cubic centimeter in space to millions of tons of atoms per cubic centimeter inside dead stars. Clearly, we should expect such densities to affect the way atoms collide with one another and thus how they absorb and emit photons.

Collisional broadening spreads out spectral lines when the atoms absorb or emit photons while they are colliding with other atoms, ions, or electrons. The collisions disturb the energy levels in the atoms, making it possible for the atoms to absorb a slightly wider range

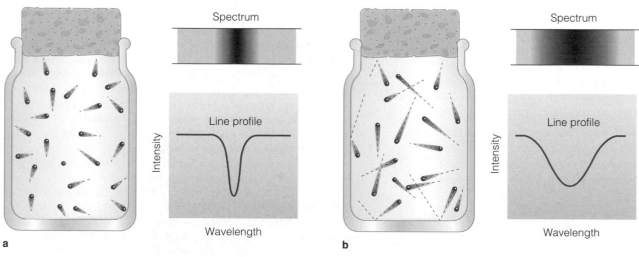

FIGURE 7-15

Doppler broadening. The atoms of a gas are in constant motion. Photons emitted by atoms moving toward the observer will have slightly shorter wavelengths, and those emitted by atoms moving away will have slightly longer wavelengths. This broadens the spectral line. If the gas is cool (a), the atoms do not move very fast, the Doppler shifts are small, and the line is narrow. If the gas is hot (b), the atoms move faster, the Doppler shifts are larger, and the line is broader.

Density

Density is a measure of the amount of matter in a given volume. Density is expressed as mass per volume, such as grams per cubic centimeter. The density of water, for example, is about 1 g/cm^3.

To get a feel for density, imagine holding a brick in one hand and a similar-sized block of styrofoam in the other hand. We can easily tell that the brick contains more matter than the styrofoam block even though both are the same size. The brick weighs more than the styrofoam, but it isn't really the weight that we should consider. Rather, we should think about the mass of the two objects. In space, where they have no weight, the brick and the styrofoam would still have mass, and we could tell by moving them about that the brick

contains more mass than the styrofoam. For example, imagine tapping each object gently against your ear. The massive brick would be easy to distinguish from the low-mass styrofoam block even in weightlessness.

When we think of density, we divide mass by volume, and our minds make that comparison at an instinctive level. We can sense the density of an object just by handling it. Gift shops sometimes sell imitation rocks made of styrofoam as humorous gifts. "Rocks" made of styrofoam seem odd when we handle them because our brain expects rocks to be dense.

Density is a common parameter in science because it is a general property of materials. Lead, for example, has a

density of about 7 g/cm^3, and rock has a density of 3 to 4 g/cm^3. Water and ice have densities of about 1 g/cm^3. If we knew that a small moon had a density of 1.5 g/cm^3, we could immediately draw some conclusions about what kinds of materials it might be made of—ice and a little rock, but not much lead. The density of an object is a basic clue to its composition.

Density also determines how materials behave. The low-density gases in a nebula behave in one way, but the same gases under high density inside a star behave in a different way. Thus, astronomers must consider the density of materials when they describe astronomical objects.

of wavelengths. Thus, the spectral lines are wider. Because atoms in a dense gas collide more often than atoms in a low-density gas, collisional broadening depends on the density of the gas, but temperature is also a factor. Atoms in a hot gas travel faster and collide more often and more violently than atoms in a cool gas. Once again, the physics of the interaction of light and matter gives us a tool to understand starlight. In later chapters, we will see how collisional broadening affects the widths of spectral lines that are formed in normal stars like the sun, in giant stars, and in gas in interstellar space.

REVIEW CRITICAL INQUIRY

Why are helium lines weak and calcium lines strong in the visible spectrum of the sun?

To analyze this problem, we must recall that the ability of an atom or ion to absorb light depends on temperature. Helium is quite abundant in the sun, but the surface of the sun is too cool to excite helium atoms and enable them to easily absorb visible-wavelength photons. On the other hand, calcium is easily ionized, and calcium atoms that have lost an electron are very good absorbers of photons at temperatures like those at the sun's surface. Thus, calcium lines are strong in the solar spectrum even though calcium ions are rare, and helium lines are weak even though helium atoms are common.

If we are to analyze a star's spectrum and consider the strength of spectral lines, we must take into account the temperature of the star. Why don't we need to take temperature into account when we use the Doppler effect to measure a star's radial velocity?

The spectra of the stars are filled with clues about the gas emitting the light. Much of astronomy is based on unraveling these clues. We will begin in the following chapter by studying the nearest star—the sun.

SUMMARY

Stars emit black body radiation from the dense gases of their photospheres, and as that radiation passes through the less dense gas of the star's atmosphere, the atoms absorb photons of certain wavelengths to produce dark lines in the spectrum.

A heated solid, liquid, or dense gas emits black body radiation that contains all wavelengths and thus produces a continuous spectrum. Black body radiation is most intense at the wavelength of maximum, λ_{max}, which depends on the temperature of the radiating body. Hot objects emit mostly short-wavelength radiation, whereas cool objects emit mostly long-wavelength radiation. This effect gives us clues to the temperatures of stars—hot stars appear blue and cool stars red.

A low-density gas excited to emit radiation will produce an emission (or bright-line) spectrum. Light from a source of a continuous spectrum passing through a low-density gas will produce an absorption (or dark-line) spectrum. These three kinds of spectra are described by Kirchhoff's laws.

The lines in spectra are produced by the electrons that surround the nucleus of the atom. The electrons may occupy only certain permitted energy levels, and photons may be absorbed or emitted when electrons move from one energy level to another. Because each kind of atom and ion has a different set of energy levels, each kind of atom and ion can absorb or emit only photons of a certain wavelength. The hydrogen atom can produce the Lyman series lines in the ultraviolet, the Balmer series in the visual and near ultraviolet, the Paschen series in the infrared, and many more.

In cool stars, the Balmer lines are weak because most atoms are not excited out of the ground state. In hot stars, the Balmer lines are weak because most atoms are excited to higher orbits or ionized. Only at medium temperatures are the Balmer lines strong. We can use this effect as a thermometer for determining the temperature of a star. In its simplest form, this amounts to classifying the stars' spectra in the spectral sequence O, B, A, F, G, K, M.

Stellar spectra can tell us the chemical compositions of the stars, but we must be careful to consider the temperature of the star in our analysis. In general, 90 percent of the atoms in a star are hydrogen.

When a source of radiation is approaching us, we observe shorter wavelengths, a blue shift, and when it is receding, we observe longer wavelengths, a red shift. This Doppler effect makes it possible for the astronomer to measure a star's radial velocity, that part of its velocity directed toward or away from the earth.

The widths of spectral lines are affected by a number of processes. With no outside influences, a spectral line has a natural width that is very small. Doppler broadening, due to the thermal motion of the gas atoms, depends on the temperature of the gas. Collisional broadening occurs when the atoms emitting or absorbing photons collide with other atoms, ions, or electrons. The collisions disturb the atomic energy levels and slightly alter the wavelengths the atoms can absorb or emit. In a dense, hot gas, where the gas atoms collide often, collisional broadening is important.

NEW TERMS

temperature

Kelvin temperature scale

absolute zero

thermal energy

electron

black body radiation

wavelength of maximum intensity (λ_{max})

color index

nucleus

proton

neutron

isotope

ionization

ion

molecule

Coulomb force

binding energy

quantum mechanics

permitted orbit

energy level

excited atom

ground state

continuous spectrum

absorption spectrum (dark-line spectrum)

absorption line

emission spectrum (bright-line spectrum)

emission line

Kirchhoff's laws

transition

Lyman series

Balmer series

Paschen series

spectral class or type

spectral sequence

L dwarf

T dwarf

Doppler effect

blue shift

red shift

radial velocity (V_r)

transverse velocity

line profile

Doppler broadening

collisional broadening

density

REVIEW QUESTIONS

Ace◐Astronomy™ Assess your understanding of this chapter's topics with additional quizzing and animations at **http:// astronomy.brookscole.com/seeds8e**

1. Why can a good blacksmith judge the temperature of a piece of heated iron by its color?

2. Why do hot stars look bluer than cool stars?

3. Use black body radiation to explain why you could hold a cigarette between your lips and light it with a match, but you couldn't hold the cigarette in your lips while you light it by sticking the tip into a bonfire. The match and bonfire have about the same temperature. (This is just one of the hazards of smoking.)

4. What is the difference between a neutral atom, an ion, and an excited atom?

5. How do the energy levels in an atom determine which wavelength photons it can absorb or emit?

6. What kind of spectrum would you expect to record if you observed molten lava? In practice, to view molten lava you must look through gases boiling out of the lava. What kind of spectrum might you see in that case?

7. Why do the strengths of the Balmer lines depend on the temperature of the star?

8. Why does a stellar spectrum tell us about the surface layers but not the deeper layers of the star?

9. Explain the similarities among Figure 7-8, Figure 7-9, Figure 7-10, and Table 7-1.

10. Why do we not see TiO bands in the spectra of hot stars?

11. In observing a star's spectrum, we note that the lines of a certain element are not present. Are we safe in concluding that the star does not contain this element? Why or why not?

12. If a star moves exactly perpendicular to a line connecting it to Earth, will the Doppler effect change its spectrum? Why or why not?

13. Why would we expect a star that rotates very rapidly to have broad spectral lines? What else could broaden the lines?

DISCUSSION QUESTIONS

1. In what ways is our model of an atom a scientific model? How can we use it when it is not a completely correct description of an atom?

2. Can you think of classification systems we commonly use to simplify what would otherwise be very complex measurements? Consider foods, movies, cars, grades, clothes, and so on.

PROBLEMS

1. Human body temperature is about 310 K (98.6°F). At what wavelength do humans radiate the most energy? What kind of radiation do we emit?

2. If a star has a surface temperature of 20,000 K, at what wavelength will it radiate the most energy?

3. Infrared observations of a star show that it is most intense at a wavelength of 2000 nm. What is the temperature of the star's surface?

4. If we double the temperature of a black body, by what factor will the total energy radiated per second per square meter increase?

5. If one star has a temperature of 6000 K and another star has a temperature of 7000 K, how much more energy per second will the hotter star radiate from each square meter of its surface?

6. Transition A produces light with a wavelength of 500 nm. Transition B involves twice as much energy as A. What wavelength light does it produce?

7. Determine the temperatures of the following stars based on their spectra. Use Figure 7-8.

    a. medium-strength Balmer lines, strong helium lines

    b. medium-strength Balmer lines, weak ionized calcium lines

    c. TiO bands very strong

    d. very weak Balmer lines, strong ionized calcium lines

8. To which spectral classes do the stars in Problem 7 belong?

9. In a laboratory, the Balmer beta line has a wavelength of 486.1 nm. If the line appears in a star's spectrum at 486.3 nm, what is the star's radial velocity? Is it approaching or receding?

10. The highest-velocity stars an astronomer might observe have velocities of about 400 km/s. What change in wavelength would this cause in the Balmer gamma line? (*Hint:* Wavelengths are given on page 133).

CRITICAL INQUIRIES FOR THE WEB

1. To what extent is it possible to tell the spectral type of a star based on observation of its color? Search for color index data on the stars in Table A-9 (Appendix A) to see how B-V indices correlate with spectral type. Observational follow-up: Is it possible to rate star colors in such a way as to determine spectral type through visual observation? (Search the Web for indications that others have tried this.)

2. Is it possible to tell the direction of motion of the sun in space by observing radial velocities of other stars? Search the Internet for stellar radial velocity data for stars in a variety of directions on the celestial sphere and look for trends that might indicate our motion in space.

EXPLORING *THESKY*

1. Locate the following stars, click on them, and determine their spectral types: Antares in Scorpius, Betelgeuse in Orion, Aldebaran in Taurus, Sirius in Canis Major, Rigel in Orion.

2. How are spectral types correlated with the colors of stars? (*Hint:* Locate Orion and choose **Spectral Colors** under the **View** menu.)

 Visit the Seeds *Foundations of Astronomy* companion Web site for critical thinking exercises, articles, and additional readings from InfoTrac College Edition, Brooks/Cole's online student library.

THE SUN—OUR STAR

All cannot live on the piazza, but everyone may enjoy the sun.

Italian Proverb

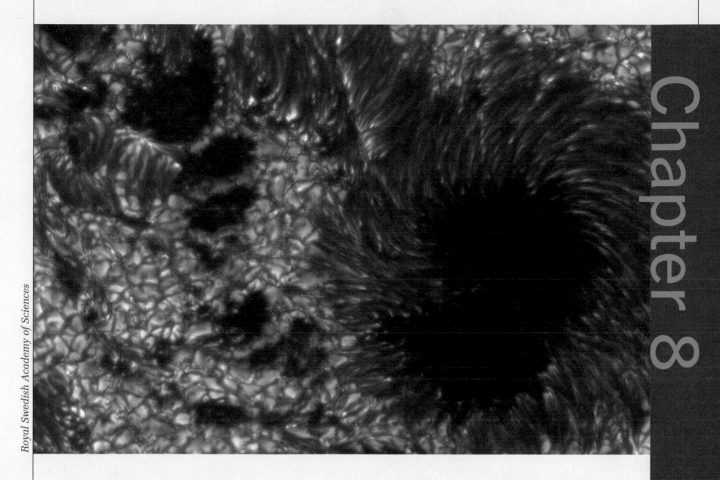

Royal Swedish Academy of Sciences

Chapter 8

GUIDEPOST

The preceding chapter described how we can get information from a spectrum. In this chapter, we apply these techniques to the sun, to learn about its complexities.

This chapter gives us our first close look at how scientists work, how they use evidence and hypothesis to understand nature. Here we will follow carefully developed logical arguments to understand our sun.

Most important, this chapter gives us our first detailed look at a star. The chapters that follow will discuss the many kinds of stars that fill the heavens, but this chapter shows us that each of them is both complex and beautiful; each is a sun.

VIRTUAL LABORATORIES

SOLAR WIND AND COSMIC RAYS

HELIOSEISMOLOGY

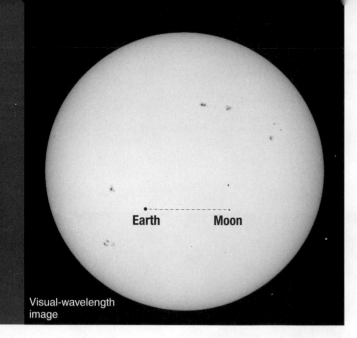

Visual-wavelength image

An image of the sun in visible light shows a few sunspots. The Earth–moon system is added for scale. *(Daniel Good)*

THE SUN

Data File One

Average distance from Earth	1.00 AU (1.495979×10^8 km)
Maximum distance from Earth	1.0167 AU (1.5210×10^8 km)
Minimum distance from Earth	0.9833 AU (1.4710×10^8 km)
Average angular diameter seen from Earth	0.53° (32 minutes of arc)
Period of rotation	25.38 days at equator
Radius	6.9599×10^5 km (109 $R_\oplus$)*
Mass	1.989×10^{30} kg (333,000 $M_\oplus$)*
Average density	1.409 g/cm³
Escape velocity at surface	617.7 km/s
Luminosity	3.826×10^{26} J/s
Surface temperature	5800 K
Central temperature	15×10^6 K
Spectral type	G2 V
Apparent visual magnitude	−26.74
Absolute visual magnitude	4.83

*In astronomy the symbols ⊙ and ⊕ represent the sun and Earth, respectively.

A wit once remarked that solar astronomers would know a lot more about the sun if it were farther away. This comment contains a grain of truth; the sun is only a humdrum star, and there are billions like it in the sky, but the sun is the only one close enough to show surface detail. Solar astronomers can see so much detail in the swirling currents of gas and arching bridges of magnetic force that present theories seem inadequate to describe it. Yet the sun is not a complicated object. It is just a star.

In their general properties, stars are very simple. They are great balls of hot gas held together by their own gravity. Their gravity would make them collapse into small, dense bodies were they not so hot. The tremendously hot gas inside stars has such a high pressure that the stars would surely explode were it not for their own confining gravity. Like soap bubbles, stars are simple structures balanced between opposing forces that individually would destroy them. Thus, we study the sun as a close-up example of a star.

Another reason to study the sun is that life on Earth depends critically on the sun. Very small changes in the sun's luminosity can alter Earth's climate, and a slightly larger change might make Earth uninhabitable. In addition, we get nearly all our energy from the sun— oil and coal are merely stored sunlight—and our pleasant climate is maintained by energy from the sun. In fact, the sun's atmosphere of very thin gas reaches out past Earth's orbit, and thus any change in the sun, such as an eruption or a magnetic storm, can have a direct effect on Earth.

Finally, we study the sun because it is beautiful. Our analysis of sunlight will reveal that the sun is both powerful and delicate. Thus, we study the sun not only because it is *a star,* not only because it is *our star,* but because it is the sun.

8-1 THE SOLAR ATMOSPHERE

The sun is 109 times Earth's diameter and 333,000 times Earth's mass. This seems dramatic, but look at ▌Data File One and notice the sun's density. It is only a little bit more dense than water. So, although the sun is very large and very massive, it must be a gas from its surface to its center. When we look at the sun we see only the outer layers of this vast sphere of gas. In fact, these outer layers, the solar atmosphere, extend high above the visible surface of the sun.

Ace◐Astronomy™ The AceAstronomy icon throughout the text indicates an opportunity for you to test yourself on key concepts and to explore animations and interactions on the AceAstronomy Web site at **http://astronomy.brookscole .com/seeds8e**

HEAT FLOW IN THE SUN

Simple logic tells us that energy in the form of heat is flowing outward from the sun's interior. The solar spectrum reveals that the temperature of the sun's surface is about 5800 K. At that temperature, every square centimeter of the sun's surface must be radiating more energy than a 6000-watt lightbulb. With all that energy radiating into space, the sun's surface would cool rapidly if energy did not flow up from the interior to keep the surface hot.

Not until the 1930s did astronomers understand how the sun makes its energy. Nuclear reactions occur in the core of the sun and generate energy, which flows upward as heat and keeps the surface hot. We will discuss these nuclear reactions in detail later in this chapter, but here we must recognize the importance of the energy flowing outward through the sun's surface.

As we study the atmosphere of the sun, we will find many phenomena that are driven by the energy flowing outward. Like a pot of boiling soup on a hot stove, the surface of the sun is in constant activity as the heat flows up from below.

THE PHOTOSPHERE

The photosphere, the sun's visible surface, seems smooth and featureless, marked occasionally by **sunspots,** dark spots on the solar disk. Recall that Galileo was the first to see sunspots. We will discuss sunspots later in this chapter; but, for now, we must consider the sun's smooth face, the photosphere.

The photosphere is not a solid surface. Remember that the sun is gaseous from its outer atmosphere right down to its center. The photosphere is the thin layer of gas from which we receive most of the sun's light. It is less than 500 km deep and has an average temperature of about 5800 K. If the sun magically shrank to the size of a bowling ball, the photosphere would be no thicker than a layer of tissue paper wrapped around the ball (▌ Figure 8-1a). For comparison, the chromosphere lies above the photosphere and is only a few times thicker in extent, but the corona, beginning above the chromosphere, extends far above the visible surface (Figure 8-1b).

Below the photosphere, the gas is denser and hotter and therefore radiates plenty of light, but that light cannot escape from the sun because of the outer layers of gas. Thus, we cannot detect light from these deeper layers. Above the photosphere, the gas is less dense and so is unable to radiate much light. The photosphere is the layer in the sun's atmosphere that is dense enough to emit plenty of light but not so dense that the light can't escape.

One reason the photosphere is so shallow is related to the hydrogen atom. Because the temperature of the photosphere is sufficient to ionize some atoms, there are a large number of free electrons in the gas. Neutral hydrogen atoms can add an extra electron and become

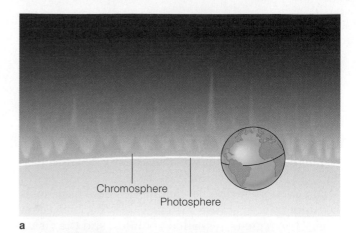

Chromosphere
Photosphere

a

b

FIGURE 8-1

(a) A cross section at the edge of the sun shows the relative thickness of the photosphere and chromosphere. Earth is shown for scale. On this scale, the disk of the sun would be more than 1.5 m (5 feet) in diameter. (b) The corona extends from the top of the chromosphere to great height above the photosphere. This photograph, made during a total solar eclipse, shows only the inner part of the corona. *(Daniel Good)*

an H^- (H-minus) ion, but this extra electron is held so loosely that almost any photon has energy enough to free it. In the process, of course, the photon is absorbed. Thus, the H^- ions are very good absorbers of photons and make the gas of the photosphere very opaque. Light from below cannot escape easily, and we see a well-defined surface—the thin photosphere.

Although the photosphere appears to be substantial, it is really a very-low-density gas. Even in the deepest and densest layers visible, the photosphere is 3400 times less dense than the air we breathe. To find gases as dense as the air we breathe, we would have to descend about 7×10^4 km below the photosphere, about 10 percent of the way to the sun's center. With a fantastically efficient insulation system, we could fly a spaceship right through the photosphere.

The spectrum of the sun is an absorption spectrum, and that can tell us a great deal about the photosphere. We know from Kirchhoff's third law that an absorption spectrum is produced when a source of a continuous spectrum is viewed through a gas. In the case of the photosphere, the deeper layers are dense enough to produce a continuous spectrum, but atoms in the photosphere absorb photons of specific wavelengths, producing the absorption lines we see.

In good photographs, the photosphere has a mottled appearance because it is made up of dark-edged regions. The regions are called granules, and the visual pattern is called **granulation** (❚ Figure 8-2a). Each granule is about the size of Texas and lasts for only 10 to 20 minutes before fading away. Faded granules are continuously replaced by new granules. Spectra of these granules show that the centers are a few hundred degrees hotter than the edges, and Doppler shifts reveal that the centers are rising and the edges are sinking at speeds of about 0.4 km/second.

From this evidence, astronomers recognize granulation as the surface effects of convection just below the photosphere. **Convection** occurs when hot fluid rises and cool fluid sinks, as when, for example, a convection current of hot gas rises above a candle flame. You can create convection in a liquid by adding a bit of cool nondairy creamer to an unstirred cup of hot coffee. The cool creamer sinks, warms, rises, cools, sinks again, and so on, creating small regions on the surface of the coffee that mark the tops of convection currents. Viewed from above, these regions look much like solar granules. Rising currents of hot gas heat small regions of the photosphere, which, being slightly hotter, emit more black body radiation and look brighter. The cool sinking gas of the edges emits less light and thus looks darker (Figure 8-2b). Thus the granulation reveals that energy flows upward from below by convection and heats the photosphere.

Spectroscopic studies of the solar surface have revealed another kind of granulation. **Supergranules** are regions about 30,000 km in diameter (about 2.3 times Earth's diameter) and include about 300 granules. These supergranules are regions of very slowly rising currents that last a day or two. They may be the surface traces of larger currents of rising gas deeper under the photosphere.

The edge, or **limb,** of the solar disk is dimmer than the center (see the figure in Data File One). This **limb darkening** is caused by the absorption of light in the photosphere. When we look at the center of the solar disk, we are looking directly down into the sun, and we see deep, hot, bright layers in the photosphere. But when we look near the limb of the solar disk, we are looking at a steep angle and cannot see as deeply. The photons we see come from shallower, cooler, dimmer layers in the photosphere. Limb darkening proves that the temperature in the photosphere decreases with height, as we would expect if energy is flowing up from below.

THE CHROMOSPHERE

Above the photosphere lies the chromosphere. Solar astronomers define the lower edge of the chromosphere as lying just above the visible surface of the sun with its upper regions blending gradually with the corona. We can think of the chromosphere as being an irregular layer with a depth on average less than Earth's diameter (see Figure 8-1). Because the chromosphere is roughly 1000 times fainter than the photosphere, we can see it with our unaided eyes only during a total solar eclipse when the moon covers the brilliant photosphere. Then, the chromosphere flashes into view as a thin line of pink just above the photosphere. The word *chromosphere* comes from the Greek word *chroma,* meaning "color." The pink color is produced by the combined light of three bright emission lines—the red, blue, and violet Balmer lines of hydrogen.

Astronomers know a great deal about the chromosphere from its spectrum. The chromosphere produces an emission spectrum, and Kirchhoff's second law tells us the chromosphere must be an excited, low-density gas. The density is about 10^8 times less dense than the air we breathe.

From spectra, astronomers can tell that atoms in the lower chromosphere are ionized, and atoms in the

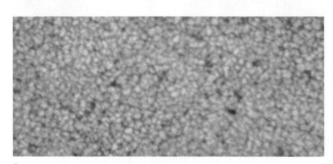

a

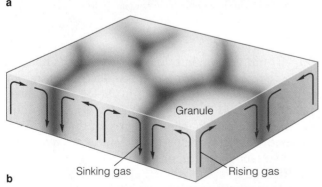

Granule

Sinking gas   Rising gas

b

FIGURE 8-2

(a) A visible-light photo of the sun's surface shows granulation. *(AURA/ NOAO/NSF)* (b) This model explains granulation as the tops of rising convection currents just below the photosphere. Heat flows upward as rising currents of hot gas and downward as sinking currents of cool gas. The rising currents heat the solar surface in small regions that we see as granules.

higher layers of the chromosphere are even more highly ionized. That is, they have lost more electrons. From this, astronomers can find the temperature in different parts of the chromosphere. Just above the photosphere the temperature falls to a minimum of about 4500 K, and then rises rapidly (▌Figure 8-3). The region where the temperature increases fastest is called the **transition region** because it makes the transition from the lower temperatures of the photosphere and chromosphere to the extremely high temperatures of the corona.

Solar astronomers often image the sun at specific wavelengths. For example, hydrogen atoms in the layers of gas above the photosphere absorb photons at the wavelength of the Balmer alpha line very efficiently, and these photons cannot escape easily from the photosphere. An image formed using only these wavelengths, often called a **filtergram,** reveals structure above the photosphere in the chromosphere. In a similar way, an image recorded in the far ultraviolet or in the X-ray part of the spectrum reveals other structures in the solar atmosphere. This chapter contains a number of such solar images.

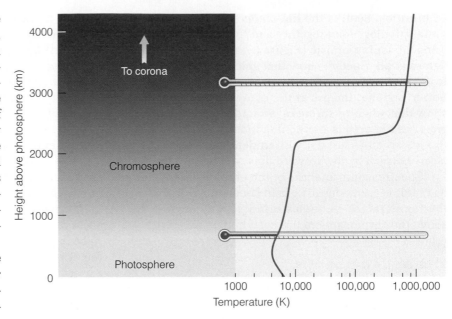

FIGURE 8-3

If we could place thermometers in the sun's atmosphere, we could construct a graph of temperature versus height as shown at the right. The curve shows that the temperature at the photosphere is about 5800 K, falls slightly in the lower chromosphere, and then rises rapidly.

The most common features of the chromosphere are shown in ▌Figure 8-4. **Filaments** are long dark features silhouetted against the brighter surface. We will see later in this chapter that filaments are related to solar activity. The **spicules** are flamelike jets of gas rising upward into the chromosphere and lasting 5 to

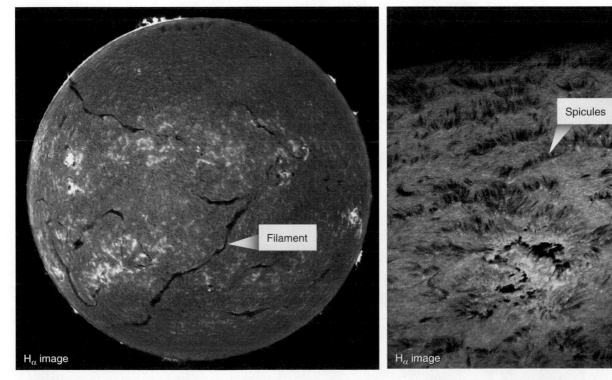

FIGURE 8-4

H_α filtergrams reveal complex structure in the chromosphere including long, dark filaments and spicules springing from the edges of supergranuals twice the diameter of Earth. *(NOAA/SEL/USAF; © 1971 NOAO/NSO)*

15 minutes. Seen at the limb (edge) of the sun's disk, these spicules blend together and look like flames covering a burning prairie (Figure 8-1a), but they are not flames at all. Spectra show that spicules are cooler gas from the lower chromosphere extending upward into hotter regions. Images at the center of the solar disk show that spicules spring up around the edge of supergranules like weeds around flagstones (Figure 8-4b). Although spicules are not well understood yet, they are clearly driven by the outward flow of energy in the sun.

Spectroscopic analysis of the chromosphere alerts us that it is a low-density gas in constant motion where the temperature increases rapidly with height. Just above the chromosphere lies even hotter gas.

THE SOLAR CORONA

The outermost part of the sun's atmosphere is called the corona, after the Greek word for crown. The corona is so dim that it is not visible in Earth's daytime sky because of the glare of scattered light from the brilliant photosphere. During a total solar eclipse, however, when the moon covers the photosphere, we can see the innermost parts of the corona, as shown in Figure 8-1b.

Specialized telescopes called **coronagraphs** can block the light of the photosphere with a mask and image the corona. Such telescopes have traditionally been located on high mountaintops where the air is thin and scatters little sunlight. Now coronagraphs are carried by satellites in the vacuum of space where scattered light is not a problem. Such coronagraphs can trace the corona out beyond 20 solar radii, almost 10 percent of the way to Earth (█ Figure 8-5).

The spectrum of the corona can tell us a great deal about the coronal gases and simultaneously illustrate how astronomers can analyze a spectrum. Some of the light from the outer corona produces a spectrum with absorption lines the same as the sun's spectrum. This light is just sunlight reflected from dust particles in the corona. In contrast, some of the light from the corona produces a continuous spectrum that lacks absorption lines, and that happens when sunlight from the photosphere is scattered off of free electrons in the ionized coronal gas. We might expect this light to reproduce the absorption spectrum of the photosphere, but it does not because the coronal gas has a temperature over 1 million K, and the electrons travel very fast. When photons scatter off of these high-speed elec-

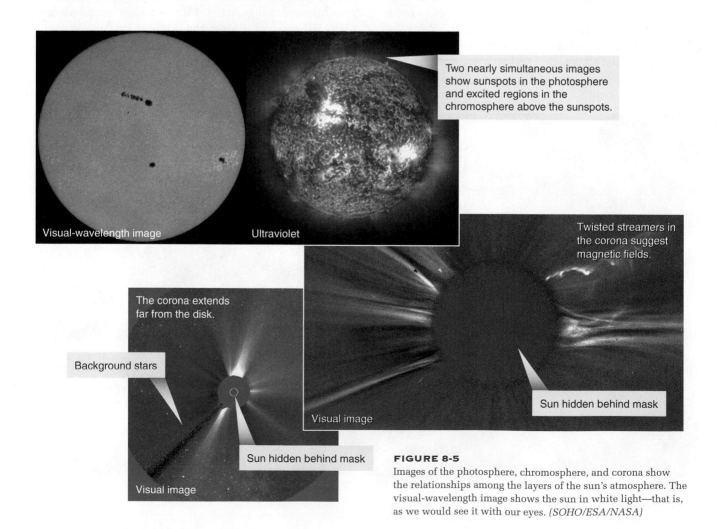

Two nearly simultaneous images show sunspots in the photosphere and excited regions in the chromosphere above the sunspots.

Visual-wavelength image

Ultraviolet

Twisted streamers in the corona suggest magnetic fields.

The corona extends far from the disk.

Background stars

Sun hidden behind mask

Visual image

Sun hidden behind mask

Visual image

Sun hidden behind mask

FIGURE 8-5

Images of the photosphere, chromosphere, and corona show the relationships among the layers of the sun's atmosphere. The visual-wavelength image shows the sun in white light—that is, as we would see it with our eyes. *(SOHO/ESA/NASA)*

trons, the photons suffer random Doppler shifts that smear out any absorption lines to produce a continuous spectrum.

Superimposed on the corona's continuous spectrum are emission lines of highly ionized gases. In the lower corona, the atoms are not as highly ionized as they are at higher altitudes, and this tells us that the temperature of the corona rises with altitude. Just above the chromosphere, the temperature is about 500,000 K; but in the outer corona the temperature can be as high as 2 million K or more.

The spectrum of the corona tells us that it is exceedingly hot gas, but it is not very bright. Its density is very low, only 10^6 atoms/cm^3 in its lower regions. That is about a trillion times less dense than the air we breath. In its outer layers the corona contains only 1 to 10 atoms/cm^3, better than the best vacuum on Earth. Because of this low density, the hot gas does not emit much radiation.

Astronomers have wondered for years how the corona and chromosphere can be so hot. Heat flows from hot regions to cool regions, so how can the heat from the photosphere, with a temperature of only 5800 K, flow out into the much hotter chromosphere and corona? Observations made by the SOHO satellite have mapped a **magnetic carpet** of looped magnetic fields extending up through the photosphere (Figure 8-6). Turbulence below the surface may be whipping these fields about and churning the low-density gases of the chromosphere and corona. That could heat the gas. In this instance, energy appears to flow outward as the agitation of the magnetic fields.

Gas from the solar atmosphere follows the magnetic fields pointing outward and flows away from the sun in a breeze called the **solar wind.** Like an extension of the corona, the low-density gases of the solar wind blow past Earth at 300 to 800 km/s with gusts as high as 1000 km/s. Thus Earth is bathed in the corona's hot breath.

As the solar wind carries gas away from the sun, the sun loses mass. This is only a minor loss for an object as massive as the sun. The sun loses about 10^7 tons per year, but that is only 10^{-14} of a solar mass per year. Even in its entire life as a stable star, the sun will lose only a tiny fraction of its mass through the solar wind.

Do other stars have chromospheres, coronae, and stellar winds like the sun? Ultraviolet and X-ray observations suggest the answer is yes. The spectra of many stars contain emission lines in the far ultraviolet that could only have formed in the low-density, high-temperature gases of a chromosphere and corona. Also, many stars are sources of X rays, which appear to have been produced by coronae. Thus, observational evidence gives us good reason to believe that the sun, for all its complexity, is a typical star.

FIGURE 8-6

This extreme ultraviolet image of a section of the sun's lower corona has been given a green color. White and black areas are regions of opposite magnetic polarity. Computer models show the location of the magnetic fields that connect these regions. The largest loops shown here could encircle Earth. The entire surface of the sun is covered by this magnetic carpet. *(Stanford-Lockheed Institute for Space Research, Palo Alto, CA, and NASA GSFC)*

HELIOSEISMOLOGY

It seems that the light we receive from the sun can tell us nothing about the layers below the photosphere, but solar astronomers have found a way to explore the sun's interior by **helioseismology,** the study of the modes of vibration of the sun. Just as geologists can study Earth's interior by observing how sound waves produced by earthquakes are reflected and transmitted by the layers of Earth's interior, so too can solar astronomers explore the sun's interior by studying how vibrations travel through the sun.

The waves are believed to arise as mechanical vibrations in the convective zone just below the photosphere. Helioseismologists refer to these waves as acoustic vibrations, but we might call them sound. As the waves travel down into the sun, they are bent back up toward the surface by the increasing temperature. Short-wavelength waves penetrate less deeply and travel shorter distances than longer-wavelength waves, as shown in Figure 8-7a. When the waves reach the surface, they cause sections of the photosphere to move up and down by small amounts—roughly plus or minus 15 km. This covers the surface of the sun with a pattern of rising and falling regions that can be mapped using the Doppler effect (Figure 8-7b).

The sun can oscillate in about 10 million different ways, and each mode of oscillation has its own characteristic wavelength and its own unique pattern on the solar surface. The waves producing different modes penetrate to different depths where conditions can weaken or strengthen a wave. By discovering which wavelength waves are actually present, solar astronomers can determine the temperature, density, pressure, composition, and motion at different depths inside the sun.

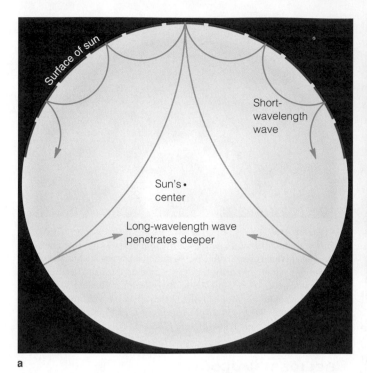

a

b

FIGURE 8-7

Helioseismology: (a) This cross section through the sun's equator shows how waves with short wavelengths do not penetrate as deeply as longer-wavelength waves. Where waves reflect off the underside of the solar surface, they cause a pattern of rising and falling regions (red and blue in this diagram) that can be observed by the Doppler effect. (b) This model shows only one of the 10 million possible patterns that are superimposed on the sun's surface. *(AURA/NOAO/NSF)*

Of course, with 10 million possible wavelengths, the observations and analysis are difficult. A single wave produces a complicated pattern of motion on the solar surface (Figure 8-7b). Large amounts of data are necessary, so helioseismologists have used a network of telescopes around the world operated by the Global Oscillation Network Group (GONG). The sun never sets on GONG. The network can observe the sun continuously for weeks at a time. The SOHO satellite in space has also observed solar oscillations continuously and can detect motions as slow as 1 mm/s (0.002 mph). Solar astronomers can then use high-speed computers to separate the different patterns on the solar surface and measure the strength of the waves at many different wavelengths.

Helioseismology sounds almost magical, but we can understand it better if we think of a duck pond. If we stood at the shore of a duck pond and looked down at the water, we would see ripples arriving from all parts of the pond. Because every duck on the pond contributes to the ripples, we could, in principle, study the ripples near the shore and draw a map showing the position and velocity of every duck on the pond. Of course, it would be difficult to untangle the different ripples, but all of the information would be there, lapping at the rocks at our feet.

Helioseismology has allowed astronomers to map the temperature, density, and rate of rotation inside the sun. They have been able to detect great currents of gas flowing below the photosphere and the emergence of sunspots before they appear in the photosphere. Helio-

seismology can even locate sunspots on the back side of the sun, sunspots that are not yet visible from Earth.

REVIEW CRITICAL INQUIRY

How deeply into the sun can we see?

This is a simple question, but it has a very interesting answer. When we look into the layers of the sun, our sight does not really penetrate into the sun. Rather, our eyes record photons that have escaped from the sun and traveled outward through the layers of the sun's atmosphere. If we observe at a wavelength at the center of a dark absorption line, then the photosphere and lower chromosphere are opaque, photons can't escape to our eyes, and the only photons we can see come from the upper chromosphere. What we see are the details of the upper chromosphere. On the other hand, if we observe at a wavelength that is not easily absorbed (a wavelength between spectral lines), the atmosphere is more transparent, and photons from deep inside the photosphere can escape to our eyes. There is a limit, however, set by the H^- ion, a hydrogen atom with an extra electron. At a certain depth, there is so much of this ion that the sun's atmosphere is opaque for almost all wavelengths, few photons can escape, and we can't see deeper.

By choosing the proper wavelength, solar astronomers can observe to different depths. But the corona is so thin and the gas below the photosphere so dense that this method doesn't work in these regions. How can we observe the corona and the deeper layers of the sun?

So far we have thought of the sun as a static, unchanging ball of gas with energy flowing outward from the interior and through the atmospheric layers. In fact, the sun is a highly variable body whose appearance is constantly changing. It is now time to think of the active sun.

8-2 SOLAR ACTIVITY

The sun is unquiet. It is home to slowly changing spots larger than Earth and vast eruptions that dwarf our imaginations. All of these seemingly different forms of solar activity have one thing in common—magnetic fields. The weather on the sun is magnetic.

SUNSPOTS AND ACTIVE REGIONS

Solar activity is often visible with even a small telescope, but we should exercise great caution in observing the sun. Sunlight is very intense, and when it enters our eye it is absorbed and converted into heat. Equally dangerous is the infrared radiation in sunlight. Our eyes can't detect the infrared, but it is converted to heat in our eyes and can burn and scar the retina. Looking at the sun with the unaided eye can cause permanent damage, and it is even more dangerous to look at the sun through a telescope, which concentrates the sunlight. ▌Figure 8-8a shows a safe way to observe the sun with a small telescope.

When we observe the sun at visual wavelengths, we see the gases of the photosphere. Dark spots on the photosphere are called sunspots, and observations over a period of days will show the sun rotating and the sunspots growing larger or shrinking away (Figure 8-8b).

If we look carefully at sunspots, we see that they consist of a dark center called the umbra and an outer border called the penumbra (▌Figure 8-9). Sunspots come in a wide range of sizes, but we might observe that the average spot is about 40 seconds of arc in diameter. The small-angle formula (Chapter 3) tells us that an average sunspot is about twice the diameter of Earth. Such spots last a week or so and tend to occur in pairs or in groups of one or two major spots accompanied by smaller spots. A large group of sunspots is called an **active region,** may contain up to 100 spots, and could last as long as two months.

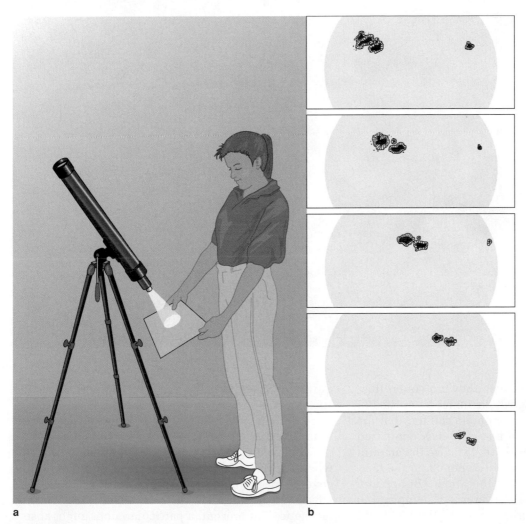

a   b

FIGURE 8-8
Looking through a telescope at the sun is dangerous, but you can always view the sun safely with a small telescope by projecting its image on a white screen (a). If you sketch the location and structure of sunspots on successive days (b), you will see the rotation of the sun and gradual changes in the size and structure of sunspots.

FIGURE 8-9

(a) Sunspots often occur in pairs or in complex groups. This large sunspot group is a number of times larger than Earth's diameter. By analogy with the parts of a shadow, the dark center of the sunspot is called an umbra, and the brighter outer border is called the penumbra. Of course, sunspots are not shadows but are slightly cooler regions of the solar surface. *(SOHO/ ESA/NASA)* (b) This extremely high-resolution image shows detailed flow patterns in the penumbra as well as granulation on the solar surace. *(Royal Swedish Academy of Sciences)*

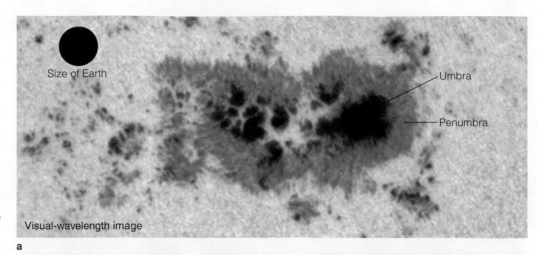

Size of Earth

Umbra

Penumbra

Visual-wavelength image

a

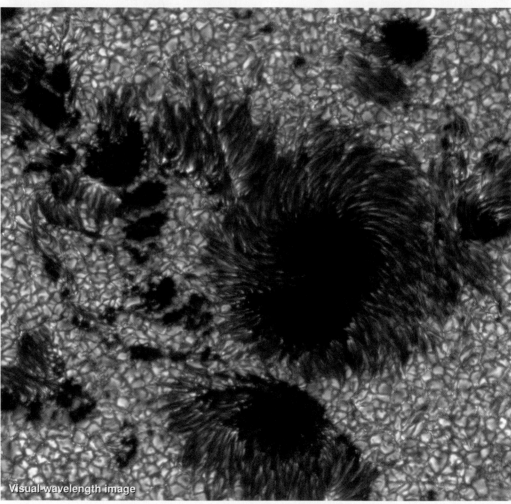

Visual-wavelength image

Sunspots look dark because they are cooler than the photosphere. By noting which atoms are excited in the spectra of sunspots, astronomers can tell that the umbra has a temperature of about 4240 K compared with 5800 K for the photosphere. Because the amount of black body radiation emitted depends on the temperature to the fourth power (the Stefan–Boltzmann law in Chapter 7), this small difference in temperature makes a big difference in brightness, and the spot looks quite dark compared to the photosphere. In fact, a sunspot umbra emits quite a bit of light. If the sun were removed and only an average-size sunspot were left behind, it would glow brighter than the full moon.

A clue to the origin of sunspots appeared in 1908 when American astronomer George Ellery Hale discovered magnetic fields in sunspots. A magnetic field affects the permitted energy levels in an atom. With no magnetic field present, a particular atom might absorb

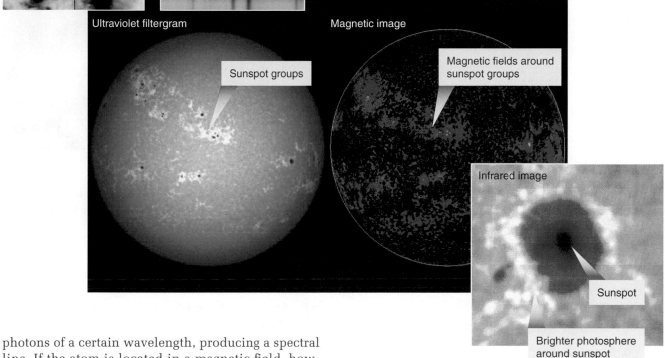

Slit allows light from sunspot to enter spectrograph.

Visual

Spectral line split by Zeeman effect

FIGURE 8-10
The Zeeman effect splits spectral lines into three or more components. The amount of Zeeman splitting allows astronomers to map the north and south magnetic fields on the sun and show that the sunspot groups are highly magnetic. Infrared images show that energy that would have flowed upward through a sunspot is deflected and emerges around the sunspot. *(Zeeman: AURA/NOAO/NSF; Magnetic map: J. Harvey/NSO and HAO/NCAR; IR image: Dan Gezari, NASA/Goddard)*

Ultraviolet filtergram

Magnetic image

Magnetic fields around sunspot groups

Sunspot groups

Infrared image

Sunspot

Brighter photosphere around sunspot

photons of a certain wavelength, producing a spectral line. If the atom is located in a magnetic field, however, the energy levels are split into multiple levels, and the single spectral line could appear as three or more spectral lines at slightly different wavelengths. This is known as the **Zeeman effect** (▌ Figure 8-10). The separation of the spectral lines depends on the strength of the magnetic field, so Hale could measure the strength of magnetic fields on the sun by looking for this effect. He found that the field in a sunspot is about 1000 times stronger than the sun's average field. Modern solar telescopes can use the Zeeman effect to map the magnetic field across the sun's surface. The magnetic map in Figure 8-10 shows that the magnetic field is strongest at the location of active regions. This suggests that the powerful magnetic field causes a sunspot by inhibiting circulation. Ionized gas is made up of electrically charged particles, and such particles cannot move freely in a magnetic field. Astronomers often say the magnetic field is "frozen into" the ionized gas. This simply means that the gas and magnetic field are locked together. Rising currents of hot gas just under the photosphere might be slowed by the strong magnetic fields in sunspots, causing a decrease in the temperature and producing a dark spot. The energy unable to emerge through the sunspot is evidently deflected and emerges in the photosphere around the sunspot.

The magnetic field that emerges through sunspots extends up into the chromosphere and corona above active regions, and high-temperature gas becomes trapped in these magnetic fields. At far-ultraviolet wavelengths, this trapped gas is much brighter than the cool photosphere and outlines the arched magnetic fields, giving us further evidence that active regions are dominated by magnetic fields (▌ Figure 8-11).

Do other stars have sunspots, or rather "starspots," on their surfaces? This is a difficult question, because, except for the sun, the stars are so far away that no surface detail is visible. Some stars, however, vary in brightness in ways that suggest they are mottled by randomly placed, dark spots. As the star rotates, its total brightness changes slightly, depending on the number of spots facing in our direction. This explains the variation of the RS Canum Venaticorum stars, whose spots may cover as much as 25 percent of the surface.

Also, some stars show spectral features that suggest the presence of magnetic fields and starspots. Ultraviolet observations reveal stars whose spectra contain emission lines commonly produced by the regions

FIGURE 8-11

(a) At visible wavelengths we see a sunspot group. (b) At far-ultraviolet wavelengths, the TRACE satellite can detect the hot gas trapped in the magnetic fields arching above the sunspot group. *(NASA/ TRACE)* Notice the resemblance in shape to the magnetic field around a bar magnet as revealed by iron fillings. *(M. Seeds)*

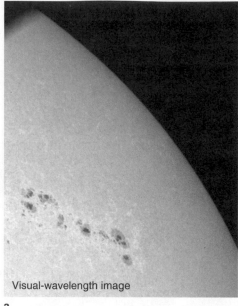

Visual-wavelength image

a

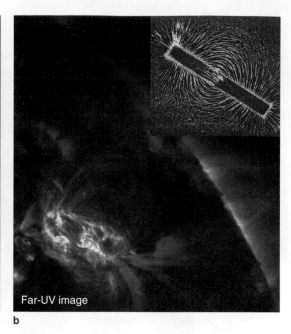

Far-UV image

b

around spots on the sun. This suggests that these stars, too, have spots. No star except the sun can be resolved into a disk, so spots cannot be seen directly. Nevertheless, spectroscopic analysis has allowed astronomers to detect spots on the surfaces of certain stars (❙ Figure 8-12). Such results tell us that the sunspots we see on our sun are not unusual.

THE SUNSPOT CYCLE

The total number of sunspots visible on our sun is not constant. In 1843, the German amateur astronomer Heinrich Schwabe noticed that the number of sunspots

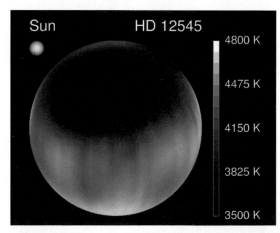

FIGURE 8-12

A very large, dark star spot lies near the axis of rotation of the star known as HD 12545. A bright spot on the star was located roughly opposite the dark spot, suggesting a magnetic field produced the spots. The size and shape of the spots on this star change over a period of a month or so. This map was constructed by observing the changing Doppler shifts in the shapes of spectral lines as the star rotated over a period of 24 days. The sun is shown for scale. *(K. Strassmeier, Vienna, AURA/NOAO/NSF)*

varies in a period of about 11 years. This is now known as the sunspot cycle (❙ Figure 8-13). At sunspot maximum, there are often as many as 100 spots visible at any one time, but at sunspot minimum there are only a few small spots. A sunspot minimum occurred in the mid-1990s with a maximum in early 2001.

At the beginning of each sunspot cycle, the spots begin to appear in the sun's middle latitudes about 35° above and below the sun's equator. As the cycle proceeds, the spots appear at lower latitudes until, near the end of the cycle, they are appearing within 5° of the sun's equator. If we plot the latitude of the appearance of sunspots versus time, the diagram takes on the shape of butterfly wings as shown in Figure 8-13. Such diagrams are known as **Maunder butterfly diagrams,** named after E. Walter Maunder of the Royal Greenwich Observatory, who first published such a diagram in 1922.

THE SUN'S MAGNETIC CYCLE

Sunspots are magnetic phenomena, so the 11-year cycle of sunspots must be caused by cyclical changes in the sun's magnetic field. To explore that idea, we must begin with the sun's rotation.

The sun does not rotate as a rigid body. It is a gas from its outermost layers down to its center, so some parts of the sun rotate faster than other parts. The equatorial region of the photosphere rotates faster than do regions at higher latitudes (❙ Figure 8-14a). At the equator, the photosphere rotates once every 25 days, but at a latitude of 45° one rotation takes 27.8 days. Helioseismology can map the rotation throughout the interior (Figure 8-14b). This phenomenon is called **differential rotation,** and it is clearly linked with the magnetic cycle.

Astronomers believe that the sun's overall magnetic field is produced by its rotation and the outward

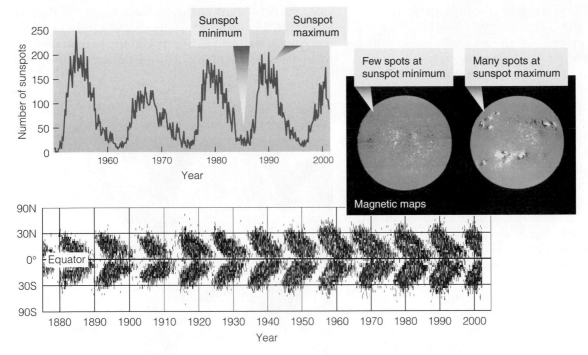

FIGURE 8-13

The number of sunspots varies in an 11-year cycle. At times of sunspot minimum, very few spots are visible on the sun, but at times of sunspot maximum, there are many spots. Early in each cycle, the spots appear at higher latitudes on the sun (farther from the sun's equator) as is shown in the Maunder butterfly diagram at the bottom of this figure. Later in the cycle, the spots appear closer to the sun's equator, producing a pattern in the graph much like butterfly wings. *(Magnetic maps: AURA/NOAO/ NSF; Butterfly diagram: HAO/SMM/NASA/MSFC and D. Hathaway)*

flow of energy. The gases of the sun are highly ionized, so they are very good conductors of electricity. When an electrical conductor rotates rapidly and is stirred by convection, it can convert some of the energy flowing outward as convection into a magnetic field. This process is called the **dynamo effect,** and it is believed to produce Earth's magnetic field as well. Helioseismologists have found evidence that the sun's magnetic field is generated in convection currents deep under the photosphere. Once the magnetic field is created, the convection and differential rotation stretch, twist, and tangle the field to produce a magnetic cycle.

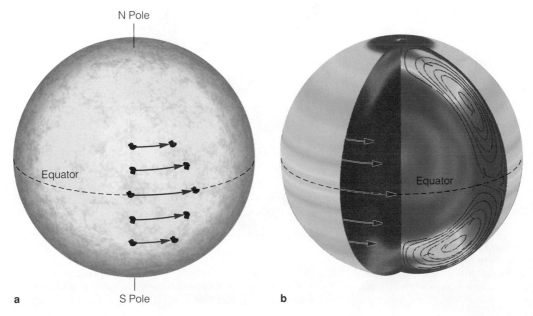

FIGURE 8-14

(a) In general, the photosphere of the sun rotates faster at the equator than at higher latitudes. If we started five sunspots in a row, they would not stay lined up as the sun rotates. (b) Detailed analysis of the sun's rotation from helioseismology reveals regions of slow rotation (blue) and rapid rotation (red). Such studies show that the interior of the sun rotates differentially and that currents similar to the trade winds in Earth's atmosphere flow through the sun. *(NASA/SOI)*

The magnetic behavior of sunspots gives us an insight into how the magnetic cycle works. Sunspots tend to occur in pairs, and the magnetic field around the pair resembles that around a bar magnet with one end magnetic north and the other end magnetic south. At any one time, sunspot pairs south of the sun's equator have reversed polarity compared to those north of the sun's equator. ▌Figure 8-15 illustrates this by showing sunspot pairs south of the sun's equator with magnetic south poles leading and sunspots north of the sun's equator with magnetic north poles leading. At the end of an 11-year sunspot cycle, the new spots appear with reversed magnetic polarity.

This magnetic cycle is not fully understood, but the **Babcock model** (named for its inventor) explains the magnetic cycle as a progressive tangling of the solar magnetic field. Because the electrons in an ionized gas are free to move, the gas is a very good conductor of electricity, and any magnetic field in the gas is "frozen" into the gas. If the gas moves, the magnetic field must move with it. Thus the sun's magnetic field is frozen into its gases, and the differential rotation wraps this field around the sun like a long string caught on a hubcap. Rising and sinking gas currents twist the field into ropelike tubes, which tend to float upward. Where these magnetic tubes burst through the sun's surface, sunspot pairs occur (▌Figure 8-16).

The Babcock model explains the reversal of the sun's magnetic field from cycle to cycle. As the magnetic field becomes tangled, adjacent regions of the sun's surface are dominated by magnetic fields that point in different directions. After about 11 years of tangling, the field becomes so complex that adjacent regions of the

solar surface begin changing their magnetic field to agree with neighboring regions. Quickly the entire field rearranges itself into a simpler pattern, and differential rotation begins winding it up to start a new cycle. But the

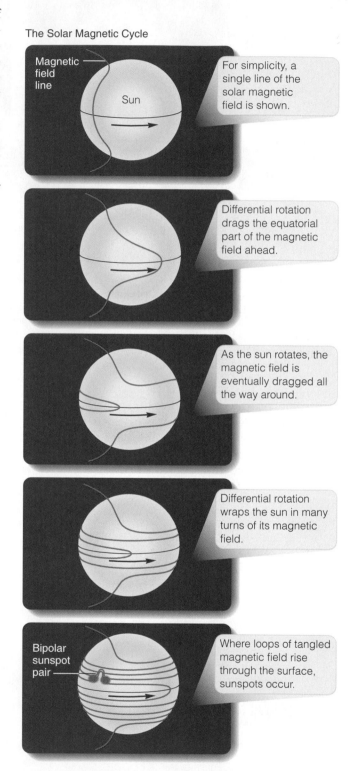

The Solar Magnetic Cycle

Magnetic field line

Sun

For simplicity, a single line of the solar magnetic field is shown.

Differential rotation drags the equatorial part of the magnetic field ahead.

As the sun rotates, the magnetic field is eventually dragged all the way around.

Differential rotation wraps the sun in many turns of its magnetic field.

Bipolar sunspot pair

Where loops of tangled magnetic field rise through the surface, sunspots occur.

FIGURE 8-16

The Babcock model of the solar magnetic cycle explains the sunspot cycle as a consequence of the sun's differential rotation gradually winding up the magnetic field.

S N

S N

Leading spot is magnetic north.

Rotation →

N S

N S

Leading spot is magnetic south.

FIGURE 8-15

The magnetic polarity of sunspot groups in the sun's southern hemisphere is the reverse of that in the northern hemisphere. When a new sunspot cycle begins, the magnetic polarity is reversed—magnetic north becomes magnetic south and vice versa. Compare this diagram with the magnetic maps in Figures 8-10 and 8-13.

Building Confidence by Confirmation and Consolidation

While many textbooks describe science as the process of testing hypotheses by observation and experiment, you should not think that every astronomer approaches the telescope with the expectation of making an observation that will disprove long-held beliefs and trigger a revolution in science. Then what is the daily grind of science really about?

First, many observations and experiments merely confirm already tested hypotheses. The biologist knows that all worker bees in a hive are sisters, but a careful study of the DNA from different workers further confirms that hypothesis. By repeatedly confirming a hypothesis, scientists build confidence in the

hypothesis and may be able to extend it to a wider application. Of course, there is always the chance that a new observation or experiment will disprove the hypothesis, but that is usually very unlikely. Much of the daily grind of science is confirmation.

Another aspect of routine science is consolidation, the linking of a hypothesis to other well-studied phenomena. Chemists may understand certain kinds of carbon molecules shaped like rings, but by repeated study they find a carbon molecule shaped like a hollow sphere. To consolidate their findings, they must show that the chemical bonding in the two molecules follows the same rules

and that the molecules have certain properties in common. No hypothesis is overthrown, but the chemists consolidate their knowledge and understand carbon molecules better.

The Babcock model of the solar magnetic cycle is an astronomical example of the scientific process. Solar astronomers know that the model explains some solar features but has shortcomings. Although most astronomers don't expect to discard the entire model, they work through confirmation and consolidation to better understand how the solar magnetic cycle works and how it is related to cycles in other stars.

---

newly organized field is reversed, and the next sunspot cycle begins with magnetic north replaced by magnetic south. Thus the complete magnetic cycle is 22 years long, and the sunspot cycle is 11 years long.

This magnetic cycle may even explain the Maunder butterfly diagrams. As a sunspot cycle begins, the twisted tubes of magnetic force first begin to float upward and produce sunspot pairs at higher latitude. Consequently the first spots in a cycle appear further north and south of the equator. Later in the cycle, when the field is more tightly wound, the tubes of magnetic force arch up through the surface closer to the equator. Thus the later sunspot pairs in a cycle appear closer to the equator.

Notice the power of a scientific model. The Babcock model may in fact be incorrect in some or all details, but it gives us a framework on which to organize all of the complex solar activity. Even though our models of the sky (see Chapter 2) and the atom (see Chapter 7) were only partially correct, they served as organizing themes to guide our thinking. Similarly, although the precise details of the solar magnetic cycle are not yet understood, the Babcock model gives us a general picture of the behavior of the sun's magnetic field (Window on Science 8-1).

MAGNETIC CYCLES ON OTHER STARS

Because we believe that the sun is a representative star, we should expect other stars to have similar cycles of starspots. We can't see individual spots from Earth, of course, but certain features in stellar spectra are associated with magnetic fields. Regions of strong magnetic

fields on the solar surface emit strongly at the central wavelengths of the two strongest lines of ionized calcium. This calcium emission appears in the spectra of other sunlike stars and tells us that these stars, too, have strong magnetic fields on their surfaces. These stars presumably have starspots as well.

In 1966, astronomers began measuring the strength of this H and K emission in the spectra of stars. The stars to be observed were selected to be similar to the sun. With temperatures ranging from 1000 K hotter than the sun to 3000 K cooler, these stars were considered most likely to have sunlike magnetic activity on their surfaces.

The observations show that the strength of the emissions in the spectra of these stars varies from year to year. The calcium emission averaged over the sun's disk varies with the sunspot cycle, and similar periodic variations can be seen in the spectra of the stars studied (▮ Figure 8-17). The star 107 Piscium, for instance, appears to have a starspot cycle lasting 9 years. Thus, we can be sure that stars like the sun do have magnetic fields and are subject to magnetic cycles.

The rotation periods of these stars are also apparent from the observations. If a star has a major region of magnetic activity and rotates every 20 days, we see the emission appear for 10 days and then disappear for 10 days as the rotation of the star carries the region to the far side of the star.

These observations confirm our belief that the sun is an average sort of star, that is, not peculiar. Most other stars like our sun have magnetic fields and starspots and go through magnetic cycles. More important, these studies are helping us understand our sun better.

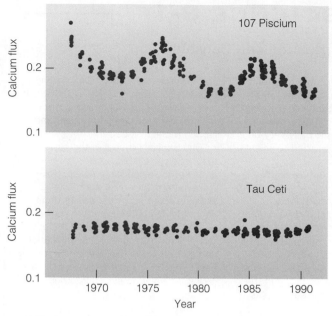

FIGURE 8-17

The average amount of emission in the lines of calcium (calcium flux) is related to magnetic activity. In the sun, this emission is stronger when sunspot activity is higher. The sunlike star 107 Piscium appears to have a magnetic cycle, while the star Tau Ceti does not. *(Adapted from data by Baliunas and Saar)*

CHROMOSPHERIC AND CORONAL ACTIVITY

The solar magnetic fields extend high into the chromosphere and corona where they produce beautiful and powerful phenomena. Study ▌ "Magnetic Solar Phenomena" on pages 162 and 163 and notice three important points.

First, notice that all solar activity is magnetic. We do not experience such events on Earth because Earth's magnetic field is weak, and our atmosphere is not ionized and so is free to move independent of the magnetic field. On the sun, however, the weather is a magnetic phenomenon.

Second, tremendous energy can be stored in arches of magnetic field. We see these near the limb of the sun as prominences, and, seen from above in filtergrams, we seem them as filaments (Figure 8-4). When that stored energy is released, it can trigger powerful eruptions, and, although these eruptions occur far from Earth, they can affect us in dramatic ways.

The third thing to notice is how solar astronomers use every part of the electromagnetic spectrum to study the sun. Some things are easily visible to our eyes, but many phenomena on the sun can only be studied at nonvisible wavelengths. To understand our sun, we need every piece of evidence we can find.

THE SOLAR CONSTANT

Even a small change in the sun's energy output could produce dramatic changes in Earth's climate. The con-

tinued existence of our civilization and our species depends on the constancy of our sun, but we know very little about the variation of the sun's energy output.

The energy production of the sun can be measured by adding up all of the energy falling on 1 square meter of Earth's surface during 1 second. Of course, some correction for the absorption of Earth's atmosphere is necessary, and we must count all wavelengths from X rays to radio waves. The result, which is called the **solar constant,** amounts to about 1360 joules per square meter per second. A change in the solar constant of only 1 percent could change Earth's average temperature by 1 to 2°C (about 1.8 to 3.6°F). For comparison, during the last ice age Earth's average temperature was about 5°C cooler than it is now.

Some of the best measurements of the solar constant were made by instruments aboard the Solar Maximum Mission satellite. These have shown variations in the energy received from the sun of about 0.1 percent that lasted for days or weeks. Superimposed on that random variation is a long-term decrease of about 0.018 percent per year that has been confirmed by observations made by sounding rockets and balloons and by the NIMBUS 7 satellite. This long-term decrease may be related to a cycle of activity on the sun with a period longer than the 22-year magnetic cycle.

Small, random fluctuations will not affect our climate, but a long-term decrease over a decade or more could cause worldwide cooling. History contains some evidence that the solar constant may have varied in the past. The "Little Ice Age" was a period of unusually cool weather in Europe and America that lasted from about 1500 to 1850. The average temperature worldwide was about 1 K cooler than it is now. Although many things affect climate, this cooling could have been caused by a decrease in the solar constant of only 1 percent.

This period of cool weather included an era of few sunspots now known as the **Maunder minimum*** (▌ Figure 8-18). Between 1645 and 1715, the sun was unusually quiet. Very large sunspots can sometimes be seen at sunrise or sunset when the sun is dimmed, but there is no record of any naked-eye sighting of sunspots during this period. Historical records show that astronomers were using the newly invented telescope to observe the sun, but they only observed a few sunspots. Reports of total solar eclipses during this period make no mention of the corona or chromosphere, and there is almost no record of auroral displays. The Maunder minimum seems to have been a period of reduced solar activity that resulted in a cooling of Earth's climate.

In contrast, a period called the Grand Maximum, lasting from about 1100 to about 1250 AD, saw a warming of Earth's climate. The Vikings explored Greenland

---

*Ironically, the Maunder minimum coincides with the reign of Louis XIV of France, the "Sun King."

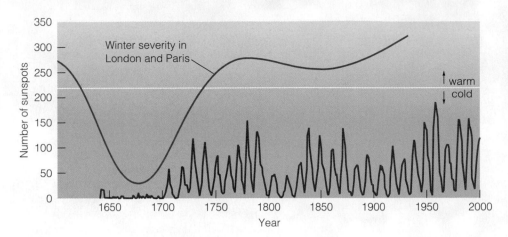

FIGURE 8-18
Historical records show that there were very few sunspots during the late 17th century and that Earth's climate was significantly cooler then. Evidence such as this suggests a link between solar activity and the solar constant, a link that has been confirmed by precise measurements of the solar constant made by instruments on satellites above Earth's atmosphere.

and parts of North America but were forced to abandon their settlements when the climate grew colder.

Other minima and maxima have been found in climate data taken from studies of tree rings. Evidently, solar activity can increase the solar constant very slightly and affect Earth's climate in dramatic ways. The future of our civilization on Earth may depend on our learning to understand the solar constant.

REVIEW CRITICAL INQUIRY

What kind of activity would the sun have if it didn't rotate differentially?

This is a really difficult question because we can see only one star close up and thus have no other examples. Nevertheless, we can make an educated guess by thinking about the Babcock model. If the sun didn't rotate differentially, with its equator traveling faster than higher latitudes, then the magnetic field might not get twisted up, and there might not be a solar cycle. Twisted tubes of magnetic field might not form and rise through the photosphere to produce prominences and flares, although convection might tangle the magnetic field and produce some activity. Is the magnetic activity that heats the chromosphere and corona driven by differential rotation or by convection? It is hard to guess, but without differential rotation, the sun might not have a strong magnetic field and high-temperature gas above its photosphere.

This is very speculative, but sometimes in the critical analysis of ideas it helps to imagine a change in a single important factor and try to understand what might happen. For example, what do you think the sun would be like if it had no convection inside?

The sun is beautiful and complex, but the evidence tells us that all of the powerful activity on its surface is driven by the energy flowing up from its interior. It is time to descend into the sun and ask how that energy is made.

8-3 NUCLEAR FUSION IN THE SUN

Astronomers often use the wrong words to describe energy generation in the sun and stars. Astronomers will say, "The star ignites hydrogen burning." We use the word *ignite* to mean *catch on fire,* and we use *burn* to mean *on fire.* What goes on inside stars isn't really burning in the usual sense.

We must also use the word *atom* with care. The interior of the sun is so hot the gas is totally ionized. That is, the electrons are not attached to the atomic nuclei, and the gas is an atomic soup of rapidly moving particles colliding with each other at high velocity. When we discuss nuclear reactions inside stars, we refer to atomic nuclei and not to atoms.

How exactly can the nucleus of an atom yield energy? The answer lies in the forces that hold the nuclei together.

NUCLEAR BINDING ENERGY

The sun generates its energy by breaking and reconnecting the bonds between the particles *inside* atomic nuclei. This is quite different from the way we generate energy by burning wood in a fireplace. The process of burning wood extracts energy by breaking and reconnecting chemical bonds between atoms in the wood. Chemical bonds are formed by the electrons in atoms, and we saw in Chapter 7 that the electrons are bound to the atoms by the electromagnetic force. Thus chemical energy originates in the electromagnetic force.

There are only four known forces in nature: the force of gravity, the electromagnetic force, the **weak force,** and the **strong force.** The weak force is involved in the radioactive decay of certain kinds of nuclear particles, and the strong force binds together atomic nuclei. Thus nuclear energy comes from the strong force.

Nuclear power plants on Earth generate energy through **nuclear fission** reactions that split uranium nuclei into less massive fragments. A uranium nucleus contains a total of 235 protons and neutrons, and it splits into a range of fragments containing roughly half

Magnetic Solar Phenomena

Magnetic phenomena in the chromosphere and corona, like magnetic weather, result as constantly changing magnetic fields on the sun trap ionized gas to produce beautiful arches and powerful outbursts. Some of this solar activity can affect Earth's magnetic field and atmosphere.

A **prominence** is composed of ionized gas trapped in a magnetic arch rising up through the photosphere and chromosphere into the lower corona. Seen during total solar eclipses at the edge of the solar disk, prominences look pink because of the three Balmer emission lines. The image below shows the arch shape suggestive of magnetic fields. Seen from above against the sun's bright surface, prominences form dark filaments, as in Figure 8-4.

Sacramento Peak Observatory

H-alpha filtergram

This ultraviolet image of the solar surface was made by the NASA TRACE spacecraft. It shows hot gas trapped in magnetic arches extending above active regions. At visual wavelengths, we would see sunspot groups in these active regions.

Quiescent prominences may hang in the lower corona for many days, whereas eruptive prominences burst upward in hours. The eruptive prominences below are many Earth diameters long.

Far-UV image

The gas in prominences may be 60,000 to 80,000 K, quite cold compared with the low-density gas in the corona, which may be as hot as a million Kelvin.

Trace/NASA

SOHO, EIT, ESA and NASA

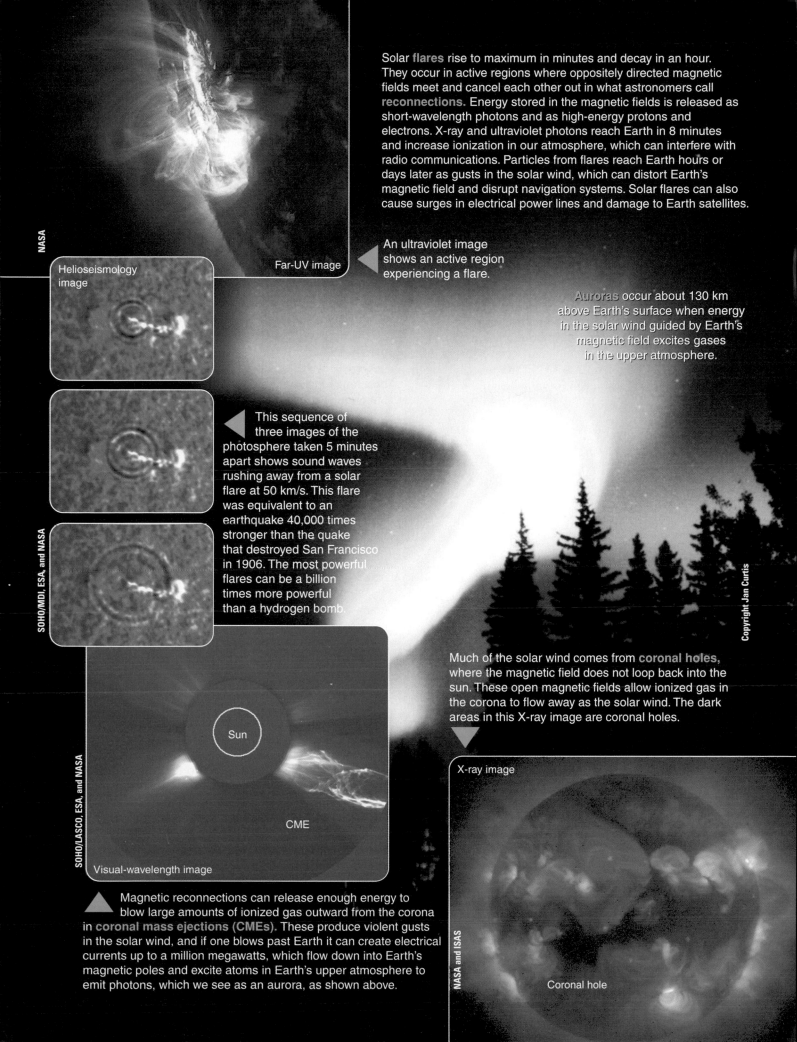

Solar **flares** rise to maximum in minutes and decay in an hour. They occur in active regions where oppositely directed magnetic fields meet and cancel each other out in what astronomers call **reconnections.** Energy stored in the magnetic fields is released as short-wavelength photons and as high-energy protons and electrons. X-ray and ultraviolet photons reach Earth in 8 minutes and increase ionization in our atmosphere, which can interfere with radio communications. Particles from flares reach Earth hours or days later as gusts in the solar wind, which can distort Earth's magnetic field and disrupt navigation systems. Solar flares can also cause surges in electrical power lines and damage to Earth satellites.

Far-UV image

An ultraviolet image shows an active region experiencing a flare.

Helioseismology image

Auroras occur about 130 km above Earth's surface when energy in the solar wind guided by Earth's magnetic field excites gases in the upper atmosphere.

This sequence of three images of the photosphere taken 5 minutes apart shows sound waves rushing away from a solar flare at 50 km/s. This flare was equivalent to an earthquake 40,000 times stronger than the quake that destroyed San Francisco in 1906. The most powerful flares can be a billion times more powerful than a hydrogen bomb.

SOHO/MDI, ESA, and NASA

Copyright Jan Curtis

Much of the solar wind comes from **coronal holes,** where the magnetic field does not loop back into the sun. These open magnetic fields allow ionized gas in the corona to flow away as the solar wind. The dark areas in this X-ray image are coronal holes.

X-ray image

Sun

CME

SOHO/LASCO, ESA, and NASA

Visual-wavelength image

Magnetic reconnections can release enough energy to blow large amounts of ionized gas outward from the corona in **coronal mass ejections (CMEs).** These produce violent gusts in the solar wind, and if one blows past Earth it can create electrical currents up to a million megawatts, which flow down into Earth's magnetic poles and excite atoms in Earth's upper atmosphere to emit photons, which we see as an aurora, as shown above.

NASA and ISAS

Coronal hole

as many particles. Because the fragments produced are more tightly bound than the uranium nuclei, binding energy is released during uranium fission.

Stars make energy in **nuclear fusion** reactions that combine light nuclei into heavier nuclei. The most common reaction, including that in the sun, fuses hydrogen nuclei (single protons) into helium nuclei (two protons and two neutrons). Because the nuclei produced are more tightly bound than the original nuclei, energy is released. Notice in ▌ Figure 8-19 that both fusion and fission reactions move downward in the diagram toward more tightly bound nuclei. Thus both produce energy by releasing the binding energy of atomic nuclei.

HYDROGEN FUSION

Not until the 1930s did astronomers realize how the sun generates energy. The sun fuses together four hydrogen nuclei to make one helium nucleus. Because one helium nucleus has 0.7 percent less mass than four hydrogen nuclei, it seems that some mass vanishes in the process:

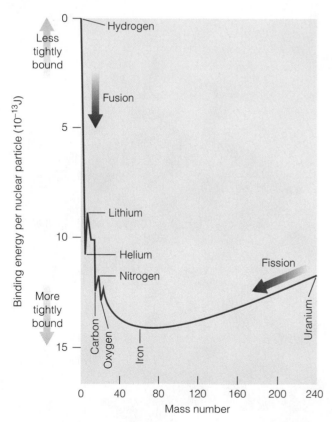

FIGURE 8-19

The red line in this graph shows the binding energy per particle, the energy that holds particles inside an atomic nucleus. The horizontal axis shows the atomic mass number of each element, the number of protons and neutrons in the nucleus. Both fission and fusion nuclear reactions move downward in the diagram (arrows) toward more tightly bound nuclei. Iron has the most tightly bound nucleus, so no nuclear reactions can begin with iron and release energy.

$$\begin{array}{r} \text{4 hydrogen nuclei} = 6.693 \times 10^{-27} \text{ kg} \\ \underline{\text{1 helium nucleus} = 6.645 \times 10^{-27} \text{ kg}} \\ \text{difference in mass} = 0.048 \times 10^{-27} \text{ kg} \end{array}$$

However, this mass does not actually vanish; it merely changes form. The equation $E = m_0 c^2$ reminds us that mass and energy are related, and under certain circumstances mass may become energy and vice versa. Thus, the 0.048×10^{-27} kg does not vanish but merely becomes energy. To see how much, we use Einstein's equation:

$$\begin{aligned} E &= m_0 c^2 \\ &= (0.048 \times 10^{-27} \text{ kg})(3 \times 10^8 \text{ m/s})^2 \\ &= 0.43 \times 10^{-11} \text{ J} \end{aligned}$$

This is a very small amount of energy, hardly enough to raise a housefly one-thousandth of an inch. Because one reaction produces such a small amount of energy, it is obvious that many reactions are necessary to supply the energy needs of a star. The sun, for example, needs 10^{38} reactions per second, transforming 5 million tons of mass into energy every second, just to stay hot enough to resist its own gravity.

It seems from this that nuclear fusion is very powerful, especially when we calculate that the fusion of a milligram of hydrogen (roughly the mass of a match head) produces as much energy as burning 30 gallons of gasoline. However, the nuclear reactions in the sun are spread through a large volume in its core, and any single gram of matter produces little energy. A person of normal mass eating a normal diet produces about 4000 times more heat per gram than the matter in the core of the sun. The sun produces a lot of energy because it contains a lot of grams of matter in its core.

Two nuclei in the sun's core can fuse only if they come close together, but atomic nuclei have positive charges and repel each other with an electrostatic force. This repulsion between the positively charged nuclei has been called the **Coulomb barrier.** To overcome this barrier, the atomic nuclei must collide at high velocity. If the particles in a material are moving at high velocities, we say that the material is hot. Thus, atomic fusion reactions can only occur if the gas is very hot—at least 10^7 K.

Fusion reactions in the sun also require that the gas be very dense—denser than solid lead. We know that the sun requires 10^{38} reactions per second to manufacture sufficient energy. But fusion occurs in only a small percentage of all collisions, so the sun requires many collisions between nuclei each second. Only where the gas is very dense are there enough collisions to meet the sun's energy needs.

Nuclear fusion requires high temperature and high density. Only in the central regions of the sun is the gas hot enough and dense enough for hyrodgen to fuse. Thus all of the energy that flows outward through the sun's surface is created in the sun's core. Sunlight is nuclear power.

We can symbolize the fusion reactions in the sun with a simple nuclear reaction:

$$4 \,^1\text{H} \rightarrow \,^4\text{He} + \text{energy}$$

In this equation, ^1H represents a proton, the nucleus of the hydrogen atom, and ^4He represents the nucleus of a helium atom. The superscripts indicate the approximate weight of the nuclei (the number of protons plus the number of neutrons). The actual steps in the process are more complicated than this convenient summary suggests. Instead of waiting for four hydrogen nuclei to collide simultaneously, a highly unlikely event, the process can proceed step by step in a chain of reactions—the proton–proton chain.

The **proton–proton chain** is a series of three nuclear reactions that builds a helium nucleus by adding together protons. This process is efficient at temperatures above 10,000,000 K. The sun, for example, manufactures over 90 percent of its energy in this way.

The three steps in the proton–proton chain entail these reactions:

$$^1\text{H} + \,^1\text{H} \rightarrow \,^2\text{H} + e^+ + \nu$$
$$^2\text{H} + \,^1\text{H} \rightarrow \,^3\text{He} + \gamma$$
$$^3\text{He} + \,^3\text{He} \rightarrow \,^4\text{He} + \,^1\text{H} + \,^1\text{H}$$

In the first step, two hydrogen nuclei (two protons) combine to form a heavy hydrogen nucleus called **deuterium,** emitting a particle called a positron, e^+ (a positively charged electron), and a **neutrino,** ν (a subatomic particle having an extremely low mass and a velocity nearly equal to the velocity of light). In the second reaction, the heavy hydrogen nucleus absorbs another proton and, with the emission of a gamma ray, γ, becomes a lightweight helium nucleus. Finally, two light helium nuclei combine to form a common helium nucleus and two hydrogen nuclei. Because the last reaction needs two ^3He nuclei, the first and second reactions must occur twice (▌Figure 8-20). The net result of this chain reaction is the transformation of four hydrogen nuclei into one helium nucleus plus energy.

The energy appears in the form of gamma rays, positrons, the energy of motion of the particles, and neutrinos. The gamma rays are photons that are absorbed by the surrounding gas before they can travel more than a fraction of a millimeter. This heats the gas and helps maintain the pressure. The positrons produced in the first reaction combine with free electrons, and both particles vanish, converting their mass into gamma rays. Thus, the positrons

also help keep the center of the star hot. In addition, when fusion produces new nuclei, they fly apart at high velocity. This energy of motion helps raise the temperature of the gas. The neutrinos, however, resemble photons except that they almost never interact with other particles. The average neutrino could pass unhindered through a lead wall 1 ly thick. Thus, the neutrinos do not help heat the gas but race out of the star at the speed of light, carrying away roughly 2 percent of the energy produced.

It is time to ask the critical question that lies at the heart of science. What is the evidence to support our theoretical explanation of how the sun makes its energy? The search for that evidence introduces us to one of the great problems of modern astronomy.

THE SOLAR NEUTRINO PROBLEM

The center of a star seems forever hidden from us, but the sun is transparent to neutrinos because these subatomic particles almost never interact with normal matter. Nuclear reactions in the sun's core produce floods of neutrinos that rush out of the sun and off into space. If we could detect these neutrinos, we could probe the sun's interior.

Because neutrinos almost never interact with atoms, we never feel the flood of over 10^{12} solar neutrinos that flow through our bodies every second. Even at night, neutrinos from the sun rush through Earth as if it weren't there, up through our beds, through us, and onward into space. Obviously we are lucky to be transparent to neutrinos, but it means that neutrinos are extremely hard to detect. Certain nuclear reactions, however, can be triggered by a neutrino of the

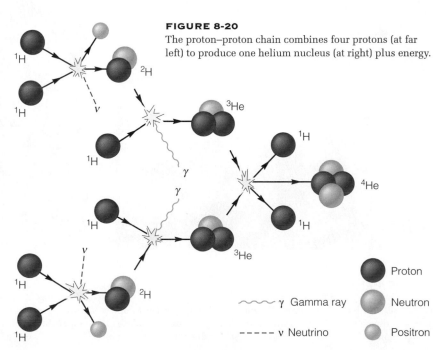

FIGURE 8-20
The proton–proton chain combines four protons (at far left) to produce one helium nucleus (at right) plus energy.

^1H ^1H ^2H ν ^1H γ ^3He ^1H ^4He γ ^1H ^1H ν ^3He ^1H ^2H ^1H

● Proton
〰〰 γ Gamma ray ○ Neutron
----- ν Neutrino ○ Positron

right energy; and in the late 1960s, chemist Raymond Davis, Jr., began using such a reaction to detect solar neutrinos.

Davis filled a 100,000-gallon tank with the cleaning fluid perchloroethylene (C_2Cl_4). Theory predicts that about once a day, a solar neutrino will convert a chlorine atom in the tank into radioactive argon, which can be detected later by its radioactive decay. To protect the detector from cosmic rays from space, the tank was buried nearly a mile deep in a South Dakota gold mine (❙ Figure 8-21a). Of course, the mile of rock overhead has no effect on the neutrinos.

The result of the Davis experiment startled astronomers. The cleaning fluid detected too few neutrinos—not one neutrino per day as predicted by models of the sun, but about one every three days. The experiment was refined, tested, and calibrated for three decades; but it did not find the missing neutrinos. Other detectors were built, and they too counted too few neutrinos coming from the sun.

The missing solar neutrinos were one of the great mysteries of modern astronomy. Some scientists argued that astronomers didn't correctly understand how the sun and stars make their energy, but other scientists wondered if there was something about neutrinos that could explain the problem. Astronomers have great confidence in their theories of the sun's interior and helioseismology confirmed those theories, so astronomers did not abandon their theories immediately (Window on Science 8-2).

As the 21st century began, scientists were able to solve the mystery. Physicists know of three kinds of neutrinos, which they call flavors. The Davis experiment could detect (or taste) only one flavor, electron-neutrinos. Theory hinted that the electron-neutrinos produced in the core of the sun might oscillate among the three flavors as they rushed out through the sun. Observations begun in 2000 confirm this theory (Figure 8-21b). Some of the electron-neutrinos produced in the sun transform into tau- and muon-neutrinos, which most detectors cannot count.

This solution to the solar neutrino problem is exciting because neutrinos can't oscillate unless they have mass. Neutrinos were long thought to be massless, but if they have even a small mass, they are so common their gravity could affect the evolution of the universe

a

b

FIGURE 8-21
(a) The Davis solar neutrino experiment used cleaning fluid and could detect only one of the three flavors of neutrinos. *(Brookhaven National Laboratory)* (b) The Sudbury Neutrino Observatory is a 12-meter-diameter globe containing water rich in deuterium in place of hydrogen. Buried 6800 feet deep in an Ontario mine, it can detect all three flavors of neutrinos and confirms that neutrinos oscillate. *(Photo courtesy of SNO)*

WINDOW ON SCIENCE 8-2

Avoiding Hasty Judgments: Scientific Faith

Scientists like to claim that every scientific belief is based on evidence, that every theory has been tested, and that the moment a theory fails a test, it is discarded or revised. The truth is much more complicated than that, and the solar neutrino problem is a good illustration. If the detection of solar neutrinos contradicted the theory of stellar structure, why wasn't the theory abandoned?

While scientists do indeed have tremendous respect for evidence, they also have a faith in theories that have been tested successfully many times. If a theory has been tested and confirmed over and over, they may even begin to call it a natural law, and that means scientists have great faith in its truth. That is, they have confidence that the theory or law is a good description of how nature works.

Nevertheless, it is not unusual for an experiment or an observation to contradict well-established theories. In many cases, the experiments and observations are simple mistakes, or they have not been interpreted correctly. Scientists resist abandoning a well-tested theory even when an observation continues to contradict it. If confidence in the theory is stronger than the evidence, scientists begin by testing the evidence. Can it be right? Do we understand it correctly? Of course, if the evidence cannot be impeached and it continues to contradict the theory, scientists must eventually abandon or modify the theory no matter how many times it has previously been tested and confirmed. This ultimate reliance on evidence is the distinguishing characteristic of science.

It is only human nature to hang on to the principles you have come to trust, and that confidence in well-tested scientific principles helps scientists avoid rushing to faulty judgments. For example, claims for perpetual motion machines occasionally crop up in the news, but the world's scientists don't instantly abandon the laws of energy and motion pending an analysis of the latest claim. Of course, if such a claim did prove true, the entire structure of scientific knowledge would come crashing down, but because the known laws of energy and motion have been well tested and no perpetual motion machine has ever been successful, scientists know which way to bet. Like the keel on a ship, confidence in well-tested theories and laws keeps the scientific boat from rocking before every little breeze.

Some forms of faith must be absolute and unshakable—religious faith, for example. But scientific faith may be better described as scientific confidence because it must be open to change. If, ultimately, a single experiment or observation conclusively contradicts our most cherished law of nature, we as scientists must abandon that law and find a new way to understand nature. We can see this scientific confidence at work in many controversies, from the origin of the human race, to the meaning of IQ measurements; but one of the best examples is the solar neutrino problem. While astronomers struggled to understand the origin of solar neutrinos, they continued to have confidence that they did indeed understand how stars make energy.

---

as a whole—something we will discuss in Chapter 18. The detection of neutrino oscillation excites astronomers for another reason. It confirms the theories that describe the interior of the sun and stars.

REVIEW CRITICAL INQUIRY

Why does nuclear fusion require that the gas be very hot?
Only under certain conditions do the nuclei of atoms fuse together to form a new nucleus. Inside a star, the gas is ionized, which means the electrons have been stripped off the atoms, and the nuclei are bare and have a positive charge. For hydrogen fusion, the nuclei are single protons. These atomic nuclei repel each other because of their positive charges, so they must collide with each other violently to overcome that repulsion and get close enough together to fuse. If the atoms in a gas are moving rapidly, we say it has a high temperature, and so nuclear fusion requires that the gas have a very high temperature. If the gas is cooler than about 10 million K, hydrogen can't fuse because the protons don't collide violently enough to overcome the repulsion of their positive charges.

It is easy to see why nuclear fusion in the sun requires high temperature, but why does it require high density?

The sun is beautiful and complex, with great eruptions, prominences, active regions, and dark spots sweeping across its surface like magnetic weather. All of this activity is driven by the outward flow of energy generated by hydrogen fusion in the core. Presumably other stars are like the sun, but other stars are so far away we can't see their surfaces. Through the largest telescopes, they look like nothing more than points of light. How can we learn about the stars? We begin to answer that question in the next chapter.

SUMMARY

The atmosphere of the sun consists of three layers: photosphere, chromosphere, and corona. The photosphere, or visible surface, is a thin layer of low-density gas from which visible photons most easily escape. It is marked by granulation, a pattern produced by circulation below the photosphere.

The chromosphere is visible to the unaided eye during total solar eclipses when the moon covers the bright photosphere and the chromosphere flashes into view as a bright pink layer of gas. The spectrum shows that the upper chromosphere is composed of ionized gas much hotter than the photosphere. Images recorded at wavelengths absorbed by specific atoms reveal that the chromosphere is filled with large jets called spicules.

The corona is the sun's outermost layer. The inner corona is visible during solar eclipses, but it is best studied with special telescopes called coronagraphs located on high mountains or in space. Spectra show that it is composed of extremely hot gas extending very far from the sun. The corona and chromosphere appear to be heated by interactions with the sun's magnetic field.

Sunspots are the most prominent example of solar activity. A sunspot appears to have a dark center, called the umbra, and a slightly lighter border, called the penumbra. A sunspot seems dark because it is slightly cooler than the rest of the photosphere. The average sunspot is about twice the size of Earth and contains magnetic fields about 1000 times stronger than the sun's average field. Sunspots are thought to form because the magnetic field inhibits convection. The average number of sunspots visible varies with a period of about 11 years and appears to be related to the solar magnetic cycle.

Alternate sunspot cycles have reversed magnetic polarity, and this has been explained by the Babcock model of the magnetic cycle. In this model, the differential rotation of the sun winds up the magnetic field. Tangles in the field rise to the surface and cause active regions containing sunspots. When the field becomes strongly tangled, it reorders itself into a simpler but reversed field, and the cycle starts over.

Prominences and flares are other examples of solar activity. Prominences occur in the chromosphere; their arched shape shows that they are formed of ionized gas trapped in the magnetic field. In filtergrams of the chromosphere, we see prominences as filaments silhouetted against the brighter surface. Flares, too, seem to be related to the magnetic field. They are sudden eruptions of X-ray, ultraviolet, and visible radiation and high-energy particles that occur among the twisted magnetic fields around sunspot groups.

Activity in the corona is also guided by the magnetic field. The corona seems to be composed of streamers of thin, hot gas escaping from the magnetic field. In some regions of the corona, the magnetic field does not loop back to the sun, and the gas escapes unimpeded. These regions are called coronal holes and are believed to be the source of the solar wind.

The sun generates its energy near its center where the temperature and density are high enough for nuclear fusion reactions that combine hydrogen to make helium. Observations of too few neutrinos coming from the sun's core have been difficult to understand, but neutrino oscillation now explains the missing neutrinos.

NEW TERMS

sunspot	transition region
granulation	filtergram
convection	filament
supergranule	spicule
limb	coronagraph
limb darkening	magnetic carpet

solar wind	coronal hole
helioseismology	coronal mass ejection (CME)
active region	solar constant
Zeeman effect	Maunder minimum
Maunder butterfly diagram	weak force
differential rotation	strong force
dynamo effect	nuclear fission
Babcock model	nuclear fusion
prominence	Coulomb barrier
flare	proton–proton chain
reconnection	deuterium
aurora	neutrino

REVIEW QUESTIONS

Ace ◟Astronomy™ Assess your understanding of this chapter's topics with additional quizzing and animations at **http:// astronomy.brookscole.com/seeds8e**

1. Why can't we see deeper than the photosphere?
2. What evidence do we have that granulation is caused by convection?
3. How are granules and supergranules related? How do they differ?
4. How can astronomers detect structure in the chromosphere?
5. What evidence do we have that the corona has a very high temperature?
6. What heats the chromosphere and corona to high temperature?
7. How are astronomers able to explore the layers of the sun below the photosphere?
8. What evidence do we have that sunspots are magnetic?
9. How does the Babcock model explain the sunspot cycle?
10. What does the spectrum of a prominence tell us? What does its shape tell us?
11. How can solar flares affect Earth?
12. Why does nuclear fusion require high temperatures?
13. Why does nuclear fusion in the sun occur only near the center?
14. How can astronomers detect neutrinos from the sun?
15. How can neutrino oscillation explain the solar neutrino problem?

DISCUSSION QUESTIONS

1. What energy sources on Earth cannot be thought of as stored sunlight?
2. What would the spectrum of an auroral display look like? Why?

3. What observations would you make if you were ordered to set up a system that could warn astronauts in orbit of dangerous solar flares? Such a warning system exists.

PROBLEMS

1. The radius of the sun is 0.7 million km. What percentage of the radius is taken up by the chromosphere?

2. The smallest detail visible with ground-based solar telescopes is about 1 second of arc. How large a region does this represent on the sun? (*Hint:* Use the small-angle formula.)

3. What is the angular diameter of a star like the sun located 5 ly from Earth? Is the Hubble Space telescope able to detect detail on the surface of such a star?

4. If a sunspot has a temperature of 4240 K and the solar surface has a temperature of 5800 K, how many times brighter is the surface compared to the sunspot? (*Hint:* Use the Stefan–Boltzmann law, Chapter 7.)

5. A solar flare can release 10^{25} J. How many megatons of TNT would be equivalent? (*Hint:* A 1-megaton bomb produces about 4×10^{15} J.)

6. The United States consumes about 2.5×10^{19} J of energy in all forms in a year. How many years could we run the United States on the energy released by the solar flare in Problem 5?

7. Neglecting energy absorbed or reflected by our atmosphere, the solar energy hitting 1 square meter of Earth's surface is 1360 J/s (the solar constant). How long does it take a baseball diamond (90 ft on a side) to receive 1 megaton of solar energy? (*Hint:* See Problem 5.)

8. How much energy is produced when the sun converts 1 kg of mass into energy?

9. How much energy is produced when the sun converts 1 kg of hydrogen into helium? (*Hint:* How does this problem differ from Problem 8?)

10. A 1-megaton nuclear weapon produces about 4×10^{15} J of energy. How much mass must vanish when a 5-megaton weapon explodes?

CRITICAL INQUIRIES FOR THE WEB

1. Do disturbances in one layer of the solar atmosphere produce effects in other layers? We have seen that filtergrams are useful in identifying the layers of the solar atmosphere and the structures within them. Visit a Web site that provides daily solar images. Then, choose today's date (or one near it) and examine the sun in several wavelengths to explore the relation between disturbances in various layers.

2. Explore the Web to find out how auroral activity is affected as solar activity rises and falls through the solar cycle. What changes in auroral visibility occur during this cycle? In what other ways can the increased activity associated with a solar maximum affect Earth?

3. Explore the Web to find photos and observations of auroras. From what places on Earth are auroras most often seen?

4. What can you find on the Web about Earth-based efforts to generate energy through nuclear fusion? How do nuclear fusion power experiments attempt to trigger and control nuclear fusion? So-called "cold fusion" has been abandoned as a false trail. How did it resemble nuclear fusion?

EXPLORING *THESKY*

1. Locate the six photos of the sun provided in *TheSky*, and attempt to draw in the sun's equator in each photo. (*Hint:* In the sun's information box, choose **More Information** and then **Multimedia.** What features are visible in these images that help us recognize the orientation of the sun's equator?)

 Visit the Seeds *Foundations of Astronomy* companion Web site for critical thinking exercises, articles, and additional readings from InfoTrac College Edition, Brooks/Cole's online student library.

THE FAMILY OF STARS

Ice is the silent language of the peak;
and fire the silent language of the star.

Conrad Aiken, And in the Human Heart

Chapter 9

GUIDEPOST

Science is based on measurement, but measurement in astronomy is very difficult. Even with the powerful modern telescopes described in Chapter 6, it is impossible to measure directly simple parameters such as the diameter of a star. This chapter shows how we can use the simple observations that are possible, combined with the basic laws of physics, to discover the properties of stars.

With this chapter, we leave our sun behind and begin our study of the billions of stars that dot the sky. In a sense, the star is the basic building block of the universe. If we hope to understand what the universe is, what our sun is, what our Earth is, and what we are, we must understand the stars.

In this chapter we will find out what stars are like. In the chapters that follow, we will trace the life stories of the stars from their births to their deaths.

VIRTUAL LABORATORIES

BINARY STARS

ASTRONOMICAL DISTANCE SCALES

THE SPECTRAL SEQUENCE AND THE H–R DIAGRAM

If the silent language of the stars is fire, then the silent language of fire is light. To understand the family relations among the stars, we must learn to speak the language of light.

Unfortunately, the stars are unimaginably remote. The nearest star is the sun, only 150 million km (93 million miles) away, so close that light takes only about 8 minutes to reach Earth. The next nearest star is nearly 300,000 times farther away, a distance so great that the starlight takes 4 years to reach Earth. The distances from Earth to the stars are so large we must express them using the light-year, the distance light travels in a year—about 9.5 trillion kilometers or about 5.9 trillion miles.

Finding out what a star is like is quite difficult. When we look at the stars, we look across vast distances and see only bright points of light. To understand stars, we must analyze starlight with great care (▌Figure 9-1). We will concentrate in this chapter on finding out three things about stars—how much energy they emit, how big they are, and how much mass they contain. These three properties, combined with stellar temperatures—already discussed in Chapter 7—will give us an insight into the family relations among stars.

Although we begin with three things to learn about stars, we immediately meet a detour. To find out how much energy a star emits, we must know how far away it is. A fantastically luminous star might look very dim if it were far away. Only when we know the distances to the stars can we decide how much light they are actually emitting. Our short detour will provide us with a method of measuring stellar distances.

After we reach our goals, we will put our data together to find out what the average star is like. Which types of stars are most common? Which are rare? By the time we finish this chapter, we will have a good idea what the family of the stars is like.

9-1 MEASURING THE DISTANCES TO STARS

Distance is the most important and the most difficult measurement in astronomy, and astronomers have found many different ways to estimate the distance to stars. Yet each of those ways depends on a direct geometrical method that is much like the method surveyors would use to measure the distance across a river they cannot cross. We will begin by reviewing this method, and then we will apply it to stars.

Ace✹Astronomy™ The AceAstronomy icon throughout the text indicates an opportunity for you to test yourself on key concepts and to explore animations and interactions on the AceAstronomy Web site at **http://astronomy.brookscole .com/seeds8e**

Infrared image

FIGURE 9-1

The modern quest to understand the universe is manifest in this photo of the Eta Carinae Nebula, roughly 7500 light-years from Earth. Only by the careful analysis of starlight have astronomers learned that the stars in the nebula are only 3 million years old and that the brightest star at lower left contains 100 times more matter than the sun. *(2MASS Sky Survey and IPAC)*

THE SURVEYOR'S METHOD

To measure the distance across a river, a team of surveyors begins by driving two stakes into the ground. The distance between the stakes is the baseline of the measurement. The surveyors then choose a landmark on the opposite side of the river, a tree perhaps, thus establishing a large triangle marked by the two stakes and the tree. Using their surveyor's instruments, they sight the tree from the two ends of the baseline and measure the two angles on their side of the river.

Knowing two angles of this large triangle and the length of the side between them, the surveyors can then find the distance across the river by simple trigonometry. Another way to find the distance is to construct a scale drawing. For example, if the baseline is 50 m and the angles are 66° and 71°, we can draw a line 50 mm long to represent the baseline. Using a protractor, we can construct angles of 66° and 71° at each end of the baseline, and then, as shown in ▌Figure 9-2, extend the two sides until they meet at *C*. Point *C* on our drawing is the location of the tree. Measuring the height of our triangle, we find it to be 64 mm and thus conclude that the distance from the baseline to the tree is 64 m.

The more distant an object is, the longer the baseline we must use to measure the distance to the object. We could use a baseline 50 m long to find the distance across a river, but to measure the distance to a mountain on the horizon, we might need a baseline 1000 m long.

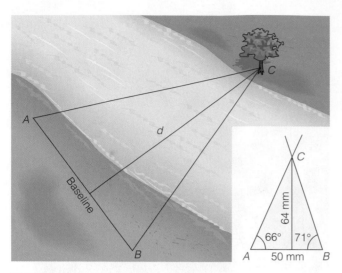

FIGURE 9-2

You can find the distance *d* across a river by measuring the baseline and the angles *A* and *B* and then constructing a scale drawing of the triangle.

Modern surveyors use computers for these problems and not simple drawings. The point is not how the problem is solved, but that it can be solved. Surveyors use simple triangulation to find the distance across a river.

THE ASTRONOMER'S METHOD

To find the distance to a star, we must use a very long baseline, the diameter of Earth's orbit. If we take a photograph of a nearby star and then wait 6 months, Earth will have moved halfway around its orbit. We can then take another photograph of the star. This second photograph is taken at a point in space 2 AU (astronomical units) from the point where the first photograph was taken. Thus our baseline equals the diameter of Earth's orbit, or 2 AU.

We then have two photographs of the same part of the sky taken from slightly different locations in space. When we examine the photographs, we will discover that the star is not in exactly the same place in the two photographs. This apparent shift in the position of the star is called *parallax*.

Parallax is the apparent change in the position of an object due to a change in the location of the observer. We saw in Chapter 4 that parallax is an everyday experience. Our thumb, held at arm's length, appears to shift position against a distant background when we look with first one eye and then with the other. In this case, the baseline is the distance between our eyes, and the parallax is the angle through which our thumb appears to move when we change eyes. The farther away we hold our thumb, the smaller the parallax. If we know the length of the baseline and measure the parallax, we can calculate the distance from our eyes to our thumb.

Because the stars are so distant, their parallaxes are very small angles, usually expressed in seconds of arc. The quantity that astronomers call **stellar parallax (*p*)** is half the total shift of the star, as shown in ▌Figure 9-3. Astronomers measure the parallax, and surveyors measure the angles at the ends of the baseline, but both measurements tell us the same thing—the shape of the triangle and thus the distance to the object in question.

The distance to a star with parallax *p* is given by the simple formula

$$d = \frac{1}{p}$$

where the parallax is measured in seconds of arc and the distance is measured in a unit of distance invented by astronomers, the **parsec (pc),*** defined as the distance to a star with a parallax of 1 second of arc. One parsec turns out to equal 206,265 AU, or 3.26 ly. This makes it very easy to calculate the distance to a star given the parallax. For example, the star Altair has a parallax of 0.20 second of arc. Then the distance to Altair in parsecs is 1 divided by 0.20, which equals 5 pc. To convert to light-years, we multiply 5 pc by 3.26 and discover that Altair is 16.3 ly away.

Measuring the small angle *p* is very difficult. The nearest star to the sun, α Centauri, has a parallax of

---

*The parsec is used throughout astronomy because it simplifies the calculation of distance. However, there are instances when the light-year is also convenient. Consequently, the chapters that follow use either parsecs or light-years as convenience and custom dictate.

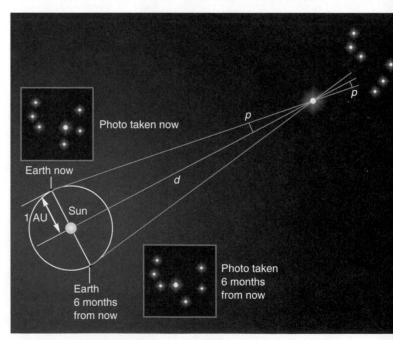

FIGURE 9-3

We can measure the parallax of a nearby star by photographing it from two points along Earth's orbit. For example, we might photograph it now and again 6 months from now. Half of the star's total change in position from one photograph to the other is its stellar parallax *p*.

only 0.76 second of arc, and more distant stars have even smaller parallaxes. To see how small these angles are, hold a piece of paper edgewise at arm's length. The thickness of the paper covers an angle of about 30 seconds of arc. You can see that the parallax of a star, smaller than 1 second of arc, must be very difficult to measure accurately.

The blurring caused by Earth's atmosphere smears star images into tiny disks about one second of arc in diameter, and that makes it difficult to measure parallax. Even averaging many observations, astronomers on Earth cannot measure parallax with an uncertainty smaller than about 0.002 seconds of arc. To limit their uncertainty to 10 percent or better, ground-based astronomers can't measure parallaxes smaller than about 0.02 seconds of arc, which corresponds to a distance of 50 pc. That means ground-based parallax measurements are limited to only the closest stars. Since the first stellar parallax was measured in 1838, ground-based astronomers have been able to measure accurate parallaxes for only about 10,000 stars.

In 1989, the European Space Agency launched the satellite Hipparcos to measure stellar parallaxes from orbit above the blurring effects of Earth's atmosphere. The little satellite observed for 4 years, and the data were reduced by the most sophisticated computers to produce two parallax catalogs in 1997. One catalog contains 120,000 stars with parallaxes 20 times more accurate than ground-based measurements. The other catalog contains over a million stars with parallaxes as accurate as ground-based parallaxes. The Hipparcos data have given astronmers new insights into the nature of stars.

Before we drop the subject of parallax measurement, we should mention a related observation that reveals the motions of the stars through space.

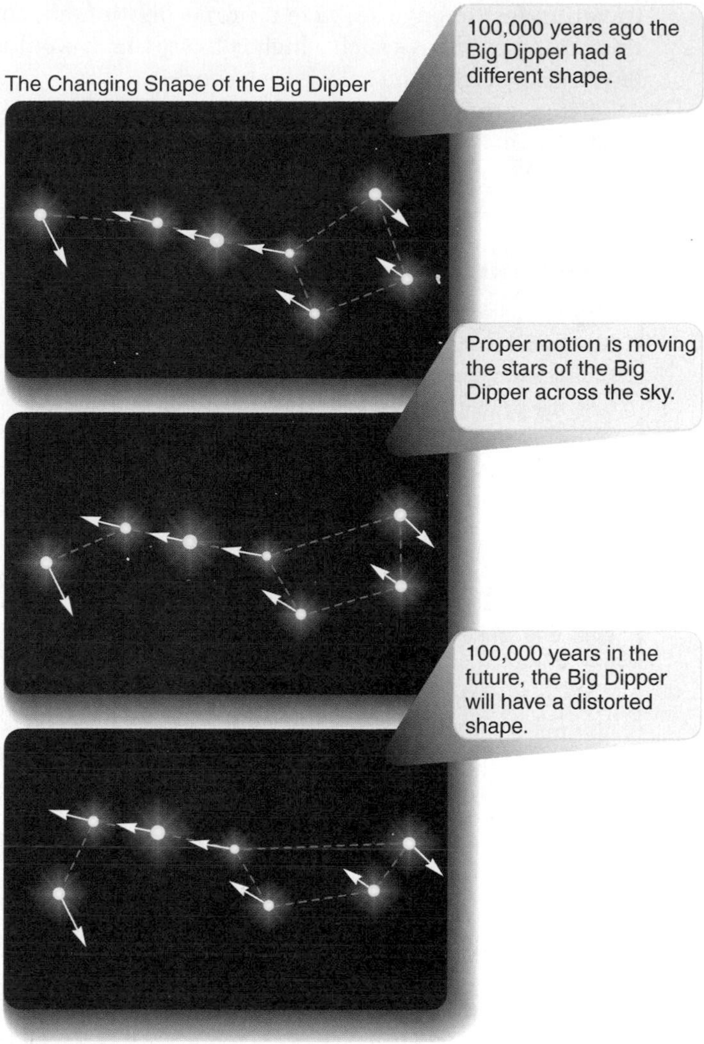

PROPER MOTION

All the stars in the sky, including our sun, are moving along orbits around the center of our galaxy. We don't notice that motion over periods of years, but over the centuries it can significantly distort the shape of constellations (▌Figure 9-4).

If we photograph a small area of the sky on two dates separated by 10 years or more, we can notice that some of the stars in the photograph have moved very slightly against the background stars. This motion, expressed in units of seconds of arc per year, is the **proper motion** of the stars. For example, the star Altair (α Aquilae) has a proper motion of 0.662 second of arc per year, but Albireo (β Cygni) has a proper motion of only 0.002 second of arc per year.

Why should one star have a larger proper motion than another star? For one thing, a star might be moving almost directly toward or away from us, and thus its position on the sky would change very slowly. That

The Changing Shape of the Big Dipper

100,000 years ago the Big Dipper had a different shape.

Proper motion is moving the stars of the Big Dipper across the sky.

100,000 years in the future, the Big Dipper will have a distorted shape.

FIGURE 9-4

"Proper motion" refers to the slow movement of the stars across the sky.

is, its transverse velocity (Chapter 7) could be quite low, resulting in a small proper motion.

Another reason why a star might have a small proper motion is that it could be quite far away from us. Then even a large transverse velocity would not produce a large proper motion. Thus Albireo, at a distance of 116 pc, has a smaller proper motion, and Altair, at a distance of only 4.9 pc, has a larger proper motion.

We can use proper motion to look for nearby stars. If we see a star with a small (or zero) proper motion, it is probably a distant star. We cannot be sure, because it could be a nearby star traveling almost directly toward or away from us. But usually stars with small proper motions are distant. You have probably seen this effect if you watch birds. Distant geese move slowly across the sky, but a nearby bird flits quickly across our field of view. Similarly, stars with large proper motions are usually nearby stars. In fact, stars with proper motions above a certain limit must be nearby. A distant star traveling fast enough to have a proper motion above this limit would have such a high velocity that it would have escaped from the galaxy long ago.

Thus, proper motions can give us statistical clues to distance. We can locate stars that are probably nearby by looking for stars with large proper motions.

REVIEW CRITICAL INQUIRY

Why are parallax measurements made from space better than parallax measurements made from Earth?

At first we might suppose that a satellite in orbit can measure the parallax of the stars better because the satellite is closer to the stars, but when we recall the immense distances to the stars, we see that being in space doesn't really put the satellite significantly closer to the stars. Rather, the reason for increased accuracy is that the satellite is above Earth's atmosphere. When we try to measure parallax, the turbulence in Earth's atmosphere blurs the star images and smears them out into blobs roughly 1 second of arc in diameter. It isn't possible to measure the position of these fuzzy images accurately. We can't measure parallax smaller than about 0.02 seconds of arc. A parallax of 0.02 second of arc corresponds to a distance of 50 pc, so ground-based astronomers can't measure parallax accurately beyond that distance.

A satellite in orbit, however, is above Earth's atmosphere, so the only blurring in the star images is that produced by diffraction in the optics. In other words, the star images are very sharp, and a satellite in orbit can measure the positions of stars and thus their parallaxes to high accuracy.

If a satellite can measure parallaxes as small as 0.001 second of arc, then how far are the most distant stars it can measure?

Having found a way to locate nearby stars and measure their distances, we are ready to discuss the first of the three stellar properties—brightness. Our goal is to find out how much energy stars emit.

9-2 INTRINSIC BRIGHTNESS

Our eyes tells us that some stars look brighter than others, and in Chapter 2 we used the scale of apparent magnitudes to refer to stellar brightness. The faintest stars we can see with the naked eye are about sixth magnitude. Brighter stars have magnitudes represented by smaller numbers and the brightest stars we see in the sky have negative magnitudes, such as Sirius, whose apparent magnitude is −1.47.

The scale of apparent magnitudes only tells us how bright stars look, however, and we want to know their true, or intrinsic, brightness. *Intrinsic* means "belonging to the thing." When we refer to the intrinsic brightness of a star, we mean a measure of the total amount of light the star emits. Apparent magnitudes are deceptive, however. An intrinsically very bright star might appear faint if it were far away. To find the true brightness of stars, we must correct the apparent magnitudes for the influence of distance.

BRIGHTNESS AND DISTANCE

If we see lights at night, it is difficult to determine which are less powerful but nearby and which are highly luminous but farther away (■ Figure 9-5). We face the same problem when we look at stars, and to resolve that problem, we must think carefully about how brightness depends on distance.

When we look at a bright light, our eyes respond to the visual-wavelength energy falling on the eye's retina, which tells us how bright the object looks. Thus, the brightness is related to the flux of energy entering our eye. **Flux** is the energy in joules (J) per second falling on one square meter. Recall that a joule is about as much energy as is released when an apple falls from a table onto the floor. We commonly refer to 1 joule per second as 1 watt. The light flux entering our eye is directly related to the intensity of the light. The more flux entering our eye, the brighter the light looks.

If we placed a screen 1 meter square near a lightbulb, a certain amount of flux would fall on the screen. If we moved the screen twice as far from the bulb, the light that previously fell on the screen would be spread

ACTIVE FIGURE 9-5
To judge the true brightness of a light source, we need to know how far away it is. With no clues to distance, the distant headlight on a truck might look as bright as the nearby headlight on a bicycle.

Ace☉Astronomy™ Go to AceAstronomy and click Active Figures to find "Brightness and Distance." You can take control of this figure.

Observer

to cover an area four times larger, and the screen would receive only one-fourth as much light. If we tripled the distance to the screen, it would receive only one-ninth as much light. Thus, the flux we receive from a light source is inversely proportional to the square of the distance to the source. This is known as the inverse square relation (Figure 5-6). (We first encountered the inverse square relation in Chapter 5, where it was applied to the strength of gravity.)

Now we understand how the brightness of a star depends on its distance. If we know the apparent magnitude of a star and its distance from Earth, we can use the inverse square law to correct for distance and learn the intrinsic brightness of the star. Astronomers do that using a special kind of magnitude scale as described in the next section.

ABSOLUTE VISUAL MAGNITUDE

Judging the intrinsic brightness of stars would be easier if they were all at the same distance. Although astronomers can't move stars about and line them up at some standard distance, they can use the inverse square law to calculate how bright a star of known distance would appear at any other distance. Using this method, they refer to the intrinsic brightness of a star as its **absolute visual magnitude** (M_v)—the apparent visual magnitude the star would have if it were 10 pc away.

The symbol for absolute visual magnitude is an uppercase M with a subscript v. The symbol for apparent visual magnitude is a lowercase m with a subscript v. The subscript tells us that the visual magnitude system is based only on the wavelengths of light we can see. Other magnitude systems are based on other parts of the electromagnetic spectrum such as the infrared, ultraviolet, and so on. Yet another magnitude system refers to the total energy emitted at all wavelengths. We will limit our discussion to visual magnitudes.

The intrinsically brightest stars known have absolute magnitudes of about −8, and the faintest about +19. The nearest star to the sun, α Centauri, is only 1.4 pc away, and its apparent magnitude is 0.0, indicating that it looks bright in the sky. However, its absolute magnitude is 4.39, telling us it is not intrinsically very bright. Because we know the distance to the sun and can measure its apparent magnitude, we can find its absolute magnitude—about 4.78. If the sun were only 10 pc away from us, it would look no brighter than the faintest star in the handle of the Little Dipper.

CALCULATING ABSOLUTE VISUAL MAGNITUDE

How can astronomers find the absolute visual magnitude of a star? This question leads us to one of the most common formulas in astronomy, a formula that relates a star's magnitude to its distance.

The **magnitude–distance formula** relates the apparent magnitude m_v, the absolute magnitude M_v, and the distance d in parsecs:

$$m_v - M_v = -5 + 5\log_{10}(d)$$

If we know any two of the parameters in this formula, we can easily calculate the third. If we are interested in knowing how we might find the absolute magnitude of a star, then we must consider an example in which we know the distance and apparent magnitude of a star. Suppose a star has a distance of 50 pc and an apparent magnitude of 4.5. A pocket calculator tells us that the log of 50 is 1.70, and $-5 + 5 \times 1.70$ equals 3.5, so we know that the absolute magnitude is 3.5 magnitudes brighter than the apparent magnitude. The absolute magnitude is thus 1.0, since 4.5 minus 3.5 is 1.0. (Remember that smaller numbers mean brighter magnitudes.) If this star were 10 pc away, it would be a first-magnitude star.

Astronomers also use the magnitude–distance formula to calculate the distance to a star if the apparent and absolute magnitudes are known. For that purpose, it is handy to rewrite the formula in the following form:

$$d = 10^{(m_v - M_v + 5)/5}$$

If we knew that a star had an apparent magnitude of 7 and an absolute magnitude of 2, then $m_v - M_v$ is 5 magnitudes, and the distance would be 10^2 or 100 parsecs.

The magnitude difference $m_v - M_v$ is known as the **distance modulus,** a measure of how far away the star is. The larger the distance modulus, the more distant the star. We can use the magnitude–distance formula to construct a table of distance and distance modulus (▐ Table 9-1).

TABLE 9-1
Distance Moduli

$m_v - M_v$	d(pc)
0	10
1	16
2	25
3	40
4	63
5	100
6	160
7	250
8	400
9	630
10	1000
⋮	⋮
15	10,000
⋮	⋮
20	100,000
⋮	⋮

The magnitude–distance formula may seem awkward at first, but a pocket calculator makes it easy to use. It is important because it performs a critical function in astronomy: It allows us to convert observations of distance and apparent magnitude into absolute magnitude, a measure of the true brightness of the star. Once we know the absolute magnitude, we can go one step further and figure out the total amount of energy a star is radiating into space.

LUMINOSITY

The **luminosity (L)** of a star is the total amount of energy the star radiates in 1 second—not just visible light, but all wavelengths. To find a star's luminosity, we begin with its absolute visual magnitude, make a small correction, and compare the star with the sun.

The correction we must make adjusts for the radiation emitted at wavelengths we cannot see. Absolute visual magnitude includes only visible light. The absolute magnitudes of hot stars and cool stars will underestimate their total luminosities because those stars radiate significant amounts of radiation in the ultraviolet or infrared parts of the spectrum. We can correct for the missing radiation because the amount of missing energy depends only on the star's temperature. For hot and cool stars, the correction can be large, but for medium-temperature stars like the sun, the correction is small. Adding the proper correction to the absolute visual magnitude changes it into the **absolute bolometric magnitude**—the absolute magnitude the star would have if we could see all wavelengths.

Once we know a star's absolute bolometric magnitude, we can find its luminosity by comparing it with the sun. The absolute bolometric magnitude of the sun is +4.7. For every magnitude a star is brighter than 4.7, it is 2.512 times more luminous than the sun. (Recall from Chapter 2 that a difference of 1 magnitude corresponds to an intensity ratio of 2.512.) Thus, a star with an absolute bolometric magnitude of 2.7 is 2 magnitudes brighter than the sun, which means it is 6.3 times more luminous (6.3 is approximately 2.512 × 2.512).

Arcturus, for example, has an absolute bolometric magnitude of −0.3. That makes it 5 magnitudes brighter than the sun. A difference of 5 magnitudes is defined to be a factor of 100 in brightness, so the luminosity of Arcturus is 100 times the sun's luminosity, or 100 $L_\odot$.

The symbol $L_\odot$ represents the luminosity of the sun, a number we can calculate in a direct way. Because we know how much solar energy hits 1 square meter in 1 second just above Earth's atmosphere (the solar constant defined in the previous chapter) and how far it is from Earth to the sun, it is a simple matter to calculate how much energy the sun must radiate in all directions to provide Earth with the energy it receives per second (see Problem 9 at the end of this chapter). We find that the luminosity of the sun is about 4×10^{26} joules/s.

Let's review: If we can measure the parallax of a star, we can find its distance, calculate its absolute visual magnitude, correct for the light we can't see to find the absolute bolometric magnitude, and then find the luminosity in terms of the sun. If we want to know the luminosity in joules per second, we need only multiply by the sun's luminosity.

When we look at the night sky, the stars look much the same; yet our study of distances and luminosities has revealed an astonishing fact. Some stars are almost a million times more luminous than the sun, and some are nearly a million times fainter. Clearly, the family of stars is filled with interesting characters.

REVIEW CRITICAL INQUIRY

Why do hot stars have an absolute bolometric magnitude that is quite different from their absolute visual magnitude? Recall that the absolute visual magnitude, M_V, is the magnitude we would see if the star were 10 pc away. But our eyes can't see all wavelengths, so a very hot star, which radiates much of its energy in the ultraviolet, would not look very bright to our eyes. The absolute bolometric magnitude includes all wavelengths, so hot stars have absolute bolometric magnitudes much brighter (smaller numbers) than their absolute visual magnitudes because of the ultraviolet energy they radiate. If we want to calculate the luminosities of hot stars, we must use absolute bolometric magnitudes. Otherwise, we'll leave out the ultraviolet radiation they emit, and the luminosities we calculate will be too small.

Stars like the sun radiate most of their energy at visible wavelengths, and we can see that energy, so there isn't a big difference between the absolute bolometric and absolute visual magnitudes for these medium-temperature stars. But what about the cool stars? Explain their absolute bolometric magnitudes.

Although we had to make a detour in order to find the distances to the stars, we have reached our first goal. We have found a way to discover the energy emitted by stars. The range of stellar luminosities is very large. The most luminous stars have luminosities of about 10^5 $L_\odot$, and the least luminous are roughly 10^{-4} $L_\odot$. Knowing the luminosities of stars helps us toward our second goal, finding their diameters.

9-3 THE DIAMETERS OF STARS

One fundamental property of stars is their diameter. Are they all the same size, or are some larger and some smaller? We know little about stars until we know their diameters. We certainly can't see their diameters through a telescope; the stars appear much too small for us to resolve their disks and measure their diameters. But there is a way to find out how big stars really are. If we know their temperature and luminosity, we can find their diameter. That relationship will introduce us to the most important diagram in astronomy, where we will discover the family relations among the stars.

LUMINOSITY, RADIUS, AND TEMPERATURE

The luminosity and temperature of a star can tell us its diameter if we understand the two factors that affect a star's luminosity: surface area and temperature. For example, you can eat dinner by candlelight because the candle flame has a small surface area, and although it is very hot, it cannot radiate much heat; it has a low luminosity. However, if the candle flame were 12 ft tall, it would have a very large surface area from which to radiate, and although it might be no hotter than a normal candle flame, its luminosity would drive you from the table.

In a similar way, a hot star may not be very luminous if it has a small surface area. It could be highly luminous if it were larger, and even a cool star could be very luminous if it were very large and so had a large surface area from which to radiate. Thus, we see that both temperature and surface area help determine the luminosity of a star.

Stars are spheres, and the surface area of a sphere is $4\pi R^2$. If we measure the radius in meters, this is the number of square meters on the surface of a star. Each square meter radiates like a black body, and the total energy given off each second is σT^4. The luminosity of the star is the surface area multiplied by the energy radiated per square meter:

$$L = 4\pi R^2 \sigma T^4$$

If we divide by the same quantities for the sun, we can cancel out the constants and get a simple formula for the luminosity of a star in terms of its radius and temperature:

$$\frac{L}{L_\odot} = \left(\frac{R}{R_\odot}\right)^2 \left(\frac{T}{T_\odot}\right)^4$$

Here the symbol $\odot$ stands for the sun, and the formula tells us that the luminosity in terms of the sun equals the radius in terms of the sun squared times the temperature in terms of the sun raised to the fourth power.

Suppose a star is 10 times the sun's radius but only half as hot. How luminous would it be?

$$\frac{L}{L_\odot} = \left(\frac{10}{1}\right)^2 \left(\frac{1}{2}\right)^4 = \frac{100}{1} \frac{1}{16} = 6.25$$

The formula tells us that it would be 6.25 times the sun's luminosity.

How can we use this to tell us the diameters of the stars? If we see a cool star that is very luminous, we know it must be very large, and if we see a hot star that is not very luminous, we know it must be very small.

The formula allows us to calculate these sizes. For instance, suppose that a star is 40 times the luminosity of the sun and twice as hot. If we put these numbers into our formula, we get:

$$\frac{40}{1} = \left(\frac{R}{R_\odot}\right)^2 \left(\frac{2}{1}\right)^4$$

Solving for the radius, we get:

$$\left(\frac{R}{R_\odot}\right)^2 = \frac{40}{2^4} = \frac{40}{16} = 2.5$$

So the radius is:

$$\frac{R}{R_\odot} = \sqrt{2.5} = 1.58$$

The star is 1.58 times larger in radius than the sun.

Because a star's luminosity depends on its surface area and its temperature, we can sort the stars by diameter, and thus discover which are large and which are small, if we can sort them by luminosity and temperature. Astronomers use a special diagram for that sorting.

THE H–R DIAGRAM

The **Hertzsprung–Russell (H–R) diagram,** named after its originators, Ejnar Hertzsprung and Henry Norris Russell, is a graph that separates the effects of temperature and surface area on stellar luminosities and enables us to sort the stars according to their diameters. Before we discuss the details of the H–R diagram (as it is often called), let us look at a similar diagram we might use to sort automobiles.

We can plot a diagram such as ▌Figure 9-6 to show horsepower versus weight for various makes of cars. And in so doing, we find that in general the more a car weighs, the more horsepower it has. Most cars fall somewhere along the sequence of cars, running from heavy, high-powered cars to light, low-powered models. We might call this the main sequence of cars. But some cars have much more horsepower than normal for their weight—the sport or racing models—and the economy models have less power than normal for cars of the same weight. Just as this diagram helps us understand the different kinds of autos, the H–R diagram helps us understand the kinds of stars.

Now we are ready to plot an H–R diagram. We put the luminosity of the stars on the vertical axis and the temperature on the horizontal axis as in the H–R diagram in ▌Figure 9-7. A star is represented by a point in the graph that tells us the luminosity of the star and its surface temperature (Window on Science 9-1). A point near the top represents a very luminous star, and a point near the bottom represents a low-luminosity star. A point toward the right represents a cool star, and a point toward the left represents a hot star.

Just as in our plot of cars, our graph separates the stars into groups. Roughly 90 percent of normal stars fall somewhere along the **main sequence** running from

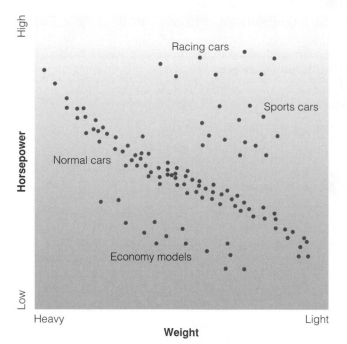

FIGURE 9-6
We could analyze automobiles by plotting their horsepower versus their weight and thus reveal relationships between various models. Most would lie somewhere along the main sequence of "normal" cars.

upper left to lower right. The sun is a main-sequence star. Stars not on the main sequence fall into a few separate groups on the H–R diagram, and we will see in the next section that stars in these groups have similar sizes. That is why the H–R diagram is so important. It sorts stars by size.

Before we discuss the different groups of stars in the H–R diagram, we should note that astronomers plot H–R diagrams in a number of different ways. However they are plotted, the diagrams always show the same information—brightness versus temperature. ▌Figure 9-8 shows two common forms of the H–R diagram.

Astronomers use H–R diagrams so often that they usually skip the words "the point that represents the star." Rather they will say that a star is located in a certain place in the diagram. Of course, they mean the point that represents the luminosity and temperature of the star and not the star itself.

Notice that the location of a star in the H–R diagram has nothing to do with the location of the star in space. Furthermore, a star may move in the H–R diagram as it ages and its luminosity and temperature change, but such motion in the diagram has nothing to do with the star's motion in space.

GIANTS, SUPERGIANTS, AND DWARFS

The dominant feature of an H–R diagram is the main sequence running from upper left to lower right. It makes sense that the hottest stars are the most luminous, but we have learned that both temperature and

size determine a star's luminosity. That alerts us that the H–R diagram contains information about the diameters of stars. In general, stars at the top are larger and stars at the bottom are smaller.

The relation among luminosity, radius, and temperature is a precise mathematical relationship that can be used to draw lines of constant radius across an H–R diagram. ▌Figure 9-9 is an H–R diagram on which slanting dashed lines show the location of stars of certain radii. For example, locate the dashed line labeled $1\ R_\odot$. That line passes through the point marked "Sun" and represents the location of any star whose radius equals that of the sun. Of course, the line slants down to the right because cooler stars are always fainter than hotter stars of the same size.

Now we can identify certain relationships within the family of stars. The stars called **giants** lie at the right above the main sequence. Although these stars are cool, they are luminous because they are 10 to 100 times larger than the sun. The **supergiants** lie near the top of the H–R diagram and are 10 to 1000 times the size of the sun (▌Figure 9-10). Betelgeuse in Orion is a supergiant. If it magically replaced the sun at the center of our solar system, it would swallow up Mercury, Venus, Earth, Mars, and would just reach Jupiter. μ Cephei is believed to be over three times larger than Betelgeuse. It would reach nearly to the orbit of Uranus.

At the bottom of the H–R diagram lie the economy models, stars that are very low in luminosity because they are very small. At the bottom end of the main sequence, the **red dwarfs** are not only small, but they are

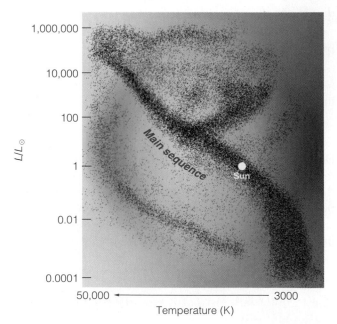

FIGURE 9-7
In this H–R diagram, the luminosity of stars relative to the sun is plotted on the vertical axis, and the temperature of their surface on the horizontal axis. Notice how the background colors are shaded to illustrate temperature. Most stars fall along a main sequence running from hot, luminous stars, to cool, faint stars. *(Adapted from an ESO graphic)*

WINDOW ON SCIENCE 9-1

Exponential Scales

Graphs are important in all branches of science, and many graphs have scales that are exponential (related to the power of some base, such as 10^2, 10^3, 10^4, and so on). Because we are accustomed to thinking about graphs with linear scales, exponential scales can be misleading. H–R diagrams are good examples of graphs with exponential scales, but such scales are common in all the sciences, including chemistry, biology, and geology. They also appear in economics, government, and other fields that deal with quantitative data.

In an exponential scale, each step (tick mark) along the axis corresponds to a constant factor. For example, suppose we plotted a diagram in which the verti-cal axis showed weight on an exponential scale. At the very bottom of the graph, we could plot the weight of a cat, and above that our own weight. Near the top of the graph, we could plot the weight of a bull elephant, and just above that the weight of a whale. In fact, just above that we could plot the weight of all the whales on Earth plus all the elephants, people, and cats. An exponential scale compresses the high numbers and spreads out the low numbers. Thus, our weight on the graph would seem much greater than that of a cat and surprisingly close to that of an elephant. Exponential scales are very helpful for compressing a wide range of numbers into a single graph, but they can be misleading.

For an astronomical example, notice that the vertical axis of the H–R diagram in Figure 9-7 has tick marks separated by a constant factor of 100 in stellar luminosity. If a star is just one tick mark higher than another star, it is *100 times* more luminous.

To see how misleading this can be, try drawing an H–R diagram with a linear scale on the vertical axis. Put zero luminosity at the bottom, and make each tick mark equal to an increase of 100,000 $L_\odot$, with 1,000,000 $L_\odot$ at the top. You will discover that it is difficult to plot all of the different kinds of stars, but their relative luminosities will be much more clearly illustrated.

---

also cool, and that means they can't radiate much energy; they have very low luminosities. In contrast, the **white dwarfs** lie in the lower left of the H–R diagram, and, although some are very hot, they are so small they can't be very bright. They are all roughly the size of Earth and can't radiate much energy from their small surface areas.

The H–R diagram tells us that there is a great range in the sizes of stars. The largest are 100,000 times larger than the smallest. The distribution of stars in the H–R

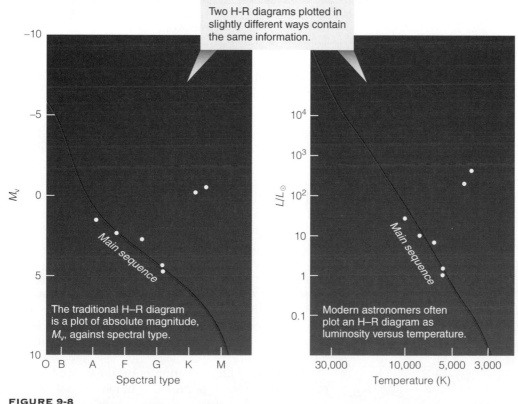

FIGURE 9-8

Here the astronomer has plotted the same data in two different ways. Both graphs are H–R diagrams, and they reveal the same information about the stars. Of the seven stars, five are main-sequence stars and two are not.

diagram according to size is a clue to how stars are born and how they die. We will follow those clues in the chapters that follow, but, before we can do that, we must gather more information about the diverse family of stars.

LUMINOSITY CLASSIFICATION

We can tell from a star's spectrum what kind of star it is. Recall from Chapter 7 that collisional broadening can make spectral lines wider when the density of a gas is relatively high and the atoms collide often.

Main-sequence stars are relatively small and have dense atmospheres in which the gas atoms collide often and distort their electron energy levels. Thus, lines in the spectra of main-sequence stars are broad. On the other hand, giant stars are larger, their atmospheres are less dense, and the atoms disturb one another relatively little. The widths of spectral lines are partially determined by the density of the gas.

Some spectral lines are particularly sensitive to density. If the atoms collide often in a dense gas, their energy levels become distorted and the spectral lines are broadened. Hydrogen Balmer lines are an example. In the spectrum of a main-sequence star, the Balmer lines are broad because the star's atmosphere is dense and the hydrogen atoms collide often. In the spectrum of a giant star, the lines are narrower (▌ Figure 9-11), because the giant star's atmosphere is less dense and

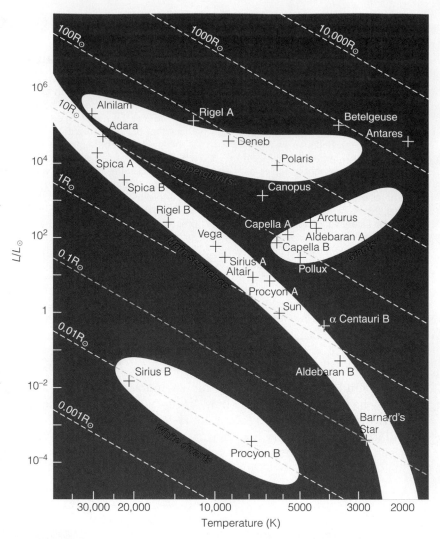

FIGURE 9-9

An H–R diagram showing the luminosity and temperature of many well-known stars. The dashed lines are lines of constant radius. (Individual stars that orbit each other are designated A and B, as in Spica A and Spica B.)

the hydrogen atoms collide less often. In the spectrum of a supergiant star, the Balmer lines are very narrow.

Thus we can look at a star's spectrum and tell roughly how big it is. These are called **luminosity classes,** because the size of the star is the dominating factor in determining luminosity. Supergiants, for example, are very luminous because they are very large.

The luminosity classes are represented by the Roman numerals I through V, with supergiants further subdivided into types Ia and Ib, as follows:

Luminosity Classes

Ia	Bright supergiant
Ib	Supergiant
II	Bright giant
III	Giant
IV	Subgiant
V	Main-sequence star

We can distinguish between the bright supergiants (Ia) such as Rigel (β Orionis) and the regular supergiants (Ib) such as Polaris, the North Star. The star Adhara (ε Canis Majoris) is a bright giant (II), Capella (α Aurigae) is a giant (III), and Altair (α Aquilae) is a subgiant (IV). The sun is a main-sequence star (V). The luminosity class appears after the spectral type, as in G2 V for the sun. White dwarfs don't enter into this classification, because their spectra are peculiar.

We can plot the positions of the luminosity classes on the H–R diagram (▌ Figure 9-12). Remember that these are rather broad classifications and that the lines on the diagram are only approximate. A star of luminosity class III may lie slightly above or below the line labeled III.

Luminosity classification is subtle and not too accurate, but it is an important tool in modern astronomy. As we will see in the next section, luminosity classification, combined with the H–R diagram, gives us a way

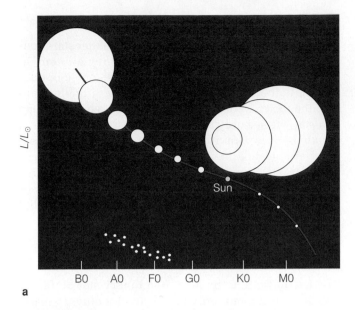

FIGURE 9-10

(a) This H–R diagram shows the relative sizes of stars. Giant stars are 10 to 100 times larger than the sun, and white dwarfs are about the size of Earth. (The dots representing white dwarfs here are much too large.) Supergiants are too large for this diagram. (b) To visualize the size of the largest stars, imagine that the sun is the size of one of your eyeballs. Then the largest supergiants would be the size of a hot air balloon.

b

to find the distance to stars that are too far away to have measurable parallaxes.

SPECTROSCOPIC PARALLAX

We can measure the stellar parallax of nearby stars, but most stars are too distant to have measurable parallaxes. We can find the distances to these stars if we can photograph their spectra and determine their luminosity classes in a process called **spectroscopic parallax**—the estimation of the distance to a star from its spectral type, luminosity class, and apparent magnitude. Spectroscopic parallax is not an actual measure of parallax, but it does tell us the distance to the star.

The method of spectroscopic parallax depends on the H–R diagram. If we photograph the spectrum of a star, we can determine its spectral class, and that tells us its horizontal location in the H–R diagram. We can also determine its luminosity class by looking at the widths of its spectral lines, and that tells us the star's vertical location in the diagram. Once we plot the point that represents the star in the H–R diagram, we can read off its absolute magnitude. As we have seen earlier in this chapter, we can find the distance to a star by comparing its apparent and absolute magnitudes.

For example, Spica is classified B1 V, and its apparent magnitude is +1. We can plot this star in an H–R diagram such as that in Figure 9-12, where we would find that it should have an absolute magnitude of about −3. Therefore, its distance modulus is 1 minus (−3), or 4, and the distance (from Table 9-1) is about 63 pc.

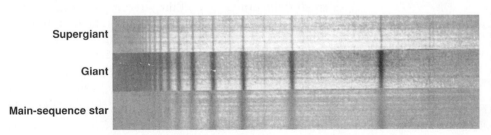

FIGURE 9-11

Differences in widths and strengths of spectral lines distinguish the spectra of supergiants, giants, and main-sequence stars. *(Adapted from H. A. Abt, A. B. Meinel, W. W. Morgan, and J. W. Tap-scott,* An Atlas of Low-Dispersion Grating Stellar Spectra, *Kitt Peak National Observatory, 1968)*

What evidence do we have that giant stars really are bigger than the sun?

We can find stars of the same spectral type as the sun that are clearly more luminous than the sun. Capella, for example, is a G star with an absolute magnitude of 0. Because it is a G star, it must have about the same temperature as the sun, but its absolute magnitude is 5 magnitudes brighter than the sun's. A magnitude difference of 5 magnitudes corresponds to an intensity ratio of 100, so Capella must be about 100 times more luminous than the sun. If it has the same surface temperature as the sun but is 100 times more luminous, then it must have a surface area 100 times greater than the sun's. Because the surface area of a sphere is proportional to the square of the radius, Capella must be 10 times larger in radius. That is clear observational evidence that Capella is a giant star.

In Figure 9-9, we see that Procyon B is a white dwarf only slightly warmer than the sun but about 10,000 times less luminous than the sun. Use the same logical procedure followed above to demonstrate that the white dwarfs must be small stars.

The first two goals of this chapter have been achieved; we know how to find the luminosities and diameters of stars. We have discovered that there are four different kinds of stars—main-sequence stars, giants, supergiants, and white dwarfs—and that these types differ in their intrinsic properties, luminosity and diameter. But this discovery raises an entirely new question. How massive are these stars?

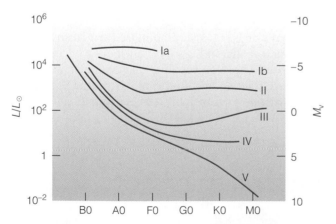

FIGURE 9-12
The approximate location of the luminosity classes on the H–R diagram.

9-4 THE MASSES OF STARS

Our third goal is to find out how much matter stars contain, that is, to know their mass. Do they all contain about the same mass as our sun, or are some more massive and others less? Unfortunately, it's difficult to determine the mass of a star. Looking through a telescope at a star, we see only a point of light that tells nothing about the star. Gravity is the key. Matter produces a gravitational field, and we can figure out how much matter a star contains if we watch an object move through the star's gravitational field. To find the masses of stars, we must study **binary stars,** pairs of stars that orbit each other.

BINARY STARS IN GENERAL

The key to finding the mass of a binary star is understanding orbital motion, which provides clues to mass.

In Chapter 5, we illustrated orbital motion by imagining a cannonball fired from a high mountain (see page 84). If Earth's gravity didn't act on the cannonball, it would follow a straight-line path and leave Earth forever. Because Earth's gravity pulls it away from its straight-line path, the cannonball follows a curved path around Earth—an orbit. When two stars orbit each other, their mutual gravitation pulls them away from straight-line paths and makes them follow closed orbits around a point between the stars.

Each star in a binary system moves in its own orbit around the system's center of mass, the balance point of the system. If the stars were connected by a massless rod and placed in a uniform gravitational field such as that near Earth's surface, the system would balance at its center of mass like a child's seesaw (see page 85). If one star were more massive than its companion, then the massive star would be closer to the center of mass and would travel in a smaller orbit, while the lower-mass star would whip around in a larger orbit (❚ Figure 9-13). The ratio of the masses of the stars M_A/M_B equals r_B/r_A, the inverse of the ratio of the radii of the orbits. If one star has an orbit twice as large as the other star's orbit, then it must be half as massive. Getting the ratio of the masses is easy, but that doesn't tell us the individual masses of the stars, which is what we really want to know. That takes further analysis.

To find the mass of a binary star system, we must know the size of the orbits and the orbital period—the length of time the stars take to complete one orbit. The smaller the orbits are and the shorter the orbital period is, the stronger the stars' gravity must be to hold each other in orbit. For example, if two stars whirl rapidly around each other in small orbits, then their gravity must be very strong to prevent their flying apart. Such stars would have to be very massive. From the size of the orbits and the orbital period, we can figure out the masses of the stars from Newton's laws.

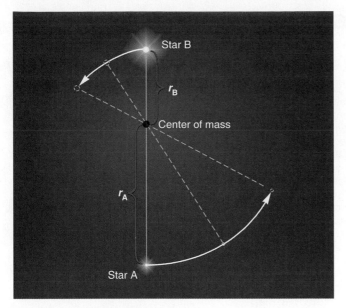

ACTIVE FIGURE 9-13

As stars in a binary star system revolve around each other, the line connecting them always passes through the center of mass, and the more massive star is always closer to the center of mass.

Ace⊘Astronomy™ Go to AceAstronomy and click Active Figures to see "Center of Mass." Experiment to see how mass ratio affects the size of the orbits.

CALCULATING THE MASSES OF BINARY STARS

According to Newton's laws of motion and gravity, the total mass of two stars orbiting each other is related to the average distance a between them and their orbital period P. If the masses are M_A and M_B, then:

$$M_A + M_B = \frac{a^3}{P^2}$$

In this formula, we will measure a in AU, P in years, and the mass in solar masses.

Notice that this formula is related to Kepler's third law of planetary motion (see Table 4-1). Almost all of the mass of the solar system is in the sun. If we apply this formula to any planet in our solar system, the total mass is 1 solar mass. Then the formula becomes $P^2 = a^3$, which is Kepler's third law.

In other star systems, the total mass is not necessarily 1 solar mass, and this gives us a way to find the masses of binary stars. If we can find the average distance in AU between the two stars and their orbital period in years, the sum of the masses of the two stars is just a^3/P^2.

Example A: If we observe a binary system with a period of 32 years and an average separation of 16 AU, what is the total mass? *Solution:* The total mass equals $16^3/32^2$, which equals 4 solar masses.

Example B: Let's call the two stars in the previous example A and B. Suppose star A is 12 AU away from the center of mass, and star B is 4 AU away. What are the individual masses? *Solution:* The ratio of the masses must be 12:4, which is the same as 3:1. What two numbers add up to 4 and have the ratio of 3:1? In this case, the answer is easy. Star B must be 3 solar masses and Star A must be 1 solar mass.

Actually, figuring out the mass of a binary star system is not as easy as it might seem from our discussion. The orbits of the two stars may be elliptical, and, although the orbits lie in the same plane, that plane can be tipped at an unknown angle to our line of sight, further distorting the shapes of the orbits. Astronomers must find ways to correct for these distortions. In addition, astronomers analyzing binary systems must find the distances to the stars so they can estimate the true size of the orbits in astronomical units. Notice that finding the masses of binary stars requires a number of steps to get from what we can observe to what we really want to know, the masses. Constructing such sequences of steps is an important part of science (Window on Science 9-2).

Although there are many different kinds of binary stars, three types are especially important for determining stellar masses. We will discuss these separately in the next sections.

VISUAL BINARY SYSTEMS

In a **visual binary system,** the two stars are separately visible in the telescope. Only a pair of stars with large orbits can be separated visually, for if the orbits are small, the star images blend together in the telescope and we see only a single point of light. Because visual binary systems must have large orbits, they also have long orbital periods. Some take hundreds or even thousands of years to complete a single orbit.

Astronomers study visual binary systems by measuring the position of the two stars directly at the telescope or in images. In either case, the astronomers need measurements over many years to map the orbits. The first frame of ❚ Figure 9-14 shows a photograph of the bright star Sirius, which is a visual binary system made up of the bright star Sirius A and its white dwarf companion Sirius B. The photo was taken in 1960. Successive frames in Figure 9-14 show the motion of the two stars as observed since 1960 and the orbits the stars follow. The orbital period is 50 years.

Binary systems are common; more than half of all stars are members of binary star systems. Few, however, can be analyzed completely. Many visual binaries are so far apart that their periods are much too long for practical mapping of their orbits. Other binary systems are so close together they are not visible as separate stars.

Learning About Nature Through Chains of Inference

Scientists can rarely observe the things they really want to know, so they must construct chains of inference. We can't observe the mass of stars directly, so we must find a way to use what we can observe, orbital period and angular separation, and step by step figure out the parameters we need to reach our goal. In this case, the goal is the masses of the two stars. We follow the chain of inference from the observable parameters to the unobservable quantities we want to know.

Chains of inference can be mathematical, as when geologists use earthquakes to calculate the temperature and density of Earth's interior. There is no way to drill a hole to Earth's center and lower a thermometer or recover a sample, so the internal temperature and density are not directly observable. Nevertheless, the speed of vibrations from distant earthquakes depends on the temperature and density of the rock they pass through. Geologists can't measure the speed of the vibrations deep inside Earth; but they can measure the delay in the arrival time at different locations on the surface, and that allows them to work their way back to the speed and, finally, the temperature and density.

Chains of inference can also be non-mathematical. Biologists studying the migration of whales can't follow individual whales for years at a time, but they can observe feeding and mating in different locations; take into consideration food sources, ocean currents, and water temperatures; and construct a chain of inference that leads back to the seasonal migration pattern for whales.

This chapter contains a number of chains of inference because it describes how we know what stars are like. Almost all sciences use chains of inference. If you can link the observable parameters step by step to the final conclusions, you have gained a strong insight into the nature of science.

SPECTROSCOPIC BINARY SYSTEMS

If the stars in a binary system are close together, the telescope, limited by diffraction and by seeing, shows us a single point of light. Only by looking at a spectrum, which is formed by light from both stars and contains spectral lines from both, can we tell that there are two stars present and not one. Such a system is a **spectroscopic binary system.**

❚ Figure 9-15 shows a pair of stars orbiting each other in identical circular orbits. From the diagram we can guess that the two stars have equal masses, and we notice that the circular orbit appears elliptical because we see it nearly edge-on. Of course, if this were a spectroscopic binary system, we would not see the separate stars. Nevertheless, the Doppler shift would tell us there were two stars orbiting each other. In the first frame of the figure, star A is approaching us while star B recedes from us. In the spectrum, we see a spectral line from star A blue shifted while the same spectral line from star B is red shifted. As we watch the two stars revolve around their orbits, they alternately approach and recede from us, and we see their spectral lines Doppler shifted first toward the blue and then toward the red. In a real spectroscopic binary, we can't see the individual stars, but the sight of pairs of spectral lines moving back and forth across each other would alert us that we were observing a spectroscopic binary system (❚ Figure 9-16).

We can find the orbital period of a spectroscopic binary system by waiting to see how long it takes for the spectral lines to return to their starting positions. We can measure the size of the Doppler shifts to find the orbital velocities of the two stars. If we multiply velocity times orbital period, we can find the circumference of the orbit, and from that we can find the radius of the orbit. Of course, if we know the orbital period and the size of the orbit we can calculate the mass. One important detail is missing, however. We don't know how much the orbits are inclined to our line of sight.

We can find the inclination of a visual binary system because we can see the shape of the orbits. In a spectroscopic binary system, however, we cannot see the individual stars, so we can't find the inclination or untip the orbits. Recall that the Doppler effect only tells us the radial velocity, the part of the velocity directed toward or away from us. Because we cannot find the inclination, we cannot correct these radial velocities to find the true orbital velocities. Consequently, we cannot find the true masses. All we can find from a spectroscopic binary system is a lower limit to the masses.

More than half of all stars are in binary systems, and most of those are spectroscopic binary systems. Many of the familiar stars in the sky are actually pairs of stars orbiting each other (❚ Figure 9-17).

We might wonder what happens when the orbits of a spectroscopic binary system lie exactly edge-on to Earth. The result is the most useful kind of binary system.

ECLIPSING BINARY SYSTEMS

As we said earlier, the orbits of the two stars in a binary system always lie in a single plane. If that plane is nearly edge-on to Earth, then the stars can cross in front of each other as seen from Earth. Imagine a model of a binary star system in which a cardboard disk represents the orbital plane and balls represent the stars, as in ❚ Figure 9-18. If we see the model from the edge, then the balls that represent the stars can move in front of

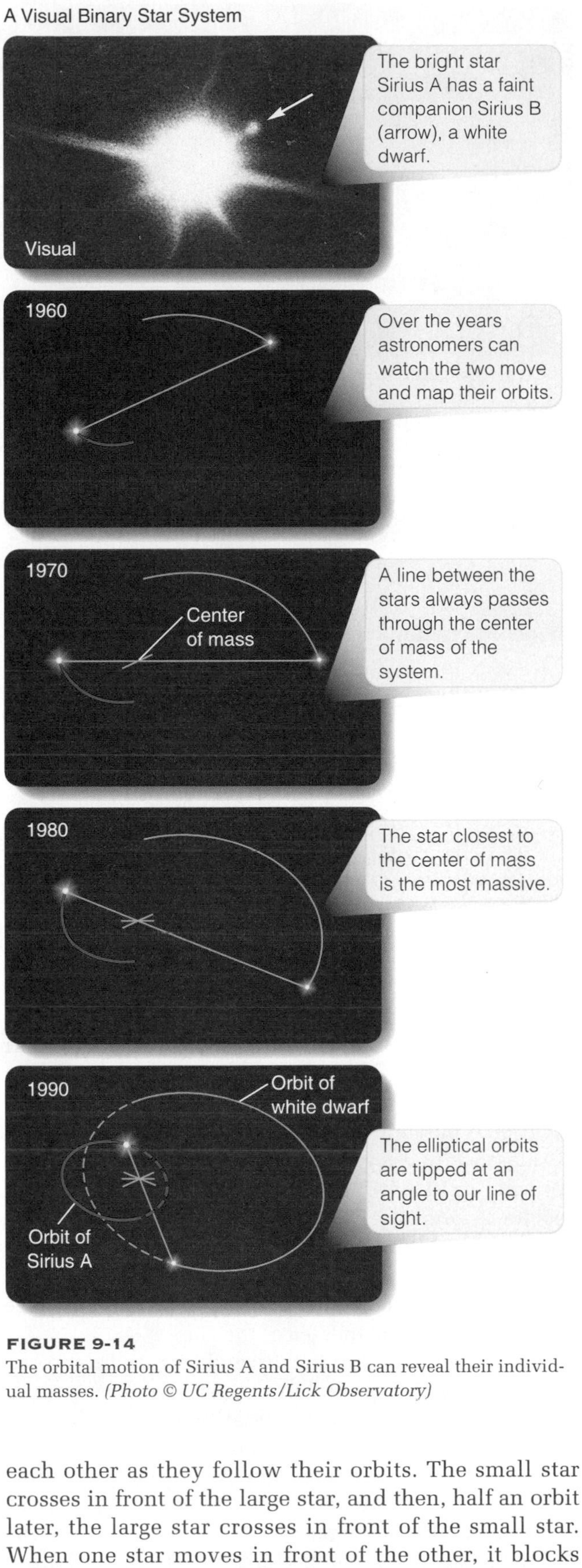

A Visual Binary Star System

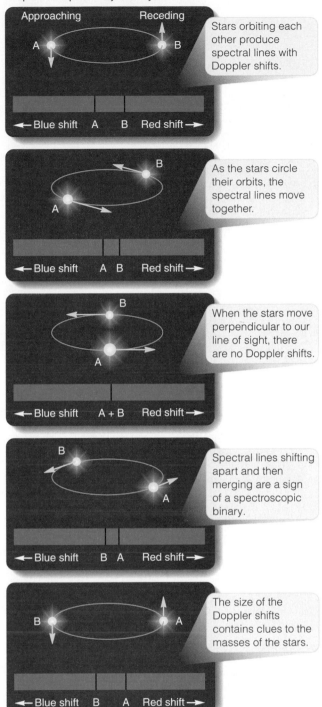

Visual

The bright star Sirius A has a faint companion Sirius B (arrow), a white dwarf.

1960

Over the years astronomers can watch the two move and map their orbits.

1970

Center of mass

A line between the stars always passes through the center of mass of the system.

1980

The star closest to the center of mass is the most massive.

1990

Orbit of white dwarf

Orbit of Sirius A

The elliptical orbits are tipped at an angle to our line of sight.

FIGURE 9-14

The orbital motion of Sirius A and Sirius B can reveal their individual masses. *(Photo © UC Regents/Lick Observatory)*

A Spectroscopic Binary Star System

Approaching   Receding

A   B

Stars orbiting each other produce spectral lines with Doppler shifts.

← Blue shift   A   B   Red shift →

B

A

As the stars circle their orbits, the spectral lines move together.

← Blue shift   A B   Red shift →

B

A

When the stars move perpendicular to our line of sight, there are no Doppler shifts.

← Blue shift   A + B   Red shift →

B

A

Spectral lines shifting apart and then merging are a sign of a spectroscopic binary.

← Blue shift   B   A   Red shift →

B   A

The size of the Doppler shifts contains clues to the masses of the stars.

← Blue shift   B   A   Red shift →

FIGURE 9-15

From Earth, a spectroscopic binary looks like a single point of light, but the Doppler shifts in its spectrum reveal the orbital motion of the two stars.

each other as they follow their orbits. The small star crosses in front of the large star, and then, half an orbit later, the large star crosses in front of the small star. When one star moves in front of the other, it blocks some of the light, and we say the star is eclipsed. Such a system is called an **eclipsing binary system.**

Seen from Earth, the two stars are not visible separately. The system looks like a single point of light. But when one star moves in front of the other star, part of the light is blocked, and the total brightness of the point of light decreases. ▌Figure 9-19 shows a smaller

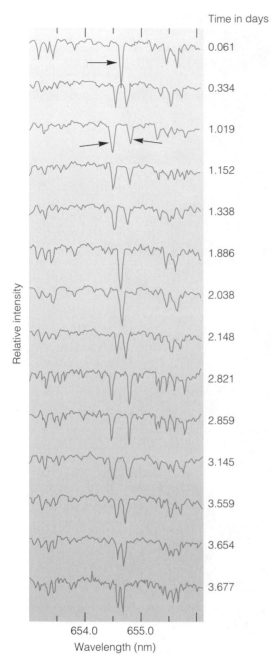

FIGURE 9-16

Fourteen spectra of the star HD80715 are shown here as graphs of intensity versus wavelength. A single spectral line (arrow in top spectrum) splits into a pair of spectral lines (arrows in third spectrum), which then merge and split apart again. These changing Doppler shifts reveal that HD80715 is a spectroscopic binary. *(Adapted from data courtesy of Samuel C. Barden and Harold L. Nations)*

star moving in an orbit around a larger star, first eclipsing the larger star and then being eclipsed as it moves behind. The resulting variation in the brightness of the system is shown as a graph of brightness versus time, a **light curve.**

The light curves of eclipsing binary systems contain tremendous amounts of information about the stars, but the curves can be difficult to analyze. Figure 9-19 shows

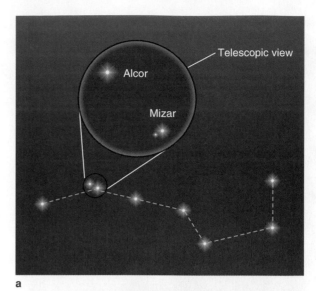

FIGURE 9-17

(a) At the bend of the handle of the Big Dipper lies a pair of stars, Mizar and Alcor. Through a telescope we discover that Mizar has a fainter companion and so is a member of a visual binary system. (b) Spectra of Mizar recorded at different times show that it is itself a spectroscopic binary system rather than a single star. In fact, both the faint companion to Mizar and the nearby star Alcor are also spectroscopic binary systems. *(The Observatories of the Carnegie Institution of Washington)*

an idealized system. ▌Figure 9-20 shows the light curve of a real system in which the stars have dark spots on their surfaces and are so close to each other that their shapes are distorted.

Once the light curve of an eclipsing binary system has been accurately observed, we can construct a chain of inference that leads to the masses of the two stars. We can find the orbital period easily, and we can get spectra showing the Doppler shifts of the two stars. We can find the orbital velocity because we don't have to untip the orbits; we know they are nearly edge-on or there would not be eclipses. Then we can find the size of the orbits and the masses of the stars.

Earlier we used luminosity and temperature to calculate the radii of stars, but eclipsing binary systems give us a way to measure the sizes of stars directly. From the light curve we can tell how long it took for the small star to cross the large star. Multiplying this time interval by the orbital velocity of the small star gives us the diameter of the larger star. We can also determine the diameter of the small star by noting how long it took to disappear behind the edge of the large star. For

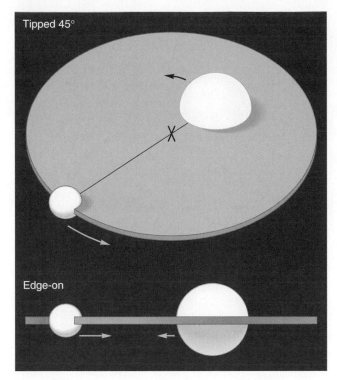

Tipped 45°

Edge-on

FIGURE 9-18
Imagine a model of a binary system with balls for stars and a disk of cardboard for the plane of the orbits. Only if we view the system edge-on do we see the stars cross in front of each other.

example, if it took 300 seconds for the small star to disappear while traveling 500 km/s relative to the large star, then it must be 150,000 km in diameter.

Of course, there are complications due to the inclination and eccentricity of orbits, but often these effects can be taken into account, and the system can tell us not only the masses of its stars but also their diameters.

Algol (β Persei) is one of the best-known eclipsing binaries, because its eclipses are visible to the naked eye. Normally, its magnitude is about 2.15, but its brightness drops to 3.4 in eclipses that occur every 68.8 hours. Although the nature of the star was not recognized until 1783, its periodic dimming was probably known to the ancients. *Algol* comes from the Arabic for "the demon's head," and it is associated in constellation mythology with the severed head of Medusa, the sight of whose serpentine locks turned mortals to stone (❚ Figure 9-21). Indeed, in some accounts, Algol is the winking eye of the demon.

From the study of binary stars, astronomers have found that the masses of stars range from roughly 0.1 solar mass at the low end to somewhere between 60 and 100 solar masses at the high end. The most massive star ever found in a binary system is an O3 star with a mass of 56.9 solar masses. A few other stars are believed to be more massive, but they do not lie in binary systems, so astronomers can only estimate their masses.

An Eclipsing Binary Star System

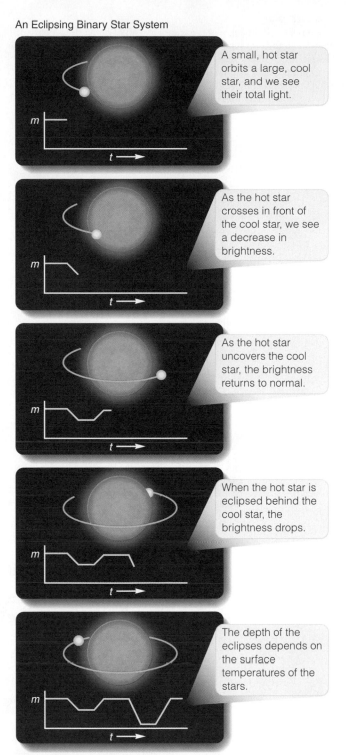

A small, hot star orbits a large, cool star, and we see their total light.

As the hot star crosses in front of the cool star, we see a decrease in brightness.

As the hot star uncovers the cool star, the brightness returns to normal.

When the hot star is eclipsed behind the cool star, the brightness drops.

The depth of the eclipses depends on the surface temperatures of the stars.

FIGURE 9-19
From Earth, an eclipsing binary looks like a single point of light, but changes in brightness reveal that two stars are eclipsing each other. Doppler shifts in the spectrum combined with the light curve, shown here as magnitude versus time, can reveal the size and mass of the individual stars.

FIGURE 9-20

The observed light curve of the binary star VW Cephei (lower curve) shows that the two stars are so close together their gravity distorts their shapes. Slight distortions in the light curve reveal the presence of dark spots at specific places on the star's surface. The upper curve shows what the light curve would look like if there were no spots. *(Graphics created with Binary Maker 2.0)*

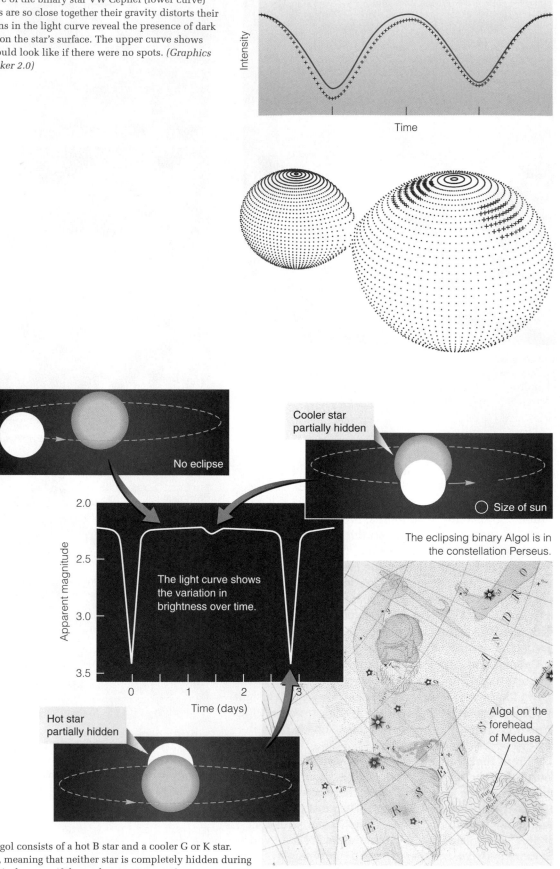

FIGURE 9-21

The eclipsing binary Algol consists of a hot B star and a cooler G or K star. The eclipses are partial, meaning that neither star is completely hidden during eclipses. The orbit here is drawn as if the cooler star were stationary.

WINDOW ON SCIENCE 9-3

Information About Nature: Basic Scientific Data

In a simple sense, science is the process by which we look at data and search for relationships that tell us how nature works. But, as a result, science sometimes requires large amounts of data. For example, astronomers need to know the masses and luminosities of many stars before they can begin looking for relationships.

Compiling basic data is one of the common forms of scientific study. This work may not seem very exciting to an outsider, but scientists often love their work not so much because they want to know nature's secrets but because they love the process of studying nature. Using a microscope or a telescope is fun. Gathering plants in the rain forest or geological samples from a cliff face can

be tremendously exciting and satisfying. It is sometimes the process of science that is most rewarding, and enjoyment of the process can lead scientists to gather significant amounts of information.

Solving a single binary star system to find the masses of the stars does not tell an astronomer a great deal about nature, but solving a binary is like solving a puzzle. It is fun and it is satisfying. Over the years, many astronomers have added their results to the growing data file on stellar masses. We can now analyze that data to search for relationships between the masses of stars and other parameters such as diameter and luminosity.

The history of science is filled with stories of hardworking scientists who compiled large amounts of data that

later scientists used to make important discoveries. For example, Tycho Brahe spent 20 years recording the positions of the stars and planets (Chapter 4). He must have loved his work to spend 20 years on his windy island, but he did not live to use his data. It was his successor, Johannes Kepler, who used Tycho Brahe's data to discover the laws of planetary motion.

Whatever science you study, you will encounter accumulations of measurements and observations that have been compiled over the years, including everything from the hardness of rocks to the attention span of infants. Determining these basic scientific data is as much a part of science as is testing a hypothesis.

REVIEW CRITICAL INQUIRY

When we look at the light curve for an eclipsing binary system with total eclipses, how can we tell which star is hotter? If we assume that the two stars in an eclipsing binary system are not the same size, then we can refer to them as the larger star and the smaller star. When the smaller star moves behind the larger star, we lose the light coming from the total area of the small star. And when the smaller star moves in front of the larger star, it blocks off light from the same amount of area on the larger star. In both cases, the same amount of area, the same number of square meters, is hidden from our sight. Then the amount of light lost during an eclipse depends only on the temperature of the hidden surface because temperature is what determines how much a single square meter can radiate per second. When the surface of the hotter star is hidden, the brightness will fall dramatically, but when the surface of the cooler star is hidden, the brightness will not fall as much. So we can look at the light curve and point to the deeper of the two eclipses and say, "That is where the hotter star is behind the cooler star."

But eclipsing binaries also tell us the diameters of the stars. How could we look at the light curve of an eclipsing binary with total eclipses and find the ratio of the diameters?

Binary stars give us a powerful tool with which to discover the masses of the stars. With that information, we are now ready to assemble our data and begin trying to understand the secrets of the family of stars.

9-5 A SURVEY OF THE STARS

We have achieved the three goals we set for ourselves at the start of this chapter. We know how to find the luminosities, diameters, and masses of stars. Now we can put that data (Window on Science 9-3) together to paint a family portrait of the stars; like most family portraits, both similarities and differences are important clues to the history of the family. As we begin trying to understand how stars are born and how they die, we ask a simple question: What is the average star like? Answering that question is both challenging and illuminating.

MASS, LUMINOSITY, AND DENSITY

The H–R diagram is filled with patterns that give us clues as to how stars are born, how they age, and how they die. When we add our data, we see traces of those patterns.

If we label an H–R diagram with the masses of the plotted stars, as is done in ▌ Figure 9-22, we discover that the main-sequence stars are ordered by mass. The most massive main-sequence stars are the hot stars. As we run our eye down the main sequence, we find lower-mass stars; and the lowest-mass stars are the coolest, faintest main-sequence stars.

Stars that do not lie on the main sequence are not in order according to mass. Some giants and supergiants are massive, while others are no more massive than our sun. All white dwarfs have about the same

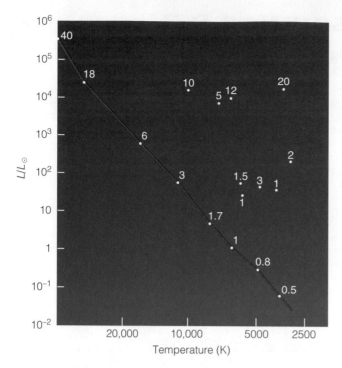

FIGURE 9-22

The masses of the plotted stars are labeled on this H–R diagram. Notice that the masses of main-sequence stars decrease from top to bottom but that masses of giants and supergiants are not arranged in any ordered pattern.

mass, somewhere in the narrow range of 0.5 to about 1 solar mass.

Because of the systematic ordering of mass along the main sequence, these main-sequence stars obey a **mass–luminosity relation**—the more massive a star is, the more luminous it is (❙ Figure 9-23). In fact, the mass–luminosity relation can be expressed as a simple formula:

$$L = M^{3.5}$$

That is, a star's luminosity (in terms of the sun's luminosity) equals its mass (in solar masses) raised to the 3.5 power. For example, a star of 4 solar masses has a luminosity of approximately $4^{3.5}$, or $4 \times 4 \times 4 \times \sqrt{4}$. This equals 64×2, or 128. So a 4-solar-mass star will have a luminosity about 128 times the luminosity of the sun. This is only an approximate equation, as shown by the yellow line in Figure 9-23.

Notice how large the range in luminosity is. The range of masses extends from about 0.08 solar mass to about 50 solar masses—a factor of 600. But the range of luminosities extends from about 10^{-6} to about 10^6 solar luminosities—a factor of 10^{12}. Clearly, a small difference in mass causes a large difference in luminosity.

Giants and supergiants do not follow the mass–luminosity relation very closely, and white dwarfs not at all. In the next chapters, the mass–luminosity relation will help us understand how stars generate their energy.

Though mass alone does not reveal any pattern among giants, supergiants, and white dwarfs, density does. Once we know a star's mass and diameter, we can calculate its average density by dividing its mass by its volume. Stars are not uniform in density but are most dense at their centers and least dense near their surface. The center of the sun, for instance, is about 100 times as dense as water; its density near the visible surface is about 3400 times less dense than Earth's atmosphere at sea level. A star's average density is intermediate between its central and surface densities. The sun's average density is approximately 1 g/cm^3—about the density of water.

Main-sequence stars have average densities similar to the sun's, but giant stars, being large, have low average densities, ranging from 0.1 to 0.01 g/cm^3. The enormous supergiants have still lower densities, ranging from 0.001 to 0.000001 g/cm^3. These densities are thinner than the air we breathe; and, if we could insulate ourselves from the heat, we could fly an airplane through these stars. Only near the center would we be in any danger, for there the material is very dense—about 3,000,000 g/cm^3.

The white dwarfs have masses about equal to the sun's, but are very small, only about the size of Earth. Thus, the matter is compressed to densities of 3,000,000 g/cm^3 or more. On Earth, a teaspoonful of this material would weigh about 15 tons.

Density divides stars into three groups. Most stars are main-sequence stars with densities like the sun's. Giants and supergiants are very-low-density stars, and white dwarfs are high-density stars. We will see in later chapters that these densities reflect different stages in the evolution of stars.

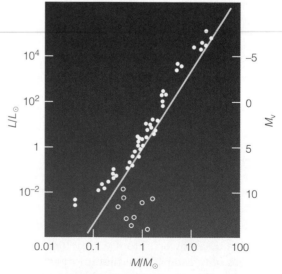

FIGURE 9-23

The mass–luminosity relation shows that the more massive a main-sequence star is, the more luminous it is. The open circles represent white dwarfs, which do not obey the relation. The yellow line represents the equation $L = M^{3.5}$.

SURVEYING THE STARS

If we want to know what the average person thinks about a certain subject, we take a survey. If we want to know what the average star is like, we must survey the stars. Such surveys reveal important relationships among the family of stars.

Not many decades ago, surveying large numbers of stars was an exhausting task, but modern computers have changed that. Specially designed telescopes controlled by computers can make millions of observations per night, and high-speed computers can compile and analyze these vast surveys and create easy-to-use data bases. Chapter 6 mentioned the Sloan Digital Sky Survey and the 2Mass infrared survey. Those and other surveys produce mountains of data that astronomers can mine searching for relationships within the family of stars.

What could we learn about stars from a survey of the stars near the sun? Because we believe the sun is in a typical place in the universe, such a survey could reveal general characteristics of the stars and might reveal unexpected processes in the formation and evolution of stars. Study "■ The Family of Stars" on pages 192 and 193 and notice three important points.

First, taking a survey is difficult because we must be sure we get an honest sample. If we don't survey enough stars or if we don't notice the faintest stars, we can get biased results. The second important point is that most stars are faint, and the most luminous stars are rare. The most common kinds of stars are the lower-main-sequence red dwarfs and the white dwarfs. Also notice that luminous stars, although they are rare, are easily visible even at great distances. Many of the brighter stars in the sky are highly luminous stars that we see even though they lie far away.

The night sky is a beautiful carpet of stars, but they are not all the same. Some are giants and supergiants, and some are dwarfs. The family of the stars is rich in its diversity.

REVIEW CRITICAL INQUIRY

Why is it difficult to be sure how common O and B stars are?

In our survey of the stars, the number of O and B stars per million cubic parsecs is difficult to determine because these stars are quite rare. We know that O and B stars are not common because there are none near us in space. If we take our survey by counting stars within 63 pc of the sun, a distance that would include a million cubic parsecs, we find no O or B stars that close. To find these stars we must look to great distances, and at such distances we can't measure their distance accurately. Thus, the chain of inference we use to find the number of sunlike stars in space does not work as well for these more luminous stars.

Now consider the chain of inference for the most common stars. Why is it difficult to be sure just how common these stars are?

In this chapter we set out to measure the basic properties of stars. Once we found the distance to the stars, we were able to find their luminosities, diameters, and masses—rather mundane data. But we have now discovered a puzzling situation. The largest and most luminous stars are so rare we might joke that they hardly exist, and the average stars are such dinky low-mass things they are hard to see even if they are near us in space. Why does nature make stars in this peculiar way? To answer that question, we must explore the birth, life, and death of stars. We begin that quest in the next chapter.

SUMMARY

Our goal in this chapter was to characterize the stars by finding their luminosities, diameters, and masses. Before we could begin, we needed to find the distances to stars. Only by first knowing the distance could we find the other properties of the stars.

We can measure the distance to the nearer stars by observing their parallaxes. The more distant stars are so far away that their parallaxes are unmeasurably small. To find the distances to these stars, we must use spectroscopic parallax. Stellar distances are commonly expressed in parsecs. One parsec is 206,265 AU—the distance to an imaginary star whose parallax is 1 second of arc.

Once we know the distance to a star, we can find its intrinsic brightness, expressed as its absolute magnitude or its luminosity. A star's absolute magnitude is the apparent magnitude we would see if the star were only 10 pc away. The luminosity of the star is the total energy radiated in 1 second, usually expressed in terms of the luminosity of the sun.

The H–R diagram plots stars according to their intrinsic brightness and their surface temperature. In the diagram, roughly 90 percent of all normal stars fall on the main sequence, the more massive being hotter, larger, and more luminous. The giants and supergiants, however, are much larger and lie above the main sequence. They are more luminous than main-sequence stars of the same temperature. Some of the white dwarfs are hot stars, but they fall below the main sequence because they are so small.

The large size of the giants and supergiants means their atmospheres have low densities and their spectra have sharper spectral lines than the spectra of main-sequence stars. In fact, it is possible to assign stars to luminosity classes by the widths of their spectral lines. Class V stars are main-sequence stars with broad spectral lines. Giant stars (III) have sharper lines, and supergiants (I) have extremely sharp spectral lines.

The only direct way we can find the mass of a star is by studying binary stars. When two stars orbit a common center of mass, we can find their masses by observing the period and sizes of their orbits.

Given the mass and diameter of a star, we can find its average density. On the main sequence, the stars are about as dense as the sun, but the giants and supergiants are very low-density stars. Some are much thinner than air. The white dwarfs, lying below the main sequence, are tremendously dense.

The Family of Stars

What is the most common kind of star? Are some rare? Are some common? To answer those questions we must survey the stars. To do so we must know their spectral class, their luminosity class, and their distance. Our census of the family of stars produces some surprising demographic results.

A sphere 62 pc in radius encloses a million cubic parsecs.

62 pc

Earth

We could survey the stars by observing every star within 62 pc of Earth. Such a survey would tell us how many stars of each type are found within a volume of a million cubic parsecs.

Our survey faces two problems.

1. The most luminous stars are so rare we find few in our survey region. There are no O stars at all within 62 pc of Earth.

2. Lower-main-sequence M stars, called red dwarfs, and white dwarfs are so faint they are hard to locate even when they are only a few parsecs from Earth. Finding every one of these stars in our survey sphere is a difficult task.

In this chart showing most of the constellation Canis Major, stars are represented as dots with colors assigned according to spectral class. The brightest stars in the sky tend to be the rare, highly luminous stars, which look bright even though they are far away. Most stars are of very low luminosity, so nearby stars tend to be very faint red dwarfs.

- O and B
- A
- F
- G
- K
- M

Red dwarf
15 pc

o² Canis Majoris
B3Ia   790 pc

Red dwarf
17 pc

δ Canis Majoris
F8Ia   550 pc

σ Canis Majoris
M0Iab   370 pc

η Canis Majoris
B5Ia   980 pc

ε Canis Majoris
B2II   130 pc

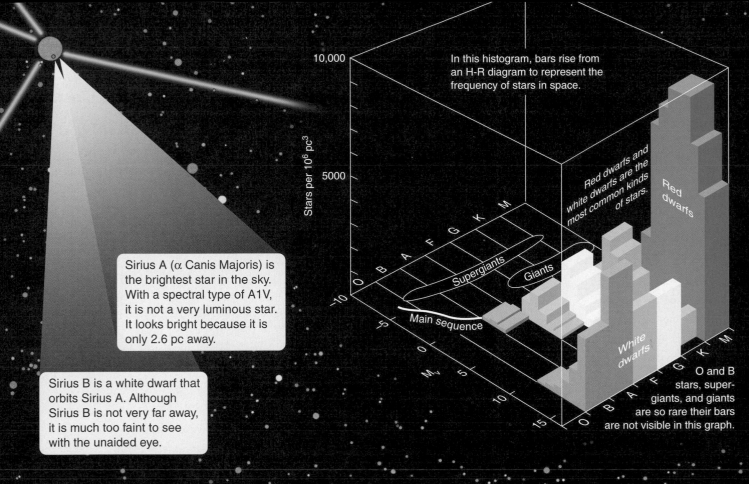

In this histogram, bars rise from an H-R diagram to represent the frequency of stars in space.

Stars per 10⁶ pc³

Red dwarfs and white dwarfs are the most common kinds of stars.

Red dwarfs

Supergiants

Giants

Main sequence

White dwarfs

O and B stars, super-giants, and giants are so rare their bars are not visible in this graph.

Sirius A (α Canis Majoris) is the brightest star in the sky. With a spectral type of A1V, it is not a very luminous star. It looks bright because it is only 2.6 pc away.

Sirius B is a white dwarf that orbits Sirius A. Although Sirius B is not very far away, it is much too faint to see with the unaided eye.

The universe is filled with uncountable stars, but their family relationships tell us how they were born and how they will die.

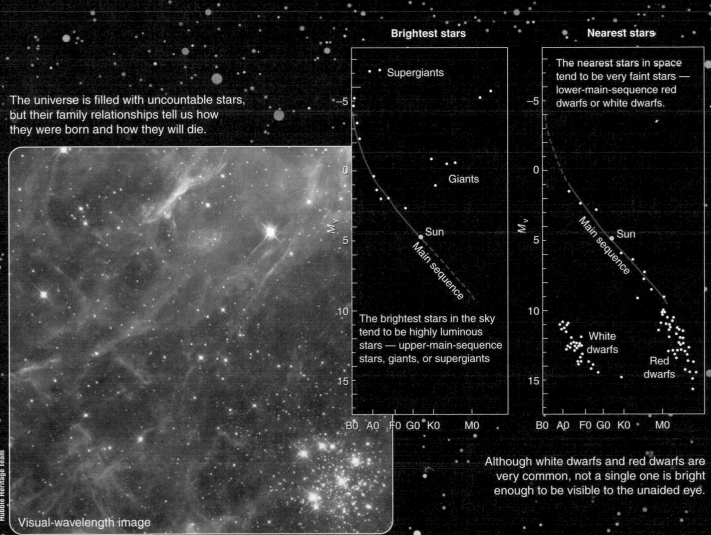

Brightest stars

Supergiants

Giants

Sun

Main sequence

The brightest stars in the sky tend to be highly luminous stars — upper-main-sequence stars, giants, or supergiants

Nearest stars

The nearest stars in space tend to be very faint stars — lower-main-sequence red dwarfs or white dwarfs.

Main sequence

Sun

White dwarfs

Red dwarfs

Although white dwarfs and red dwarfs are very common, not a single one is bright enough to be visible to the unaided eye.

Visual-wavelength image

Hubble Heritage Team

The mass–luminosity relation says that the more massive a star is, the more luminous it is. Main-sequence stars follow this rule closely, the most massive being the upper-main-sequence stars and the least massive the lower-main-sequence stars. Giants and supergiants do not follow the relation precisely, and white dwarfs not at all.

A survey in the neighborhood of the sun shows us that lower-main-sequence stars are the most common type. The hot stars of the upper main sequence are very rare. Giants and supergiants are also rare, but white dwarfs are quite common, although they are faint and hard to find.

NEW TERMS

stellar parallax (p)

parsec (pc)

proper motion

flux

absolute visual magnitude (M_v)

magnitude–distance formula

distance modulus ($m_v - M_v$)

luminosity (L)

absolute bolometric magnitude

H–R (Hertzsprung–Russell) diagram

main sequence

giants

supergiants

red dwarf

white dwarf

luminosity class

spectroscopic parallax

binary stars

visual binary system

spectroscopic binary system

eclipsing binary system

light curve

mass–luminosity relation

REVIEW QUESTIONS

Ace⊗Astronomy™ Assess your understanding of this chapter's topics with additional quizzing and animations at **http:// astronomy.brookscole.com/seeds8e**

1. Why are parallax measurements made from Earth-based telescopes limited to the nearest stars?

2. How would having an observatory on Mars help astronomers measure parallax more accurately?

3. How did having an observatory in orbit above Earth's atmosphere help astronomers measure parallax more accurately?

4. For which stars does absolute visual magnitude differ least from absolute bolometric magnitude? Why?

5. How can a cool star be more luminous than a hot star? Give some examples.

6. How can we be certain that the giant stars are actually larger than the sun?

7. Describe the steps in using the methods of spectroscopic parallax. Do we really measure a parallax?

8. Give the approximate radii and intrinsic brightnesses of stars in the following classes: G2 V, G2 III, G2 Ia.

9. Why can we find the masses of the stars in visual binary systems and in eclipsing binary systems, but not in spectroscopic binary systems?

10. What is the mass–luminosity relation?

11. What measurements must we make to find out how common different kinds of stars are?

DISCUSSION QUESTIONS

1. If someone asked you to compile a list of the nearest stars to the sun based on your own observations, what measurements would you make, and how would you analyze them to detect nearby stars?

2. The sun is sometimes described as an average star. What is the average star really like?

PROBLEMS

1. If a star has a parallax of 0.050 second of arc, what is its distance in parsecs? in light-years? in AU?

2. If a star has a parallax of 0.016 second of arc and an apparent magnitude of 6, how far away is it, and what is its absolute magnitude?

3. Complete the following table:

m_v	M_v	d (pc)	P (seconds of arc)
—	7	10	—
11	—	1000	—
—	−2	—	0.025
4	—	—	0.040

4. If a main-sequence star has a luminosity of 400 $L_\odot$, what is its spectral type? (*Hint:* See Figure 9-12).

5. If a star has an apparent magnitude equal to its absolute magnitude, how far away is it in parsecs? in light-years?

6. If a star has an absolute bolometric magnitude that is eight magnitudes brighter than the sun, what is the star's luminosity?

7. If a star has an absolute bolometric magnitude that is one magnitude fainter than the sun, what is the star's luminosity?

8. An O8 V star has an apparent magnitude of +1. Use the method of spectroscopic parallax to find the distance to the star.

9. Find the luminosity of the sun, given the radius of Earth's orbit and the solar constant (Chapter 8). Make your calculation in two steps. First, use $4\pi R^2$ to calculate the surface area in square meters of a sphere surrounding the sun with a radius of 1 AU. Second, multiply by the solar constant to find the total solar energy passing through the

sphere in 1 second. That is the luminosity of the sun. Compare your result with that in Data File One.

10. In the following table, which star is brightest in apparent magnitude? most luminous in absolute magnitude? largest? farthest away?

Star	Spectral Type	m_v
a	G2 V	5
b	B1 V	8
c	G2 Ib	10
d	M5 III	19
e	White dwarf	15

11. What is the total mass of a visual binary system if its average separation is 8 AU and its period is 20 years?

12. Measure the wavelengths of the iron lines in Figure 9-16 and plot a graph showing the radial velocities of the two stars in this system. What is the period? Assuming that the orbit is circular and edge-on, what is the total mass? What are the individual masses?

13. If the eclipsing binary in Figure 9-19 has a period of 32 days, an orbital velocity of 153 km/s, and an orbit that is nearly edge-on, what is the circumference of the orbit? the radius of the orbit? the mass of the system?

14. If the orbital velocity of the eclipsing binary in Figure 9-19 is 153 km/s and the smaller star becomes completely eclipsed in $2\frac{1}{2}$ hours, what is its diameter?

15. What is the luminosity of a 4-solar-mass main-sequence star? of a 9-solar-mass main-sequence star? of a 7-solar-mass main-sequence star?

CRITICAL INQUIRIES FOR THE WEB

1. The Hertzsprung–Russell diagram was named for two famous astronomers. Who were they? What did they do to earn such an honor?

2. Algol is a famous binary star. What can you find out about the mythology of Algol?

3. An entire class of binary stars is called the Algol binary stars. How would you characterize such stars?

EXPLORING *THESKY*

1. Locate the following stars and determine their apparent magnitude, parallax, distance in parsecs and in light-years, and spectral classification. (*Hint:* To center on an object, use **Find** under the **Edit** menu and type the object's name followed by a period.)

   Sirius
   Aldebaran
   Vega
   Deneb
   Betelgeuse
   Antares
   Altair

2. Use the spectral type and parallax of the stars above to estimate their distance. Compare with the distances given in *TheSky*.

3. Take a survey of the stars. Center on Orion and adjust the field until it is about 100° wide. Click on the 10 brightest stars and record the spectral types. Now zoom in until only a few dozen stars are in the frame. Click on the 10 faintest stars and record their spectral types. Is there a difference between the brightest and faintest stars? Is the result what you expected? Are certain kinds of stars missing from this computer data base?

4. Repeat Activity 3 for a region centered on the Big Dipper.

 Visit the Seeds *Foundations of Astronomy* companion Web site for critical thinking exercises, articles, and additional readings from InfoTrac College Edition, Brooks/Cole's online student library.

when he shall die,
Take him and cut him out in little stars,
And he will make the face of heaven so fine
That all the world will be in love with night,
And pay no worship to the garish sun.

Shakespeare, Romeo and Juliet *III, ii, 21*

Chapter 10

The Hubble Heritage Team STScI/AURA/NASA

GUIDEPOST

In a discussion of bread baking, we might begin with a chapter on wheat and flour. In our discussion of the birth and death of stars, the theme of the next five chapters, we begin with a chapter about the gas and dust between the stars. It is the flour from which nature bakes stars.

This chapter clearly illustrates how astronomers use the interaction of light and matter to learn about nature on the astronomical scale. That tool, which we developed in Chapter 7, "Starlight and Atoms," is powerfully employed here, especially when we include observations at many different wavelengths.

We also see in this chapter the interplay of observation and theory. Neither is useful alone, but together they are a powerful method for studying nature, a method generally known as science.

Juliet loved Romeo so much she compared him to the beauty of the stars, but had she known what was between the stars, she might have compared him to that instead. True, the gas and dust between the stars are mostly dark and cold, but where they are illuminated by stars they create beautiful nebulae, and where they are densest they give birth to beautiful stars. If there is beauty in vast extent and sweeping power, then the **interstellar medium,** the gas and dust between the stars, could steal worship from the garish stars.

We are interested in the interstellar medium because it is the matter from which stars are born. We will discover both large and small clouds of gas and dust, and both hot and cold gas, and we will find that the densest clouds can contract under the influence of their own gravity and create new stars. Clearly, before we can understand how stars are born, we must understand what fills the spaces between the stars.

Another reason we are interested in studying the interstellar medium is that it further illustrates the importance of the interaction of light and matter. Our study of starlight and atoms in Chapter 7 will help us understand the vast but nearly invisible clouds of gas and dust between the stars.

In Chapter 1, we saw that we live in a disk-shaped galaxy, the Milky Way (Figure 1-11). The interstellar medium is confined mostly within a few hundred parsecs of the plane of the disk of the galaxy; if we journeyed out of the galaxy, we would find conditions to be quite different. In our study of the interstellar medium, we confine ourselves to the disk of our galaxy in the neighborhood of the sun—probably a typical part of interstellar space.

To detect the interstellar medium, we must use all the tools of the astronomer. Observations at radio, infrared, visual, ultraviolet, and X-ray wavelengths reveal the properties of the interstellar medium. We will analyze images and spectra, consider the nature of atoms, and utilize the physics of light in order to unlock the secrets hidden in the material between the stars.

10-1 VISIBLE-WAVELENGTH OBSERVATIONS

A quick glance at the night sky does not reveal any matter between the stars, but evidence of its existence is clear when we examine photographs and spectra. In fact, if we look carefully, one important piece of evidence is visible to the naked eye in a familiar constellation.

NEBULAE

On a cold, clear winter night, Orion hangs high in the southern sky, a large constellation composed of brilliant stars. If you look carefully at Orion's sword, you will see that one of the stars is a hazy cloud (Figure 2-4). A small telescope reveals even more such clouds of gas and dust. Astronomers refer to these clouds as **nebulae** (singular, nebula), from the Latin word for mist or cloud.

Study ▌ "Three Kinds of Nebulae" on pages 198 and 199 and notice three important points. First, very hot stars can excite clouds of gas and dust to emit light, and that tells us the clouds contain mostly hydrogen gas at very low densities. Second, where slightly cooler stars illuminate clouds we get evidence that the dust in the clouds is made up of very small particles. The third thing to notice is that some dense clouds of gas and dust are only detectable where they are silhouetted against background regions filled with stars or bright nebulae.

A detailed analysis of the spectra of nebulae can tell us even more. Certain lines in the spectra of emission nebulae are called **forbidden lines** because they are almost never seen in excited gas on Earth. Two good examples are the strong green lines at 495.9 nm and 500.7 nm produced by oxygen atoms that have lost two electrons. (Following our convention for naming ions, twice-ionized oxygen is OIII.) The oxygen ions can become excited by collision with a high-energy photon or a rapidly moving ion or electron, and the atom can emit various-wavelength photons as its electron cascades back down to lower energy levels. However, transitions between certain energy levels are so unlikely they are called "forbidden." If an electron enters such an energy level, it will remain there for a relatively long time before it can decay further and emit the appropriate photon. For a normal transition, the electron might wait 10^{-8} to 10^{-7} second. If it enters one of these **metastable levels,** the electron may be stuck there for as long as an hour before it becomes likely to fall to a lower level and emit the proper photon.

This explains why we don't see these forbidden lines in laboratories on Earth. The atoms in a dense gas collide with each other so often that there isn't time for an electron in a metastable level to decay and emit a photon. Such electrons get excited back up to higher levels before they can drop downward and emit a photon. In an emission nebula, the gas has a very low density, and an atom could go for an hour or more between collisions, giving an electron in a metastable level time to decay to a lower level and emit a photon at a supposedly forbidden wavelength. The forbidden lines clearly show that nebulae have very low densities. This is a dramatic example of how astronomers can use knowledge of atomic physics to understand astronomical objects.

Three Kinds of Nebulae

At visual wavelengths, astronomers see three kinds of nebulae. The differences between these clouds of gas and dust give us insights into the interstellar medium.

Emission nebulae are also called **HII regions**, following the custom of naming gas with a roman numeral to show its state of ionization. HI is neutral hydrogen, and HII is ionized.

In an HII region, the ionized nuclei and free electrons are mixed. When a nucleus captures an electron, the electron falls down through the atomic energy levels, emitting photons at specific wavelengths. Spectra tell us that the nebulae have compositions much that like of the sun – mostly hydrogen.

Emission nebulae have densities of 100 to 1000 atoms per cubic centimeter, equivalent to the best vacuums produced in laboratories on Earth.

Emission nebulae are produced when a hot star excites the gas near it to produce an emission spectrum. The star must be hotter than about B1 (25,000 K). Cooler stars do not emit enough ultraviolet radiation to ionize the gas. Emission nebulae have a distinctive pink color produced by the blending of the red, blue, and violet Balmer lines.

ESO

Visual-wavelength image

A **reflection nebula** is produced when starlight scatters from a dusty nebula. Consequently, the spectrum of a reflection nebula is just the reflected absorption spectrum of starlight. Gas is surely present in a reflection nebula, but it is not excited to emit photons.

▼

Reflection nebulae NGC 1973, 1975, and 1977 lie just north of the Orion nebula. The pink tints are produced by ionized gases deep in the nebulae.

▼ Reflection nebulae look blue for the same reason the sky looks blue. Short wavelengths scatter more easily than long wavelengths.

Sunlight enters Earth's atmosphere

Blue photons are scattered more easily than longer wavelengths.

Blue photons enter our eyes from all directions, making the sky look blue.

Visual-wavelength image   AATB

Ace ◐ Astronomy™

Go to AceAstronomy and click Active Figures to see "Scattering in Earth's Atmosphere." Take control of this diagram.

The blue color of reflection nebulae tells us that the dust particles must be very small in order to preferentially scatter the blue photons. Interstellar dust grains must have diameters ranging from 0.001 mm down to 1 nm or so.

The hottest star in the Pleiades star cluster is Merope, a B3 star. It is not hot enough to ionize the gas so we see a reflection nebula rather than an emission nebula.

Merope

Visual-wavelength image

Caltech

A dusty reflection nebula is located very close to the star Merope just out of the image to upper right. The glare from the star is caused by internal reflections in the telescope, but the wispy nature of the nebula is real. The intense light from the star is pushing the dust particles away and may destroy the little nebula over the next few thousand years.

Merope

Visual   NASA

Reflection

Emission

Trifid Nebula

The Milky Way in Sagittarius contains two nebulae that dramatically demonstrate the difference between emission and reflection nebulae.

Emission

Visual   Lagoon Nebula

Daniel Good

Dark Nebula Barnard 86

Star Cluster NGC6520

Visual   AATB

Dark nebulae are dense clouds of gas and dust that obstruct our view of more distant stars. Some are generally round, but others are twisted and distorted, suggesting that even when there are no nearby stars to ionize the gas or produce a reflection nebula, there are breezes and currents pushing through the interstellar medium.

Northern Coalsack

Cygnus

Milky Way

Great Rift

The Horsehead Nebula in Orion is a dark nebula silhouetted against a more distant emission nebula.

Visual

Hubble Heritage Team

Large dark nebulae obstruct our view of more distant stars and form holes and rifts along the Milky Way. The Great Rift extends from Cygnus to Sagittarius.

The three kinds of nebulae introduce us to the interstellar medium. To learn more, we have to think carefully about light passing through this gas and dust.

EXTINCTION AND REDDENING

The dust in space is called **interstellar dust.** Its presence is dramatically evident in dark nebulae, but simple observations tell us that the interstellar dust is spread throughout space, making up roughly 1 percent of the mass of the interstellar medium.

One way we know that dust is present in the interstellar medium is that it makes distant stars appear fainter than they would if space were perfectly transparent. This phenomenon is called **interstellar extinction,** and in the neighborhood of the sun it amounts to about 2 magnitudes per thousand parsecs. That is, if a star lies 1000 pc from Earth, it will look about 2 magnitudes fainter than it would if space were perfectly empty. If it were 2000 pc away, it would look about 4 magnitudes dimmer, and so on. This is a dramatic effect, and it shows that the interstellar medium is not confined to a few nebulae scattered here and there. The spaces between the stars are far from empty.

Another way we can detect the presence of dust is through the effect it has on the colors of stars. An O star should be blue, but some stars with the spectrum of an O star seem much redder than they should be. Termed **interstellar reddening,** this effect is caused by dust particles scattering light. As we saw in the case of the reflection nebulae, the dust particles are small, with diameters roughly equal to the wavelength of light, and they scatter shorter wavelengths better than longer wavelengths. That means they scatter blue photons more often than red photons. Thus, the light from a distant star has lost some of its blue photons because of scattering, and consequently the star looks redder (❚ Figure 10-1).

As we discussed earlier, this scattering of blue light is what makes the sky blue, but it is also what makes distant city lights look yellow. If you view the lights of a city at night from a high-flying aircraft or a distant mountaintop, the lights will look yellow. As you descend toward the city, the lights will become bluer. The light from the city is reddened by microscopic particles in the air. If the particles are especially dense, we call them smog.

Astronomers can measure the amount of reddening by comparing two stars of the same spectral type, one of which is dimmed more than the other. The more obscured star will look redder. If we plot the difference in brightness between the two stars as a function of wavelength, we get a curve that shows the reddening. That is, it shows how the starlight is dimmed at different wavelengths (❚ Figure 10-2). In general, the light is dimmed in proportion to the reciprocal of the wavelength, a pattern typical of scattering from small dust particles. Laboratory measurements show that the high extinction at about 220 nm is caused by a form of carbon, so we must conclude that some of the dust particles are carbon. Other evidence suggests that some grains contain silicates and metals and may have coatings of water ice, frozen ammonia, or carbon-based molecules.

INTERSTELLAR ABSORPTION LINES

If we look at the spectra of distant stars, we can see dramatic evidence of an interstellar medium. Of course, we see spectral lines produced by the gas in the atmospheres of the stars, but we also see sharp spectral lines produced by the gas in the interstellar medium. These **interstellar absorption lines** give us a new way to study the gas between the stars.

Astronomers can recognize interstellar absorption lines in three ways: by their ionization, by their widths, and by their multiple components. Some stellar spectra contain absorption lines that just don't belong because they represent the wrong ionization state. For example, if we look at the spectrum of a very hot star such as an O star, we would expect to see no lines of once-ionized calcium (CaII) because that ion cannot exist at the high temperatures in the atmosphere of such a hot star. But many O-star spectra contain lines of CaII, so we must conclude that these lines were produced not in the star but in the interstellar medium.

In addition, the widths of the interstellar lines give away their identity. In the hot atmosphere of a star, the atoms move rapidly, and the Doppler effect broadens spectral lines. Also, the atoms collide with each other often enough to blur the electron energy levels, and that also broadens the spectral lines. We called these two effects Doppler broadening and collisional broadening in Chapter 7. These effects make the lines in the spectrum of a main-sequence star quite broad, but even in the atmosphere of a giant or supergiant star, the gas atoms move rapidly and collide often enough to broaden the spectral lines. Interstellar lines, on the other hand, are exceedingly sharp. This tells us that the interstellar matter is cold and has a low density. If it were hot, Doppler broadening would smear out the lines due to the motions of individual atoms. If the gas were dense, collisional broadening would produce wider lines. The exceedingly narrow widths of the interstellar absorption lines are typical of cold, low-density gas.

Another revealing characteristic of the interstellar lines is that they are often split into two or more components. These multiple components have slightly different wavelengths and appear to have been produced when the light from the star passed through different clouds of gas on its way to Earth. Because the clouds of gas have slightly different radial velocities, they produce absorption lines with slightly Doppler-shifted wavelengths. ❚ Figure 10-3 illustrates all three of these characteristics of interstellar absorption lines—the wrong ionization, narrow widths, and multiple components.

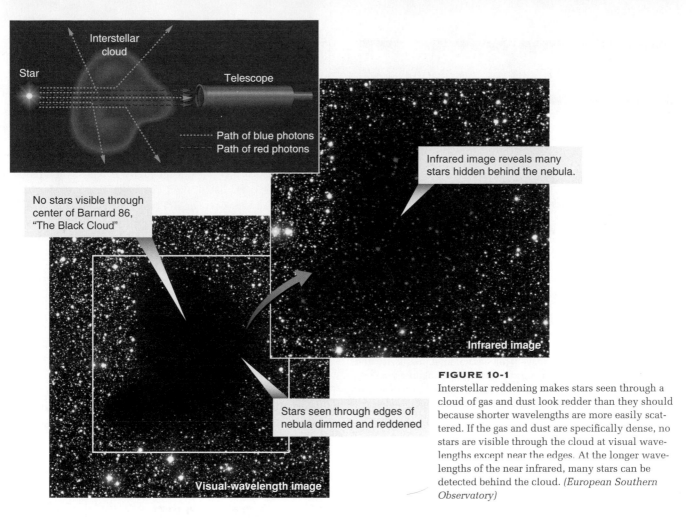

No stars visible through center of Barnard 86, "The Black Cloud"

Infrared image reveals many stars hidden behind the nebula.

Stars seen through edges of nebula dimmed and reddened

Visual-wavelength image

Infrared image

FIGURE 10-1

Interstellar reddening makes stars seen through a cloud of gas and dust look redder than they should because shorter wavelengths are more easily scattered. If the gas and dust are specifically dense, no stars are visible through the cloud at visual wavelengths except near the edges. At the longer wavelengths of the near infrared, many stars can be detected behind the cloud. *(European Southern Observatory)*

Astronomers disagree as to the structure of the interstellar medium, and the boundaries and characteristics of clouds are ill defined. Nevertheless, astronomers find it convenient to categorize clouds into a few main varieties. Studies of interstellar absorption lines reveal clouds of neutral gas (and presumably dust) with densities of 10 to a few hundred atoms/cm^3. Because these clouds are not ionized, they are called **HI clouds.** These clouds must be 50 to 150 pc in diameter and have masses of a few solar masses. The gas temperature is only about 100 K. Starlight seems to pass through six to ten of these clouds for every 1000 pc near the plane of our galaxy. While it is easy to think of these clouds as more or less spherical blobs of gas, observations show that they are usually twisted into long filaments, flattened into thin sheets, or tangled into chaotic shapes—further evidence that the interstellar medium is not static and motionless.

Between these HI clouds of neutral gas lies a hot **intercloud medium** with a temperature of a few thousand K and a density of only about 0.1 atom/cm^3. This intercloud medium consists of ionized hydrogen (HII). The gas is partially ionized by the ultraviolet photons in starlight. If it is to be in equilibrium with the HI clouds (if the HI clouds are not expanding or contracting), the pressure in both regions should be about the same.

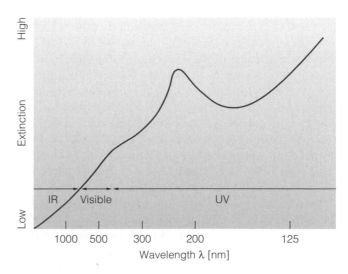

FIGURE 10-2

Interstellar extinction, the dimming of starlight by dust between the stars, depends strongly on wavelength. Infrared radiation is only slightly affected, but ultraviolet light is strongly scattered. The strong extinction at about 220 nm is caused by a certain form of carbon dust in the interstellar medium.

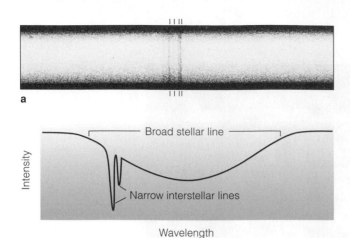

a

b

FIGURE 10-3

Interstellar absorption lines can be recognized in three ways. (a) The B0 supergiant ∈ Orionis is much too hot to show spectral lines of once-ionized calcium (CaII), yet this short segment of its spectrum reveals narrow, multiple lines of CaII (tick marks) that must have been produced in the interstellar medium. *(The Observatories of the Carnegie Institution of Washington)* (b) Spectral lines produced in the atmospheres of stars are much broader than the spectral lines produced in the interstellar medium. *(Adapted from a diagram by Binnendijk)* In both (a) and (b) the multiple interstellar lines are produced by separate interstellar clouds with slightly different radial velocities.

The pressure in a gas depends on its density and its temperature (Parameters of Science 10-1). In the interstellar medium, the HI clouds are cool but dense, and the HII clouds of the intercloud medium are hot but low in density. Thus the two regions could have similar pressures even though their temperatures are quite different. Nevertheless, observations reveal some regions where the pressures are not equal, and that is poorly understood.

How can the intercloud medium be ionized when it is not close to hot stars? To understand how, imagine that we are atoms floating in the intercloud medium. Ultraviolet photons from distant stars are not common, but they do whiz by now and then. Soon we become ionized by absorbing one of these photons and losing an electron. In a denser gas, we would quickly find another electron, capture it, and become neutral again, but the interstellar medium has such a low density that we must wait a very long time to find an electron. Thus, the atoms of the intercloud medium spend almost all their time in an ionized state because of the low density.

Studies of interstellar absorption lines give us further clues about the composition of the interstellar gas. It is much like the composition of the sun. Hydrogen is most abundant, with helium second. Light elements such as carbon, nitrogen, and oxygen are as common as in the sun, but elements such as iron, calcium, and titanium are less abundant than they are in the sun. Most likely, these elements are missing from the gas because they have condensed to form the dust.

Observations at visible wavelengths can give us important information about the interstellar medium, but in order to paint a complete picture of the matter between the stars, we must observe at other wavelengths.

10-2 LONG- AND SHORT-WAVELENGTH OBSERVATIONS

The interstellar medium is mostly invisible to our eyes, so we must observe at the longer wavelengths of infrared and radio radiation and at the shorter wavelengths of ultraviolet and X-ray radiation to further explore the interstellar medium.

21-CM OBSERVATIONS

Cold, neutral hydrogen floating in space can emit electromagnetic radiation with a wavelength of 21 cm. This **21-cm radiation** allows radio astronomers to map the distribution of neutral hydrogen (HI) throughout our galaxy.

The existence of the 21-cm radiation was predicted theoretically in the 1940s by H. C. van de Hulst, but it was not detected until 1951. We can understand how a theoretical prediction could be made by thinking of the structure of a hydrogen atom. A hydrogen atom consists of a proton and an electron, and physicists know that both of these particles must spin like tiny tops. They can never stop. Because these particles have an electrostatic charge, their rapid spin is the same as the circulation of an electric current through a coil of wire.

Pressure

One of the most fundamental parameters in science is **pressure,** a measure of force per unit area. Doctors measure blood pressure and astronomers measure gas pressure, but they are really measuring the same fundamental parameter.

Pressure is expressed in the units of force per unit area. When we inflate the tires on a car, we use the unit pounds per square inch. A typical pressure might be 34 lb/in.^{2}. It is important to note that this is not the total force pushing out on the inside of the tire but only the force exerted on a single square inch. When you stand, your weight exerts a force on the floor, and the pressure under your shoes is your weight divided by the surface area of your shoes' soles. A typical pressure might be only 4 lb/in.^{2}. If you step on someone's toe, that is the pressure you exert. Of course, if you were wearing ice skates, your weight would be spread over a much smaller area, the area of the bottom of the blade, and you might exert a pressure of 150 lb/in.^{2} or more. We must be careful not to step on someone's toe while we are wearing ice skates. The pressure would be dangerously high.

Astronomers are most commonly interested in the behavior of matter when it is a gas, and pressure in a gas arises when atoms or molecules collide. Consider, for example, how the gas molecules colliding with the inside of a balloon exert an outward force on the rubber and keep the balloon inflated. If the gas is hot, the atoms or molecules move rapidly, and the resulting pressure is higher than for a cooler gas. If the gas is dense, there will be many gas particles colliding with the inside of the balloon, and the pressure will be higher than for a lower-density gas. Thus, pressure depends on both the temperature and the density of the gas.

Notice that pressure and density are related, but they are not at all the same thing. Density is a measure of the amount of matter in a given volume, and pressure is a measure of the force that matter exerts on its surroundings. A very-low-density gas and a very-high-density gas might have the same pressure if they had different temperatures.

In daily life, we think of pressure when we inflate an automobile tire, but pressure is common in nearly all of the sciences. Astronomers must consider pressure in thinking about the gas inside stars and the thin gas between the stars.

It creates a magnetic field. Because the charge on the proton is positive and the charge on the electron is negative, the magnetic fields are reversed when the particles spin in the same direction.

The magnetic fields of the spinning proton and electron affect the binding energy that holds the electron to the proton in a hydrogen atom. We have all played with small magnets and noticed that they repel each other in one orientation and attract each other if we turn one around. In the same way, the small magnetic fields produced by the spinning proton and electron can repel or attract each other. In one orientation the electron is slightly less tightly bound, and in the other orientation it is slightly more tightly bound (▌ Figure 10-4). The energy difference is very small, but it makes a big difference in astronomy.

Now we can follow van de Hulst's logic and predict the existence of 21-cm radiation. Because an electron in the ground state of a hydrogen atom could spin in either of two directions—the same way as the proton or the opposite way—the ground state must really be two energy levels separated by the tiny amount of energy produced by the spinning particles. If an electron is spinning such that it is in the higher of the two states, it can spontaneously flip over and spin in the other direction. Then the atom must drop to the lower of the two energy levels, and the excess energy is radiated away as a photon. The energy difference is small, so the photon has a long wavelength—21 cm (▌ Figure 10-5).

The 21-cm radiation was predicted theoretically but it could not be detected in laboratory experiments. Of the two closely spaced energy levels, the upper one

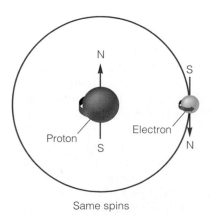

Same spins

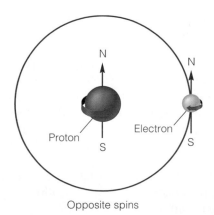

Opposite spins

FIGURE 10-4
Both the proton and electron in a neutral hydrogen atom spin and consequently have small magnetic fields. When they spin in the same direction, their magnetic fields are reversed, and when they spin in opposite directions, their magnetic fields are aligned. As explained in the text, this allows cold, neutral hydrogen in space to emit radio photons with a wavelength of 21 cm.

FIGURE 10-5

The lowest energy level of the hydrogen atom, the ground state, is actually two closely spaced energy levels that differ because the proton and electon spin. In this diagram, we would need a magnifying glass to distinguish the energy levels that make up the ground state. When an atom decays from the upper energy level to the lower energy level, it emits a photon with a wavelength of 21 cm.

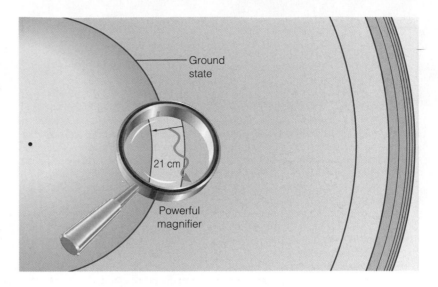

is metastable. Once an electron gets caught in the upper energy level, it will, on average, stay there for 11 million years before spontaneously dropping to the lower energy level and emitting a photon of 21-cm wavelength. The atoms in a gas in a laboratory jar will collide with each other millions of times a second, so none of those atoms can remain undisturbed with its electron in the upper energy level long enough to produce a photon of 21-cm radiation. Atoms in space, however, collide much less often, so a few do manage to emit 21-cm radiation. Thus, the existence of 21-cm radiation was confirmed observationally by radio astronomers who detected it coming from clouds of neutral hydrogen in space.

The 21-cm observations give us a way to map the cold, neutral hydrogen that fills much of our galaxy. Astronomers can locate individual clouds and measure their motion by observing the Doppler shifts in the 21-cm radiation (❚ Figure 10-6).

Of course, ionized hydrogen lacks an electron, so it can't emit 21-cm radiation. Further, hydrogen atoms locked in molecules are also unable to emit 21-cm wavelength photons. We must find other ways to study these parts of the interstellar medium.

MOLECULES IN SPACE

Radio telescopes can also detect radiation from various molecules in the interstellar medium. A molecule can store energy in a number of different ways. For example, it can rotate at different rates, or the atoms in a molecule can vibrate as if they were linked together by small springs. If a molecule suffers a collision or absorbs a photon, it can be excited to vibrate and rotate in some higher energy state. Quickly, however, it will return to a lower energy state and radiate the excess energy as a photon. Because these energy levels are closely spaced, the emitted photons typically have low energies, and we detect them in the radio or far-infrared part of the electromagnetic spectrum. Just as neutral hydrogen radiates at a specific wavelength of 21 cm, many natural molecules radiate at their own unique wavelengths.

Unfortunately for astronomers, molecules of hydrogen (H_2) are difficult to detect. In the far ultraviolet, molecular hydrogen can be detected by the photons it absorbs, but these observations must be made by specialized telescopes in space. Mapping the location of molecular hydrogen would be easier if the molecules emitted radio-wavelength photons, but a molecule containing two identical atoms does not radiate in the radio part of the spectrum. However, clouds of gas dense enough to form molecular hydrogen also form tiny amounts of other molecules, and many of them are good emitters of radio energy. Nearly 100 different molecules have been detected (❚ Table 10-1). Some are quite complex, and it is not clear how they form. Most astronomers believe that the atoms meet and bond to form molecules on the surfaces of dust grains. Some of these molecules have not yet been synthesized on Earth, but others are common, such as N_2O (nitrous oxide), also known as laughing gas. Ethyl alcohol, which some humans drink, has also been detected. Although this is a very rare molecule compared to molecular hydrogen, an interstellar cloud can contain ethyl alcohol in amounts equivalent to 10^{28} fifths of whisky (about 100 Earth masses).

One of the most important of the interstellar molecules may be carbon monoxide (CO). It is one of the poisonous gases that comes out of the tailpipes of cars, but it is also a very good emitter of radio energy at a wavelength of 2.6 mm. An interstellar cloud may contain only 1 CO molecule for every 10,000 molecules of hydrogen, but the CO can be detected while the molecular hydrogen cannot. Thus, radio astronomers can map the interstellar clouds by searching for the CO radio emission; where they detect CO, they can be certain that molecular hydrogen is common.

Another important constituent of the interstellar medium is OH. Because this combination of an oxygen

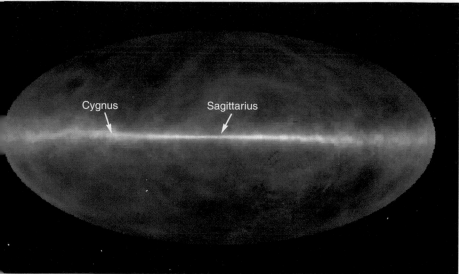

a

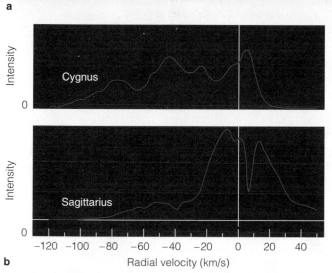

b

FIGURE 10-6

(a) This radio map of the entire sky shows the distribution of neutral hydrogen. The Milky Way is the bright band running from left to right. Notice how wispy and irregular the hydrogen is. *(J. Dickey, UMn, F. Lockman, NRAO, SkyView)* (b) These 21-cm radio observations contain many peaks, each produced by a separate cloud of neutral hydrogen with its own radial velocity. *(Adapted from observations by Burton)*

atom and a hydrogen atom is not electrically neutral, it is called a radical instead of a molecule. The OH radical forms in interstellar clouds and is a good emitter of radio energy.

These molecules are very fragile, and a high-energy photon such as an ultraviolet photon has enough energy to break the molecule into separate atoms. Thus, the molecules cannot exist outside the dense clouds. Only deep inside the densest clouds, where dust absorbs and scatters the short-wavelength, high-energy photons, can the molecules survive. The very fact that these molecules are detected tells us that some of these **molecular clouds** must be very dense.

The molecules, especially CO, are such good radiators of energy that they cool the clouds to low temperatures. Thermal energy is present in the cloud as motion among the atoms and molecules. When a molecule collides with an atom or another molecule, some of the energy of motion can be stored in the rotation and vibration of a molecule. When the molecule emits that energy as a radio or infrared photon, the energy escapes from the cloud. In this way, molecular radiation can cool the interior of the cloud and keep it very cold.

The largest of these cool, dense clouds are called **giant molecular clouds.** They are 15 to 60 pc across and may contain 100 to 1,000,000 solar masses. The internal temperature is a frigid 10 K. Although we detect these clouds by their carbon monoxide molecular emission, remember that the gas is mostly hydrogen.

Giant molecular clouds are the nests of star birth. Deep inside these great clouds, gravity can pull the matter inward and create new stars. That is a story we will discuss in detail in the next chapter. For now, we must study the dirty part of the interstellar medium— the dust. To do this, we must trade our radio telescope for an infrared telescope.

TABLE 10-1
Selected Molecules Detected in the Interstellar Medium

H_2	molecular hydrogen	H_2S	hydrogen sulfide
C_2	diatomic carbon	N_2O	nitrous oxide
CN	cyanogen	H_2CO	formaldehyde
CO	carbon monoxide	C_2H_2	acetylene
NO	nitric oxide	NH_3	ammonia
OH	hydroxyl	HCO_2H	formic acid
NaCl	common table salt	CH_4	methane
HCN	hydrogen cyanide	CH_3OH	methyl alcohol
H_2O	water	CH_3CH_2OH	ethyl alcohol

INFRARED RADIATION FROM DUST

The dust in the interstellar medium makes up only about 1 percent of the mass, and at a temperature of 100 K (−143°C) or less it is very cold. Nevertheless, it is easy to detect at infrared wavelengths. To see how such cold dust can radiate a lot of energy, consider a simple experiment with the dust in an imaginary giant molecular cloud.

For the sake of quick calculation, let us assume that our giant molecular cloud has a mass of about 10^5 solar masses. Only 1 percent of that mass is dust, so all of the dust in the cloud amounts to a mass of 10^3 solar masses. Imagine that we could collect all of the dust into a single sphere. It would be only about 10 times the diameter of the sun, and its surface area would be about 100 times that of the sun. But suppose we left the dust as separate specks, each 0.0005 mm in diameter, a typical size. Then the cloud would contain about 10^{43} dust specks, and the total surface area of the dust would be about 10^{29} times that of the sun. When matter is finely divided into dust, it has a very large surface area, and even though it is much colder than the sun, its vast surface area can radiate tremendous amounts of infrared radiation.

In 1983, the Infrared Astronomy Satellite mapped the sky at far-infrared wavelengths and found the galaxy filled with infrared radiation from dust. Most of this dust is confined to the region near the plane of our galaxy, and, as we would expect, the dust is thickest where the gas is thickest. Much of the dust is distributed in wispy clouds that became known as the **infrared cirrus** because of their overall resemblance to cirrus clouds in Earth's atmosphere. The infrared cirrus consists of dusty clouds of interstellar matter with temperatures of about 30 K (▌ Figure 10-7a). Studies of CO emission from molecular clouds show that at least some of the infrared cirrus is associated with molecular clouds within a few hundred parsecs of the sun. This suggests that the gas and dust in the molecular clouds is not uniform but rather very patchy.

As we mentioned before, the interstellar dust appears to be composed of carbon, silicates, and iron, elements that are underabundant in the gas. Some grains appear to contain water ice contaminated with ammonia, methane, and other compounds.

Infrared observations can help us understand the interstellar medium, but X-ray and ultraviolet observations complete the picture by allowing us to detect a part of the gas between the stars that is otherwise invisible.

X RAYS FROM THE INTERSTELLAR MEDIUM

The interstellar medium seems very cold, so we would hardly expect to detect X rays. These high-energy photons are commonly produced by high temperatures. Nevertheless, X-ray telescopes can help us understand the interstellar medium.

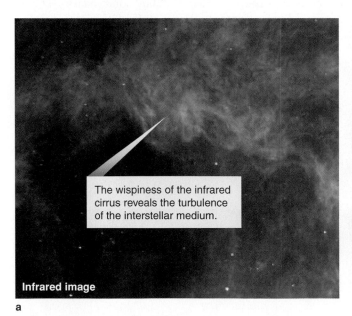

The wispiness of the infrared cirrus reveals the turbulence of the interstellar medium.

Infrared image

a

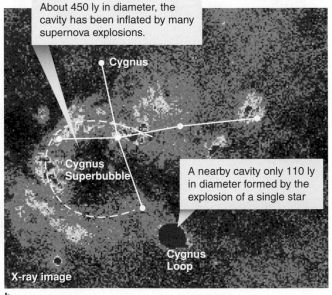

About 450 ly in diameter, the cavity has been inflated by many supernova explosions.

Cygnus

Cygnus Superbubble

A nearby cavity only 110 ly in diameter formed by the explosion of a single star

Cygnus Loop

X-ray image

b

FIGURE 10-7
Infrared and X-ray observations tell us about different components of the interstellar medium. (a) This far-infrared image shows the infrared cirrus, wispy clouds of very cold dusty gas spread all across the sky. *(NASA/IPAC, Courtesy Deborah Levine)* (b) This ROSAT X-ray image shows the Cygnus Superbubble, a cavity filled with gas from exploding stars. The gas is heated and emits X rays where the expanding bubble pushes into the surrounding gas. *(Courtesy Steve Snowden and Max-Planck Institute for Extraterrestrial Physics, Germany)*

X-ray telescopes above Earth's atmosphere have detected X rays from a part of the interstellar medium with a very high temperature. This gas has been called the **coronal gas** because it has temperatures of 10^6 K or higher, as does the sun's corona. Of course, the coronal gas in the interstellar medium is not related in any way to the actual corona of the sun.

We might expect the coronal gas to have a very high pressure because it is so hot, but it has a very low density. It contains only 0.0004 to 0.003 particles/cm³. That is, we would have to search through a few hundred to a few thousand cubic centimeters of coronal gas to find a single particle—an ionized atom or an electron. Because of its very low density, the gas pressure in the coronal gas can be roughly the same as in the HI clouds and the intercloud medium. Nevertheless, exceptions are known. The cloud of interstellar matter near the sun has a pressure that is more than 20 times less than the pressure of neighboring coronal gas. Such differences in pressure are not well understood.

The coronal gas appears to originate when a massive star explodes violently in what astronomers call a supernova. Although rare, such explosions blast large amounts of very hot gas outward in expanding shells. Gas flowing away from very hot, young stars must add to the coronal gas. In some cases, neighboring shells of coronal gas may expand into each other and merge to form larger volumes, but an earlier theory that coronal gas fills a large network of tunnels and shells throughout interstellar space seems to be contradicted by evidence. Probably about 20 percent of the space between the stars is filled with isolated pockets of coronal gas.

The Cygnus superbubble appears to be related to the coronal gas. Located in Cygnus, it is a very large shell of hot gas about 450 pc in diameter (Figure 10-7b). The energy needed to create such a shell is equivalent to hundreds of supernova explosions. The bubble may have developed as a large cluster of stars was born and grew old and the most massive stars died in supernova explosions. Several of these large bubbles are known.

ULTRAVIOLET OBSERVATIONS OF THE INTERSTELLAR MEDIUM

We can divide the ultraviolet spectrum into the near ultraviolet, with wavelengths only slightly shorter than visible light, and the far ultraviolet, with much shorter wavelengths. Only a few decades ago, astronomers believed that far-ultraviolet photons could not travel far through the interstellar medium because they would be absorbed so easily by neutral hydrogen atoms. These atoms of neutral hydrogen are good absorbers of far-ultraviolet photons because the photons have enough energy to ionize the atoms. In fact, even one atom of HI per cubic centimeter would make the interstellar medium opaque to far-ultraviolet photons, and we would be unable to see more than a light-year into space with a far-ultraviolet telescope.

When the Extreme Ultraviolet Explorer (EUVE) satellite was put into Earth orbit in 1992, it discovered that the interstellar medium was only partly cloudy. While some regions were filled with clouds of neutral hydrogen and so were opaque, other regions were filled with hot, ionized hydrogen that was transparent to far-ultraviolet photons. This confirms the description of the interstellar medium produced by X-ray observations.

The far-ultraviolet observations tell us that the sun is located just inside a large region of hot, ionized hydrogen, while only a few light-years from the sun lies a cool, neutral hydrogen cloud that is opaque to far-ultraviolet photons. This **local bubble or void** of gas in which the sun is located appears to be linked to other hot, transparent regions. Thus, ultraviolet observations, combined with X-ray observations, suggest that the apparently cold and empty regions between the stars are filled with a complex, evolving mixture of hot and cold gas.

REVIEW CRITICAL INQUIRY

If hydrogen is the most common molecule, why do astronomers depend on the CO molecule to map molecular clouds?

Although hydrogen is the most common atom in the universe and molecular hydrogen the most common molecule, a molecule of hydrogen does not radiate in the radio part of the spectrum. Consequently, radio astronomers cannot detect it. But the much less common CO (carbon monoxide) molecule is a very efficient radiator of radio energy, so radio astronomers use it as a tracer of molecular clouds. Wherever radio telescopes reveal a great cloud of CO, we can be confident that most of the gas is molecular hydrogen.

We can also be confident that the molecular clouds contain dust, because it is the dust that protects the molecules in the cloud from the ultraviolet radiation that would otherwise break down the molecules into atoms.

Dust doesn't radiate longer-wavelength radio energy, so we must study it in the infrared. Although the dust in a molecular cloud is very cold and makes up a small percentage of the total mass, it is a very good radiator of infrared radiation. How can a small amount of cold dust radiate vast amounts of infrared radiation?

From one end of the spectrum to the other, we have used every wavelength to study the interstellar medium. Now we will try to put our data in order and create a model of the gas and dust between the stars. That is the first step toward understanding how stars are born.

TABLE 10-2

Four Components of the Interstellar Medium

Component	Temperature (K)	Density (atoms/cm³)	Gas
HI clouds	50–150	1–1000	Neutral hydrogen; other atoms ionized
Intercloud medium (HII)	10^3–10^4	0.01	Partially ionized
Coronal gas	10^5–10^6	10^{-4}–10^{-3}	Highly ionized
Molecular clouds	20–50	10^3–10^5	Molecules

10-3 A MODEL OF THE INTERSTELLAR MEDIUM

When we look at bright nebulae like the Great Nebula in Orion, we see a part of the interstellar medium, but what we see is only a very special region that happens to be close enough to hot, bright stars to become excited. Most of the interstellar medium is invisible to our eyes, so we must gather the evidence—the observational facts (Window on Science 10-1)—and use them to develop a model of the interstellar medium.

We can divide the interstellar medium into four basic components (❚ Table 10-2). Our problem is to describe these components and explain how they interact and evolve. We will discover that their evolution is intimately connected to the process of star formation and death.

FOUR COMPONENTS OF THE INTERSTELLAR MEDIUM

The interstellar medium is not at all uniform. Rather, it is lumpy, and the lumps differ dramatically in temperature and density.

HI clouds are cool, with temperatures of 50 to 150 K and densities of ten to a few hundreds of atoms per cubic centimeter. These clouds are only a few parsecs in diameter and contain a few solar masses.

Between the cool HI clouds lies the warm intercloud medium of HII, with temperatures of a few thousand Kelvin and densities of 0.01 atoms/cm³. The intercloud medium is in approximate equilibrium with the HI clouds in that the hot, low-density gas has about the same pressure as the colder, denser gas in the HI clouds.

Molecular clouds are especially dense. Molecules cannot survive if they are exposed to ultraviolet photons in starlight, so they can form only in the densest clouds, where dust absorbs and scatters ultraviolet photons. The molecular clouds can be very large, with diameters up to 60 pc and masses up to a million solar masses, but they are also very cold. Strong evidence suggests that stars are born when giant molecular clouds

contract under the influence of their own gravity, a subject we will explore in detail in the next chapter.

Pushing through this stew of interstellar clouds are regions of coronal gas. Most of this very hot gas is probably produced in supernova explosions, although some may be gas flowing away from very hot stars. With temperatures up to a million degrees and densities as low as 10^{-4} atoms/cm³, the coronal gas occupies a large part of the interstellar medium.

Astronomers believe that the HI clouds make up about 25 percent of the interstellar mass, and the intercloud medium about 50 percent. The coronal gas contributes only 5 percent of the mass, although it seems to occupy about one-fifth of the volume. The giant molecular clouds amount to about 25 percent of the mass. If we add up these percentages, we get slightly more than 100 percent, which illustrates the uncertainty inherent in our model of the interstellar medium.

As the stars move through the interstellar medium, they meet no resistance, but they do have a dramatic effect on the gas and dust. The Pleiades, for example, is a relatively young star cluster that is moving rapidly through space. As it moves through the interstellar medium, it is leaving behind a trail like that left by a boat in water (❚ Figure 10-8). This wake has been detected in IRAS images and is apparently produced by the ultraviolet radiation from the stars in the cluster. None of the stars is hot enough to ionize the gas, but there are a number of relatively hot stars; the ultraviolet radiation from those stars heats, but does not quite ionize, the gas. The heated gas then expands and forms a wake showing the path of the cluster through the interstellar medium.

The Pleiades clearly illustrates the close relationship between the stars and the interstellar medium. In fact, we can now outline a cycle that links the stars to the gas and dust between them.

THE INTERSTELLAR CYCLE

The story of the interstellar medium is closely linked to star formation. We will see in the next chapter that stars form in giant molecular clouds. As soon as a group

Understanding Science: Separating Facts from Theories

The fundamental work of science is testing theories by comparing them with facts. As we think about science, we need to distinguish clearly between facts and theories. The facts are the evidence against which we test the theories.

Scientific facts are those observations or experimental results of which we are confident. An astronomer makes observations of stars, and a botanist collects samples of related plants. A fact could be a precise measurement, such as the mass of a star expressed as a specific number, or it could be a simple observation, such as that a certain butterfly no longer visits a certain mountain valley. In each case, the scientist is gathering facts.

A theory, however, is a conjecture as to how nature works. If we are uncertain of the theory, we might call it a hypothesis. In any case, these conjectures are not facts; they are attempts to explain how nature works. In a sense, a theory or hypothesis is a story that scientists have made up to explain how nature works in some specific case. These stories can be wonderfully detailed and ingenious, but without evidence they are nothing more than hunches.

When one of these stories is tested against the facts and confirmed, we have more confidence that the story is more or less right. The more a story is tested successfully, the more confidence we have in it. The facts represent reality, and every theory or hypothesis must be repeatedly tested against reality.

We can't test one theory against another theory. Theories are not evidence; they are conjectures. If we were allowed to test one theory against another theory, we might fall into the trap of circular reasoning. "Elves make the flowers bloom. I know that elves exist because the flowers bloom." That is using two theories to confirm each other, and it leads us to nonsense. Only facts can be evidence.

Nor can we use a theory to deduce facts. We can use a theory to make predictions, which we can then test against facts, but the predictions themselves can never be certainties, so they can't be facts. The only way to arrive at facts is to consult nature and make direct measurements or observations.

As we study different problems in astronomy, we must carefully distinguish between facts and theories. Facts are the basic building blocks of all science, and they can come only from the careful study of nature. As Galileo would say, we must read the book of nature.

---

of stars forms, the hottest stars begin ionizing the gas to produce emission nebulae. The pressure of the starlight and the gas flowing away from the hot, young stars pushes the interstellar cloud outward and may disrupt the cloud entirely.

The composition of the interstellar dust suggests that it is formed mostly in the atmospheres of cool stars. There the temperatures are low enough for some atoms to condense into specks of solid matter, much as soot can condense in a candle flame. The pressure of the starlight can push these dust specks out of the star and thus replenish the interstellar medium. Other stars that eject mass into space, such as supernovae, probably also add to the supply of interstellar dust.

The most massive stars die quickly in supernova explosions, and those tremendously violent events blast high-temperature gas outward that further disrupts the interstellar medium (Figure 10-9). Much of the motion we see in interstellar clouds and their twisted shapes are probably produced by these supernova explosions and the hot coronal gas they produce. It seems that our galaxy produces about two supernova explosions per century (although most are not visible from Earth because of obscuration by dust) and that these explosions keep the interstellar medium stirred and create the vast regions filled with coronal gas. In fact, the sun lies inside such a region of high-temperature, low-density gas called the local bubble or void. With a typical diameter

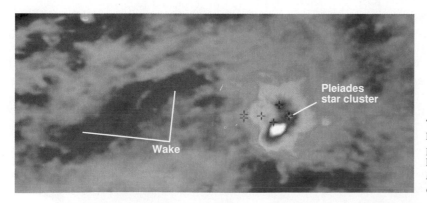

FIGURE 10-8

This infrared image shows the brighter stars of the Pleiades as small crosses. (Compare with photo on page 199.) The motion of the cluster from left to right has left a wake in the interstellar medium. *(Courtesy Richard E. White; image rendered by Duncan Chesley of American Image, Inc.)*

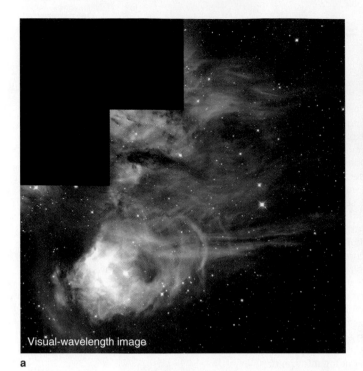

FIGURE 10-9

(a) The nebula N44C is part of a larger region where young, hot, massive stars have formed and where at least one supernova has blown a bubble filled with hot gas. *(The Hubble Heritage Team STScI/ AURA/NASA)* (b) This image of the entire N44 region combines a visual-wavelength image (white) with an X-ray image (red). X rays reveal regions of very high temperature gas that has been ejected from supernova explosions. *(R. Chris Smith, CTIO and Y.-H. Chu, UICU)*

of a few hundred parsecs, the local bubble may be a cavity in the interstellar medium inflated by a supernova explosion within the last million years or so.

While supernovae and hot stars keep the interstellar medium in motion, the natural gravitation of gas clouds and collisions between clouds may gradually build more massive clouds. In the most massive clouds, the dust protects the interior from ultraviolet photons, and molecules can form. These giant molecular clouds eventually give birth to new stars, and the cycle begins all over again.

The Trifid Nebula (❙ Figure 10-10) is a dramatic illustration of this cycle. Measuring over 12 pc in diameter, the nebula is illuminated by a number of hot, young stars that have apparently formed recently from the gas. Near the stars, the gas is ionized and glows as a pink-red HII region, but farther from the stars, where the ultraviolet radiation is weaker, the gas is not ionized. Nevertheless, dust in the nebula scatters blue light, so this un-ionized part of the nebula is visible as a blue reflection nebula. Dark lanes of obscuring dust cross the face of the nebula as if to remind us again of the importance of dust in the interstellar cycle.

REVIEW CRITICAL INQUIRY

How can the coronal gas make up only a few percent of the mass but 20 percent of the volume of the interstellar medium?

The solution to this puzzle is the extremely low density of the coronal gas, roughly one particle for every thousand cubic centimeters. The coronal gas is the least dense part of the interstellar medium. It is a much better vacuum than any laboratory vacuum on Earth. It doesn't take much mass in coronal gas to occupy a large volume. In contrast, molecular clouds are the densest part, and they contain roughly a quarter of the mass, although they occupy only a small part of the volume.

The density and temperature of the different components of the interstellar medium determine how we can observe them. How do we observe the coronal gas and the molecular clouds?

Our study of the interstellar medium is incomplete for two reasons. First, astronomers don't yet understand all of its components or how those components interact. Second, we can't fully understand the interstellar medium until we understand how stars are born and how they die. We begin that story in the next chapter.

FIGURE 10-10

The Trifid Nebula embodies an important part of the interstellar cycle. A dense cloud of gas has given birth to stars, and the hottest are ionizing the gas to produce the pink emission nebula, while dust scatters light to produce the blue reflection nebula. As the nebula is disrupted by the newborn stars, the gas and dust merge with the interstellar medium. *(AURA/NOAO/NSF and Nigel Sharp)*

SUMMARY

The interstellar medium, the gas and dust between the stars, is confined near the plane of our Milky Way Galaxy. We see clear evidence of an interstellar medium when we look at an emission nebula, a cloud of gas near one or more hot stars whose ultraviolet radiation ionizes the hydrogen and makes the nebula glow like a giant neon sign. The red, blue, and violet Balmer lines blend together to produce the characteristic pink-red glow of ionized hydrogen.

A reflection nebula is produced by gas and dust illuminated by a star that is not hot enough to ionize the gas. Rather, the dust scatters the starlight to produce a reflection of the stellar absorption spectrum. Because shorter-wavelength photons scatter more than longer-wavelength photons, reflection nebulae look blue. The daytime sky looks blue for the same reason.

A dark nebula is a cloud of gas and dust that is visible because it blocks the light of distant stars. We see such nebulae as dark shapes.

Further evidence of an interstellar medium is the extinction, or dimming, of the light of distant stars, and interstellar reddening. Light from distant stars suffers scattering by dust particles in the interstellar medium, and blue light is scattered more than red light. This makes distant stars look redder than their spectral types suggest. The dependence of this extinction on wavelength tells us that the scattering dust particles are very small. The dust is made of carbon, silicates, iron, and ice.

Interstellar absorption lines in the spectra of distant stars are very narrow. The interstellar gas is cold and has a very low density, and this makes the interstellar lines much narrower than the spectral lines produced in stars. Multiple interstellar lines tell us that the light has passed through more than one interstellar cloud on its way to Earth. Radio observations at a wavelength of 21 cm also reveal multiple clouds of neutral hydrogen. These neutral clouds drift through a warmer but lower-density intercloud medium. Radio telescopes tuned to other wavelengths have detected nearly 100 different molecules in the interstellar medium, most of them found in giant molecular clouds.

The Infrared Astronomy Satellite has detected dust in the interstellar medium in the form of the infrared cirrus. X-ray and far-ultraviolet observations have detected very hot coronal gas produced by supernova explosions.

The four main components of the interstellar medium are the small neutral HI clouds, the warm intercloud medium, coronal gas, and molecular clouds. Stars are born in the dense molecular clouds, and the energy from hot stars and supernova explosions causes currents in the interstellar medium and creates the coronal gas.

NEW TERMS

interstellar medium

nebula

emission nebula

HII region

reflection nebula

dark nebula

forbidden line

metastable level

interstellar dust

interstellar extinction

interstellar reddening

interstellar absorption lines

HI clouds

intercloud medium

pressure

21-cm radiation

molecular cloud

giant molecular clouds

infrared cirrus

coronal gas

local bubble or void

REVIEW QUESTIONS

Ace◐Astronomy™ Assess your understanding of this chapter's topics with additional quizzing and animations at **http:// astronomy.brookscole.com/seeds8e**

1. What evidence do we have that the spaces between the stars are not totally empty?

2. What evidence do we have that the interstellar medium contains both gas and dust?

3. How do the spectra of HII regions differ from the spectra of reflection nebulae? Why?

4. Why are interstellar lines so narrow? Why do some spectral lines forbidden in spectra on Earth appear in spectra of interstellar clouds and nebulae? What does that tell us?

5. How is the blue color of a reflection nebula related to the blue color of the daytime sky?

6. Why do distant stars look redder than their spectral types suggest?

7. If starlight on its way to Earth passed through a cloud of interstellar gas that was hot instead of very cold, would you expect the interstellar absorption lines to be broader or narrower than usual? Why?

8. How can the HI clouds and the intercloud medium have similar pressures when their temperatures are so different?

9. What does the shape of the 21-cm radio emission line of neutral hydrogen tell us about the interstellar medium?

10. What produces the coronal gas?

DISCUSSION QUESTIONS

1. When we see distant streetlights through smog, they look dimmer and redder than they do normally. But when we see the same streetlights through fog or falling snow, they look dimmer but not redder. Use your knowledge of the interstellar medium to discuss the relative sizes of the particles in smog, fog, and snowstorms compared to the wavelength of light.

2. If you could see a few stars through a dark nebula, how would you expect their spectra and colors to differ from similar stars just in front of the dark nebula?

PROBLEMS

1. A small nebula has a diameter of 20 seconds of arc and a distance of 1000 pc from Earth. What is the diameter of the nebula in parsecs? in meters?

2. The dust in a molecular cloud has a temperature of about 50 K. At what wavelength does it emit the maximum energy? (*Hint:* Consider black body radiation, Chapter 7.)

3. Extinction dims starlight by about 1.9 magnitudes per 1000 pc. What fraction of photons survives a trip of 1000 pc? (*Hint:* Consider the definition of the magnitude scale in Chapter 2.)

4. If the total extinction through a dark nebula is 10 magnitudes, what fraction of photons makes it through the cloud? (*Hint:* See Problem 3.)

5. The density of air in a child's balloon 20 cm in diameter is roughly the same as the density of air at sea level, 10^{19} particles/cm³. To how large a diameter would you have to expand the balloon to make the gas inside the same density as the interstellar medium, about 1 particle/cm³? (*Hint:* The volume of a sphere is $\frac{4}{3}\pi R^3$.)

6. If a giant molecular cloud has a diameter of 30 pc and drifts relative to neighboring clouds at 20 km/s, how long will it take to travel its own diameter?

7. An HI cloud is 4 pc in diameter and has a density of 100 hydrogen atoms/cm³. What is its total mass in kilograms? (*Hints:* The volume of a sphere is $\frac{4}{3}\pi R^3$, and the mass of a hydrogen atom is 1.67×10^{-27} kg.)

8. Find the mass in kilograms of a giant molecular cloud that is 30 pc in diameter and has a density of 300 hydrogen molecules/cm³. (*Hint:* See Problem 7.)

9. At what wavelength does the coronal gas radiate most strongly? (*Hint:* Consider black body radiation, Chapter 7.)

CRITICAL INQUIRIES FOR THE WEB

1. What causes the varied colors in images of gaseous nebulae that grace textbooks and Web sites? Search the Web for a color image of a nebula in the Messier catalog. Describe the structure and colors you see in terms of the concepts discussed in this chapter. (Be careful to distinguish real color from false color when answering this question. See Window on Science 13-1 for more information about false-color images.)

2. An interesting highlight in the history of astronomy is the discovery of "Nebulium." Search the Internet for information on this once-mysterious source of nebular spectral lines. How can astronomers today explain these lines in terms of known elements?

EXPLORING *THESKY*

1. The following nebulae are all star-formation regions. What kind of nebulae are they? (*Hint:* To center on an object, use **Find** under the **Edit** menu. Choose **Messier Objects** and pick from the list.)

   M42, M20, M8, M17

2. Locate M8 in *TheSky,* zoom in, and identify other nebulae in the region. Study the photo of NGC6559. What kind of nebula is it?

3. Locate M42, M20, and M8. Zoom in and identify the spectral type of the central star(s). What common characteristic do these stars share? Why?

 Visit the Seeds *Foundations of Astronomy* companion Web site for critical thinking exercises, articles, and additional readings from InfoTrac College Edition, Brooks/Cole's online student library.

THE FORMATION OF STARS

Jim he allowed [the stars] was made, but I allowed they happened.
Jim said the moon could'a laid them; well, that looked kind of reasonable,
so I didn't say nothing against it, because I've seen a frog lay most as many,
so of course it could be done.

Mark Twain, The Adventures of Huckleberry Finn

N. Walborn and J. Maiz-Apellániz, STScI, R. Barbá, La Plata Observatory, and NASA

GUIDEPOST

Previous chapters have used the basic principles of physics as a way to deduce things about stars and the interstellar medium. All of the data we have amassed will now help us understand the life stories of the stars in this chapter and those that follow.

In this chapter, we use the laws of physics in a new way. We develop theories and models based on physics that help us understand how stars work. For instance, what stops a contracting star and gives it stability? We can understand this phenomenon because we understand some of the basic laws of physics.

Throughout this chapter and the chapters that follow, we search for evidence. What observational facts confirm or contradict our theories? That is the basis of all science, and it must be part of any critical analysis of what we know and how we know it.

Stars exist because of gravity. They form because gravity makes clouds of gas contract, and they generate nuclear energy because gravity squeezes their cores to unearthly densities and temperatures. In the end, stars die because they exhaust their fuel supply and can no longer withstand the force of their own gravity.

A star can remain stable only by maintaining great pressure in its interior. Gravity tries to make it contract. However, if the internal temperature is high enough, pressure pushes outward just enough to balance gravity. Thus, a star is a battlefield where pressure and gravity struggle for dominance.

Only by generating tremendous amounts of energy can a star maintain the gravity–pressure balance. The sun, for example, generates 6×10^{13} times more energy per second than all of the coal, oil, natural gas, and nuclear power plants on Earth. Like the sun, most stars generate their power by fusion reactions that consume hydrogen and produce helium. However, we will discover that not all stars fuse their hydrogen in the same way as does the sun.

In this chapter, we will see how gravity creates stars from the thin gas of space and how nuclear reactions inside stars generate energy. We will see how the flow of that energy outward toward the surface of the star balances gravity and makes the stars stable. In the next two chapters, we will follow the life story of stars from the beginning of their stable lives to their ultimate deaths.

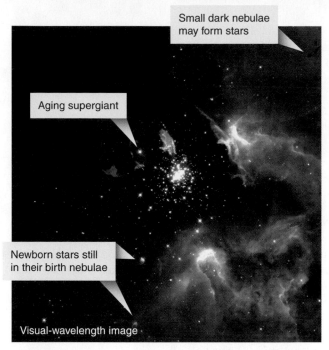

FIGURE 11-1

This young star cluster contains hot, luminous O stars, which are blowing away the nebula from which the cluster formed. Yet star formation is continuing in the nebula as it is compressed. Young in years, the supergiant at upper left has already begun its death throes as it has ejected a ring of gas around its equator and clouds of gas from its poles. The entire cycle of star birth and star death is illustrated in this true color image. *(W. Brandner, JPL/IPAC, E.K. Grebel, Univ. of Washington, Y. Chu, Univ. of Illinois, and NASA)*

11-1 MAKING STARS FROM THE INTERSTELLAR MEDIUM

Stars have been forming continuously since our galaxy took shape over 10 billion years ago. We know this for two reasons. First, the sun is only about 5 billion years old, a relative newcomer compared to the older stars in our galaxy. Second, we can see hot, blue stars such as Spica (α Virginis), a B1 main-sequence star. As we will see in the next chapter, such massive stars have very short lives. In fact, a star like Spica can last only 10 million years and thus must have formed recently.

The key to understanding star formation is the correlation between young stars and clouds of gas. Where we find the youngest groups of stars, we also find large clouds of gas illuminated by the hottest and brightest of the new stars (▌Figure 11-1). This leads us to suspect that stars form from such clouds, much as raindrops condense from the water vapor in a thundercloud. Indeed,

the giant molecular clouds discussed in the preceding chapter can give birth to entire clusters of new stars.

The central problem for our discussion of star formation is how these large, low-density, cold clouds of gas become comparatively small, high-density, hot stars. Gravity is the key.

STAR BIRTH IN GIANT MOLECULAR CLOUDS

The giant molecular clouds are the sites of active star formation, yet they are very unlike stars. With a typical diameter of 50 pc and a typical mass exceeding 10^5 solar masses, a giant molecular cloud is vastly larger than a star. Also, the gas in a giant molecular cloud is about 10^{20} times less dense than a star and has temperatures of only a few degrees Kelvin. These clouds can form stars because gravity can force some small regions of the clouds to contract to high density and high temperature.

Radio observations show that at least some giant molecular clouds develop dense cores that are only 0.1 pc in radius and that contain roughly 1 solar mass. A single giant molecular cloud may contain many of these dense cores and thus can give birth to star clusters containing hundreds of stars. However, both theory and observations suggest that many giant molecular clouds cannot begin the formation of dense cores spontaneously. At least four factors resist the contraction of a gas cloud, and gravity must overcome those four factors before star formation can begin.

First, thermal energy in the gas is present as motion among the atoms and molecules. Even at temperatures of 10 K, the average hydrogen molecule moves at about 0.35 km/s (almost 800 mph). This thermal motion would make the cloud drift apart if gravity were too weak to hold it together.

The interstellar magnetic field is the second factor that gravity must overcome to make the cloud contract. Neutral atoms and molecules are unaffected by a magnetic field, but ions, having an electric charge, cannot move freely through a magnetic field. Although the gas in a molecular cloud is mostly neutral, there are some ions, and thus a magnetic field can exert a force on the gas. The magnetic field present all through our galaxy averages only about 10^{-4} times as strong as that on Earth, but it can act like an internal spring to resist the contraction of the gas cloud.

The third factor is rotation. Everything in the universe rotates to some extent. As the gas cloud begins to contract, it spins more and more rapidly as it conserves angular momentum, just as ice skaters spin faster as they pull in their arms (Figure 5-7). This rotation can become so rapid that it resists further contraction of the cloud.

The turbulence in the interstellar medium is the fourth thing that could prevent a cloud from contracting. Our discussion of the interstellar medium in the previous chapter revealed that nebulae are often twisted and distorted by moving currents in the interstellar medium. This turbulence could make it difficult for a large molecular cloud to contract.

Given these four resistive factors, it seems surprising that any giant molecular clouds can contract, form dense cores, and eventually form stars. In some cases, the gas in giant molecular clouds can gradually recombine with free electrons and become less ionized. Neutral gas is free to "slip past" the magnetic field and contract. This gradual process has been observed to form dense cores inside isolated molecular clouds.

In contrast, both theory and observation suggest that many giant molecular clouds are triggered to form stars by a passing **shock wave,** the astronomical equivalent of a sonic boom (▌Figure 11-2). During such a triggering event, a few regions of the large cloud can be compressed to such high densities that the resistive factors can no longer oppose gravity, and star formation begins.

At least four different processes can produce shock waves that trigger star formation. Supernova explosions (Chapter 13) can produce powerful shock waves that rush through the interstellar medium. Also, the ignition of very hot stars can ionize nearby gas and drive it away to produce a shock wave where it pushes into

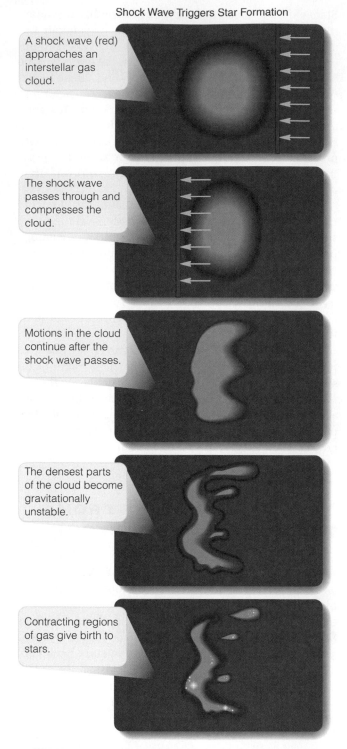

Shock Wave Triggers Star Formation

A shock wave (red) approaches an interstellar gas cloud.

The shock wave passes through and compresses the cloud.

Motions in the cloud continue after the shock wave passes.

The densest parts of the cloud become gravitationally unstable.

Contracting regions of gas give birth to stars.

FIGURE 11-2

In this summary of a computer model, an interstellar gas cloud is triggered into star formation by a passing shock wave. The events summarized here might span about 6 million years.

the colder, denser interstellar matter. A third trigger is the collision of molecular clouds. Because the clouds are large, they are likely to run into each other occasionally; and because they contain magnetic fields, they cannot pass through each other. A collision between such clouds can compress parts of the clouds and trigger star formation. The fourth trigger is the spiral pattern of our Milky Way Galaxy (see Figure 1-11). One theory suggests that the spiral arms are shock waves that travel around the galaxy like the moving hands of a clock (Chapter 15). As a cloud passes through a spiral arm, the cloud could be compressed, and star formation could begin.

Of course, a single giant molecular cloud containing a million solar masses does not contract to form a single humongous star. The cloud fragments and the densest parts form a number of dense cores. Exactly why a cloud fragments isn't fully understood, but the rotation, magnetic field, and turbulence probably play important roles. In any case, a giant cloud of gas typically contracts to form a number of newborn stars.

HEATING BY CONTRACTION

We have explained how clouds of interstellar gas can become dense enough to make stars, but we have not explained how the gas can become hot enough. The answer, once again, is gravity.

To consider the formation of a star, we now shift our attention to a single dense core of gas destined to become a star. Once a small cloud of gas begins to contract, gravity draws the atoms toward the center, and the atoms gather speed as they fall. In fact, astronomers refer to this early stage in the formation of a star as **free-fall contraction.** Whereas the atoms may have had low velocities to start with, by the time they have fallen most of the way to the center of the cloud, they are traveling at high velocities. We have defined thermal energy as the agitation of the particles in a gas, so this increase in velocity is a step toward heating the gas. But we can't say that the gas is hot simply because all of the atoms are moving rapidly. The air in the cabin of a jet airplane is traveling rapidly, but it isn't hot because all of the atoms are moving in generally the same direction with the plane. To convert the high velocity of the infalling atoms into thermal energy, the motion must become randomized, and that happens when the atoms begin to collide with one another as they fall into the central region of the cloud. The jumbled, random motion of the atoms represents thermal energy, and the temperature of the gas increases.

This is an important principle in astronomy. Whenever a cloud of gas contracts, gravitational energy is converted into thermal energy, and the gas grows hotter. Whenever a gas cloud expands, thermal energy is converted into gravitational energy, and the gas cools. This principle applies not only to clouds of interstellar gas, but also to contracting and expanding stars, as we will see in the following chapters.

Our study of gas clouds has shown us how nature can begin the contraction of dense cores in giant molecular clouds and how contraction can heat the gas. Now we will construct a detailed story of the transformation from gas cloud to star.

PROTOSTARS

To follow the story of star formation farther, we must concentrate on a single fragment of a collapsing cloud as it contracts, heats up, and begins to behave like a star. Although the term **protostar** is used rather loosely by astronomers, we will define it here to be a prestellar object that is hot enough to radiate infrared radiation but not hot enough to generate energy by nuclear fusion.

A protostar begins life as a contracting cloud of gas and dust falling inward under the influence of gravity (Figure 11-3). As the cloud contracts, it develops a higher-density region at the center and a low-density envelope. Mass continues to flow inward from the outer parts of the cloud. That is, the cloud contracts from the inside out, with the protostar taking shape deep inside an enveloping cloud of cold, dusty gas. These clouds have been called **cocoons** because they hide the forming protostar from our view as it takes shape.

The contraction of the cloud cannot lead directly to the formation of a small protostar, because the cloud has some net rotation. Everything rotates one way or another. The rotation of the cloud may be undetectable at first, but its rotation must grow more and more pronounced as it contracts and conserves its angular momentum. The rapidly spinning core of the cloud must

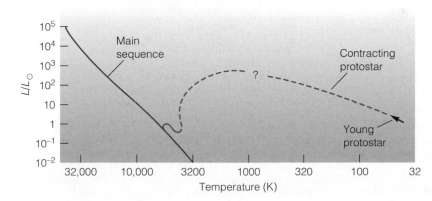

FIGURE 11-3

A contracting protostar begins as a very cold, very faint fragment of a gas cloud as shown in this H–R diagram extended to low temperatures. Contraction heats the protostar, but it remains hidden inside its dust cloud. The behavior of dust in the cloud makes the exact behavior of a protostar difficult to model as it approaches the main sequence.

Formation of a Protostellar Disk

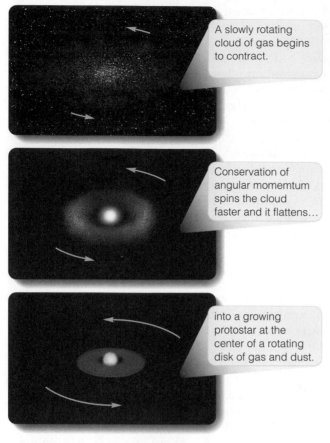

A slowly rotating cloud of gas begins to contract.

Conservation of angular momemtum spins the cloud faster and it flattens...

into a growing protostar at the center of a rotating disk of gas and dust.

FIGURE 11-4

The rotation of a contracting gas cloud forces it to flatten into a disk, and the protostar grows at the center.

flatten into a spinning disk like a blob of pizza dough spun into the air. Gas that has lost its angular momentum through collisions can sink directly to the center of the cloud, where the protostar begins to grow, surrounded by the disk. As more gas falls inward, it passes through the disk, giving up much of its angular momentum before it sinks into the protostar (❙ Figure 11-4).

Basic physics predicts that protostars should form at the centers of spinning disks of gas, which are called **protostellar disks.** That is important to us because astronomers believe that planets form within these disks. Our Earth formed in a disk around the protosun 4.6 billion years ago. As we will see in Chapter 19, the evidence is very strong that planetary systems form in these protostellar disks.

If we could see a protostar without its cocoon, it would be a very luminous and cool red star in the upper right part of the H–R diagram. The dust cocoon, however, absorbs almost all of the visible radiation and, growing warm, reradiates the energy as infrared radiation. Thus we should not expect to see protostars at visible wavelengths, but they should be emitting infrared radiation.

When the protostar becomes hot enough, it can drive away the surrounding gas and dust and become

visible. The location in the H–R diagram where protostars first emerge from their cocoons and become visible is called the **birth line.** Once a star crosses the birth line, it continues to contract toward the main sequence with a speed that depends on its mass. More massive stars have stronger gravity and contract more rapidly (❙ Figure 11-5). The sun took about 30 million years to reach the main sequence, but a 30-solar-mass star takes only 30,000 years. A 0.2-solar-mass star needs about 1 billion years to contract from a gas cloud to the main sequence.

The theory of star formation takes us into an unearthly realm filled with unfamiliar processes and objects. We might think it was nothing more than a fairy tale if observations did not support the theory.

EVIDENCE OF STAR FORMATION

In astronomy, evidence means observations. We must ask what observations confirm our theories of star formation. Unfortunately, a protostar is not easy to observe. The protostar stage is less than 0.1 percent of a star's lifetime, so although that is a long time in human terms, we cannot expect to find many stars in the protostar stage. Furthermore, protostars form deep inside clouds of dusty gas that absorb any light the protostar might emit. We must depend on observations at infrared wavelengths to search for hidden protostars.

Study ❙ "Observational Evidence of Star Formation" on pages 220 and 221 and notice three important points.

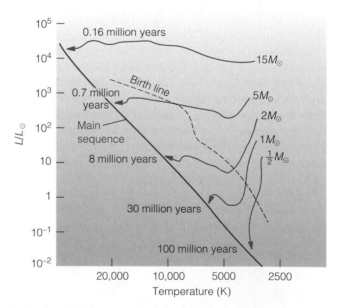

FIGURE 11-5

The more massive a protostar is, the faster it contracts. A 1-$M_\odot$ star requires 30 million years to reach the main sequence. (Recall that $M_\odot$ means "solar mass.") The dashed line is the birth line, where contracting protostars first become visible as they dissipate their surrounding clouds of gas and dust. Compare with Figure 11-3, which shows the evolution of a protostar of about 1-$M_\odot$ as a dashed line up to the birth line and as a solid line from the birth line to the main sequence. *(Illustration design by author)*

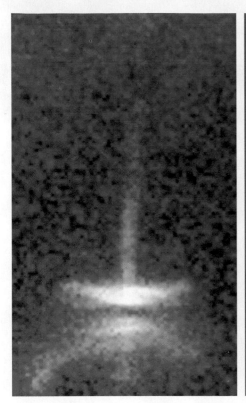

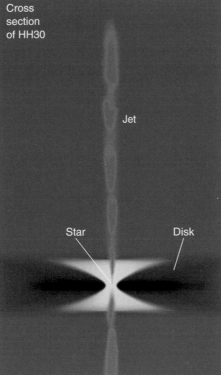

Cross section of HH30

Jet

Star   Disk

FIGURE 11-6

In the object HH30, a newly formed star lies at the center of a dense disk of dusty gas that is narrow near the star and thicker farther away. Although the star is hidden from us by the edge-on dusty disk, the star illuminates the inner surface of the disk. Interactions between the infalling material in the disk and the spinning star eject jets of gas along the axis of rotation. *(C. Burrows, STScI & ESA, WFPC 2 Investigation Definition Team, NASA)*

First, we can be sure that star formation is going on right now because we can find regions containing stars so young they must have formed recently. They lie between the birth line and the main sequence. Second, notice that regions of star formation are rich in gas and dust and contain infrared protostars and stars still contracting toward the main sequence. The third thing to notice is how observations give us insight into the process by which stars form. For example, the observation of jets coming from hidden protostars alerts us that protostars form surrounded by disks of gas and dust.

In some cases, these dark disks of gas and dust are clearly visible around newborn stars (▌ Figure 11-6). Such disks are exciting not only because they are evidence of star formation, but because they are the regions where planets form. We will discuss these disks later in this chapter and again when we discuss the origin of our solar system.

An **association** is a widely distributed star cluster that is not held together by its own gravity—its stars wander away as the association ages. These associations consist of young stars, because the stars wander apart so quickly. The constellation Orion, a known region of star formation, is filled with T Tauri stars in a **T association.** T Tauri stars are relatively low-mass objects ranging from

0.75 to 3 solar masses, but the stars in **O associations,** extended groups of O stars, are more massive. It is not clear why some gas clouds give birth to compact star clusters held together by their own gravity and others give birth to larger associations not bound together by gravity.

Not only can we locate evidence of star formation, but we can find evidence that star formation can stimulate more star formation. If a gas cloud produces massive stars, those massive stars can ionize the gas nearby and drive it away. Where that gas pushes into surrounding gas, it can compress the gas and trigger more star formation (▌ Figure 11-7). Another way massive stars can trigger star formation is through the explosion of a massive star in a supernova, a subject we will explore in detail in Chapter 13. Such an explosion can drive a shock wave through surrounding gas and trigger more star formation. Like a grass fire spreading through the interstellar medium, star formation can reignite itself if it creates massive stars (▌ Figure 11-8). Of course, lower-mass stars also form from such star formation, but they are not hot enough, nor do they explode as supernovae, so they alone cannot trigger further star formation.

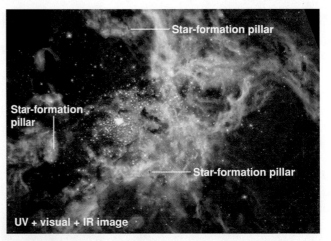

Star-formation pillar

Star-formation pillar

Star-formation pillar

UV + visual + IR image

FIGURE 11-7

The hot, massive stars in the star cluster 30 Doradus are pouring out intense ultraviolet radiation and powerful winds of hot gas that are pushing back and compressing the surrounding nebula. Slightly denser regions in the nebula protect the gas behind them to form pillars like those in the Eagle Nebula that point back at the star cluster. These pillars are each a few light-years long. The dense gas at the tips of the pillars is being compressed and may form more stars within the next few million years. *(N. Walborn and J. Maiz-Apellániz, STScI, R. Barbá, La Plata Observatory, and NASA)*

Observational Evidence of Star Formation

At visible, infrared, and radio wavelengths, astronomers find many objects in the sky that tell us how stars form.

Nebulae containing young stars usually contain **T Tauri** stars. These stars fluctuate irregularly in brightness, and many are bright in the infrared, suggesting they are surrounded by dust clouds and in some cases by dust disks. Doppler shifts show that gas is flowing away from many T Tauri stars. Located near the birth line in the H–R diagram, the T Tauri stars appear to be newborn stars just blowing away their dust cocoons. T Tauri stars appear to have ages ranging from 100,000 years to 100,000,000 years.

Spectra of T Tauri stars show signs of an active chromosphere as we might expect from young, rapidly rotating stars with powerful dynamos and strong magnetic fields.

The nebula around the star S Monocerotis is bright with hot stars. Such stars live short lives of only a few million years, so they must have formed recently. Such regions of young stars are common. The entire constellation of Orion is filled with young stars and clouds of gas and dust.

Visual-wavelength image

AATB

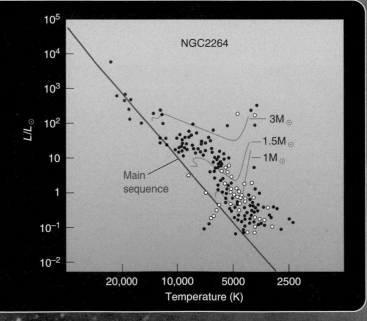

The star cluster NGC2264, imbedded in the nebula on this page, is only a few million years old. Lower-mass stars have not yet reached the main sequence, and the cluster contains many T Tauri stars (open circles), which are found above and to the right of the main sequence, near the birth line.

Bok globules, named after astronomer Bart Bok, are dense, dusty clouds seen silhouetted against glowing nebulae. Only a light-year or so in diameter, they contain from 10 to 1000 solar masses, and infrared observations show that most are very cold at their centers. Only a few have warm interiors and appear to be contracting to form protostars.

Visual

NASA

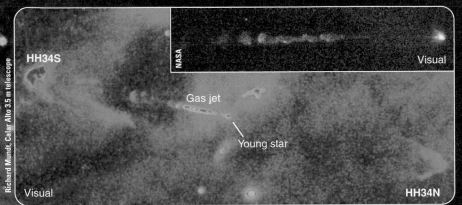

HH34S

Gas jet

Young star

NASA

Visual

HH34N

Visual

Richard Mundt, Calar Alto 3.5 m telescope

At the center of this image, a newborn star is emitting powerful jets to left and right. Where the jets strike the interstellar medium, they produce Herbig–Haro objects. The inset shows how irregular the jet is. Such jets can be over a light year long and contain gas traveling at 100 km/s or more.

Herbig–Haro objects, named after the two astronomers who first described them, are small nebulae that fluctuate in brightness. They appear to be produced by flickering jets from newborn stars exciting the interstellar medium.

Dusty disk

Jet

Herbig–Haro object

Jet

Herbig–Haro object

Matter flowing into a protostar swirls through a thick disk and, by a process believed to involve magnetic fields, ejects high-energy jets in opposite directions. Observation of these **bipolar flows** is evidence that protostars are surrounded by disks because only disks could focus the flows into jets.

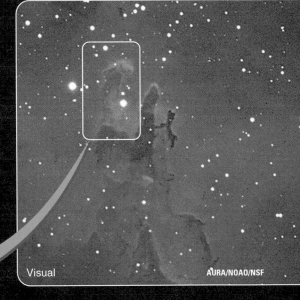

The Eagle Nebula, shown at the right in a ground-based photograph, is a region of known star formation. Turn the page a quarter turn to the right to see the eagle flying with a salmon in its talons.

Visual

AURA/NOAO/NSF

NASA

EGGs

EGGs

Visual

This Hubble Space Telescope image of part of the Eagle Nebula shows that radiation from bright stars out of the field to the upper right is evaporating the dust and driving away the gas to expose small globules of denser gas and dust. Infrared observations show that about 15 percent of these objects have formed protostars, and at least one seems to contain a newborn star (arrow). In part because these objects were first found in the Eagle Nebula, astronomers have enjoyed calling them EGGs — evaporating gaseous globules. They are evidently stars exposed in the act of forming.

The study of star formation is an exciting part of modern astronomy, and astronomers are using every wavelength to search for more information about the birth of stars. Although much remains to be discovered, the general outline is clear. Gravity creates stars by forcing interstellar clouds to contract and grow hotter.

REVIEW CRITICAL INQUIRY

What evidence do we have that stars are forming right now?

First, we should note that some extremely luminous stars can't live very long, so when we see such stars, such as the hot, blue stars in Orion, we know they must have formed in the last few million years. Star formation must be a continuous process, or we would not see any of these short-lived stars. But we have more direct evidence when we look at T Tauri stars, which lie above the main sequence and are often associated with gas and dust. In fact, we see entire associations of T Tauri stars as well as other associations of short-lived O stars. Many regions of gas and dust contain objects visible in the infrared that appear to be actual protostars. Furthermore, we can find cases where jets flowing outward from protostars create Herbig–Haro objects. In a few cases, such as the Eagle Nebula, the dense cores of gas and dust that are the forerunners of star formation are exposed by the evaporation of the nebula.

There seems to be no doubt that star formation is an ongoing process, and we can even discuss how it happens. For example, what evidence do we have that protostars are often surrounded by disks of gas and dust?

FIGURE 11-8

Generations of stars appear in the two star clusters in this Hubble Space Telescope image mosaic. The large cluster of yellow stars is about 50 million years old, but the bright white stars scattered through the image are massive young stars in a cluster only about 4 million years old. The younger cluster lies 200 ly beyond the older cluster, suggesting that star formation began 50 million years ago and created the nearer cluster. Supernova explosions in that cluster could have compressed gas clouds nearby and triggered the formation of the slightly more distant younger cluster. *(R. Gilmozzi, STScI/ESA, Shawn Ewald, JPL, and NASA)*

Gravity pulls the gas and dust together to make new stars, but this raises a natural question. If all this matter is falling inward to make a star, why does the contracting star stop? How do contracting protostars turn themselves into stable main-sequence stars? The answer is that the contracting stars begin to generate their own energy.

11-2 THE SOURCE OF STELLAR ENERGY

Stars are born when gravity pulls matter together, but what stops the contraction? Contracting stars somehow begin to make large amounts of energy, and somehow that energy stops the contraction and stabilizes the stars. In this section, we see how stars make energy; in the next section, we will see how contracting stars reach stability. This story will lead our imaginations into a region where our bodies can never go—the heart of a star.

A REVIEW OF THE PROTON–PROTON CHAIN

In Chapter 8 we visited the center of the sun and discovered that it manufactured energy through hydrogen fusion using a series of nuclear reactions called the proton–proton chain. The reaction must begin with the fusion of two protons. Protons are the nuclei of hydrogen atoms and have positive charges and repel each other with a force called the Coulomb barrier. Consequently, the proton–proton chain cannot occur if the gas temperature is lower than about 10 million Kelvin. High-velocity collisions are required to overcome the Coulomb barrier, and high velocity means high temperature.

We also discovered that the gas must be dense if the proton–proton chain is to produce significant energy. The fusion of two protons is unlikely, so many collisions are necessary to produce a few fusion reactions. Also, a single cycle through the proton–proton chain produces only a tiny amount of energy, so a vast number of fusion reactions are needed to supply the energy the sun needs. There will be a large number of fusion reactions only if the gas density is high.

Our study of the sun revealed that the proton–proton chain can produce energy only near the sun's center, where the temperature and density are high. In fact, only about 30 percent of the sun's mass is actually hot enough to fuse.

We might expect other stars to fuse hydrogen the same way the sun does, and we would be right for most stars. Some stars, however, can fuse hydrogen using a different recipe, and that makes a big difference to the structure of those stars.

THE CNO CYCLE

Main-sequence stars more massive than 1.1 solar mass fuse hydrogen into helium using the **CNO cycle,** a hydrogen fusion process that uses carbon, nitrogen, and oxygen as steppingstones.

Look carefully at ▌Figure 11-9 and notice the steps in the CNO cycle. The cycle begins with a carbon-12 nucleus absorbing a proton and becoming nitrogen-13, which decays to become carbon-13. The carbon-13 nucleus absorbs a second proton and becomes nitrogen-14, which absorbs a third proton and becomes oxygen-15. The oxygen-15 decays to become nitrogen-15, which absorbs a fourth proton, ejects a helium nucleus, and becomes carbon-12. Notice that carbon-12 begins the cycle and ends the cycle, and, along the way, four protons are combined to make a helium nucleus. This CNO cycle has the same outcome as the proton–proton chain, but it is different in an important way.

The CNO cycle begins with a carbon nucleus combining with a proton, a bare hydrogen nucleus. Because a carbon nucleus has a positive charge six times higher than a proton, the Coulomb barrier is high, and temperatures higher than 16 million Kelvin are required to make the CNO cycle work.

The difference between the proton–proton chain and the CNO cycle is their dependence on temperature. The CNO cycle is so critically sensitive to temperature that it produces almost no energy at all at temperatures below 16 million Kelvin. The proton–proton chain can make energy at these lower temperatures, so the sun makes nearly all of its energy using the proton–proton chain. At slightly higher temperatures, the proton–proton chain makes slightly more energy, but the CNO cycle produces a flood of energy. Thus stars with higher-temperature cores make nearly all of their energy using the CNO cycle, while stars with cooler cores make their energy using the proton–proton chain.

The CNO cycle and the proton–proton chain both fuse hydrogen to make helium, but their sensitivity to temperature is dramatic. We will see later in this chapter how this affects the internal structure of the stars.

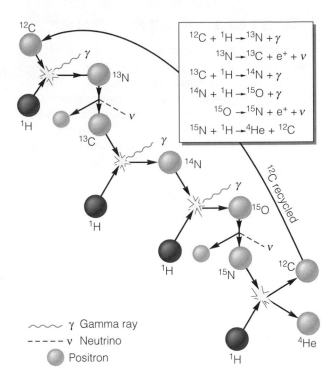

$$^{12}C + {}^{1}H \rightarrow {}^{13}N + \gamma$$
$$^{13}N \rightarrow {}^{13}C + e^{+} + \nu$$
$$^{13}C + {}^{1}H \rightarrow {}^{14}N + \gamma$$
$$^{14}N + {}^{1}H \rightarrow {}^{15}O + \gamma$$
$$^{15}O \rightarrow {}^{15}N + e^{+} + \nu$$
$$^{15}N + {}^{1}H \rightarrow {}^{4}He + {}^{12}C$$

〜〜 γ Gamma ray
- - - - - ν Neutrino
◯ Positron

FIGURE 11-9
The CNO cycle uses ^{12}C as a catalyst to combine four hydrogen nuclei (^{1}H) to make one helium nucleus (^{4}He) plus energy. The carbon nucleus reappears at the end of the process, ready to start the cycle over.

┤ **REVIEW** CRITICAL INQUIRY ├

Why does the CNO cycle require higher temperatures than the proton–proton cycle?
Nuclear fusion happens when atomic nuclei collide and fuse to form a new nucleus. Inside a star, the gas is ionized, so the nuclei have no orbiting electrons, and thus the nuclei have positive charges. Two objects with positive charges will repel each other in what we have called the Coulomb barrier. To start fusion, the nuclei must collide at high enough velocity to overcome that barrier. Heavier atomic nuclei such as carbon have higher positive charges, so the Coulomb barrier is higher. That requires that the gas be hotter so the collisions will be violent enough to overcome the barrier.

The CNO cycle and the proton–proton chain are strikingly similar in some ways but different in other ways. Describe these similarities and differences.

Because the law of gravity and the rules of nuclear fusion determine how stars work, we can understand what the inside of a star is like. That is, we can describe the internal structure of a star.

11-3 STELLAR STRUCTURE

Gravity makes stars contract, and when the density and temperature at the centers of the stars are high enough, nuclear fusion begins making energy. Why do the stars stop contracting? Somehow, the energy generated at the center of the star flows outward to the surface, and that flow of energy stops the contraction. To understand the stable structure of a star, we must first understand how energy flows through the star and how that energy flow affects the entire star.

ENERGY TRANSPORT

The sun and other stars generate nuclear energy in their deep interiors, and that energy must flow outward to their surfaces to replace the energy radiated into space as light and heat. If this outward flow of energy in the sun were suddenly shut off, the sun's surface would gradually cool and dim, and the sun would begin to shrink. Thus, stars can exist only as long as energy can move from their cores to their surfaces. In the material of which stars are made, energy can move by conduction, radiation, or convection.

Conduction is the most familiar form of heat flow. If you hold the bowl of a spoon in a candle flame, the handle of the spoon grows warmer. Heat, in the form of the motion of the particles in the metal, is conducted from particle to particle up the handle, until the particles under your fingers begin to move faster and you sense heat (▌ Figure 11-10). Thus, conduction requires close contact between the particles. Because matter in most stars is gaseous, conduction is an unimportant means of energy flow. Conduction is significant only in peculiar stars that have tremendous internal densities.

The transport of energy by radiation is another familiar experience. Put your hand beside a candle flame, and you can feel the heat. What you actually feel are infrared photons radiated by the flame (Figure 11-10). Because photons are packets of energy, your hand grows warm as it absorbs them. Radiation is the principal means of energy transport in the sun's interior. Photons are absorbed and reemitted in random directions over and over as they work their way outward. The process of absorption and reemission breaks the high-energy photons common near the sun's center into large numbers of low-energy photons in the cooler outer layers. The energy of a single high-energy photon at the center of the sun takes about 1 million years to reach the sun's surface, where it emerges as roughly 2500 photons of visible light.

The flow of energy by radiation depends on how difficult it is for the photons to move through the gas. If the gas is cool and dense, the photons are more likely to be absorbed or scattered, and thus the radiation does not penetrate the gas very well. We call such a gas opaque. In a hotter, lower-density gas, the photons can

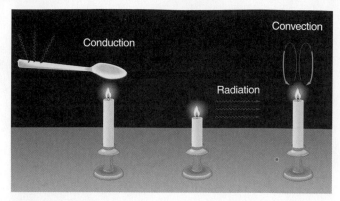

ACTIVE FIGURE 11-10

The three modes by which energy may be transported from the flame of a candle, as shown here, are the three modes of energy transport within a star.

Ace⊙Astronomy™ Go to AceAstronomy and click Active Figures to see "Conduction, Convection, and Radiation." Take control of this diagram.

penetrate more easily; such a gas is less opaque. The **opacity** of a gas—its resistance to the flow of radiation—depends strongly on its temperature.

If the opacity of a gas is high, radiation cannot flow through it easily. Energy moving outward from the center of the sun first moves through the hotter gas of the deep interior. Because the gas is so hot, it is transparent, and the energy moves as radiation. But the outer portions of the sun are cooler and therefore more opaque. Like water behind a dam, energy builds up, raising the temperature until the gas begins to churn. Hot gas, being less dense, rises, and cool gas, being denser, sinks. This is convection, the third way energy can move in a star.

Convection is a common experience; the wisp of smoke rising above a candle flame travels upward in a small convection current (Figure 11-10). If you hold your hand above the flame, you can feel the rising current of hot gas. In stars, energy may be carried upward by rising currents of hot gas hundreds or thousands of miles in diameter.

Convection is important in stars because it both carries energy and mixes the gas. Convection currents flowing through the layers of a star tend to homogenize the gas, giving it a uniform composition throughout the convective zone. As you might expect, this mixing affects the fuel supply of the nuclear reactions, just as the stirring of a campfire makes it burn more efficiently.

The convection in the sun stirs a zone just below the visible surface (▌ Figure 11-11). We see the tops of these rising currents of hot gas as granulation in the solar photosphere (Figure 8-2). That is, we can see the upper layers of the solar convection zone. Below that, the energy moves as radiation, and the gas does not get mixed. Deep in the interior of the sun, there is no convection. Hydrogen nuclei are fusing to form helium, and nothing removes the helium ashes or brings fresh

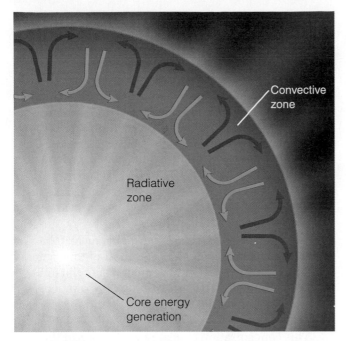

ACTIVE FIGURE 11-11

A cross section of the sun. Near the center, nuclear fusion reactions generate high temperatures. Energy flows outward through the radiation zone as photons. In the cooler, more opaque outer layers, the energy is carried outward by rising convection currents of hot gas (red) and sinking currents of cooler gas (blue).

Ace⊙Astronomy™ Go to AceAstronomy and click Active Figures to see "Inside the Sun." Watch energy flow outward from the core to the surface.

hydrogen down into the core. Thus, the sun is like a great pot of mashed potatoes burning at the bottom and stirred only slightly at the top. We will see later that most stars have interiors like the sun's.

WHAT SUPPORTS THE SUN?

From its surface to its interior, the sun is gaseous. On Earth, a puff of gas, vapor from a smokestack perhaps, dissipates rapidly, driven by the motion of the air and by the random motions of the atoms in the gas. However, the sun differs from a puff of gas in a very important characteristic—its mass. The sun is over 300,000 times more massive than Earth, and that mass produces a tremendous gravitational field that draws the sun's gases into a sphere.

With such a strong gravitational field, it might seem as if the gases of the sun should compress into a tiny ball, but a second force balances gravity and prevents the sun from shrinking. The gas of which the sun is made is quite hot; and, because it is hot, it has a high pressure. The pressure pushes outward; indeed, without the restraint of the sun's gravity, it would blow the sun apart. Thus, the sun is balanced between two forces—gravity trying to squeeze it tighter and gas pressure trying to make it expand.

To discuss the forces inside the sun, we can imagine that the sun's interior is divided into concentric shells like those in an onion (❚ Figure 11-12). We can then discuss the temperature, density, pressure, and so on in each shell. Keep in mind, however, that these helpful shells do not really exist. The sun and stars are not composed of separable layers. The layers are only a convenience in our discussion.

The gravity–pressure balance that supports the sun is a fundamental part of stellar structure known as the law of **hydrostatic equilibrium.** It says that, in a stable star like the sun, the weight of the material pressing downward on a layer must be balanced by the pressure of the gas in that layer. *Hydro* implies that we are discussing a fluid—the gases of the star. *Static* implies that the fluid is stable—neither expanding nor contracting.

The law of hydrostatic equilibrium can prove to us that the interior of the sun must be very hot. Near the sun's surface, there is little weight pressing down on the gas, so the pressure must be low, implying a low temperature. But as we go deeper into the sun, the weight becomes larger, so the pressure, and therefore the temperature, must also increase (❚ Figure 11-13). Near the sun's surface, the pressure is only a few times Earth's atmospheric pressure, and the temperature is only about 5800 K. At the center of the sun, the weight pressing down equals a pressure of about 3×10^{14} times Earth's atmospheric pressure. To support that weight, the gas temperature has to be almost 15 million K.

Of course, the interior of the sun is kept hot by the nuclear reactions occurring at the core, and the outward flow of energy keeps each layer in the sun hot enough to support the weight pressing down from above. In a sense, the sun is supported by the flow of energy from its center to its surface. Turn off that energy, and

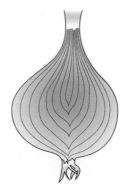

FIGURE 11-12

To discuss the structure of a star, it is helpful to divide its interior into concentric shells much like the layers in an onion. This model is, of course, only an aid to our imaginations. Stars are not really divided into separable layers.

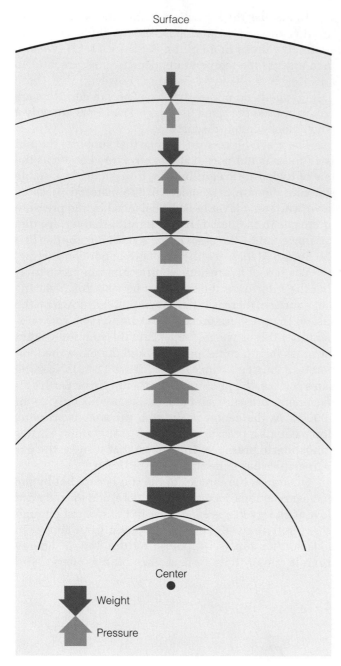

Surface

Weight

Pressure

Center

FIGURE 11-13

The law of hydrostatic equilibrium says the pressure in each layer must balance the weight on that layer. As a result, pressure and temperature must increase from the surface of a star to its center.

gravity would gradually force the sun to collapse into its center.

INSIDE STARS

In Chapter 9, we discovered that the stars on the main sequence are ordered according to mass, with the upper-main-sequence stars being most massive and the lower-main-sequence stars being least massive. Combining this with what we know about hydrogen fusion tells us that there are two kinds of main-sequence stars: upper-

main-sequence stars, which fuse hydrogen on the CNO cycle, and lower-main-sequence stars, which fuse hydrogen on the proton–proton chain. Viewed from the outside, these stars differ only in size, temperature, and luminosity, but inside they are quite different.

The upper-main-sequence stars are more massive and thus must have higher central temperatures to withstand their own gravity. These high central temperatures permit the star to fuse hydrogen on the CNO cycle, and that affects the internal structure of the star.

The CNO cycle is very temperature-sensitive. If the central temperature of the sun rose by 10 percent, energy production by the proton–proton chain would rise by about 46 percent, but energy production by the CNO cycle would shoot up 350 percent. This means that the more massive stars generate almost all of their energy in a tiny region at their very centers where the temperature is highest. A 10-solar-mass star, for instance, generates 50 percent of its energy in its central 2 percent of mass.

This concentration of energy production at the very center of the star causes a "traffic jam" as the energy tries to flow away from the center. Transport of energy by radiation can't drain away the energy fast enough, and the central core of the star churns in convection as hot gas rises upward and cooler gas sinks downward. Farther from the center, the traffic jam is less severe, and the energy can flow outward as radiation. Thus, massive stars have convective cores at their centers and radiative envelopes extending from their cores to their surfaces (❚ Figure 11-14).

Main-sequence stars less massive than about 1.1 solar masses cannot get hot enough to fuse much hydrogen on the CNO cycle. They generate nearly all of their energy by the proton–proton chain, which is not as sensitive to temperature, and thus the energy generation occurs in a larger region in the star's core. The sun, for example, generates 50 percent of its energy in a region that contains 11 percent of its mass. Because the energy generation is not concentrated at the very center of the star, no traffic jam develops, and the energy flows outward as radiation. Only near the surface, where the gas is cooler and therefore more opaque, does convection stir the gas. Consequently, the less massive stars on the main sequence have radiative cores and convective envelopes.

The lowest-mass stars have a slightly different kind of structure. For stars less than about 0.4 solar mass, the gas is relatively cool compared to the inside of more massive stars, and the radiation cannot flow outward easily. Thus, the entire bulk of these low-mass stars is stirred by convection.

We have answered the questions of how the energy in stars is made and how it gets from the core to the surface. We are now ready to answer the question: What makes stars stable?

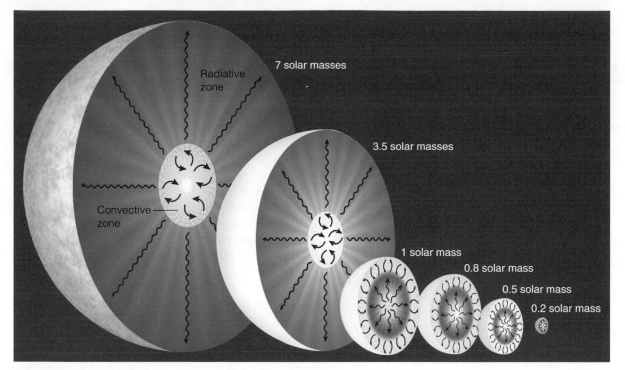

FIGURE 11-14

Inside stars. The more massive stars have small convective cores and radiative envelopes. Stars like the sun have radiative cores and convective envelopes. The lowest-mass stars are convective throughout. *(Illustration design by author)*

THE PRESSURE–TEMPERATURE THERMOSTAT

We can now understand what stops a newborn star from contracting into a tiny ball. Once nuclear fusion begins, energy flows outward, heats the layers of gas, and stops the contraction. This raises an interesting question. How does the star manage to make just the right amount of energy? The answer involves the relationship between pressure and temperature.

In a star, the nuclear reactions generate just enough energy to balance the inward pull of gravity. Consider what would happen if the reactions began to produce too much energy. The star balances gravity by generating energy, so the extra energy flowing out of the star would force it to expand. The expansion would lower the central temperature and density and slow the nuclear reactions until the star regained stability. Thus the star has a built-in regulator that keeps the nuclear reactions from occurring too rapidly.

The same thermostat keeps the reactions from dying down. Suppose the nuclear reactions began making too little energy. Then the star would contract slightly, increasing the central temperature and density and increasing the nuclear energy generation.

The stability of a star depends on this relation between pressure and temperature. If an increase or decrease in temperature produces a corresponding change in pressure, the thermostat functions correctly, and the star is stable. We will see in the next chapter how the thermostat accounts for the mass–luminosity relation. In Chapter 13, we will see what happens to a star when the thermostat breaks down completely and the nuclear fires burn unregulated.

──────── **REVIEW** CRITICAL INQUIRY ────────

What would happen if the sun stopped generating energy?
Stars are supported by the outward flow of energy generated by nuclear fusion in their interiors. That energy keeps each layer of the star just hot enough for the gas pressure to support the weight of the layers above. Each layer in the star must be in hydrostatic equilibrium; that is, the inward weight must be balanced by outward pressure. If the sun stopped making energy in its interior, nothing would happen at first, but over many thousands of years the loss of energy from its surface would reduce the sun's ability to withstand its own gravity, and it would begin to contract. We wouldn't notice much for 100,000 years or so, but eventually the sun would lose its battle with gravity.

Stars are elegant in their simplicity. Nothing more than a cloud of gas held together by gravity and warmed by nuclear fusion, a star can achieve stability by balancing its weight through the generation of nuclear energy. But how does the star manage to make exactly the right amount of energy to support its weight?

We have traced the birth of stars from the first instability in the interstellar medium to the final equilibrium of nuclear fusion and gravity. We have constructed a complete story of star formation, but we should demand evidence to support our theories. One of the best places to search for evidence of star formation is in the Great Nebula in Orion. It is just part of a great storm of star birth sweeping through Orion.

11-4 THE ORION NEBULA

On a clear winter night, you can see with your naked eye the Great Nebula of Orion as a fuzzy wisp in Orion's sword. With binoculars or a small telescope it is striking, and through a large telescope it is breathtaking. At the center lie four brilliant blue-white stars known as the Trapezium, the brightest of a cluster of a few hundred stars. Surrounding the stars are the glowing filaments of a nebula more than 8 pc across. Like a great thundercloud illuminated from within, the churning currents of gas and dust suggest immense power. The significance of the Orion Nebula lies hidden, figuratively and literally, beyond the visible nebula. The region is ripe with star formation.

EVIDENCE OF YOUNG STARS

We should not be surprised to find star formation in Orion. The constellation is a brilliant landmark in the winter sky because it is marked by hot, blue stars. These stars are bright in our sky, not because they are nearby, but because they are tremendously luminous. These O and B stars cannot live more than a few million years, so we can conclude that they must have been born recently. Furthermore, the constellation contains large numbers of T Tauri stars, which are known to be young. Orion is rich with young stars.

The history of star formation in the constellation of Orion is written in its stars. The stars at Orion's west shoulder are about 12 million years old, while the stars of Orion's belt are about 8 million years old. The stars of the Trapezium at the center of the Great Nebula are no older than 2 million years. Apparently, star formation began near the west shoulder, and the massive stars that formed there triggered the formation of the stars we see in Orion's belt. That star formation may have triggered the formation of the stars we see in the Great Nebula. Like a grass fire, star formation has swept across Orion from northwest to southeast.

Study ▌ "Star Formation in the Orion Nebula" on pages 230 and 231 and notice four points. First, the nebula we see is only a small part of a vast, dusty molecular cloud. We see the nebula because the stars born within it have ionized the gas and driven it outward, breaking out of the molecular cloud. Also notice that a single very hot star is almost entirely responsible for ionizing the gas. The third point to notice is that infrared observations reveal clear evidence of active star formation deeper in the molecular cloud just to the northwest of the Trapezium. Finally, notice that many stars visible in the Orion Nebula are surrounded by disks of gas and dust. Such disks do not last long and are clear evidence that the stars are very young.

In the next million years, the familiar outline of the Great Nebula will change, and a new nebula may begin to form as the protostars in the molecular cloud ionize the gas, drive it away, and become visible. Centers of star formation may develop and then dissipate as massive stars are born and force the gas to expand. If enough massive stars are born, they can blow the entire molecular cloud apart and bring the successive generations of star formation to a final conclusion. The Great Nebula in Orion and its invisible molecular cloud are a beautiful and dramatic example of the continuing cycles of star formation.

REVIEW CRITICAL INQUIRY

What did Orion look like to the ancient Egyptians, to the first humans, and to the dinosaurs?

The Egyptian civilization had its beginning only a few thousand years ago, and that is not very long in terms of the history of Orion. The stars we see in the constellation are hot and young, but they are a few million years old, so the Egyptians saw the same constellation we see. (They called it Osiris.) Even the Orion Nebula hasn't changed very much in a few thousand years, and Egyptians may have admired it in the dark skies along the Nile.

Our oldest human ancestors lived about 3 million years ago, and that was about the time that the youngest stars in Orion were forming. Thus, they may have looked up and seen some of the stars we see, but some stars have formed since that time. Also, the Great Nebula is excited by the Trapezium stars, and they are not more than a few million years old, so our early ancestors probably didn't see the Great Nebula.

The dinosaurs saw something quite different. The last of the dinosaurs died about 65 million years ago, long before the birth of the brightest stars in the constellation we see. The dinosaurs, had they had the brains to appreciate the view, might have seen bright stars along the Milky Way, but they didn't see Orion. All of the stars in the sky are moving through space, and the sun is orbiting the center of our galaxy. Over many millions of years, the stars move appreciable distances across the sky. The night sky above the dinosaurs contained totally different star patterns.

The Orion Nebula is the product of a giant molecular cloud, but such a cloud can't continue spawning new stars forever. What processes limit star formation in a molecular cloud?

The ancient Aztecs of central Mexico told the story of how the stars, known as the Four Hundred Southerners, were scattered across the sky when they lost a cosmic battle with their brother, the great war god Huitzilopochtli. Modern astronomy tells a less colorful story of the birth of stars, but the modern story is supported by evidence and leads us to ask further questions. If stars are born, then how do they die? We will begin that story in the next chapter.

SUMMARY

The largest and densest clouds in the interstellar medium are the giant molecular clouds. Triggered by the compression of a passing shock wave, the densest regions of such a cloud may begin to contract under the pull of their own gravity and eventually fragment to form a cluster or association of stars.

A contracting protostar begins as a very large, cool object, but the acceleration of the atoms as they fall inward causes the temperature of the gas to rise. Rotation in the gas causes some of the material to settle into a rapidly rotating disk around the growing protostar. Although the protostar is hot and luminous, it is hidden inside the dusty remains of the cloud from which it formed and thus remains invisible until the cloud is driven away by the increasing luminosity of the protostar.

Protostars can be detected at infrared wavelengths because the longer-infrared-wavelength photons can escape from the dusty clouds. Bipolar flows are ejected along the axis of rotation of the protostar and its disk. Where these jets strike the surrounding gas, they can excite small nebulae called Herbig–Haro objects. T Tauri stars appear to be very young stars just shedding their cocoons of gas and dust. Very young star clusters contain T Tauri stars, and some are born in very large, loosely bound groups called T associations.

As a protostar's center grows hot enough to fuse hydrogen into helium, the star settles onto the main sequence to begin its long, stable, hydrogen-fusion life. The sun is typical of these stars. It generates energy in its interior through the proton–proton chain, and that energy flows outward from the core as radiation and, in the outermost layers, as convection. All lower-mass stars such as the sun generate energy by the proton–proton chain, but more massive stars have hotter interiors and use the CNO cycle.

Because of the temperature sensitivity of the CNO cycle, most of the energy generated by stars more massive than about 1.1 solar masses is produced very near the star's center. The resulting traffic jam as this energy flows outward causes the core to be convective, whereas the outer part of the star transports energy by radiation. In stars less massive than about 1.1 solar masses, the energy is generated on the proton–proton chain in a larger volume of the core. Thus, there is no traffic jam, and these stars have radiative cores and convective envelopes.

In all stable stars, the nuclear reactions are regulated by the pressure–temperature thermostat. Thus, stars cannot generate less energy or more energy than that needed to support them against their own gravity.

The Orion Nebula is an example of a star-forming region. The bright stars we now see are ionizing part of the gas and forcing it to expand, and this gas is pushing into cooler regions of the cloud. Radio and infrared observations show that the compressed regions of the gas cloud are giving birth to a new generation of stars.

NEW TERMS

shock wave	bipolar flow
free-fall contraction	association
protostar	T association
cocoon	O association
protostellar disk	CNO (carbon–nitrogen–oxygen) cycle
birth line	opacity
T Tauri star	hydrostatic equilibrium
Bok globule	
Herbig–Haro object	

REVIEW QUESTIONS

Ace◯Astronomy™ Assess your understanding of this chapter's topics with additional quizzing and animations at **http://astronomy.brookscole.com/seeds8e**

1. What factors resist the contraction of a cloud of interstellar matter?

2. Explain four different ways a giant molecular cloud can be triggered to contract.

3. What evidence do we have that (a) star formation is a continuing process? (b) protostars really exist? (c) the Orion region is actively forming stars?

4. How does a contracting protostar convert gravitational energy into thermal energy?

5. How does the geometry of bipolar flows and Herbig–Haro objects support our hypothesis that protostars are surrounded by rotating disks?

6. How does the CNO cycle differ from the proton–proton chain? How is it similar?

7. How does the extreme temperature sensitivity of the CNO cycle affect the structure of stars?

8. How does energy get from the core of a star, where it is generated, to the surface, where it is radiated into space?

9. Describe the principle of hydrostatic equilibrium as it relates to the internal structure of a star.

10. How does the pressure–temperature thermostat control the nuclear reactions inside stars?

Star Formation in the Orion Nebula

The visible Orion Nebula shown below is a pocket of ionized gas on the near side of a vast, dusty molecular cloud that fills much of the southern part of the constellation Orion. The molecular cloud can be mapped by radio telescopes. To scale, the cloud would be many times larger than this page. As the stars of the Trapezium were born in the cloud, their radiation has ionized the gas and pushed it away. Where the expanding nebula pushes into the larger molecular cloud, it is compressing the gas and may be triggering the formation of the protostars that can be detected at infrared wavelengths within the molecular cloud.

Side view of Orion Nebula

Hot Trapezium stars

Protostars

To Earth

Expanding ionized hydrogen

Molecular cloud

Daniel Good

A few hundred stars lie within the nebula, but only the four brightest, those in the Trapezium, are easy to see with a small telescope. A fifth star, at the narrow end of the Trapezium, may be visible on nights of good seeing.

Trapezium

The near-infrared image below reveals about 50 low-mass, very cool stars that must have formed recently.

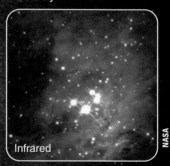

Infrared

NASA

The cluster of stars in the nebula is less than 2 million years old. This must mean the nebula is similarly young.

Visual-wavelength image

Roughly 1000 young stars with hot chromospheres appear in this X-ray image of the Orion Nebula.

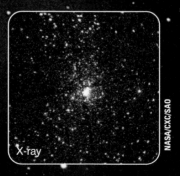

X-ray

NASA/CXC/SAO

Photons with enough energy to ionize H

Energy radiated by O6 star

Energy radiated by B1 star

Energy

0   100   200   300
Wavelength (nanometers)

Of all the stars in the Orion Nebula, only one is hot enough to ionize the gas. Only photons with wavelengths shorter than 91.2 nm can ionize hydrogen. The second-hottest stars in the nebula are B1 stars, and they emit little of this ionizing radiation. The hottest star, however, is an O6 star 30 times the mass of the sun. At a temperature of 40,000 K, it emits plenty of photons with wavelengths short enough to ionize hydrogen. Remove that one star, and the nebula would turn off its emission.

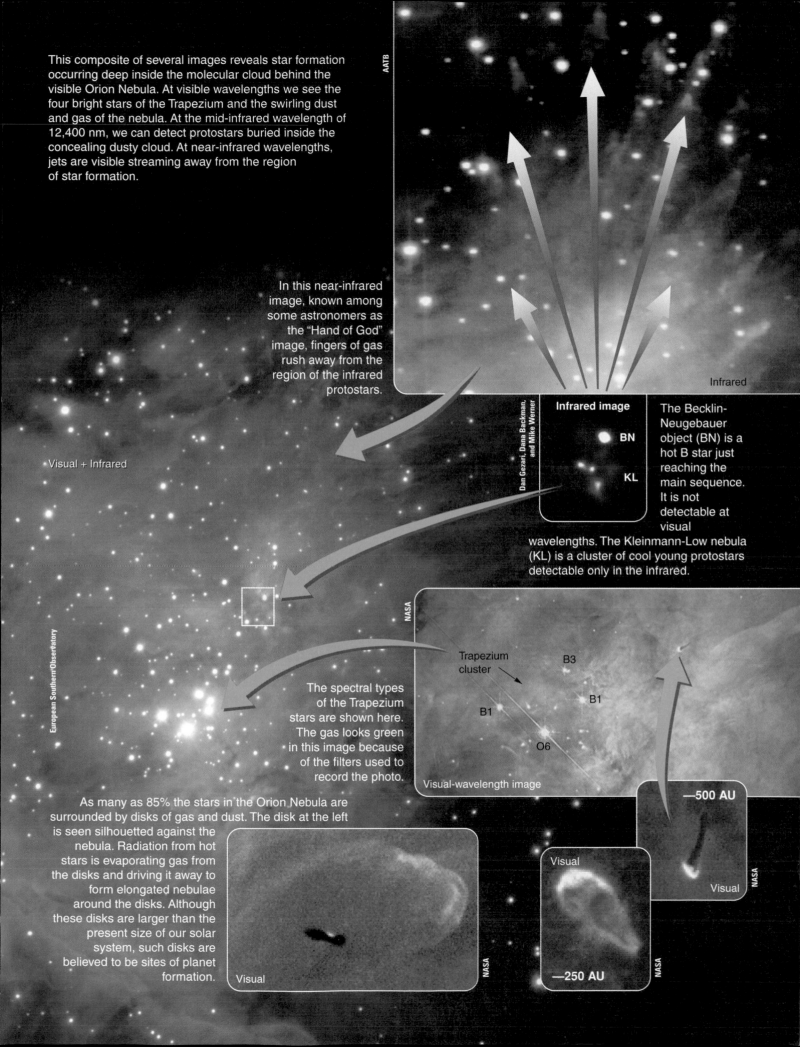

This composite of several images reveals star formation occurring deep inside the molecular cloud behind the visible Orion Nebula. At visible wavelengths we see the four bright stars of the Trapezium and the swirling dust and gas of the nebula. At the mid-infrared wavelength of 12,400 nm, we can detect protostars buried inside the concealing dusty cloud. At near-infrared wavelengths, jets are visible streaming away from the region of star formation.

AATB

In this near-infrared image, known among some astronomers as the "Hand of God" image, fingers of gas rush away from the region of the infrared protostars.

Infrared

Visual + Infrared

European Southern Observatory

Dan Gezari, Dana Backman, and Mike Werner

Infrared image

BN

KL

The Becklin-Neugebauer object (BN) is a hot B star just reaching the main sequence. It is not detectable at visual wavelengths. The Kleinmann-Low nebula (KL) is a cluster of cool young protostars detectable only in the infrared.

NASA

Trapezium cluster

B3

B1

B1

O6

The spectral types of the Trapezium stars are shown here. The gas looks green in this image because of the filters used to record the photo.

Visual-wavelength image

—500 AU

Visual

Visual

NASA

As many as 85% the stars in the Orion Nebula are surrounded by disks of gas and dust. The disk at the left is seen silhouetted against the nebula. Radiation from hot stars is evaporating gas from the disks and driving it away to form elongated nebulae around the disks. Although these disks are larger than the present size of our solar system, such disks are believed to be sites of planet formation.

Visual

—250 AU

NASA

NASA

DISCUSSION QUESTIONS

1. Ancient astronomers, philosophers, and poets assumed that the stars were eternal and unchanging. Is there any observation they could have made or any line of reasoning that could have led them to conclude that stars don't live forever?

2. How does hydrostatic equilibrium relate to hot-air ballooning?

PROBLEMS

1. The ring around the bright star in Figure 11-1 has a radius of about 7 seconds of arc. If the cluster is 20,000 ly from Earth, what is the radius of the ring in light-years?

2. If a giant molecular cloud is 50 pc in diameter and a shock wave can sweep through it in 2 million years, how fast is the shock wave going in kilometers per second?

3. If a giant molecular cloud has a mass of 10^{35} kg, and it converts 1 percent of its mass into stars during a single encounter with a shock wave, how many stars can it make? Assume the stars each contain 1 solar mass.

4. If a protostellar disk is 200 AU in radius, and the disk plus the forming star contain 2 solar masses, what is the orbital velocity at the outer edge of the disk in kilometers per second?

5. If a contracting protostar is five times the radius of the sun and has a temperature of only 2000 K, how luminous will it be? (*Hint:* See Chapter 9.)

6. The gas in a bipolar flow can travel as fast as 100 km/s. If the length of the jet is 1 ly, how long does it take for a blob of gas to travel from the protostar to the end of the jet?

7. If a T Tauri star is the same temperature as the sun but is ten times more luminous, what is its radius? (*Hint:* See Chapter 9.)

8. Circle all of the ^{1}H and ^{4}He nuclei in Figure 11-9 and explain how the CNO cycle can be summarized by $4\ ^1\text{H} \rightarrow\ ^4\text{He} + \text{energy}$.

9. How much energy is produced when the CNO cycle converts 1 kg of mass into energy? Is your answer different if the mass is fused by the proton–proton chain?

10. If the Orion Nebula is 8 pc in diameter and has a density of about 600 hydrogen atoms/cm^3, what is its total mass? (*Hint:* The volume of a sphere is $\frac{4}{3}\pi R^3$.)

11. The hottest star in the Orion Nebula has a surface temperature of 40,000 K. At what wavelength does it radiate the most energy? (*Hint:* See Chapter 7.)

CRITICAL INQUIRIES FOR THE WEB

1. Use the Web to supply additional details concerning the evolution of protostars and T Tauri stars.

2. Astronomers continue to study the Orion Nebula and star formation in the molecular cloud behind the nebula. What is the latest news from Orion?

3. What is BM Orionis? Where is it located, and what does it do? Why might it be interesting to observe with even a small telescope?

EXPLORING *THESKY*

1. The following nebulae are all star formation regions. What kind of nebulae are they? (*Hint:* To center on an object, use **Find** under **Edit.** Choose **Messier Objects** and pick from the list.)

    M42, M20, M8, M17

2. Locate M8 in *TheSky,* zoom in, and identify other nebulae in the region. Study the photo of NGC6559. What kind of nebula is it?

Visit the Seeds *Foundations of Astronomy* companion Web site for critical thinking exercises, articles, and additional readings from InfoTrac College Edition, Brooks/Cole's online student library.

STELLAR EVOLUTION

We should be unwise to trust scientific inference very far
when it becomes divorced from opportunity for observational test.

Sir Arthur Eddington, The Internal Constitution of the Stars

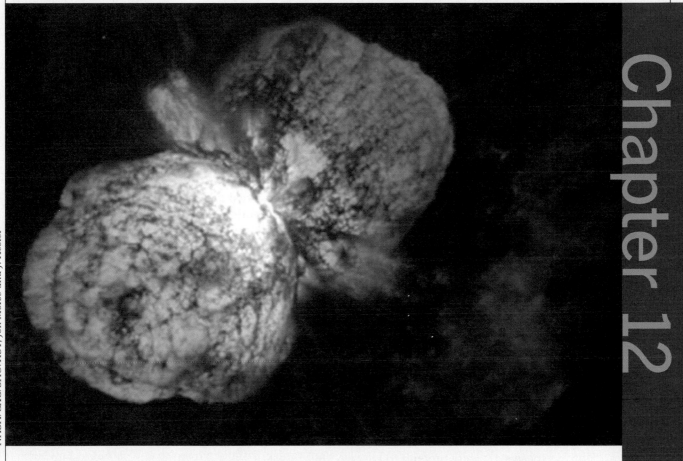

NASA/CXC/SAO/HST; Jon Morse and J. Hester

Chapter 12

GUIDEPOST

This chapter is the heart of any discussion of astronomy. Previous chapters showed how astronomers make observations with telescopes and how they analyze their observations to find the luminosity, diameter, and mass of stars. All of that aims at understanding what stars are.

This is the middle of three chapters that tell the story of stars. The preceding chapter told us how stars form, and the next chapter tells us how stars die. This chapter is the heart of the story—how stars live.

As always, we accept nothing at face value. We expect theory to be supported by evidence. We expect carefully constructed models to help us understand the structure inside stars. In short, we exercise our critical faculties and analyze the story of stellar evolution rather than merely accepting it.

After this chapter, we will know how stars work, and we will be ready to study the rest of the universe, from galaxies that contain billions of stars to the planets that form around individual stars.

The stars are going out. Although stars enjoy long, stable lives on the main sequence, they suffer the first pangs of stellar age as they gradually consume their hydrogen fuel. Once hydrogen is exhausted at their centers, they swell rapidly into giant stars 10 to 1000 times the size of the sun. For a short time, a few million to a billion years, they are majestic beacons visible across thousands of parsecs.

Although we say a star swells rapidly and spends a short time as a giant, these changes happen slowly compared with a human life. How can we be sure stars really do age? Stellar evolution and the formation of giant stars might be just an astronomer's daydream.

In this chapter, we see the full interplay of theory and evidence. We use the basic laws of physics, combined with our knowledge of the nature of stars, to create a theory to describe how main-sequence stars maintain their equilibrium and how such stars change when they exhaust their nuclear fuels. Theory alone, however, is never enough. At each step, we compare these theories with the evidence. We will discover that clusters of stars can make the slow evolution of stars visible even during our short lifetimes, and we will find that some evolving stars can become unstable and pulsate like beating hearts. Those pulsating stars tell us more about how stars evolve.

From beginning to end, this chapter tells the story of the evolution of stars from main sequence to giant. We will trace the passage of stars through this phase of their existence and see how they generate their energy and why they swell so large. We begin with the main-sequence stars.

12-1 MAIN-SEQUENCE STARS

If Shakespeare were alive today, he would probably have something sarcastic to say about modern astronomers. "How do these stargazers pretend to know the hearts of the eternal stars?" he might ask. In fact, one of the greatest triumphs of modern astronomy is the discovery that the stars are not eternal and that mere humans can indeed probe their centers. We can know the conditions at the center of a star because stars are fundamentally very simple objects. We begin by considering how we can use stellar models to understand the interior of main-sequence stars; later we will use these same models to understand what happens to stars when they grow old.

STELLAR MODELS

Astronomers describe the insides of stars by imagining that the stars are divided into concentric shells as shown in Figure 11-12. Of course, stars are not made like onions, but defining these shells mathematically helps astronomers discuss the conditions at different levels inside the star—what astronomers call the "structure" of the star.

The structure of a star is described by four simple laws of physics, two of which we have already discussed. In Chapter 11, we met the law of hydrostatic equilibrium, which says that the weight pressing down on a shell in a star must be balanced by the pressure pushing outward in that shell. That is, at every level in a star, the weight of the overlying matter must be supported by the pressure. Of course, that means that there can be no empty shells; an empty gap could exert no pressure and would be unable to support the weight of the material above. Chapter 11 also introduced us to the law of energy transport, which describes how energy flows from hot to cool regions by radiation, convection, or conduction.

To these two laws we add two basic laws of nature. The **conservation of mass law** says that the total mass of the star must equal the sum of the masses in its shells. Of course, no shell can be empty, and there is no such thing as negative mass. The **conservation of energy law** says that the amount of energy flowing out the top of a shell in the star must be equal to the amount of energy coming in at the bottom of the shell plus whatever energy is generated within the shell. Of course, that means that the energy leaving the surface of the star, its luminosity, must equal the sum of the energy generated in all of the shells inside the star. This is like saying that the total number of new cars driving out of a factory must equal the sum of cars manufactured on each assembly line. No car can vanish into nothing or appear from nothing. Energy in a star may not vanish without a trace or appear out of nowhere.

The laws of stellar structure, described in general terms in ▌ Table 12-1, can be written as mathematical equations. By solving those equations in a special way, astronomers can build a mathematical model of the inside of a star.

If we wanted to build a model of a star, we would have to divide the star into about 100 concentric shells and then write down the four equations of stellar structure for each shell. We would then have 400 equations that would have 400 unknowns, namely, the temperature, density, mass, and energy flow in each shell. Solving 400 equations simultaneously is not easy, and the first such solutions, done by hand before the invention of electronic computers, took months of work. Now a properly programmed computer can solve the equations in a few seconds and print a table of numbers that represent the conditions in each shell of the star. Such a table is a **stellar model**.

TABLE 12-1
The Four Laws of Stellar Structure

1. Hydrostatic equilibrium	The weight on each layer is balanced by the pressure in that layer.
2. Energy transport	Energy moves from hot to cool by radiation, convection, or conduction.
3. Conservation of mass	Total mass equals the sum of the shell masses. No gaps are allowed.
4. Conservation of energy	Total luminosity equals the sum of the energies generated in each shell.

The table shown in ▍Figure 12-1 is a model of the sun. The bottom line, for radius equal to 0.00, represents the center of the sun, and the top line, for radius equal to 1.00, represents the surface. The other lines tell us the temperature and density in each shell, the mass inside each shell, and the fraction of the sun's luminosity flowing outward through the shell. We can use the table to understand the sun. The bottom line tells us the temperature at the center of the sun is about 15 million Kelvin. At such a high temperature, the gas is highly transparent, and energy flows as radiation. Nearer the surface, the temperature is lower, the gas is more opaque, and the energy is carried by convection.

Notice that stellar models are quantitative; that is, properties have specific numerical values. Earlier in this book, we examined models that were qualitative—

our model of the sun's magnetic cycle, for instance. Both kinds of models are useful, but the quantitative model gives us deeper insights into how nature works, reflecting the power of mathematics as a precise way of thinking (Window on Science 12-1).

Stellar models also let us look into a star's past and future. In fact, we can use models as time machines to follow the evolution of stars over billions of years. To look into a star's future, for instance, we use a stellar model to determine how fast the star uses its fuel in each shell. As the fuel is consumed, the chemical composition of the gas changes and the amount of energy generated declines. By calculating the rate of these changes, we can predict what the star will look like at any point in the future.

Although this sounds simple, it is actually a highly challenging problem involving nuclear and atomic physics, thermodynamics, and sophisticated computational methods. Only since the 1950s have electronic computers made the rapid calculation of stellar models possible, and the advance of astronomy since then has been heavily influenced by the use of such models to study the structure and evolution of stars. Our summary of star formation in this chapter is based on thousands of stellar models. We will continue to rely on theoretical models as we study main-sequence stars in the next section and the deaths of stars in the next chapter.

WHY THERE IS A MAIN SEQUENCE

We can now understand why there is a main sequence. The clue is the law of hydrostatic equilibrium, and we can use models of stars to analyze this clue.

There is a main sequence because the centers of contracting protostars eventually grow hot enough to

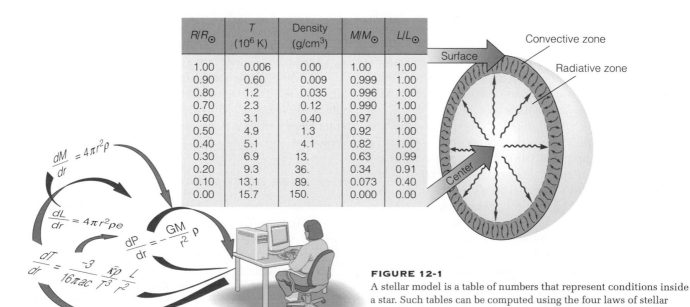

$R/R_{\odot}$	T (10^6 K)	Density (g/cm^3)	$M/M_{\odot}$	$L/L_{\odot}$
1.00	0.006	0.00	1.00	1.00
0.90	0.60	0.009	0.999	1.00
0.80	1.2	0.035	0.996	1.00
0.70	2.3	0.12	0.990	1.00
0.60	3.1	0.40	0.97	1.00
0.50	4.9	1.3	0.92	1.00
0.40	5.1	4.1	0.82	1.00
0.30	6.9	13.	0.63	0.99
0.20	9.3	36.	0.34	0.91
0.10	13.1	89.	0.073	0.40
0.00	15.7	150.	0.000	0.00

$$\frac{dM}{dr} = 4\pi r^2 \rho$$

$$\frac{dL}{dr} = 4\pi r^2 \rho e$$

$$\frac{dP}{dr} = -\frac{GM}{r^2}\rho$$

$$\frac{dT}{dr} = \frac{-3}{16\pi ac}\frac{\bar{\kappa}\rho}{T^3}\frac{L}{r^2}$$

FIGURE 12-1

A stellar model is a table of numbers that represent conditions inside a star. Such tables can be computed using the four laws of stellar structure, shown here in mathematical form. The table in this figure describes the sun. *(Illustration design by author)*

begin hydrogen fusion, and that stops their contraction. They reach equilibrium somewhere along a line in the H–R diagram that we call the main sequence. Hot stars are more luminous and cool stars are less luminous, as we would expect. There is, however, a mystery about the main sequence that we can now solve. Why does the luminosity of a star depend on its mass?

In Chapter 9, we used binary stars to find the masses of stars, and we discovered that the masses of main-sequence stars are ordered along the main sequence. The least massive stars are at the bottom, and the most massive stars are at the top. Further, we discovered a direct relationship between the mass of a star and its luminosity—the mass–luminosity relation. This is one of the most fundamental observations in astronomy, and stellar models can help us understand why there must be a mass–luminosity relation, and that tells us why there has to be a main sequence.

The keys to the mass–luminosity relation are the law of hydrostatic equilibrium, which says that pressure must balance weight, and the pressure–temperature thermostat, which regulates energy production. We have seen that a star's internal pressure stays high because the generation of thermonuclear energy keeps its interior hot. Because more massive stars have more weight pressing down on the inner layers, their interiors must have high pressures and thus must be hot. For example, the temperature at the center of a 15-solar-mass star is about 34,000,000 K, more than twice the central temperature of the sun.

Because massive stars have hotter cores, their nuclear reactions burn more fiercely. That is, their pressure–temperature thermostat is set higher. The nuclear fuel at the center of a 15-solar-mass star fuses over 3000 times more rapidly than the fuel at the center of the sun. The rapid reactions in the cores of massive stars produce more energy, but that energy cannot remain in the core. The energy must flow outward toward the cooler surface, and in doing so, the flowing energy heats each level in the star and enables it to support the weight pressing inward. When all that energy reaches the surface, it radiates into space and makes the star highly luminous. Thus there must be a mass–luminosity relation because each star must support its weight by generating nuclear energy.

The main sequence is elegant in its simplicity. It exists because stars balance their weight by fusing hydrogen in their cores. To understand the main-sequence stars even better, we can look at the ends of the main sequence.

THE ENDS OF THE MAIN SEQUENCE

Stellar structure gives us a way to study the extreme ends of the main sequence, the most massive and least massive stars. The first are rare, but the latter are common. Nevertheless, both are very difficult to study.

We can identify two reasons why there is an upper limit to the mass of stars. First, it appears that gas clouds contracting to form a star can fragment and form two or more stars. If a gas cloud contains more mass, it is more likely to fragment, so we don't find very massive stars because those gas clouds broke into smaller fragments and formed multiple-star systems.

But second, stellar models tell us that stars of roughly 100 solar masses are unstable at formation. To support the tremendous weight in such stars, the internal gas must be very hot, and that means it must emit floods of radiation that flow outward through the star and blow gas away from the star's surface in powerful stellar winds. This mass loss from very massive stars could reduce a 60-solar-mass star to less than 30 solar masses in a million years.

When we compare our theory with observation, we discover that it is difficult to find truly massive stars. Most of the O and B stars we see in the sky have masses of 10 to 25 solar masses. Our survey of stars at the end of Chapter 9 revealed that the stars at the upper end of the main sequence are very rare, so we must search to great distances to find just a few. Nevertheless, a few stars are known that are thought to be very massive, and their spectra contain blue-shifted emission lines. Kirchhoff's laws tell us that emission lines come from excited low-density gas, and the blue shift must be caused by a Doppler shift in gas coming toward us. These stars are losing mass.

❚ Figure 12-2 shows a famous star, Eta Carinae, that is believed to be a massive binary in which the stars contain 60 solar masses and 70 solar masses. They may have formed with about 100 solar masses each, but mass loss is reducing their mass. An eruption 150 years ago made Eta Carinae the second brightest star in the sky and ejected the two expanding lobes of dusty gas. Although the star has faded, it is still very active, and

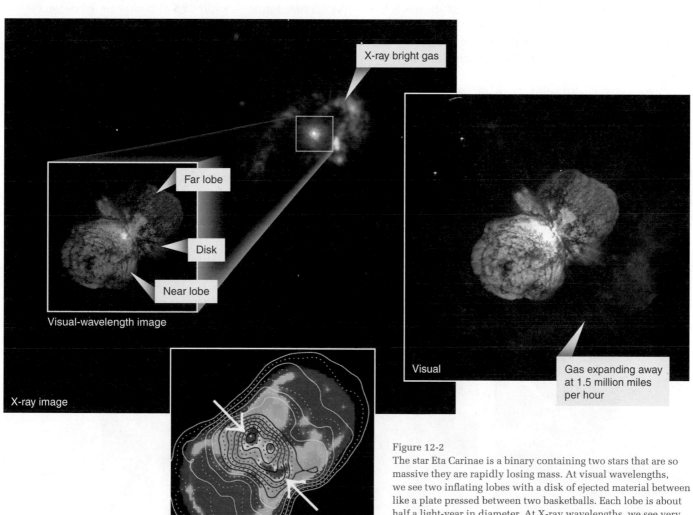

Figure 12-2
The star Eta Carinae is a binary containing two stars that are so massive they are rapidly losing mass. At visual wavelengths, we see two inflating lobes with a disk of ejected material between like a plate pressed between two basketballs. Each lobe is about half a light-year in diameter. At X-ray wavelengths, we see very hot gas excited by collision with high-speed gas ejected from the stars. An infrared image reveals a 15-solar-mass torus (doughnut) of gas and dust squeezing the outflowing gas into the two lobes. (NASA/CXC/SAO/HST; Jon Morse and J. Hester; IR image: ISO, Courtesy ESA)

more recent eruptions have ejected jets and an equatorial disk of gas and dust. Whatever their status, these massive stars are clearly unstable.

The lower end of the main sequence is difficult to study, not because the stars are rare but because they are dim. If a red dwarf from the lower end of the main sequence replaced the sun, it would shine only a few times brighter than the full moon. Such stars are difficult to find even when they are only a few light-years away.

Stellar models predict that stars less massive than 0.08 solar masses cannot get hot enough to ignite hydrogen fusion. These **brown dwarfs** should cool as they convert their gravitational energy into heat and radiate it away. A star of 0.08 solar masses fusing hydrogen has a surface temperature of about 2500 K, but brown dwarfs should have temperatures of 1000 K or so, giving them a color even ruddier than red dwarfs—thus the term "brown dwarf."

Brown dwarfs were difficult to find because they are so faint, but large surveys and infrared studies have turned up lots. Some are located in binary systems with normal stars (█ Figure 12-3a), but large numbers are free-floating objects without stellar companions (Figure 12-3b).

Brown dwarfs are clearly different from normal stars. Some brown dwarfs such as Gliese 229B in Figure 12-3 have methane bands in their spectra, and that means they must be quite cool. Methane molecules would be broken up at the temperatures of true stars. Color variations suggest that some brown dwarfs may be cool enough to have weather patterns.

The discovery of brown dwarfs has created a controversy over what we call them. Are they failed stars or are they planets? To discuss this we use a new unit of mass related to Jupiter, the giant planet in our solar system. Jupiter is one thousand times less massive than the sun, and astronomers refer to the mass of brown dwarfs in Jupiter masses. Thus a brown dwarf must be less than about 80 Jupiter masses. Generally, astronomers think of stars as bodies that generate energy by nuclear fusion, and a star less than 80 Jupiter masses can't heat its center to the 2.7 million Kelvin temperature needed to start hydrogen fusion. Then brown dwarfs aren't stars. But astronomers tend to think of a planet as a nonluminous body that orbits a star, so can a brown dwarf floating free in space be called a planet?

Brown dwarfs more massive than 13 Jupiter masses can fuse deuterium, a heavy isotope of hydrogen, and some astronomers draw the line between stars and planets at 13 Jupiter masses. However, deuterium fusion does not generate much energy and does not stop the star from cooling rapidly. Not everyone agrees that this is a good dividing line.

Another approach to the controversy is to note how these objects form. A star forms from the contraction of a gas cloud, but astronomers believe planets form from the accumulation of solid bits of matter in disk-shaped nebulae around stars (a subject we will discuss in Chapter 19). In that case, free-floating brown dwarfs can't be planets.

This is a controversy over the definition of words, but it illustrates our uncertainty concerning the lower end of the main sequence. Perhaps the study of brown dwarfs will eventually help us better understand what stars and planets are and how they form.

THE LIFE OF A MAIN-SEQUENCE STAR

A normal main-sequence star supports its weight by fusing hydrogen into helium, but its supply of hydrogen is limited. As it consumes its hydrogen, the chemical composition in its core changes, and the star evolves. Mathematical models of stars allow astronomers to follow that evolution.

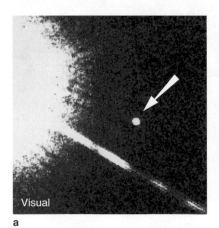
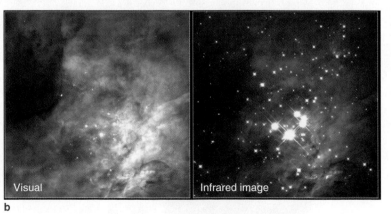

a   b

FIGURE 12-3
(a) Gliese 229B (arrow) is a brown dwarf that is a binary companion to the bright star at the left. The bright ray of light is caused by scattered light in the telescope. *(Kulkarni, Caltech, Golimowski, JHU, NASA)* (b) Too dim to see at visual wavelengths, about 50 free-floating brown dwarfs appear in an infrared image of the center of the Orion nebula. The bright stars of the Trapezium are located at the center. *(STScI and NASA)*

Hydrogen fusion combines four nuclei into one. Thus, as a main-sequence star consumes its hydrogen, the total number of nuclei in its interior decreases. Each newly made helium nucleus can exert the same pressure as a hydrogen nucleus, but because the gas has fewer nuclei, its total pressure is less. This unbalances the gravity–pressure stability, and gravity squeezes the core of the star more tightly. As the core contracts, its temperature and density increase, and the nuclear reactions burn faster, releasing more energy and making the star more luminous. This additional energy flowing outward through the envelope forces the outer layers to expand and cool, so the star becomes slightly larger, brighter, and cooler.

As a result of these gradual changes in main-sequence stars, the main sequence is not a sharp line across the H–R diagram but rather a band (blue in ▌Figure 12-4). A star begins its stable life fusing hydrogen and falls on the lower edge of this band, the **zero-age main sequence (ZAMS).** As it combines hydrogen nuclei to make helium nuclei, the star slowly changes. In the H–R diagram, the point that represents the star's luminosity and surface temperature moves upward and to the right, eventually reaching the upper edge of the main sequence, just as the star exhausts nearly all of the hydrogen in its center. Thus, we find main-sequence stars plotted throughout this band at various stages of their main-sequence lives.

These gradual changes in the sun will spell trouble for Earth. When the sun began its main-sequence life about 5 billion years ago, it was only about 75 percent of its present luminosity. This, by the way, makes it difficult to explain how Earth has remained at roughly its present temperature for at least 3 billion years. Some experts suggest that Earth's atmosphere has gradually changed and thus compensated for the increasing luminosity of the sun.

By the time the sun leaves the main sequence in a few billion years, it will have twice its present luminosity. This will raise the average temperature on Earth by at least 19°C (34°F). As this happens over the next billion years, the polar caps will melt, the oceans will evaporate, and much of the atmosphere will vanish into space. Clearly, the future of Earth as the home of life is limited by the evolution of the sun.

Once a star leaves the main sequence, it evolves rapidly and dies. The average star spends 90 percent of its life fusing hydrogen on the main sequence. This explains why 90 percent of all normal stars are main-sequence stars. We are most likely to see a star during that long, stable period when it is on the main sequence.

The number of years a star spends on the main sequence depends on its mass (▌Table 12-2). Massive stars use fuel rapidly and live short lives, but low-mass stars conserve their fuel and shine for billions of years. For example, a 25-solar-mass star will exhaust its hydrogen and die in only about 7 million years. This means, for one thing, that life is very unlikely to develop on planets orbiting massive stars. These stars do not live long enough for life to get started and evolve into complex creatures. We will discuss this problem in detail in Chapter 26.

Very-low-mass stars, the red dwarfs, use their fuel so slowly they should survive for 200 to 300 billion years. Because the universe seems to be only 10 to 20 billion years old, red dwarfs must still be in their infancy. None of them should have exhausted their hydrogen fuel yet.

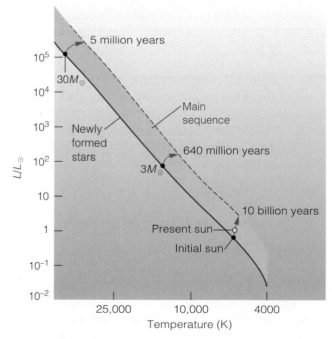

FIGURE 12-4

The main sequence is not a line but a band (shaded). Newly formed stars begin fusing hydrogen on the lower edge, the zero-age main sequence (ZAMS). As hydrogen changes their composition, the stars move across the band. Notice that lower-mass stars take longer to cross the band of the main sequence than stars of higher mass.

TABLE 12-2 Main-Sequence Stars

Spectral Type	Mass (sun = 1)	Luminosity (sun = 1)	Approximate Years on Main Sequence
O5	40	405,000	1×10^6
B0	15	13,000	11×10^6
A0	3.5	80	440×10^6
F0	1.7	6.4	3×10^9
G0	1.1	1.4	8×10^9
K0	0.8	0.46	17×10^9
M0	0.5	0.08	56×10^9

Nature makes more low-mass stars than massive stars, but this fact is not sufficient to explain the vast numbers of low-mass stars that fill the sky. An additional factor is the stellar lifetimes. Because low-mass stars live long lives, there are more of them in the sky than massive stars. Look at page 193 and notice how much more common the lower-main-sequence stars are than the massive O and B stars. The main-sequence K and M stars are so faint they are difficult to locate, but they are very common. The O and B stars are luminous and easy to locate, but because of their fleeting lives, there are never more than a few on the main sequence at any one time.

THE LIFE EXPECTANCIES OF STARS

To understand how nature makes stars and how stars evolve, we must be able to estimate how long they can survive. In general, massive stars live short lives and lower-mass stars live long lives, but we can make more accurate estimates by using simple stellar models. In fact, we can calculate the approximate life expectancy of a star from its mass.

Because main-sequence stars consume their fuel at a constant rate, we can estimate the amount of time a star spends on the main sequence—its life expectancy T—by dividing the amount of fuel by the rate of fuel consumption. This is a common calculation. If we drive a truck that carries 20 gallons of fuel and uses 5 gallons of fuel per hour, we know the truck can run for 4 hours.

The amount of fuel a star has is proportional to its mass, and the rate at which it burns its fuel is proportional to its luminosity; thus, we could make a first estimate of the star's life expectancy by dividing mass by luminosity. A 2-solar-mass star is about 11 times more luminous than the sun and should live about 2/11, or 18 percent as long as the sun. We can, however, make the calculation even easier if we remember that the mass–luminosity relation tells us that the luminosity of a star equals $M^{3.5}$. The life expectancy then is:

$$T = \frac{M}{L} = \frac{M}{M^{3.5}} \quad \text{or} \quad T = \frac{1}{M^{2.5}}$$

This means that we can estimate the life expectancy of a star by dividing 1 by the star's mass raised to the 2.5 power. If we express the mass in solar masses, the life expectancy will be in solar lifetimes.

For example, how long can a 4-solar-mass star live?

$$T = \frac{1}{4^{2.5}} = \frac{1}{4 \cdot 4 \cdot \sqrt{4}} = \frac{1}{32} \text{ solar lifetimes}$$

Detailed studies of models of the sun show that the sun, presently 5 billion years old, can last another 5 billion years. Thus, a solar lifetime is approximately 10 billion years, and a 4-solar-mass star will last about (10 billion)/32 or about 310 million years.

Our estimation of stellar life expectancies is very approximate. For example, our model ignores mass loss, which may affect the life expectancies of very luminous and very faint stars. Nevertheless, it serves to illustrate a very important point. Stars that are only slightly more massive than the sun have dramatically shorter lifetimes on the main sequence.

| REVIEW CRITICAL INQUIRY |

Why are main-sequence O stars so rare compared with stars like the sun (G stars)?

In Chapter 9, we discussed the family of stars and noticed that space contains only about 0.036 main-sequence O stars per million cubic parsecs, but that there are about 2000 main-sequence G stars per million cubic parsecs (see page 193). Dividing, we discover that main-sequence G stars are over 50,000 times more common than O stars. Our discussion of the lives of main-sequence stars gives us a way to analyze these numbers. O stars live very short lifetimes. Our simple formula predicts a 20-solar-mass O star has a life expectancy of about 6 million years. The sun, however, will live a total of about 10 billion years or about 1600 times longer than an O star. Assuming nature makes equal numbers of different kinds of stars, we would expect to see about 1600 times more main-sequence G stars than O stars. But we see over 50,000 times more, so there must be another factor involved. Nature must be making vastly more lower-mass stars than massive stars. By comparing theory and observation, our analysis has revealed something new about how nature makes stars.

Now we can understand why O stars are so rare. What factors must you take into consideration to explain why main-sequence M stars are even more common than G stars?

When a star finally exhausts its hydrogen fuel, it can no longer resist the pull of its own gravity. The contraction that began when it was a protostar resumes and begins the process that leads to its death. It can delay its end by fusing other fuels, but, as we will discover in Chapter 13, nothing can steal gravity's final victory.

12-2 POST-MAIN-SEQUENCE EVOLUTION

In earlier chapters, we asked three questions: Why are giant stars so large? Why are they so uncommon? And why do they have such low densities? Now we are ready to answer those questions by discussing the evolution of stars after they leave the main sequence.

EXPANSION INTO A GIANT

To understand how stars evolve, we must recognize that they are not well mixed; that is, their interiors are not stirred. The centers of stars like the sun are radiative, meaning the energy moves as radiation and not as circulating currents of heated gas. The gas does not move in such stars, and thus they are not mixed at all. More massive stars have convective cores that mix the central regions (Figure 11-14), but these regions are not very large, and thus, for the most part, these stars, too, are not mixed. (The lowest-mass stars are an exception that we will examine in the next chapter.)

In this respect, stars are like a campfire that is not stirred; the ashes accumulate at the center, and the fuel in the outer parts never gets used. Nuclear fusion consumes hydrogen nuclei and produces helium nuclei, the "ashes," at the star's center. Nothing mixes the interior of the star, so the helium nuclei remain where they are in the center of the star, and the hydrogen in the outer parts of the star is not mixed down to the center where it can be fused.

The helium ashes that accumulate in the star's core cannot fuse into heavier elements because the temperature is too low. As a result, the helium accumulates, the hydrogen is used up, and the core becomes an inert ball of helium. As this happens, the energy production in the core falls, and the weight of the outer layers forces the core to contract.

Although the contracting helium core cannot generate nuclear energy, it does grow hotter because it converts gravitational energy into thermal energy (see Chapter 11). The rising temperature heats the unprocessed hydrogen just outside the core, hydrogen that was never before hot enough to fuse. When the temperature of the surrounding hydrogen becomes high enough, it ignites in a shell of fusing hydrogen. Like a grass fire burning outward from an exhausted campfire, the hydrogen-fusing shell burns outward, leaving helium ash behind and increasing the mass of the helium core.

At this stage in its evolution, the star overproduces energy; that is, it produces more energy than it needs to balance its own gravity. The helium core, having no nuclear energy sources, must contract, and that contraction converts gravitational energy into thermal energy—the contraction heats the helium core. Some of that heat leaks outward through the star. At the same time, the hydrogen-fusing shell produces energy as the contracting core brings fresh hydrogen closer to the center of the star and heats it to high temperature. The result is a flood of energy flowing outward through the outer layers of the star, forcing them to puff up and swelling the star into a giant (Figure 12-5).

The expansion of the envelope dramatically changes the star's location in the H–R diagram. Just as contraction heats a star, expansion cools it. As the outer layers of gas expand, energy is absorbed in lifting and expanding the gas. The loss of that energy lowers the temperature of the gas. Consequently the point that represents the star in the H–R diagram moves quickly to the right (in less than a million years for a star of 5 solar

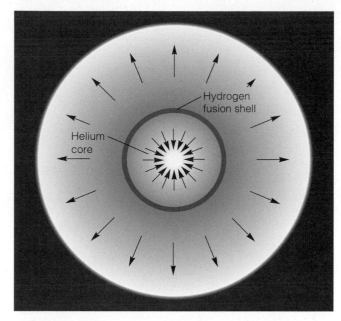

FIGURE 12-5

When a star runs out of hydrogen at its center, it ignites a hydrogen-fusing shell. The helium core contracts and heats, while the envelope expands and cools. (For a scale drawing, see Figure 12-8.)

masses). A massive star moves to the right across the top of the H–R diagram and becomes a supergiant, while a medium-mass star like the sun becomes a red giant (Figure 12-6). As the radius of a giant star continues to increase, the enlarging surface area makes the star more luminous, moving its point upward in the H–R diagram. Aldebaran, the glowing red eye of Taurus the bull, is such a red giant, with a diameter 25 times that of the sun, but with a surface temperature only half that of the sun.

More massive stars cross higher in the H–R diagram and become supergiants, stars even larger than giants. Betelgeuse (α Orionis) is a very cool, red supergiant over 800 times the diameter of the sun. Rigel (β Orionis) is a supergiant 50 times larger than the sun. It may seem odd to say that Rigel, a blue star, has expanded and cooled. At a temperature of 12,000 K, Rigel looks quite blue to our eyes. When it was on the main sequence, Rigel had a much hotter surface, but it would not have looked much bluer because we cannot see the ultraviolet radiation that the hottest stars emit. Any star as hot or hotter than Rigel will look blue to our eyes, as shown in Figure 7-2. Thus, blue Rigel is indeed a star that has expanded and cooled.

Our discussion of post-main-sequence evolution explains a number of things we noticed in earlier chapters. Giants and supergiants are large because they have expanded, and they have very low densities for the same reason. Also, the giant and supergiant region of the H–R diagram contains a jumble of different mass stars because all main-sequence stars funnel through this part of the diagram as they die. (See Figure 9-22.)

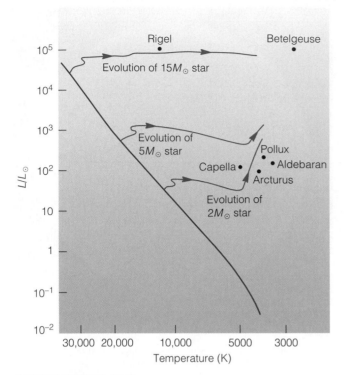

The evolution of a massive star moves the point that represents it in the H–R diagram to the right of the main sequence into the region of the supergiants such as Rigel and Betelgeuse. The evolution of a medium-mass star moves its point in the H–R diagram into the region of the giants such as those shown here.

Ace⊛Astronomy™ Go to AceAstronomy and click Active Figures to see "Evolution of Stars" and compare the evolution of a massive star and a low-mass star.

DEGENERATE MATTER

Although the hydrogen-fusing shell can force the envelope of the star to expand, it cannot stop the contraction of the helium core. Because the core has no energy source, gravity squeezes it tighter, and it becomes very small. If we represent the helium core of a 5-solar-mass star with a quarter, the outer envelope of the star would be about the size of a baseball diamond. Yet the core would contain about 12 percent of the star's mass compressed to very high density. When gas is compressed to such extreme densities, it begins to behave in astonishing ways that can alter the evolution of the star. Thus, to follow the story of stellar evolution, we must consider the behavior of gas at extremely high densities.

Normally, the pressure in a gas depends on its temperature. The hotter the gas is, the faster its particles move and the more pressure it exerts. But the gas inside a star is ionized, so there are two kinds of particles: atomic nuclei and free electrons. If the gas is compressed to very high densities, for example, in the core of a giant star, the difference between these two kinds of particles is important.

If the density is very high, the particles of the gas are forced close together, and two laws of quantum me-

chanics become important. First, quantum mechanics says that the moving electrons confined in the star's core can have only certain amounts of energy, just as the electron in an atom can occupy only certain energy levels (see Chapter 7). We can think of these permitted energies as the rungs of a ladder. An electron can occupy any rung, but not the spaces between.

The second quantum-mechanical law (called the Pauli Exclusion Principle) says that two identical electrons cannot occupy the same energy level. Because electrons spin in one direction or the other, two electrons can occupy an energy level if they spin in opposite directions. That level is then completely filled, and a third electron cannot enter because, whichever way it spins, it will be identical to one or the other of the two electrons already in the level. Thus, no more than two electrons can occupy the same energy level.

A low-density gas has few electrons per cubic centimeter, so there are plenty of energy levels available. If a gas becomes very dense, however, nearly all of the lower energy levels may be occupied (❙ Figure 12-7). In such a gas, a moving electron cannot slow down, because slowing down would decrease its energy, and there are no open energy levels for it to drop down to. It can speed up only if it can absorb enough energy to leap to the top of the energy ladder, where there are empty energy levels.

When a gas is so dense that the electrons are not free to change their energy, astronomers call it **degenerate matter**. Although it is a gas, it has two peculiar properties that can affect the star. First, the degenerate gas resists compression. To compress the gas, we must push against the moving electrons, and changing their motion means changing their energy. That requires tremendous effort, because we must boost them to the

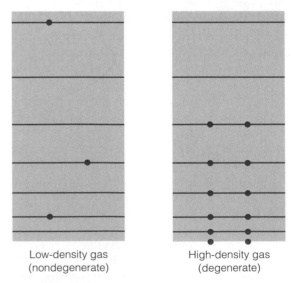

FIGURE 12-7
Electron energy levels are arranged like rungs on a ladder. In a low-density gas, many levels are open, but in a degenerate gas, all lower energy levels are filled.

Causes in Nature: The Very Small and the Very Large

One of the most interesting lessons of science is that the behavior of very small things often determines the structure and behavior of very large things. In degenerate matter, the quantum-mechanical behavior of electrons helps determine the evolution of giant stars. Such links between the very small and the very large are common in astronomy, and we will see in later chapters how the nature of certain subatomic particles may determine the fate of the entire universe.

This is a second reason why astronomers study atoms. Recall that the first reason is that atoms interact with light and astronomers must understand that interaction in order to analyze the light. The second reason is the link between the very small and the very large. It would be impossible to understand the evolution of the largest stars if we failed to understand how degenerate electrons behave. Similarly, astronomers studying galaxies must understand how microscopic dust in space reddens starlight, and astronomers studying planets must know how molecules form crystals.

As you study any science, look for the way very small things affect very large things. Sociologists and psychologists know that the mass behavior of very large groups of people can depend on the behavior of a few key individuals. Biologists study the visible consequences of atomic bonds in molecules we call genes, and meteorologists know that tiny changes in temperature in one part of the world can affect worldwide climate weeks or months later.

Nature is the sum of many tiny parts, and science is the study of nature. Thus, one way to do science is to search out the tiny causes that determine the way nature behaves on the grandest scales.

---

top of the energy ladder. Thus, degenerate matter, though still a gas, takes on the consistency of hardened steel.

Second, the pressure of degenerate gas does not depend on temperature but rather on the speed of the electrons, which cannot be changed without tremendous effort. The temperature, however, depends on the motion of all the particles in the gas, both electrons and nuclei. If we add heat to the gas, most of that energy goes to speed up the motions of the nuclei, and only a few electrons can absorb enough energy to reach the empty energy levels at the top of the energy ladder. Thus, changing the temperature of the gas has almost no effect on the pressure.

These two properties of degenerate matter become important when stars end their main-sequence lives and approach their final collapse (Window on Science 12-2). Eventually, many stars collapse into white dwarfs, and we will discover that these tiny stars are made of degenerate matter. But long before that, the cores of many giant stars become so dense that they are degenerate, a situation that can produce a cosmic bomb when helium begins to fuse.

HELIUM FUSION

Hydrogen fusion leaves behind helium ash, which cannot fuse at the relatively low temperatures of hydrogen fusion. Helium nuclei have a positive charge twice that of a hydrogen nucleus (a proton), which means the Coulomb barrier is higher. At the temperature of hydrogen fusion, the particles move too slowly and collide too gently to produce helium fusion. Thus the helium that accumulates is an ash that cannot fuse until the temperature rises significantly.

As the star becomes a giant star, fusing hydrogen in a shell, the inner core of helium ash contracts and grows hotter. It may even become degenerate. Finally, as the temperature approaches 100,000,000 K, helium nuclei begin to fuse together to make carbon.

We can summarize the helium-fusing process in two steps:

$$^4\text{He} + {}^4\text{He} \rightarrow {}^8\text{Be} + \gamma$$
$$^8\text{Be} + {}^4\text{He} \rightarrow {}^{12}\text{C} + \gamma$$

This process is complicated by the fact that beryllium-8 is very unstable and may break up into two helium nuclei before it can absorb another helium nucleus. Three helium nuclei can also form carbon directly, but such a triple collision is unlikely. Because a helium nucleus is called an alpha particle, astronomers often refer to helium fusion as the **triple alpha process.**

Some stars begin helium fusion gradually, but stars in a certain mass range begin helium fusion with an explosion called the **helium flash.** This explosion is caused by the density of the helium, which can reach 1,000,000 g/cm^3. On Earth, a teaspoon of this material would weigh as much as a large truck. At these densities, the gas is degenerate, and its pressure no longer depends on temperature. Thus, the pressure–temperature thermostat that controls the nuclear fusion reactions no longer works.

When the helium ignites, it generates energy, which raises the temperature. Because the pressure–temperature thermostat is not working in the degenerate gas, the core does not respond to the higher temperature by expanding. Rather, the higher temperature forces the reactions to go faster, which makes more energy, which raises the temperature, which makes the reactions go faster, and so on. The ignition of helium fusion in a degenerate gas results in a runaway explosion so violent that for a few minutes the helium core generates more energy than an entire galaxy. At its

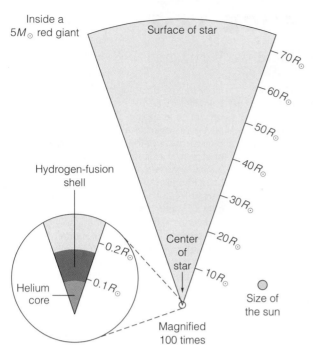

FIGURE 12-8

As a star becomes a giant, its envelope (yellow) expands to enormous size, while its core contracts and becomes very small. *(Illustration design by the author)*

peak, the core generates 10^{14} times more energy per second than the sun.

Although the helium flash is sudden and powerful, it does not destroy the star. In fact, if you were observing a giant star as it experienced a helium flash, you would see no outward evidence of an eruption. The helium core is quite small (❙ Figure 12-8), and all of the energy of the explosion is absorbed by the distended envelope. Also, the helium flash is a very short-lived event. In a matter of minutes, the core of the star becomes so hot it is no longer degenerate, the pressure–temperature thermostat brings the helium fusion under control, and the star proceeds to fuse helium steadily in its core. We will discuss this post-helium-flash evolution later.

Not all stars experience a helium flash. Stars less massive than about 0.4 solar mass can never get hot enough to ignite helium, and stars more massive than about 3 solar masses ignite helium before their cores become degenerate (❙ Figure 12-9). In such stars, pressure depends on temperature, so the pressure–temperature thermostat keeps the helium fusion under control.

If the helium flash occurs only in some stars and is a very short-lived event that is not visible from outside the star, why should we worry about it? The answer is that it limits the reliability of the mathematical models astronomers use to study stellar evolution. The medium-mass stars, which experience the helium flash, are those that we must consider most carefully in studying stellar evolution. Massive stars are not very common; low-mass stars evolve so slowly that we cannot see much evidence of low-mass stellar evolution. Thus, studies of stellar evolution must concentrate on medium-mass stars, which do experience the helium flash. But the helium flash occurs so rapidly and so violently that computer programs cannot follow the changes in the star's internal structure in detail. To follow the evolution of medium-mass stars like the sun past the helium flash, astronomers must make assumptions about the way the helium flash affects stars' internal structures.

Post-helium-flash evolution is generally understood. As the core temperature rises, the gas ceases to be degenerate, and helium fusion proceeds under the control of the pressure–temperature thermostat. Throughout these events, the hydrogen-fusion shell continues to produce energy, but the new helium-fusion energy produced in the core makes the core expand. That expansion absorbs energy previously used to support the outer layers of the star. As a result, the outer layers contract, and the surface of the star grows slightly hotter. In the H–R diagram, the point that represents the star initially moves down toward lower luminosity and to the left toward higher temperature. Later, as the star stabilizes fusing helium in its core, its luminosity recovers, but its surface grows hotter.

Helium fusion produces carbon and oxygen as "ash," and these accumulate at the center of the star. They do not fuse because the core is not hot enough. Thus, the helium in the core is gradually converted into carbon–oxygen ashes. As this happens, the core contracts, grows hotter, and ignites a helium-fusion shell. At this stage, the star has two shells producing energy. A hydrogen-fusion shell eats its way outward, leaving behind helium ash; a helium-fusion shell eats its way

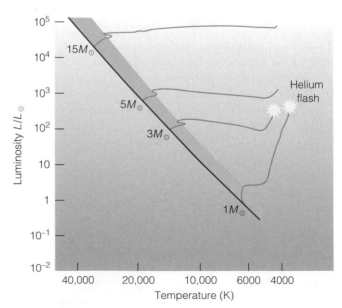

FIGURE 12-9

Stars like the sun suffer a helium flash, but more massive stars begin helium fusion without a helium flash. The lower-mass stars cannot get hot enough to ignite helium. The zero-age main sequence is shown here in red, with the band of the main sequence in blue.

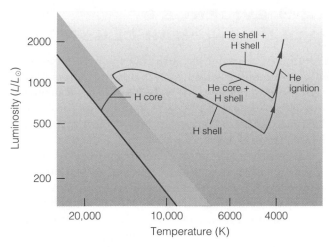

FIGURE 12-10

The evolution of a 5-$M_\odot$ star illustrates stages in energy generation. From the zero-age main sequence (red) across the band of the main sequence (blue), the star makes energy in a hydrogen-fusion core. Hydrogen-shell fusion begins as the star leaves the main sequence and continues after helium fusion ignites in the core. When the core exhausts its helium, the star makes energy with a helium-fusion shell around the core and a hydrogen-fusion shell outside the helium-fusion shell.

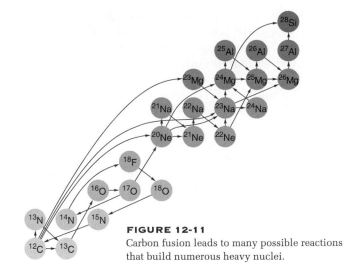

FIGURE 12-11
Carbon fusion leads to many possible reactions that build numerous heavy nuclei.

outward into the helium, leaving behind carbon–oxygen ash. Unable to generate energy in its carbon–oxygen core, the star expands its outer layers and its point in the H–R diagram moves back toward the right, completing a loop (▌ Figure 12-10).

FUSING ELEMENTS HEAVIER THAN HELIUM

Modeling the evolution of stars after helium exhaustion requires great sophistication. The inert core of carbon–oxygen ash contracts and becomes hotter. Stars more massive than about 4 solar masses can heat their cores to temperatures as high as 600,000,000 K, hot enough to fuse carbon. Subsequently, the stars can fuse oxygen, silicon, and other heavy elements.

Carbon fusion in a star begins a complex network of reactions illustrated in ▌ Figure 12-11, where each circle represents a possible nucleus and each arrow represents a different nuclear reaction. Nuclei can react by capturing a proton, capturing a neutron, capturing a helium nucleus, or by combining directly with other nuclei. Also, unstable nuclei can decay by ejecting an electron, ejecting a positron, ejecting a helium nucleus or by splitting into fragments. Astronomers who model the structure and evolution of these stars must use sophisticated nuclear physics to compute the energy production and changes in chemical composition as the stars fuse heavier elements.

These stars develop complex internal structure as they fuse heavier and heavier elements in concentric shells. Like great spherical cakes with many layers, the stars may be fusing a number of different elements simultaneously in concentric shells.

Eventually, all stars face collapse. Less massive stars cannot get hot enough to ignite the heavier nuclei. The more massive stars can ignite heavy nuclei, but these stars consume their fuels at a tremendous rate, so they run through their available fuel quickly (▌ Table 12-3). No matter what the star's mass, it will eventually run out of usable fuels, and gravity will win the struggle with pressure. We will explore the ultimate deaths of stars in the next chapter.

TABLE 12-3
Nuclear Reactions in Massive Stars

Nuclear Fuel	Nuclear Products	Minimum Ignition Temperature	Main-Sequence Mass Needed to Ignite Fusion	Duration of Fusion in a 25-$M_\odot$ Star
H	He	4×10^6 K	0.1 $M_\odot$	7×10^6 yr
He	C, O	120×10^6 K	0.4 $M_\odot$	0.5×10^6 yr
C	Ne, Na, Mg, O	600×10^6 K	4 $M_\odot$	600 yr
Ne	O, Mg	1.2×10^9 K	~8 $M_\odot$	1 yr
O	Si, S, P	1.5×10^9 K	~8 $M_\odot$	~0.5 yr
Si	Ni to Fe	2.7×10^9 K	~8 $M_\odot$	~1 day

The time a star spends as a giant or supergiant is small compared with its life on the main sequence. The sun, for example, will spend about 10 billion years as a main-sequence star but only about a billion years as a giant. The more massive stars pass through the giant stage even more rapidly. Because of the short time a star spends as a giant, we are unlikely to see many such stars. This illustrates an important principle in astronomy: the shorter the time a given evolutionary stage takes, the less likely we are to see stars in that particular stage. Thus, we see a great many main-sequence stars, but few giants (see page 193).

REVIEW CRITICAL INQUIRY

What uncertainties limit the accuracy of our story of stellar evolution?

Mathematical models of stars apply four basic laws of physics to describe the inside of a star. Modern astronomers have used such models to study stellar evolution, a process that takes billions of years and thus is unobservable for humans. But two things limit the accuracy of the models. First, some events in the history of a star happen so fast that computers can't predict the results in a reasonable amount of time. In fact, some astronomers estimate that a computer following the evolution of a model star undergoing the helium flash would have to recompute the model in such short time steps that the computer program would take millions of years to predict events that occur inside the star in just seconds. Thus, truly rapid changes such as explosions in stars make the models uncertain.

Second, we have seen how the general properties of a giant star can depend on the subatomic properties of matter. Not all of those subatomic properties are well understood. For example, the solar neutrino problem hints that physicists might not understand exactly how neutrinos and other subatomic particles behave, and if that is true, then stellar models may be slightly different from the real stars they are supposed to represent.

Can you think of any other physical processes that have been omitted from stellar models? What features of the sun's surface suggest motions and forces that are ignored in most stellar models?

We cannot see the stars evolving, so we must test our theoretical story of stellar evolution against observation. To paraphrase the quotation at the beginning of this chapter, we shouldn't believe any theory until we compare it with reality—seeing is believing. Thus, in the rest of this chapter, we will search for evidence to support the theory of stellar evolution. At the same time, we will discover some interesting consequences of the fact that the stars are not eternal.

12-3 EVIDENCE OF EVOLUTION: STAR CLUSTERS

The theory of stellar evolution is so complex and involves so many assumptions that astronomers would have little confidence in it were it not for H–R diagrams of clusters of stars. By observing the properties of star clusters of different ages, we can see clear evidence of the evolution of stars.

To grasp the difficulty of understanding stellar evolution, consider an analogy suggested by William Herschel. Suppose a visitor who had never seen a tree wandered through a forest for an hour looking at the falling seeds, rising sprouts, young saplings, mature trees, and fallen logs. Could such an observer understand the life cycle of the trees? Astronomers face the same problem when they try to understand the life story of the stars.

We do not live long enough to see stars evolve. We see only a momentary glimpse of the universe, a snapshot in which all the stages in the life cycle of the stars are represented. Unscrambling these stages and putting them in order is an almost impossible task. Only by looking at selected groups of stars—star clusters—can we see the pattern.

OBSERVING STAR CLUSTERS

Just as Sherlock Holmes studies peculiar dust on a lamp shade as a clue to a crime, astronomers look at star clusters and say, "Aha!" Each star cluster freezes a moment in the evolution of the cluster and makes the evolution of the stars visible.

Study ❚ "Star Cluster H–R Diagrams" on pages 248 and 249 and notice three important points. First, there are two kinds of star clusters, but they are similar in the way their stars evolve. We will discuss these clusters in more detail in a later chapter. Second, notice that we can estimate the age of a star cluster by observing the distribution of the points that represent its stars in the H–R diagram. Finally, we notice that the shape of a star cluster's H–R diagram is governed by the evolutionary path the stars take. By comparing clusters of different ages, we can visualize how stars evolve almost as if we were watching a film of a star cluster evolving over billions of years.

Were it not for star clusters, astronomers would have little confidence in the theories of stellar evolution. Star clusters make that evolution visible and assure astronomers that they really do understand how stars are born, live, and die.

THE EVOLUTION OF STAR CLUSTERS

What we know about star formation and stellar evolution can help us understand star clusters and how they evolve.

A star cluster is formed when a cloud of gas contracts, fragments, and forms a group of stars. Some groups of stars, which we have called associations, are so widely scattered the group is not held together by its own gravity. The stars in an association wander away from each other rather quickly. Some groups of stars are more compact, and even after the stars become luminous and the gas and dust are blown away, gravity holds the group together as a star cluster.

Open clusters are not as old or as massive as the globular clusters, and that helps us understand their appearance. Close encounters between stars are rare in an open cluster, and such clusters have an irregular appearance. The globular clusters appear to be nearly perfect globes because encounters between their stars are more common and because the globular clusters have had time to evenly distribute the energy of motion among all of the stars.

As a star cluster ages, some stars traveling a bit faster than others can escape. Globular clusters are compact and massive and have survived for 11 billion years or more. A star cluster with only a few stars widely distributed may evaporate completely as its stars escape one by one. Our sun probably formed in a star cluster 5 billion years ago, but there is no way to trace it back to its family home even if that star cluster still exists.

REVIEW CRITICAL INQUIRY

Why do open clusters contain only a small number of giant stars but many main-sequence stars?

When we look at a star cluster, we see a snapshot of the cluster that freezes it in time. In the H–R diagram of an open cluster, we see a main sequence containing many stars, but typically we see significantly fewer giant stars. Those giant stars were once main-sequence stars in the cluster, but they exhausted the hydrogen fuel in their cores and expanded to become giants. A star like the sun can fuse hydrogen on the main sequence for many billions of years but can remain a giant for only a billion years or less. More massive stars evolve even faster. Thus, at any given moment, we see most of the stars in a star cluster on the main sequence and only a few, caught in the snapshot, as giant stars.

Star clusters are important in astronomy because their H–R diagrams reveal the progress of stellar evolution. How do they reveal the age of the star cluster?

Star clusters give us dramatic confirmation that stars evolve and also reveal some of the details of stellar structure and evolution. Certain pulsating giant stars give us further information about how stars work.

12-4 EVIDENCE OF EVOLUTION: VARIABLE STARS

For millennia, poets and astronomers believed that the stars were constant and unchanging. Modern theory tells us that the stars are slowly evolving as they consume their fuels. But can we really be sure the stars are changing and evolving? Certain stars that pulsate and change their brightness give us evidence of stellar evolution and illustrate once more the importance of energy flowing outward through the layers of the stars.

A **variable star** is any star that changes its brightness in a periodic way. Variable stars include many different kinds of stars. Some variable stars are eclipsing binaries (Chapter 9), but many others are stars that expand and contract, heat and cool, and grow more luminous and less luminous because of internal characteristics. These stars are often called **intrinsic variables** to distinguish them from eclipsing binaries. Of these intrinsic variables, one particular kind is centrally important to modern astronomy even though the first star of that class was discovered centuries ago.

CEPHEID AND RR LYRAE VARIABLE STARS

In 1784, the deaf and mute English astronomer John Goodricke, then 19 years old, discovered that the star δ Cephei is variable, changing its brightness by almost a magnitude over a period of 5.37 days (Figure 12-12a). Goodricke died at the age of 21, but his discovery ensures his place in the history of astronomy. Since 1784, hundreds of stars like δ Cephei have been found, and they are now known as **Cepheid variable stars.**

Cepheid variables are supergiant or bright giant stars of spectral type F or G. The fastest complete a cycle—bright to faint to bright—in about 2 days, whereas the slowest take as long as 60 days. A plot of a variable's magnitude versus time, a light curve, shows a typical rapid rise to maximum brightness and a slower decline (Figure 12-12b). Some Cepheids change their brightness by only 0.5 magnitude (about 10 percent), while others change by more than a magnitude.

Supergiants and bright giants are rare, and not all are Cepheids. Nevertheless, some familiar stars are Cepheids. Polaris, the North Star, is a Cepheid with a period of 3.9696 days and a very low amplitude.

The **RR Lyrae variable stars** are related to the Cepheids. The RR Lyrae stars pulsate with periods of less than a day and are fainter than the faintest Cepheids.

Studies of Cepheids and RR Lyrae stars reveal two related facts of great importance. First, there is a **period–luminosity relation** that connects the period of pulsation of a Cepheid to its luminosity. The longer-period Cepheids are about 40,000 times more luminous than the sun, while the shortest-period Cepheids are only a few hundred times more luminous than the sun. Second, there are two types of Cepheids. Type I Cepheids,

Star Cluster H–R Diagrams

Star clusters are beautiful, but they are also keys to understanding the evolution of stars. All of the stars in a star cluster formed at about the same time from the same cloud of gas. That means the stars have the same age and the same initial chemical composition. The differences we see among stars in a star cluster must arise from differences in mass, and that makes stellar evolution visible.

An **open cluster** is a collection of 10 to 1000 stars in a region about 25 pc in diameter. Some open clusters are quite small and some are large, but they all have an open, transparent appearance because the stars are not crowded together.

**Open Cluster
The Jewel Box**

In a star cluster each star follows its orbit around the center of mass of the cluster.

Visual-wavelength image

**Globular Cluster
47 Tucanae**

A **globular cluster** can contain 10^5 to 10^6 stars in a region only 10 to 30 pc in diameter. The term "globular cluster" comes from the word "globe," although globular cluster is pronounced like "glob of butter." These clusters are nearly spherical, and the stars are much closer together than the stars in an open cluster. Globular clusters are discussed in more detail in Chapter 15.

Visual-wavelength image

Astronomers can construct an H–R diagram for a star cluster by plotting a point to represent the luminosity and temperature of each star.

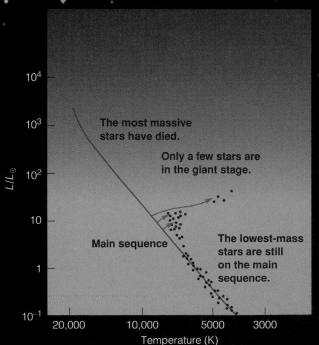

The most massive stars have died.

Only a few stars are in the giant stage.

Main sequence

The lowest-mass stars are still on the main sequence.

The H–R diagram of a star cluster can make the evolution of stars visible. The key is to remember that all of the stars in the star cluster have the same age but differ in mass. The H–R diagram of a star cluster gives us a snapshot of the evolutionary state of the stars at the time we happen to be alive. The diagram here represents a medium-age star cluster in which the most massive stars have died and a few medium mass stars are in the act of evolving off of the main sequence to become giants.

As a star cluster ages, its main sequence grows shorter like a candle burning down. We can judge the age of a star cluster by looking at the turnoff point, the point on the main sequence where stars evolve to the right to become giants. Stars at the **turnoff point** have lived out their lives and are about to die. Consequently, the life expectancy of the stars at the turnoff point equals the age of the cluster.

Ace ✪ Astronomy™

Go to AceAstronomy and click Active Figures to see "Cluster Turnoff" and notice how the shape of a cluster's H–R diagram changes with time.

From theoretical models of stars, we could construct a film to show how the H–R diagram of a star cluster changes as it ages. We can then compare theory (left) with observation (right) to understand how stars evolve. Note that the time step for each frame in this film increases by a factor of 10.

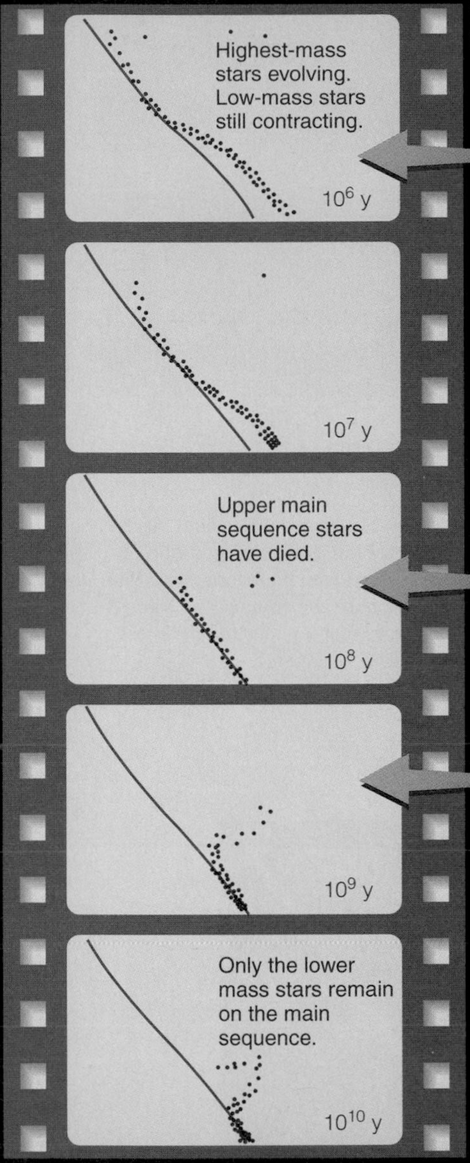

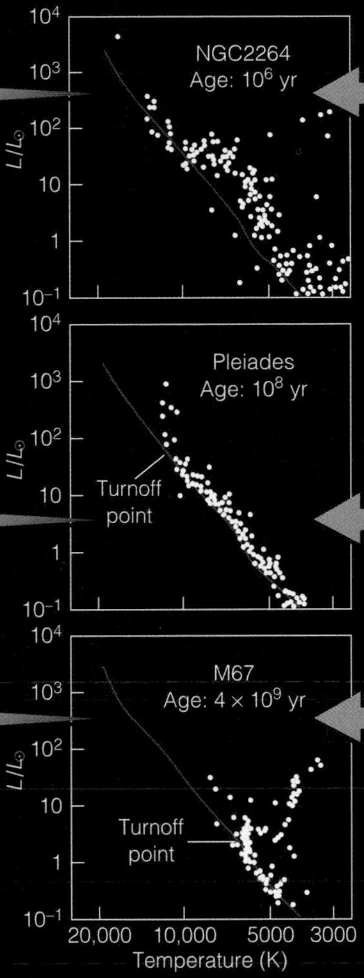

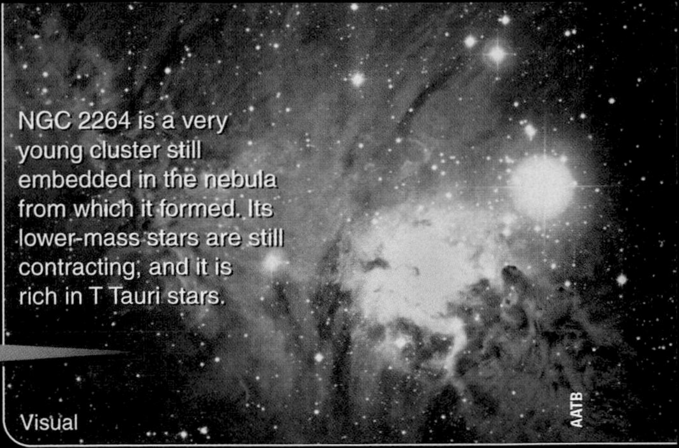

NGC 2264 is a very young cluster still embedded in the nebula from which it formed. Its lower-mass stars are still contracting, and it is rich in T Tauri stars.

Visual

AATB

Highest-mass stars evolving. Low-mass stars still contracting.

10^6 y

10^4

10^3

10^2

10^1

1

10^{-1}

$L/L_\odot$

NGC2264
Age: 10^6 yr

10^7 y

10^8 y

Upper main sequence stars have died.

Pleiades
Age: 10^8 yr

Turnoff point

10^4

10^3

10^2

10^1

1

10^{-1}

$L/L_\odot$

The nebula around the Pleiades is produced by gas and dust through which the cluster is passing. Its original nebula dissipated long ago.

Visual

Caltech

10^9 y

M67
Age: 4×10^9 yr

Turnoff point

10^4

10^3

10^2

10^1

1

10^{-1}

$L/L_\odot$

Only the lower mass stars remain on the main sequence.

10^{10} y

20,000   10,000   5000   3000
Temperature (K)

M67 is an old open cluster. In photographs, such clusters have a uniform appearance because they lack hot, bright stars. Compare with the Jewel Box on the opposite page.

Visual

Nigel Sharp, Mark Hanna/AURA/NOAO/NSF

Globular cluster H–R diagrams as shown below, resemble the last frame in our film, which tells us that globular clusters are very old.

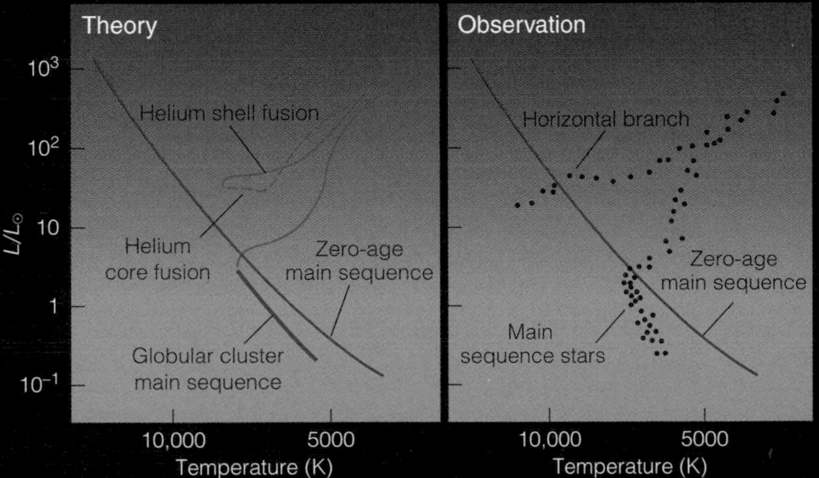

Theory

Helium shell fusion

Helium core fusion

Zero-age main sequence

Globular cluster main sequence

10^3

10^2

10

1

10^{-1}

$L/L_\odot$

10,000   5000
Temperature (K)

Observation

Horizontal branch

Zero-age main sequence

Main sequence stars

10,000   5000
Temperature (K)

The H–R diagrams of globular clusters have very faint turnoff points showing that they are very old clusters. The best analysis suggests these clusters are about 11 billion years old.

The **horizontal branch** stars are giants fusing helium in their cores and then in shells. The shape of the horizontal branch outlines the evolution of these stars.

The main-sequence stars in globular clusters are fainter and bluer than the zero-age main sequence. Spectra reveal that globular cluster stars are poor in elements heavier than helium, and that means their gases are less opaque. That means energy can flow outward more easily, which makes the stars slightly smaller and hotter. Again the shape of star cluster H–R diagrams illustrates principles of stellar evolution.

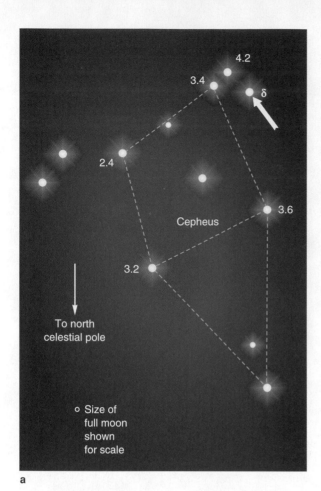

a

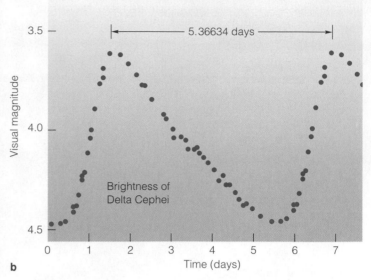

b

FIGURE 12-12

(a) The star δ Cephei (upper right) changes its brightness from about magnitude 3.6 at brightest to about magnitude 4.5 at faintest. The magnitudes of a few stars in the constellation Cepheus are given here for comparison. The constellation is shown as it appears on evenings in the autumn. (b) A graph of the brightness of δ Cephei versus time shows that it varies in brightness with a period slightly longer than 5 days.

including δ Cephei itself, have chemical compositions like that of the sun, but type II Cepheids and the RR Lyrae stars are poor in elements heavier than helium. These two facts can be summarized in a graph of absolute magnitude versus period (▌ Figure 12-13). The two types of Cepheids are clearly separated, and the fainter RR Lyrae stars lie at the lower left.

Cepheid and RR Lyrae variable stars are critically important in astronomy because they can be used to find the distance to star clusters and galaxies. We will discuss this use of variable stars in later chapters; here we are interested in them as peculiar giant stars that are unstable.

PULSATING STARS

Why should a star pulsate? Why should there be a period–luminosity relation? Why are there two types of Cepheids? The answers to these questions help us understand the insides of stars. To find the answers, we start with the evolution of stars in the H–R diagram.

After a star leaves the main sequence, it fuses hydrogen in a shell, and the point that represents it in the H–R diagram moves to the right as the star becomes a cool giant. When it ignites helium in its core, the point moves back to the left. Soon, however, the helium core

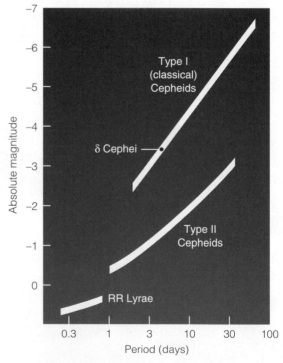

FIGURE 12-13

The Cepheid variable stars obey a period–luminosity relation: The longer-period stars are more luminous. The Cepheids are divided into two types. Type I Cepheids have chemical compositions like that of the sun, but type II Cepheids are poor in elements heavier than helium. The RR Lyrae stars are variable stars with lower luminosity and shorter periods.

is exhausted, and helium fusion continues only in a shell, forcing the star to expand again, and the point in the H–R diagram moves back to the right. Because giant stars evolve through these stages, their size and temperature change in complicated ways. Certain combinations of size and temperature make the star unstable, and it pulsates as an intrinsic variable star. In the H–R diagram, the region where size and temperature lead to pulsation is called the **instability strip** (▌Figure 12-14).

Why do some stars pulsate? The answer has to do with the flow of energy out of the star's interior. We have seen that a star supports its own weight by generating energy and that the flow of that energy outward prevents the star from contracting. The star is balanced between the outward flow of energy and the inward force of gravity. We could make a stable star pulsate if we squeezed it and then released it. The star would rebound outward, and it might oscillate for a few cycles, but the oscillation would eventually run down because of friction. The energy of the moving gas would be converted to heat and radiated away, and the star would return to stability. To make a star pulsate continuously, some process must drive the oscillation, just as a spring in a wind-up clock is needed to keep it ticking.

Studies using stellar models show that variable stars in the instability strip pulsate like beating hearts because of an energy-absorbing layer in their outer envelopes. This layer is the region where helium is partially ionized. Above this layer the temperature is too low to ionize helium, and below the layer it is hot enough to ionize all of the helium. Like a spring, the helium ionization zone can absorb energy when it is compressed and release it when the zone expands. That is enough to keep the star pulsating.

We can follow this pulsation in our imaginations and see how the helium ionization zone makes a star pulsate. As the outer layers of a pulsating star expand, the ionization zone expands, and the ionized helium becomes less ionized and releases stored energy, which forces the expansion to go even faster. The surface of the star overshoots its equilibrium position—it expands too far—and eventually falls back. As the surface layers contract, the ionization zone is compressed and becomes more ionized, which absorbs energy. Robbed of some of their energy, the layers of the star can't support the weight, and they compress further. This allows the contraction to go even faster, and the in-falling layers overshoot the equilibrium point until the pressure inside the star slows them to a stop and makes them expand again. Thus, the surface of these stars is not in hydrostatic equilibrium but rather pulsates in and out.

Cepheids change their radius by 5 to 10 percent as they pulsate, and this motion can be observed as Doppler shifts in their spectra. When they expand, the surface layers approach, and we see a blue shift. When they contract, the surface layers recede, and we see a

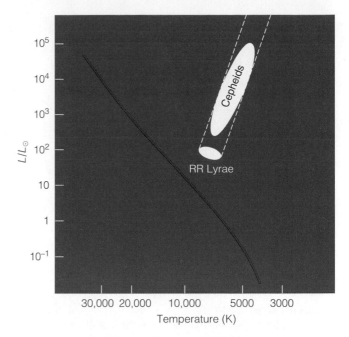

FIGURE 12-14
Both the Cepheid and RR Lyrae variable stars lie in the instability strip (dashed lines) because their variability is caused by the same mechanism.

red shift. Although a 10 percent change in radius seems like a large change, it affects only the outer layers of the star. The center of the star is much too dense to be affected.

Now we can understand why there is an instability strip. Only stars in the instability strip have their helium ionization zone at the right depth to act as a spring and drive the pulsation. A star will pulsate if the zone lies at the right depth below the surface. In cool stars, the zone is too deep and cannot overcome the weight of the layers above it. That is, the spring is overloaded and can't expand. In hot stars, the helium ionization zone lies near the surface, and there is little weight above it to compress it. That is, the spring never gets squeezed. The stars in the instability strip have the right temperature and radius for the helium ionization zone to fall exactly where it is most effective, and thus those stars pulsate.

We can also understand why there is a period–luminosity relation. The more massive stars are larger and more luminous. When they cross the instability strip, they cross higher in the H–R diagram. But the larger a star is, the more slowly it pulsates—just as large bells go "bong" and little bells go "ding." Pulsation period depends on size, which depends on mass, and mass also determines luminosity. So the most luminous Cepheids at the top of the instability strip are also the largest and pulsate the slowest (▌Figure 12-15). There has to be a period–luminosity relation because period and luminosity both depend on mass.

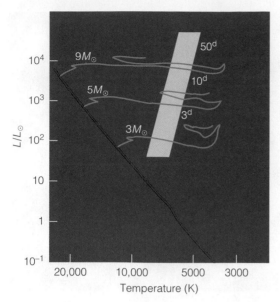

FIGURE 12-15
The instability strip (shaded) is a region of the H–R diagram in which stars are unstable and pulsate as variable stars. Massive stars cross the strip at high luminosities, and because of their large radius they pulsate with longer periods, such as 50 days. Lower-mass stars cross the instability strip at lower luminosities, and because of their smaller radius they pulsate with shorter periods, such as 3 days. This accounts for the period–luminosity relation.

We also know enough about stars to understand why there are two types of Cepheid variable stars. Type I Cepheids have chemical abundances roughly like those of the sun, but type II Cepheids are poor in elements heavier than helium, and that means the gases in type II Cepheids are not as opaque as the gases in type I Cepheids. The all-important energy flowing outward can escape more easily in type II Cepheids, and they reach a slightly different equilibrium. For stars of the same period of pulsation, type II Cepheids are less luminous (Figure 12-13).

PERIOD CHANGES IN VARIABLE STARS

Some Cepheids have been observed to change their periods of pulsation, and those changes seem to arise from the evolution of the stars.

The evolution of a star may carry it through the instability strip a number of times, and each time it can become a variable star. If we could watch a star as it entered the instability strip, we could see it begin to pulsate. Similarly, if we could watch a star leave the instability strip, we could watch it stop pulsating. Such an observation would confirm our belief that the stars are evolving. Unfortunately, stars evolve so slowly that these events are surely rare.

One famous Cepheid, RU Camelopardalis, was observed to stop pulsating temporarily. Its pulsations died away in 1966, and many astronomers assumed that it had stopped pulsating because it was leaving the

instability strip. But its pulsations resumed in 1967. Polaris may be going through a similar change now, but it does not lie near the edge of the instability strip in the H–R diagram. Thus, it may not be an example of a star evolving out of the instability strip.

Even more dramatic evidence of stellar evolution can be found in the slowly changing periods of Cepheid variables. Even a tiny change in period can become easily observable because it accumulates, just as a clock that gains a second each day becomes obviously fast over a time as long as a year. Some Cepheids have periods that are growing shorter, and others have periods that are growing longer. These changes occur because the stars are evolving and changing their radii; contraction shortens the period, and expansion lengthens the period. The star X Cygni, for example, has a period that is gradually growing shorter, and it has been "gaining" since observations began in the late 19th century (❙ Figure 12-16). These changes in the periods of pulsating stars give us dramatic evidence that the stars are evolving.

Some kinds of variable stars lie outside the instability strip because their pulsation is driven by different mechanisms. It is sufficient here for us to discuss the Cepheids. They are common, they illustrate important ideas about stellar structure, and their changing periods make stellar evolution clearly evident. Also, we will use Cepheids in later chapters to study the distances to galaxies and the fate of the universe.

REVIEW CRITICAL INQUIRY

How do Cepheid variable stars give us evidence that stars are evolving?

If we observe Cepheids for a few years, we may notice that some pulsations are slowing down and others are speeding up. This is a very small change in the period of pulsation, but if we observe the stars for long periods of time, the change in their pulsation causes them to gain or lose just as a clock can gain or lose time over many weeks. These changes in period are caused by the expansion or contraction of the stars as they evolve. The period of pulsation depends on the size of the star, so pulsations of a star that is expanding should slow down slightly, and those of a star that is contracting should speed up slightly. When we observe these small changes in the pulsation periods of Cepheid variable stars, we are seeing direct evidence that the stars are evolving.

If we wanted to conduct this research project, what observations would we need to make, and how would we analyze them to detect changes in period? Think about a clock. How would we determine whether a clock was ticking a tiny bit too fast or too slow?

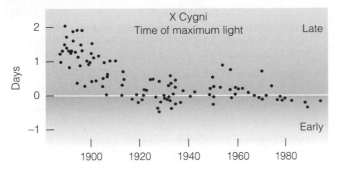

FIGURE 12-16
Like a clock running slow, the Cepheid variable star X Cygni has been reaching maximum light late for most of this century. As it evolves to the left across the instability strip, it is slowly contracting, and its period is growing shorter.

This chapter has discussed the evolution of stars, a process that occurs so slowly we cannot expect to see much happen in a single human lifetime. Thus, the study of stellar evolution depends heavily on theory and its verification by evidence from observations of real stars. We have seen that star clusters make the evolution of stars visible in the H–R diagram and that the pulsation of the intrinsic variable stars can give us insight into how giant stars evolve through the instability strip. We have not yet considered, however, the most interesting question of stellar evolution. What happens to a star when it uses up the last of its nuclear fuel? Clearly, stars must die—the question is how. We will explore that question in the next chapter.

SUMMARY

Almost everything we know about the internal structure of stars comes from mathematical stellar models. The models are based on four simple laws of stellar structure. Two laws say that mass and energy must be conserved and spread smoothly through the star. Another, the law of hydrostatic equilibrium, says that the star must balance the weight of its layers by its internal pressure. Yet another says that energy can flow outward only by conduction, convection, or radiation.

The mass–luminosity relation is explained by the requirement that a star support the weight of its layers by its internal pressure. The more massive a star is, the more weight it must support and the higher its internal pressure must be. To keep its pressure high, it must be hot and generate large amounts of energy. Thus, the mass of a star determines its luminosity. The massive stars are very luminous and lie along the upper main sequence. The less-massive stars are fainter and lie lower on the main sequence.

How long a star can stay on the main sequence depends on its mass. The more massive a star is, the faster it uses up its hydrogen fuel. A 25-solar-mass star will exhaust its hydrogen and die in only about 7 million years, but the sun can last for 10 billion years.

When a main-sequence star exhausts its hydrogen, it does not just wink out. Its core contracts, and it begins to fuse hydrogen in a shell around its core. The outer parts of the star—its envelope—swell, and the star becomes a giant. Because of this expansion, the surface of the star cools, and it moves toward the right in the H–R diagram. The most massive stars move across the top of the diagram as supergiants.

As the core of the star continues to contract, it finally ignites helium fusion, a process that converts helium into carbon. If the core becomes degenerate before helium ignites, the pressure of the gas does not depend on its temperature, and when helium ignites, it explodes in the helium flash. Although the helium flash is violent, the star absorbs the extra energy and quickly brings the helium-fusing reactions under control.

After helium is exhausted in the core, it can fuse in a shell around the core. Then, if the star is massive enough, it can contract and ignite other fuels, such as carbon.

Confirmation of stellar evolution comes from star clusters. Because all the stars in a cluster have about the same distance, composition, and age, we can see the effects of stellar evolution in the H–R diagram of a cluster. Massive stars evolve faster than low-mass stars, so in a given cluster the most massive stars leave the main sequence first. We can judge the age of such a cluster by looking at the turnoff point, the location on the main sequence where the stars turn off to the right and become giants. The life expectancy of a star at the turnoff point equals the age of the cluster.

There are two types of star clusters. Open clusters contain 10 to 1000 stars and have an open, transparent appearance. Globular clusters contain 10^5 to 10^6 stars densely packed into a spherical shape. The open clusters tend to be young to middle-aged, but globular clusters tend to be very old—11 billion years or more. Also, globular clusters tend to be poor in elements heavier than helium.

A simple understanding of stellar structure explains many of the properties of pulsating stars such as the Cepheid and RR Lyrae variables. These intrinsic variable stars lie in an instability strip in the H–R diagram because they contain a layer in their envelopes that stores and releases energy as they expand and contract. Stars outside the instability strip do not pulsate, because the layer is too deep or too shallow to make the stars unstable. The Cepheids obey a period–luminosity relation because more massive stars, which are more luminous, become larger and pulsate more slowly.

Some Cepheids have periods that are slowly changing, showing that the evolution of the star is changing the star's radius and thus its period of pulsation.

NEW TERMS

conservation of mass law	globular cluster
conservation of energy law	turnoff point
stellar model	horizontal branch
brown dwarf	variable star
zero-age main sequence (ZAMS)	intrinsic variable
degenerate matter	Cepheid variable star
triple alpha process	RR Lyrae variable star
helium flash	period–luminosity relation
open cluster	instability strip

REVIEW QUESTIONS

Ace⟲Astronomy™ Assess your understanding of this chapter's topics with additional quizzing and animations at **http://astronomy.brookscole.com/seeds8e**

1. Why is there a lower end to the main sequence? Why is there an upper end?

2. What is a brown dwarf?

3. Why is there a mass–luminosity relation?

4. Why does a star's life expectancy depend on mass?

5. Why do expanding stars become cooler and more luminous?

6. What causes the helium flash? Why does it make it difficult for astronomers to understand the later stages of stellar evolution?

7. How do some stars avoid the helium flash?

8. Why are giant stars so low in density?

9. Why are lower-mass stars unable to ignite more massive nuclear fuels such as carbon?

10. How can we estimate the age of a star cluster?

11. How do star clusters confirm that stars evolve?

12. How do some variable stars prove that stars are evolving?

DISCUSSION QUESTIONS

1. How do we know that the helium flash occurs if it cannot be observed? Can we accept an event as real if we can never observe it?

2. Can you think of ways that chemical differences could arise in stars in a single star cluster? Consider the mechanism that triggered their formation.

PROBLEMS

1. In the model shown in Figure 12-1, how much of the sun's mass is hotter than 13,000,000 K?

2. What is the life expectancy of a 16-solar-mass star? of a 50-solar-mass star?

3. How massive could a star be and still survive for 5 billion years?

4. If the sun expanded to a radius 100 times its present radius, what would its density be? (*Hint:* The volume of a sphere is $\frac{4}{3}\pi R^3$.)

5. If a giant star 100 times the diameter of the sun were 1 pc from us, what would its angular diameter be? (*Hint:* Use the small-angle formula, in Chapter 3.)

6. What fraction of the volume of a 5-solar-mass giant star is occupied by its helium core? (*Hints:* See Figure 12-8. The volume of a sphere is $\frac{4}{3}\pi R^3$.)

7. If the stars at the turnoff point in a star cluster have masses of about 4 solar masses, how old is the cluster?

8. If an open cluster contains 500 stars and is 25 pc in diameter, what is the average distance between the stars? (*Hints:* What share of the volume of the cluster surrounds the average star? The volume of a sphere is $\frac{4}{3}\pi R^3$.)

9. Repeat Problem 8 for a typical globular cluster containing a million stars in a sphere 25 pc in diameter.

10. If a Cepheid variable star has a period of pulsation of 2 days and its period increases by 1 second, how late will it be in reaching maximum light after 1 year? after 10 years? (*Hint:* How many cycles will it complete in a year?)

CRITICAL INQUIRIES FOR THE WEB

1. Find Hertzsprung–Russell (or color-magnitude) diagrams for five star clusters. Based on the distributions of stars on the diagrams, rank the clusters in order from oldest to youngest. At what value (spectral type, temperature, or B-V color index) does each cluster have its turnoff?

2. What is the latest news about Eta Carinae? It is a complex system and is constantly under study as astronomers try to understand what is happening to its stars.

3. Both professional and amateur astronomers study variable stars. Search the Web for the American Association of Variable Star Observers and see what information you can find. Can you find recent data and light curves for known variables like δ Cephei and X Cyg?

4. What is the latest news about brown dwarfs?

EXPLORING *THESKY*

1. Locate the open clusters below and describe their location with respect to the Milky Way Galaxy.

    M7, M41, M44, M45

2. Locate the clusters in Activity 1 and zoom in until you can see individual stars. Plot an H–R diagram for each cluster using the spectral type and apparent magnitude of the stars you see. Discuss the ages of the clusters. (*Hint:* Remember you can see only the brightest stars in each cluster.)

3. Use the H–R diagram of the Pleiades star cluster (M45) from Activity 2 to find the distance to the cluster. (*Hint:* How bright should the main-sequence stars be in absolute magnitude?)

 Visit the Seeds *Foundations of Astronomy* companion Web site for critical thinking exercises, articles, and additional readings from InfoTrac College Edition, Brooks/Cole's online student library.

THE DEATHS OF STARS

Natural laws have no pity.

Robert Heinlein, The Notebooks of Lazarus Long

Hubble Heritage Team, AURA/STScI/NASA

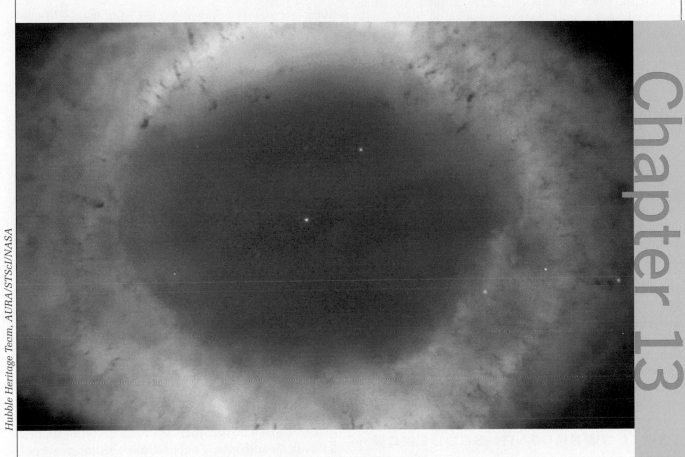

Chapter 13

GUIDEPOST

As you read this chapter, take a moment to be astonished and proud that the human race knows how stars die. Finding the masses of stars is difficult, and understanding the invisible matter between the stars takes ingenuity, but we humans have used what we know about stars and what we know about the physics of energy and matter to solve one of nature's deepest mysteries. We know how stars die.

The previous chapter made heavy use of theory in describing the internal structure of stars. We tested that theory against observations of clusters of stars and variable stars. In this chapter, we have more direct observational evidence to help us understand how stars die. We still need theory, but images and spectra of dying stars give us landmarks to guide our theoretical explorations.

This chapter leads us onward, not only to discuss the remains of dead stars—neutron stars and black holes—but it also leads us into discussions of the formation of our galaxy, the origin of the universe, the formation of planets, and the origin of life. Our world is made of atoms that were born in the deaths of stars. This chapter of endings is also a chapter of beginnings.

VIRTUAL LABORATORY

STELLAR EXPLOSIONS, NOVAE, AND SUPERNOVAE

Gravity is patient—so patient it can kill stars. Occasionally astronomers see a new star appear in the sky, grow brighter, fade away after a few weeks or a year, and vanish. We will discover that a **nova,** what seems to be a new star in the sky, is in fact produced by a very old, dying star, and that a **supernova,** a particularly luminous and long-lasting nova, is caused by the explosive death of a star. Modern astronomers find a few novae (plural of *nova*) each year, but supernovae (plural) are so rare that there are only one or two supernovae each century in our galaxy. Astronomers know that supernovae occur because they occasionally see them flare in other galaxies, and because we see in our own galaxy stars about to explode as supernovae and the shattered remains of these violent explosions (▌ Figure 13-1).

In the previous chapter, we saw that stars resist their own gravity by generating energy through nuclear fusion. The energy keeps their interiors hot, and the resulting high pressure balances gravity and prevents the star from collapsing. Stars, however, have limited fuel, and gravity is patient. When the stars exhaust their fuel, gravity wins, and the stars die.

The mass of a star is critical in determining its fate. Massive stars use up their nuclear fuel at a furious rate and die after only a few million years, while the lowest-mass stars use their fuels sparingly and may be able to live hundreds of billions of years. Massive stars can die in violent explosions that we see as supernovae, but low-mass stars die quiet deaths. To follow the evolution of stars to their graves, we will sort the stars into three categories according to their masses: low-mass red dwarfs, medium-mass sunlike stars, and massive upper-main-sequence stars. These stars lead dramatically different lives and die in different ways.

13-1 LOWER-MAIN-SEQUENCE STARS

The stars of the lower main sequence share a common characteristic: They have relatively low masses. That means that they face similar fates as they exhaust their nuclear fuels.

When a star exhausts one nuclear fuel, its interior contracts and grows hotter until the next nuclear fuel ignites. The contracting star heats up by converting gravitational energy into thermal energy, so low-mass stars cannot get very hot, and this limits the fuels they can ignite. The lowest-mass stars, for example, cannot get hot enough to ignite helium fusion.

Ace🟣Astronomy™ The AceAstronomy icon throughout the text indicates an opportunity for you to test yourself on key concepts and to explore animations and interactions on the AceAstronomy Web site at **http://astronomy.brookscole .com/seeds8e**

Structural differences divide the low-mass stars into two subgroups—very-low-mass stars and medium-mass stars such as the sun. The critical difference between the two groups is the extent of interior convection (see Figure 11-14). If the star is convective, fuel is constantly mixed, and the resulting evolution of the star is drastically altered.

RED DWARFS

Stars less massive than about 0.4 solar mass—the red dwarfs—have two advantages over more massive stars. First, they have very small masses, and thus they have very little weight to support. Their pressure–temperature thermostats are set low, and they consume their hydrogen fuel very slowly. Our discussion of the life expectancies of stars in Chapter 12 predicted that the red dwarfs would live very long lives.

The red dwarfs have a second advantage because they are totally convective. That is, they are stirred by circulating currents of hot gas rising from their interiors and cool gas sinking inward from their surfaces. This mixes the stars like pots of soup that are constantly stirred. Hydrogen is consumed, and helium accumulates uniformly throughout the stars; and that has two important consequences for red dwarfs.

Because they use up their hydrogen uniformly, they can never develop a helium core and a hydrogen fusion shell. That means they can never become giant stars. They are doomed to convert their hydrogen into helium, a fuel they are not massive enough to fuse, and then die slow, unremarkable deaths.

Because red dwarf stars are completely mixed and can fuse all of their hydrogen, their lives are prolonged even further. Models of a 0.1-solar-mass red dwarf predicted it would take 2 billion years to contract to the main sequence and 6 trillion years to use up its hydrogen fuel. Because the universe is only about 14 billion years old, none of these stars should have exhausted its fuel yet.

SUNLIKE STARS

Stars with masses between roughly 0.4 and 4 solar masses,* including the sun, ignite hydrogen and then helium and become giants, but they cannot get hot enough to ignite carbon, the next fuel in the sequence (see Table 12-3). When they reach that impasse, they can no longer maintain their stability, and their interiors contract while their envelopes expand.

To understand the fate of these stars, we must consider two concepts: mixing and expansion. The interiors of these sunlike stars are not well mixed (see Figure 11-14). As we learned in Chapter 11, stars of 1.1 solar masses or less, including the sun, have no convection

*This mass limit is uncertain, as are many of the masses stated here. The evolution of stars is highly complex, and such parameters are not precisely known.

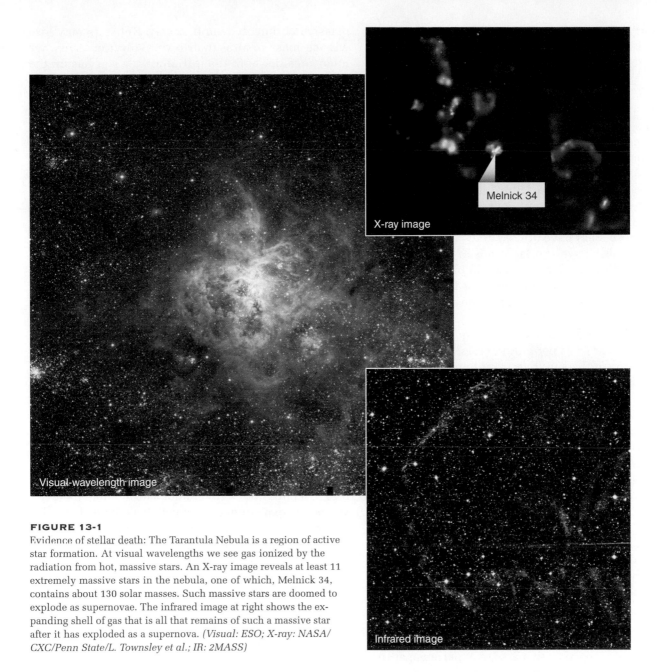

FIGURE 13-1

Evidence of stellar death: The Tarantula Nebula is a region of active star formation. At visual wavelengths we see gas ionized by the radiation from hot, massive stars. An X-ray image reveals at least 11 extremely massive stars in the nebula, one of which, Melnick 34, contains about 130 solar masses. Such massive stars are doomed to explode as supernovae. The infrared image at right shows the expanding shell of gas that is all that remains of such a massive star after it has exploded as a supernova. *(Visual: ESO; X-ray: NASA/ CXC/Penn State/L. Townsley et al.; IR: 2MASS)*

near their centers, so they are not mixed at all. Stars more massive than 1.1 solar masses have small zones of convection at their centers, but this mixes no more than about 12 percent of the star's mass. Thus, medium-mass stars, whether they have convective cores or not, are not mixed, and the helium accumulates in an inert helium core surrounded by unprocessed hydrogen. When this core contracts, the unprocessed hydrogen just outside the core ignites in a shell and swells the star into a giant.

As a giant, the star fuses helium in its core and then in a shell surrounding a core of carbon and oxygen. This core contracts and grows hotter, but it cannot become hot enough to ignite the carbon. Thus, the carbon–oxygen core is a dead end for these medium-mass stars.

Because no nuclear reactions can begin in the carbon–oxygen core, it cannot resist the weight pressing down on it. In addition, just outside the carbon–oxygen core, the helium-fusion shell converts helium into carbon and thus increases the mass of the carbon–oxygen core, so the core must contract. The energy released by the contracting core, plus the energy generated in the helium- and hydrogen-fusing shells, flows outward and makes the envelope of the star expand.

This forces the star to become a very large giant. Its radius may become as large as the radius of Earth's orbit, and its surface becomes as cool as 2000 K. Such a star can lose large amounts of mass from its surface.

MASS LOSS FROM SUNLIKE STARS

We have clear evidence that stars like the sun lose mass. The sun itself is losing mass. The solar wind is a gentle breeze of gas blowing outward from the solar corona and carrying mass into space.

Other stars like the sun are losing mass, and the evidence lies in their spectra. Ultraviolet and X-ray spectra of many stars reveal strong emission lines that must originate in hot chromospheres and coronas like the sun's. If these stars have atmospheres like the sun's, they presumably have similar winds of hot gas. Also, the spectra of some giant stars contain blue-shifted absorption lines. The blue shifts must be Doppler shifts produced as the gas flowing out of the star comes toward us.

The solar wind is not very strong; the sun loses about 0.001 solar mass per billion years. Even over the entire lifetime of the sun, that would not significantly alter the sun's mass. However, once the sun becomes a giant star, it may lose mass more rapidly. Because the spectra of some of these stars do not show the characteristic emission, we must assume that the stars do not have hot coronas, but other processes could drive mass loss. The stars are so large that gravity is weak at their surfaces, and convection in the cool gas can drive shock waves outward and power mass loss. In addition, some giants are so cool that specks of carbon dust condense in their atmospheres, just as soot can condense in a fireplace. The pressure of the star's radiation can push this dust and any atoms that collide with the dust completely out of the star.

Another factor that can cause mass loss is periodic eruptions in the helium-fusion shell. The triple-alpha process that fuses helium into carbon is extremely sensitive to temperature, and that can make the helium-fusion shell so unstable that it may experience eruptions called **thermal pulses.** Every 200,000 years or so, the shell can erupt and suddenly produce energy equivalent to a million times the luminosity of the sun. Almost all of this energy goes into lifting the outer layers of the star and driving gas away from the surface.

Whatever drives this mass loss from giants, it can affect the mass of the star appreciably in a short time. A star expanding as its carbon–oxygen core contracts could lose an entire solar mass in only 10^5 years, which is not a long time in the evolution of a star. Thus, a star that began its existence on the main sequence with a mass of 8 solar masses might reduce its mass to only 3 solar masses in half a million years.

Stellar mass loss confuses our story of stellar evolution. We would like to say that stars more massive than a certain limit will evolve one way, and stars less massive will evolve another way. But stars may lose enough mass to alter their own evolution. Thus, we must consider both the initial mass a star has on the main sequence and the mass it retains after mass loss. Because we don't know exactly how effective mass loss is, it is difficult for us to be exact about the mass limits in our discussion of stellar evolution.

The mass loss we have discussed so far has been a gradual process, but a sunlike star with a contracting carbon–oxygen core expands so far that it can expel all of its outer layers. This surface expulsion is helped by the growing pressure of the radiation inside the star and by instabilities in the helium-fusion shell. The expanding layers are visible as small nebulae.

PLANETARY NEBULAE

When a medium-mass star like the sun becomes a distended giant, it can expel its outer atmosphere to form one of the most interesting objects in astronomy, a **planetary nebula,** so called because it looks like the greenish-blue disk of a planet such as Uranus or Neptune. In fact, a planetary nebula has nothing to do with a planet. It is composed of ionized gases expelled by a dying star.

Study ❚ "The Formation of Planetary Nebulae" on pages 260 and 261 and notice four things. First, we can understand what planetary nebulae are like by using simple observational methods we have discussed before. Second, notice the model astronomers have developed to explain planetary nebulae. The real nebulae are more complex than our simple model of a slow wind and a fast wind, but the model gives us a way to organize the observed phenomena. The third thing to notice is how oppositely directed jets (much like bipolar flows) produce many of the asymmetries seen in planetary nebulae. Finally, notice the fate of the star itself; it must contract into a white dwarf.

The medium-mass stars, including the sun, must die by forming planetary nebulae and contracting into white dwarfs. This suggests another way we can search for evidence of the deaths of medium-mass stars: We can study white dwarfs.

WHITE DWARFS

White dwarfs are among the most common stars. The survey we took in Chapter 9 revealed that white dwarfs and red dwarfs, although very faint, are the dominant kinds of stars (page 193). Now we can recognize white dwarfs as the remains of medium-mass stars that fused hydrogen and helium, failed to ignite carbon, drove away their outer layers to form planetary nebulae, and collapsed and cooled to form white dwarfs. The billions of white dwarfs in our galaxy must be the remains of medium-mass stars.

The first white dwarf discovered was the faint companion to Sirius. In that visual binary system, the bright star is Sirius A. The white dwarf, Sirius B, is 10,000 times fainter than Sirius A. The orbital motions of the stars (shown in Figure 9-14) tell us that the white dwarf's mass is 0.98 solar mass, and its blue-white color tells us that its surface is hot, about 25,000 K. Although it is very hot, it has a very low luminosity, so it must have a small surface area—in fact, its diameter is about twice Earth's. Dividing its mass by its volume reveals that it is very dense—over 3×10^6 g/cm^3. On Earth, a teaspoonful of Sirius B material would weigh more than 15 tons (▮ Figure 13-2). Thus, basic observations and simple physics lead us to the conclusion that white dwarfs are astonishingly dense.

A normal star is supported by energy flowing outward from its core, but a white dwarf cannot generate energy by nuclear fusion. It has exhausted its hydrogen and helium fuel and produced carbon. As the star collapses into a white dwarf it converts gravitational energy into thermal energy, and its interior becomes very hot, but it cannot get hot enough to ignite carbon fusion. The star contracts until it becomes degenerate. Although a tremendous amount of energy flows out of the hot interior, it is not the energy flow that supports the star. The white dwarf is supported against its own gravity by the inability of its degenerate electrons to pack into a smaller volume.

The interior of a white dwarf is mostly carbon and oxygen ions floating among a whirling storm of degenerate electrons. It is the degenerate electrons that exert the pressure to support the star's weight, but most of its mass is represented by the carbon and oxygen ions. Theory predicts that as the star cools these ions will lock together to form a crystal lattice, so there may be some truth in thinking of aging white dwarfs as great crystals of carbon and oxygen. Near the surface, where the pressure is lower, a layer of ionized gases makes up a hot atmosphere.

The tremendous surface gravity of white dwarfs—100,000 times that of Earth—affects their atmospheres in strange ways. The heavier atoms in the atmosphere tend to sink, leaving the lightest gases at the surface. We see some white dwarfs with atmospheres of almost pure hydrogen, whereas others have atmospheres of nearly pure helium. Still others, for reasons not well understood, have atmospheres that contain traces of heavier atoms. In addition, the powerful surface gravity pulls the white dwarf's atmosphere down into a very shallow layer. If Earth's atmosphere were equally shallow, people on the top floors of skyscrapers would have to wear oxygen masks.

Clearly, a white dwarf is not a normal star. It generates no nuclear energy, is almost totally degenerate, and, except for a thin layer at its surface, contains no gas. Instead of calling a white dwarf a "star," we can

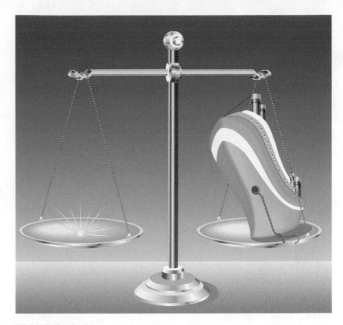

FIGURE 13-2
The degenerate matter from inside a white dwarf is so dense that a lump the size of a beach ball would, transported to Earth, weigh as much as an ocean liner.

call it a **compact object.** In the next chapter, we will discuss two other compact objects—neutron stars and black holes.

A white dwarf's future is bleak. As it radiates energy into space, its temperature gradually falls, but it cannot shrink any smaller because its degenerate electrons cannot get closer together. This degenerate matter is a very good thermal conductor, so heat flows to the surface and escapes into space, and the white dwarf gets fainter and cooler, moving downward and to the right in the H–R diagram. Because the white dwarf contains a tremendous amount of heat, it needs billions of years to radiate that heat through its small surface area. Eventually, such objects may become cold and dark, so-called **black dwarfs.** Our galaxy is not old enough to contain black dwarfs. The coolest white dwarfs in our galaxy are about the temperature of the sun.

Perhaps the most interesting thing about white dwarfs has come from mathematical models. The equations predict that if we added mass to a white dwarf, its radius would *shrink,* because added mass would increase its gravity and squeeze it tighter. If we added enough to raise its total mass to about 1.4 solar masses, its radius would shrink to zero (▮ Figure 13-3). This is called the **Chandrasekhar limit** after Subrahmanyan Chandrasekhar, the astronomer who discovered it. It seems to imply that a star more massive than 1.4 solar masses could not become a white dwarf unless it shed mass in some way.

Stars do lose mass. Young stars have strong stellar winds, and aging giants also lose mass (▮ Figure 13-4).

The Formation of Planetary Nebulae

Simple observations tell us what planetary nebulae are like. Their angular size and their distance tell us that their radii range from 0.2 to 3 light-years. The presence of emission lines in their spectra assures us they are excited, low-density gas. Doppler shifts show they are expanding at 10 to 20 km/s. If we divide radius by velocity, we find planetary nebulae are no more than 10,000 years old. Older planetary nebulae evidently become mixed into the interstellar medium.

We see about 1500 planetary nebulae in our sky. Because planetary nebulae are short-lived formations, we can conclude that they must be a common part of stellar evolution. All stars like the sun, up to a mass of about 8 solar masses, are destined to die by forming planetary nebulae.

IC3568   Visual

This nearly spherical planetary nebula has a low-luminosity outer envelope and a highly excited inner region.

H. Bond, B. Ballick, and NASA

Visual

Hubble Heritage Team, AURA/STScI/NASA

The Ring Nebula in the constellation Lyra is visible even in small telescopes. Note the hot blue star at its center and the radial texture in the gas, suggesting outward motion.

The process that produces planetary nebulae involves two stellar winds. First, as an aging giant, the star gradually blows away its outer layers in a slow breeze of low-excitation gas that is not easily visible. Once the hot interior of the star is exposed, it ejects a high-speed wind that overtakes and compresses the gas of the slow wind like a snowplow, while ultraviolet radiation from the hot remains of the central star excites the gases to glow like a giant neon sign.

Slow stellar wind from a red giant

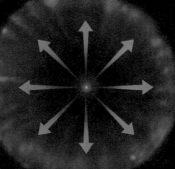

The gases of the slow wind are not easily detectable.

Fast wind from exposed interior

We see a planetary nebula where the fast wind compresses the slow wind.

Images from the Hubble Space Telescope reveal that asymmetry is the rule in planetary nebulae rather than the exception. A number of causes have been suggested. A disk of gas around a star's equator might form during the slow-wind stage and then deflect the fast wind into oppositely directed flows. Another star or planets orbiting the dying star, rapid rotation, or magnetic fields might cause these peculiar shapes. The Hour Glass Nebula seems to have formed when a fast wind overtook an equatorial disk (white in the image). The nebula Menzel 3, as do many planetary nebulae, shows evidence of multiple ejections.

Some shapes suggest bubbles being inflated in the interstellar medium (Hubble 5 and the Hour Glass Nebula) or precessing jets of gas (the Cat's Eye Nebula).

Visual + X-ray

The Cat's Eye Nebula

The purple glow in the image above is a region of X-ray bright gas with a temperature measured in millions of degrees. It is apparently driving the expansion of the nebula.

Hubble 5   **Visual**

Visual

Menzel 3   **Visual**

The Hour Glass Nebula

Visual

NGC6826

Visual

NGC7009

Some planetary nebulae, such as M2-9, are highly elongated, and it has been suggested that the Ring Nebula (opposite page) is a tubular shape pointed roughly at Earth.

Infrared   **M2-9**   **Visual**

This infrared image of the Egg Nebula reveals an irregular, thick disk from which beams of gas and dust emerge. Such beams may create many of the asymmetries in planetary nebulae.

The Egg Nebula

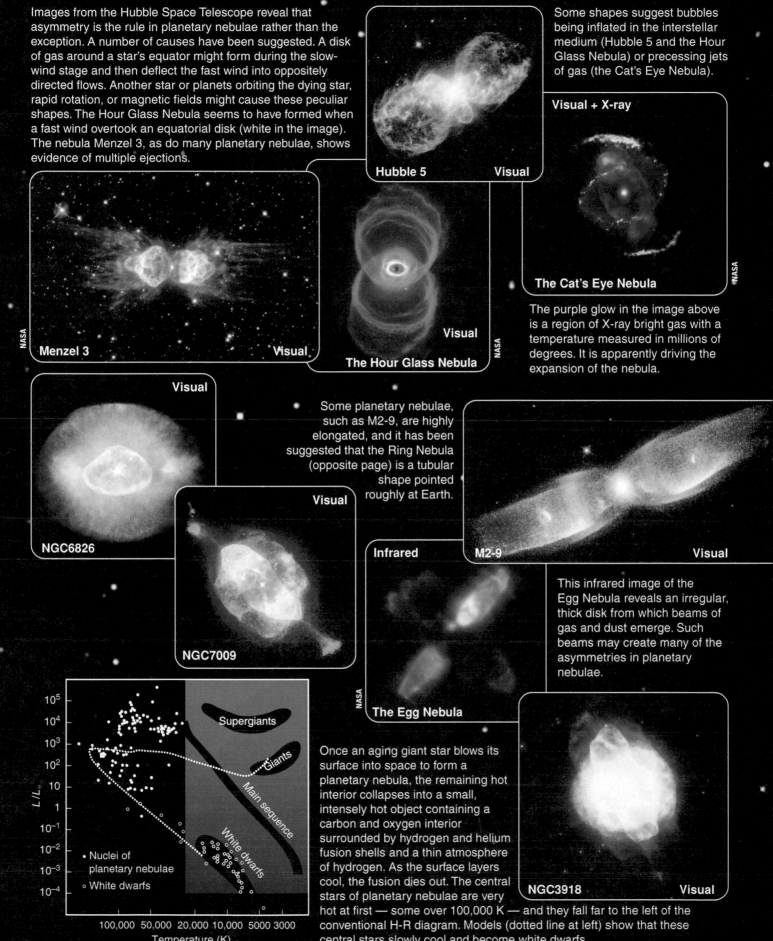

$L/L_\odot$

10^5
10^4
10^3
10^2
10
1
10^{-1}
10^{-2}
10^{-3}
10^{-4}

Supergiants

Giants

Main sequence

White dwarfs

• Nuclei of planetary nebulae
○ White dwarfs

100,000 50,000 20,000 10,000 5000 3000
Temperature (K)

Once an aging giant star blows its surface into space to form a planetary nebula, the remaining hot interior collapses into a small, intensely hot object containing a carbon and oxygen interior surrounded by hydrogen and helium fusion shells and a thin atmosphere of hydrogen. As the surface layers cool, the fusion dies out. The central stars of planetary nebulae are very hot at first — some over 100,000 K — and they fall far to the left of the conventional H-R diagram. Models (dotted line at left) show that these central stars slowly cool and become white dwarfs.

NGC3918   **Visual**

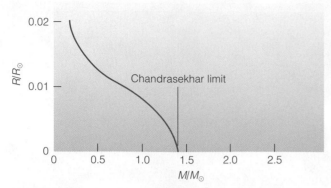

FIGURE 13-3
The more massive a white dwarf, the smaller its radius. Stars more massive than the Chandrasekhar limit of 1.4 solar masses cannot be white dwarfs.

This suggests that stars more massive than the Chandrasekhar limit can eventually die as white dwarfs if they reduce their mass. Theoretical models show an 8-solar-mass star should be able to reduce its mass to 1.4 solar masses before it collapses, and slightly more massive stars may also be able to get under the limit. Thus, a wide range of medium-mass stars eventually die as white dwarfs.

REVIEW CRITICAL INQUIRY

How will the sun die?
We have seen that lower-main-sequence stars like the sun die by producing planetary nebulae and then becoming white dwarfs. For the sun, this process will begin in a few billion years, when it exhausts the hydrogen in its core. It will then expand to become a giant star. Although the surface of the sun will become cooler, it will expand until it has such a large surface that it will be about 100 times more luminous than it is now. As a giant star, the sun will have a strong solar wind carrying gas into space and pushing back the interstellar medium. Eventually, the loss of mass will expose deeper layers in the sun. These extremely hot layers will ionize the gas around the sun and propel it outward to scoop up previously expelled gases and produce a planetary nebula. Soon the last remains of the sun will collapse inward and slowly cool to form a small, dense white dwarf. The surface will be very hot but so small that the sun will be about 100 times fainter than it is now.

The story of stellar evolution tells us that our sun will someday create a planetary nebula and become a white dwarf. At that point, what will have become of all the hydrogen the sun once contained?

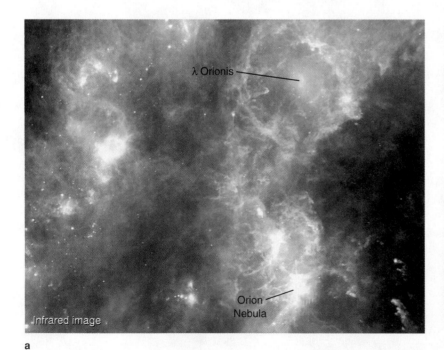

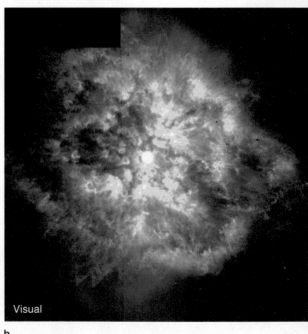

a   b

FIGURE 13-4
Many stars lose mass as they age. (a) This far-infrared map of the constellation Orion shows clouds of gas and dust heated and pushed away by hot, young stars. The Orion Nebula is the brightest area at lower right, and the large ring at upper right is expanding away from the O8 giant star λ Orionis. *(NASA/IPAC, Courtesy Deborah Levine)* (b) The massive star WR124 is ejecting mass in a violent stellar wind, forming chaotic arcs and blobs. *(Grosdidier and Moffat, University of Montreal; Joncas, Université Laval; Acker, Observatorie de Strasbourg; and NASA)*

Lower-main-sequence stars die by producing beautiful planetary nebulae and shrinking to become white dwarfs. Now we are ready to see what happens when one of these dying stars has a stellar companion.

13-2 THE EVOLUTION OF BINARY STARS

So far we have discussed the deaths of stars as if they were all single objects that never interact. But more than half of all stars are members of binary star systems. Most such binaries are far apart, and one of the stars can swell into a giant and eventually collapse without affecting the companion star. The two stars in some systems, however, are close together. When the more massive star begins to expand, it interacts with its companion star in peculiar ways.

These interacting binary stars are interesting objects themselves, but they are also important because they help us explain observed phenomena such as nova explosions. In the next chapter, we will use them to help us find black holes.

MASS TRANSFER

Binary stars can sometimes interact by transferring mass from one star to the other. To understand this process, we must understand how the gravity of the two stars controls the matter in the binary system.

Each star in a binary system controls a region of space near it, and if we could make these regions visible they would look like two teardrop-shaped lobes enclosing the stars and meeting tip to tip between the stars. These two volumes are called the **Roche lobes,** and the surface of the lobes is called the **Roche surface.** Because of the interaction of the gravity of the two stars and their rotation around their center of mass, there are five points of stability in the orbital plane called **Lagrangian points.*** ▌ Figure 13-5 illustrates the arrangement of the Roche surface around the stars and the Lagrangian points.

The Lagrangian points are important because they are locations where gas is stable in the revolving binary system. Gas at the L_2 or L_3 point tends to leak away, but gas at the L_4 or L_5 point can be trapped. (We will use the L_4 and L_5 points when we discuss planetary satellites and the orbits of asteroids in later chapters.) For our purposes here, the inner Lagrangian point L_1 located between the two stars is critical. If matter from one star can reach the inner Lagrangian point, it can

*The Lagrangian points are named after French mathematician Joseph Louis Lagrange, who solved this famous mathematical problem around the time of the French Revolution.

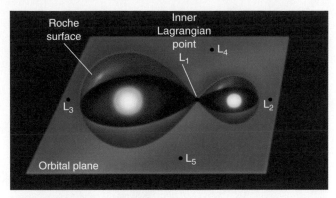

FIGURE 13-5

A pair of binary stars controls the region of space located inside the Roche surface. The Lagrangian points are locations of stability, with the inner Lagrangian point making a connection through which the two stars can transfer matter.

flow onto the other star. That is, the stars can transfer mass through the inner Lagrangian point.

In general, there are only two ways matter can escape from a star and reach the inner Lagrangian point. First, if a star has a strong stellar wind, some of the gas blowing away from the star can pass through the inner Lagrangian point and be captured by the other star. Second, if an evolving star expands so far that it fills its Roche surface, it will be forced to take on the teardrop shape of the Roche surface. We have seen that effect in our study of binary stars (Figure 9-20). If the star expands farther, matter will flow through the inner Lagrangian point (like water in a pond flowing over a dam) and will fall into the other star. Mass transfer driven by a stellar wind tends to be slow, but mass transfer driven by an expanding star can occur rapidly.

RECYCLED STELLAR EVOLUTION

Mass transfer between stars can affect their evolution in surprising ways. In fact, it provides the explanation for a problem that puzzled astronomers for many years.

In some binary systems, the less massive star has become a giant, while the more massive star is still on the main sequence. If more massive stars evolve faster than lower-mass stars, how does the low-mass star in such binaries manage to leave the main sequence first? This is called the Algol paradox, after the binary system Algol (see Figure 9-21).

Mass transfer explains how this could happen. Imagine a binary system that contains a 5-solar-mass star and a 1-solar-mass companion (▌ Figure 13-6). The two stars formed at the same time, so the more massive star will evolve faster and leave the main sequence first. When it expands into a giant, it can fill its Roche surface and transfer matter to the low-mass companion. Thus, the massive star could evolve into a lower-

The Evolution of a Binary System

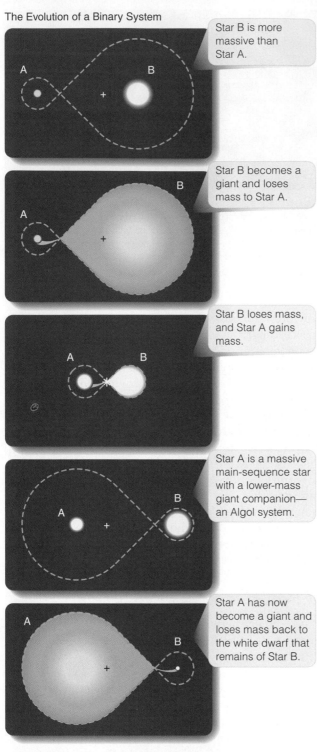

Star B is more massive than Star A.

Star B becomes a giant and loses mass to Star A.

Star B loses mass, and Star A gains mass.

Star A is a massive main-sequence star with a lower-mass giant companion—an Algol system.

Star A has now become a giant and loses mass back to the white dwarf that remains of Star B.

FIGURE 13-6
A pair of stars orbiting close to each other can exchange mass and modify their evolution.

mass star, and the companion could gain mass and become a massive star still on the main sequence. We might find a system such as Algol containing a 5-solar-mass main-sequence star and a 1-solar-mass giant.

The evolution of close binary stars could result in one of the stars having its outer layers "peeled" away. A massive star expanding to become a giant could lose its outer layers to its companion and then collapse to form a lower-mass peculiar star. A few such peculiar stars are known.

Another exotic result of the evolution of close binary systems is the merging of the stars. We see many binaries in which both stars have expanded to fill their Roche surfaces and spill mass out into space. If the stars are close enough together and expand rapidly enough, theorists believe, the two stars could merge into a single, rapidly rotating giant star. Most giants rotate slowly because they conserved angular momentum as they expanded and consequently slowed. Examples of rapidly rotating giant stars are known, however, and they are believed to be the results of merged binary stars. Inside the distended envelope of such a star, the cores of the two stars could continue to orbit each other until friction slowed them down and they sank to the center.

Mass transfer can lead to dramatic violence. The first four frames of Figure 13-6 show mass transfer producing a system like Algol. The last frame shows an additional stage in which the giant star has expelled its outer layers and has collapsed to form a white dwarf. The more massive companion has expanded, transferring matter back onto the white dwarf. Such systems can become the site of tremendous explosions. To see how this can happen, we must consider in detail how mass falls into a star.

ACCRETION DISKS

We say that matter flows through the inner Lagrangian point and falls onto the white dwarf, but the matter cannot fall directly into a white dwarf because of the conservation of angular momentum (as described in Chapter 5). Instead, it falls into a whirlpool around the white dwarf. For a common example, consider a bathtub full of water. Gentle currents in the water give it some angular momentum, but its slow circulation is not apparent until we pull the stopper. Then, as the water rushes toward the drain, conservation of angular momentum forces it to form a whirlpool. This same effect forces gas falling into a white dwarf to form a whirling disk of gas called an **accretion disk** (Figure 13-7).

Two important things happen in an accretion disk. First, the gas in the disk grows very hot due to friction and tidal forces. The disk also acts as a brake, ridding the gas of its angular momentum and allowing it to fall into the white dwarf. The temperature of the gas in the inner parts of an accretion disk can exceed 1,000,000 K, and the gas can emit X rays. In addition, the matter falling inward from the accretion disk can cause a violent explosion if it accumulates on the white dwarf.

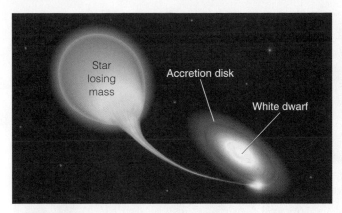

FIGURE 13-7
Matter falling into a compact object forms a whirling accretion disk. Friction and tidal forces can make the disk very hot.

NOVA EXPLOSIONS

At the beginning of this chapter we saw that the word "nova" refers to a new star that appears in the sky for a while and then fades away (■ Figure 13-8). The evidence tells us that it is not a new star at all. After a nova fades, astronomers can photograph the spectrum of the object, and they invariably find a closely spaced binary star containing a normal star and a white dwarf. A nova is evidently an explosion involving a white dwarf in a binary system.

Observational evidence can tell us how nova explosions occur. As the explosion begins, spectra show blue-shifted absorption lines, which tells us the gas is dense and coming toward us at a few thousand kilometers per second. After a few days, the spectral lines change to emission lines, which tells us the gas has thinned, but the blue shifts remain, showing that a cloud of debris had been ejected into space.

Theory gives us a way to understand nova explosions as the result of mass transfer from a normal star, through the inner Lagrangian point, through an accretion disk, onto the surface of the white dwarf. Because the matter comes from the surface of a normal star, it is rich in unfused fuel, mostly hydrogen, and when it accumulates on the surface of the white dwarf, it forms a layer of unprocessed fuel.

As the layer of fuel grows deeper, it becomes hotter and denser and eventually becomes degenerate, much like the core of a sunlike star approaching the helium flash. In such a gas, the pressure–temperature thermostat does not work, so the layer of unfused hydrogen is a thermonuclear bomb waiting to explode. By the time the white dwarf has accumulated about 100 Earth masses of hydrogen, the temperature at the base of the hydrogen layer reaches millions of degrees, and the density is 10,000 times the density of water. Suddenly, the hydrogen begins to fuse on the proton–proton chain, and the energy released drives the temperature so high that the CNO cycle begins. With no pressure–temperature thermostat to control the fusion, in seconds the temperature shoots to 100 million degrees. Quickly the temperature rises high enough to force the gas to expand, and it stops being degenerate. But it is too late. By the time the gas begins to expand, the amount of energy released is enough to blow the surface layers of the star into space as a violently expanding shell of hot gas, which we see from Earth as a nova.

The explosion of its surface hardly disturbs the white dwarf or its companion star. Mass transfer quickly resumes, and a new layer of fuel begins to accumulate. How fast the fuel builds up depends on the rate of mass transfer. Accordingly, we can expect novae to repeat each time an explosive layer accumulates. Many novae take thousands of years to build an explosive layer, but some take only a few years (■ Figure 13-9).

THE END OF EARTH

Astronomy is about us. Although we have been discussing the deaths of medium-mass stars, white dwarfs, and novae explosions, we have also been discussing the future of our planet. The sun is a medium-mass star

FIGURE 13-8
Nova Cygni 1975 near maximum at about second magnitude (top) and later when it had declined to about eleventh magnitude (bottom). *(Photo © UC Regents)*

FIGURE 13-9

Nova T Pyxidis erupts about every two decades, expelling shells of gas into space. The shells of gas are visible from ground-based telescopes, but the Hubble Space Telescope reveals much more detail. The shell consists of knots of excited gas that presumably form when a new shell collides with a previous shell. *(M. Shara and R. Williams, STScI; R. Gilmozzi, ESO; and NASA)*

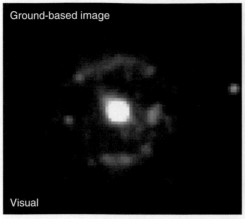

Ground-based image

Visual

Hubble Space Telescope image

Visual

and must eventually die by becoming a giant, possibly producing a planetary nebula, and collapsing into a white dwarf. That will spell the end of Earth.

Mathematical models of the sun suggest that it may survive for 5 billion years or so, but it is already growing more luminous as it fuses hydrogen into helium. In a few billion years, it will exhaust hydrogen in its core and swell into a giant star about 100 times its present radius. That giant sun will be about as large as the orbit of Earth, so that will mark the end of our world. Whether or not the expanding sun becomes large enough to totally engulf Earth, its growing luminosity will certainly evaporate our oceans, drive away our atmosphere, and even vaporize most of Earth's crust.

Ace **Astronomy**™ Go to AceAstronomy and click Active Figures to see "The Future of the Sun" and how Earth's temperature will change as the sun evolves.

While it is a giant star, the sun will lose mass into space. This mass loss is a relatively gentle process, so any cinder that might remain of Earth would not be disturbed. Of course, when the sun finally collapses into a white dwarf, the solar system will become a much colder place, but not much of Earth will remain by then.

If the collapsing sun becomes hot enough, it will ionize the expelled gas to form a planetary nebula. Some models, however, suggest that the sun is not quite massive enough. It will collapse into a hot object that will cool to become a white dwarf, but it may not become hot enough to ionize the expelled gas and "turn on" a planetary nebula. An astronomer recently expressed disappointment that the dying sun might not leave behind a beautiful planetary nebula, but that embarrassment lies a few billion years in the future.

There is no danger that the sun will explode as a nova; it has no binary companion. And, as we will see, the sun is not massive enough to die the violent death of the massive stars.

The most important lesson of astronomy is that we are part of the universe and not just observers. The atoms we are made of are destined to return to the in-

terstellar medium in just a few billion years. That's a long time, and it is possible that the human race will migrate to other planetary systems. That might save the human race, but our planet is star dust.

REVIEW CRITICAL INQUIRY

How does modern astronomy explain the Algol paradox?

When we encounter a paradox in nature, it is usually a warning that we don't understand things as well as we thought we did. The Algol paradox is a good example. The binary star Algol contains a lower-mass giant star and a more massive main-sequence star. Because the two stars must have formed together, they are the same age; and the more massive star should have evolved first and left the main sequence. But in binary systems such as Algol, the lower-mass star has left the main sequence, or so it seems.

We can understand this paradox if we add mass transfer to our theory of stellar evolution. The first star to leave the main sequence must be the more massive star; but if the stars are close together, the star that is initially more massive can fill its Roche surface and transfer mass back to its companion. The companion can grow more massive, and the giant can grow less massive. Thus, the lower-mass giant star we see in the system may have originally been more massive and may have evolved away from the main sequence, leaving its companion behind to increase in mass as it decreased.

Mass transfer can explain the Algol paradox, and it can also explain the violent explosions called novae. How does mass transfer explain why novae can explode over and over in the same binary system?

Our study of the deaths of low- and medium-mass stars has led us to discuss planetary nebulae, white dwarfs, and the cause of novae. But there are more violent fish in the cosmic sea. What causes supernovae? To find out, we must consider how massive stars die.

13-3 THE DEATHS OF MASSIVE STARS

We have seen that low- and medium-mass stars die relatively quietly as they exhaust their hydrogen and helium and then eject their surface layers to form planetary nebulae. In contrast, massive stars live spectacular lives and destroy themselves in violent explosions.

NUCLEAR FUSION IN MASSIVE STARS

Stars on the upper main sequence have too much mass to die as white dwarfs, but their evolution begins much like that of their lower-mass cousins. They consume the hydrogen in their cores and ignite hydrogen shells; as a result, they expand and become giants or, for the most massive stars, supergiants. Their cores contract and fuse helium first in the core and then in a shell, producing a carbon–oxygen core.

A massive star can lose significant mass as it ages, but if it still has a mass over 4 solar masses when its carbon–oxygen core contracts, it can reach a temperature of 600 million Kelvin and ignite carbon fusion. The fusion of carbon produces heavier nuclei such as oxygen and neon. As soon as the carbon is exhausted in the core, the core contracts, and carbon ignites in a shell. This pattern of core ignition followed by shell ignition continues with fuel after fuel, and the star develops a layered structure at its center (Figure 13-10), with a hydrogen-fusion shell above a helium-fusion shell above a carbon-fusion shell and so on. After carbon fusion, oxygen, neon, magnesium, and heavier elements fuse right up to iron.

The fusion of these nuclear fuels goes faster and faster as the massive star evolves rapidly. Recall that massive stars must consume their fuels rapidly to support their great weight, but other factors also cause the heavier fuels like carbon, oxygen, and silicon to fuse at increasing speed. For one thing, the amount of energy released per fusion reaction decreases as the mass of the fusing atom increases. To support its weight, a star

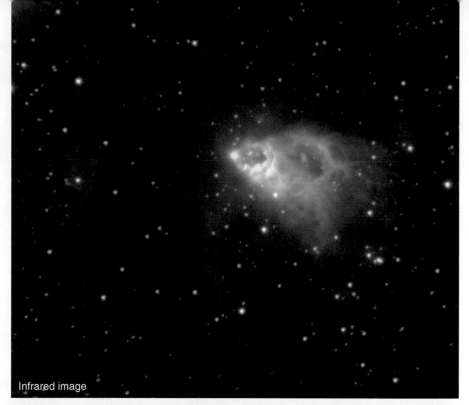

Infrared image

a

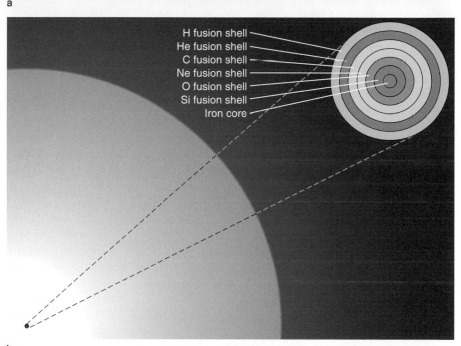

H fusion shell
He fusion shell
C fusion shell
Ne fusion shell
O fusion shell
Si fusion shell
Iron core

b

ACTIVE FIGURE 13-10

Massive stars live fast and die young. (a) Hidden deep inside the molecular cloud where it was born, the massive star AFGL2591 is rapidly expelling rings of gas to the right. Similar clouds of gas being expelled to the left are hidden behind a dusty disk surrounding the star. *(Gemini Observatory/NSF/C. Aspin)* (b) Inside an aging massive star lies an Earth-size core containing most of the mass. As nuclear fuels fuse one after the other, the core develops layers surrounding an inert iron core at the center.

Ace⊗Astronomy™ Go to AceAstronomy and click Active Figures to see "Inside Stars" and compare the interior of a sunlike star with that of a massive star.

must fuse oxygen much faster than it fused hydrogen. Also there are fewer atoms in the core of the star by the time heavy atoms begin to fuse. Four hydrogens made a helium atom, and three heliums made a carbon, so there are 12 times fewer atoms of carbon available for fusion than there were hydrogen. This means the heavy elements are used up and fusion goes very quickly in massive stars (▌Table 13-1). Hydrogen fusion can last 7 million years in a 25-solar-mass star, but that same star will fuse its oxygen in 6 months and its silicon in a day.

THE IRON CORE

Heavy-element fusion ends with iron, because nuclear reactions that use iron as a fuel cannot produce energy. Nuclear reactions can produce energy if they proceed from less tightly bound nuclei to more tightly bound nuclei. As shown in Figure 8-19, both nuclear fission and nuclear fusion produce nuclei that are more tightly bound than the starting fuel. Look again at Figure 8-19, and notice that iron is the most tightly bound nucleus of all. No nuclear reaction, fission or fusion, that starts with iron can produce a more tightly bound nucleus, and that means that iron is a dead end.

When a massive star develops an iron core, nuclear fusion cannot produce energy, and the core contracts and grows hotter. The shells around the core burn outward, fusing lighter elements and leaving behind more iron, which further increases the mass of the core. When the mass of the iron core exceeds 1.3 to 2 solar masses, the core must collapse.

As the core begins to contract, two processes can make it contract even faster. Heavy nuclei in the core can capture high-energy electrons, thus removing thermal energy from the gas. This robs the gas of some of the pressure it needs to support the crushing weight of the outer layers. Also, in more massive stars, temperatures are so high that many photons have gamma-ray wavelengths and can break more massive nuclei into less-massive nuclei. This reversal of nuclear fusion absorbs energy and allows the core to collapse even faster.

Although a massive star may live for millions of years, its iron core—about 500 km in diameter—collapses in only a few thousandths of a second. This collapse happens so rapidly that our most powerful computers are not capable of following the details. One thing is clear, however. The collapse of the iron core in a massive star triggers a star-destroying explosion—a supernova.

THE SUPERNOVA DEATHS OF MASSIVE STARS

Modern theory predicts that the collapse of a massive star can eject the outer layers of the star to produce a supernova explosion while the core of the star col-

TABLE 13-1
Heavy-Element Fusion in a 25-$M_\odot$ Star

Fuel	Time	Percentage of Lifetime
H	7,000,000 years	93.3
He	500,000 years	6.7
C	600 years	0.008
O	0.5 years	0.000007
Si	1 day	0.00000004

lapses to form a neutron star or a black hole. We will examine neutron stars and black holes in detail in the next chapter; here we concentrate on the process that triggers the supernova explosion.

A supernova explosion is rare, remote, rapid, and violent. It is an event in nature that is extremely difficult to study in person, which is why astronomers have used powerful mathematical modeling techniques and high-speed supercomputers to explore the inside of a star exploding as a supernova. Such models allow astronomers to experiment on an exploding star as if the star were in a laboratory beaker.

Mathematical models reveal that the key to the supernova explosion is the collapse of the iron core, which allows the rest of the interior of the star to fall inward, creating a tremendous "traffic jam" as all the nuclei try to fall toward the center. It is as if all the residents of Indiana suddenly tried to drive their cars into downtown Indianapolis. There would be a tremendous traffic jam, not only downtown but also in the suburbs. As more cars arrived, the traffic jam would spread outward. Similarly, as the inner core of the star falls inward, a shock wave (a traffic jam) develops and begins to move outward. Containing about 100 times more energy than necessary to destroy the star, such a shock wave was first thought to be the cause of the supernova explosion.

Computer models revealed, however, that the shock wave spreading outward through the collapsing star stalls within a few hundredths of a second. Matter flowing inward smothers the shock wave and pushes it back into the star. If this happened in a real star, it presumably would collapse without any visible explosion.

However, theory predicts that 99 percent of the energy released in the collapse will appear as neutrinos. In the sun, neutrinos zip outward unimpeded by the gas of the solar layers, but in a collapsing star the density is as high as 10^{12} g/cm^3, nearly as dense as an atomic nucleus. This gas is opaque to neutrinos, and they are partially absorbed in the gas. Not only does the tremendous burst of neutrinos remove energy from the core and allow the core to collapse even faster, but the neutrinos are absorbed outside the core and heat those layers. For some years, astronomers thought that these neutrinos could spread energy across the shock wave and, within a quarter of second or so, reaccelerate the stalled shock wave. The computer models, however, refused to explode.

The final ingredient that the models needed to make them explode was turbulent convection. When the collapse begins, the very center of the star forms a highly dense core, the beginnings of a neutron star, and the infalling material bounces off of that core. With the temperature shooting up, the bouncing material produces highly turbulent convection currents that give the stalled shock wave an outward boost (▌Figure 13-11). Within a second or so, the shock wave begins to push outward, and after just a few hours it bursts out through the surface, blasting the star apart in a supernova explosion.

The supernova we see is the brightening of the star as its distended outer layers are blasted outward. As the months pass, the cloud of gas expands, thins, and begins to fade. But the way it fades in some cases tells astronomers about the death throes of the star. Essentially all of the iron in the core of the star is destroyed when the core collapses, but the violence in the outer layers can produce densities and temperatures high enough to trigger nuclear fusion reactions that produce as much as half a solar mass of radioactive nickel-56. The nickel gradually decays to form radioactive cobalt, which decays to form normal iron. The rate at which the supernova fades matches the rate at which these radioactive elements decay. Thus the destruction of iron in the core of the star is matched by the production of iron through nuclear fusion in the expanding outer layers.

The presence of nuclear fusion in the outer layers of the supernova testifies to the violence of the explosion. A typical supernova is equivalent to the explosion of 10^{28} megatons of TNT—about 3 million solar masses of high explosive.

Collapsing massive stars can trigger cosmic violence, but astronomers have observed more than one kind of supernova.

TYPES OF SUPERNOVAE

Supernovae are rare and only a few have been seen in our galaxy, but astronomers have been able to observe supernovae occurring in other galaxies (▌Figure 13-12). From these data accumulated over decades, astronomers have noticed that there are two main types. **Type I supernovae** have spectra that contain no hydrogen lines. They reach a maximum brightness about 4 billions times more luminous than the sun, decline rapidly at first and then more slowly. **Type II supernovae** have spectra containing hydrogen lines. They reach a maximum brightness up to about 0.6 billion times more luminous than the sun, decline to a standstill and then fade rapidly. The light curves in ▌Figure 13-13 summarize the behavior of the two types of supernovae.

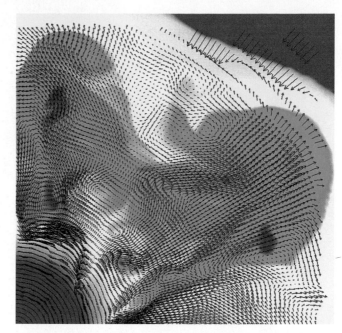

FIGURE 13-11
The inside of an exploding star is shown in this mathematical model of a supernova explosion. The image here represents a region 150 km on a side with the center of the star at the lower left corner. Black arrows indicate the motion of the gas. Turbulent material (blue) has bounced off the core and is rushing outward to meet the rest of the star, which is falling inward (red at upper right). The innermost part of the star is forming a neutron star (red at lower left). To show the entire star at this scale, this figure would have to be 140 km in diameter. *(Courtesy Adam Burrows, John Hayes, and Bruce Fryxell)*

The evidence is clear that type II supernovae occur when a massive star develops an iron core and collapses. They occur in regions of active star formation where there is plenty of gas and dust. These are the

FIGURE 13-12
This supernova (arrow) was seen in the outskirts of the galaxy NGC4526. From its spectrum and from the way its brightness faded with time, astronomers recognize it as a type Ia supernova produced by the collapse and total destruction of a white dwarf. *(High-Z Supernova Search Team, HST, NASA)*

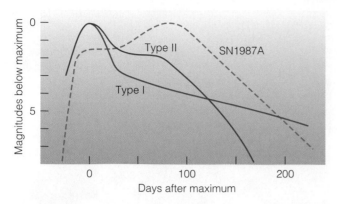

FIGURE 13-13
Type I supernovae decline rapidly at first and then more slowly, but type II supernovae pause for about 100 days before beginning a steep decline. Supernova 1987A was odd in that it did not rise directly to maximum brightness. These light curves have been adjusted to the same maximum brightness. Generally, type II supernovae are about two magnitudes fainter than type I.

regions where we would expect to find massive stars. The spectra of type II supernovae contain hydrogen lines, as we would expect from the explosion of a massive star that contains large amounts of hydrogen in its outer layers.

Type I supernovae are further classified into type Ia and type Ib according to their spectra and light curves. Type Ia supernovae are often found in regions where star formation ended long ago, and they lack hydrogen lines in their spectra, so they can't be caused by the deaths of massive stars. Rather, the evidence shows that a type Ia occurs when a white dwarf gaining mass in a binary star system exceeds the Chandrasekhar limit and collapses.

The collapse of a white dwarf is different from the collapse of a massive star because the core of the white dwarf contains usable fuel. As the collapse begins, the temperature shoots up, but the gas cannot halt the collapse because it is degenerate, and the temperature–pressure thermostat is turned off in the core. Even as carbon fusion begins, the increased temperature cannot increase the pressure and make the gas expand and slow the reactions. The core of the white dwarf is a bomb. The carbon–oxygen core fuses suddenly in violent nuclear reactions called **carbon deflagration.** The word *deflagration* means to be totally destroyed by fire, in this case by nuclear fusion. In a flicker of a stellar lifetime the entire star is consumed, and the outermost layers are blasted away in a violent explosion that at its brightest is 3 to 6 times more luminous than a type II supernova. The white dwarf is entirely destroyed. No neutron star or black hole is left behind. Of course, we see no hydrogen lines in the spectrum of a type Ia supernova because white dwarfs contain very little hydrogen.

The less common type Ib supernovae are believed to occur when a massive star in a binary system loses its hydrogen-rich outer layers to its companion star.

The remains of the massive star could develop an iron core and collapse, producing a supernova explosion that lacks hydrogen lines in the spectrum.

To summarize, a type II supernova is caused by the collapse of a massive star. A type Ia is caused by the collapse of a white dwarf. A type Ib is caused by the collapse of a massive star that has lost its outer envelope of hydrogen.

Much of our discussion so far has been based on theory, so it is time to compare our theory with observation of real supernova explosions.

OBSERVATIONS OF SUPERNOVAE

In 1054, Chinese astronomers saw a "guest star" appear in the constellation we know as Taurus, the Bull. The star quickly became so bright it was visible in the daytime. After a month's time, it slowly faded, taking almost two years to vanish from sight. When modern astronomers turned their telescopes to the location of the guest star, they found a cloud of gas about 1.35 pc in radius, expanding at 1400 km/s. Projecting the expansion back in time, they concluded that the expansion must have begun about nine centuries ago, just when the guest star made its visit. Thus, we think the nebula, now called the Crab Nebula because of its shape (■ Figure 13-14), marks the site of the 1054 supernova.

Supernovae are rare. Only a few have been seen with the naked eye in recorded history. Arab astrono-

FIGURE 13-14
The Crab Nebula, the remains of the supernova of AD 1054, is located in the constellation Taurus and is bright enough to see with a small telescope. At visual wavelengths, the nebula consists of a network of expanding filaments containing a hazy glow. *(European Southern Observatory)*

False-Color Images and Reality

Astronomers use computers to add false color to images and reveal things our eyes cannot see. Radio, infrared, ultraviolet, and X-ray radiation are not visible to our eyes, so data collected at these wavelengths can be displayed in images that are given colors visible to our eyes. Of course, the colors used are entirely arbitrary. Radio astronomers might choose to use reds and yellows to color a radio map of a cloud of gas, or they might choose greens and blues. The colors make the radio energy visible to our eyes, but the choice of colors is up to the astronomer analyzing the image.

The variation in color across an image does have meaning, however.

Radio astronomers might decide to use reds for the regions of strongest radio signal, yellows for fainter regions, and blues for the faintest. Thus, we could look at the radio map of a gas cloud and immediately pick out the parts of the cloud that were emitting the strongest radio energy.

In addition to using false color in images recorded in wavelengths our eyes cannot see, astronomers often use false color in photographs recorded at visible wavelengths. These images can reveal to our eyes subtle variations in brightness not visible in the original photographs.

False-color images are common in astronomy, but they are also used in other sciences. Doctors often analyze medical X-ray images, CAT scans, and so on by converting the images to false color. Biologists use false color to analyze microscopic photographs, and geologists use false color to study photographs of Earth recorded by a satellite at various wavelengths.

We might think of false-color images as deceptive, but they actually help us see parts of the universe that would otherwise never register on our limited eyes.

---

mers saw one in 1006, and the Chinese saw one in 1054. European astronomers observed two—one in 1572 (Tycho's supernova) and one in 1604 (Kepler's supernova). Also, the guest stars of 185, 386, 393, and 1181 may have been supernovae. For 383 years following 1604, no naked-eye supernova appeared, and then in February 1987 a star in the southern sky exploded. We will examine the great supernova of 1987 later in this section. Here we will explore supernovae in general.

Most supernovae are discovered in distant galaxies, and searching for these supernovae was once a tedious job. Robotic telescopes now observe long lists of galaxies each night searching for the appearance of a new star. Detecting supernovae early is important because astronomers need to study the early stages of the explosion in order to understand the complex processes that occur in the rapidly expanding cloud of gas.

The supernova explosion fades to obscurity in a year or two, but an expanding shell of gas marks the explosion site. The gas, originally expelled at 10,000 to 20,000 km/s, may carry away one-fifth of the mass of the star. The collision of that expanding gas with the surrounding interstellar medium can sweep up even more gas and excite it to produce a **supernova remnant,** the nebulous remains of a supernova explosion.

Supernova remnants look quite delicate and do not survive very long—a few tens of thousands of years—before they gradually mix with the interstellar medium and vanish. The Crab Nebula is a young remnant, only about 950 years old, and isn't very large, only a few parsecs in diameter (Figure 13-14). Older remnants can be larger. Some supernova remnants are detectable only at radio and X-ray wavelengths. They have become too tenuous to emit detectable light, but the collision of the expanding hot gas with the interstellar medium can

generate radio and X-ray radiation and allows us to create images of them at these nonvisible wavelengths (Window on Science 13-1). In general, supernova remnants are tenuous spheres of gas expanding into the interstellar medium (▌ Figure 13-15). We saw in Chapter 11 that the compression of the interstellar medium by expanding supernova remnants can trigger star formation.

Radio images of supernova remnants reveal that they are emitting **synchrotron radiation**—electromagnetic energy radiated by high-speed electrons spiraling through a magnetic field (▌ Figure 13-16). Low-speed electrons radiate at longer wavelengths, and high-speed electrons radiate at shorter wavelengths, so synchrotron radiation is spread over a wide range of wavelengths. The presence of synchrotron radiation coming from a nebula tells us that the nebula contains magnetic fields and energetic electrons.

Some of the energy coming from the Crab Nebula is synchrotron radiation, and it is so energetic that some of it has wavelengths as short as light. The general hazy or foggy glow of the nebula is produced by synchrotron radiation. As the electrons in a nebula emit synchrotron radiation, they slow. To emit synchrotron radiation as light, the electrons in the Crab Nebula must be traveling very fast, but over the 950-year history of the nebula, those electrons should have radiated their energy away and slowed down. That is, the synchrotron glow of the nebula can't have lasted since the supernova explosion. The observation of synchrotron radiation coming from the Crab Nebula today tells us that something must be continuously supplying energy to keep the electrons moving rapidly. We will solve this mystery in the next chapter when we locate a neutron star in the center of the Crab Nebula.

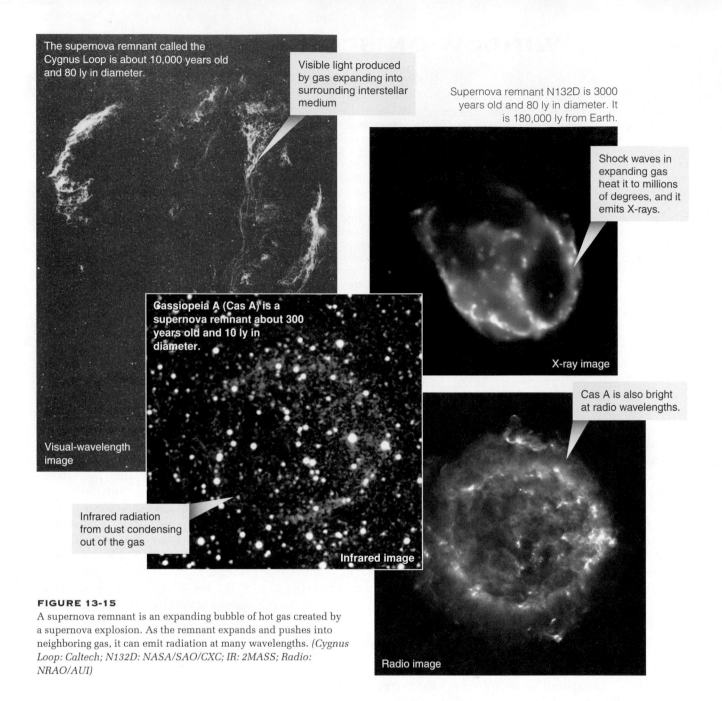

The supernova remnant called the Cygnus Loop is about 10,000 years old and 80 ly in diameter.

Visible light produced by gas expanding into surrounding interstellar medium

Supernova remnant N132D is 3000 years old and 80 ly in diameter. It is 180,000 ly from Earth.

Shock waves in expanding gas heat it to millions of degrees, and it emits X-rays.

Cassiopeia A (Cas A) is a supernova remnant about 300 years old and 10 ly in diameter.

Visual-wavelength image

X-ray image

Cas A is also bright at radio wavelengths.

Infrared radiation from dust condensing out of the gas

Infrared image

Radio image

FIGURE 13-15

A supernova remnant is an expanding bubble of hot gas created by a supernova explosion. As the remnant expands and pushes into neighboring gas, it can emit radiation at many wavelengths. *(Cygnus Loop: Caltech; N132D: NASA/SAO/CXC; IR: 2MASS; Radio: NRAO/AUI)*

THE GREAT SUPERNOVA OF 1987

Until 1987, astronomers had never looked at a bright supernova through a telescope. Nearly all supernovae are in distant galaxies and thus are very faint. The last supernova visible to the naked eye, Kepler's supernova of 1604, predated the invention of the telescope. Then, in late February 1987, the news raced around the world. Astronomers in Chile had discovered a naked-eye supernova in the Large Magellanic Cloud, a small galaxy very near our Milky Way Galaxy (❚ Figure 13-17). Because the supernova was only 20 degrees from the south celestial pole, it could be studied only from southern latitudes. It was named SN1987A to denote the first supernova discovered in 1987.

The hydrogen-rich spectrum suggested that the supernova was a type II, caused by the collapse of the core of a massive star. As the months passed, however, the light curve proved to be odd (Figure 13-13) in that it paused for a few weeks before rising to its final maximum. From photographs of the area made some years before, astronomers were able to determine that the star that exploded, cataloged as Sanduleak −69°202, was not the expected red supergiant but rather a hot, blue supergiant of only 15 solar masses and 50 solar radii, not extreme for a supergiant. Theorists now believe that the star was chemically poor in elements heavier than helium and had consequently contracted and heated up after a phase as a cool, red supergiant, during which it lost mass into space. The relatively small

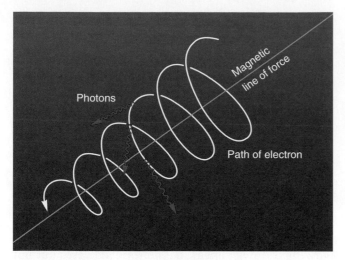

FIGURE 13-16

Synchrotron radiation is emitted when an electron spirals around a magnetic field line and radiates its energy away as photons. Only a few photons are shown here, for clarity.

FIGURE 13-17

Supernova 1987A eventually reached a peak brightness of magnitude 2.9, nearly as bright as the stars in the Big Dipper, and was easily visible to the unaided eye. An earlier photograph of the region, left, allows astronomers to identify the star that exploded (arrow). *(Anglo-Australian Observatory)*

size of the supergiant may explain the pause in the light curve. Much of the energy of the explosion went into blowing apart the smaller, denser-than-usual star and making it expand.

The brightening of the supernova after the first few weeks seems to have been caused by the decay of radioactive nickel into cobalt. Theory predicts the production of such nickel atoms in the explosion, and their decay into cobalt would release gamma rays that would heat the expanding shell of gas and make it brighter. About 0.07 solar mass of nickel was produced, about 20,000 times the mass of Earth.

The cobalt atoms are also unstable, but they decay more slowly, so it was not until sometime later, after much of the nickel had decayed, that the decay of cobalt into iron began providing energy to keep the expanding gas hot and luminous. Although these processes had been predicted, they were clearly observed in SN1987A. Gamma rays from the decay of cobalt to iron were detected, and cobalt and iron are clearly visible in the infrared spectra of the supernova.

As the supernova dimmed, photographs revealed bright rings of gas (▮ Figure 13-18). Comparison with mathematical models shows that the rings were produced by two stellar winds. When the star was a red supergiant, it expelled a slow stellar wind. Later when it became a blue supergiant it expelled a fast stellar wind. The interaction of these two winds shaped by a magnetic field and illuminated by the supernova explosion produces the rings. Compare this with the rings seen in some of the planetary nebulae shown on page 261.

Two independent observations confirm that SN1987A probably gave birth to a neutron star. Theory predicts that the collapse of a massive star's core should liberate a tremendous blast of neutrinos that leave the star hours before the shock wave from the

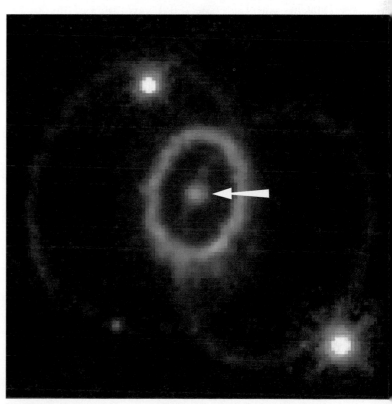

FIGURE 13-18

Supernova 1987A (arrow) is surrounded by three rings. (The bright stars are foreground stars not related to the supernova.) The rings are caused by the interaction of a fast stellar wind, an older slow stellar wind, and a magnetic field. The rings resemble the rings seen in some planetary nebulae such as the Hour Glass Nebula on page 261. *(Christopher Burrows, ESA/STScI, NASA)*

interior blows the star apart. Two detectors, one in Ohio and one in Japan, recorded a burst of neutrinos passing through Earth at 2:35:41 AM EST on February 23, 1987, about 18 hours before the supernova was seen. The detectors caught only 19 neutrinos during a 12-second interval, but recall that neutrinos hardly ever react with normal matter. The full flood of neutrinos was immense. Within a few seconds of that time, roughly 20 trillion neutrinos passed harmlessly through each human body on Earth. The detection of the neutrino blast confirms that the collapsing core gave birth to a neutron star.

The expanding gas shell of the supernova will continue to thin and cool, and eventually astronomers on Earth will be able to peer inside and see what is left of Sanduleak $-69°202$. Will we see the expected neutron star, or will theories need further revision? For the first time in almost 400 years, astronomers can observe a bright supernova to test their theories.

LOCAL SUPERNOVAE AND LIFE ON EARTH

Although supernovae are rare events, they are very powerful and could affect life on planets orbiting nearby stars. In fact, supernovae explosions long ago may have affected Earth's climate and the evolution of life.

If a supernova occurred within about 50 ly of Earth, the human race would have to abandon the surface and live below ground for at least a few decades. The burst of gamma rays and high-energy particles from the supernova explosion could kill many life forms and cause serious genetic damage in others. The only way we could avoid this radiation would be to move our population into tunnels below Earth's surface. Of course, if a supernova did occur, we would not have time to dig enough tunnels.

Even if we could survive in tunnels long enough for the radioactivity on Earth's surface to subside, we might not like Earth when we emerged. Genetic mutation induced by radioactivity could alter plant and animal life so seriously that we might not be able to support our population. After all, we humans depend almost totally on grass for food. The basic human foods—milk, butter, eggs, wheat, tomatoes, lettuce, and meat (Big Macs, in other words)—are grass and similar vegetation processed into different forms. Seafood is merely processed ocean plankton and plant life, which might also be altered or damaged by a local supernova explosion. Even if surface life survived the radiation, damage to the delicate upper layers of our atmosphere might alter the climate dramatically.

Local supernovae have been suggested as a possible cause of occasional climate changes and extinctions in Earth's past. It is quite possible that, as the sun moves through space, supernovae explode near Earth every few hundred million years. A really close call could trigger an extinction.

But do supernovae really affect Earth? In 1979, scientists studying Antarctic ice cores containing very old ice noticed that ice from the years 1181, 1320, 1572, and 1604 contains high abundances of nitrates. Bright supernovae were seen in the years 1181, 1572, and 1604. Ultraviolet radiation from those supernovae might have altered Earth's upper atmosphere and produced acid rain that was recorded in the ice as nitrates. But what of the ice from 1320? In 1999, X-ray astronomers noticed a supernova remnant in the southern sky. When they calculated its age, they found the supernova should have occurred about 1320. It isn't clear why there is no historical record of this supernova or why the nitrates appear in Antarctic ice cores but not in Greenland ice cores. In any case, this evidence may alert us that even distant supernova explosions can affect Earth in subtle ways.

A study of cores from the seafloor has revealed a deposit of the isotope iron-60 in a layer that was laid down at the time of the Pliocene-Pleistocene extinction about two million years ago. Iron-60 is produced in supernova explosions and has a half-life of only 1.5 million years. To reach earth before it decayed, this iron must have been produced in a supernova explosion no more than 100 ly away. Thus, the Pliocene-Pleistocene extinction may have been caused by a nearby supernova explosion.

Earth seems to be fairly safe at the moment. No star within 50 ly is known to be a massive giant capable of exploding as a supernova. Massive stars are rare, and most are far away. The most massive star known is 25,000 ly from Earth and poses no threat (❙ Figure 13-19). We only need to worry about nearby supernova explosions, and massive stars are so rare and so easy to locate that we can be sure that no type II supernova explosion will occur near us for the time being.

Type Ia supernovae are caused by collapsing white dwarfs, and white dwarfs are both very common and very hard to locate. There is no way to be sure that a white dwarf teetering on the edge of the Chandrasekhar limit does not lurk near us in space. We can take confidence, however, in the extreme rarity of supernova explosions. Earth faces greater dangers from human ignorance than from exploding stars.

ergy generation begins to fall, the star contracts, but because iron can't ignite there is no new energy source to stop the contraction. In seconds, the core of the star falls inward and a shock wave moves outward. Aided by a flood of neutrinos and sudden convective turbulence, the shock wave blasts the star apart, and we see it brighten as its surface gases expand into space.

Type II supernovae are easy to recognize because their spectra contain hydrogen lines. Use what you know about type Ia supernovae to explain why the spectra of these supernovae do not contain visible hydrogen lines.

The study of the deaths of stars has led us to discover astonishing objects of unbelievable density, temperature, and violence—all consequences of the victory of gravity over matter. But we have not considered the strangest circumstance of all. What happens when degenerate matter can't support the weight of the dying star, that is, when the mass of the compact object exceeds the Chandrasekhar limit? The answer is in the title of the next chapter.

FIGURE 13-19
The most massive star known is about 100 times more massive than the sun and 10 million times more luminous. Originally about 200 solar masses, it is rapidly expelling mass. At a distance of 25,000 ly, it will not be dangerous when it explodes as a supernova within the next few million years. *(Don F. Figer, UCLA, NASA)*

SUMMARY

When a star's central hydrogen-fusing reactions cease, its core contracts and heats up, igniting a hydrogen-fusing shell and swelling the star into a cool giant. The contraction of the star's core ignites helium first in the core and later in a shell. If the star is massive enough, it can eventually fuse carbon and other elements.

How a star evolves depends on its mass. Stars less massive than about 0.4 solar mass are completely mixed and have very little hydrogen left when they die. They cannot ignite a hydrogen shell or a helium core, so they cannot become giants. Their fuel will last for trillions of years. Our universe is not old enough for any of these stars to have died.

Medium-mass stars between about 0.4 and 4 solar masses become giants and fuse helium but cannot fuse carbon. They swell very large as their carbon–oxygen cores contract and eventually eject their surface layers to form planetary nebulae. Such nebulae are expanding gas ionized by the high-temperature core of the star left at the center. These planetary nebula nuclei eventually collapse to form white dwarfs.

Because stars lose mass through stellar winds, the mass of a star can decrease as it evolves. Thus, it seems possible for a star with a main-sequence mass as great as eight solar masses to lose enough mass to produce a planetary nebula and a white dwarf. No white dwarf can exist if it is more massive than the Chandrasekhar limit of 1.4 solar masses, so the most massive stars cannot collapse into white dwarfs. Presumably, they collapse to form other compact objects—neutron stars or black holes.

Evolving stars in close binary systems can transfer matter from one star to the other, producing an accretion disk around the star gaining mass. This can dramatically alter the evolution of the stars. If the star gaining mass is a white dwarf, the added fuel on its surface can explode as a nova.

The most massive stars can get hot enough to ignite carbon and other nuclear fuels. Eventually, they develop iron cores, which cannot release energy through fusion reactions, and collapse to produce a supernova explosion as a powerful shock wave, aided by a blast of neutrinos and violent turbulence, rushes out of the core. The expanding shell of gas produces a supernova remnant, and the remains of the core of the star collapse to form a neutron star or black hole. Supernovae caused by the collapse of a massive star are called type II supernovae.

A type Ia supernova is produced by the collapse of a white dwarf that has gained enough matter from a binary companion to exceed the Chandrasekhar limit. The collapse of the white dwarf causes rapid nuclear fusion, called carbon deflagration, and the entire star is destroyed. Unlike the spectra of type II supernovae, the spectrum of a type Ia supernova contains no hydrogen lines.

NEW TERMS

nova	Roche surface
supernova	Lagrangian point
thermal pulse	accretion disk
planetary nebula	supernova (type I)
compact object	supernova (type II)
black dwarf	carbon deflagration
Chandrasekhar limit	supernova remnant
Roche lobe	synchrotron radiation

REVIEW QUESTIONS

Ace⊘Astronomy™ Assess your understanding of this chapter's topics with additional quizzing and animations at **http:// astronomy.brookscole.com/seeds8e**

1. Why can't the lowest-mass stars become giants?

2. Presumably, all the white dwarfs we see in our galaxy were produced by sunlike stars of medium mass. Why couldn't any of the white dwarfs we see have been produced by the deaths of the lowest-mass stars?

3. What leads us to believe that stars can lose mass?

4. What kind of spectrum does the gas in a planetary nebula produce? Where does it get the energy to radiate?

5. The coolest stars we see at the center of planetary nebulae are about 25,000 K. Why don't we see planetary nebulae containing cooler central stars? (*Hint:* What kind of photons excite the gas in a planetary nebula?)

6. As white dwarfs cool, they move toward the lower right in the H–R diagram (Figure 9-9 and page 261), maintaining constant radius. Why don't they contract as they cool?

7. All white dwarfs are about the same mass. Why?

8. Why have no white dwarfs cooled to form black dwarfs in our galaxy?

9. What happens to a star when it becomes a giant if it has a close binary companion?

10. Why do novae repeat but supernovae do not?

11. Why can't massive stars generate energy from iron fusion?

12. How can we use the spectrum to tell the difference between a type I supernova and a type II supernova? Why does this difference arise?

13. What processes produce type I and type II supernovae?

14. Why do supernova remnants emit X rays?

DISCUSSION QUESTIONS

1. How certain can we be that the sun will never explode as a supernova? What does it mean when a scientist uses the word "certainty"?

2. If we lived for a billion years, we might notice a star that exploded occasionally as a nova and finally destroyed itself in a supernova explosion. What could account for this evolution?

PROBLEMS

1. Use the formula in Chapter 12 to compute the life expectancy of a 0.4-solar-mass star. Why might this be an underestimate if the star is fully mixed?

2. The Ring Nebula in Lyra is a planetary nebula with an angular diameter of 72 seconds of arc and a distance of 5000 ly. What is its linear diameter? (*Hint:* Use the small-angle formula.)

3. If the Ring Nebula is a light-year in diameter and is expanding at a velocity of 15 km/s, typical of planetary nebulae, how old is it? (*Hint:* 1 ly = 9×10^{12} km, and 1 y = 3.15×10^7 s.)

4. Suppose that a planetary nebula is 1 pc in diameter and the Doppler shifts in its spectrum show that it is expanding at 30 km/s. How old is it? (*Hint:* See Problem 3.)

5. A planetary nebula photographed 20 years ago and photographed today has increased its radius by 0.6 second of arc. If Doppler shifts in its spectrum show that it is expanding at a velocity of 20 km/s, how far away is it? (*Hint:* First figure out how many parsecs it has increased in radius in 20 years. Then use the small-angle formula.)

6. The Crab Nebula is now 1.35 pc in radius and is expanding at 1400 km/s. About when did the supernova occur?

7. Doppler shifts in the spectra of the Cygnus Loop show that it is now expanding at 70 km/s. If it is increasing its radius by 0.03 second of arc per year, how far away is it? (*Hint:* How many kilometers will its radius increase in 1 year? Use the small-angle formula.)

8. The Cygnus Loop is now 2.6° in diameter and lies about 500 pc distant. If it is 10,000 years old, what was its average velocity of expansion? (*Hint:* Find its radius in parsecs first.)

9. The supernova remnant Cassiopeia A is expanding in radius at a rate of about 0.5 second of arc per year. Doppler shifts show that the velocity of expansion is about 5700 km/s. How far away is the nebula?

10. Cassiopeia A has a radius of about 2.5 minutes of arc. If it is expanding at 0.5 second of arc per year, when did the supernova explosion occur? (There is no record of a supernova being seen at that time.)

CRITICAL INQUIRIES FOR THE WEB

1. As seen on pages 260 and 261, planetary nebulae show an incredible diversity in appearance. Browse the Web for images and information on these dying stars, and discuss why we see such a range of shapes of planetary nebulae.

2. Naked-eye supernovae in our galaxy are rare, but astronomers have noted supernovae in other galaxies for years. Search the Web for summaries of observations of recent supernovae. Are similar numbers of type I and type II being seen? Compare the number of supernovae seen during the last few years with that of two decades ago. Why are we finding so many more supernovae in recent years than in the past?

3. The word "nova" is used in many ways, so you must be thoughtful to search out photos and observations of recently discovered novae. Limit your search to astronomy categories, and also try pages about variable stars.

4. What is U Scorpii?

EXPLORING *THESKY*

1. Locate the planetary nebulae M57, M97, and M27. How does their shape distinguish them from the star-formation nebulae such as M42 and M8? (*Hint:* To find an object, use **Find** under **Edit.** Choose **Messier Objects** and pick from the list.)

2. The Crab Nebula is M1. Locate it, zoom in, measure its angular size in seconds of arc, and compute its diameter assuming it is about 6000 ly from Earth.

3. Locate the supernova remnant called the Cygnus Loop just south of ε Cygni. How big is this object in angular diameter compared to the full moon? (*Hint:* Under the **View** menu, choose **Labels** and **Setup.** Check **Bayer Designation,** go to Cygnus, and zoom in on ε Cygni until the Cygnus Loop appears.)

 Visit the Seeds *Foundations of Astronomy* companion Web site for critical thinking exercises, articles, and additional readings from InfoTrac College Edition, Brooks/Cole's online student library.

NEUTRON STARS AND BLACK HOLES

Almost anything is easier to get into than out of.

Agnes Allen

Chapter 14

ESO and NASA

GUIDEPOST

The preceding chapters have traced the story of stars from their birth as clouds of gas in the interstellar medium to their final collapse. This chapter finishes the story by discussing the kinds of objects that remain after a massive star dies.

How strange and wonderful that we humans can talk about places in the universe where gravity is so strong it bends space, slows time, and curves light back on itself! To carry on these discussions, astronomers have learned to use the language of relativity.

Throughout this chapter, remember that our generalized discussions are made possible by astronomers studying general relativity in all its mathematical sophistication. That is, our understanding rests on a rich foundation of theory.

This chapter ends the story of individual stars. The next three chapters, however, extend that story to include the giant communities in which stars live—the galaxies.

VIRTUAL LABORATORIES

 NEUTRON STARS AND PULSARS

 GENERAL RELATIVITY AND BLACK HOLES

Gravity always wins. In a star's struggle to withstand its own gravity, the star must lose and gravity must win. Gravity ensures that this star's last remains must eventually reach one of three final states—white dwarf, neutron star, or black hole. These objects, often called compact objects, are small, high-density monuments to the power of gravity. Every star faces an ultimate collapse into such an object.

We discussed white dwarfs in the preceding chapter. In this chapter, we complete our story of stellar death by discussing the other two forms a dead star can take—neutron stars and black holes. In many ways, we learn more about normal stars by studying the remains of dead stars.

In this chapter, we must compare evidence and hypothesis with great care. Theory predicts the existence of these objects; but, by their nature, they are difficult to detect. Is theory right? Do these objects really exist? To confirm the theories, astronomers have searched for real objects that could be identified as having the properties predicted by theory. That is, they have tried to find real neutron stars and real black holes. No medieval knight ever rode off on a more difficult quest, but modern astronomers have found objects that do seem to be neutron stars and black holes. We can judge their success by critically analyzing the evidence to see if it does confirm the theory.

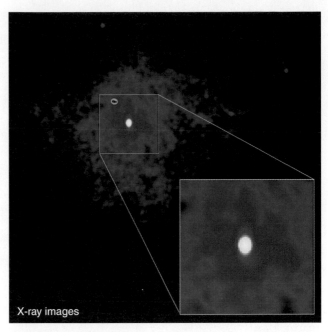

X-ray images

FIGURE 14-1

A supernova explosion seen in 1181 AD left behind an expanding supernova remnant. The Chandra X-Ray Observatory has imaged the nebula in X rays and finds a tiny hot object within—a neutron star. *(NASA/SAO/CXC/P. Slane et al.)*

14-1 NEUTRON STARS

Theory predicts that the death of a massive star could leave behind a **neutron star,** an object containing a little over one solar mass compressed to a radius of about 10 km and heated to high temperatures (▌ Figure 14-1). Modern astronomers have found many objects that appear to be neutron stars, but we must tell their story with care in order to separate theory from observation.

Theory predicts that a neutron star should spin a number of times a second, be nearly as hot at its surface as the inside of the sun, and have a magnetic field a trillion times stronger than Earth's. Two questions should occur to us immediately. First, how could any theory predict such a wondrously unbelievable star? And second, do such neutron stars really exist?

THEORETICAL PREDICTION OF NEUTRON STARS

The neutron was discovered in the laboratory in 1932, and its properties suggested something fantastic. Neutrons spin in much the way that electrons do, which

Ace ◯ Astronomy™ The AceAstronomy icon throughout the text indicates an opportunity for you to test yourself on key concepts and to explore animations and interactions on the AceAstronomy Web site at **http://astronomy.brookscole .com/seeds8e**

means that neutrons must obey the Pauli exclusion principle. In that case, if neutrons are packed together tightly enough, they can become degenerate just as electrons do. White dwarfs are supported by degenerate electrons, and, in 1932, the Russian physicist Lev Landau predicted that neutron stars might exist supported by degenerate neutrons. Of course, the inside of a neutron star would have to be much denser than the inside of a white dwarf.

Only two years later, in 1934, Walter Baade and Fritz Zwicky provided a pedigree for neutron stars. They suggested that some of the most luminous novae in the historical record were not true novae but were caused by the collapse of a massive star in an even larger explosion they called a supernova. What was left of the core of the star, they proposed, was a small, high-density neutron star.

Atomic physics gives us an explanation of how the collapsing core of a massive star could form a neutron star. If the collapsing core is more massive than the Chandrasekhar limit of 1.4 solar masses, then it cannot reach stability as a white dwarf. The weight is too great to be supported by degenerate electrons. The collapse of the core continues, and the atomic nuclei are broken apart by gamma rays. Almost instantly, the increasing density forces the freed protons to combine with electrons and become neutrons by the emission of neutrinos:

$$e + p \rightarrow n + \nu$$

The burst of neutrinos, as we saw in the previous chapter, helps blow the star apart; and the core of the star, in

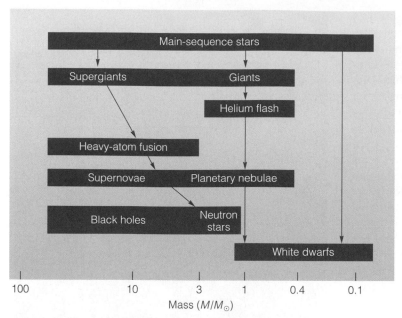

FIGURE 14-2

How a star evolves depends on its mass. The lowest-mass stars will someday become white dwarfs, while stars like the sun become giants, produce planetary nebulae, and then collapse into white dwarfs. Mass loss can move stars to the right in this diagram, so a star of 8 or less solar masses may be able to shed enough mass to die as a planetary nebula and a white dwarf. A more massive supergiant may be able to eject enough mass to leave behind a neutron star instead of a black hole. *(Illustration design by author)*

a fraction of a second, is transformed into a contracting ball of neutrons left behind by the supernova explosion as a neutron star.

Our discussion of stellar evolution in the previous chapter revealed that a star of 8 solar masses or less could lose enough mass to die as a planetary nebula leaving behind a white dwarf. More massive stars up to a limit of about 15 solar masses will lose mass rapidly, but they cannot reduce their mass fast enough and will apparently face death as a supernova explosion leaving behind a neutron star (Figure 14-2). (We will see later in this chapter that the most massive stars probably leave behind black holes.)

Theoretical calculations predict that a neutron star will be only 10 or so kilometers in radius (Figure 14-3) and will have a density of about 10^{14} g/cm^3. On Earth, a sugar-cube-sized lump of this material would

weigh 100 million tons. This is roughly the density of the atomic nucleus, and we can think of a neutron star as matter with all of the empty space squeezed out of it.

How massive can a neutron star be? That is a critical question, and a difficult one to answer because we don't know the strength of pure neutron material. Scientists can't make such matter in the laboratory, so its properties must be predicted theoretically. The most widely accepted calculations suggest that a neutron star cannot be more massive than 2 to 3 solar masses. If a neutron star were more massive than that, the degenerate neutrons would not be able to support the weight, and the object would collapse (presumably into a black hole).

Simple physics, the physics we have used in previous chapters to discuss normal stars, predicts that neutron stars should be hot, spin rapidly, and have strong magnetic fields. We have seen that contraction heats the gas in a star. As the gas particles fall inward, they pick up speed, and when they collide, their high speeds become thermal energy. The sudden collapse of the core of a massive star to a radius of 10 km should heat it to millions of degrees. Furthermore, neutron stars should cool slowly because the heat can escape only from the surface, and neutron stars are so small they have little surface from which to radiate. Thus, basic theory predicts that neutron stars should be very hot.

The conservation of angular momentum predicts that neutron stars should spin rapidly. All stars rotate to some extent because they form from swirling clouds of interstellar matter. As such a star collapses, it must rotate faster because it conserves angular momentum.

FIGURE 14-3

A tennis ball and a road map illustrate the relative size of a neutron star. Such an object, containing slightly more than the mass of the sun, would fit with room to spare inside the beltway around Washington, DC. *(Photo by author)*

Recall that we see this happen when ice skaters spin slowly with their arms extended and then speed up as they pull their arms closer to their bodies (see Figure 5-7). In the same way, a collapsing star must spin faster as it pulls its matter closer to its axis of rotation. If the sun collapsed to a radius of 10 km, its period of rotation would decrease from 25 days to about 0.001 second. We might expect the collapsed core of a massive star to rotate 10 or 100 times a second.

Basic theory also predicts that a neutron star should have a powerful magnetic field. Whatever magnetic field a star has is frozen into the star. The gas of the star is ionized, and that means the magnetic field cannot move easily through the gas. When the star collapses, the magnetic field is carried along and squeezed into a smaller area, which could make the field a billion times stronger. Because some stars have magnetic fields over 1000 times stronger than the sun's, we might expect a neutron star to have a magnetic field as much as a trillion times stronger than the sun's. For comparison, that is about 10 million times stronger than any magnetic field ever produced in the laboratory.

Basic physics predicted the properties of neutron stars, but it also predicted that such objects should be difficult to observe. Neutron stars are very hot, so from our understanding of black body radiation we can predict they will radiate most of their energy in the X-ray part of the spectrum, radiation that could not be observed in the 1940s and 1950s because astronomers could not put their telescopes above Earth's atmosphere. Also, the small surface areas of neutron stars mean that they will be faint objects. Thus astronomers of the mid-20th century were not surprised that none of the newly predicted neutron stars were found. Neutron stars were, at that point, entirely theoretical objects. Progress came not from theory but from observation.

THE DISCOVERY OF PULSARS

In November 1967, Jocelyn Bell, a graduate student at Cambridge University in England, found a peculiar pattern on the paper chart from a radio telescope. Unlike other radio signals from celestial bodies, this was a series of regular pulses (‖ Figure 14-4). At first she and the leader of the project, Anthony Hewish, thought the signal was interference, but they found it day after day in the same place in the sky. Clearly, it was celestial in origin.

Another possibility, that it came from a distant civilization, led them to consider naming it LGM for Little Green Men. But within a few weeks, the team found three more objects in other parts of the sky pulsing with different periods. The objects were clearly natural, and the team dropped the name LGM in favor of **pulsar**—a contraction of *pulsing star*. The pulsing radio source Bell had observed with her radio telescope was the first known pulsar.

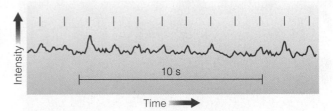

FIGURE 14-4

The 1967 detection of regularly spaced pulses in the output of a radio telescope led to the discovery of pulsars. This record of the radio signal from the first pulsar, CP1919, contains regularly spaced pulses (marked by ticks). The period is 1.33730119 seconds.

As more pulsars were found, astronomers argued over their nature. Periods ranged from 0.033 to 3.75 seconds and were nearly as exact as an atomic clock. Months of observation showed that many of the periods were slowly growing longer by a few billionths of a second per day. Thus, whatever produced the regular pulses had to be highly precise, nearly as exact as an atomic clock, but it had to gradually slow down.

It was easy to eliminate possibilities. Pulsars could not be stars. A normal star, even a small white dwarf, is too big to pulse that fast. Nor could a star with a hot spot on its surface spin fast enough to produce the pulses. Even a small white dwarf would fly apart if it spun 30 times a second.

The pulses themselves gave the astronomers a clue. The pulses lasted only about 0.001 second. This places an upper limit on the size of the object producing the pulse. If a white dwarf blinked on and then off in that interval, we would not see a 0.001-second pulse. The near side of the white dwarf would be about 6000 km closer to us, and light from the near side would arrive 0.022 seconds before the light from the bulk of the white dwarf. Thus its short blink would be smeared out into a longer pulse. This is an important principle in astronomy—an object cannot change its brightness appreciably in an interval shorter than the time light takes to cross its diameter. If pulses from pulsars are no longer than 0.001 second, then the objects cannot be larger than 300 km (190 miles) in diameter.

Only a neutron star is small enough to be a pulsar. In fact, a neutron star is so small, it can't vibrate slowly enough, but it can spin as fast as 1000 times a second without flying apart. The missing link between pulsars and neutron stars was found in late 1968, when astronomers discovered a pulsar at the heart of the Crab Nebula (‖ Figure 14-5). The Crab Nebula is a supernova remnant, and theory predicts that some supernovae leave behind a neutron star.

The short pulses and the discovery of the pulsar in the Crab Nebula are strong hints that pulsars are neutron stars. If we combine theory and observation, we can devise a model of a pulsar.

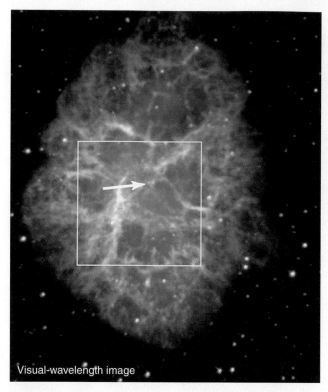

FIGURE 14-5

The pulsar at the center of the Crab Nebula (arrow) is detectable in visual-wavelength photographs. The star to the left of the pulsar lies much closer to Earth and is not in the Crab Nebula. The white box outlines the area imaged on page 285. *(Caltech)*

A MODEL PULSAR

Scientists often work by building a model of a natural phenomenon—not a physical model made of plastic and glue, but an intellectual conception of how nature works in a specific instance. The model may be limited and incomplete, but it helps us organize our understanding of pulsars.

The modern model of a pulsar has been called the **lighthouse model** and is shown in ▌ "The Lighthouse Model of a Pulsar" on pages 284 and 285. Notice four important points. First, a pulsar does not pulse but rather emits beams of radiation that sweep around the sky as the neutron star rotates. Second, notice that the mechanism that produces the beams involves extremely high energies and is not fully understood. The third thing to notice is that we tend to see only those pulsars whose beams sweep over Earth. Finally, notice how modern space telescopes observing at various wavelengths can help us confirm and refine our model.

Neutron stars are not simple objects, and modern astronomers need both general relativity and quantum mechanics to try to understand them. Nevertheless, the life story of pulsars can be understood in terms of the lighthouse model.

The model of a pulsar as a spinning neutron star won the support of many astronomers because of two properties of pulsars. First, many pulsars are slowing down. Their periods are increasing by a few billionths of a second each day—a change radio astronomers can measure using atomic clocks. Evidently, the spinning neutron star is converting some of its energy of rotation into various kinds of electromagnetic energy and a powerful outflow of high-speed particles called a **pulsar wind** (▌ Figure 14-6). About 99.9 percent of the energy released by the slowing of the neutron star is carried away by the pulsar wind, and only 0.1 percent goes into producing the radio beams we detect. The energy that keeps the Crab Nebula glowing nearly 1000 years after the explosion is coming from the rotational energy of the neutron star and the pulsar wind that the neutron star produces.

The second property of pulsars that supports the neutron-star model of a pulsar is the **glitch**—a sudden increase in the pulse rate seen in some pulsars (▌ Figure 14-7). Two theories have been proposed to explain these changes, and both depend on the internal structure of spinning neutron stars.

One theory suggests that "starquakes" occur on the surface of a neutron star and produce glitches. To understand this theory we must note that theoretical models of neutron stars suggest that their interiors are fluids composed mostly of neutrons. Near the surface, where the pressure is lower, matter can exist as a rigid crust composed mostly of iron nuclei about 10^{16} times stronger than steel. This crust might be a few hundred meters thick. The rapidly spinning neutron star would be slightly flattened, but as it slowed, its gravity would squeeze it until the crust broke in a "starquake"—a neutron-star earthquake. When the crust breaks, the neutron star could become slightly less flattened, and the conservation of angular momentum would cause the neutron star to spin slightly faster. From Earth we would see a sudden increase in the pulse rate.

The starquake theory has a drawback; starquakes should occur every few thousand years on any given neutron star, but glitches are more common. The Vela pulsar has glitched 13 times in 25 years. A newer theory may explain most glitches. Because the liquid in the core of a neutron star can circulate with almost no friction, vortexes should develop in the fluid. Like swarms of frictionless tornadoes, they should store large amounts of angular momentum. Calculations applying quantum mechanics to these vortices show that as the neutron star slows, swarms of vortices should suddenly transfer their angular momentum to the crust. That would make the crust spin faster and increase the pulse rate. That is, it would produce a glitch. This theory has been confirmed in laboratory experiments with rotating fluids.

A rare glitch may be caused by a starquake, but most are probably caused by changes in internal vortices. Notice, in any case, that both explanations depend on the pulsar being a spinning neutron star. Con-

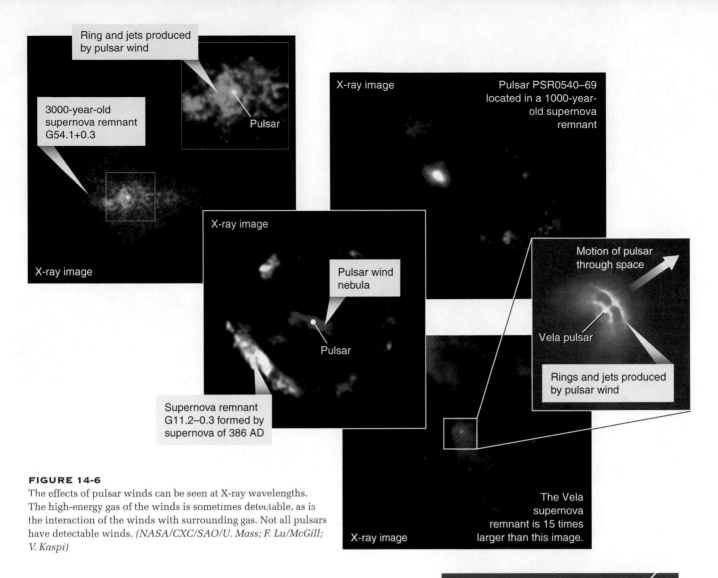

FIGURE 14-6

The effects of pulsar winds can be seen at X-ray wavelengths. The high-energy gas of the winds is sometimes detectable, as is the interaction of the winds with surrounding gas. Not all pulsars have detectable winds. *(NASA/CXC/SAO/U. Mass; F. Lu/McGill; V. Kaspi)*

sequently, glitches give astronomers further confidence that pulsars are spinning neutron stars.

Since the discovery of the first pulsar, over 1000 have been found, and there are probably about 100 million in our galaxy. Of course, we see only those pulsars whose beams sweep over Earth. Most neutron stars are invisible from Earth. Nevertheless, astronomers can describe how neutron stars are born, how they age, and how they die.

THE EVOLUTION OF PULSARS

When a pulsar first forms, it is spinning quickly, perhaps nearly 100 times a second. The energy it radiates into space comes from its energy of rotation, so as it blows away its pulsar wind and blasts beams of radiation outward, its rotation slows. The average pulsar is apparently only a few million years old, and the oldest is about 10 million years old. Presumably, older neutron stars rotate too slowly to generate detectable radio beams.

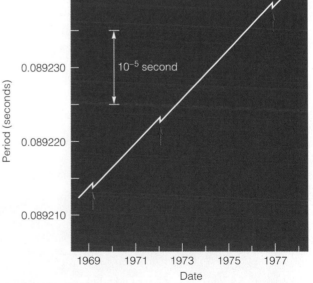

FIGURE 14-7

Soon after pulsars were discovered, radio astronomers accumulated enough data to show that pulsars were very gradually slowing down. That is, their pulses were growing longer. Some pulsars, such as the Vela pulsar whose data are shown here, experience glitches in which the pulsar suddenly speeds up only to resume its more leisurely decline.

The Lighthouse Model of a Pulsar

Astronomers think of pulsars not as pulsing objects, but rather as objects emitting beams. As they spin, the beams sweep around the sky; when a beam sweeps over us, we detect a pulse of radiation. Understanding the details of this lighthouse model is a challenge, but the implications are clear. Although a neutron star is only a few kilometers in radius, it can produce powerful beams. Also, we tend to notice only those pulsars whose beams happen to sweep over us.

In this artist's conception, gas trapped in the neutron star's magnetic field is excited to emit light and outline the otherwise invisible magnetic field.

Beams of electromagnetic radiation would probably be invisible unless they excited local gas to glow.

What color should an artist use to paint a neutron star? With a temperature of a million degrees, the surface emits most of its electromagnetic radiation at X-ray wavelengths. Nevertheless, it would probably look blue-white to our eyes.

How a neutron star can emit beams is one of the challenging problems of modern astronomy, but astronomers have a general idea. A neutron star contains a powerful magnetic field and spins very rapidly. The spinning magnetic field generates a tremendously powerful electric field, and the field causes the production of electron–positron pairs. As these charged particles are accelerated through the magnetic field, they emit photons in the direction of their motion, which produce powerful beams of electromagnetic radiation emerging from the magnetic poles.

Neutron Star Rotation with Beams

As in the case of Earth, the magnetic axis of a neutron star could be inclined to its rotational axis.

The rotation of the neutron star will sweep its beams around like beams from a lighthouse.

While a beam points roughly toward Earth, we detect a pulse.

While neither beam is pointed toward us, we detect no energy.

Beams may not be as exactly symmetric as in this model.

Ace Astronomy™

Go to AceAstronomy and click Active Figures to see "Neutron Star" and adjust the inclination of the neutron star's magnetic field to produce pulses.

NASA

Crab pulsar

Visual-wavelength image

The hazy glow of the Crab Nebula is produced by synchrotron radiation. In the nearly 10 centuries since the supernova, the high-speed electrons should have radiated their energy away, and the synchrotron radiation should have faded. Evidently, the Crab pulsar powers the nebula. The Hubble Space Telescope image above shows the region boxed in Figure 14-5. Circular wisps excited by the pulsar change and flicker from day to day.

Red and yellow in this image show the radio and visual extent of the Crab Nebula. Blue, an X-ray image, reveals that the central pulsar has flung off rings of highly ionized gas in a disk up to a light-year in radius. The pulsar is ejecting jets of high-energy particles perpendicular to the disk.

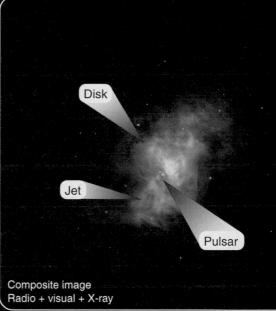

Disk

Jet

Pulsar

NASA

Composite image
Radio + visual + X-ray

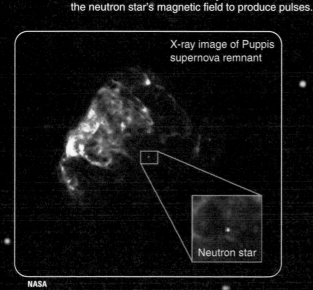

X-ray image of Puppis supernova remnant

Neutron star

NASA

Hubble Space Telescope visual-wavelength image

Neutron star

NASA

If a pulsar's beams do not sweep over us, we detect no pulses and the neutron star is difficult to find. A few such objects are known, however. The Puppis A supernova remnant is about 4000 years old and contains a point source of X rays believed to be a neutron star. The isolated neutron star in the right-hand image has a temperature of 700,000 K.

FIGURE 14-8

High-speed images of the Crab Nebula pulsar (arrow) show it pulsing at visual wavelengths and at X-ray wavelengths. *(© AURA, Inc., NOAO, KPNO)* The period of pulsation is 33 milliseconds, and each cycle includes two pulses as its two beams of unequal intensity sweep over Earth. *(Courtesy F. R. Harnden, Jr., from* The Astrophysical Journal, *published by the University of Chicago Press; © 1984 The American Astronomical Society)*

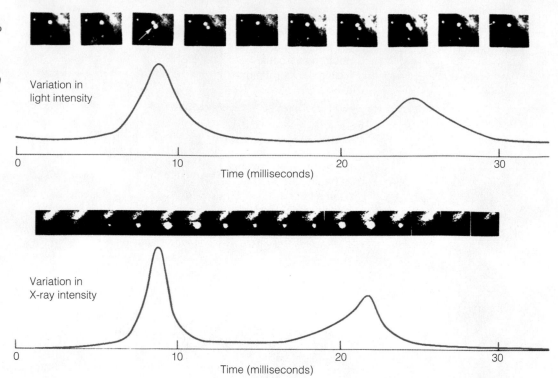

From this we can expect that a young neutron star should emit powerful beams of radiation. The Crab Nebula gives us an example of such a system. Only about 950 years old, the Crab pulsar is so powerful it emits photons of radio, infrared, visible, X-ray, and gamma-ray wavelengths (❚ Figure 14-8). Careful measurements of its brightness with high-speed instruments show that it blinks twice for every rotation. One beam sweeps almost directly over us, and we detect a strong pulse. Half a rotation later, the edge of the other beam sweeps past us, and we detect a weaker pulse.

We would expect only the most energetic pulsars to produce short-wavelength photons and thus pulse at visible wavelengths. The Crab Nebula pulsar is young and powerful, and it produces visible pulses, and so does a pulsar called the Vela pulsar (located in the Southern Hemisphere constellation Vela). The Vela pulsar is fast, pulsing about 11 times a second, and, like the Crab Nebula pulsar, is located inside a supernova remnant. Its age is estimated at about 20,000 to 30,000 years, young in terms of the average pulsar. Thus we suspect that pulsars are capable of producing optical pulses only when they are young.

We might expect to find all pulsars inside supernova remnants, but the statistics must be examined with care. Not every supernova remnant contains a pulsar, and not every pulsar is located inside a supernova remnant. Many supernova remnants probably contain pulsars whose beams never sweep over Earth, and it is difficult to detect such pulsars. Also, some pulsars move through space at high velocity, which suggests that supernova explosions can occur asymmetrically, perhaps because of the violent turbulence in the exploding core.

Such a violent, off-center explosion could give the neutron star a high velocity through space (❚ Figure 14-9). Also, some supernovae probably occur in binary systems and fling the two stars apart at high velocity. In any case, pulsars are known to have such high velocities that many probably escape the disk of our galaxy. We should not be surprised that many neutron stars quickly leave their supernova remnant behind. And finally, we must remember that a pulsar can remain detectable for 10 million years or so, but a supernova remnant cannot survive more than about 50,000 years before it is mixed into the interstellar medium. Consequently, we might expect that most pulsars are not located in supernova remnants, and that most supernova remnants do not contain pulsars.

Careful analysis, however, shows that there are more empty supernova remnants than we should expect. Where are these missing pulsars? Searches suggest that many neutron stars may not be emitting powerful radio beacons. If they are born with very strong magnetic fields, they may have slowed down rapidly and may be rotating so slowly they do not emit beams of radio energy. These objects would have magnetic fields 1000 times stronger than those of a typical pulsar and have been dubbed **magnetars.** We will see more evidence that such magnetars do exist. This line of research is ongoing, but it suggests that we have more to learn about the evolution of pulsars.

The explosion of Supernova 1987A in February 1987 apparently formed a neutron star. We can draw this conclusion because a burst of neutrinos was detected passing through Earth, and theory predicts that the collapse of a massive star's core into a neutron star would

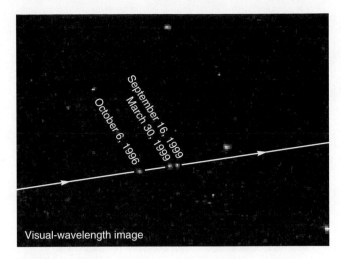

FIGURE 14-9

Many neutron stars have high velocities through space. Here the neutron star known as RX J185635-3754 was photographed on three different dates as it rushes past background stars. *(NASA and F. M. Walter)*

produce such a burst of neutrinos. At first the neutron star would be hidden at the center of the expanding shells of gas ejected into space, but as the gas expands and thins we might be able to see it. Or, if its beams don't sweep over Earth, we might be able to detect it from its X-ray and gamma-ray emission. Although, years after the explosion, no neutron star has been detected, astronomers continue to watch the site, hoping to see the youngest pulsar known.

One reason pulsars are so fascinating is the extreme conditions we find in spinning neutron stars. To see natural processes of even greater violence, we have only to look at pulsars in binary systems.

BINARY PULSARS

Over a thousand pulsars are now known, and some are located in binary systems. These pulsars are of special interest because we can learn more about the neutron star by studying the orbital motions of the binary. Also, in some cases, mass can flow from the companion star onto the neutron star, and that produces high temperatures and X rays.

The first binary pulsar was discovered in 1974 when astronomers Joseph Taylor and Russell Hulse noticed that the pulse period of the pulsar PSR 1913+16 was changing. The period first grew longer and then grew shorter in a cycle that took 7.75 hours. Thinking of the Doppler shifts seen in spectroscopic binaries, the radio astronomers realized that the pulsar had to be in a binary system with an orbital period of 7.75 hours. When the orbital motion of the pulsar carries it away from Earth, we see the pulse period slightly lengthened, just as the wavelength of light emitted by a receding source is lengthened. That is, we see a red shift. Then, when the pulsar rounds its orbit and approaches

Earth, we see the pulse period slightly shortened—a blue shift. From these changing Doppler shifts, the astronomers could calculate the radial velocity of the pulsar around its orbit just as if it were a spectroscopic binary star (Chapter 9). The resulting graph of radial velocity versus time could be analyzed to find the shape of the pulsar's orbit (▌ Figure 14-10). When Taylor and Hulse analyzed PSR 1913+16, they discovered that the binary system consisted of two neutron stars separated by a distance roughly equal to the radius of our sun.

Yet another surprise was hidden in the motion of PSR 1913+16. In 1916, Einstein's general theory of relativity described gravity as a curvature of space-time. Einstein realized that any rapid change in a gravitational field should spread outward at the speed of light as **gravitational radiation.** Gravity waves have not been detected, but Taylor and Hulse were able to show that the orbital period of the binary pulsar was slowly grow-

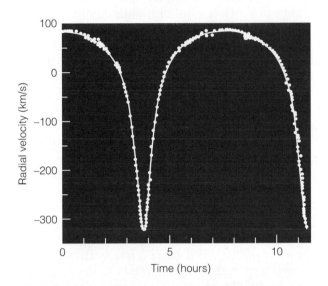

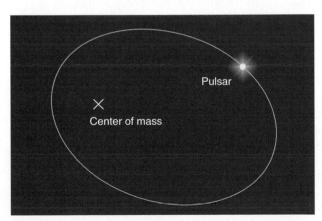

FIGURE 14-10

The radial velocity of pulsar PSR 1913+16 can be found from the Doppler shifts in its period of pulsation. The shape of the radial velocity curve is typical for a spectroscopic binary with an elliptical orbit and allows astronomers to determine the pulsar's orbit. In the plotted orbit, the center of mass does not appear to be at a focus of the elliptical orbit, because the orbit is inclined. *(Adapted from data by Joseph Taylor and Russell Hulse)*

ing shorter because the stars are gradually spiraling toward each other. They are radiating orbital energy away as gravitational radiation. Taylor and Hulse won the Nobel prize in 1993 for their work with binary pulsars.

Dozens of binary pulsars have been found; by analyzing the Doppler shifts in their pulse periods, astronomers can estimate the mass of the neutron stars. Typical masses are about 1.35 solar masses, in good agreement with models of neutron stars.

Binary pulsars can emit strong gravitational waves because the neutron stars contain large amounts of mass in a small volume. This also means that binary pulsars can be sites of tremendous violence because of the strength of gravity at the surface of a neutron star. An astronaut stepping onto the surface of a neutron star would be instantly smushed into a layer of matter only 1 atom thick. Matter falling onto a neutron star can release titanic amounts of energy. If you dropped a single marshmallow onto the surface of a neutron star from a distance of 1 AU, it would hit with an impact equivalent to a 3-megaton nuclear warhead. In general, a particle falling from a large distance to the surface of a neutron star will release energy equivalent to $0.2\ mc^2$, where m is the particle's mass at rest. Even a small amount of matter flowing from a companion star to a neutron star can generate high temperatures and release X rays and gamma rays.

As an example of such an active system, we can examine Hercules X-1. It emits pulses of X rays with a period of about 1.2 seconds, but every 1.7 days the pulses vanish for a few hours (■ Figure 14-11). Astronomers can understand this system by comparing it to an eclipsing binary star. Hercules X-1 seems to contain a 2-solar-mass star with a temperature of 7000 K and a neutron star that orbit each other with a period of 1.7 days. Matter flowing from the normal star into an accretion disk around the neutron star can reach temperatures of millions of degrees and emit a powerful X-ray glow. Interactions with the neutron star's magnetic field can produce beams of X rays that sweep around with the rotating neuron star. We receive a pulse of X rays every time a beam points our way. The X rays shut off every 1.7 days when the neutron star is eclipsed behind the normal star. The X rays from the neutron star and its accretion disk heat the near side of the normal star to about 20,000 K. As the system rotates, we alternately see the hot side of the star and then the cool side, and its brightness at visible wavelengths varies. Hercules X-1 is a complex system and is still not well understood, but our quick analysis serves to illustrate how complex and powerful such binary systems are during mass transfer.

THE FASTEST PULSARS

Our discussion of pulsars suggests that newborn pulsars should blink rapidly and old pulsars should blink

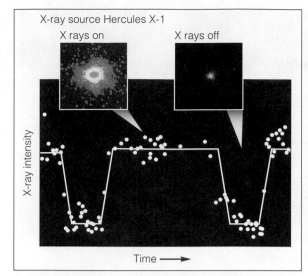

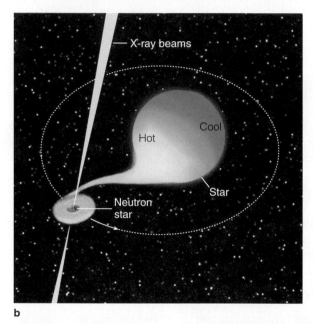

FIGURE 14-11

Sometimes the X-ray pulses from Hercules X-1 are on and sometimes they are off. A graph of X-ray intensity versus time looks like the light curve of an eclipsing binary. *(Insets: J. Trümper, Max-Planck Institute)* (b) In Hercules X-1, matter flows from a star into an accretion disk around a neutron star producing X rays, which heat the near side of the star to 20,000 K compared with only 7000 K on the far side. X rays turn off when the neutron star is eclipsed behind the star.

slowly, but the handful that blink the fastest may be quite old. One of the fastest known pulsars is cataloged as PSR 1937+21 in the constellation Vulpecula. It pulses 642 times a second and is slowing down only slightly. The energy stored in the rotation of a neutron star at this rate is equal to the total energy of a supernova explosion, so it seemed difficult at first to explain this pulsar. It now appears that PSR 1937+21 is an old neutron star that has gained mass and rotational energy from a companion in the binary system. Like water hitting a mill wheel, the matter falling on the neutron star

The Impossibility of Proof in Science

No scientific theory or hypothesis can be proved correct. We can test a theory over and over by performing experiments or making observations, but we can never prove that the theory is absolutely true. It is always possible that we have misunderstood the theory or the evidence, and the next observation we make might disprove the theory.

For example, we might propose the theory that the sun is mostly iron. We might test the theory by looking at the iron lines in the solar spectrum, and the strength of the iron lines would suggest that our theory is right. Although our observation has confirmed our theory, it has not proven the theory is right. We might confirm the theory many times before we realized that iron absorbs photons much more efficiently than hydrogen. Although the hydrogen lines are weak in the sun's spectrum, they tell us that most of the atoms in the sun are hydrogen and not iron.

The nature of scientific thinking can lead to two kinds of mistakes. Sometimes nonscientists will say, "You scientists just want to tear everything down—you don't believe in anything." Scientists test a theory over and over to test its worth. If a theory survives many tests, scientists begin to have confidence it is true.

The second error nonscientists make is to say, "You scientists are never sure of anything." Again, the scientist knows that no theory can be proven correct. That the sun will rise tomorrow is very likely, and scientists have great confidence in that theory. But in the end it is still a theory.

People will say of an idea they dislike, "That is only a theory," as if a theory were simply a random guess. In fact, a theory can be a well-tested truth in which all scientists have great confidence. Yet we can never prove that any theory is absolutely true.

---

has spun it up to 642 rotations per second. With its weak magnetic field, it slows down very slowly and will continue to spin for a very long time.

A number of other very fast pulsars have been found. They are known generally as **millisecond pulsars** because their pulse periods are almost as short as a millisecond (0.001 s). This produces some fascinating physics because the pulse period of a pulsar equals the rotation period of a neutron star. If a neutron star 10 km in radius spins 642 times a second, as does PSR 1937+21, then the period is 0.0016 second, and the equator of the neutron star must be traveling about 40,000 km/s. That is fast enough to flatten the neutron star into an ellipsoidal shape and is nearly fast enough to break it up.

All scientists should be made honorary citizens of Missouri, the "Show Me" state, because scientists demand evidence. The hypothesis that the millisecond pulsars are spun up by mass transfer from a companion star is quite reasonable, but astronomers demand evidence, and evidence has been found. For example, the pulsar PSR J1740-5340 has a period of 42 milliseconds and is orbiting with a bloated red star that is losing mass to the neutron star. This appears to be a pulsar in the act of being spun up to high speed. For another example, consider the X-ray source XTE J1751-305, a pulsar with a period of only 2.3 milliseconds. X-ray observations show that it is in the act of gaining mass from a companion star. The orbital period is only 42 minutes, and the mass of the companion star is only 0.014 solar masses. The evidence suggests this neutron star has devoured all but the last morsel of its binary partner.

Although some millisecond pulsars have binary companions, some are solitary neutron stars. A pulsar known as the Black Widow may explain how a fast pulsar can lack a companion. The Black Widow has a period of 1.6 milliseconds, meaning it is spinning 622 times per second, and it orbits with a low-mass companion. Presumably the neutron star was spun up by mass flowing from the companion, but spectra show that the blast of radiation and high-energy particles from the neutron star is now boiling away the surface of the companion. The Black Widow has eaten its fill and is now evaporating the remains of its companion. It will soon be a solitary millisecond pulsar.

"Show me," say scientists, and in the case of neutron stars, the evidence seems very strong. Of course, we can never prove a theory is absolutely true (Window on Science 14-1), but the evidence for neutron stars is so strong that astronomers have great confidence that they really do exist. Other theories that describe how they emit beams of radiation and how they form and evolve are less certain, but continuing observations at many wavelengths are expanding our understanding of these last embers of massive stars. In fact, observations of one pulsar have turned up objects no one predicted.

PULSAR PLANETS

Finding planets orbiting stars other than the sun is very difficult, and only about a hundred are known. Oddly, the first such planets were found orbiting a neutron star.

Because a pulsar's period is so precise, astronomers can detect tiny variations by comparison with atomic clocks. When astronomers checked pulsar PSR 1257+12, they found variations in the period of pulsation much like those caused by the orbital motion of the binary pulsar (❚ Figure 14-12a). However, in the case of PSR 1257+12, the variations were much smaller; and, when they were interpreted as Doppler shifts, it became evident that the pulsar was being orbited by at least two objects with planetlike masses of 4.3 and 3.9 Earth masses. The gravitational tugs of the planets make the pulsar wobble about the center of mass of the system by no more than 800 km, and that produces the tiny changes in period (Figure 14-12b).

Astronomers greeted this discovery with both enthusiasm and skepticism. As usual, they looked for ways to test the hypothesis. Simple gravitational theory predicts that the planets should interact and slightly modify each other's orbit. When the data were analyzed, that interaction was found, further confirming the hypothesis that the variations in the period of the pulsar are caused by planets. In fact, further data revealed the presence of a third planet of about the mass of Earth's moon, and a fourth planet with a mass of about 100 Earth masses is now believed to follow a much larger orbit. This illustrates the astonishing precision of studies based on pulsar timing. At least one other pulsar is known where small shifts in pulse timing imply the presence of an orbiting planet.

Astronomers wonder how a neutron star can have planets. The planets that orbit PSR 1257+12 are very close to the pulsar; the inner three orbit at 0.19 AU, 0.36 AU, and 0.47 AU—closer to the pulsar than Venus is to the sun. Any planets that orbit a star would be lost or vaporized when the star exploded. Furthermore, a star about to explode as a supernova would be a large giant or a supergiant, and planets only a few AU distant would be inside such a large star and could not survive. It seems more likely that these planets are the remains of a stellar companion that was devoured by the neutron star. In fact, the pulsar is very fast (162 pulses per second), suggesting that it was spun up in a binary system.

Another pulsar planet has been found orbiting in a binary system containing a neutron star and a white dwarf. Because this system is located in a very old star cluster and contains a white dwarf, astronomers suspect that the planet may be very old. Planets probably orbit other neutron stars, and small shifts in the timing of the pulses may eventually reveal their presence.

We can imagine what these worlds might be like. Planets formed from the remains of dying stars might be rich in heavy elements, but truly ancient planets might be poor in such elements. We can imagine visiting these worlds, landing on their surfaces, and hiking across their valleys and mountains. Above us, the neutron star would glitter in the sky, a tiny point of light.

REVIEW CRITICAL INQUIRY

How can a neutron star be found at X-ray wavelengths?
First, we should remember that a neutron star is very hot because of the heat released when it contracts to a radius of 10 km. It could easily have a surface temperature of 1,000,000 K, and Wien's law (Chapter 7) tells us that such an object will radiate most intensely at a very short wavelength typical of X rays. However, we know that the total luminosity of a star depends on its surface temperature and

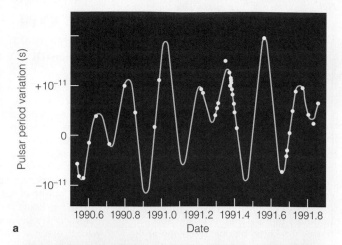

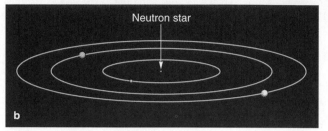

FIGURE 14-12

(a) The dots in this graph are observations showing that the period of pulsar PSR 1257+12 varies from its average value by a fraction of a billionth of a second. The blue line shows the variation that would be produced by planets orbiting the pulsar. (b) As the planets orbit the pulsar, they cause it to wobble by less than 800 km, a distance that is invisibly small in this diagram. *(Adapted from data by Alexander Wolszczan)*

its surface area, and a neutron star is so small it can't radiate much energy. X-ray telescopes have found such neutron stars, but they are not easy to locate.

There is, however, a second way a neutron star can radiate X rays. If a normal star in a binary system loses mass to a neutron star companion, the inflowing matter will hit with so much energy that it will be heated to very high temperatures. It may form a very hot accretion disk that can radiate intense X rays easily detectable by X-ray telescopes orbiting above Earth's atmosphere.

If you discovered a pulsar, what observations would you make to determine whether it was young or old, single or a member of a binary system, alone or accompanied by planets?

Perhaps the strangest planets in the universe are those orbiting pulsars. But however strange a pulsar planet may be, we can imagine going to one. Our next topic, in contrast, seems beyond the reach of even our imaginations.

14-2 BLACK HOLES

We have now studied white dwarfs and neutron stars, two of the three end states of dying stars. Now we turn to the third—black holes.

Although the physics of black holes is difficult to discuss without sophisticated mathematics, simple logic is sufficient to predict that they should exist. Our problem is to consider their predicted properties and try to confirm that they exist. What objects observed in the heavens could be real black holes? More difficult than the search for neutron stars, the quest for black holes has nevertheless met with success.

To begin our discussion of black holes, we must consider a simple question. How fast must an object travel to escape from the surface of a celestial body? The answer will lead us to black holes.

ESCAPE VELOCITY

Suppose we threw a baseball straight up. How fast must we throw it if it is not to come down? Of course, gravity would pull back on the ball, slowing it, but if the ball were traveling fast enough to start with, it would never come to a stop and fall back. Such a ball would escape from Earth. The escape velocity is the initial velocity an object needs to escape from a celestial body (▌ Figure 14-13). (See Chapter 5.)

Whether we are discussing a baseball leaving Earth or a photon leaving a collapsing star, the escape velocity depends on two things: the mass of the celestial body and the distance from the center of mass to the escaping object. If the celestial body has a large mass, its gravity is strong, and we need a high velocity to escape; but if we begin our journey farther from the center of mass, the velocity needed is less. For example, to escape from Earth, a spaceship would have to leave Earth's surface at 11 km/s (25,000 mph), but if we could launch spaceships from the top of a tower 1000 miles high, the escape velocity would be only 8.8 km/s (20,000 mph). If we could make an object massive enough or small enough, its escape velocity could be greater than the speed of light. Relativity tells us that nothing can travel faster than the speed of light, so even photons, which have no mass, would be unable to escape. Such a small, massive object could never be seen because light could not leave it.

Long before Einstein and relativity, the Rev. John Mitchell, a British gentleman astronomer, realized that Newton's laws of gravity and motion had peculiar consequences. In 1783, Mitchell pointed out that an object 500 times the radius of the sun but of the same density would have an escape velocity greater than the speed of light. Then, "all light emitted from such a body would be made to return towards it." Mitchell didn't know it, but he was talking about a black hole.

FIGURE 14-13

Escape velocity, the velocity needed to escape from a celestial body, depends on mass. The escape velocity at the surface of a very small, low-mass body would be so low we could jump into space. Earth's escape velocity is much larger, about 11 km/s (25,000 mph).

SCHWARZSCHILD BLACK HOLES

If the core of a star collapses and contains more than about three solar masses, no known force can stop it. The object cannot stop collapsing when it reaches the size of a white dwarf because degenerate electrons cannot support the weight. It cannot stop when it reaches the size of a neutron star because degenerate neutrons cannot support the weight. No force remains to stop the object from collapsing to zero radius.

As an object collapses, its density and the strength of its surface gravity increase; and if an object collapses to zero radius, its density and gravity become infinite. Mathematicians call such a point a **singularity;** but in physical terms it is difficult to imagine an object of zero radius. Some theorists believe that a singularity is impossible and that when we better understand the laws of physics, we will discover that the collapse halts before the radius zero. Astronomically, it seems to make little difference.

If the contracting core of a star becomes small enough, the escape velocity in the region around it is so large that no light can escape. We can receive no information about the object or about the region of space near it, and we refer to this region as a **black hole.** If the core of an exploding star collapsed into a black hole, the core would vanish without a trace. Thus a supernova remnant that lacks a central pulsar may harbor a black hole (▌Figure 14-14).

The boundary of the black hole is called the **event horizon,** because any event that takes place inside the event horizon is invisible to an outside observer. To see how a black hole can exist, we must consider general relativity.

In 1916, Albert Einstein published a mathematical theory of space and time that became known as the general theory of relativity. Einstein treated space and time as a single entity—space-time. His equations showed that gravity could be described as a curvature of space-time, and almost immediately the astronomer Karl Schwarzschild found a way to solve the equations to describe the gravitational field around a single, nonrotating, electrically neutral lump of matter. That solution contained the first general relativistic description of a black hole; nonrotating, electrically neutral black holes are now known as Schwarzschild black holes. In most cases in astronomy, we can use the Schwarzschild

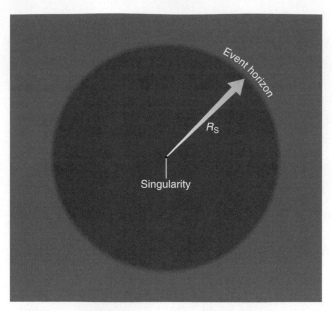

ACTIVE FIGURE 14-15

A black hole forms when an object collapses to a small size (perhaps to a singularity) and the escape velocity in its neighborhood is so great that light cannot escape. The boundary of this region is called the event horizon because any event that occurs inside is invisible to outside observers. The radius of the region is R_s, the Schwarzschild radius.

Ace Astronomy™ Go to AceAstronomy and click Active Figures to see "Schwarzschild Radius." Take control of this diagram.

solution to think about black holes. We will see later in this chapter what difference rotation makes.

Schwarzschild's solution shows that if matter is packed into a small enough volume, then space-time curves back on itself. Objects can still follow paths that lead into the black hole, but no path leads out, so nothing can escape, not even light. Thus, the inside of the black hole is totally beyond the view of an outside observer. The event horizon is the boundary between the isolated volume of space-time and the rest of the universe, and the radius of the event horizon is called the **Schwarzschild radius, R_S**—the radius within which an object must shrink to become a black hole (▌Figure 14-15).

Although Schwarzschild's work was highly mathematical, his conclusion is quite simple. The Schwarzschild radius (in meters) depends only on the mass of the object (in kilograms):

$$R_S = \frac{2GM}{c^2}$$

In this simple formula, G is the gravitational constant, M is the mass, and c is the speed of light. A bit of arithmetic shows that a 1-solar-mass black hole will have a Schwarzschild radius of 3 km, a 10-solar-mass black hole will have a Schwarzschild radius of 30 km, and so on (▌Table 14-1). Even a very massive black hole would not be very large.

Every object with mass has a Schwarzschild radius, but not every object is a black hole. For example, Earth

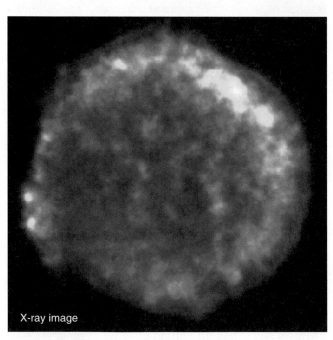

X-ray image

FIGURE 14-14

Some supernova remnants contain no neutron star perhaps because the supernova formed a black hole instead. However, this supernova remnant, formed by Tycho Brahe's supernova of 1572, has the chemical composition typical of the remains of a type Ia supernova. Such supernovae are believed to destroy the white dwarf entirely and do not leave behind a neutron star or a black hole. *(John P. Hughes, Rutgers University)*

TABLE 14-1
The Schwarzschild Radius

Object	Mass ($M_\odot$)	Radius
Star	10	30 km
Star	3	9 km
Star	2	6 km
Sun	1	3 km
Earth	0.000003	0.9 cm

has a Schwarzschild radius of about 1 cm, but it could become a black hole only if we squeezed it inside that radius. Fortunately, Earth will not collapse spontaneously into a black hole because its mass is less than the critical mass of about 3 solar masses. Only exhausted stellar cores more massive than this can form black holes under the sole influence of their own gravity. In this chapter, we are interested in black holes that might originate from the deaths of stars. These black holes would have masses larger than 3 solar masses. In later chapters, we will encounter black holes whose masses might exceed a million solar masses.

Do not think of black holes as giant vacuum cleaners that will pull in everything in the universe. A black hole is just a gravitational field, and at a reasonably large distance its gravity is no greater than that of a normal object of similar mass. If the sun were replaced by a 1-solar-mass black hole, the orbits of the planets would not change at all. The gravity of a black hole becomes extreme only when we approach close to it. The universe contains many black holes. So long as we and other objects stay at a safe distance from the black holes, they have no catastrophic effects.

BLACK HOLES HAVE NO HAIR

Theorists who study black holes are fond of saying, "Black holes have no hair." By that they mean that once matter forms a black hole, it loses almost all of its normal properties. A black hole made of a collapsed star will be indistinguishable from a black hole made from peanut butter and fake-fur mittens. Once the matter is inside the event horizon, it retains only three properties—mass, angular momentum, and electrical charge.

The Schwarzschild black hole is represented by a solution to Einstein's equations for the special case where the object has only mass. Schwarzschild black holes do not rotate or have charge. The solutions for rotating or charged black holes (or for rotating, charged black holes) are more difficult and have been found in only the last few decades. Generally, rotating, charged black holes are similar to Schwarzschild black holes.

It seems that astronomers need not worry about charged black holes because stars, whose collapse pre-

sumably forms black holes, cannot have large electrostatic charges. Suppose that we could give the sun a large positive charge. It would begin to repel protons in its corona and attract electrons and would soon return to neutral charge. Thus, we should expect stars and black holes to be electrically neutral.

But everything in the universe seems to rotate, and collapsing stars spin rapidly as they conserve angular momentum. Thus, we should probably expect black holes to have angular momentum. In 1963, New Zealand mathematician Roy P. Kerr found a solution to Einstein's equations that describes a rotating black hole. This is now known as the **Kerr black hole.**

The mass of a black hole curves neighboring space-time, and the Kerr solution shows that the rotation of a black hole drags space-time around with it. The **ergosphere** is a region outside the event horizon in which space-time rotates with the rotating black hole so powerfully that nothing could avoid being dragged along. No one has ever approached a rotating black hole, so we have no idea what it might feel like to enter a region where space-time was moving past us.

The word *ergosphere* comes from the Greek word *ergo,* meaning "work," because the rotating space-time in the ergosphere can do work on a particle; that is, the particle can gain energy. In particular, the Kerr solution shows that a particle that enters the ergosphere can break into two pieces, one falling into the black hole and the other escaping with more energy than it had when it entered. Thus, energy can be extracted from a rotating black hole, and, as a result, the black hole slows its rotation very slightly.

The Kerr solution is a fascinating bit of theoretical physics, but it has an important application in astronomy. Almost certainly black holes rotate, and matter falling into black holes must pass through the ergosphere. Thus, we may eventually find situations where energy is extracted from rotating black holes.

A LEAP INTO A BLACK HOLE

Before we can search for real black holes, we must understand what theory predicts about the appearance of a black hole. To explore that idea, we can imagine that we leap, feet first, into a Schwarzschild black hole.

If we were to leap into a black hole of a few solar masses from a distance of an astronomical unit, the gravitational pull would not be very large, and we would fall slowly at first. Of course, the longer we fell and the closer we came to the center, the faster we would travel. Our wristwatches would tell us that we fell for about 65 days before we reached the event horizon.

Our friends who stayed behind would see something different. They would see us falling more slowly as we came closer to the event horizon because, as explained by general relativity, clocks slow down in curved space-time. This is known as **time dilation.** In fact, our friends would never actually see us cross the event horizon. To them we would fall more and more slowly until we seemed hardly to move. Generations later, our descendants could focus their telescopes on us and see us still inching closer to the event horizon. We, however, would have sensed no slowdown and would conclude that we had crossed the event horizon after only about 65 days.

Other relativistic effects would make it difficult to see us. As light travels out of a gravitational field, it loses energy, and its wavelength grows longer. This is known as a **gravitational red shift** (▌Table 14-2). Light leaving us and traveling away from the black hole would suffer a larger and larger gravitational red shift. In addition, as our inward velocity grew higher and higher, a relativistic effect would cause more and more of the light leaving us to be emitted in the forward direction. This would reduce the amount of light leaving us and traveling outward, making us even more difficult to see. Although we would notice none of these effects as we fell toward the black hole, our friends would need to observe at longer wavelengths and with larger telescopes to detect us.

While these relativistic effects seem merely peculiar, other effects would be quite unpleasant. Imagine again that we are falling feet first toward the event horizon of a black hole. We would feel our feet, which would be closer to the black hole, being pulled in more strongly than our heads. This is a tidal force, and at first it would be minor. But as we fell closer, the tidal force would become very large. Another tidal force would compress us as our left side and our right side both fell toward the center of the black hole. For any black hole with a mass like that of a star, the tidal forces would crush us laterally and stretch us longitudinally long before we reached the event horizon (▌Figure 14-16). The friction from such severe distortions of our bodies would heat us to millions of degrees, and we would emit X rays and gamma rays. (Needless to say, this would render us inoperative as thoughtful observers.)

Some years ago a popular book suggested that we could travel through the universe by jumping into a black hole in one place and popping out of another somewhere far across space. That might make for good science fiction, but tidal forces would make it an unpopular form of transportation even if it worked.

Our imaginary leap into a black hole is not frivolous. We now know how to find a black hole: Look for a strong source of X rays. It may be a black hole into which matter is falling.

THE SEARCH FOR BLACK HOLES

Do black holes really exist? Beginning in the 1970s, astronomers searched for observational evidence that their theories were correct. They tried to find one or more objects that were obviously black holes. That very

TABLE 14-2	
The Gravitational Red Shift	
Object	Red Shift (percent)
Sun	0.0002
White dwarf	0.01
Neutron star	20
Black hole event horizon	Infinite

difficult search is a good illustration of how the unwritten rules of science help us understand nature (Window on Science 14-2).

A black hole alone is totally invisible because nothing can escape from the event horizon. But a black hole into which matter is flowing would be a source of X rays. Of course, X rays can't escape from inside the

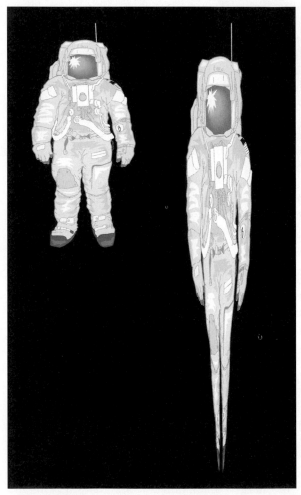

FIGURE 14-16
Leaping feet first into a black hole, a person of normal proportions (left) would be distorted by tidal forces (right) long before reaching the event horizon around a typical black hole of stellar mass. Tidal forces would stretch the body lengthwise while compressing it laterally. Friction from this distortion would heat the body to high temperatures.

WINDOW ON SCIENCE 14-2

Natural Checks on Fraud in Science

Fraud is actually quite rare in science. The nature of science makes fraud difficult, and the way scientists publish their research makes it almost impossible. In fact, we can think of science as a set of unwritten rules of behavior that have evolved to prevent scientists from lying to one another or to themselves, even by accident.

Suppose for a moment that we wanted to commit scientific fraud. We would have to invent data supposedly obtained from experiment or observation. We might invent X-ray data supposedly obtained by observing an X-ray binary star. Or if we were interested in theory, we might invent a fraudulent mathematical calculation of the physics going on in an X-ray binary. We might get away with it for a short time, but one of the most important rules in science is that good results must be reproducible. Other people must be able to repeat our observations, experiments, and calcula-

tions. In fact, most scientists routinely repeat other scientists' work as a way of getting started on a research topic. As soon as someone tries to repeat our fraudulent research, we will be caught. The more important a scientific result is, the sooner other scientists will repeat it, so we don't have much of a chance of getting away with scientific fraud. In this way, science is self-correcting.

Even if we could invent some convincing scientific research, we would probably have difficulty publishing it. When a scientist submits an article to a scientific journal, it is subject to peer review. That is, the editor of the journal sends the article to one or two other experts in the field for comment and suggestions. These reviewers often make helpful suggestions, but they may also point out errors that have to be fixed before the journal can publish the article. In some cases, an article may be so flawed the editor will refuse to publish it

at all. If we submitted our fraudulent research on X-ray binaries, the reviewers would almost certainly notice things wrong with it, and it would never get into print.

Scientists know the rules, and they use them. If someone makes a big discovery and is interviewed by the press, scientists will begin asking, "Has this work been published in a peer-reviewed journal yet?" That is, they want to know if other experts have checked the work. Until research is published, it isn't really official, and most scientists would treat the results with care.

Fraud isn't impossible in science. Some cases have even been in the national news. Big grant money is a terrible temptation. But in science, "the truth will out." Because of the way scientists reproduce research and because of the way research is published, scientific fraud is quite rare.

---

event horizon, but X rays emitted by the heated matter flowing into the black hole could escape if the X rays were emitted before the matter crossed the event horizon. An isolated black hole will not have much matter flowing into it, but a black hole in a binary system might receive a steady flow of matter transferred from the companion star. Thus, we can search for black holes by searching among X-ray binaries.

Some X-ray binaries, such as Hercules X-1, contain a neutron star, and they will emit X rays much as would a binary containing a black hole. We can tell the difference between a neutron star and a black hole in an X-ray binary in two ways. If the compact object emits pulses, we know it is a neutron star. Otherwise, we must depend on the mass of the object. If the compact object has a mass greater than 3 solar masses, the object can't be a neutron star, and we can conclude that it must be a black hole.

The first X-ray binary suspected of harboring a black hole was Cygnus X-1, the first X-ray object discovered in Cygnus. It contains a supergiant B0 star and a compact object orbiting each other with a period of 5.6 days. Matter flows from the B0 star as a strong stellar wind, and some of that matter enters a hot accretion disk around the compact object (Figure 14-17). The accretion disk is about 5 times larger in diameter than the orbit of Earth's moon, and the inner few hundred kilometers of the disk have a temperature of about 2 million Kelvin—hot enough to radiate X rays. The com-

pact object is invisible, but Doppler shifts in the spectrum reveal the motion of the B0 star around the center of mass of the binary. By making assumptions about the mass of the B0 star and the geometry of the orbit,

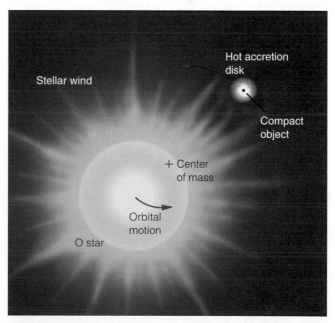

FIGURE 14-17
The X-ray source Cygnus X-1 is a supergiant O star and a compact object orbiting each other. Gas from the O star's stellar wind flows into the hot accretion disk, and the X rays we detect come from the disk.

TABLE 14-3
Nine Black-Hole Candidates

Object	Location	Companion Star	Orbital Period	Mass of Compact Object
Cygnus X-1	Cygnus	B0 Supergiant	5.6 days	>3.8 $M_\odot$
LMC X-3	Dorado	B3 main-sequence	1.7 days	6.7 ± 1.1 $M_\odot$
A0620-00	Monocerotis	K main-sequence	7.75 hours	10 ± 5 $M_\odot$
V404 Cygni	Cygnus	K main-sequence	6.47 days	12 ± 2 $M_\odot$
J1655-40	Scorpius	F main-sequence	2.61 days	6.9 ± 1 $M_\odot$
QZ Vul	Vulpecula	K main-sequence	8 hours	10 ± 4 $M_\odot$
4U 1543-47	Lupus	A main-sequence	1.123 days	2.7–7.5 $M_\odot$
V4641 Sgr	Sagittarius	B supergiant	2.81678 days	8.7–11.7 $M_\odot$
XTE J1118+480	Ursa Major	K main-sequence	0.170113 days	>6 $M_\odot$

astronomers can calculate the mass of the compact object—roughly 5 solar masses, well above the maximum for a neutron star.

To confirm that black holes exist, astronomers needed to find a conclusive example, an object that couldn't be anything else. Cygnus X-1 didn't quite succeed. Astronomers testing the hypothesis that it contained a black hole were quick to point out that the B0 star might not be a normal star. If it were of lower mass, the compact object might not be as massive as it seemed. Also, there could be a third star in the system confusing the analysis. The compact object in the Cygnus X-1 binary system cannot be shown conclusively to be more massive than 3 solar masses. Most astronomers believe it is a black hole, but the loophole remains, and more conclusive examples have been found.

As X-ray telescopes have found more X-ray objects, the list of black hole candidates has grown to a few dozen. A few of these objects, such as the first two in ▮ Table 14-3, contain massive stellar companions, either giants, supergiants, or massive main-sequence stars. This makes such systems difficult to analyze, because such massive companions dominate the system. A good example of this sort is LMC X-3 (LMC refers to the Large Magellanic Cloud, a small galaxy near our own). The compact object in LMC X-3 has a mass of about 10 solar masses. One reason astronomers think the compact object is so massive is that it distorts the shape of the B main-sequence star into an egg shape; and, as the system rotates, the light from the B star varies because we see the side of the egg and then the end. By analyzing the light change, astronomers can determine the shape of the B star and from that find the mass of the compact object.

Many of the black hole candidates are binary systems in which the normal star is a lower-mass main-sequence star. Such systems do not remain X-ray sources continuously, but suffer X-ray nova outbursts as matter flows rapidly into the accretion disk. A year or so after

an outburst, the flow has stopped, the accretion disk has dimmed, and astronomers can detect the spectrum of the main-sequence star. The spectral type and Doppler motions of the main-sequence companion reveal the mass of the compact object with only limited uncertainty. The lower-mass stellar companion makes these X-ray nova systems easier to analyze than are systems with massive companions.

A number of examples of these X-ray nova systems are shown in Table 14-3. A0620-00 is an old nova that erupted again in 1975. It contains an ordinary main-sequence K star and a compact object that orbit each other with a period of 7.75 hours. From the orbital motion and the distortion of the K star, astronomers conclude the compact object must have a mass between 5 and 15 solar masses. Three of the best understood examples are V404 Cygni; J1655-40, also known as Nova Scorpii 1994; and QZ Vul. The compact objects in these systems seem much too massive to be anything but black holes.

The growing list of X-ray binaries with compact objects exceeding 3 solar masses has convinced astronomers that black holes really do exist. The problem now is to understand how these objects interact with matter flowing into them through accretion disks to produce X rays, gamma rays, and jets of matter.

Ace◐Astronomy™ Go to AceAstronomy and click Active Figures to see "End States of Stars." Compare the deaths of high-, medium-, and low-mass stars.

┤ **REVIEW** CRITICAL INQUIRY ├

How can a black hole emit X rays?

Once a bit of matter falling into a black hole crosses the event horizon, no light or other electromagnetic radiation it emits can escape from the black hole. The matter becomes lost to our view. But if it emitted radiation before it crossed the event horizon, that radiation could escape, and we could detect it. Furthermore, the powerful gravitational field near

Astronomers are confident that black holes really exist. Now the problem is to understand how compact objects can explain the exotic objects we see in the sky.

14-3 COMPACT OBJECTS WITH DISKS AND JETS

Neutron stars and black holes seem to be exotic objects, but they generate equally exotic phenomena. By studying those phenomena, we can learn more about the strange objects.

X-RAY BURSTERS

Beginning in the 1970s, X-ray telescopes revealed that some objects emit irregularly spaced bursts of X rays. Typically, bursts that follow a long quiet period are especially large (❚ Figure 14-18), and this suggests that some mechanism is accumulating energy that is released by the bursts. The longer the quiet phase, the more energy accumulates.

These **X-ray bursters** are thought to be binaries containing neutron stars. The X-ray burster 4U 1820−30 in the globular cluster NGC6624, for example, appears to be a neutron star pulling mass away from a white dwarf (❚ Figure 14-19). Matter from the white dwarf first flows into an accretion disk and then falls to the surface of the neutron star where the impact heats the gas. Hydrogen fuses on the surface of the star, leaving helium ash to accumulate. When the helium reaches a depth of about a meter, it fuses explosively into carbon and produces an X-ray burst. The rapid increase in brightness and the total amount of energy produced fit well with theoretical models of an explosion on the

surface of an object as small as a neutron star. Of course, mass continues to accumulate, producing burst after burst. Notice the similarity with the mechanism that produces nova explosions on the surfaces of white dwarfs.

Dozens of X-ray bursters are known, and astronomers suspect that they occur in binary systems where matter flows through accretion disks and falls into neutron stars.

ACCRETION DISK OBSERVATIONS

When material falls into a neutron star or black hole it forms an accretion disk, and high-speed observations have been able to reveal some of the processes that go on in such disks.

When a blob of matter gets caught in an accretion disk, it can orbit so fast its orbital period is measured in thousandths of a second. Because the blob of matter looses energy by friction, it spirals inward, and its orbital period grows shorter. This process can be detected as a quick flicker as the disk emits a short series of pulses of electromagnetic radiation. Disks around some neutron stars produce pulses with a period as short as 0.00075 seconds, as blobs of material orbit only a dozen kilometers from the center of the neutron star. Because the pulse trains last only a few cycles, and because the period of the pulses grows rapidly shorter, they are called **quasi-periodic oscillations (QPOs).** Do not confuse these rapid flickering pulses with the regular pulses emitted by a pulsar.

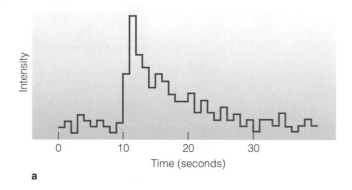

a

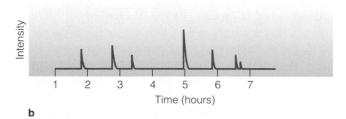

b

FIGURE 14-18

(a) X-ray bursters emit bursts of X rays that rise to full intensity in a few seconds and then fade in about 20 seconds. (b) The longer the interval since the preceding burst, the brighter the next burst will be.

FIGURE 14-19

(a) At visible wavelengths, the center of star cluster NGC6624 is crowded with stars. (b) In the ultraviolet, one object stands out, the X-ray burster 4U 1820−30 consisting of a neutron star orbiting a white dwarf. (c) An artist's conception shows matter flowing from the white dwarf into an accretion disk around the neutron star. *(a and b, Ivan King and NASA/ESA)*

Visual-wavelength image

a

X-ray source 4U 1820–30

UV image

b

c

QPOs can be observed in J1550-564, a star shedding mass into an accretion disk around a black hole. The flow of mass is irregular and causes X-ray flares. Short flickering strings of pulses are seen with periods as short as 0.003 second, and the period decreases rapidly. This suggests we are seeing material at some boundary in the accretion disk moving rapidly inward toward orbits of shorter and shorter period.

Cyg X-1 appears to be a star shedding mass into a black hole of about 10 solar masses. Pulse trains have been detected with periods of 0.018 seconds. The pulses grow dimmer and faster, which seems to mean the material is spiraling into the black hole. But no impact is seen, as we would expect of material falling into a neutron star. Evidently, the pulses fade away because the material approaches the event horizon and a powerful gravitational red shift stretches the photons to such long wavelengths they cannot be detected.

Further evidence of an event horizon was found in a detailed study of 12 X-ray novae. Six of the novae consisted of binary systems containing neutron stars, and bursts of energy were detected, produced by in-falling material striking the surfaces of the neutron stars. In contrast, six of the novae were systems containing black holes, and no bursts were seen from impacts on a surface. Apparently, in these systems, the material spiraled inward and disappeared across event horizons (▌Figure 14-20).

Because space-time is so strongly curved near a compact object, very small orbits are unstable. Once material spirals inward to that smallest possible orbit, it must quickly fall into the object. Yet the Rossi X-ray Timing Explorer satellite found an X-ray binary containing a black hole in which a QPO had a period of 0.002 second. The smallest possible orbit around the black hole in that system has a period of a little over

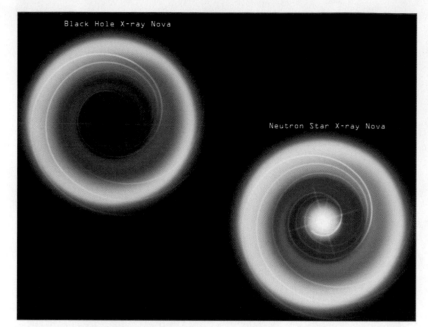

FIGURE 14-20
Gas spiraling into an accretion disk grows hot, and as it nears the central object, a strong gravitational red shift makes it appear redder and dimmer. Systems containing a neutron star emit bursts of energy when the gas hits the surface of the neutron star, but such bursts are not seen for systems containing black holes. In those systems, the matter vanishes across the event horizon. *(NASA/CXC/SAO)*

0.003 second. This appears to be evidence that the spinning black hole is dragging space-time with it as theory predicts and making smaller, shorter period orbits stable. Astronomers are searching for more of these very-short-period QPOs, because they confirm the Kerr solution's prediction that spinning black holes drag space-time.

High-speed observations at many wavelengths are allowing astronomers to follow the rapid processes that occur inside accretion disks. Those same accretion disks can eject high-energy jets extending out into space.

JETS OF ENERGY FROM COMPACT OBJECTS

Our first impression of a black hole might suggest that it is impossible to get any energy out of such an object. In later chapters, we will meet galaxies that extract vast amounts of energy from massive black holes, so we should pause here to see how a compact object can produce energy.

Whether a compact object is a black hole or a neutron star, it has a strong gravitational field. Any matter flowing into that field is accelerated inward; and, because it must conserve angular momentum, it flows into an accretion disk made so hot by friction that the inner regions can emit X rays and gamma rays. Thermal and magnetic processes can cause the inner parts of the accretion disk to eject high-speed jets of matter in opposite directions along the axis of rotation (▌ Figure 14-21). This process is similar to the bipolar outflows ejected by protostars, but it is much more powerful. We have seen in the X-ray image on page 285 that the Crab Nebula pulsar is ejecting jets of highly excited gas. The Vela pulsar does the same (Figure 14-6). Systems containing black holes can also eject jets. The black hole can-

didate J1655-40 is observed at radio wavelengths to be sporadically ejecting oppositely directed jets at 92 percent the speed of light.

One of the most powerful examples of this process is an X-ray binary called SS433. Its optical spectrum shows sets of spectral lines that are Doppler-shifted by about one-fourth the speed of light, with one set shifted to the red and one set shifted to the blue. Furthermore, the two sets of lines shift back and forth across each other with a period of 164 days. News media reported that astronomers had discovered an object that was both approaching and receding at a fantastic speed, but astronomers recognized the Doppler shifts as evidence of oppositely directed jets.

Apparently, SS433 is a binary system in which a compact object thought to be a neutron star or possibly a black hole pulls matter from its companion star and forms an extremely hot accretion disk. Jets of high-temperature gas blast away in beams aimed in opposite directions. As the disk precesses, it sweeps these beams around the sky once every 164 days, and we see light from gas trapped in both beams. One beam produces a red shift, and the other produces a blue shift (▌ Figure 14-22). SS433 is a prototype that illustrates how the gravitational field around a compact object can produce powerful beams of radiation and matter. We will meet this phenomenon again when we study peculiar galaxies.

GAMMA-RAY BURSTS

The cold war has an odd connection to neutron stars and black holes. In 1963, a nuclear test ban treaty was signed; and, by 1968, the United States was able to put a series of Vela satellites in orbit to watch for nuclear tests that were violations of the treaty. A nuclear detonation emits gamma rays, so the Vela satellites were

FIGURE 14-21

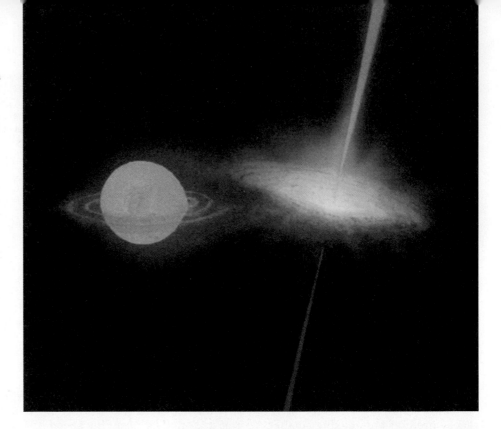

In this artist's impression, matter from a normal star at left flows into an accretion disk around a compact object at right. Processes in the spinning disk eject gas and radiation in jets perpendicular to the disk. The jet inclined slightly toward us has a blue shift, and the jet inclined slightly away from us has a red shift. (NASA)

designed to watch for bursts of gamma rays coming from Earth. The experts were startled when the satellites began detecting about one gamma-ray burst a day coming from space. When those data were finally declassified, astronomers realized that the bursts might be coming from neutron stars or black holes. These objects are now known as **gamma-ray bursters.**

The Compton Gamma Ray Observatory reached orbit in 1991 and immediately began reporting gamma-ray bursts at the rate of a few a day. The intensity of the gamma rays rises to a maximum in seconds and then fades away quickly; a burst is usually over in seconds or minutes. Furthermore, the Compton Observatory discovered that the gamma-ray bursts were coming from all over the sky and not from any particular region. This helped astronomers sort out the different theories.

Some theories proposed that the gamma-ray bursts were being produced among the stars in our galaxy, but the Compton Observatory observation that the bursts were coming from all over the sky eliminated that possibility. If the gamma-ray bursts were produced among stars in our galaxy, we would expect to see them most often from along the Milky Way, where many stars are located. The observation of bursts occurring all over the sky meant that bursts were probably coming from distant galaxies.

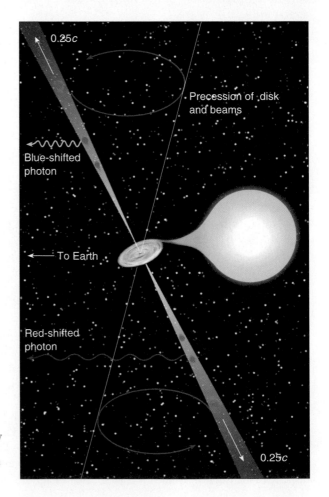

FIGURE 14-22

The generally accepted model of SS433 includes a compact object with a very hot accretion disk producing beams of radiation with embedded blobs of gas traveling at one-fourth the velocity of light. The precession of the disk swings the beams around a conical path every 164 days.

Gamma-ray bursts are hard to study because they occur without warning and fade so quickly; but, in 1997, the Italian-Dutch satellite BeppoSAX began giving Earth's astronomers immediate notice when a gamma-ray burst occurred, along with an accurate position in the sky. When astronomers quickly looked at those positions at visible wavelengths, they could see the fading glow of the explosion (▌ Figure 14-23). The spectra of these glowing clouds of gas reveal that most gamma-ray bursts are in very distant galaxies.

Given the intensity of a gamma-ray burst and its distance, astronomers can calculate its luminosity. If the gamma-ray bursts occur at very great distances, out among the galaxies, then they must be very powerful—at least as powerful as the most violent supernova explosions. Or perhaps the eruption blasts gamma rays out in beams. In that case, the explosion doesn't have to be quite as violent and could still be detected at very great distance if a beam were pointed at Earth. But that would mean that we don't see most of these events because their beams do not happen to be pointed toward us. In that case, the eruptions would have to be much more common than they appear.

Some bursts of lower-energy (soft) gamma rays repeat and are therefore known as **soft gamma-ray repeaters (SGRs).** Only a few are known, and they lie along the Milky Way. Astronomers suspect that they occur on young neutron stars that have magnetic fields as much as 1000 times stronger than the average neutron stars. We have discussed these magnetars earlier. Evidently, such a highly magnetic neutron star can produce bursts of gamma rays when shifts in the magnetic field break the crust of the neutron stars. One of these objects produced a burst of gamma rays that reached Earth on August 27, 1998, and temporarily ionized Earth's upper atmosphere, disrupting radio communication worldwide.

Bursts of higher-energy gamma rays are spread uniformly across the sky and do not appear to repeat, so they can't be explained by magnetars in our galaxy. These gamma-ray bursts must be produced by violent explosions occurring in distant galaxies. Two theories have been proposed. One suggestion is that gamma-ray bursts occur when two neutron stars orbiting each other lose energy by gravitational radiation and fall together. Theoretical models show that the merger of two neutron stars would produce a tremendous eruption lasting only seconds as the neutron stars ripped each other apart with tides and vanished into a newborn black hole. Such an event could produce bursts of gamma rays possibly in beams.

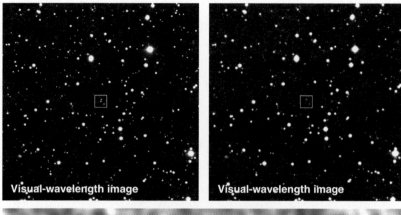

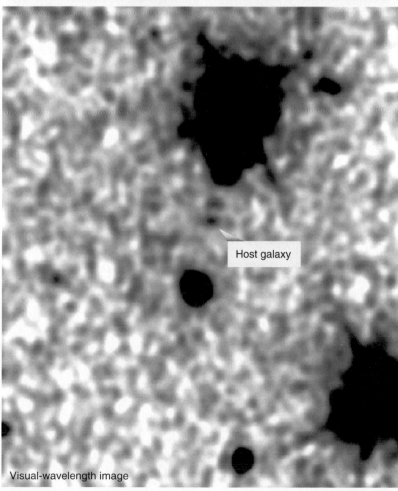

Visual-wavelength image

FIGURE 14-23

Alerted by gamma-ray detectors on satellites, observers used one of the VLT 8.2-meter telescopes on a mountaintop in Chile to image the location of a gamma-ray burst only hours after the burst. The image at top left shows that fading glow of the eruption. The image at top right, recorded 13 years before, reveals no trace of an object at the location of the gamma-ray burst. The Hubble Space Telescope image at bottom was recorded a year later and reveals a very faint galaxy at the location of the gamma-ray burst. *(ESO and NASA)*

The second theory proposes that gamma-ray bursts are produced when an extremely massive star exhausts its nuclear fuels. For the most massive stars, the resulting supernova explosion can be smothered by the massive outer layers of the star, and the star may collapse directly into a black hole. These have been called **hypernovae** or **collapsars**. Although the hypernova may produce no supernova explosion, it can emit jets of gamma rays because its rotation slows the collapse of its equatorial regions. The rotating star would collapse first along its poles, and moments later the equatorial regions would fall in. Models suggest this could produce powerful bursts of gamma rays shooting out of the poles of rotation. These hypernovae are thought to be much more rare than normal supernovae because they are produced only by the most massive stars.

The growing evidence seems to favor the hypernovae theory over the merging neutron star theory, but this may be a case where similar phenomena are produced by different processes. Binary neutron stars are common, and mergers should occur occasionally. We may eventually find that many gamma-ray bursts are produced by hypernovae but that some are produced by neutron star mergers. The study of gamma-ray bursts is still new and may hold further surprises.

Incidentally, if a gamma-ray burst occurred only 1600 ly from Earth, the distance to a nearby binary pulsar, the gamma-ray burst would shower the Earth with radiation equivalent to a 10,000-megaton nuclear blast. (The largest bombs ever made were a few megatons.) The gamma rays could create enough nitric oxide in the atmosphere to produce intense acid rain and could destroy the ozone layer and expose life on Earth to deadly levels of solar ultraviolet radiation. Gamma-ray bursts may occur relatively near the Earth as often as every few 100 million years and could be one of the causes of the mass extinctions that show up in the fossil record.

It may seem odd that such rare events as merging neutron stars and hypernovae are blamed for something so common that gamma-ray telescopes observe one or more every day. But we must remember that these events are so powerful they can be detected over very great distances. There may be 30,000 neutron star binaries in each galaxy, so mergers must occur now and then in any galaxy. Massive stars explode as hypernovae only rarely in any one galaxy, but we can detect these explosions among a vast number of galaxies. Thus the 1000 or so gamma-ray bursts that the Compton Observatory detected each year appear to be coming from the final deaths of stars far across space.

---| **REVIEW** CRITICAL INQUIRY |---

Why do fluctuations from accretion disks have such short periods?

When we see an X-ray flicker coming from an accretion disk, we are seeing a blob of hot gas orbiting the neutron star or black hole at a very small radius. Earth takes a year to orbit the sun once, but it is very far from the sun. A blob of gas in an accretion disk around a neutron star is orbiting an object with a mass of about a solar mass, but it may have an orbit only a few dozen kilometers in radius. The orbital period must be very small, as we would expect from Kepler's third law.

The flickers we see coming from accretion disks tell us about fantastic processes going on there. Why do these trains of pulses fade away?

Neutron stars and black holes give rise to violent processes and emit high-energy radiation, but they are only single stars. When we turn our attention to galaxies, we study communities containing billions of stars. The processes there are much more powerful.

SUMMARY

If the remains of a star collapse with a mass greater than the Chandrasekhar limit of 1.4 solar masses, the object cannot reach stability as a white dwarf. It must collapse to the neutron star stage with a radius of about 10 km and a density equal to that of an atomic nucleus. Such a neutron star can be supported by the pressure of its degenerate neutrons. But if the mass is greater than 2 to 3 solar masses, the degenerate neutrons cannot stop the collapse, and the object must become a black hole.

Theory predicts that a neutron star should rotate very fast, be very hot, and have a strong magnetic field. Such objects have been identified as pulsars, sources of pulsed radio energy. Pulsars are evidently spinning neutron stars that emit beams of radiation from their magnetic poles. As they spin, they sweep the beams around the sky; and, if the beams sweep over Earth, we detect pulses. This is known as the lighthouse model. The spinning neutron star slows as it radiates energy into space, mostly as a pulsar wind, and glitches in its pulsation appear to be caused by starquakes in the rigid crust of the neutron star or by changes in the circulation of the fluid interior.

Dozens of pulsars have been found in binary systems. One such binary pulsar is losing orbital energy at the rate predicted by the theory of gravitational waves. In some binary systems, such as Hercules X-1, mass flows into a hot accretion disk around the neutron star and causes the emission of X rays. In other systems, called X-ray bursters, the accumulation of fuel on the neutron star causes periodic outbursts of X rays.

If a collapsing star has a mass greater than 2 to 3 solar masses, it must contract to a very small size—perhaps to a singularity, an object of zero radius. Near such an object, gravity is so strong not even light can escape, and we term the region a black hole. The surface of this region, called the event horizon, marks the boundary of the black hole. The Schwarzschild radius is the radius of this event horizon, amounting to only a few kilometers for black holes of stellar mass.

If we were to leap into a black hole, our friends who stayed behind would see two relativistic effects. They would see our clock slow relative to their own clock because of time dilation. Also, they would see our light red-shifted to longer wavelengths. We would not notice these effects, but we would feel powerful tidal forces that would deform and heat our mass until we grew hot enough to emit X rays. Any X rays we emitted before crossing the event horizon could escape.

To search for black holes, we must look for binary star systems in which mass flows into a compact object and emits X rays. If the mass of the compact object is greater than about 3 solar masses, then the object is presumably a black hole. A number of such objects have been located.

Accretion disks produce rapid, violent phenomena. X-ray bursters are caused by matter flowing through an accretion disk onto a neutron star. Flickering in the light from accretion disks is caused by the rapid orbital motion of matter in the disk. In many cases, the rapid rotation of the disk and the inflow of matter produce high-energy jets perpendicular to the disk. Soft gamma-ray repeaters appear to be produced by magnetars, highly magnetized neutron stars. True gamma-ray bursters appear to be produced by the merger of two neutron stars or by hypernovae.

NEW TERMS

neutron star

pulsar

lighthouse model

pulsar wind

glitch

magnetar

gravitational radiation

millisecond pulsar

singularity

black hole

event horizon

Schwarzschild radius (R_S)

Kerr black hole

ergosphere

time dilation

gravitational red shift

X-ray burster

quasi-periodic oscillations (QPOs)

gamma-ray burster

soft gamma-ray repeater (SGR)

hypernova

collapsar

REVIEW QUESTIONS

1. Why is there an upper limit to the mass of neutron stars? Why is that upper limit not well known?

2. Explain in detail why we expect neutron stars to be hot, spin fast, and have strong magnetic fields.

3. Why can't we use visual-wavelength telescopes to locate neutron stars?

4. Why does the short length of pulsar pulses eliminate normal stars as possible pulsars?

5. What do we mean when we say "Every pulsar is a neutron star, but not every neutron star is a pulsar"?

6. According to our model of a pulsar, if a neutron star formed with no magnetic field at all, could it be a pulsar? Why or why not?

7. Why did astronomers first assume that the millisecond pulsar was very young?

8. Why do we suspect that only very fast pulsars can emit visible pulses?

9. If the sun were replaced by a 1-solar-mass black hole, how would Earth's orbit change?

10. If the sun has a Schwarzschild radius, why isn't it a black hole?

11. How can a black hole emit X rays?

12. What do we mean when we say "Black holes have no hair"?

13. What evidence do we have that black holes really exist?

14. How can mass transfer into a compact object produce jets of high-speed gas?

15. How does an X-ray burster resemble a nova?

16. Discuss the possible causes of gamma-ray bursters.

DISCUSSION QUESTIONS

1. Has the existence of neutron stars been sufficiently tested to be called a theory, or should it be called a hypothesis? What about the existence of black holes?

2. Why would you expect an accretion disk around a star the size of the sun to be cooler than an accretion disk around a compact object?

3. In this chapter, we imagined what would happen if we jumped into a Schwarzschild black hole. From what you have read, what do you think would happen to you if you jumped into a Kerr black hole?

PROBLEMS

1. If a neutron star has a radius of 10 km and rotates 642 times a second, what is the speed of the surface at the neutron star's equator in terms of the speed of light?

2. Suppose that a neutron star has a radius of 10 km and a temperature of 1,000,000 K. How luminous is it? (*Hint:* See Chapter 9.)

3. A neutron star and a white dwarf have been found orbiting each other with a period of 11 minutes. If their masses are typical, what is their average separation? Compare the separation with the radius of the sun, 7×10^5 km. (*Hint:* See Chapter 9.)

4. If the accretion disk around a neutron star has a radius of 2×10^5 km, what is the orbital velocity of a particle at its outer edge? (*Hint:* Use circular velocity, Chapter 5.)

5. What is the escape velocity from the surface of a typical neutron star? How does that compare with the speed of light? (*Hint:* See Chapter 5.)

6. If Earth's moon were replaced by a typical neutron star, what would the angular diameter of the neutron star be as seen from Earth? (*Hint:* Use the small-angle formula, Chapter 3.)

7. If the inner accretion disk around a black hole has a temperature of 1,000,000 K, at what wavelength will it radiate the most energy? What part of the spectrum is this in? (*Hint:* Use Wien's law, Chapter 7.)

8. What is the orbital period of a bit of matter in an accretion disk 2×10^5 km from a 10-solar-mass black hole? (*Hint:* Use circular velocity, Chapter 5.)

9. If SS433 consists of a 20-solar-mass star and a neutron star orbiting each other every 13.1 days, what is their average separation? (*Hint:* See Chapter 9.)

CRITICAL INQUIRIES FOR THE WEB

1. Imagine that you are on a mission to explore one of the pulsar planets noted in this chapter. What would you find there? Look on the Web for information about pulsars and the known pulsar planets, and describe what you might encounter on such a mission.

2. What would you experience if you were to pilot a spacecraft near a black hole? Visit related Internet sites to determine what the gravitational effects and general environment would be like. Also, use the Internet to find the limits of human tolerance to strong gravitational forces. (*Hint:* Look for information about astronaut training and find out how many Gs a human can withstand.) Using these sources, give a brief account of what your voyage would be like.

3. Search for information about gamma-ray bursters. What is the latest news in this developing story?

EXPLORING *THESKY*

1. The Crab Nebula pulsar is located in the Crab Nebula, also known as M1. Locate the nebula, zoom in, and compare its shape and size with Figure 14-4. (*TheSky* does not show the pulsar.)

 Visit the Seeds *Foundations of Astronomy* companion Web site for critical thinking exercises, articles, and additional readings from InfoTrac College Edition, Brooks/Cole's online student library.

THE MILKY WAY GALAXY

Jane was watching Mrs. Corry splashing the glue on the sky and Mary Poppins sticking on the stars. . . .
"What *I* want to know," said Jane, "is this: Are the stars gold paper or is the gold paper stars?"
There was no reply to her question and she did not expect one. She knew that only someone
very much wiser than Michael could give her the right answer.

P. L. Travers, Mary Poppins

2Mass

Chapter 15

GUIDEPOST

This chapter plays three parts in our cosmic drama. First, it introduces the concept of a galaxy. Second, it discusses our home, the Milky Way Galaxy, a natural object of our curiosity. Third, it elaborates our story of stars by introducing us to galaxies, the communities in which stars exist.

Science is based on the interaction of theory and evidence, and this chapter will show a number of examples of astronomers using evidence to test theo-

ries. If the theories seem incomplete and the evidence contradictory, we should not be disappointed. Rather, we must conclude that the adventure of discovery is not yet over.

We struggle to understand our own galaxy as an example. We will extend the concept of the galaxy in Chapters 16 and 17 on normal and peculiar galaxies. We will then apply our understanding of galaxies in Chapter 18 to the study of the universe as a whole.

P. L. Travers wrote her story about Mary Poppins, the supercalifragilisticexpialidocious nanny, in 1934, and at that time no one could answer the question that Jane asked in the quotation given on the previous page. But by the 1960s, astronomers could assure her that the gold paper was stars, or rather that the atomic elements of which gold paper is made were created inside generation after generation of stars. How the stars made the elements is one of the keys to understand our Milky Way Galaxy.

It isn't obvious that we live in a galaxy. We are inside, and we see nearby stars scattered all over the sky. From a dark site far from city lights, we can see great clouds of stars in our galaxy making a faint band of light circling the sky (■ Figure 15-1a). The ancient Greeks named that band *galaxies kuklos,* the "milky circle." The Romans changed the name to *via lactia,* "milky road" or "milky way." We will see in this chapter how the evidence tells us that we live inside a great wheel of stars, a galaxy.

Almost every object you see in the sky is part of our Milky Way Galaxy. Exceptions include the **Magellanic Clouds,** small irregular galaxies located near our galaxy in space. They are visible as faint patches of haze in the southern sky. Another exception is the Andromeda Galaxy, just visible to our unaided eyes as a faint patch of light in the constellation Andromeda. (Consult the star charts at the end of this book to locate the Andromeda

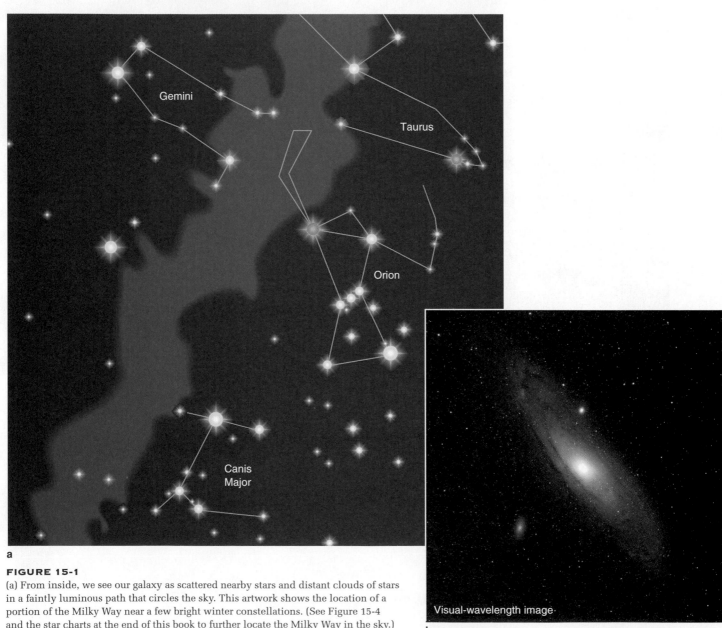

Visual-wavelength image

a

b

FIGURE 15-1
(a) From inside, we see our galaxy as scattered nearby stars and distant clouds of stars in a faintly luminous path that circles the sky. This artwork shows the location of a portion of the Milky Way near a few bright winter constellations. (See Figure 15-4 and the star charts at the end of this book to further locate the Milky Way in the sky.) (b) This photograph of the Andromeda Galaxy, a spiral galaxy about 2.3 million ly from us, shows us what our Milky Way Galaxy would look like if we could view it from such a great distance. *(AURA/NOAO/NSF)*

Galaxy.) We will see evidence that our galaxy looks much like the disk-shaped Andromeda Galaxy (Figure 15-1b).

The largest telescopes can detect an estimated 100 billion extremely faint, hazy objects which astronomers, drawing on the Greek word for milk, call galaxies. They appear to be other systems of stars similar to our own. Our goal in this chapter is to learn about our Milky Way Galaxy—how it formed and evolved and why it takes the shape it does. As always, we begin with observations, the evidence upon which all astronomy is based.

15-1 THE NATURE OF THE MILKY WAY GALAXY

The secret to studying any science lies in the simple question, "How do we know?" Because all scientific knowledge is based on evidence and logical arguments about natural processes, we need to know facts. But we also need to understand how we know those facts. Thus, we set the stage in the next section with a few facts, but our real interest is in how we know what our galaxy is like.

THE STRUCTURE OF OUR GALAXY

We live in a disk-shaped galaxy with beautiful spiral arms reaching from its center to its edge. Astronomers commonly give the diameter of the galaxy as 25,000 pc or 75,000 ly, with the sun located about two thirds of the way from the center to the edge (█ Figure 15-2). The diameter of our galaxy is not known very accurately, as you can see from the two numbers given above. Convert 25,000 pc into light-years, and you get significantly more than 75,000 ly. That these two numbers don't quite match is a warning that we don't know the size of our galaxy to better than about 10 percent.

The disk of the galaxy contains most of the stars and nearly all of the gas and dust. Although most of the stars in the disk are middle- to lower-main-sequence stars like the sun, a few are giants, and fewer still are brilliant O and B stars. These hot, luminous stars light up the disk and make it bright. It is difficult to judge just how big the disk is, because it is filled with gas and dust that dim starlight and make it difficult to see distant stars in the disk. Also, large dusty clouds block our view in many directions within the disk. In most directions in the plane of the disk, we can't see farther than a few **kiloparsecs.** (A kiloparsec is 1000 pc.)

If we look up or down out of the disk, we look away from the dust and gas, and we can see out into the **halo** of our galaxy, a spherical cloud of stars and star clusters that contains almost no gas and dust. The halo completely encloses the disk of the galaxy. The size of the halo is uncertain, but it is at least a few times larger than the disk.

75,000 ly

Sun Nuclear bulge Disk

Halo

Globular cluster

ACTIVE FIGURE 15-2

An artist's conception of our Milky Way Galaxy, seen face-on and edge-on. Note the position of the sun and the distribution of globular clusters in the halo. Only the inner halo is shown here.

Ace⌇Astronomy™ Go to AceAstronomy and click Active Figures to see "Explorable Milky Way" and go deeper into this diagram.

Around the center of our galaxy lies the **nuclear bulge,** a flattened cloud of stars about 3 kpc in radius. It contains little gas and dust but billions of stars similar to those found in the halo.

This quick portrait of our galaxy is just a collection of bare facts. It raises two important questions that will guide us through the rest of this chapter. How do we know what our galaxy is like? How did our galaxy get this way?

FIRST STUDIES OF THE GALAXY

How do astronomers know what our galaxy is like? Humanity has known of the Milky Way, the hazy path of light around the sky, since antiquity, but not until Galileo looked at the sky with his telescope in 1610 did anyone know the Milky Way was made of stars. Little more was known for two centuries after Galileo lived.

By the mid-18th century, astronomers generally understood that the stars were other suns, but they had little understanding of how the stars were distributed in space. One of the first people to study this problem was the English astronomer Sir William Herschel (1738–1822). He and his sister Caroline (1750–1848) mapped the stars in three dimensions by assuming that the sun was located inside a great cloud of stars and that they could see to the edges of the cloud in any direction. They believed that by counting the number of stars visible in different directions, they could gauge the relative distance to the edge of the cloud. If their telescope revealed few stars in one direction, they assumed that the edge was not far away, and if they saw many stars in another direction, they assumed that the edge was very distant. Calling their method "star gauges," they counted stars in 683 directions in the sky and outlined a model of the star cloud. Their data showed that the cloud was a great disk with the sun near the center, and using the technology of their day as an analogy, they called it the "grindstone model" (❙ Figure 15-3).

This model of the star system explains the most obvious feature of the Milky Way—it is a glowing path that circles the sky. The grindstone model reproduces this as a disk of stars with the sun near the center. Unfortunately, there was no way for the Herschels to find the diameter of the disk.

Generations of astronomers studied the size of our star system, culminating with Jacobus C. Kapteyn (1851–1922). In the early 20th century, he analyzed the brightness, number, and motion of stars and concluded that our star system is a disk about 10 kpc in diameter and about 2 kpc thick, with the sun near the center.

As we saw in the previous section, modern astronomers believe our galaxy is much larger and that

the sun is not at the center. Why did these first studies of the galaxy underestimate its size and mistakenly place the sun near the center? The studies were misled because astronomers assumed that space was empty and that they could see to great distances. Today, we know that the disk of our galaxy is filled with gas and dust that dim stars and block our view. We can see only a small region near us in the disk of our galaxy. How we discovered the true size of our galaxy and came to understand the nature of a galaxy is one of the great adventures of human knowledge. It began about a century ago when a man studying star clusters and a woman studying variable stars unlocked one of nature's biggest secrets—that our star system is a galaxy.

DISCOVERING THE GALAXY

It seems odd to say that astronomers discovered something that is all around us, but until early in the 20th century, no one knew that the star system we live in is a galaxy. That began to change in the second decade of the 20th century, when a young astronomer named Harlow Shapley (1885–1972) discovered how big our star system really is. Besides being one of the turning points of modern astronomy, Shapley's study illustrates one of the most common techniques in astronomy. If we want to know how we know things in astronomy, then Shapley's story is well worth tracing in detail.

Shapley began by noticing that open star clusters lie along the Milky Way but that more than half of all globular clusters lie in or near the constellation Sagittarius (❙ Figure 15-4). The globular clusters seemed to be distributed in a great cloud whose center was located in the direction of Sagittarius. Shapley assumed that the orbital motion of these clusters was controlled by the gravitation of the entire star system, and thus the

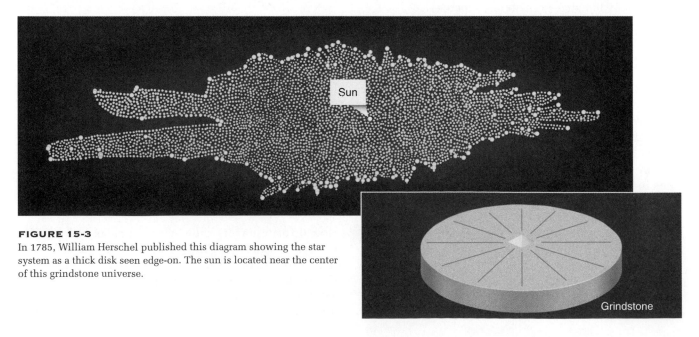

FIGURE 15-3
In 1785, William Herschel published this diagram showing the star system as a thick disk seen edge-on. The sun is located near the center of this grindstone universe.

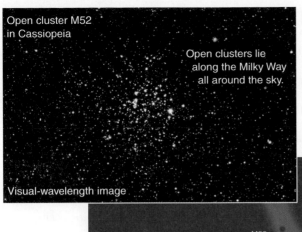

Open cluster M52 in Cassiopeia

Open clusters lie along the Milky Way all around the sky.

Visual-wavelength image

FIGURE 15-4

Nearly half of cataloged globular clusters (red dots) are located in or near Sagittarius and Scorpius. A few of the brighter globular clusters, labeled with their catalog designations, are visible in binoculars or small telescopes. Constellations are shown as they appear above the southern horizon on a summer night as seen from latitude 40° N, typical for most of the United States. *(M52: NOAO/AURA/NSF; M19: Doug Williams, N.A. Sharp/ NOAO/AURA/NSF)*

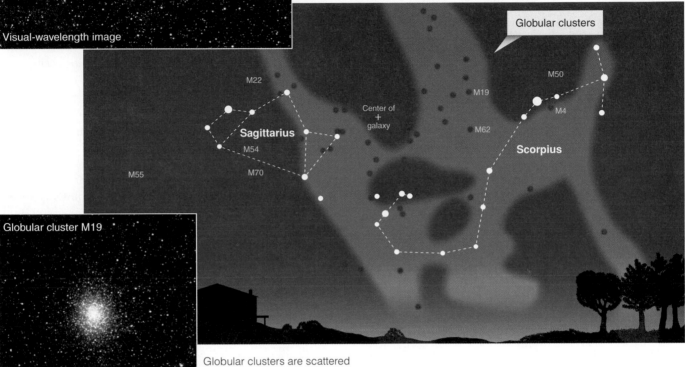

Globular cluster M19

Visual-wavelength image

Globular clusters are scattered over the entire sky but are strongly concentrated toward Sagittarius.

center of the star system could not be near the sun but must lie somewhere toward Sagittarius.

To find the distance to the center of the star system and thus the size of the star system as a whole, Shapley needed to find the distance to the star clusters, but that was difficult to do. The clusters are much too far away to have measurable parallaxes. They do, however, contain variable stars (Chapter 12), and those were the lampposts Shapley needed to find the distances to the clusters.

Shapley knew of the work of Henrietta S. Leavitt (1868–1921), who in 1912 had shown that a Cepheid variable star's period of pulsation was related to its brightness. Leavitt worked on stars whose distances were unknown, so she was unable to find their absolute magnitudes or luminosities, but astronomers realized that the Cepheids could be a powerful tool in astronomy if their true luminosities could be discovered.

Cepheids are giant stars and so are relatively rare. None lies close enough to Earth to have measurable par-

allax. Their proper motions—slow movement across the sky—can be measured, and the more distant a star is, the smaller its proper motion tends to be. Thus proper motions contain clues to distance. Ejnar Hertzsprung (codiscoverer of the H–R diagram) made an early attempt to determine the absolute magnitudes of the Cepheids, but Shapley went further. He found 11 Cepheids with measured proper motions and used a statistical process to find their average distance and thus their average absolute magnitude. That meant he could erase Leavitt's apparent magnitudes from the period–luminosity diagram (see Figure 12-13) and write in absolute magnitudes. All of the stars in the diagram were thus calibrated as lampposts that astronomers could use to find distances (Window on Science 15-1), including the RR Lyrae stars that Shapley saw in the globular clusters.

Finding distance using Cepheid variable stars is so important in astronomy we should pause to illustrate the process. Suppose we studied a star cluster and discovered that it contained a type I Cepheid with a period

The Calibration of One Parameter to Find Another

Astronomers often say that Shapley "calibrated" the Cepheids for the determination of distance, meaning that he did all the detailed background work so that the Cepheids could be used to find distances. Other astronomers could use the Cepheids without repeating the detailed calibration.

Calibration is actually very common in science. Chemists, for instance, have carefully calibrated the colors of certain compounds against acidity. They can quickly measure the acidity of a solution by dipping into it a slip of paper containing the indicator compound and looking at the color. They don't have to repeat the careful calibration every time they measure acidity.

Engineers in steel mills have calibrated the color of molten steel against its temperature. They can use a hand-held device that measures the color of a ladle of molten steel and then look up the temperature in a table. They don't have to repeat the calibration every time. Astronomers have made the same kind of color–temperature calibration for stars.

As you read about any science, notice how calibrations are used to simplify common measurements. But notice, too, how important it is to get the calibration right. An error in calibration can throw off every measurement made with that calibration. Some of the biggest errors in science have been errors of calibration.

---

of 10 days and an average apparent magnitude of 10.5. How far away is the cluster? From the period–luminosity diagram (Figure 12-13), we see that the absolute magnitude of the star must be about -3. Then the distance modulus is:

$$m - M_v = 10.5 - (-3) = 13.5$$

We could use this distance modulus and a table such as Table 9-1 to estimate the distance to the cluster as somewhere between 1000 and 16,000 pc. We can be more accurate if we solve the magnitude–distance formula (Chapter 9) for distance:

$$d = 10^{(m_v - M_v + 5)/5}$$

Now we can substitute the apparent magnitude and absolute magnitude, and we discover that the distance is equal to $10^{3.7}$. A pocket calculator tells us that the distance is about 5000 pc. This is one of the most common calculations in astronomy.

Once Shapley had calibrated the period–luminosity relation, he could use it to find the distance to any cluster in which he could identify variable stars. He could measure the apparent magnitudes of the variable stars in a cluster from a photographic plate, and he could find the periods of the variable stars from a series of photographic plates. Because the periods were very short, he knew the variable stars he saw in the clusters were the lowest-luminosity variable stars (known today as RR Lyrae stars). That allowed him to read their absolute magnitude off of the period–luminosity diagram. Once he knew the apparent magnitude and the absolute magnitude, he could calculate the distance to the star cluster.

This worked well for the nearer globular clusters, but the variable stars in the more distant clusters are too faint to detect. Shapley estimated the distances to these more distant clusters by calibrating the diameters of the clusters. For the clusters whose distance he knew, he could use their angular diameters and the small-angle formula to calculate linear diameters in parsecs. He found that the nearby clusters are about 25 pc in diameter, which he assumed is the average diameter of all globular clusters. He then used the angular diame-

ters of the more distant clusters to find their distances. This is another illustration of calibration in astronomy.

Shapley later wrote that it was late at night when he finally plotted the direction and distance to the globular clusters and found that, just as he had supposed, they formed a great swarm whose center lay many thousands of light-years in the direction of Sagittarius. Evidently the center of the star system was not near the sun but rather was far away in Sagittarius (❙ Figure 15-5). He called the only other person in the building, a cleaning lady, and the two stood looking at his graph as he explained that they were the only two people on Earth who understood that humanity lives, not at the center of a small star system, but in the suburbs of a vast wheel of stars.

Why did astronomers before Shapley think we lived near the center of a small star system? Space is filled with gas and dust that dim our view of distant stars. When we look toward the band of the Milky Way, we can see only the neighborhood near the sun. Most of the star system is invisible, and, like travelers in a fog, we seem to be at the center of a small region. Shapley was able to see the globular clusters at greater distances because they lay outside the plane of the star system and are not dimmed very much by the gas and dust.

Building on Shapley's work, other astronomers began to suspect that some of the faint patches of light visible through telescopes were other star systems like our own. We will continue this story in the next chapter when we discuss galaxies in general. It is sufficient here to note that within a few years astronomers had found evidence that the faint patches of light were other galaxies much like our own Milky Way Galaxy.

Shapley's study of star clusters led to the discovery that we live in a galaxy and that the universe is filled with similar galaxies. Like Copernicus, Shapley moved us from the center to the suburbs. Nevertheless, our home galaxy is dramatically beautiful. To get a true impression of the great galaxy in which we live, we must conduct a careful analysis.

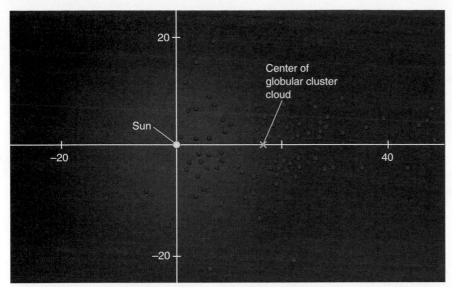

a

FIGURE 15-5

(a) Shapley's study of globular clusters showed that they were not centered on the sun, at the origin of this graph, but rather formed a great cloud centered far away in the direction of Sagittarius. Distances on this graph are given in thousands of parsecs. (b) Looking toward Sagittarius, we see nothing to suggest this is the center of the galaxy. Gas and dust block our view. Only the distribution of globular clusters told Shapely the sun lay far from the center of the star system. *(Daniel Good)*

b

AN ANALYSIS OF THE GALAXY

We want to know how we know, and we are now ready to explore our modern knowledge of the galaxy. We can survey the galaxy's components, understand how we have learned about them, and begin to understand what they tell us about the nature of our home galaxy.

We can divide our galaxy into two main components whose characteristics will help us understand how our galaxy formed. The **disk component** of the galaxy contains most of the stars, gas, and dust that orbit within the disk. In contrast, the **spherical component** includes the nuclear bulge and the halo.

In addition to stars, the disk of the galaxy contains vast amounts of gas and dust—the interstellar medium. We saw in Chapter 10 how we can learn about this material. It contains mostly hydrogen gas and helium gas and is irregularly distributed as dense clouds separated by less dense regions. The densest of these clouds are the giant molecular clouds within which stars form (Chapter 11), and thus nearly all star formation in our galaxy takes place in the disk. The most massive of these newborn stars are the most luminous, the O and B stars. They light up the disk and make it glow brightly. Some of these stars are hot enough to ionize the nearby gas to produce HII regions, and thus the disk contains all of the emission nebulae in our galaxy.

Although we can survey the location and distance of the stars (Chapter 9), we cannot cite a single number for the thickness of the disk because it lacks sharp boundaries. Stars become less crowded farther from the central plane of the galaxy. Also, the thickness of the disk depends on the kind of object we study. Stars like the sun with ages of a few billion years lie within about 500 pc above and below the central plane. But the youngest stars, including the O and B stars, and the gas and dust from which these young stars are forming are confined to a disk only about 100 pc thick (Figure 15-6). With a diameter of 25 kpc, the disk is thinner than a thin pizza crust.

From the shape and size of our galaxy, astronomers conclude that we live in a spiral galaxy (Figure 15-7). The most dramatic features of these disk galaxies are the spiral arms—long spiral patterns of bright stars, HII regions, star clusters, and clouds of gas and dust. The sun is located on the inside edge of one of these spiral arms. We will see later in this chapter how we can observe the spiral arms in our own galaxy.

We see two kinds of star groupings in the disk of our galaxy. Associations are groups of 10 to a few hundred stars so widely scattered in space that their mutual gravity cannot hold the association together. From the turnoff points in their H–R diagrams (Chapter 12) we can tell they are very young groups of stars. We see them moving together through space (Figure 15-8) because they formed from a single gas cloud and haven't had time to wander apart. Two kinds of association—O and B associations and T Tauri associations (Chapter 11)—are located along the spiral arms.

The second kind of cluster in the disk is the open cluster (Figure 15-4), a group of 100 to a few thousand stars in a region about 25 pc in diameter. Because they have more stars in less space than associations, open clusters are more firmly bound by gravity. Although they lose stars occasionally, they can survive for a long time, and the turnoff points in their H–R diagrams give ages from a few million to a few billion years.

In contrast to the disk, the halo of our galaxy contains almost no gas and dust and only a thin scattering of stars and globular clusters. Because the halo contains no dense gas clouds, it cannot make new stars. Halo stars are old, cool, lower-main-sequence stars, red giants, and white dwarfs. It is difficult to judge the extent of the halo, but it could extend far beyond the edge of the visible disk and may be slightly flattened like a thick hamburger bun.

The halo contains roughly 200 globular clusters, each of which contains 50,000 to a million stars in a sphere about 25 pc in diameter (Figure 15-4). Because they contain so many stars in such a small region, the clusters are very stable and have survived for billions of years. From the turnoff points in their H–R diagrams, we know they are about 11 billion years old.

The nuclear bulge at the center of the galaxy is the most crowded part of the spherical component. Visible above and below the plane of the galaxy and through gaps in the obscuring dust, the nuclear bulge contains stars that are old and cool like the stars in the halo.

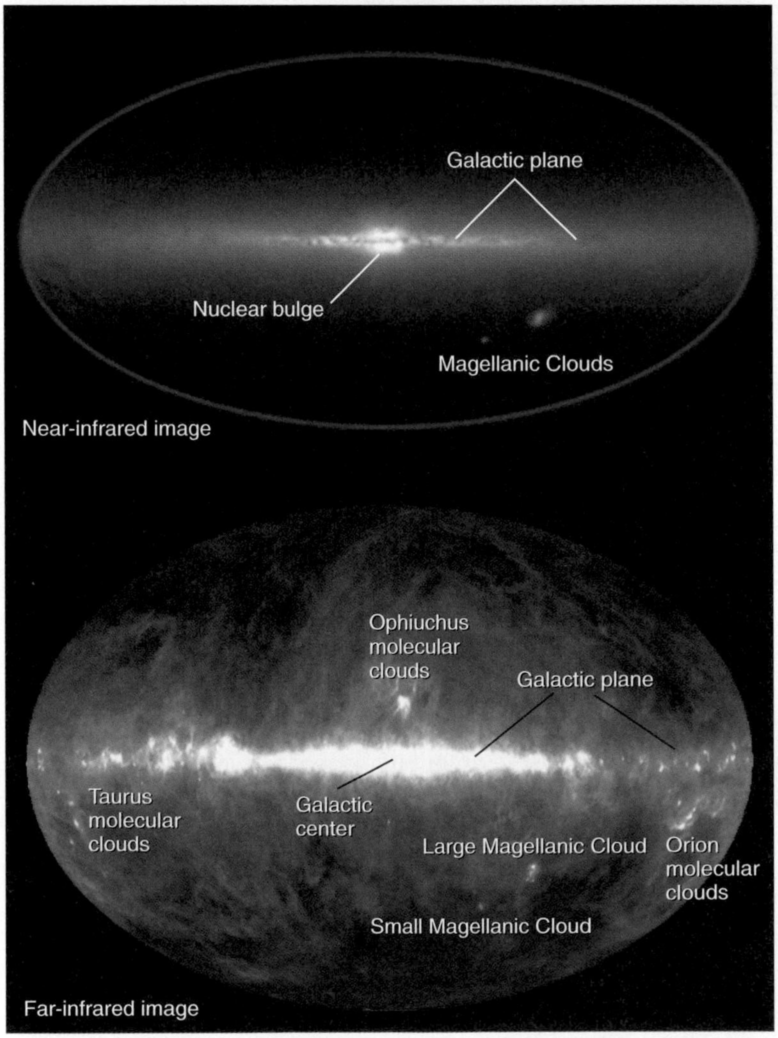

FIGURE 15-6

In these infrared images, the entire sky has been projected onto ovals with the center of the galaxy at the center. The Milky Way extends from left to right. In the near infrared, the nuclear bulge is prominent, and dust clouds block our view. At longer wavelengths, the dust emits significant energy and glows brightly. *(Near-IR: 2Mass; Far-IR: DIRBE Image courtesy Henry Freudenreich)*

Our analysis of the components of our galaxy leave us wondering about one critical question. How much matter does our galaxy contain? To answer that question, we must watch our galaxy rotate.

THE MASS OF THE GALAXY

To find the mass of an object, astronomers must observe it in orbital motion as, for example, in a binary star system. We don't live long enough to see our galaxy rotate significantly, but astronomers can observe the radial velocities, proper motions, and distances of stars and then calculate their orbits. The results can tell us the mass of the galaxy and give us hints about its origin.

Stars in the disk of the galaxy follow nearly circular orbits in the plane of the disk (❙ Figure 15-9a). By observing the radial velocities of other galaxies in various directions, astronomers can conclude that the sun is moving at about 220 km/s in the direction of Cygnus as it orbits the center of our galaxy in a nearly circular path. Given the distance to the center of our galaxy, 8.5 kpc, we can conclude that the sun has an orbital period of about 240 million years.

Halo stars and globular clusters, however, follow highly elongated orbits tipped steeply to the plane of the disk (Figure 15-9b). Although we show these orbits as elliptical, they are more like rosettes because of the influence of the thick bulge. Where halo stars pass through the disk, they are cutting across the regular traffic of disk stars and seem to have very high velocities. Thus, they are called **high-velocity stars.** The dramatic difference between the motion of halo stars and disk stars will be important evidence when we discuss the formation of the galaxy in the next section.

The orbital motion of the sun can tell us the mass of the galaxy. When we studied binary stars in Chapter 9, we saw that the total mass (in solar masses) of a binary star system equals the cube of the separation of the stars a (in AU) divided by the square of the period P (in years):

$$M = \frac{a^3}{P^2}$$

The Sombrero Galaxy

Visual-wavelength image

a

NGC 2997

Visual-wavelength image

b

FIGURE 15-7
Our disk galaxy resembles the spiral galaxies visible through large telescopes. (a) Seen edge-on, spiral galaxies have nuclear bulges and dust-filled disks. *(Todd Boroson/NOAO/AURA/NSF)* (b) Seen face-on, the spiral arms are dramatically outlined by hot O and B stars, emission nebulae, and dust. *(ESO)*

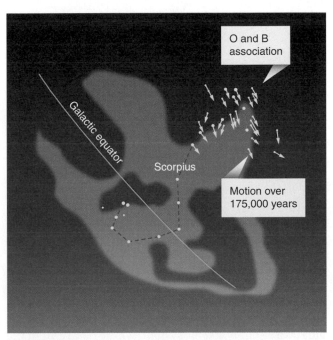

FIGURE 15-8
Many of the stars in the constellation Scorpius are members of an O and B association that has formed recently from a single cloud of gas. As the stars orbit the center of our galaxy, they are moving together southwest along the Milky Way.

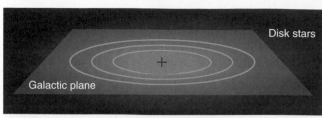

a

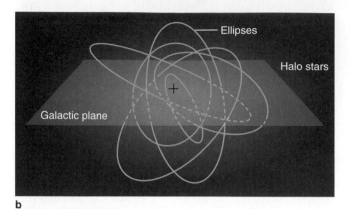

b

FIGURE 15-9

(a) Stars in the galactic disk have nearly circular orbits that lie in the plane of the galaxy. (b) Stars in the halo have randomly oriented, highly elongated orbits.

The radius of the sun's orbit around the galaxy is about 8500 pc, and each parsec contains 206,265 AU. Multiplying, we find that the radius of the sun's orbit is 1.75×10^9 AU. The orbital period is 240 million years, so the mass is:

$$M = \frac{(1.75 \times 10^9)^3}{(240 \times 10^6)^2} = 0.93 \times 10^{11}\ M_\odot$$

This is only a rough estimate because it does not include any mass outside the orbit of the sun. Adding an estimate for the mass outside the sun's orbit gives us a total mass for our galaxy of about 4×10^{11} solar masses.

The rotation of our galaxy is actually much more interesting than the description above. Stars at different distances from the center revolve around the center of the galaxy with different periods, so three stars near each other will draw apart as time passes (▌Figure 15-10). This motion is called differential rotation. (Recall that we defined differential rotation in Chapter 8.) To fully describe the rotation of our galaxy, we must graph orbital velocity versus radius—a **rotation curve** (▌Figure 15-11).

If all of the mass of the galaxy were concentrated near its center, then we would expect to see orbital velocities fall as we moved away from the center. This is what we see in our solar system, where nearly all of the mass is concentrated in the sun, and it is called **Keplerian motion** (a reference to Kepler's laws). The best observations of the rotation curve of the Milky Way

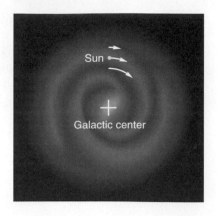

FIGURE 15-10

The differential rotation of the galaxy means that stars at different distances from the center have different orbital periods. In this example, the star just inside the sun's orbit has a shorter period and gains on the sun, while the star outside falls behind.

show orbital velocities constant or rising in the outer disk rather than falling. These higher orbital velocities indicate that the larger orbits enclose more mass and seem to tell us that our galaxy is more massive than it appears to be.

The evidence tells astronomers that the extra mass lies in an extended halo sometimes called a **galactic corona.** It may extend up to 10 times farther than the edge of the visible disk and could contain a trillion solar masses. Much of this mass is invisible, so astronomers conclude it is not emitting or absorbing light and refer to it as **dark matter.** At least some of the mass in the galactic corona is made up of low-luminosity stars and white dwarfs, but much of the mass must be some other form of matter. We will return to the problem of the dark matter in the following two chapters. It is one of the fundamental problems of modern astronomy.

REVIEW CRITICAL INQUIRY

If gas and dust block our view, how do we know how big our galaxy is?

The gas and dust in our galaxy block our view only in the plane of the galaxy. When we look away from the plane of the galaxy, we look out of the gas and dust and can see to great distances. The globular clusters are scattered through the halo with a strong concentration in the direction of Sagittarius. When we look above or below the plane of the galaxy, we can see those clusters, and careful observations reveal variable stars in the clusters. By using a modern calibration of the period–luminosity diagram, we can find the distance to those clusters. If we assume that the distribution of the clusters is controlled by the gravitation of the galaxy as a whole, then we can find the distance to the center of the galaxy by finding the center of the distribution of globular clusters.

In the critical analysis of ideas, it is important to ask "How do we know?", and that is especially true in science. For example, what measurements and assumptions tell us the total mass of our galaxy?

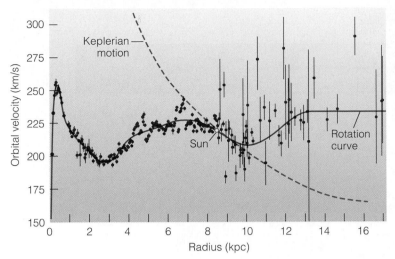

FIGURE 15-11

The rotation curve of our galaxy is plotted here as orbital velocity versus radius. Data points show measurements made by radio telescopes with error bars showing the uncertainty in the measurements. The fitted rotation curve (solid line) does not decline outside the sun's orbit as we would expect for Keplerian motion (dashed line). Rather, the curve is approximately flat at great distances, suggesting that the galaxy contains significant mass outside the orbit of the sun. *(Adapted from a diagram by Françoise Combes)*

The differences between the disk and spherical components extend even to their chemical composition. As we will see in the next section, those differences hold clues to the origin of our galaxy.

15-2 THE ORIGIN OF THE MILKY WAY

Just as dinosaurs left behind fossilized footprints, our galaxy has left behind a fossil footprint of its youth. The stars of the spherical component are old and must have formed long ago when the galaxy was very young.

STELLAR POPULATIONS

In the 1940s, astronomers realized that there are two families of stars in the galaxy. They form and evolve in similar ways, but they differ in the abundance of atoms that are heavier than helium—atoms that astronomers call **metals.** (Note that this is not the way the word "metal" is used by nonastronomers.) **Population I stars** are metal rich, containing 2 to 3 percent metals, whereas **population II stars** are metal poor, containing only about 0.1 percent metals or less. The difference is dramatically evident in spectra (❚ Figure 15-12).

Population I stars belong to the disk component of the galaxy and are sometimes called disk population stars. They have nearly circular orbits in the plane of the galaxy and are relatively young stars that formed within the last few billion years. The sun is a population I star, as are the type I Cepheid variables discussed in Chapter 12.

Population II stars belong to the spherical component of the galaxy and are sometimes called the halo population stars. These stars have randomly tipped, elliptical orbits and are old stars. The metal-poor globular clusters are part of the halo population, as are the RR Lyrae and type II Cepheids.

Since the discovery of stellar populations, astronomers have realized that there is a gradation between populations. Extreme population I stars are found only in the spiral arms. Slightly less metal-rich population I stars, called intermediate population I stars, are located throughout the disk. The sun is such a star. Stars even less metal rich, such as stars in the nuclear bulge, belong to the intermediate population II. The most metal-poor stars are those in the halo, including those in globular clusters. These are extreme population II stars.

Why do the disk and halo stars have different metal abundances? The answer to this question is the key to our understanding of the history of our galaxy.

The different populations of stars differ in other ways such as type of orbit and age (❚ Table 15-1). These are clues to the formation of our galaxy. To follow these clues we must consider the origin of the chemical elements.

THE ELEMENT-BUILDING PROCESS

We can use our understanding of stellar structure and stellar evolution (Chapters 11 to 13) to understand the origin of the chemical elements heavier than helium—elements that astronomers call metals. When the universe began, it contained only hydrogen and helium atoms. (We will review the evidence that supports this statement in Chapter 18.) All of the remaining chemical elements have been produced by **nucleosynthesis,** the process that fuses hydrogen and helium inside stars to make the heavier elements (Window on Science 15-2).

When our galaxy formed sometime after the beginning of the universe, the gas contained about 90 percent hydrogen atoms, 10 percent helium atoms, and almost no heavier atoms. Thus, the first stars to form were metal poor. Succeeding generations of stars manufactured heavier atoms, and the metal abundance gradually increased.

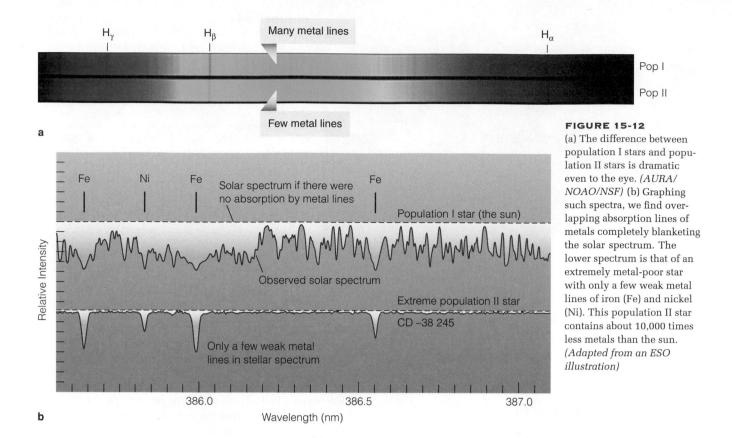

FIGURE 15-12

(a) The difference between population I stars and population II stars is dramatic even to the eye. *(AURA/NOAO/NSF)* (b) Graphing such spectra, we find overlapping absorption lines of metals completely blanketing the solar spectrum. The lower spectrum is that of an extremely metal-poor star with only a few weak metal lines of iron (Fe) and nickel (Ni). This population II star contains about 10,000 times less metals than the sun. *(Adapted from an ESO illustration)*

The most massive stars fuse helium into carbon, nitrogen, oxygen, and heavier elements up to iron (Chapter 11), and when those stars die in supernova explosions, traces of those elements are spread back into the interstellar medium. Furthermore, the supernova explosion itself can fuse nuclei into atoms heavier than iron. These are rare atoms, such as gold, silver, uranium, and platinum, but they, too, get spread back into space during supernova explosions.

Elements heavier than helium are quite rare in the universe. Traditionally, astronomers graph the abundance of the elements using an exponential scale, but if we replot the data using a linear scale we see how rare these atoms are (Figure 15-13).

When we look at population II stars, such as those in the halo, we are looking at the survivors of the early generations of stars in our galaxy. Those stars formed from gas that was metal poor, and thus their spectra show only weak metal lines. The survivors of these early generations are low-mass, long-lived stars that don't make heavy elements. Furthermore, any elements they might have made are locked in their cores. Thus, the spectrum we see tells us the composition of the gas from which they formed and that the gas was metal poor. Population I stars, such as the sun, formed more recently, after the interstellar medium had been enriched in metals, and their spectra show stronger metal lines. Stars forming now show the strongest metal lines.

TABLE 15-1
Stellar Populations

	Population I		Population II	
	Extreme	*Intermediate*	*Intermediate*	*Extreme*
Location	Spiral arms	Disk	Nuclear bulge	Halo
Metals (%)	3	1.6	0.8	Less than 0.8
Shape of orbit	Circular	Slightly elliptical	Moderately elliptical	Highly elliptical
Average age (yr)	100 million and younger	0.2–10 billion	2–10 billion	10–13 billion

WINDOW ON SCIENCE 15-2

Understanding Nature as Processes

One of the most common organizing themes in science is the concept of a process—a sequence of events that leads to some result or condition. Scientists try to discover the natural processes that shape the universe. Telling the stories of these processes explains why the universe is the way it is. At the same time, recognizing these processes and learning to tell their stories simplify much of science.

For example, astronomers try to assemble the sequence of events that led to the formation of the chemical elements. We know that the first stars formed from nearly pure hydrogen and helium and that they made certain heavier elements in their cores. Supernova

explosions made other elements and mixed the elements back into the interstellar medium. New stars formed from this gas, and generation after generation of stars added to the heavier elements that we are made of. Telling that story explains much of the science of the stars.

As you study any science, be alert for processes as organizing themes. Science, at first glance, seems to be nothing but facts, but one of the ways to organize the facts is to fit them into the story of a process. If you study biology and learn the process by which plant cells know whether to become roots, stems, or leaves, you will have organized a great deal of factual information into a story

that is not only easy to remember but also makes sense.

When you see a process in science, ask yourself a few basic questions. What conditions prevailed at the beginning of the process? What sequence of steps occurred? Can some steps occur simultaneously, or must one step occur before another can occur? What is the final state that this process produces? Most important, what evidence do we have that nature really works this way?

Not everything in science is a process, but identifying a process will help you remember and understand a great many details and principles.

GALACTIC FOUNTAINS

Supernova remnants are rich in metals and eventually mix back into the interstellar medium. In fact, supernovae continuously stir the interstellar medium. But a larger-scale process may be even more efficient at spreading newly created metals throughout the disk of the galaxy where they can be incorporated into the newly forming stars.

In Chapter 10, we saw how neighboring supernova remnants can merge to form a superbubble of hot gas (Figure 10-7b). A supernova remnant may be only a few tens of parsecs in diameter, but a superbubble can be more than ten times bigger. The denser gas of the galac-

tic disk tends to confine an expanding superbubble, but the disk is only a few hundred parsecs thick. If a superbubble breaks out of the denser gas of the galactic disk, it could spew hot gas high above the plane of the galaxy in a **galactic fountain** (▌Figure 15-14). As this gas cools and falls back into the disk, it could spread metals through the galaxy.

Galactic fountains have not yet been observed directly, but isolated clouds of hot gas have been found high in the halo, and cool clouds have been found falling toward the disk. It seems very likely that this process contributes to spreading metals through the disk. Even without direct observational confirmation

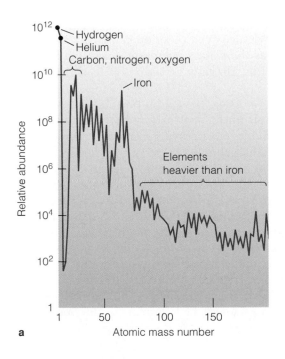

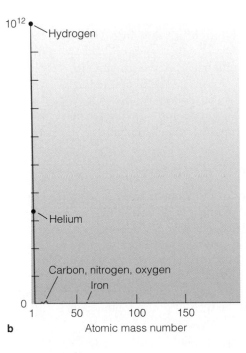

FIGURE 15-13
The abundance of the elements in the universe. (a) When the elements are plotted on an exponential scale, we see that elements heavier than iron are about a million times less common than iron and that all elements heavier than helium (the metals) are quite rare. (b) The same data plotted on a linear scale provide a more realistic impression of how rare the metals are. Carbon, nitrogen, and oxygen make small peaks near atomic mass 15, and iron is just visible in the graph.

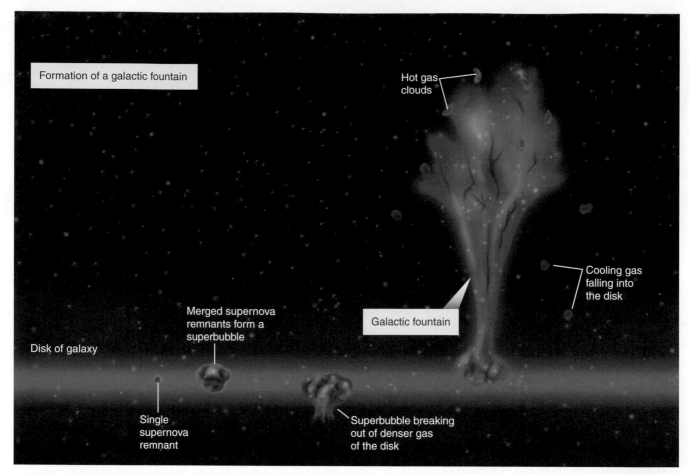

FIGURE 15-14
In this edge-on view of the galactic disk, we see that superbubbles, formed of multiple supernova remnants, can be roughly as large as the thickness of the disk. Once a superbubble breaks out of the denser gas of the disk, it is believed to produce a galactic fountain. The metal-enriched gas could flow out into the halo where it would cool and fall back to enrich the interstellar medium in metals.

of galactic fountains, we can be sure that supernovae create metals and mix them back into the interstellar medium where they can be incorporated into newly forming stars.

Because the metal abundance of newborn stars increases as the galaxy ages, we can use metal abundance as an index to the ages of the components of our galaxy. The halo is old, and the disk is younger. Such ages can help us understand the formation of the galaxy.

THE AGE OF THE MILKY WAY

Because we know how to find the age of star clusters, we can estimate the age of our galaxy. Nevertheless, uncertainties make our easy answer hard to interpret.

The oldest open clusters are 9 to 10 billion years old. These ages come from the turnoff point in the H–R diagram (see Chapter 12), but finding the age of an old cluster is difficult because old clusters change so slowly. Also, the exact location of the turnoff point depends on chemical composition, which differs slightly among clusters. Finally, open clusters are not strongly bound by their gravity, so even older open clusters may have dissipated as their stars wandered away. Thus, open clusters tell us the galactic disk is at least 10 billion years old.

Globular clusters have faint turnoff points in their H–R diagrams and are clearly old, but finding these ages is difficult. Clusters differ slightly in chemical composition, which must be accounted for in calculating the stellar models from which ages are determined. Also, to find the age of a cluster, astronomers must know the distance to the cluster. Precise parallaxes from the Hipparcos satellite have allowed astronomers to more precisely calibrate the Cepheid variable stars, and careful studies with the newest large telescopes have refined the chemical compositions and better defined the H–R diagrams of globular clusters. The best analysis of all of the data suggests that the average globular clusters are about 11 billion years old. Some globular clusters are younger than that, and some are older. Studies of the oldest globular clusters suggest that the halo of our galaxy is at least 13 billion years old.

Both populations and clusters tell us that the disk is younger than the halo. We can combine these ages with the process of nucleosynthesis to tell the story of our galaxy.

THE HISTORY OF THE MILKY WAY GALAXY

In the 1950s, astronomers began to develop a hypothesis to explain the formation of our galaxy. Recent observations, however, are forcing a reevaluation of that traditional hypothesis.

The traditional hypothesis says that the galaxy formed from a single large cloud of gas roughly 14 billion years ago. As gravity pulled the gas together, the cloud began to fragment into smaller clouds, and, because the gas was turbulent, the smaller clouds had random velocities. Stars and star clusters that formed from these fragments went into orbits that had a wide range of shapes; a few were circular, but most were elliptical, and some were extremely elliptical. The orbits were also inclined at different angles, so the orbits produced a spherical cloud of stars. Of course, these first stars were metal poor because no stars had existed earlier to enrich the gas with metals. This is the way the traditional hypothesis explains the formation of the spherical component of the galaxy.

The second stage of this hypothesis accounts for the disk component. The contracting gas cloud was roughly spherical at first, but the turbulent motions canceled out, as do eddies in recently stirred coffee, leaving the cloud with uniform rotation. A rotating, low-density cloud of gas cannot remain spherical. A star is spherical because its high internal pressure balances its gravity; but, in a low-density cloud, the pressure cannot support the weight. Like a blob of pizza dough spun in the air, the cloud must flatten into a disk (Figure 15-15).

This contraction into a disk took billions of years, with the metal abundance gradually increasing as generations of stars were born from the gradually flattening gas cloud. The stars and globular clusters of the halo were left behind by the cloud when it was spherical, and subsequent generations of stars formed in flatter distributions. The gas distribution in the galaxy now is so flat that the youngest stars are confined to a disk only about 100 parsecs thick. These stars are metal rich and have nearly circular orbits.

This traditional hypothesis accounts for many of the Milky Way's properties. Advances in technology, however, have improved astronomical observation, and, beginning in the 1980s, contradictions arose. For example, many halo stars have metal abundances similar to those of globular clusters and may have escaped

Origin of the Halo and Disk

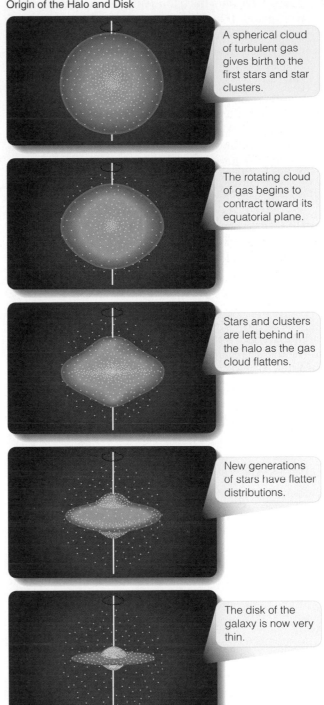

A spherical cloud of turbulent gas gives birth to the first stars and star clusters.

The rotating cloud of gas begins to contract toward its equatorial plane.

Stars and clusters are left behind in the halo as the gas cloud flattens.

New generations of stars have flatter distributions.

The disk of the galaxy is now very thin.

FIGURE 15-15
The traditional theory for the origin of our galaxy begins with a spherical gas cloud that flattens into a disk.

from such clusters, but some stars are even more metal poor. One star contains 200,000 times less metals than the sun. These stars may be older than the globular clusters. Also, some of the oldest stars in our galaxy are in the central bulge, not in the halo. Furthermore, not all globular clusters have the same age, and the younger clusters seem to be in the outer halo. The traditional hypothesis says that the halo formed first and that the clusters within it should have a uniform age.

Ages are a key problem for the traditional hypothesis. The globular clusters are about 10 to 13 billion years old, but the disk seems much younger. The oldest open star clusters in the disk are only 9 to 10 billion years old. Also, white dwarfs tend to accumulate over the eons, and there aren't as many white dwarfs in the disk as there should be if the disk were 10 to 13 billion years old. A successful hypothesis should account for these ages.

Another problem is that the oldest stars are metal poor but not metal free. There must have been at least a few massive stars to create these metals before the formation of the oldest stars we see in the halo.

Can we modify the traditional hypothesis to explain these observations? Perhaps the galaxy began with the contraction of a gas cloud to form the central bulge and the later accumulation of the halo from gas clouds that had been slightly enriched in metals by an early generation of massive stars. This would explain the age of the central bulge and the metals in the oldest stars.

The disk could have formed later as the gas that was already in the galaxy flattened and as more gas fell into the galaxy and settled into the disk. Perhaps entire galaxies were captured by the growing Milky Way Galaxy. In 2003, astronomers announced the discovery of a ring containing millions of stars circling our galaxy about twice as far from the center as the sun (❚ Figure 15-16). The ring may be the remains of a small galaxy that was absorbed by the Milky Way roughly 10 billion years ago. (We will see dramatic evidence in the next chapter that such galaxy mergers do occur.)

If the young Milky Way Galaxy absorbed a few small but partially evolved galaxies, then some of the globular clusters we see in the halo may be hitchhikers. This would explain the range of globular cluster ages. In fact, some theories hold that our galaxy formed by a rapid merger of a number of smaller pregalaxy clouds of gas and stars with later additions of in-falling gas and small merging galaxies.

The problem of the formation of our galaxy is frustrating because the theories are incomplete. We prefer certainty, but astronomers are still gathering observations and testing hypotheses. We see an older theory that has proven to be inadequate to explain the observations, and we see astronomers attempting to refine the observations and devise new theories. The metal abundances and ages of the stars in our galaxy seem to be important clues, but metal abundance and age do not tell the whole story. Astronomer Bernard Pagel was thinking of this when he said, "Cats and dogs may have the same age and metallicity, but they are still cats and dogs."

FIGURE 15-16

Surveys of many faint stars have uncovered a ring of stars that encircles the Milky Way Galaxy. One astronomer described it as the "smoking gun" to show that our galaxy has absorbed at least one smaller galaxy in its past. Similar but less dramatic streams of stars have been found throughout the Milky Way Galaxy's halo, suggesting that such mergers have occurred repeatedly. *(Rensselaer Polytechnic Institute and the Sloan Digital Sky Survey [www.sdss.org])*

REVIEW CRITICAL INQUIRY

Why do metal-poor stars have the most elongated orbits? Of course, the metal abundance of a star cannot affect its orbit, so our analysis must not confuse cause and effect with the relationship between these two factors. Both chemical composition and orbital shape depend on a third factor—age. The oldest stars are metal poor because they formed before there had been many supernova explosions to create and scatter metals into the interstellar medium. Those stars formed long ago when the galaxy was young and motions were not organized into a disk, and they tended to take up randomly shaped orbits, many of which are quite elongated. Thus, today, we see the most metal-poor stars following the most elongated orbits.

Nevertheless, even the oldest stars we can find in our galaxy contain some metals. They are metal poor, not metal free. How does that constrain our theories?

The story of our galaxy is an astronomical work in progress, but that is not the only mystery hidden in the Milky Way. The next two sections discuss two special problems that astronomers are trying to understand— the nature of spiral arms and the secret of the galactic nucleus.

UV + visual-wavelength image

a

FIGURE 15-17

(a) Images of other galaxies, such as NGC1232 shown here, show the spiral arms illuminated by clouds of young stars. The hot O and B stars in these clouds give the arms a blue tint. *(European Southern Observatory)* (b) Segments of spiral arms near the sun appear when we plot the location of O and B associations and the ionized nebulae that often surround such associations. These groups of massive stars have very short lives, so we find them only in regions of recent star formation. Dust blocks our view, and we can see only a few kiloparsecs in the plane of the Milky Way, but in the region we can see near the sun, O and B associations trace out segments of three spiral arms as diagrammed here.

b

15-3 SPIRAL ARMS

The most striking feature of galaxies like the Milky Way is the system of spiral arms that wind outward through the disk. These arms contain swarms of hot, blue stars, clouds of dust and gas, and young star clusters. These young objects hint that the spiral arms involve star formation. As we try to understand the spiral arms, we face two problems. First, how can we be sure our galaxy has spiral arms when our view is obscured by gas and dust? Second, why doesn't the differential rotation of the galaxy destroy the arms? The solution to both problems involves star formation.

TRACING THE SPIRAL ARMS

Studies of other galaxies show us that spiral arms contain hot, blue stars (▌ Figure 15-17a). Thus, one way to study the spiral arms of our own galaxy is to locate these stars. Fortunately, this is not difficult, since O and B stars are often found in associations and, being very bright, are easy to detect across great distances. Unfortunately, at these great distances their parallax is too small to measure, so their distances must be found by other means, usually by spectroscopic parallax (see Chapter 9).

O and B associations in the sky are not located randomly but reveal parts of three spiral arms near the sun, which have been named for the prominent constellations through which they pass (Figure 15-17b). If we could penetrate the gas and dust, we could locate other O and B associations and trace the spiral arms farther; but, like a traveler in a fog, we see only the region near us.

Objects used to map spiral arms are called **spiral tracers.** O and B associations are good spiral tracers because they are bright and easy to see at great distances. Other tracers include young open clusters, clouds of hydrogen ionized by hot stars (emission nebulae), and certain higher-mass variable stars.

Notice that all spiral tracers are young objects. O stars, for example, live for only a few million years. If their orbital velocity is about 250 km/s, they cannot have moved more than about 500 pc since they formed. This is less than the width of a spiral arm. Because they don't live long enough to move away from the spiral arms, they must have formed there.

The youth of spiral tracers gives us an important clue about spiral arms. Somehow they are associated with star formation. Before we can follow this clue, however, we must extend our map of spiral arms to show the entire galaxy.

RADIO MAPS OF SPIRAL ARMS

The dust that blocks our view at visual wavelengths is transparent at radio wavelengths because radio waves are much longer than the diameter of the dust particles. When we point a radio telescope at a section of the Milky Way, we receive 21-cm radio signals coming from cool hydrogen in spiral arms at various distances across the galaxy. Fortunately, the signals can be unscrambled by measuring the Doppler shifts of the 21-cm radiation, except in the direction toward the nucleus. There, the orbital motions of gas clouds are perpendicular to the line of sight, and all of the radial velocities are zero. Thus, the radio map shown in ▌ Figure 15-18 reveals spiral arms throughout the disk of the galaxy, except in the wedge-shaped region toward the center.

The analysis of the 21-cm radial velocity data requires that astronomers estimate the orbital velocity of the clouds at different distances from the center of the galaxy. Because these velocities are not precisely known and because turbulent motions in the gas distort the radial velocities, we can't trace a perfect spiral pattern in Figure 15-18. We can be confident our galaxy has spiral arms, but we can't see the overall pattern.

Radio astronomers can also use the strong radio emission from carbon monoxide (CO) to map the location of giant molecular clouds in the plane of the galaxy. Recall from Chapter 11 that these clouds are sites of active star formation. Maps constructed from such observations reveal that the giant molecular clouds are located in large structures that resemble segments of spiral arms (▌ Figure 15-19).

Radio maps combined with studies at infrared and visible wavelengths paint a picture of our galaxy. Clearly we live in a spiral galaxy, but the spiral pattern appears slightly irregular with branches and spurs and gaps. The stars we see in Orion, for example, appear to be in a detached segment of a spiral arm, a spur. Some astronomers argue that our galaxy has a four-armed spiral pattern, but most believe it has a classic two-armed pattern with branches and spurs. The nuclear bulge of our galaxy does not seem to be a sphere but rather is an elongated bar pointing partially away from us. The spiral arms may spring from a ring that encloses this bar. In the next chapter we will see that such structures are common in other galaxies. Thus we can construct a best guess at what our galaxy might look like for comparison with other galaxies (▌ Figure 15-20).

The most important feature in the radio maps is easy to overlook—spiral arms are regions of higher gas density. Spiral tracers told us that the arms contain young objects, and we suspected active star formation. Now radio maps confirm our suspicion by telling us that the material needed to make stars is abundant in spiral arms.

THE DENSITY WAVE THEORY

Just what are spiral arms? We can be sure they are not physically connected structures like bands of magnetic field holding the gas in place. If they were, the strong

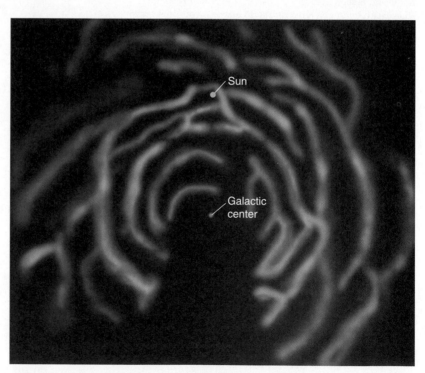

FIGURE 15-18

This map of our galaxy was made using 21-cm radio observations, and it confirms that our galaxy has spiral arms. Branches and spurs in the spiral arms are visible. The conical region around and beyond the center of the galaxy is the region where 21-cm radiation from gas clouds at different distances cannot be unscrambled. *(Adapted from a radio map by Gart Westerhout)*

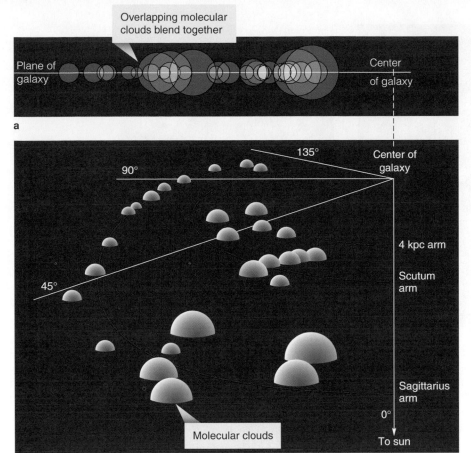

a

Overlapping molecular clouds blend together

Plane of galaxy

Center of galaxy

b

135°

90°

45°

Center of galaxy

4 kpc arm

Scutum arm

Sagittarius arm

0°

To sun

Molecular clouds

FIGURE 15-19

(a) At the wavelength emitted by the CO molecule we find many molecular clouds along the Milky Way, but they overlap in a confusing jumble. By using a model of the rotation of the galaxy and the radial velocities of the clouds, radio astronomers can find the distances to each cloud and use them to map spiral arms. (b) This map represents the view from a point 2 kpc directly above the sun. The molecular clouds, shown as hemispheres extending above the plane of our galaxy, are located along spiral arms. Angles in this diagram are galactic longitudes. *(Adapted from a diagram by T. M. Dame, B. G. Elmegreen, R. S. Cohen, and P. Thaddeus)*

differential rotation of the galaxy would destroy them within a billion years. They would get wound up and torn apart like paper streamers caught on the wheel of a speeding car. Yet spiral arms are common in galaxies (▌ Figure 15-21) and must last billions of years.

Astronomers believe that spiral arms are dynamically stable—they retain the same appearance even though the gas, dust, and stars in them are constantly changing. To see how this works, think of the traffic jam behind a slow-moving truck. Seen from an airplane, the

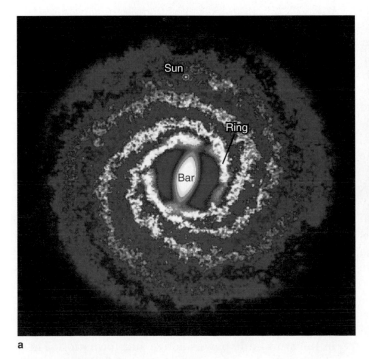

Sun

Ring

Bar

a

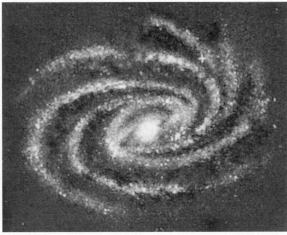

b

FIGURE 15-20

(a) This reconstruction of our galaxy is based on far-infrared observations of dust. The model includes a central bar surrounded by a ring from which spring four spiral arms. *(Courtesy Henry Freudenreich)* (b) This painting, based on radio and optical studies of the galaxy, includes a bar-shaped nucleus and a number of irregular arms with branches and spurs. *(Painting by M. A. Seeds, based on a study by G. De Vaucouleurs and W. D. Pence)*

FIGURE 15-21

Typical of spiral galaxies, M83 has a two-armed spiral pattern with prominent dust lanes and bright O and B stars along the arms. The galaxy also contains branches and spurs. *(European Southern Observatory)*

The Spiral Density Wave

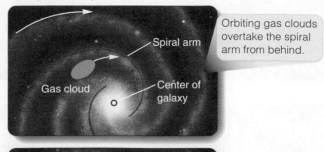

Orbiting gas clouds overtake the spiral arm from behind.

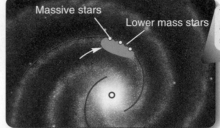

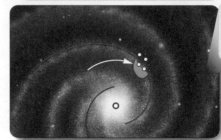

The compression of a gas cloud triggers star formation.

Massive stars are highly luminous and light up the spiral arm.

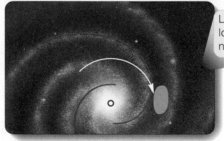

The most massive stars die quickly.

Low-mass stars live long lives but are not highly luminous.

FIGURE 15-22

According to the density wave theory, star formation occurs as gas clouds pass through spiral arms.

traffic jam would be stable, moving slowly down the highway. But any given car would approach from behind, slow down, work its way to the front of the jam, pass the truck, and resume speed. The individual cars in the jam are constantly changing, but the traffic jam itself is dynamically stable.

In the **density wave theory,** the spiral arms are dynamically stable regions of compression that move slowly around the galaxy, just as the truck moves slowly down the highway. Gas clouds, moving at orbital velocity around the galaxy, overtake the slow-moving arms from behind and slam into the gas already in the arms. The sudden compression of the gas can trigger the collapse of the gas clouds and the formation of new stars (see Chapter 11). The newly formed stars and the remaining gas eventually move on through the arm and emerge from the front of the slow-moving arm to resume their travels around the galaxy (▌ Figure 15-22). Thus, whenever we look we see a spiral arm, although the stars, gas, and dust are constantly changing.

Science depends on evidence, so we should ask what evidence we have that spiral arms are caused by spiral density waves. For one thing, the stars in spiral arms fit the theory well. We find stars of all masses forming in spiral arms, but the O and B stars are the brightest. They live such short lifetimes that they die before they can move out of the spiral arm. Thus, the presence of these stars along with gas and dust assures us that spiral arms are sites of star formation. Lower-mass stars, such as the sun, also form in spiral arms. But because they live so long, they can leave the arms and circle the galaxy many times. The sun probably formed

as part of a cluster in a spiral arm roughly 5 billion years ago, left that cluster, and has circled the galaxy about 20 times, passing through spiral arms many times.

Giant molecular clouds and thick dust in the arms (▌ Figure 15-23) are evidence that material is compressed in the arms and spawns active star formation. These clouds can convert up to 30 percent of their mass into

FIGURE 15-23

This chance alignment of a small galaxy in front of a larger, more distant galaxy illustrates the nature of spiral arms. The small galaxy's spiral arms are bright with O and B stars where they are silhouetted against the darkness of space; but, seen against the glare of the distant galaxy, the arms are revealed to contain large amounts of dusty gas. *(NASA and The Hubble Heritage Team)*

stars. The Orion complex and the Ophiuchus dark cloud are just two of these regions filled with infrared protostars. Such a cloud can form stars for 10 to 100 million years before it is disrupted by heat and light from the most massive new stars. The cloud needs about this long to pass through a spiral arm.

The evidence seems to fit the spiral density wave theory well, but the theory has two problems. First, how is the complicated spiral disturbance started and sustained? Spiral density waves should slowly die out in a billion years or so. Something must regenerate the spiral wave. Mathematical models show that the galaxy is naturally unstable to certain disturbances, just as a guitar string is unstable to certain vibrations. Any sudden disturbance—the rumble of a passing truck, for example—can set the string vibrating at its natural frequencies. Similarly, minor fluctuations in the galaxy's disk, such as supernova explosions, might regenerate the waves.

Another suggestion is that gravitational disturbances can generate a spiral density wave. Observations show that the center of our galaxy is not a sphere but a bar. The rotation of such a bar could disturb the disk of the galaxy and stimulate the formation of a spiral density wave. A close encounter or collision with a nearby galaxy could also produce a spiral pattern.

The second problem for the density wave theory involves the spurs and branches in the arms of our own and other galaxies. Computer models of density waves produce regular, two-armed spiral patterns. Some galaxies, called grand-design galaxies, do indeed have symmetric two-armed patterns, but others do not. Some galaxies have a great many short spiral segments, giving them a fluffy appearance. These galaxies have been termed **flocculent,** meaning "woolly" (❚ Figure 15-24). Our galaxy is probably an intermediate galaxy. How can we explain these variations in the spiral pattern? Perhaps the answer lies in a process that sustains star formation once it begins.

STAR FORMATION IN SPIRAL ARMS

Star formation is a critical process in the creation of the spiral patterns that we see. It makes spiral arms visible and may shape the spiral pattern itself.

The spiral density wave creates spiral arms by the gravitational attraction of the stars and gas flowing through the arms. Even if there were no star formation at all, rotating disk galaxies could form spiral arms. But without star formation to make young, hot, luminous stars, the spiral arms would be difficult to see. It is the star formation that lights up the spiral arms and makes them so prominent. Thus, star formation helps determine what we see when we look at a spiral galaxy.

But star formation can also control the shape of spiral patterns if the birth of stars in a cloud of gas can renew itself and continue making more new stars. Consider a newly formed star cluster with a single massive star. The intense radiation from that hot star can compress nearby parts of the gas cloud and trigger further star formation (❚ Figure 15-25). Also, massive stars evolve so quickly that their lifetimes are only an instant in the history of a galaxy. When they explode as supernovae, the expanding gasses can compress neighboring clouds of gas and trigger more star formation. Examples of such a **self-sustaining star formation** have been found. The Orion complex consisting of the Great Nebula in Orion and the star formation buried deep in the dark interstellar clouds behind the nebula is such a region.

Self-sustaining star formation can produce growing clumps of new stars, and the differential rotation of the galaxy can drag the inner edge ahead and let the outer edge lag behind to produce a cloud of star formation shaped like a segment of a spiral arm, a spur (❚ Figure 15-26). Mathematical models of galaxies filled with such segments of spiral arms do have a spiral appearance, but they lack the bold, two-armed spiral that astronomers refer to as the grand-design spiral pattern (❚ Figure 15-27). Rather, astronomers suspect that self-sustaining star formation can produce the branches and spurs so prominent in flocculent galaxies, but only the spiral density wave can generate the beautiful two-armed spiral patterns.

Our discussion of star formation in spiral structure illustrates the importance of natural processes. The spiral density wave creates graceful arms, but it is the star formation in the arms that makes them stand out so prominently. Self-sustaining star formation can act in some galaxies to modify the spiral arms and produce branches and spurs. In some galaxies, it can make the

a

b   Visual-wavelength images

FIGURE 15-24

(a) Grand-design spiral galaxies, such as M100, have two dominant spiral arms even if they have minor spurs and branches. (b) Flocculent or woolly galaxies have a spiral appearance but lack two distinct spiral arms, as in the case of NGC300 shown here. *(Anglo-Australian Telescope Board)*

spiral pattern flocculent. By searching out and understanding the details of such natural processes, we can begin to understand the overall structure and evolution of the universe around us.

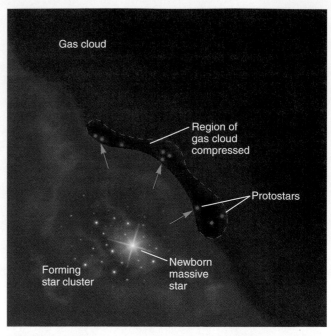

FIGURE 15-25

The energetic birth of a massive star compresses nearby gas clouds, triggering the gravitational contraction of the gas to form new stars.

REVIEW CRITICAL INQUIRY

Why can't we use solar-type stars as spiral tracers?

Sometimes the timing of events is the critical factor in an analysis. In this case, we must think about the evolution of stars and their orbital periods around the galaxy. Stars like the sun live about 10 billion years, but the sun's orbital period around the galaxy is 240 million years. The sun almost certainly formed when a gas cloud passed through a spiral arm, but since then the sun has circled the galaxy many times and has passed through spiral arms often. Thus, the sun's present location has nothing to do with the location of any spiral arms. An O star, however, lives only a few million years. It is born in a spiral arm and lives out its entire lifetime before it can leave the spiral arm. We find short-lived stars such as O stars only in spiral arms, but G stars are found all around the galaxy.

The spiral arms of our galaxy must make it beautiful in photographs taken from a distance, but we are trapped inside it. How do we know that the spiral arms we can trace out near us actually extend across the disk of our galaxy?

Spiral arms are a puzzle of great beauty, but the center of our galaxy hides a puzzle of great power.

15-4 THE NUCLEUS

The most mysterious region of our galaxy is its very center, the nucleus. At visual wavelengths, this region is totally hidden from us by gas and dust that dim the

Self-Sustaining Star Formation

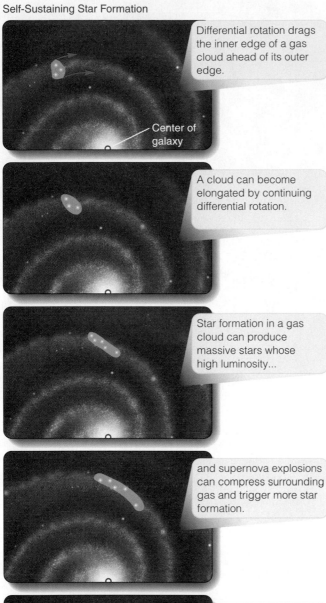

Differential rotation drags the inner edge of a gas cloud ahead of its outer edge.

Center of galaxy

A cloud can become elongated by continuing differential rotation.

Star formation in a gas cloud can produce massive stars whose high luminosity...

and supernova explosions can compress surrounding gas and trigger more star formation.

If star formation continues long enough, a cloud can be elongated into a spiral segment.

FIGURE 15-26

Continuing star formation may be able to produce long clouds of young stars that look like segments of spiral arms.

Enhanced visual image

FIGURE 15-27

The beautiful symmetry of the grand-design spiral pattern is clear in this image of the Whirlpool Galaxy, M51. Young star clusters containing hot, bright stars and ionized hydrogen (pink regions) are located along the arms, while dark lanes of dust mark the inner edges of the arms. *(NASA and The Hubble Heritage Team. STScI/AURA)*

light by 30 magnitudes. If a trillion (10^{12}) photons of light left the center of the galaxy on a journey to Earth, only one would make it through the gas and dust. Thus, visual-wavelength photos tell us nothing about the nucleus. Observations at radio and infrared wavelengths paint a picture of tremendously crowded stars orbiting the center at high velocity. To understand what is happening at the center of our galaxy, we must carefully compare observations and theories.

OBSERVATIONS

If we observe the Milky Way on a dark night, we might notice a slight thickening in the direction of the constellation Sagittarius, but nothing specifically identifies this as the direction of the heart of the galaxy. Even Shapley's study of globular clusters identified the center only approximately. When radio astronomers turned their telescopes toward Sagittarius, they found a complex collection of radio sources, with one, **Sagittarius A*** (abbreviated Sgr A* and usually pronounced "sadge A-star"), lying at the expected location of the galactic core.

Observations show that Sgr A* is less than a few astronomical units in diameter but is a powerful source of radio energy. The tremendous amount of infrared radiation coming from the central area appears to be produced by crowded stars and by dust warmed by those stars. But what could be as small as Sgr A* and produce so much energy?

Study ▌"Sagittarius A*" on pages 328 and 329 and notice three important points. First, observations at visual wavelengths are useless in the study of Sgr A*. Only

Sagittarius A*

The constellation of Sagittarius is so filled with stars and with gas and dust we can see nothing at visual wavelengths of the center of our galaxy. At infrared wavelengths, as shown at bottom left, we see the glow of the cool stars, gas, and dust at the center of our galaxy.

The image below is a wide-field radio image of the center of our galaxy. Many of the features are supernova remnants (SNR), and a few are clouds of star formation. Peculiar features such as threads, the Arc, and the Snake may be gas trapped in magnetic fields. At the center lies Sagittarius A, the center of our galaxy.

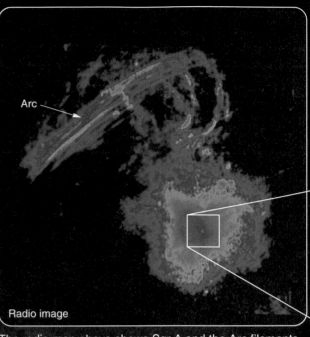

Arc

Radio image

NRL

The radio map above shows Sgr A and the Arc filaments, 50 parsecs long. The image was made with the VLA radio telescope. The contents of the white box are shown on the opposite page.

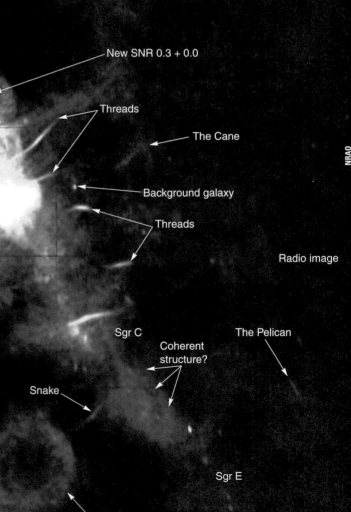

Sgr D HII

Sgr D SNR

SNR 0.9 + 0.1

Apparent angular size of the moon for comparison

Sgr B2

Sgr B1

New SNR 0.3 + 0.0

Threads

The Cane

Arc

Background galaxy

Sgr A

Threads

Infrared photons with wavelengths longer than 4 microns (4000 nm) come almost entirely from warm interstellar dust. The radiation at these wavelengths coming from Sagittarius is intense, and that tells us that the region contains lots of dust and is crowded with stars that warm the dust.

Sgr C

Coherent structure?

The Pelican

Snake

Sgr E

SNR 359.1 − 00.5

Radio image

NRAO

Infrared image

2MASS

This high-resolution radio image of Sgr A (the white boxed area on the opposite page) reveals a spiral swirl of gas around an intense radio source known as Sgr A*, the presumed central object in our galaxy. About 3 pc across, this spiral lies in a low-density cavity inside a larger disk of neutral gas. The arms of the spiral are thought to be streams of matter flowing into Sgr A* from the inner edge of the larger disk (drawing at right).

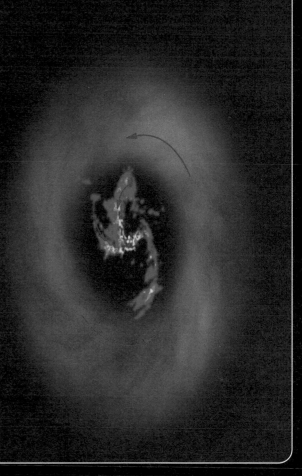

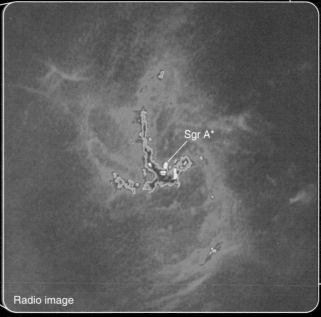

Sgr A*

Radio image

N. Killeen and Kwok-Yung Lo

Evidence of a Black Hole at the Center of Our Galaxy

Star S2 near Sgr A*

ESO

Only one in a trillion photons of visible light makes it through the gas and dust from the center of our galaxy to our eyes. But one in ten infrared photons survives. The infrared image at left shows individual stars in the central light-year around Sgr A*.

March 1992

Orbit of star S2

Sept. 2002

Sgr A*

1 light-day

The star S2 has been observed since early 1992, and, using active optics and large telescopes, astronomers have been able to observe the star orbiting Sgr A*. Crosses show the uncertainty in observations.

The orbital period of S2 is 15.2 years, and the semimajor axis of its orbit is 950 AU. The orbital motion of S2 combined with the motion of other stars in the region suggest a mass for Sgr A* of 2.6 million solar masses.

Our solar system is half a light-day in diameter.

At its closest, S2 comes within 17 light-hours of Sgr A*. Alternative theories that Sgr A* is a cluster of stars, neutron stars, or black holes are eliminated. Only a black hole could contain so much mass in so small a region.

A black hole with a mass of 2.6 million solar masses would have an event horizon no bigger than the white dot shown in this orbital diagram.

The Chandra X-ray telescope has imaged Sgr A* as shown here.

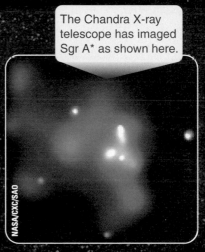

NASA/CXC/SAO

The evidence of a massive black hole at the center of our galaxy seems conclusive. It is much too massive to be the remains of a dead star, however, and astronomers conclude that it probably formed as the galaxy first took shape.

A slow dribble of only 0.0002 solar masses of gas flowing into the black hole could produce a hot accretion disk and release enough energy to power Sgr A*. A sudden increase as when a star falls in could produce an eruption.

at the longer wavelengths of infrared or radio waves, or at the very short wavelengths of X rays, can we image the features near the center of our galaxy. A second thing to notice is the evidence of star formation in the gas clouds near the center. Not only do we see clouds of gas forming new stars, but we also see supernova remnants produced by the deaths of massive stars, and those massive stars must have been born recently. Third, notice the evidence that Sgr A* is a supermassive black hole into which gas is flowing.

A supermassive black hole is an exciting idea, but we must notice the difference between adequacy and necessity. A supermassive black hole is adequate to explain the observations, but could there be other explanations? For example, astronomers have suggested that gas flowing inward could trigger tremendous bursts of star formation. Such theories have been considered and tested against the evidence, but none appears to be adequate to explain the observations. So far, the only theory that seems adequate is that our galaxy is home to a supermassive black hole. That is, a central black hole appears to be not only adequate but necessary to explain the observations. In later chapters we will see that such supermassive black holes are found at the centers of many galaxies.

REVIEW CRITICAL INQUIRY

Why do we think the center of our galaxy contains a large mass?

The best way to measure the mass of an astronomical object is to watch something orbit around it. We can then use Kepler's third law to find the mass inside the orbit. Because of gas and dust, we can't see to the center of our galaxy at visual wavelengths, but infrared observations can detect individual stars orbiting Sgr A*. The star S2 has been particularly well observed, but a number of stars can be followed as they orbit the center. The sizes and periods of these orbits, interpreted using Kepler's laws, tell us that Sgr A* contains roughly 2.6 million solar masses.

In addition to these orbital motions, the large amount of infrared radiation at wavelengths longer than 4 microns (4000 nm) also implies that vast numbers of stars are crowded into the center. Why?

Our observations of the Milky Way Galaxy suggest it is a place of great beauty and great power. Is it normal? Are all galaxies like ours? To answer that question we must survey other galaxies, which we begin in the next chapter.

SUMMARY

Our galaxy is a typical spiral galaxy with a disk about 25,000 pc in diameter. The sun is located about two-thirds of the way from the center to the edge. The disk contains nearly all of the gas and dust in the galaxy and most of the star formation. It is illuminated by the brilliant O and B stars. The nuclear bulge at the center of the galaxy is about 3000 pc in radius and contains stars similar to the large spherical halo of stars and globular clusters.

Gas and dust in the disk block our view, so until the 20th century, astronomers assumed that we were located at the center of a grindstone-shaped star cloud. Harlow Shapley was able to calibrate variable stars to find the distance to globular clusters, and he concluded that our galaxy was much bigger and that we were not located in the center. Soon after, other astronomers concluded that ours was only one of many galaxies in the universe.

The disk component of our galaxy contains metal-rich population I stars moving in nearly circular orbits in the plane of the disk. The spherical component, made up of the nuclear bulge and the halo, contains metal-poor population II stars moving in randomly tipped and highly elongated orbits. Observations of the rotation curve of our galaxy reveal that orbital velocities do not fall in the outer disk, suggesting that the galaxy contains large amounts of unseen mass commonly thought to be located in an extended halo.

The distribution of populations through the galaxy suggests a hypothesis for the formation of our galaxy from a single cloud of gas and dust that gradually contracted into a disk shape. Evidence suggests that the story is more complicated, however, and many astronomers now suspect that the nuclear bulge and halo formed first and the disk formed later as more gas fell into the galaxy. The galaxy has evidently absorbed smaller galaxies as well.

The very youngest stars lie along spiral arms within the disk. The most massive live such short lives that they don't have time to move from their place of birth in the spiral arms. Maps of the spiral tracers and cold hydrogen clouds reveal the spiral pattern of our galaxy. The spiral arms are also outlined by giant molecular cloud complexes. These molecular clouds are sites of star formation.

The spiral density wave theory suggests that spiral arms are regions of compression that move through the disk. When an orbiting gas cloud smashes into the compression wave, the gas cloud forms stars. Self-sustaining star formation is an important process that may modify the arms as the birth of massive stars triggers the formation of more stars by compressing neighboring clouds.

Dust hides the nucleus of our galaxy at visible wavelengths, but radio and infrared observations can penetrate the dust. Such observations reveal that the central region is crowded with stars orbiting at high velocity. This tells us that the central few light years must contain millions of solar masses. At the very center lies the radio source Sagittarius A*, an object no larger than a few AU in diameter. Theorists suspect that the object is a supermassive black hole that is drawing in matter through an accretion disk.

NEW TERMS

Magellanic Clouds

kiloparsec (kpc)

halo

nuclear bulge

disk component

spherical component

high-velocity star

rotation curve

Keplerian motion

galactic corona

dark matter

metals

population I star

population II star

nucleosynthesis

galactic fountain

spiral tracers

density wave theory

flocculent

self-sustaining star
  formation

Sagittarius A*

REVIEW QUESTIONS

Ace ✏Astronomy™ Assess your understanding of this chapter's topics with additional quizzing and animations at **http:// astronomy.brookscole.com/seeds8e**

1. Why is it difficult to specify the dimensions of the disk and halo?

2. Why didn't astronomers before Shapley realize how large the galaxy is?

3. What evidence do we have that our galaxy has a galactic corona?

4. Explain why some star clusters lose stars more slowly than others.

5. Contrast the motion of the disk stars and that of the halo stars. Why do their orbits differ?

6. Why do high-velocity stars have lower metal abundance than the sun?

7. Why are metals less abundant in older stars than in younger stars?

8. Why are all spiral tracers young?

9. Why couldn't spiral arms be physically connected structures? What would happen to them?

10. Why does self-sustaining star formation produce clouds of stars that look like segments of spiral arms?

11. Describe the kinds of observations we can make to study the galactic nucleus.

12. What evidence do we have that the nucleus of the galaxy contains a supermassive energy source that is very small in size?

DISCUSSION QUESTIONS

1. How would this chapter be different if interstellar dust did not scatter light?

2. Why doesn't the Milky Way circle the sky along the celestial equator or the ecliptic?

PROBLEMS

1. Make a scale sketch of our galaxy in cross section. Include the disk, sun, nucleus, halo, and some globular clusters. Try to draw the globular clusters to scale.

2. Because of dust, we can see only about 5 kpc into the disk of the galaxy. What percentage of the galactic disk can we see? (*Hint:* Consider the area of the entire disk and the area we can see.)

3. If the fastest passenger aircraft can fly 0.45 km/s (1000 mph), how long would it take to reach the sun? the galactic center? (*Hint:* 1 pc = 3×10^{13} km.)

4. If a typical halo star has an orbital velocity of 250 km/s, how long does it take to pass through the disk of the galaxy? Assume that the disk is 1000 pc thick.

5. If the RR Lyrae stars in a globular cluster have apparent magnitudes of 14, how far away is the cluster? (*Hint:* See Figure 12-13.)

6. If intersellar dust makes an RR Lyrae variable star look 1 magnitude fainter than it should, by how much will we overestimate its distance? (*Hint:* Use the magnitude–distance formula or Table 9-1.)

7. If a globular cluster is 10 minutes of arc in diameter and 8.5 kpc away, what is its diameter? (*Hint:* Use the small-angle formula.)

8. If we assume that a globular cluster 4 minutes of arc in diameter is actually 25 pc in diameter, how far away is it? (*Hint:* Use the small-angle formula.)

9. If the sun is 5 billion years old, how many times has it orbited the galaxy?

10. If the true distance to the center of the galaxy is found to be 7 kpc and the orbital velocity of the sun is 220 km/s, what is the minimum mass of the galaxy? (*Hint:* Use Kepler's third law.)

11. What temperature would interstellar dust have to have to radiate most strongly at 100 μm? (*Hints:* 1 μm = 1000 nm. Use Wien's law, Chapter 7.)

12. Infrared radiation from the center of our galaxy with a wavelength of about 2 μm (2×10^{-6} m) comes mainly from cool stars. Use this wavelength as λ_{max} and find the temperature of the stars.

13. If an object at the center of our galaxy has a linear diameter of 10 AU, what will its angular diameter be as seen from Earth? (*Hint:* Use the small-angle formula, Chapter 3.)

CRITICAL INQUIRIES FOR THE WEB

1. Henrietta Leavitt discovered the period–luminosity relation for Cepheids while working on the staff at Harvard College Observatory under Edward Pickering. She was one of several women "computers" on staff there a century ago. Search the Web for information on Leavitt, her colleagues, and their work at Harvard. List three of the women employed by Pickering and note their contributions to astronomy. What was life like for a woman in astronomy at the beginning of the 20th century?

2. What if we lived near the center of the galaxy? Search the Web for research and information on the distribution of material near the center of our galaxy. Based on what you find, speculate as to how the sky would appear from a planet associated with a star near the galactic center.

3. Who was Wilhelm Heinrich Walter Baade, and why did he win the Bruce Medal? What other winners of the Bruce Medal do you recognize?

EXPLORING *THESKY*

1. Locate Sagittarius and examine the shape of the Milky Way there and the profusion of globular clusters. (*Hint:* To turn on Messier object labels, use **Labels** and **Setup** under the **View** menu.)

2. Locate the following globular clusters: M3, M4, M5, M10, M12, M13, M15, M22, M55, M92. Where are they located in the sky? (*Hint:* Use **Find** under the **Edit** menu.)

3. Compare the distribution of globular clusters with that of open clusters. (*Hint:* Use **Filters** under the **View** menu to turn off everything but globular clusters, the Milky Way, the Galactic Equator, and Constellation Boundaries. Use the thumbwheel at the bottom of the sky window to rotate the sky. Now repeat with globular clusters off and open clusters on.)

 Visit the Seeds *Foundations of Astronomy* companion Web site for critical thinking exercises, articles, and additional readings from InfoTrac College Edition, Brooks/Cole's online student library.

GALAXIES

A hypothesis or theory is clear, decisive, and positive, but it is believed
by no one but the man who created it. Experimental findings, on the other hand,
are messy, inexact things which are believed by everyone
except the man who did that work.

Harlow Shapley, Through Rugged Ways to the Stars

Bill Schoening/AURA/NOAO/NSF

Chapter 16

GUIDEPOST

The preceding chapter was about our Milky Way Galaxy, an important object to us but only one of the many billions of galaxies visible in the sky. We can no more understand galaxies by understanding a single example, the Milky Way, than we could understand humanity by understanding a single person. This chapter expands our horizon to discuss the different kinds of galaxies and their complex histories.

We take two lessons from this chapter. First, galaxies are not solitary beasts; they collide and interact with each other. Second, most of the matter in the universe is invisible. The galaxies we see are only the tip of a cosmic iceberg.

We will carry the lessons of this chapter into the next, where we will discuss violently active galaxies, and on into Chapter 18, where we discuss the universe as a whole.

VIRTUAL LABORATORIES

ASTRONOMICAL DISTANCE SCALES

THE HUBBLE LAW

Science fiction heroes flit effortlessly between the stars, but almost none voyages between the galaxies. As we leave our home galaxy, the Milky Way, behind, we voyage out into the vast depths of the universe, out among the galaxies, space so deep it is unexplored even in fiction.

Less than a century ago, astronomers did not understand that there were galaxies. Nineteenth-century telescopes revealed faint nebulae scattered among the stars, and some were spiral. Astronomers argued about the nature of these nebulae, but it was not until the 1920s that astronomers understood that some were other galaxies much like our own; and it was not until recent decades that astronomical telescopes could reveal the tremendous beauty and intricacy of the galaxies (∥ Figure 16-1).

In this chapter, we will try to understand how galaxies form and evolve. We will discover that their complex shapes reflect long periods of peaceful star formation and episodes of cosmic violence.

Before we can build theories, however, we must gather some basic data concerning galaxies. How many different kinds of galaxies are there? How big are they?

Ace ◈ Astronomy™ The AceAstronomy icon throughout the text indicates an opportunity for you to test yourself on key concepts and to explore animations and interactions on the AceAstronomy Web site at **http://astronomy.brookscole .com/seeds8e**

How massive are they? That is, we must characterize the family of galaxies just as we characterized the family of stars in Chapter 9.

16-1 THE FAMILY OF GALAXIES

Like leaves on the forest floor, galaxies carpet the sky. Pick any spot on the sky away from the dust and gas of the Milky Way, and you are looking deep into space. Photons that have traveled for billions of years enter your eye, but they are too few to register on your retina. Only the largest telescopes can gather enough light to detect the most distant galaxies.

When astronomers picked a seemingly empty spot on the sky near the Big Dipper and used the Hubble Space Telescope to record an amazing time exposure of 10 days' duration, they found thousands of galaxies crowded into the image. This image, now known as a Hubble Deep Field, contains a few relatively nearby galaxies and others that are over 10 billion light-years away (∥ Figure 16-2). The entire sky appears to be as thickly carpeted with galaxies as is the Hubble Deep Field.

Less than a century ago, humanity did not know that it lived in a galaxy or that the universe was filled

NGC4414 60 million ly

Young blue stars illuminate spiral arms.

Visual-wavelength image

M83 12 million ly

Dust clouds glow red in this infrared image.

Infrared image

ESO510-G13 150 million ly

Dusty disk of galaxy warped by interaction with another galaxy

New stars forming in dust clouds.

Visual-wavelength image

FIGURE 16-1
A century ago, photos of galaxies looked like spiral clouds of haze. Modern images of these relatively nearby galaxies reveal newborn stars and clouds of gas and dust. *(NGC4414 and ESO 510-G13: Hubble Heritage Team, Aura/STScI/NASA; M83: ESO)*

Location of
Hubble Deep Field

Big
Dipper

Visual-wavelegth image

FIGURE 16-2

An apparently empty spot on the sky only 1/30 the diameter of the full moon contains over 1500 galaxies in this extremely long time exposure known as the Hubble Deep Field. Presumably the entire sky is similarly filled with galaxies. *(R. Williams and the Hubble Deep Field Team, STScI, NASA)*

with other galaxies. The discovery of galaxies is one of the great stories of astronomy.

THE DISCOVERY OF GALAXIES

Galaxies are faint objects, so they were not noticed until telescopes had grown large enough to gather significant amounts of light. In 1845, William Parsons, third Earl of Rosse in Ireland, built a telescope 72 in. in

diameter. It was, for some years, the largest telescope in the world. Before the invention of astronomical photography (in the late 1800s), Parsons had to view the faint nebulae directly at the eyepiece and sketch their shapes. He noticed that some have a spiral shape, and they became known as **spiral nebulae.** Parsons immediately concluded that the spiral nebulae were great spirals of stars. The German philosopher Immanuel Kant in 1755 had proposed that the universe was filled with

Searching for Clues Through Classification in Science

Classification is one of the most basic and most powerful of scientific tools. It is often a way to begin studying a topic, and it often leads to dramatic insights. Charles Darwin, for example, sailed around the world with a scientific expedition aboard the ship HMS *Beagle*. Everywhere he went, he studied the living things he saw and tried to classify them. For example, he studied the tortoises, mockingbirds, and finches he saw on the Galápagos Islands. His classifications of these and other animals eventually led him to think about evolution and natural selection. Scientists in many fields depend on carefully designed systems of classification.

Classification is an everyday mode of thought. We all use classifications when we order lunch, buy shoes, catch a bus, and so on. In science, classifications reveal the relationships between different kinds of objects and at the same time save the scientist valuable time by bringing order out of seeming disorder. An economist can classify different kinds of businesses and does not have to analyze each of the millions of businesses as if it were unique. Classifications of minerals, plants, psychological learning styles, modes of transportation, or sandwiches bring order to the world and help us deal with information.

Astronomers use classifications of galaxies, stars, moons, telescopes, and more. Whenever you encounter a scientific discussion, look for the classifications on which it is based. Classifications are the orderly framework on which much of science is built.

---

great wheels of stars that he called **island universes.** Parsons adopted that term for the spiral nebulae.

Not everyone agreed that the spiral nebulae were island universes. An alternative point of view was that we lived in a great star system in an otherwise empty universe. Sir William Herschel had counted stars in different directions and had shown that the star system was shaped approximately like a grindstone (Chapter 15). Beyond the edge of this grindstone universe, space was supposed to be a limitless void. The spiral nebulae, in this view, were vortexes of gas or faint stars within the star system.

The nature of the spiral nebulae could not be resolved in the 19th century because the telescopes were not large enough and the photographic plates were not sensitive enough. The spiral nebulae remained foggy swirls on the best photographs, and astronomers continued to disagree on their true nature. In April 1920, two astronomers debated the issue at the National Academy of Science in Washington, DC. Harlow Shapley of Mt. Wilson Observatory had recently shown that the Milky Way was a much larger star system than had been thought. He argued that the spiral nebulae were nearby nebulae within the star system. Heber D. Curtis of Lick Observatory argued that the spiral nebulae were island universes. Historians of science mark this **Shapley–Curtis Debate** as a turning point in modern astronomy, but it was inconclusive. Shapley and Curtis cited the right evidence but drew incorrect conclusions. The disagreement was finally resolved, as is often the case in astronomy, with a bigger telescope.

On December 30, 1924, Mt. Wilson astronomer Edwin Hubble (namesake of the Hubble Space Telescope) announced that he had taken photographic plates of a few bright galaxies with the new 100-in. telescope. Not only could he detect individual stars, but he could identify some of them as Cepheid variables with apparent magnitudes of about 18. For the Cepheids, which are supergiants, to look that faint, they had to be very dis-

tant, and thus the spiral nebulae were external to the Milky Way star system. They were galaxies.

Although it was spiral galaxies that first attracted the attention of astronomers, Hubble quickly realized that there were other kinds of galaxies. Not all galaxies had spiral shapes.

THE SHAPES OF GALAXIES

Look at the galaxies in the Hubble Deep Field, and you will see galaxies of different shapes. Some are spiral, some are elliptical, and some are irregular. Astronomers classify galaxies according to their shape using a system developed by Edwin Hubble in the 1920s. Such systems of classification are a fundamental technique in science (Window on Science 16-1).

Study ▌ "Galaxy Classification" on pages 338 and 339 and notice two important points. First, the amount of gas and dust in a galaxy strongly influences its appearance. Galaxies rich in gas and dust have active star formation and contain hot, bright stars. Such galaxies tend to be bluer and contain emission nebulae. Galaxies that are poor in gas and dust contain few or none of these highly luminous stars, so those galaxies look redder and have a much more uniform appearance.

Second, notice the wide range in the shapes of galaxies. Some look like spherical- or football-shaped clouds of stars, but others are thin disks. Still others are so irregular they look like scrambled galaxy.

Spiral galaxies are clearly disk shaped, but the true, three-dimensional shape of elliptical galaxies isn't obvious from images. Some are spherical, but the more elongated elliptical galaxies could be shaped like flattened spheres (bun shaped) or like elongated spheres (football shaped). Some may even have three different diameters; they are longer than they are thick and thicker than they are wide, like a flattened loaf of bread. Statistical studies suggest that all of these shapes are represented among elliptical galaxies.

Traditionally astronomers have classified galaxies according to their appearance in visual-wavelength images, but infrared images reveal the location of the vast majority of stars—stars that are not as luminous as the hot stars that produce most of the light. These studies suggest that the structure of many galaxies is dominated by the gravitational fields of large masses of cool stars. For example, some spiral galaxies that appear to lack bars have massive bars evident in infrared images. As we think about galaxies we must remember that the classification of galaxies tells us where the light comes from in a galaxy and not necessarily where most of the mass is located.

Why do galaxies look so different? Later in this chapter we will try to understand what determines the stellar content and shape of a galaxy. Our basic classification system will be an important tool.

HOW MANY GALAXIES?

Looking at the Hubble Deep Field forces even the most sophisticated astronomer to ask what seems to be a silly question. Just how many galaxies are there? The answer is impressive.

Two Hubble Deep Fields have been imaged, one in the northern hemisphere and one in the southern hemisphere, and they are both filled with galaxies. A new effort called GOODS (Great Observatories Origins Deep Survey) is continuing such studies using an improved visual-wavelength camera on the Hubble Space Telescope, X-ray observations by the Chandra spacecraft, and infrared observations by the Spitzer spacecraft. All of these studies are revealing tremendous numbers of distant galaxies in the small regions of the sky being studied. If the entire sky is similarly covered with galaxies, there must be over 100 billion galaxies visible with existing telescopes. Surely there are other galaxies too distant or too faint to see.

We might also wonder what proportion of the galaxies is elliptical, spiral, and irregular, but that is a difficult question to answer. In catalogs of galaxies, about 70 percent are spiral, but that is because spiral galaxies are gaudy and easy to notice. Spiral galaxies contain hot, bright stars and clouds of ionized gas. Most ellipticals are fainter and harder to notice. Small galaxies such as dwarf ellipticals and dwarf irregulars are very common, but they are hard to detect. From careful studies we can conclude that ellipticals are more common than spirals, and that irregulars make up only about 25 percent of all galaxies.

tains lots of young stars, and a few of those young stars will be massive, hot, luminous O and B stars. They will give the galaxy a distinct blue tint. In contrast, a galaxy that contains little gas and dust will probably contain few young stars. It will lack O and B stars, and the most luminous stars in such a galaxy will be red giants. They will give the galaxy a red tint. Because the light from a galaxy is a blend of the light from billions of stars, the colors are only tints. Nevertheless, the most luminous stars in a galaxy determine the color. From this we can conclude that elliptical galaxies tend to be red and the disks of spiral galaxies tend to be blue.

Of course, the halo of a spiral galaxy would be both dimmer and redder than the disk. Why?

If we want to create a theory to explain the shapes of galaxies, we need answers to some basic questions. How big are galaxies? How luminous are they? Are they all the same mass, or are some more massive than others? To answer those questions, we must discover the basic properties of galaxies. We begin that task in the next section.

16-2 MEASURING THE PROPERTIES OF GALAXIES

Looking beyond the edge of our Milky Way Galaxy, astronomers find many billions of galaxies. Great clusters of galaxies, some containing thousands of galaxies, fill space as far as we can see. We can recognize the different types of galaxies, but what are the properties of these star systems? What are the diameters, luminosities, and masses of galaxies? Just as in our study of stellar characteristics (Chapter 9), the first step in our study of galaxies is to find out how far away they are. Once we know a galaxy's distance, its size and luminosity are relatively easy to find. Later in this section, we will see that finding mass is more difficult.

DISTANCE

The distances to galaxies are so large that it is not convenient to measure them in light-years, parsecs, or even kiloparsecs. Instead, we will use the unit **megaparsec (Mpc),** or 1 million pc. One Mpc equals 3.26 million ly, or approximately 2×10^{19} miles.

To find the distance to a galaxy, we must search among its stars, nebulae, and star clusters for familiar objects whose luminosity or diameter we know. Such objects are called **distance indicators** because we can use them to find the distance to a galaxy. Most distance indicators are objects whose brightness is known, and astronomers often refer to them as **standard candles.** If we can find a standard candle in a galaxy we can judge its distance.

Galaxy Classification

The most commonly used system of galaxy classification is based on the appearance of galaxies in photographs made at visual wavelengths.

AURA/NOAO/NSF

Elliptical galaxies are round or elliptical, contain no visible gas and dust, and lack hot, bright stars. They are classified with a numerical index ranging from 1 to 7; E0s are round, and E7s are highly elliptical. The index is calculated from the largest and smallest diameter of the galaxy used in the following formula and rounded to the nearest integer.

$$\frac{10(a-b)}{a}$$

Outline of an E6 galaxy

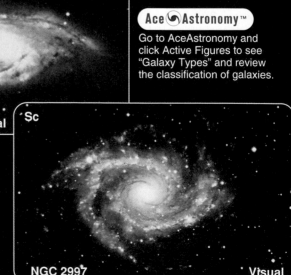

Visual-wavelength image

M87 is a giant elliptical galaxy classified E1. It is a number of times larger in diameter than our own galaxy and is surrounded by a swarm of over 500 globular clusters.

Anglo-Australian Telescope Board

Visual

The Leo 1 dwarf elliptical galaxy is not many times bigger than a globular cluster.

Spiral galaxies contain a disk and spiral arms. We do not see their halo stars, but presumably all spiral galaxies have halos. Spirals contain gas and dust and hot, bright O and B stars. The presence of short-lived O and B stars alerts us that star formation is occurring in these galaxies. Sa galaxies have larger nuclei, less gas and dust, and fewer hot, bright stars. Sc galaxies have small nuclei, lots of gas and dust, and many hot, bright stars. Sb galaxies are intermediate.

Anglo-Australian Telescope Board

Sa

Visual   NGC 3623

Anglo-Australian Telescope Board

BAR

NGC 1365   Visual

Sb

NGC 3627   Visual

Ace Astronomy™
Go to AceAstronomy and click Active Figures to see "Galaxy Types" and review the classification of galaxies.

Sc

NGC 2997   Visual

Roughly 2/3 of all spiral galaxies are **barred spiral galaxies** classified SBa, SBb, and SBc. They have an elongated nucleus with spiral arms springing from the ends of the bar. Our own galaxy is a barred spiral.

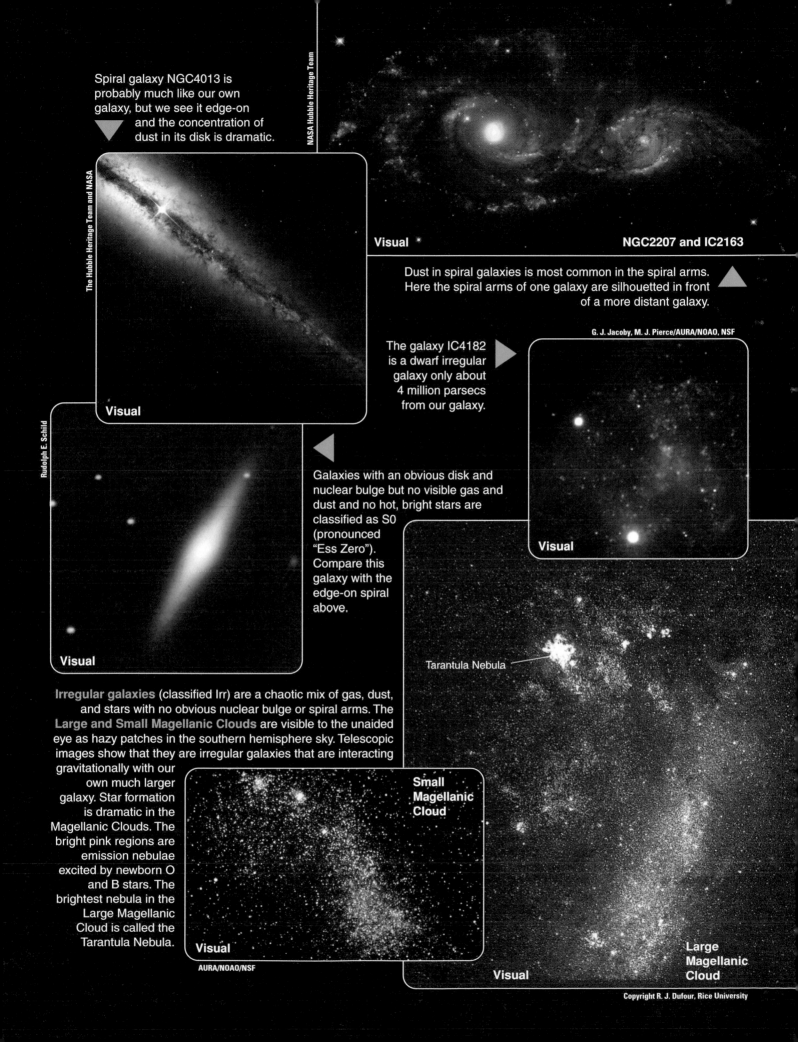

Spiral galaxy NGC4013 is probably much like our own galaxy, but we see it edge-on and the concentration of dust in its disk is dramatic.

NASA Hubble Heritage Team

The Hubble Heritage Team and NASA

Visual

Visual

NGC2207 and IC2163

Dust in spiral galaxies is most common in the spiral arms. Here the spiral arms of one galaxy are silhouetted in front of a more distant galaxy.

The galaxy IC4182 is a dwarf irregular galaxy only about 4 million parsecs from our galaxy.

G. J. Jacoby, M. J. Pierce/AURA/NOAO, NSF

Visual

Rudolph E. Schild

Galaxies with an obvious disk and nuclear bulge but no visible gas and dust and no hot, bright stars are classified as S0 (pronounced "Ess Zero"). Compare this galaxy with the edge-on spiral above.

Visual

Tarantula Nebula

Irregular galaxies (classified Irr) are a chaotic mix of gas, dust, and stars with no obvious nuclear bulge or spiral arms. The **Large and Small Magellanic Clouds** are visible to the unaided eye as hazy patches in the southern hemisphere sky. Telescopic images show that they are irregular galaxies that are interacting gravitationally with our own much larger galaxy. Star formation is dramatic in the Magellanic Clouds. The bright pink regions are emission nebulae excited by newborn O and B stars. The brightest nebula in the Large Magellanic Cloud is called the Tarantula Nebula.

Small Magellanic Cloud

Visual

AURA/NOAO/NSF

Large Magellanic Cloud

Visual

Copyright R. J. Dufour, Rice University

April 23

May 4

May 9

May 16

May 20

May 31

Visual-wavelength image

FIGURE 16-3
The vast majority of spiral galaxies are too distant for Earth-based telescopes to detect Cepheid variable stars. The Hubble Space Telescope, however, can locate Cepheids in some of these galaxies, as it has in the bright spiral galaxy M100. From a series of images taken on different dates, astronomers can locate Cepheids (inset), determine the period of pulsation, and measure the average apparent brightness. They can then deduce the distance to the galaxy—51 million ly for M100. *(J. Trauger, JPL; Wendy Freedman, Observatories of the Carnegie Institution of Washington; and NASA)*

Because their period is related to their luminosity (see Figure 12-13), Cepheid variable stars are reliable distance indicators. If we know the period of the star's variation, we can use the period–luminosity diagram to learn its absolute magnitude. By comparing its absolute and apparent magnitudes, we can find its distance. ▌Figure 16-3 shows a galaxy in which the Hubble Space Telescope detected Cepheids.

Even with the Hubble Space Telescope, Cepheids are not visible in galaxies much beyond 80 million ly (25 Mpc), so astronomers must search for less common but brighter distance indicators and calibrate them using nearby galaxies containing visible Cepheids. For example, by studying nearby galaxies with distances known from Cepheids, astronomers have found that the brightest globular clusters have absolute magnitudes of about −10. If we find globular clusters in a more distant galaxy, we can assume that the brightest have similar absolute magnitudes and thus calculate the distance.

Another distance indicator is the cloud of ionized hydrogen called an HII region (Chapter 10) that forms around a very hot star. Astronomers have attempted to calibrate these, but they have not been as reliable as Cepheids because some are larger and brighter than others. Also, the size of HII regions seems to depend on the kind of galaxy they occupy, and so astronomers have not been able to calibrate them as accurately as other distance indicators.

Planetary nebulae have proven to be very useful distance indicators, which is surprising considering how faint their central star is. These central stars are small and hot. Their small surface area limits the amount of energy they can radiate, and their high temperature means they radiate most of their energy in the ultraviolet. To our eyes, these stars look faint, but the gas of the surrounding planetary nebula absorbs the ultraviolet radiation and reradiates it as visible emission lines, such as the oxygen emission line at a wavelength of 500.7 nm. Images of galaxies recorded at such wavelengths reveal planetary nebulae glowing like brightly colored paper lanterns illuminated from within. Astronomers have been able to calibrate the planetary nebulae for distance measurements by observing the absolute magnitude of planetary nebulae in nearby galaxies such as the Andromeda Galaxy, galaxies whose distances are known from other distance indicators. When planetary nebulae are detected in a more distant galaxy, a quick comparison of their apparent and absolute magnitudes reveals the distance to the galaxy. Notice once again the importance of calibration (Window on Science 15-1).

When a supernova explodes in a distant galaxy, astronomers rush to observe it. Type Ia supernovae, those produced by the collapse of a white dwarf, have been calibrated. By locating Cepheid variable stars in nearby galaxies that have been home to type Ia supernovae, astronomers have learned that these type of supernovae all reach the same absolute magnitude at maximum. Look at the image of the small galaxy IC4182 on page 339. It is an important link in this calibration because it is nearby and its Cepheids are visible and because astronomers saw a type Ia supernova there in 1937. When astronomers see a type Ia supernova in a more distant galaxy, they observe the apparent magnitude of the supernova at maximum and compare that with absolute magnitude these supernovae reach at maximum to find the distance to the galaxy. As we will see in Chapter 18, this is a critical calibration in modern astronomy.

At the greatest distances, astronomers must calibrate the total luminosity of the galaxies themselves. For example, studies of nearby galaxies show that an average galaxy like our Milky Way Galaxy has a luminosity about 16 billion times the sun's. If we see a similar galaxy far away, we can measure its apparent magnitude and calculate its distance. Of course, it is important to

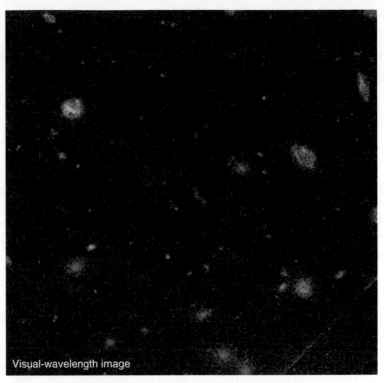

Visual-wavelength image

FIGURE 16-4

Only the general characteristics of the galaxies are recognizable in this cluster of galaxies 4 to 6 billion light-years away. The bright blue galaxy at upper left is probably a foreground galaxy much closer than the cluster. The larger galaxies in this cluster are about the size of our Milky Way Galaxy. The diagonal line at lower right was produced by an artificial Earth satellite. *(K. Ratnatunga, R. Griffiths, Carnegie Mellon University, and NASA)*

recognize the different types of galaxies, and that is difficult to do at great distances (█ Figure 16-4). Averaging the distances to the brightest galaxies in a cluster can reduce the uncertainty in this method.

Notice how astronomers use calibration (Window on Science 15-1) to build a **distance scale** reaching from the nearest galaxies to the most distant visible galaxies. Often astronomers refer to this as the distance pyramid or the distance ladder because each step depends on the steps below it. Of course, the foundation of the distance scale rests on the Cepheid variable stars and our understanding of the luminosities of the stars in the H–R diagram, which ultimately rest on measurements of the parallax of stars.

The most distant visible galaxies are roughly 10 billion ly (3000 Mpc) away, and at such distances we see an effect akin to time travel. When we look at a galaxy millions of light-years away, we do not see it as it is now but as it was millions of years ago when its light began the journey toward Earth. Thus, when we look at a distant galaxy, we look back into the past by an amount called the **look-back time,** a time in years equal to the distance to the galaxy in light-years.

You may have experienced look-back time if you have ever made a long-distance phone call carried by satellite or watched a TV newscaster interview someone on the other side of the world via satellite. A half-second delay occurs as a radio signal carries a question 23,000 miles out to a satellite then back to Earth and then carries the answer out to the satellite and again back to Earth. That half-second look-back delay can make people hesitate on long-distance phone calls and produces a seemingly awkward delay in intercontinental TV interviews.

The look-back time to nearby objects is usually not significant. The look-back time across a football field is a tiny fraction of a second. The look-back time to the moon is 1.3 seconds, to the sun only 8 minutes, and to the nearest star about 4 years. The Andromeda Galaxy has a look-back time of about 2 million years, a mere eye blink in the lifetime of a galaxy. But when we look at more distant galaxies, the look-back time becomes an appreciable part of the age of the universe. We will see evidence in Chapter 18 that the universe began 14 billion years ago. Thus, when we look at the most distant visible galaxies, we are looking back 10 billion years to a time when the universe may have been significantly different. This effect will be important in our discussions in this and the next chapter.

THE HUBBLE LAW

Although astronomers find it difficult to measure the distance to a galaxy, they often estimate such distances using a simple relationship that was first noticed at about the same time astronomers were beginning to understand the nature of galaxies.

In 1913, V. M. Slipher at Lowell Observatory reported on the spectra of faint, nebulous objects in the sky. Their spectra seemed to be composed of a mixture of stellar spectra: Some had Doppler shifts that suggested rotation; most had red shifts as if they were receding; and the faintest had the largest red shifts. Within two decades, astronomers concluded that the faint objects were galaxies similar to our own Milky Way and that the galaxies are indeed receding from us in a general expansion.

In 1929, Edwin Hubble and Milton Humason published a graph that plotted the velocity of recession versus distance for a number of galaxies. The points in the graph fell along a straight line (▌Figure 16-5). The straight-line relation from Hubble's diagram can be written as the simple equation

$$V = Hd$$

That is, the velocity of recession V equals the distance d in millions of parsecs times the constant H. This relation between red shift and distance is known as the **Hubble law,** and the constant H is known as the **Hubble constant.**

Modern attempts to measure H have been difficult because of the uncertainty in the distances to galaxies.

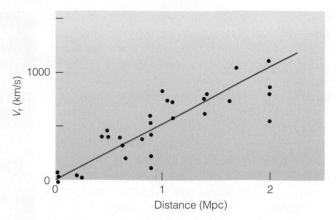

FIGURE 16-5
Edwin Hubble's first diagram of the velocities and distances of galaxies did not probe very deeply into space. It did show, however, that the galaxies are receding from one another.

During the last decades of the 20th century, astronomers were caught up in a great controversy over the value of H. Different teams measuring distance differently got answers that ranged from 50 to 100 km/s/Mpc.* The controversy was settled by better observations made with the Hubble Space Telescope and with new-generation giant telescopes on Earth. The value of H appears to be about 70 km/s/Mpc.

The Hubble law is important because it is commonly interpreted to show that the universe is expanding. In Chapter 18, we will discuss this expansion, but here we will use the Hubble law as a practical way to estimate the distance to a galaxy.

Simply stated, a galaxy's distance equals its velocity of recession divided by H. For example, if a galaxy has a radial velocity of 700 km/s, and H is 70 km/s/Mpc, then the distance to the galaxy is 700 divided by 70, or about 10 Mpc. This makes it relatively easy to estimate the distances to galaxies, because a large telescope can photograph the spectrum of a galaxy and determine its velocity of recession even when it is too distant to have visible distance indicators.

DIAMETER AND LUMINOSITY

The distance to a galaxy is the key to finding its diameter and its luminosity. We can easily photograph a galaxy and measure its angular diameter in seconds of arc. If we know the distance, we can use the small-angle formula (Chapter 3) to find its linear diameter. Also, if we measure the apparent magnitude of a galaxy, we can use the distance to find its absolute magnitude and thus its luminosity (Chapter 9).

---

*H has the units of a velocity divided by a distance. These are usually written as km/s/Mpc, meaning km/s per Mpc.

FIGURE 16-6

The small galaxy cluster known as Hickson Compact Group 87 contains three spiral galaxies and one elliptical. That means the four galaxies lie at the same distance from Earth, and the differences in size among the galaxies are real. Some are much bigger than others. The edge-on spiral at bottom is very large compared to the small spiral at the center. *(Hubble Heritage Team, STScI/AURA/NASA)*

The results of such observations show that galaxies differ dramatically in size and luminosity. Irregular galaxies tend to be small, 1 to 25 percent the size of our galaxy, and of low luminosity. Although they are common, they are easy to overlook. Our Milky Way Galaxy is large and luminous compared with most spiral galaxies, though we know of a few spiral galaxies that are even larger and more luminous—nearly four times larger in diameter and about 10 times more luminous. Elliptical galaxies cover a wide range of diameters and luminosities. The largest, called giant ellipticals, are five times the size of our Milky Way. But many elliptical galaxies are very small, dwarf ellipticals, only 1 percent the diameter of our galaxy.

To put galaxies in perspective, we can use an analogy. If our galaxy were an 18-wheeler, the smallest dwarf galaxies would be the size of pocket-size toy cars, and the largest giant ellipticals would be the size of jumbo jets. Even among spiral galaxies that looked similar, the sizes could range from that of a go-cart to that of a small yacht (❚ Figure 16-6).

Clearly, the diameter and luminosity of a galaxy do not determine its type. Some small galaxies are irregular, and some are elliptical. Some large galaxies are spiral, and some are elliptical. We need more data before we can build a theory for the origin and evolution of galaxies.

Of the three basic parameters that describe a galaxy, we have found two—diameter and luminosity. The third, as was the case for stars, is more difficult to discover.

MASS

Although the mass of a galaxy is difficult to determine, it is an important quantity. It tells us how much matter the galaxy contains, which gives us clues to the galaxy's origin and evolution. In this section, we will examine two fundamental ways to find the masses of galaxies.

One way to find the mass of a galaxy is to watch it rotate. We know the stars in the outer parts of the galaxy are in orbit, so we can use Newton's laws to find the mass. All we need to know is the radius of the galaxy in AU and the orbital period of the stars in years at the galaxy's outer edge. Then we can use Newton's version of Kepler's third law to find the mass just as we found the mass of binary stars in Chapter 9.

Recall from the previous section that it is easy to find the size of a galaxy if we know the distance. We just measure its angular size and use the small-angle formula (Chapter 3).

We can find the orbital velocity of the stars from the Doppler effect. If we focused the image of the galaxy on the slit of a spectrograph, we would see a bright spectrum formed by the bright nucleus of the galaxy, but we would also see fainter emission lines produced by ionized gas in the disk of the galaxy. Because the galaxy rotates, one side moves away from us and one side moves toward us, so the emission lines would be red shifted on one side of the galaxy and blue shifted on the other side. We could measure those changes in wavelength, use the Doppler formula to find the velocities, and plot a diagram showing the velocity of rotation at different distances from the center of the galaxy—a diagram called a **rotation curve.** The artwork in ❚ Figure 16-7a shows the process of creating a rotation curve, and Figure 16-7b shows a real galaxy, its spectrum, and its rotation curve.

Of course, if we know the orbital velocity and the radius of a star's orbit around a galaxy, we can calculate its orbital period. The circumference of the orbit (2π times the radius) divided by the orbital velocity equals the orbital period. So knowing the velocity and size of an orbit easily tells us the orbital period, and from that we can find the mass.

The rotation curve of a galaxy contains all the information we need to find the mass of the galaxy, and the method is known as the **rotation curve method.** This is the most accurate way to find the mass of a galaxy, but it works only for the nearer galaxies, whose rotation curves can be observed. More distant galaxies appear so small we cannot measure the radial velocity at different points along the galaxy. Also, recent studies of our own galaxy and others show that the outer parts of the rotation curve do not decline to lower

FIGURE 16-7

(a) In the artwork in the upper half of this diagram, the astronomer has placed the image of the galaxy over a narrow slit so that light from the galaxy can enter the spectograph and produce a spectrum. A very short segment of the spectrum shows an emission line red shifted on the receding side of the rotating galaxy and blue shifted on the approaching side. Converting these Doppler shifts into velocities, the astronomer can plot the galaxy's rotation curve (right). (b) Real data are shown in the bottom half of this diagram. Galaxy NGC2998 is shown over the spectrograph slit, and the segment of the spectrum includes three emission lines. *(Courtesy Vera Rubin)*

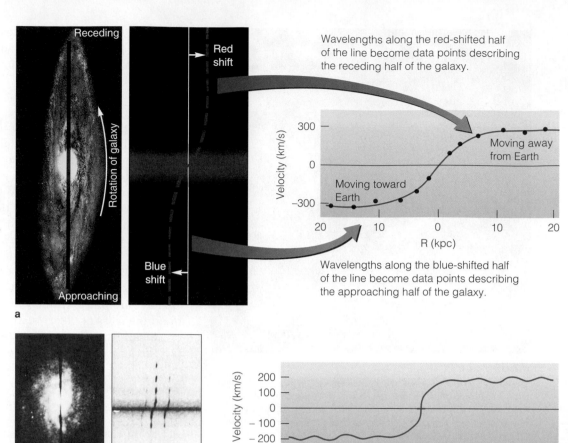

velocities (Figure 15-11). This shows that the outermost visible parts of some galaxies do not travel more slowly and tells us that the galaxies contain large amounts of mass outside this radius, perhaps in extended galactic coronas like the one that seems to surround our own galaxy (see Chapter 15). Because the rotation curve method can be applied only to nearby galaxies and because it cannot determine the masses of galactic coronas, we must look at the second way to find the masses of galaxies.

The **cluster method** of finding galactic mass depends on the motions of galaxies within a cluster. If we measure the radial velocities of many galaxies in a cluster, we find that some velocities are larger than others because of the orbital motions of the individual galaxies in the cluster. Given the range of velocities and the size of the cluster, we can ask how massive a cluster of this size must be to hold itself together with this range of velocities. Dividing the total mass of the cluster by the number of galaxies in the cluster yields the average mass of the galaxies. This method contains the built-in assumption that the cluster is not flying apart. If it is, our result is too large. Because it seems likely that most clusters are held together by their own gravity, the method is probably valid.

A related way of measuring a galaxy's mass is called the **velocity dispersion method**. It is really a ver-

sion of the cluster method. Instead of observing the motions of galaxies in a cluster, we observe the motions of matter within a galaxy. In the spectra of some galaxies, broad spectral lines indicate that stars and gas are moving at high velocities. If we assume the galaxy is bound by its own gravity, we can ask how massive it must be to hold this moving matter within the galaxy. This method, like the one before, assumes that the system is not coming apart.

Measuring the masses of galaxies tells us two things. First the range of masses is wide—from 10^{-6} as much as that of the Milky Way Galaxy to 50 times more (∎ Table 16-1). And second, there is more to a galaxy than meets the eye.

TABLE 16-1
The Properties of Galaxies*

	Elliptical	Spiral	Irregular
Mass	0.0001–50	0.005–2	<0.0005–0.15
Diameter	0.01–5	0.2–1.5	0.05–0.25
Luminosity	0.00005–5	0.005–10	0.00005–0.1

*In units of the mass, diameter, and luminosity of the Milky Way.

SUPERMASSIVE BLACK HOLES IN GALAXIES

Rotation curves show the motions of the outer parts of a galaxy, but it is possible to detect the Doppler shifts of stars orbiting close to the centers of galaxies. Although these motions are not usually shown on rotation curves, they reveal something astonishing.

Measurements show that the stars near the centers of most galaxies are orbiting very rapidly. To hold stars in such small, short-period orbits, the centers of galaxies must contain masses of a million to a few billion solar masses. Yet no object is visible, so most astronomers believe that the nuclei of galaxies contain supermassive black holes. We saw in Chapter 15 that the Milky Way contains a supermassive black hole at its center. Evidently that is typical of galaxies.

Such supermassive black holes cannot be the remains of a dead star. Rather, these black holes must have formed as the galaxy formed or they must have accumulated over billions of years as matter sank into the centers of the galaxies. Measurements show that the masses of these supermassive black holes are related to the mass of nuclear bulges. A galaxy with a large nuclear bulge has a supermassive black hole whose mass is greater than that in a galaxy with a small nuclear bulge. Rare galaxies lacking nuclear bulges also lack supermassive black holes (▌Figure 16-8). This suggests that the supermassive black holes formed long ago as the galaxies formed. Of course, matter has continued to drain into the black holes, but they do not appear to have grown dramatically since they formed.

A billion-solar-mass black hole sounds like a lot of mass, but note that it is roughly 1 percent of the mass of a galaxy. The 2.6 million-solar-mass black hole at the center of the Milky Way Galaxy contains only a thousandth of a percent of the mass of the galaxy. In the next chapter we will discover that these supermassive black holes can produce titanic eruptions, but they represent only a small fraction of the mass of a galaxy.

DARK MATTER IN GALAXIES

Given the size and luminosity of a galaxy, we can make a rough guess as to the amount of matter it should contain. We know how much light stars produce, and we know about how much matter there is between the stars, so it is quite possible to estimate very roughly the mass

Visual-wavelength image

FIGURE 16-8

The Pinwheel Galaxy is near our Milky Way Galaxy in space, and it can be studied in detail. The velocities at the center of the Pinwheel Galaxy are low, showing that it does not contain a supermassive black hole at its center. It also lacks a nuclear bulge, and that confirms the observation that the mass of a supermassive black hole is related to the size of a galaxy's nuclear bulge. *(Bill Schoening/AURA/NOAO/NSF)*

of a galaxy from its luminosity. But when astronomers measure the masses of galaxies, they often find that the measured masses are much too large. We discovered this effect in Chapter 15 when we studied the rotation curve of our own galaxy. This seems to be true of most galaxies. Measured masses of galaxies amount to 10 to 100 times more mass than we can see. Because this mass is not visible, it is called dark matter.

X-ray observations reveal more evidence of dark matter. X-ray images of galaxy clusters show that many of them are filled with very hot, low-density gas. The amount of gas present is much too small to account for the dark matter. Rather, the gas is important because it is very hot and its rapidly moving atoms have not leaked away. Evidently the gas is held in the cluster by a strong gravitational field. To provide enough gravity to hold the hot gas, the cluster must contain much more matter than what we see. The detectable galaxies in the Coma cluster, for instance, amount to only a small fraction of the total mass of the cluster (▌Figure 16-9).

Dark matter is not an insignificant issue. Observations of galaxies and clusters of galaxies tell us that approximately 95 percent of the matter in the universe is

Visual-wavelength image

X-ray image

a

b

FIGURE 16-9

(a) The Coma cluster of galaxies contains at least 1000 galaxies and is especially rich in E and S0 galaxies. Two giant galaxies lie near its center. Only the central area of the cluster is shown in this image. If the cluster were visible in the sky, it would span 8 times the diameter of the full moon. *(L. Thompson © NOAO)* (b) In false colors, this X-ray image of the Coma cluster shows it filled and surrounded by hot gas. The box outlines the area of the cluster shown in part a. *(Simon D. M. White, Ulrich G. Briel, and J. Patrick Henry)*

dark matter; the universe we see—the kind of matter that we and the stars are made of—has been compared to the foam on an invisible ocean. We will cite further evidence of dark matter when we discuss cosmology in Chapter 18.

Dark matter is difficult to detect, and it is even harder to explain. Some astronomers have suggested that dark matter consists of low-luminosity white dwarfs and brown dwarfs scattered through the halos of galaxies. Both observation and theory support the idea that galaxies have massive extended halos, and searches for white dwarfs and brown dwarfs in the halo of our galaxy have been successful. Nevertheless, the searches have not turned up enough of these low-luminosity objects to make up all of the dark matter.

The dark matter can't be hidden in vast numbers of black holes and neutron stars, because we don't see the X rays these objects would emit. Remember, we need roughly 10 times more dark matter than visible matter, and that many black holes and neutron stars should produce X rays that are easy to detect.

Because observations seem to tell us the dark matter can't be hidden in normal objects, some theorists have suggested that the dark matter is made up of unexpected forms of matter. Until recently neutrinos were thought to be massless, but studies now suggest they have a very small mass. Thus they may represent part of the dark matter.

Dark matter remains one of the fundamental unresolved problems of modern astronomy. We will return to this problem in Chapter 18 when we try to under-

stand how dark matter affects the nature of the universe, its past, and its future.

REVIEW CRITICAL INQUIRY

Why do we have to know the distance to a galaxy to find its mass?

Often the critical analysis of a problem is complicated by interrelationships between factors. Usually a bit of care is enough to reveal those connections. To find the mass of a galaxy, we find the size of the orbits and the orbital periods of stars at the galaxy's outer edge, and then use Kepler's third law to find the mass inside the orbits. Measuring the orbital velocity of the stars as the galaxy rotates is easy if we can obtain a rotation curve from a spectrum. But we must also know the radii of the stars' orbits in meters or astronomical units. That is where the distance comes in. To find the radii of the orbits, we must know the distance to the galaxy. Then we can use the small-angle formula to convert the radius in seconds of arc into a radius in parsecs or AU. If our measurement of the distance to the galaxy isn't accurate, we will get inaccurate radii for the orbits, and we will compute an inaccurate mass for the galaxy.

Many different measurements in astronomy depend on the calibration of the distance scale. For example, what would happen to our measurements of the diameters and luminosities of the galaxies if astronomers discovered that the Cepheid variable stars were slightly more luminous than had been believed?

We have gathered the basic data to help us characterize galaxies. That is, we can describe their general characteristics. Now we turn to the more difficult and more interesting question. How did galaxies get to be the way they are?

16-3 THE EVOLUTION OF GALAXIES

Our goal in this chapter has been to build a theory to explain the evolution of galaxies. In Chapter 15 we discussed the origin of our own Milky Way Galaxy; presumably, other galaxies formed similarly. But why did some galaxies become spiral, some elliptical, and some irregular? An important clue to that mystery lies in the clustering of galaxies.

CLUSTERS OF GALAXIES

Single galaxies are rare. Most occur in clusters containing a few to a few thousand galaxies in a volume 1 to 10 Mpc across (▮ Figure 16-10). Our Milky Way Galaxy is a member of a cluster containing slightly over three dozen galaxies, and surveys have cataloged over 2700 other clusters within 4 billion ly.

Visual-wavelength image

FIGURE 16-10

The Hercules galaxy cluster is named after the constellation in which it is found. It is a small cluster containing roughly a hundred galaxies both elliptical and spiral. It lies over 360 million light-years from our galaxy. *(NOAO/AURA/NSF)*

For purposes of our study, we can sort clusters of galaxies into rich clusters and poor clusters. **Rich clusters** contain a thousand or more galaxies, many elliptical, scattered through a volume roughly 3 Mpc (10^7 ly) in diameter. Such a cluster is nearly always condensed; that is, the galaxies are concentrated toward the cluster center. And, at the center, such clusters often contain one or more giant elliptical galaxies.

The Virgo cluster is an example of a rich cluster. It contains over 2500 galaxies located about 17 Mpc (55 million ly) away. The Virgo cluster, like most rich clusters, is centrally condensed and contains the giant elliptical galaxy M87 at its center (page 338).

X-ray observations have found that many of these rich clusters are filled with a hot gas—an intracluster medium. Some of this gas has presumably been driven out of galaxies by supernovae explosions, but much of it appears to be left over from the earliest generations of galaxy formation.

Poor clusters contain fewer than a thousand (and often only a few) galaxies spread through a region that can be as large as a rich cluster. That means the galaxies are more widely separated.

Our Milky Way Galaxy is a member of a poor cluster known as the Local Group (▮ Figure 16-11a). The total number of galaxies in the Local Group is uncertain, but it probably contains more than three dozen galaxies scattered irregularly through a volume roughly 1 Mpc in diameter. Of the brighter galaxies, 15 are elliptical, 4 are spiral, and 13 are irregular.

The total number of galaxies in the Local Group is uncertain because some lie in the plane of the Milky Way Galaxy and are difficult to detect. For example, a small dwarf galaxy, known as the Sagittarius Dwarf, has been found on the far side of our own galaxy where it is almost totally hidden behind the stars clouds of Sagittarius (Figure 16-11b). Astronomers in Dwingeloo, the Netherlands, located another spiral galaxy, dubbed Dwingeloo 1, hidden behind the Milky Way (Figure 16-11c). Dwingeloo 1 and its two neighbor galaxies, Maffie I and Maffie II, are a bit too far away to be part of our Local Group, but the large spiral galaxy known as the Andromeda Galaxy shown in Figure 15-1b is always included. Our Local Group surely contains other small galaxies hidden behind the stars, gas, and dust of the Milky Way.

The Local Group illustrates the subclustering found in poor galaxy clusters. The two largest galaxies, the Milky Way and the Andromeda Galaxy, are the centers of two subclusters. The Milky Way is accompanied

FIGURE 16-11

(a) The Local Group. Our galaxy is located at the center of this diagram. The vertical lines giving distances from the plane of the Milky Way are solid above the plane and dashed below. (b) The Sagittarius Dwarf Galaxy (Sgr Dwarf) lies on the other side of our galaxy. If we could see it in the sky, it would be 17 times larger than the full moon. (c) Spiral galaxy Dwingeloo 1, as large as our own galaxy, was not discovered until 1994 because it is hidden behind the Milky Way. Note the faint spiral pattern. Nearly all individual stars here are in our galaxy. *(Courtesy Shaun Hughes and Steve Maddox, Royal Greenwich Observatory; and Ofer Lahav and Andy Loan, University of Cambridge)*

by the Magellanic Clouds and seven other dwarf galaxies. The Andromeda Galaxy is attended by more dwarf elliptical galaxies (two of which are visible in Figure 15-1b) and a small spiral, M33 (Figure 16-8). The Andromeda Galaxy and our Milky Way Galaxy, with their retinue of smaller galaxies, are moving toward each other and will almost certainly collide in a few billion years.

The critical clue hidden in the clusters of galaxies is the relative abundance of the different types of galax-

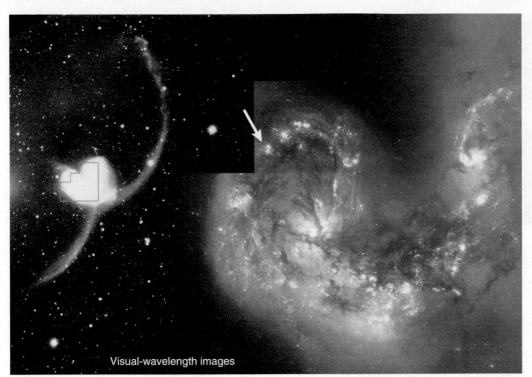

FIGURE 16-12
The colliding galaxies NGC4038 and NGC4039 are known as the Antennae because their long curving tails resemble the antennae of an insect. Earth-based photos (left) show little detail, but a Hubble Space Telescope image (right) reveals the collision of two galaxies producing thick clouds of dust and raging star formation creating roughly a thousand massive star clusters such as the one at the top (arrow). Such collisions between galaxies are common. *(Brad Whitmore, STScI, and NASA)*

Visual-wavelength images

ies. In general, rich clusters tend to contain 80 to 90 percent E and S0 galaxies and a few spirals. Poor clusters contain a larger percentage of spirals; and, among isolated galaxies, those that are not in clusters, 80 to 90 percent are spirals. Galaxies that are crowded together tend to be E or S0 rather than spiral. Somehow the environment around a galaxy helps determine its type.

Astronomers suspect that collisions between galaxies are an important process. In a rich cluster, the galaxies are crowded much closer together, and they must collide with each other more often than galaxies in a poor cluster. Could such collisions explain the excess of elliptical galaxies in the crowded rich clusters? Astronomers have discovered that galaxy smashups are both entertaining and important.

COLLIDING GALAXIES

Galaxies should collide fairly often. The average separation between galaxies is only about 20 times their diameter. Like two blindfolded elephants blundering about at random under a circus tent, galaxies should bump into each other once in a while. Stars, on the other hand, almost never collide. In the region of the galaxy near the sun, the average separation between stars is about 10^7 times their diameter. Thus, collision between two stars is about as likely as collision between two blindfolded gnats flitting about at random in a football stadium.

Large telescopes reveal hundreds of galaxies that appear to be colliding with other galaxies. One of the most famous pairs of colliding galaxies, NGC4038 and NGC4039, are called the Antennae because the long tails resemble the antennae of an insect (❙ Figure 16-12). In addition to having tails, some interacting galaxies are connected by bridges of gas, dust, and stars.

Galaxies do not need to collide directly to affect each other. If two galaxies pass near each other, their gravitational fields distort their shapes, drawing out long tails of gas, dust, and stars. Sometimes bridges of material are created that connect the interacting galaxies. These are tidal effects. The gravitation of one galaxy causes tides that distort the other galaxy and vice versa. Computer models show that even when galaxies rush past each other rapidly without interpenetrating, the distortions can be severe. Astronomers have begun to refer to this as "harassment" to distinguish it from direct collisions. One of the most famous pairs of interacting galaxies is called the Mice because of the tail-like deformities flung out as the galaxies whip around their common center of mass and tidally harass each other (❙ Figure 16-13).

A collision between galaxies can last hundreds of millions of years, but we can watch it happen in computer models. The Whirlpool Galaxy, M51, provides a good example of such a model (❙ Figure 16-14). Models show that the galaxies passed near each other but did not actually penetrate. The smaller galaxy passed behind the larger galaxy.

Unusual **ring galaxies** consist of a bright nucleus surrounded by a ring (❙ Figure 16-15). Models show that they are produced by a galaxy passing roughly perpendicularly through the disk of the larger galaxy. Indeed, many ring galaxies have nearby companions.

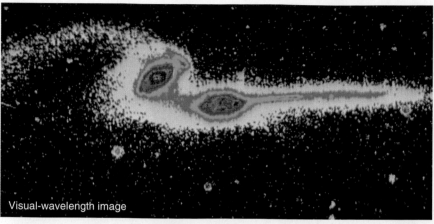

Visual-wavelength image

a

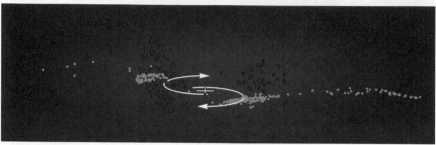

b

FIGURE 16-13

(a) The Mice are a pair of colliding galaxies with peculiar tails (shown here in a false-color image). *(NOAO)* (b) A computer model of a close encounter between normal galaxies shows how the centers of mass whip around each other, producing tails similar to the Mice. Colors of dots show the original membership of the stars and reveal that the interacting galaxies have exchanged mass. *(Illustration by Allen Beechel)*

As colliding galaxies approach each other, they carry tremendous orbital momentum. Models show that tidal forces can convert some of the orbital motion into random motion among the stars, thus robbing the galaxies of orbital momentum. Such galaxies may lose enough orbital momentum that they fall back together and merge.

A small galaxy merging with a larger galaxy will be pulled apart gradually, and its stars will spread through the larger galaxy in a process called **galactic cannibalism.** Computer simulations of mergers show that galaxies throw off shells of stars as the nuclei of the two galaxies spiral into each other, and such shells have been observed around real galaxies (▌ Figure 16-16).

Evidence of collisions often appears in the motions inside a galaxy. The tails on NGC7252 (▌ Figure 16-17a) resemble the tails on mice, and when the Hubble Space Telescope imaged the center of the galaxy it found a small, backward-spinning spiral. Evidently, NGC7252 was created about a billion years ago when two oppositely spinning galaxies collided and merged. The apparently normal galaxy M64 reveals its secrets in radio maps (Figure 16-17b). The center is rotating backward

compared to the outer galaxy—the product of a merger between counter-rotating galaxies.

Evidence of past mergers also appears in the multiple nuclei in some giant elliptical galaxies (Figure 16-17c). These galaxies are often located in crowded rich clusters of galaxies, and the extra nuclei appear to be the dense centers of smaller galaxies that have been cannibalized and only partly digested.

At infrared wavelengths, astronomers are able to image galaxies that are a hundred times more luminous than our Milky Way Galaxy, but they are deeply shrouded in dust and thus are dim at visible wavelengths. They are evidently undergoing tremendous bursts of star formation that are producing dust and warming it with starlight until the warm dust emits vast amounts of infrared radiation. The best images of these **ultra-luminous infrared galaxies** show distorted bridges and tails. Many of these galaxies are undergoing multiple collisions as three or more galaxies are merging and presumably triggering firestorms of star formation.

Our own Milky Way Galaxy is an unreformed cannibal. The ring of stars around its outer edge (Figure 15-16) and streams of misplaced population I stars in its halo are traces of smaller galaxies it has digested. It is currently pulling the Sagittarius Dwarf galaxy (Figure 16-11b) apart, and astronomers find elongated clouds of gas and stars leading from the two Magellanic Clouds back to the disk of our galaxy. They must have passed through the disk recently and will eventually fall back and be cannibalized. Perhaps the Magellanic Clouds are irregular galaxies because they have been severely harassed by the Milky Way Galaxy.

We can find further evidence of harassment among galaxies by looking at the Virgo cluster. Studies with the Hubble Space Telescope and with very large telescopes on Earth have identified free-floating stars and planetary nebulae between the galaxies in the Virgo cluster. Evidently a significant number of the stars do not lie inside the galaxies but have been freed to wander alone through the cluster as galaxies have harassed each other.

The new evidence for collisions, harassment, and cannibalism among galaxies is very strong. These are the processes we need to understand in order to tell the story of the galaxies.

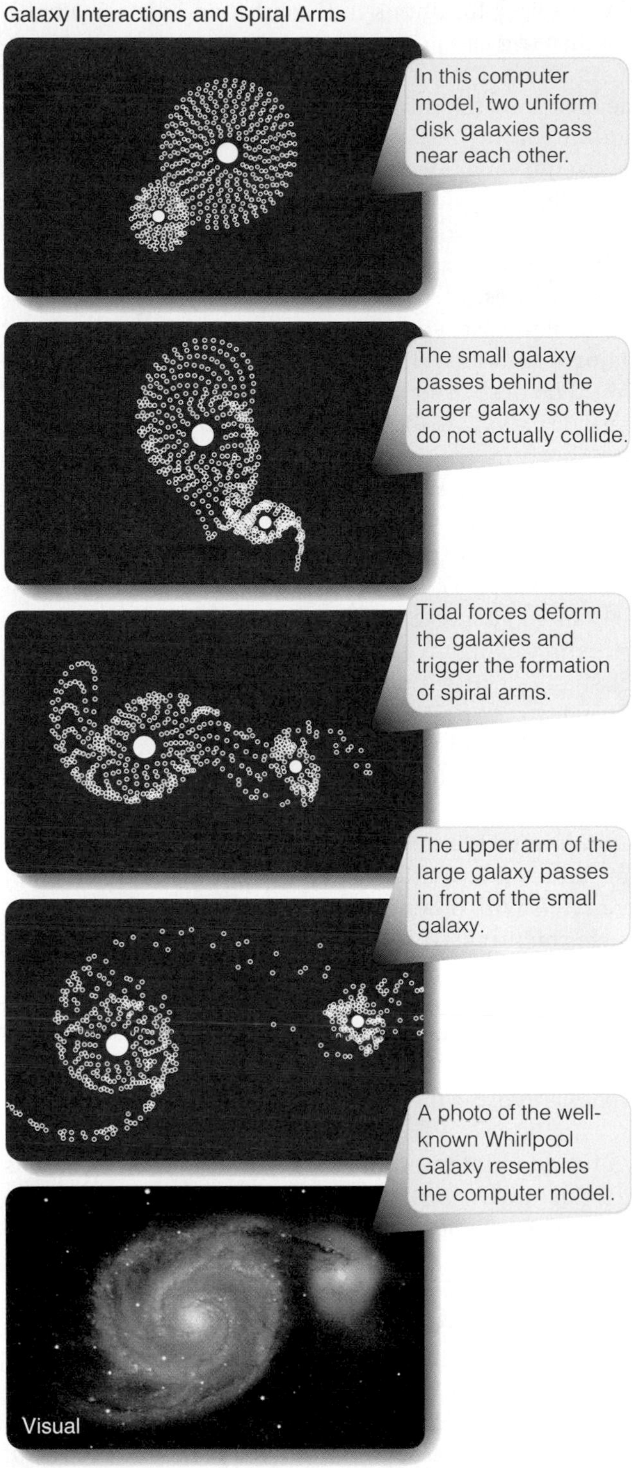

Galaxy Interactions and Spiral Arms

In this computer model, two uniform disk galaxies pass near each other.

The small galaxy passes behind the larger galaxy so they do not actually collide.

Tidal forces deform the galaxies and trigger the formation of spiral arms.

The upper arm of the large galaxy passes in front of the small galaxy.

A photo of the well-known Whirlpool Galaxy resembles the computer model.

Visual

FIGURE 16-14

An interaction between two galaxies that takes hundreds of millions of years can be followed step by step in a computer model. *(Illustration by Allen Beechel; photo: NOAO)*

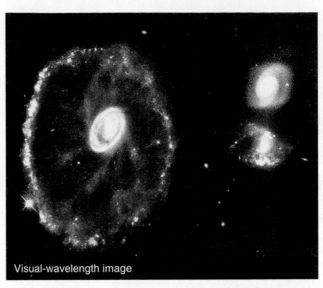

Visual-wavelength image

FIGURE 16-15

The Cartwheel Galaxy was once a normal spiral galaxy, but a few hundred thousand years ago one or the other of the two smaller galaxies at the right passed through the center of the larger galaxy. Tidal forces generated an expanding wave of star formation and a central core of newborn stars heavily reddened by dust. Such ring galaxies appear to be the fleeting consequences of bulls-eye collisions. Soon after the collision, the violent star formation exhausts itself, and normal spiral arms may re-form. This is suggested in the Cartwheel Galaxy by the faint spiral structure inside the outer ring. *(Kirk Borne, STScI and NASA)*

THE ORIGIN AND EVOLUTION OF GALAXIES

The test of any scientific understanding is whether we can put all the evidence and theory together to tell the history of the objects we study. Can we describe the origin and evolution of the galaxies? Just a few decades ago, it would have been impossible, but the evidence from space telescopes and new-generation telescopes on Earth combined with advances in computer modeling and theory allow astronomers to outline the story of the galaxies.

As we begin, we should eliminate a few older ideas immediately. It is easy to imagine that galaxies evolve from one type to another. But an elliptical galaxy cannot become a spiral galaxy or an irregular galaxy, because ellipticals contain almost no gas and dust from which to make new stars. Elliptical galaxies can't be young galaxies. We can also argue that spiral and irregular galaxies cannot evolve into elliptical galaxies because spiral and irregular galaxies contain both young and old stars. The old stars mean that spiral and irregular galaxies can't be young. The galaxy classes tell us something important, but a single galaxy does not change from one class to another any more than a cat can change into a dog.

Another old idea held that galaxies that form from rapidly rotating clouds of gas would have lots of angular momentum and would contract slowly to form disk-shaped spiral galaxies. Clouds of gas that rotated less rapidly would contract faster, form stars quickly, use up all the gas and dust, and become elliptical galaxies. The evidence clearly shows that galaxies didn't form

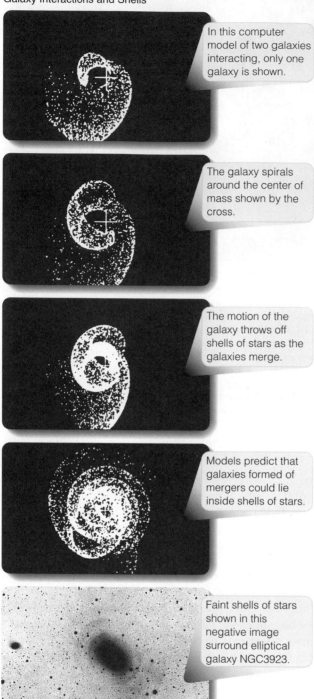

In this computer model of two galaxies interacting, only one galaxy is shown.

The galaxy spirals around the center of mass shown by the cross.

The motion of the galaxy throws off shells of stars as the galaxies merge.

Models predict that galaxies formed of mergers could lie inside shells of stars.

Faint shells of stars shown in this negative image surround elliptical galaxy NGC3923.

▶Visual

FIGURE 16-16

As two galaxies merge, they spiral around their common center of mass, which could produce shells of stars. Such faint shells are seen around some galaxies. *(Model courtesy François Schweizer and Alan Toomre; photo courtesy David Malin, Anglo-Australian Telescope Board)*

that are very luminous in the infrared because a collision has triggered a burst of star formation that is heating the dust (▌ Figure 16-18). The warm dust reradiates the energy in the infrared. Supernovae in such a galaxy can blow away any remaining gas and dust that don't get used up making stars. A few collisions and mergers could leave a galaxy with no gas and dust from which to make new stars. Astronomers now suspect that most ellipticals are formed by the merger of at least two or three galaxies.

In contrast, spirals seem never to have suffered major collisions. Their thin disks are delicate and would be destroyed by tidal forces in a collision with a massive galaxy. Also, they retain plenty of gas and dust and continue making stars. Of course, spiral galaxies can cannibalize smaller galaxies. We see evidence in our own galaxy, and some astronomers suspect that much of the halo consists of cannibalized galaxies. But our Milky Way Galaxy has never collided with a similarly large galaxy. Such a collision would trigger star formation using up gas and dust, scramble the orbits of stars, and destroy the thin disk. In a few billion years, our galaxy may merge with the approaching Andromeda Galaxy, and the final result may be an elliptical galaxy.

Barred spiral galaxies may be the product of tidal interactions. Mathematical models show that bars are not stable and eventually dissipate. It may take tidal interactions with other galaxies to regenerate the bars. Because roughly half of all spiral galaxies have bars, we can suspect that these tidal interactions are common.

Other processes can alter galaxies. The S0 galaxies retain a disk shape but lack gas and dust. They may have lost gas and dust during a starburst, but another process involves their motion through the gas in a cluster of galaxies. A galaxy orbiting through such a gas would encounter a tremendous wind blowing its gas and dust away. A disk-shaped spiral galaxy could be reduced to an S0 galaxy. This may also explain the small dwarf ellipticals. They are too small to be the product of galaxy mergers, but they have somehow lost their gas and dust. In contrast, the irregular galaxies may be small fragments ripped from larger galaxies.

Other factors must influence the evolution of galaxies. The hot gas in galaxy clusters warns us that galaxies do not form in isolation. Cooler gas clouds may fall into galaxies and add material for star formation. Also, we need to consider differences in the initial clouds of gas from which galaxies form. Astronomers are just beginning to understand which factors are important in the birth of galaxies.

THE FARTHEST GALAXIES

Observations with the largest and most sophisticated telescopes are taking us back to the age of galaxy formation. At great distances the look-back time is so great that we see the galaxies as they were long ago, and we

and evolve in isolation. Collisions and mergers dominate the history of the galaxies.

The ellipticals appear to be the product of galaxy mergers, which triggered star formation that used up the gas and dust. In fact, we see many **starburst galaxies**

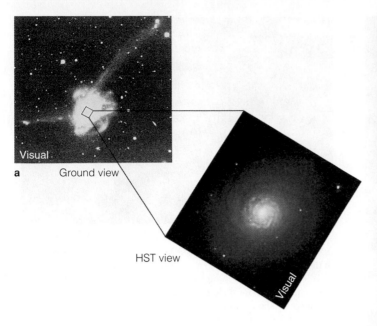

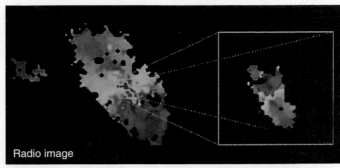

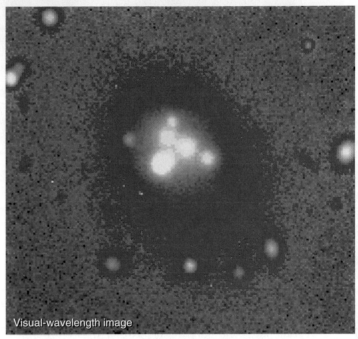

FIGURE 16-17

Evidence of mergers. (a) The twisted shape of NGC7252 suggests a collision, and a Hubble Space Telescope image reveals a small spiral of young stars spinning backward in the heart of the larger galaxy. This is thought to be the remains of two oppositely rotating galaxies that merged about a billion years ago. *(François Schweizer, Carnegie Institution of Washington, and Brad Whitmore, STScI)* (b) The apparently normal galaxy M64 is shown at left in a radio map that shows radial velocity as color. Red regions are receding from us and blue regions approaching. At the center of the galaxy, the rotation is reversed, evidently the consequence of a merger between counter-rotating galaxies. *(Robert Braun, NRAO)* (c) Giant elliptical galaxies in rich clusters sometimes have multiple nuclei, thought to be the densest parts of smaller galaxies that have been absorbed and partly digested. *(Michael J. West)*

discover that there were more spirals then and fewer ellipticals. On the whole, galaxies long ago were more compact and more irregular than they are now. We can even see that galaxies were closer together long ago; about a third of all distant galaxies are in close pairs, but only 7 percent of nearby galaxies are in pairs. The observational evidence clearly supports the hypothesis that galaxies have evolved by merger.

At great distances, astronomers see tremendous numbers of small, blue, irregularly shaped galaxies that have been called blue dwarfs. The blue color tells us they are rapidly forming stars, but the role of these blue dwarfs in the formation of galaxies is not clear. They may be clouds of gas and stars that were eventually absorbed in the formation of larger galaxies, or they may have used up their gas and dust and have faded to obscurity.

At the limits of the largest telescopes, astronomers see faint red galaxies (▌ Figure 16-19). They look red to us because of their great red shifts. The floods of ultraviolet light emitted by these star-forming galaxies have been shifted into the far red part of the spectrum. We must be seeing them at a time when the universe was

only a billion years old. These galaxies appear to be among the first to begin shining after the beginning of the universe, a story we will discuss in Chapter 18.

Visual-wavelength image

FIGURE 16-18

Starburst galaxy NGC3310 is forming stars at a prodigious rate. Several hundred star clusters, some containing as many as a million stars, outline the spiral arms in a blue glow. From the blue color of the clusters, astronomers conclude that the burst of star formation began about 100 million years ago, probably when a companion galaxy collided with NGC3310. *(Hubble Heritage Team, STScI/ AURA/NASA)*

Visual-wavelength image

FIGURE 16-19

Arrows point to three very distant red galaxies. The look-back time to these galaxies is so large we see them when the universe was only a billion years old. Such highly red-shifted galaxies are believed to be among the first to form stars and begin shining after the beginning of the universe about 14 billion years ago. *(NASA, H.-J. Yan, R. Windhorst and S. Cohen, Arizona State University)*

Before we can consider the universe as a whole, we must examine the galaxies from a different perspective. Some galaxies are suffering tremendous eruptions as their cores blast radiation and matter outward. In the next chapter, we will discover that these peculiar galaxies are closely related to collisions and mergers between galaxies.

SUMMARY

Astronomers did not realize that the spiral nebulae visible in larger telescopes were other galaxies until Edwin Hubble detected Cepheid variable stars in a few nearby galaxies. That announcement in 1924 showed that the universe was filled with galaxies much like our own Milky Way.

We can divide galaxies into three classes—elliptical, spiral, and irregular—with subclasses giving the galaxy's shape or the amount of gas and dust present. The galaxy types appear to reflect different histories of star formation. The elliptical galaxies contain little gas and dust and cannot make many new stars. The spiral and irregular galaxies contain large amounts of gas and dust and are still forming stars.

To measure the properties of galaxies, we must first find their distances. For the nearer galaxies, we can judge distances using distance indicators, objects whose luminosities are known. The best of these standard candles are the Cepheid

variable stars. Once the distances to some galaxies are known, astronomers can calibrate other distance indicators to piece together a distance scale. Some of the most-used distance indicators are globular clusters, planetary nebulae, and supernovae. For the most distant galaxies, astronomers must depend on a calibration of the luminosity of galaxies themselves.

The Hubble law shows that the radial velocity of a galaxy is proportional to its distance. Thus, we can use the Hubble law to estimate distances. The galaxy's radial velocity divided by the Hubble constant is its distance in megaparsecs.

The masses of galaxies can be measured in two ways— the rotation curve method and the cluster method. The first method is the most accurate but it is applicable only to nearby galaxies.

Many measurements of the mass of galaxies show that galaxies contain many times more mass than what we can see. This dark matter amounts to roughly 90 percent of the matter in the universe. Although some mass has been found as brown dwarfs, white dwarfs, neutrinos with mass, and so on, most of the dark matter seems to be made up of an as yet unrecognized form of matter.

The origin and evolution of galaxies is not fully understood, but it is clear that interaction between galaxies is important. Close encounters, collisions, and mergers of two galaxies can cause dramatic tidal distortions and bursts of star formation. Stripping can remove gas and dust from a galaxy. These processes appear to be common among galaxies. The Magellanic Clouds are being cannibalized by our own Milky Way Galaxy.

Normal elliptical galaxies are apparently formed by the merger of spiral galaxies. The average, bright elliptical galaxy may contain the merged remains of two or three spiral galaxies. Dwarf ellipticals are different systems and appear to be small galaxies stripped of gas and dust by their motion through the intracluster gas and by interactions with larger galaxies. According to this theory, the spiral galaxies have not experienced collisions with large galaxies very often since they were formed.

NEW TERMS

spiral nebula

island universe

Shapley–Curtis Debate

elliptical galaxy

spiral galaxy

barred spiral galaxy

irregular galaxy

megaparsec (Mpc)

distance indicator

standard candle

distance scale

look-back time

Hubble law

Hubble constant (H)

rotation curve

rotation curve method

cluster method

velocity dispersion method

rich cluster

poor cluster

ring galaxy

galactic cannibalism

ultraluminous infrared
  galaxy

starburst galaxy

REVIEW QUESTIONS

Ace⌬Astronomy™ Assess your understanding of this chapter's topics with additional quizzing and animations at **http:// astronomy.brookscole.com/seeds8e**

1. If a civilization lived on a planet in an E0 galaxy, do you think it would have a Milky Way? Why or why not?

2. Why can't the evolution of galaxies go from elliptical to spiral? from spiral to elliptical?

3. If all elliptical galaxies had three different diameters, we would never see an elliptical galaxy with a circular outline on a photograph. True or false? Explain your answer. (*Hint:* Can a football ever cast a circular shadow?)

4. What is the difference between an Sa and an Sb galaxy? between an S0 and an Sa galaxy? between an Sb and an SBb galaxy? between an E7 and an S0 galaxy?

5. Why wouldn't white dwarfs make good distance indicators?

6. Why isn't the look-back time important among nearby galaxies?

7. Explain how the rotation curve method of finding a galaxy's mass is similar to the method used to find the masses of binary stars.

8. Explain how the Hubble law permits us to estimate the distances to galaxies.

9. How can collisions affect the shape of galaxies?

10. What evidence do we have that galactic cannibalism really happens?

11. Describe the future evolution of a galaxy that we now see as a starburst galaxy. What will happen to its interstellar medium?

12. Why does the intracluster medium help determine the nature of the galaxies in a cluster?

DISCUSSION QUESTIONS

1. From what you know about star formation and the evolution of galaxies, do you think the Infrared Astronomy Satellite should have found irregular galaxies to be bright or faint in the infrared? Why or why not? What about starburst galaxies? What about elliptical galaxies?

2. Imagine that we could observe a gas cloud at such a high look-back time that it is just beginning to form one of the first galaxies. Further, suppose we discovered that the gas was metal rich. Would that support or contradict our understanding of galaxy formation?

PROBLEMS

1. If a galaxy contains a type I (classical) Cepheid with a period of 30 days and an apparent magnitude of 20, what is the distance to the galaxy?

2. If you find a galaxy that contains globular clusters that are 2 seconds of arc in diameter, how far away is the galaxy? (*Hints:* Assume that a globular cluster is 25 pc in diameter and use the small-angle formula.)

3. If a galaxy contains a supernova that at its brightest has an apparent magnitude of 17, how far away is the galaxy? (*Hint:* Assume that the absolute magnitude of the supernova is −19.)

4. If we find a galaxy that is the same size and mass as our Milky Way Galaxy, what orbital velocity would a small satellite galaxy have if it orbited 50 kpc from the center of the larger galaxy?

5. Find the orbital period of the satellite galaxy described in Problem 4.

6. If a galaxy has a radial velocity of 2000 km/s and the Hubble constant is 70 km/s/Mpc, how far away is the galaxy? (*Hint:* Use the Hubble law.)

7. If you find a galaxy that is 20 minutes of arc in diameter, and you measure its distance to be 1 Mpc, what is its diameter?

8. We have found a galaxy in which the outer stars have orbital velocities of 150 km/s. If the radius of the galaxy is 4 kpc, what is the orbital period of the outer stars? (*Hints:* 1 pc = 3.08×10^{13} km, and 1 yr = 3.15×10^7 seconds.)

9. A galaxy has been found that is 5 kpc in radius and whose outer stars orbit the center with a period of 200 million years. What is the mass of the galaxy? On what assumptions does this result depend?

10. Among the globular clusters orbiting a distant galaxy, the fastest is traveling 420 km/s and is located 11 kpc from the center of the galaxy. Assuming the globular cluster is gravitationally bound to the galaxy, what is the mass of the galaxy? (*Hint:* The galaxy had a slightly faster globular cluster, but it escaped some time ago. What is the escape velocity?)

CRITICAL INQUIRIES FOR THE WEB

1. How far out into the universe can we see Cepheid variables? Research sources on the Internet to find other galaxies whose distances have been found through observation of Cepheids. List the galaxies in which Cepheids have been identified and the distances determined from these data.

2. Who was Milton Humason, and how did he get started in astronomy?

3. In the early 1900s the nature of the "spiral nebulae" was not well understood. In 1920 a "great debate" was held between Harlow Shapley, who held that these objects were relatively nearby swirling clouds of gas, and Heber Curtis, who saw them as distant "island universes." Use the Internet to find information about the debate, outline the lines of evidence used by the two participants to present their views, and explain who was right and who was wrong.

4. Locate a Web page dedicated to the Messier Catalogue—a list of galaxies, clusters, and nebulae that is often used as a list of targets for small telescopes. Be sure that your destination includes images of the objects. For each of the galaxies in the Messier list, determine its Hubble classification. You may be given this information at the site, but examine the images to see if the features of these galaxies conform to a particular Hubble type.

EXPLORING *THESKY*

1. Locate the Andromeda Galaxy, also known as M31, and its companion galaxies. Zoom in on it and estimate its angular size compared to the full moon. (*Hint:* Use **Find** under the **Edit** menu.)

2. Take a survey of galaxies and see how many are spiral and how many are elliptical. Does your method produce a fair count or are you biased to notice only the brighter galaxies? (*Hint:* Use **Filters** under the **View** menu to turn off everything but **Galaxies** and **Mixed Deep Sky** objects. Make sure the Messier labels are switched on using the **Labels** and **Setup** under the **View** menu.)

3. Locate the Sombrero Galaxy (M104). Study the photographs and discuss this galaxy's special properties. Zoom in on it and estimate its angular size compared to the moon.

4. Study the distribution of galaxies and notice how they cluster together. Can you find the Virgo cluster? Zoom in until more galaxies appear and then scroll north to find the Coma Cluster. Zoom in on Leo to find the cluster there. What other clusters can you find?

5. Describe the galaxy located near the south celestial pole.

 Visit the Seeds *Foundations of Astronomy* companion Web site for critical thinking exercises, articles, and additional readings from InfoTrac College Edition, Brooks/Cole's online student library.

GALAXIES WITH ACTIVE NUCLEI

Virtually all scientists have the bad habit of displaying feats of virtuosity
in problems in which they can make some progress and
leave until the end the really difficult central problems.

M. S. Longair, High Energy Astrophysics

NASA

Chapter 17

GUIDEPOST

This chapter is important for two reasons. First, it draws together ideas from many previous chapters to show how nature uses the same basic rules on widely different scales. Matter flowing into a protostar, into a white dwarf, into a neutron star, or into the heart of a galaxy must obey the same laws of physics, so we see the same geometry and the same phenomena. The only difference is the level of violence.

Second, this chapter is important because the most distant objects we can see in the universe are the most luminous galaxies, and many of those are erupting in outbursts and are thus peculiar. By studying these galaxies, our attention is drawn out in space to the edge of the visible universe and back in time to the earliest stages of galaxy formation. In other words, we are led to think of the origin and evolution of the universe, the subject of the next chapter.

VIRTUAL LABORATORY

ACTIVE GALACTIC NUCLEI

Explosion? Eruption? Detonation? Colossal? Titanic? Words fail us when we try to describe the violence that astronomers find at the center of some galaxies. If we say a massive star explodes as a supernova, what word shall we use for the eruption of an entire galaxy? Astonishing it may be, but violence in the centers of galaxies is easily detected across billions of light-years, and our double goal in this chapter is to explain the origin of these eruptions and relate them to the formation and evolution of galaxies.

The adventure of studying active galaxies began when radio astronomers first noticed such objects in the 1950s. In the decades that followed, astronomers used all parts of the electromagnetic spectrum to study these galaxies and piece together the puzzle.

Finding the solution was a long and difficult journey, but modern astronomers generally agree on the basic facts. Eruptions in the hearts of active galaxies are caused by supermassive black holes. We have see evidence in the previous two chapters that our own galaxy and most other galaxies contain supermassive black holes at their centers. In this chapter, we study not only the behavior of these strange objects, but also the process by which astronomers came to understand them.

A few astronomers have objected to the quotation that opens this chapter, perhaps because it seems to belittle scientists. But it is a realistic description of how scientists attack a problem. Big problems are made of little problems, and scientists begin by solving the little problems, learning from those solutions, and moving ever closer to the center of the big problem. Perhaps no better example can be found than modern astronomy's attack on "the really difficult central problem" in the nuclei of active galaxies.

17-1 ACTIVE GALAXIES

All galaxies, including our Milky Way, emit radio energy from neutral hydrogen, from molecules, from pulsars, and so on. In the 1950s, radio astronomers discovered that some galaxies emit as much as 10 million times more radio energy than a normal galaxy. These galaxies were called **radio galaxies.** Beginning in the 1960s, astronomers could send spacecraft above Earth's atmosphere, and they discovered that many of the radio galaxies also emit powerfully at infrared, ultraviolet, and X-ray wavelengths. As a result, these galaxies are now called **active galaxies** in recognition of the many wavelengths at which they radiate.

As we will see in this section, strong evidence suggests that the energy that powers active galaxies arises

Ace◑Astronomy™ The AceAstronomy icon throughout the text indicates an opportunity for you to test yourself on key concepts and to explore animations and interactions on the AceAstronomy Web site at **http://astronomy.brookscole .com/seeds8e**

in their central nuclei. One of the great adventures of modern astronomy has been the quest to understand these **active galactic nuclei (AGN).** What are the mysterious and powerful AGN?

SEYFERT GALAXIES

In 1943, Mount Wilson astronomer Carl K. Seyfert published his study of spiral galaxies. Observing at visual wavelengths, Seyfert found that some spiral galaxies have small, highly luminous nuclei (❙ Figure 17-1b). Today, these **Seyfert galaxies** are recognized by the peculiar spectra of these bright nuclei.

The light from a normal galaxy comes from stars, and its spectrum is the combined spectra of many millions of stars. Thus, we see the principal absorption lines found in stellar spectra, mainly hydrogen Balmer lines and lines of calcium II. But the spectra of the Seyfert galaxy nuclei contain broad emission lines of highly ionized atoms. Emission lines suggest a hot, low-density gas, and ionized atoms suggest that the gas is very excited. The widths of the spectral lines suggest large Doppler shifts produced by high velocities in the nuclei; gas approaching us would produce blue-shifted spectral lines, and gas going away from us would produce red-shifted lines. The combined light would produce broad spectral lines. The velocities at the centers of Seyfert galaxies are roughly 10,000 km/s, about 30 times greater than velocities at the centers of normal galaxies.

About 2 percent of spiral galaxies appear to be Seyfert galaxies, and they are classified into two categories. Type 1 Seyfert galaxies are very luminous at X-ray and ultraviolet wavelengths and have the typical broad emission lines in their spectra. Type 2 Seyfert galaxies lack strong X-ray emission and have emission lines that are narrower than those of type 1 Seyferts but still broader than spectral lines produced by a normal galaxy.

In the 1960s, astronomers discovered that the brilliant nuclei of Seyfert galaxies fluctuate rapidly. A Seyfert nucleus can change its brightness by 50 percent in a few months. As we saw in Chapter 14, an astronomical body cannot change its brightness in a time shorter than the time it takes light to cross its diameter. If the Seyfert nucleus can change in a few months, then it cannot be larger in diameter than a few light-months. In spite of their small size, the cores of Seyfert galaxies produce tremendous amounts of energy. The brightest emit a hundred times more energy than the entire Milky Way. Something in the centers of these galaxies not much bigger than our solar system produces a galaxy's worth of energy.

The shapes of Seyferts can give us clues to their energy source. They are three times more common in interacting pairs of galaxies than in isolated galaxies. Also, about 25 percent have peculiar shapes suggesting tidal interactions with other galaxies (❙ Figure 17-2a). This statistical evidence (Window on Science 17-1) hints that Seyfert galaxies may have been triggered into activity by collisions or interactions with companions.

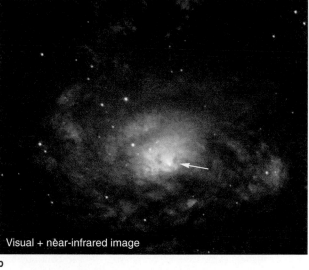

a

b

FIGURE 17-1

(a) Seyfert galaxy NGC1566 looks like a normal spiral galaxy, but short exposure images reveal that its nucleus is very small and very luminous. *(AATB)* (b) The small, bright nucleus of this Seyfert galaxy in the southern constellation Circinus is ejecting gas at high speeds in a bright, V-shaped cone located just above the nucleus (arrow). Higher above the disk, the ejected gas is visible as the bright pink regions extending to the top of the image. Presumably a similar cone of ejected gas is hidden below the disk of the galaxy. *(Andrew S. Wilson, University of Maryland; Patrick L. Shopbell, Caltech; Chris Simpson, Subaru Telescope; Thaisa Storchi-Bergmann and F. K. B. Barbosa, UFRGS, Brazil; Martin J. Ward, University of Leicester, U.K.; and NASA)*

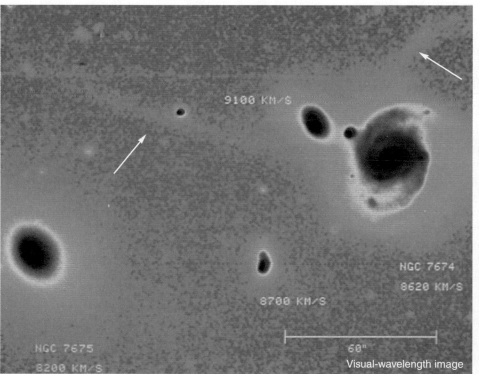

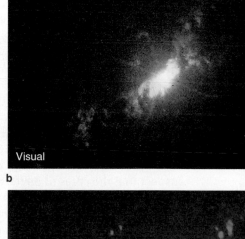

a

b

c

FIGURE 17-2

(a) Seyfert galaxy NGC7674 is distorted with tails to upper right and upper left (arrows) in this false-color image. Note the companion galaxies. *(John W. Mackenty, Institute for Astronomy, University of Hawaii)* (b) Seyfert galaxy NGC4151 has a brilliant nucleus in visible light images (top). The Space Telescope Imaging Spectrograph reveals Doppler shifts showing that part of the expelled gas has a blue shift (to the right) and is approaching us and part has a red shift (to the left) and is receding. *(John Hutchings, Dominion Astrophysical Observatory; Bruce Woodgate, GSFC/NASA; Mary Beth Kaiser, Johns Hopkins University; and the STIS Team)*

It Wouldn't Stand Up in Court: Statistical Evidence

Notice that some scientific evidence is statistical. For example, we might note that Seyfert galaxies are three times more likely to have a nearby companion than a normal galaxy is. This is statistical evidence because we can't be certain that any single Seyfert galaxy will prove to have a companion. Yet the probability is higher than if it were a normal galaxy, and that leads us to suspect that interactions between galaxies are involved. Such statistical evidence can tell us something in general about the cause of Seyfert eruptions.

Statistical evidence is common in science, but it is inadmissible in most courts of law. The American legal system is based on the principle of reasonable doubt, so most judges would not allow statistical evidence of guilt. For example, plaintiffs have had great trouble suing tobacco companies for causing their cases of lung cancer, even though there is a clear statistical link between smoking and lung cancer. The statistics tell us something in general about smoking but cannot be used to prove that any specific case of lung cancer was caused by smoking. There are other causes of lung cancer, so there is always a reasonable doubt as to the cause of any specific case.

We can use statistical evidence in science because we do not demand that the statistics demonstrate anything conclusively about any single example. To continue our example of Seyfert galaxies, we don't demand that the statistics predict that any specific galaxy has a companion. Rather, we use the statistical evidence to gain a general insight into the cause of active galactic nuclei. That is, we are trying to understand galaxies as a whole, not to convict any single galaxy.

Of course, if we surveyed only a few galaxies, our statistics might not be very good, and a critic might be justified in making the common complaint, "Oh, that's only statistics." For example, if we surveyed only four galaxies and found that three had companions, our statistics would not be very reliable. But if we surveyed 1000 galaxies, our statistics could be very good indeed, and our conclusions could be highly significant.

Thus, scientists can use statistical evidence if it passes two tests. It cannot be used to draw conclusions about specific cases, and it must be based on large enough samples so the statistics are significant. With these restrictions, statistical evidence can be a powerful scientific tool.

Some Seyferts are expelling matter in oppositely directed flows (Figure 17-2b), a geometry we have seen on smaller scales when matter flows into accretion disks around neutron stars or black holes.

Examine Figure 17-2c. This image of the core of Seyfert galaxy NGC4151 is a spectrum obtained with a special spectrograph that records an image at whatever wavelengths are bright in an object's emission spectrum. This image shows that some of the gas near the center of the galaxy has a high velocity toward us and some has a high velocity away from us. These high velocities are typical of gas in the centers of Seyfert galaxies.

All of this evidence leads modern astronomers to suspect that the cores of Seyfert galaxies contain supermassive black holes—black holes with masses as high as a billion solar masses. The gas in the centers of Seyfert galaxies is traveling so fast it would escape from a normal galaxy. Only very large central masses could exert enough gravity to hold the gas inside the nuclei, and that suggests supermassive black holes. Encounters with other galaxies could throw matter into a black hole; and, as we have seen in Chapter 14, large amounts of energy can be liberated by matter flowing into a black hole. We will expand this theory later, but first we must search for more evidence of supermassive black holes by turning our attention to galaxies that emit powerful radio signals.

DOUBLE-LOBED RADIO SOURCES

Beginning in the 1950s, radio astronomers found that some sources of radio energy in the sky consisted of pairs of radio-bright regions. When optical telescopes studied the locations of these **double-lobed radio sources,** the photographs revealed galaxies located between the two radio lobes. Apparently, the central galaxies were producing the radio lobes.

Study ▌ "Cosmic Jets and Radio Lobes" on pages 362 and 363 and notice four important points. First, the geometry suggests that radio lobes are inflated by jets of excited gas emerging from the central galaxy. This has been called the **double-exhaust model.** Second, notice how observations such as the presence of synchrotron radiation and hot spots give us evidence in support of the model. The third thing to notice is how the complex shapes of some jets and radio lobes can be explained by the motions of the active galactic nuclei. A good example of this is 3C449 (the 449th source in the *Third Cambridge Catalog of Radio Sources*) with its twisting radio lobes. Finally, notice how these jets seem to be related to matter falling into a central black hole. We have seen similar jets produced by rotating disks in the case of protostars and neutron stars; although the details are not understood, it is presumably the same process that produces all of these jets.

Study the image of NGC5128 on page 362 closely for a further clue. In visual-wavelength images, the galaxy looks like a round elliptical galaxy surrounded by a belt of dust and gas. Doppler shifts reveal that the spherical cloud of stars is rotating about an axis that lies in the plane of the dust disk. That is, the galaxy appears to have two different axes of rotation. Furthermore, bright blue stars have been identified in the dust ring, and those stars must have formed recently. All of

Scientific Arguments

An argument can be a shouting match, but another definition of the word is "a discourse intended to persuade." Scientists construct arguments as part of the business of science not because they want to persuade others they are right, but because they want to test their ideas. For example, we can construct a scientific argument to show that the radio lobes that flank some active galaxies have their origin in jets coming from the cores of the galaxies. If we do our best and our argument is not persuasive, we begin to suspect we are on the wrong track. But if our argument seems convincing, we gain confidence that we are beginning to understand radio lobes.

A scientific argument is a logical presentation of evidence and theory with interpretations and explanations that help us understand some aspect of nature. Geologists might construct an argument to explain volcanism on mid-ocean islands, and the argument could include almost anything—maps, for example, or mineral samples compared with seismic evidence and mathematical models of subsurface magma motions. Such an argument might involve the physics of radioactive decay or observations of the shapes of bird beaks on the islands. The scientists would be free to include any evidence or theory that helps persuade, but they must observe one fundamental rule of scientific argument: They must be totally honest. The purpose of a scientific argument is to test our understanding, not to win votes or sell soap. Dishonesty in a scientific argument is self-deluding, and scientists consider such dishonesty the worst possible behavior.

On the other hand, scientists are human, and they often behave no better than other humans. They sometimes defend their own theories and attack opposing theories more vigorously than the evidence warrants. They may unintentionally overlook contradictory evidence or unconsciously misinterpret data. Of course, if a scientist did these things intentionally, we would be shocked because he or she would be violating the most serious rules of honesty in science. In the struggle to understand how nature works, however, it is not surprising to see that human passions rise and that scientific debate becomes heated. That is why science is so exciting—it is a struggle to understand how nature works. Nonetheless, no matter how scientists behave in the excitement of a controversy, they all aspire to that ideal of the scientific argument—the cool, logical presentation of evidence that leads to a new understanding of nature.

Much of this book or any other scientific book consists of scientific arguments—logical presentations of evidence. As you read about any science, look for the arguments and practice organizing and presenting them to test your own understanding. The "Critical Inquiry" that ends each chapter section in this book is intended to illustrate how arguments are constructed. Use the same plan when you develop your own arguments.

---

this is evidence of a recent collision, and we will see later that collisions between galaxies can trigger eruptions in their cores.

Drawing on all of the evidence available and guided by their understanding of the behavior of light and matter, astronomers have constructed a convincing scientific argument (Window on Science 17-2) that active galactic nuclei and radio lobes are powered by violent events in the core. Now we must consider the evidence that the central body is really a black hole.

TESTING THE BLACK HOLE HYPOTHESIS

So far we have skirted the "really difficult central problem." Are the central energy sources in active galactic nuclei really supermassive black holes? The evidence takes two forms—observations of size and observations of motion.

Earlier in this chapter we saw evidence that the cores of some active galaxies are very small. The cores of Seyfert galaxies fluctuate in brightness over very short time periods, and that must mean they are very small, so small that astronomers know of nothing except a supermassive black hole that could produce so much energy in such a small region.

The motion of stars and gas near the centers of galaxies is important because it can tell us the amount of mass located there. Consider the giant elliptical galaxy M87 (page 338). It has a very small, bright nucleus and a visible jet of matter 1800 pc long racing out of its core (∎ Figure 17-3). Radio observations show that the nucleus must be no more than a light-week in diameter. This evidence is suggestive, but observations of motions around the center are decisive. A high-resolution image shows that the core lies at the center of a spinning disk and the jet lies along the axis of the disk. Only 60 ly from the center, the gas in the disk is orbiting at 750 km/s. Substituting radius and velocity into the equation for circular velocity (Chapter 5) tells us that the central mass must be roughly 2.4 billion solar masses. Only a supermassive black hole could pack so much mass into such a small region.

The evidence of rapidly rotating disks at the centers of galaxies is dramatic. Although we think of elliptical galaxies as being nearly free of dust and gas, some have visible dust and gas around their centers. Presumably this dust and gas are left over from recent mergers with smaller galaxies. By placing the slit of the spectrograph across the core of such galaxies, astronomers can sometimes detect Doppler shifts produced by rapidly rotating disks of gas; one side of the disk comes toward us and produces a blue shift, and one side goes away from us and produces a red shift (∎ Figure 17-4).

One such galaxy, NGC7052, contains a dust disk about 570 pc in radius. Spectra show that a smaller disk only 57 pc from the center rotates with an orbital

Cosmic Jets and Radio Lobes

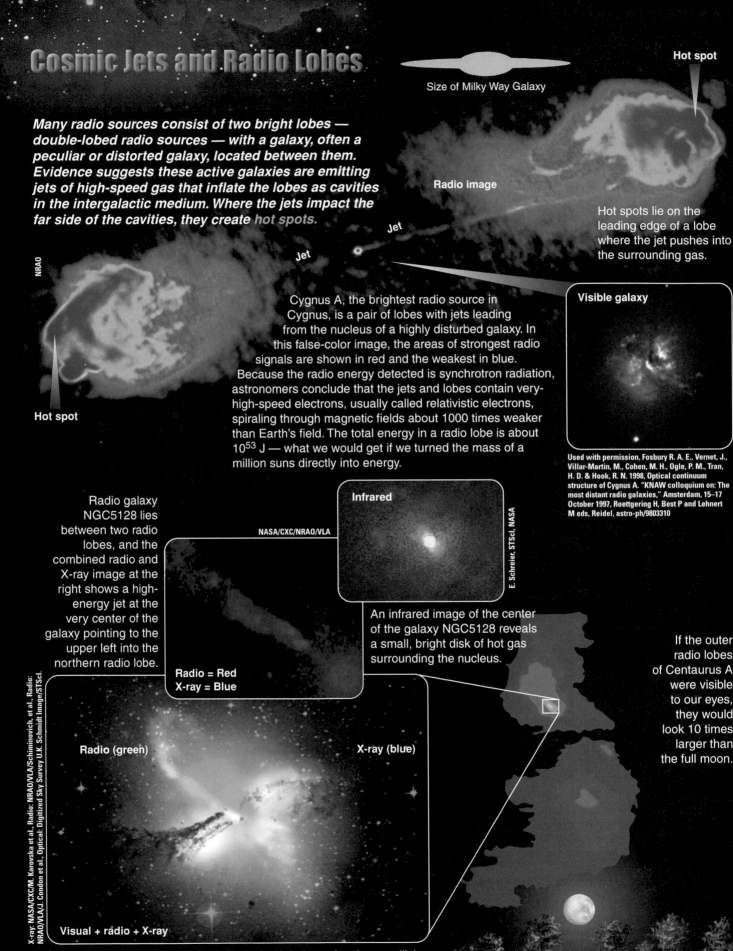

Size of Milky Way Galaxy

Hot spot

Radio image

Many radio sources consist of two bright lobes — double-lobed radio sources — with a galaxy, often a peculiar or distorted galaxy, located between them. Evidence suggests these active galaxies are emitting jets of high-speed gas that inflate the lobes as cavities in the intergalactic medium. Where the jets impact the far side of the cavities, they create hot spots.

Jet

Jet

Hot spots lie on the leading edge of a lobe where the jet pushes into the surrounding gas.

NRAO

Hot spot

Visible galaxy

Cygnus A, the brightest radio source in Cygnus, is a pair of lobes with jets leading from the nucleus of a highly disturbed galaxy. In this false-color image, the areas of strongest radio signals are shown in red and the weakest in blue. Because the radio energy detected is synchrotron radiation, astronomers conclude that the jets and lobes contain very-high-speed electrons, usually called relativistic electrons, spiraling through magnetic fields about 1000 times weaker than Earth's field. The total energy in a radio lobe is about 10^{53} J — what we would get if we turned the mass of a million suns directly into energy.

Used with permission, Fosbury R. A. E., Vernet, J., Villar-Martin, M., Cohen, M. H., Ogle, P. M., Tran, H. D. & Hook, R. N. 1998, Optical continuum structure of Cygnus A. "KNAW colloquium on: The most distant radio galaxies," Amsterdam, 15–17 October 1997, Roettgering H, Best P and Lehnert M eds, Reidel, astro-ph/9803310

Radio galaxy NGC5128 lies between two radio lobes, and the combined radio and X-ray image at the right shows a high-energy jet at the very center of the galaxy pointing to the upper left into the northern radio lobe.

Infrared

NASA/CXC/NRAO/VLA

E. Schreier, STScI, NASA

An infrared image of the center of the galaxy NGC5128 reveals a small, bright disk of hot gas surrounding the nucleus.

Radio = Red
X-ray = Blue

If the outer radio lobes of Centaurus A were visible to our eyes, they would look 10 times larger than the full moon.

X-ray: NASA/CXC/M. Karovska et al., Radio: NRAO/VLA/Schiminovich, et al., Radio: NRAO/VLA/J. Condon et al., Optical: Digitized Sky Survey U.K. Schmidt Image/STScI.

Radio (green)

X-ray (blue)

Visual + radio + X-ray

NGC5128 appears to be a giant elliptical galaxy experiencing a collision with a dusty spiral galaxy. It is known to radio astronomers as Centaurus A, an inner pair of radio lobes and a larger outer pair.

Richard A. Perley

3C449

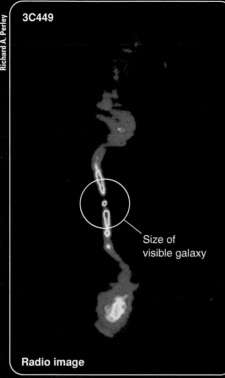

Size of visible galaxy

Radio image

The source of jets in the giant elliptical galaxy 3C449 appears to be in orbital motion about the nucleus and may be the result of a merger.

Radio galaxy 3C129 appears to be moving through the intergalactic medium as its source of jets precesses to produce kinks.

3C129

Radio image

Laurence Rudnick, University of Minnesota, and Jack Burns, University of New Mexico

Ace 🌀 Astronomy™

Go to AceAstronomy and click Active Figures to see "Jet Deflection" and take control of your own model of a moving radio galaxy.

The radio source 3C75 is produced by the interaction of two galaxies with active nuclei. The visible galaxies in these images would be many times the size of their active cores.

Direction of motion

Computer model

Vincent Icke, University of Minnesota and *The Astrophysical Journal*, published by the University of Chicago Press, © 1981 American Astronomical Society

Radio image

3C75

NRAO/AUI, F. N. Owen, C. P. O'Dea, and M. Inoue

High-energy jets appear be caused by matter flowing into a supermassive black hole in the core of an active galaxy. Conservation of angular momentum forces the matter to form a whirling accretion disk around the black hole. How that produces a jet is not entirely understood, but it appears to involve magnetic fields that are drawn into the accretion disk and tightly wrapped to eject high-temperature gas. The twisted magnetic field confines the jets in a narrow beam and causes synchrotron radiation.

Jets from active galaxies may have velocities from thousands of kilometers per second up to a large fraction of the speed of light. Compare this with the jets in bipolar flows, where the velocities are only a few hundred kilometers per second. Active-galaxy jets can be millions of light-years long. Bipolar-flow jets are typically a few light-years long. The energy is different, but the geometry is the same.

Black hole

Magnetic field lines

Accretion disk

Excited matter traveling at very high speeds tends to emit photons in the direction of travel. Consequently, a jet pointed roughly toward us will look brighter than a jet pointed more or less away. This may explain why some radio galaxies have one jet brighter than the other, as in Cygnus A shown at the top of the opposite page. It may also explain why some radio galaxies appear to have only one jet. The other jet may point generally away from us and be too faint to detect, as in the case of NGC5128, also shown on the opposite page.

Adapted from a diagram by Ann Field, NASA, STScI.

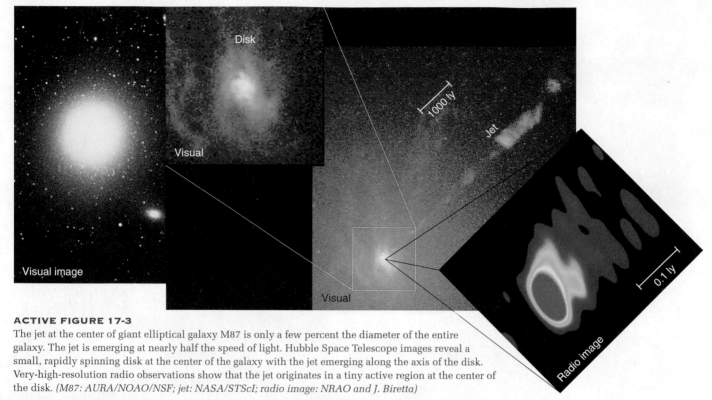

ACTIVE FIGURE 17-3

The jet at the center of giant elliptical galaxy M87 is only a few percent the diameter of the entire galaxy. The jet is emerging at nearly half the speed of light. Hubble Space Telescope images reveal a small, rapidly spinning disk at the center of the galaxy with the jet emerging along the axis of the disk. Very-high-resolution radio observations show that the jet originates in a tiny active region at the center of the disk. *(M87: AURA/NOAO/NSF; jet: NASA/STScI; radio image: NRAO and J. Biretta)*

Ace✵Astronomy™ Go to AceAstronomy and click Active Figures to see "M31 at Many Wavelengths." Different orbiting telescopes give us dramatically different views of a nearby galaxy.

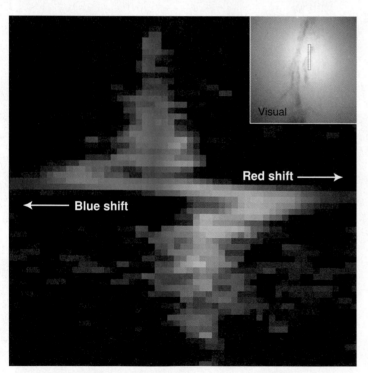

FIGURE 17-4

The elliptical galaxy M84 contains dust rings and an active core. Placing the slit of a spectrograph across the nucleus, astronomers found that spectral lines have a blue shift on one side of the nucleus and a red shift on the other side—evidence of rotation. Colors have been added to this figure to illustrate the blue shift and red shift. The actual width of the line is only about 2 nm. *(Gary Bower, Richard Green, NOAO; The STIS Instrument Definition Team; and NASA)*

velocity of 155 km/s (▌Figure 17-5). The equation for circular velocity (Chapter 5) tells us the disk must enclose a mass of about 300 million solar masses. This galaxy ejects high-speed jets of ionized gas, but the jets are not perpendicular to the dust disk. Apparently, the central black hole and its accretion disk are tipped at an angle to the dust disk, which was evidently produced by a recent merger.

While the observational evidence for supermassive black holes continues to grow, two problems remain. Astronomers are not sure how these huge black holes formed. They appear to have formed when the galaxies first took shape, but the process is unknown. Also, astronomers do not know how a spinning accretion disk produces jets of mass and electromagnetic radiation. The mechanism is believed to involve magnetic fields, but the details are not understood. Modern astronomers now work to model the complex processes that occur around these supermassive black holes.

THE SEARCH FOR A UNIFIED MODEL

When a field of research is young, scientists often find many seemingly different phenomena such as double-lobed radio galaxies, Seyfert galaxies, cosmic jets, and so on. As the research matures, the scientists begin seeing similarities and eventually are able to unify the different phenomena as different aspects of a single process. This is the real goal of science, to organize evi-

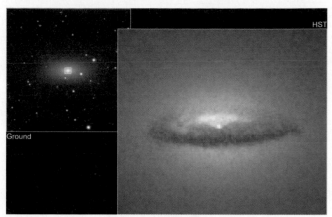

Visual-wavelength images

FIGURE 17-5

Near the center of active elliptical galaxy NGC7052, a dust disk circles a massive nucleus. Deep in a "haze" of billions of stars, the near side of the dust disk obscures more stars than the far side and looks darker. Scattering of shorter-wavelength photons in the dust gives it a red tint. Velocities near the center suggest a central black hole containing 300 million solar masses. *(Roeland P. van der Marel, STScI, Frank C. van den Bosch, University of Washington, and NASA)*

dence and theory in logical arguments that explain how nature works (Window on Science 17-2). Astronomers studying active galaxies are now struggling to find a **unified model** of active galaxy cores.

It seems clear that the cores of active galaxies contain supermassive black holes surrounded by accretion disks that are extremely hot near the black hole but cooler further out. Theory predicts that the central cavity in the disk where the black hole lurks is very narrow but that the disk there is "puffed up" and thick. This means the black hole may be hidden deep inside this narrow central well. The hot inner disk seems to be the source of the jets often seen coming out of active galaxy cores, but the process by which jets are generated is not understood. The outer part of the disk, according to calculations, is a fat, dense doughnut (a torus) of dusty gas (Figure 17-6).

What we see when we view the core of an active galaxy must depend on how this disk is tipped with respect to our line of sight. We should note, at this point, that the accretion disk may be tipped at a steep angle to the plane of a galaxy, so just because we see a galaxy face-on doesn't mean we are looking at the accretion disk face-on. Thus the important factor is not the inclination of the galaxy but the inclination of the accretion disk.

For example, in 1968 astronomers realized that an object they thought was an irregular variable star in the constellation Lacerta was actually the core of an active galaxy. The visible spectrum is featureless, but the spectrum of surrounding nebulosity is that of a giant elliptical galaxy. The object, and others like it known as **BL Lac objects** or **blazars,** are 10,000 times more luminous than the Milky Way Galaxy and fluctuate in only hours. They are believed to be the cores of active galaxies in which we happen to be looking directly into

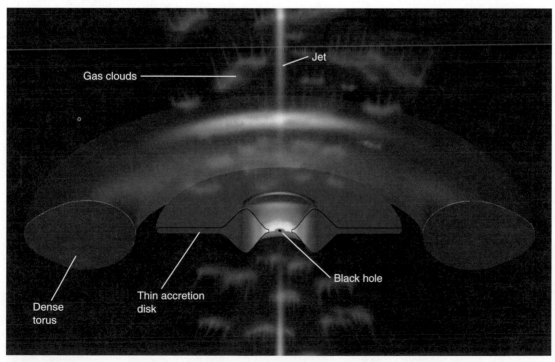

FIGURE 17-6

An accretion disk around a massive black hole. Matter flowing inward passes first through a large, opaque torus; then into a thinner, hotter disk; and finally into a hot, narrow cavity around the black hole. Gas clouds above and below the disk are excited by radiation from the hot disk. This diagram is not to scale. The central cavity may be only 0.01 pc in radius, while the outer torus may be 1000 pc in radius.

the brilliant jets. In this case, the accretion disk is apparently face-on, and we are looking directly into the central cavity around the black hole—down the dragon's throat.

If the accretion disk is tipped slightly, we may not be looking directly into the jet, but we may be able to see some of the intensely hot gas in the central cavity. This region emits broad spectral lines because the gas is orbiting at high velocities and the high Doppler shifts smear out the lines. Seyfert 1 galaxies may be explained by this phenomenon.

If we view the accretion disk from the edge, we cannot see the central area at all because the thick dusty torus blocks our view. Instead we see radiation emitted by gas lying above and below the central disk. Because this gas is farther from the center, it orbits more slowly and has smaller Doppler shifts. Thus we see narrower spectral lines. This might account for the Seyfert 2 galaxies.

This unified model is far from complete. The actual structure of accretion disks is poorly understood, as is the process by which the disks produce jets. Furthermore, the Seyfert galaxies are different from the giant elliptical galaxies that have double radio lobes. Also, our discussion of unification has ignored important factors such as differences in the rate at which mass flows inward and the influence of magnetic fields. A unified model does not explain all of the differences among active galaxies. Rather, it gives us some clues to what is happening in the cores of active galaxies.

BLACK HOLES AND GALAXY FORMATION

For decades, astronomers have confronted the central knotty problem of active galaxies: What makes a galaxy erupt? The solution to that problem seems at hand, and astronomers are going beyond that problem to explore the formation of galaxies.

The cores of active galaxies erupt when matter flows into their central supermassive black holes. A growing mass of evidence shows that active galaxies contain supermassive black holes. NGC4261, for example, is an active galaxy with a rapidly rotating core containing a 1.2 billion-solar-mass black hole (▌ Figure 17-7).

Furthermore, studies show that nearly all galaxies contain supermassive black holes, but most are dormant (▌ Figure 17-8). Evidently the cores of galaxies become active when interactions with other galaxies throw matter into their central black holes, and mathematical models of interacting galaxies show that such encounters are capable of throwing matter into the core of a galaxy (▌ Figure 17-9). The importance of collisions is shown by the prevalence of companions to Seyfert galaxies and the distorted shapes of active galaxies.

Astronomers have been able to measure the masses of a few dozen supermassive black holes, and, as discussed in the previous chapter, they have discovered that the mass of the black hole is correlated with the mass of the nuclear bulge. Galaxies with massive nuclear bulges have massive black holes at their centers. This includes both elliptical galaxies and spiral galaxies with nuclear bulges. But there is no relationship between the masses of the central black holes and the disks of galaxies. This must mean that the supermassive black holes were built by the same process that formed the bulges. Galaxy formation may have begun with matter falling together to form bulges in which a small fraction of the mass fell all the way to the centers of the bulges to form supermassive black holes. Such a process would release a tremendous amount of energy. In contrast, the disks of galaxies are very delicate and are not correlated with the mass of the supermassive black holes. Disks may have formed later as additional material fell into orbit around the central bulges.

Perhaps the history of most galaxies includes a violent youth when matter fell together rapidly and cre-

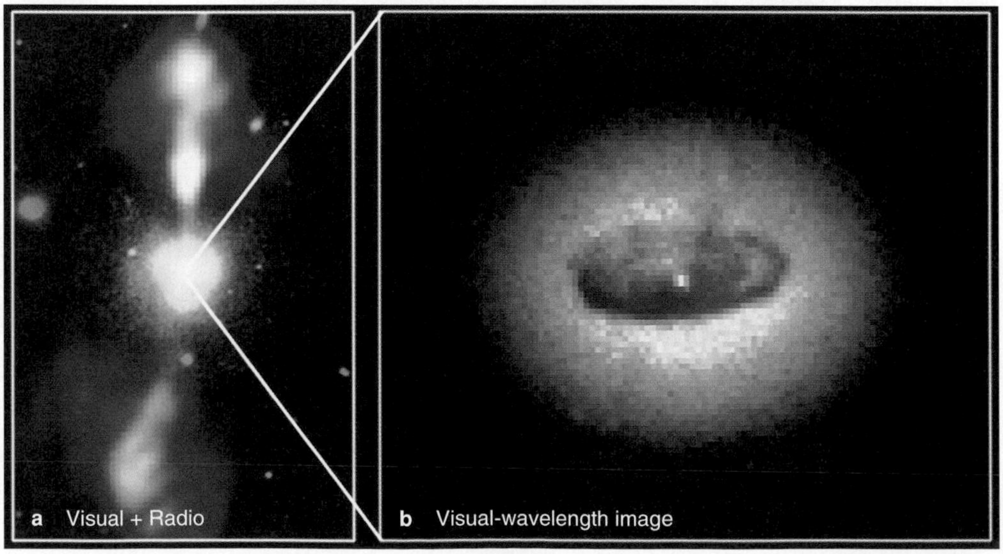

FIGURE 17-7
(a) Galaxy NGC4261 looks like a distorted fuzzy blob in an Earth-based photograph at visible wavelengths (white), but radio telescopes reveal jets leading into radio lobes (orange). (b) A Hubble Space Telescope image of the core of the galaxy reveals a bright central spot surrounded by a disk with its axis pointing at the radio lobes. *(L. Ferrarese, Johns Hopkins University, and NASA)*

a Visual + Radio

b Visual-wavelength image

Visual image

X-ray image

FIGURE 17-8

The Andromeda Galaxy, the nearest large neighbor to our own galaxy, is a normal spiral galaxy and is not active. An X-ray image of the core of the galaxy reveals a number of X-ray binaries with accretion disks as hot as 20 million Kelvin. The blue source is a central black hole containing about 30 million solar masses. The blue color signifies a lower temperature of only about 1 million Kelvin. Evidently, matter is not flowing rapidly into the black hole, and it is dormant. *(AURA/ NOAO/NSFa; NASA/CXC/SAO)*

ated short-lived outbursts and giant black holes. If so, the supermassive black holes we see now are scars left from that ancient age of violence.

REVIEW CRITICAL INQUIRY

Construct an argument to show that double-lobed radio galaxies and Seyfert galaxies have similar energy sources. Observations show that double-lobed radio sources are produced by two jets flowing out of a galaxy's core and inflating the radio lobes. The evidence of small size and very-high-speed motions in the core show that these galaxies must contain large amounts of mass in small regions. Seyfert galaxies also have small nuclei, as shown by their rapid fluctuations, and the high velocities in their cores imply that the cores must contain very large amounts of mass. Consequently both kinds of objects are suspected to contain supermassive black holes at their centers.

The preceding paragraph is a quick summary of a logical argument citing evidence and theory to reach a conclusion. Construct an argument to propose that galaxy interactions are necessary to trigger active galactic nuclei.

Observational evidence and basic theory have led us to the unified model of active galactic nuclei. Although the model seems successful, there are still difficulties, and astronomers continue to struggle to understand the details. Nevertheless, the unified model may help us understand the mysterious quasars.

17-2 QUASARS

The largest telescopes detect multitudes of faint, starlike points of light with peculiar emission spectra, objects called **quasars** (also known as *quasi-stellar objects,* or *QSOs*). Although astronomers now recognize quasars as some of the most distant visible objects in the universe, they were a mystery when they were first identified. The discovery of these objects in the 1960s and the struggle to understand them comprise a classic example of how scientists explore hypotheses, gather evidence, and build confidence in new ways to understand nature. We begin this section by telling that story, a story that will lead us to think about the depths of the universe in both space and time.

THE DISCOVERY OF QUASARS

In the early 1960s, radio interferometers (see Chapter 6) showed that a number of radio sources are much smaller than normal radio galaxies. Photographs of the location of these radio sources did not reveal a galaxy, not even a faint wisp, but rather a single starlike point of light. The first of these objects so identified was 3C48, and later the source 3C273 was added. Though these objects emitted radio signals like those from radio galaxies, they were obviously not normal radio galaxies. Even the most distant photographable galaxies look fuzzy, but these objects looked like stars. Their spectra, however, were totally unlike stellar spectra, so the objects were called quasi-stellar objects (▮ Figure 17-10).

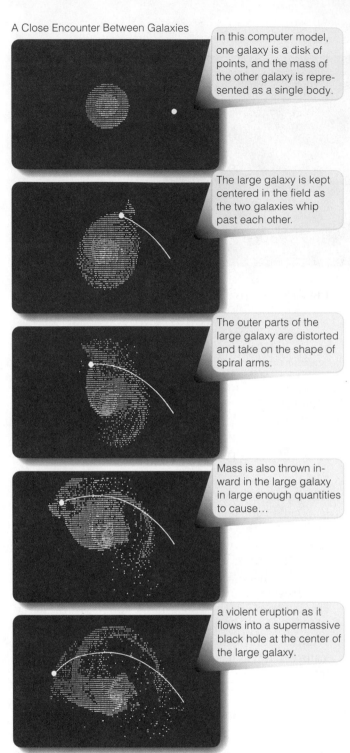

In this computer model, one galaxy is a disk of points, and the mass of the other galaxy is represented as a single body.

The large galaxy is kept centered in the field as the two galaxies whip past each other.

The outer parts of the large galaxy are distorted and take on the shape of spiral arms.

Mass is also thrown inward in the large galaxy in large enough quantities to cause…

a violent eruption as it flows into a supermassive black hole at the center of the large galaxy.

FIGURE 17-9

A computer model of the gravitational interaction between two passing galaxies shows that matter is not only thrown outward but is also thrown inward. *(G. Byrd, M. Valtonen, B. Sundelius, and L. Valtaoja)*

For a few years, the spectra of quasars were a mystery. A few unidentifiable emission lines were superimposed on a continuous spectrum. But, in 1963, Maarten Schmidt at Hale Observatories tried redshifting the hydrogen Balmer lines to see if they could

be made to agree with the lines in 3C273's spectrum. At a red shift of 15.8 percent, three lines clicked into place (❙ Figure 17-11). Other quasar spectra quickly yielded to this approach, revealing even larger red shifts.

The red shift z is the change in wavelength $\Delta\lambda$ divided by the unshifted wavelength λ_0:

$$\text{red shift} = z = \frac{\Delta\lambda}{\lambda_0}$$

Astronomers commonly use the Doppler formula to convert these red shifts into radial velocities. Thus, the large red shifts of the quasars imply large apparent velocities of recession.

Quasar spectra contain bright emission lines. The brightest are the Balmer lines of hydrogen, but other spectral lines also appear. The red shifts of quasars can be so large that spectral lines can be shifted completely out of the visible spectrum, and lines in the ultraviolet can be shifted into the visible spectrum. ❙ Figure 17-12 shows the spectra of four quasars with relatively low red shifts in which the H_α Balmer line normally seen in the red part of the spectrum is shifted into the infrared. Quasars have been found with even larger red shifts.

To understand the significance of these large red shifts and the large velocities of recession they imply, we must draw on the Hubble law. Recall that the Hubble law states that galaxies have apparent velocities of recession proportional to their distances, and thus the distance to a quasar is equal to its apparent velocity of recession divided by the Hubble constant. The large red shifts of the quasars imply that they must be at great distances.

The red shift of 3C273 is 0.158, and the red shift of 3C48 is 0.37. These are large red shifts, but not as large as the largest then known for galaxies, about 1.0. Soon, however, quasars were found with red shifts much larger than that of any known galaxy. Some quasars are evidently so far away that galaxies at those distances are very difficult to detect. Yet the quasars are easily photographed. This leads us to the conclusion that quasars must have 10 to 1000 times the luminosity of a large galaxy. Quasars must be superluminous.

Soon after quasars were discovered, astronomers detected fluctuations in brightness over times as short as a few months. Recall from our discussion of pulsars (Chapter 14) that an object cannot change its brightness appreciably in less time than it takes light to cross its diameter. The rapid fluctuations in quasars showed that they are small objects, not more than a few light-months in diameter.

Thus, by the late 1960s astronomers faced a problem: How could quasars be superluminous but also be very small? What could make 10 to 1000 times more energy than a galaxy in a region less than a light-year in diameter? Since that time, evidence has accumu-

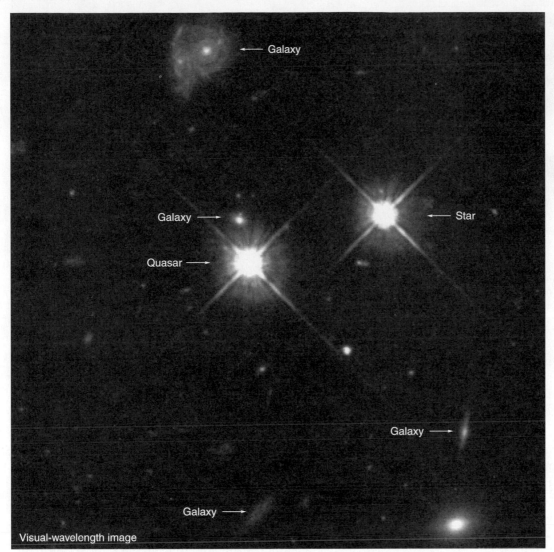

FIGURE 17-10
Quasars have starlike images clearly different from the images of distant galaxies. The spectra of quasars are unlike the spectra of stars or galaxies. (*C. Steidel, Caltech, NASA*)

Galaxy

Galaxy

Quasar

Star

Galaxy

Galaxy

Visual-wavelength image

lated that quasars are the active cores of very distant galaxies, and the rest of this chapter will discuss that evidence. The key, as in most astronomical arguments, is distance. The distance to quasars is so important to our discussion and so much larger than the distances we have considered before that we need to pay special attention to how astronomers estimate the distances to these objects.

QUASAR DISTANCES

Astronomers estimate the distance to a quasar by using the Hubble law and dividing the quasar's velocity of recession by the Hubble constant (Chapter 16). Thus, knowing the distance to a quasar depends critically on knowing its red shift and from that its velocity. Many quasars have red shifts greater than 1, and if we put such large red shifts into the Doppler formula we discussed in Chapter 7, we get radial velocities greater than the velocity of light, which is unreasonable. Quasar red shifts greater than 1 do make

sense, however, if we consider the nature of very large red shifts.

First we must note that the Doppler formula given in Chapter 7 is only an approximation good for low velocities. It is correctly called the classical Doppler formula, and it works fine if the velocity is significantly less than the velocity of light. For very high velocities, however, we need to use the **relativistic Doppler formula** derived from Einstein's theory of relativity. It relates the radial velocity V_r to the speed of light c and the red shift z:

$$\frac{V_r}{c} = \frac{(z+1)^2 - 1}{(z+1)^2 + 1}$$

The relativistic Doppler formula cannot produce a radial velocity greater than the speed of light, no matter how large z becomes. For example, suppose a quasar had a red shift of 2. The classical Doppler formula would give the unreasonable result that the radial velocity was twice the speed of light, but when we put 2 into the relativistic Doppler formula we discover that

Visual-wavelength image

a

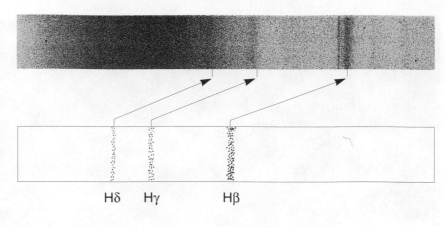

Hδ   Hγ   Hβ

b

FIGURE 17-11

(a) This image of 3C273 shows the bright quasar at the center surrounded by faint fuzz. Note the jet protruding to lower right. (b) The spectrum of 3C273 (top) contains three hydrogen Balmer lines red-shifted by 15.8 percent. The drawing shows the unshifted positions of the lines. *(Courtesy Maarten Schmidt)*

$z + 1$ is 3, and therefore $(z + 1)^2$ is 9. Then the radial velocity divided by the speed of light is:

$$\frac{V_r}{c} = \frac{(9) - 1}{(9) + 1} = \frac{8}{10} = 0.8$$

Thus, the quasar has an apparent velocity of recession of 80 percent the speed of light.

❚ Figure 17-13 compares the classical Doppler formula and the relativistic Doppler formula. At low velocities they give the same result, but at higher velocities the classical formula gives velocities that are too high. The important point for us is that no matter how large the red shift becomes, the velocity never quite equals the speed of light.

Now that we feel some comfort with quasar red shifts greater than 1, we should note that the red shifts associated with the expansion of the universe are not really Doppler shifts. The exact relation between the red shifts of distant galaxies and their expansion velocities is similar to the relativistic Doppler formula in that red shifts greater than 1 imply velocities near but not exceeding the velocity of light. But the exact relation depends critically on parameters such as the average density of the universe. Consequently, we can't use a Doppler formula to convert quasar red shifts into velocities. But we can easily conclude that the large red shifts imply very large distances, and that red shifts greater than 1 are not unreasonable. (We will discuss these red shifts produced by the expansion of the universe in the next chapter.)

If quasars are as far away as their large red shifts indicate, their look-back times are very large, and we see them as they were when the universe was significantly younger. In the next chapter, we will see that the universe is believed to be about 14 billion years old.

Thus, we see the most distant quasars as they were when the universe was only a few billion years old.

What evidence do we have that quasars really are located at these great distances? A discovery made in 1979 provides a dramatic illustration of the tremendous distances to quasars. The object cataloged as 0957+561 lies just a few degrees west of the bowl of the Big Dipper and consists of two quasars separated by only 6 seconds of arc. The spectra of these quasars are identical and even have the same red shift of 1.4136. Quasar spectra do resemble each other, but in detail they are as different as fingerprints. When two quasars so close together were discovered to have the same red shift and identical spectra, astronomers concluded that they were separate images of the same quasar.

The two images are formed by a **gravitational lens,** an effect first predicted by Einstein in 1936. The gravitational field of a galaxy between us and a distant quasar acts as a lens, bending the light from the quasar and focusing it into multiple images. A few dozen gravitational lenses are known (❚ Figure 17-14); and, in most cases, the lensing galaxy is so far away it is difficult to detect. This provides powerful evidence that quasars really are distant objects as their large red shifts imply.

For more evidence that quasars are very distant, look again at Figure 17-10. Just above the image of the quasar lies the small, oval image of an elliptical galaxy, which, evidenced by its red shift, is about 7 billion light-years away. The spectrum of the quasar contains absorption lines with the same red shift as the elliptical galaxy. Evidently the lines are produced by the thin gas in the elliptical galaxy. From this we conclude that the quasar, though very bright, is actually farther away than the distant elliptical galaxy.

About 10 percent of quasars have absorption lines as well as emission lines in their spectra. These absorp-

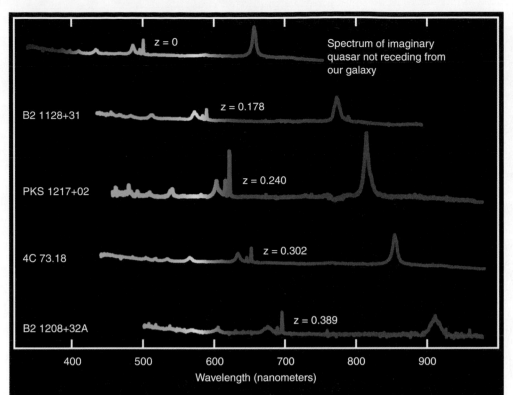

FIGURE 17-12

Spectra of four quasars are compared here with an idealized spectrum for an imaginary quasar that has no velocity of recession. The first three Balmer lines of hydrogen are visible, plus lines of other atoms, but the red shifts of the quasars move these lines to longer wavelengths. Nevertheless, the relative spacing of the spectral lines is unchanged, and astronomers can recognize the lines even with a large red shift. *(C. Pilachowski, M. Corbin, AURA/NOAO/NSF)*

tion lines have a smaller red shift than the corresponding emission lines, and, in some cases, multiple sets of absorption lines are observed, each with its individual red shift. Some of the lines are hydrogen lines, but some are lines of metals, heavier elements such as carbon, silicon, and magnesium. Apparently, these absorption lines are formed when the light from the quasar passes through objects on its way to Earth. The hydrogen lines must be formed as the light from the quasar passes through low-density clouds of hydrogen between the galaxies. The metal lines must be formed when the light passes through galaxies on its way to Earth. But the galaxies themselves are so distant that they are not visible, further evidence that quasars are very distant objects.

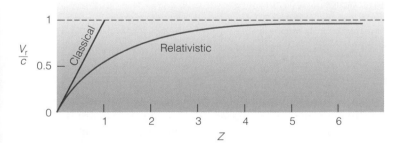

FIGURE 17-13

At high velocities, the relativistic Doppler formula must be used in place of the classical approximation. Note that no matter how large Z gets, the velocity can never equal that of light.

The quasars are very distant and must be superluminous and quite small. To understand quasars, we must examine their observed characteristics and compare that evidence with the unified model of active galaxies.

EVIDENCE OF QUASARS IN DISTANT GALAXIES

How do astronomers know that quasars are located in distant galaxies? Thousands of quasars have been found, and the largest known red shifts are a bit over 6, meaning that the objects are exceedingly far away. The lookback time to these objects is so large that we see them as they were when the universe was only 10 percent its present age. At such great distances it is difficult to detect the host galaxy because of the glare of the quasar.

Quasars with lower red shifts are not quite so far away, however, and fuzzy wisps are visible around such objects. By blocking the glare of the quasar, astronomers were able to photograph the spectrum of this quasar fuzz and found that it is the same as the spectrum of a normal galaxy with the same red shift as the quasar. For example, 3C273 appears to lie in a giant elliptical galaxy.

Moreover, some quasars have faint objects near them in the sky. The spectra of these faint objects are the same as the spectra of galaxies, and they have the same red shift as that of the quasar they accompany. This is evidence that these quasars are located in galaxies that are part of larger clusters of galaxies (❚ Figure 17-15).

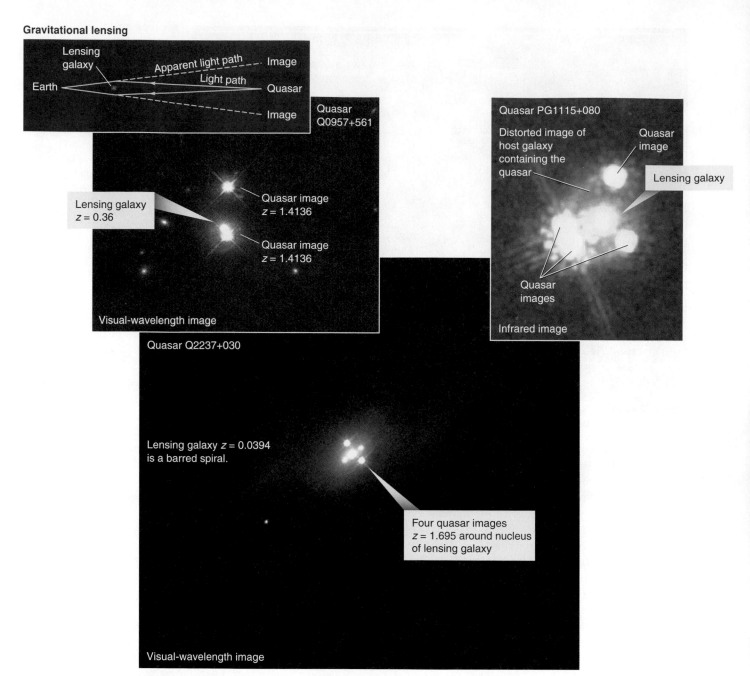

FIGURE 17-14

Gravitational lensing can occur when a relatively nearby lensing galaxy is aligned with a distant quasar. This can produce multiple images of the quasar. In the case of the quasar at upper right, the host galaxy is visible as a faint, distorted ring around the lensing galaxy. Gravitational lenses give us direct evidence that quasars are at the great distances suggested by their red shifts. *(Q0957+561: George Rhee, NASA, STScI; color composite courtesy Bill Keel; PG1115+080: Chris Impy, Univ. of Arizona and NASA; Q2237+030: Image by W. Keel from data in the NASA/ESA Hubble Space Telescope archive, originally obtained with J. Westphal as Principal Investigator)*

Some quasars have characteristics related to active galaxies. For example, quasar 3C273 seems to be ejecting a jet much like the jets seen in active galaxies. By blocking the brilliance of the quasar, astronomers have been able to image other details that suggest active galaxies (▌Figure 17-16). Some quasars even have radio lobes, which further suggests that they are related to active galaxies.

Photographs made with the largest telescopes on Earth and with the Hubble Space Telescope give us dramatic evidence. The galaxies that host quasars are in many cases clearly visible with these large telescopes. Furthermore, the galaxies are distorted or have nearby companions (▌Figure 17-17). The host galaxies appear to be involved in collisions and interactions with other galaxies, or they have the distorted shapes that we

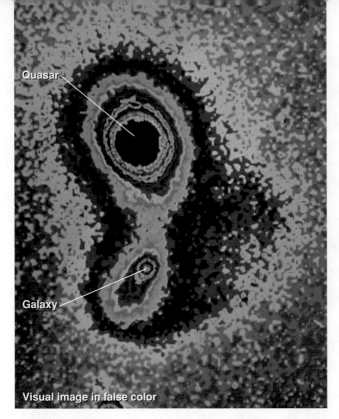

Quasar

Galaxy

Visual image in false color

FIGURE 17-15

Quasar 0351+026 is located near a faint galaxy that has the same red shift and thus, presumably, the same distance from Earth as the quasar. The extended red region around the quasar (fuzz) is typical of lower-red-shift quasars and reveals the spectrum of a normal galaxy. Thus, this image shows two interacting galaxies, one of which contains a quasar. *(AURA/NOAO/NSF)*

recognize as the result of such collisions. Because we suspect that galaxy collisions can cause the core of a galaxy to erupt, we can also suspect that quasars may have been triggered into existence in the cores of galaxies by interactions between galaxies.

SUPERLUMINAL EXPANSION

As astronomers gathered more and more evidence and grew ever more confident that quasars were located in distant galaxies, they hit a snag. A few quasars were found that seem to be expanding faster than the velocity of light, a behavior called **superluminal expansion** that is forbidden by the theory of relativity. Thus we must consider this peculiar phenomenon before we can go further with our discussion of quasars.

Radio astronomers discovered superluminal expansion by using very long baseline interferometry (VLBI)—the interconnection of radio telescopes in different parts of the world. Such a network of antennae can resolve details about 0.002 second of arc in diameter. At the distance of quasar 3C273, for example, this corresponds to a few parsecs.

For a few quasars these radio maps show small blobs near the quasars, and maps made over a few years show these blobs moving away from the quasar. Quasar 3C273 has a blob that is moving about 0.0008 second of arc farther from the quasar each year (∥ Figure 17-18). If we believe that the quasar's red shift of 0.16 arises from the expansion of the universe, the quasar is about 960 Mpc distant, and the small-angle formula tells us that the blob is moving away from the quasar at a speed of 3.8 pc per year. This is the same as 12 ly per year, which is impossible. The theory of relativity clearly shows that nothing can travel faster than the velocity of light.

This superluminal expansion has been found in a number of quasars, and it is also seen in blazars. For quasars the apparent velocities of expansion are usually greater than five times the velocity of light, and for blazars the velocity is usually less than five times the velocity of light. As we have seen earlier in this chapter, we suspect that blazars are active galaxy cores in which one of the jets is pointed almost directly at us. This is an important clue.

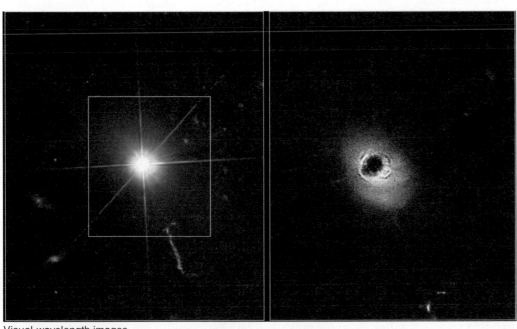

Visual-wavelength images

FIGURE 17-16

Quasar 3C273 is so bright it drowns out details around it (left). By blocking the brilliant light of the quasar (right), the Hubble Space Telescope can detect curving arms of stars in the galaxy that contains the quasar and can detect the inner parts of the jet extending to lower right. *(J. Bahcall, IAS; A. Marte, H. Ford, JHU; M. Clampin, G. Hartig [STScI]; G. Illingworth, UCO/Lick Observatory; ACS Science Team NASA and ESA)*

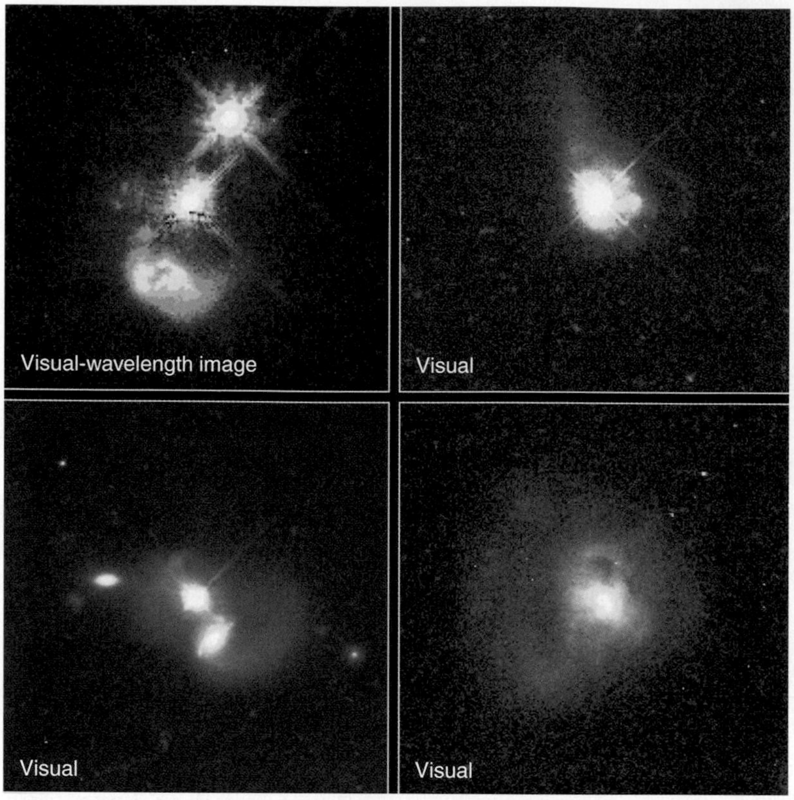
Visual-wavelength image
Visual
Visual
Visual

FIGURE 17-17
The bright object at the center of each of these images is a quasar. Fainter objects near the quasar are galaxies distorted by collisions. Compare the upper left ring-shaped galaxy with Figure 16-15 and compare the tail in the lower left image with the Antennae galaxies in Figure 16-12. *(J. Bahcall, Institute for Advanced Study, Mike Disney, University of Wales, NASA)*

One way to solve the problem is to assume that the quasar is local, by which astronomers mean nearby. If 3C273 were only 10 Mpc away, then the blob is moving at a velocity of less than 15 percent the velocity of light. But, of course, all the other evidence points to quasars being very far away.

The clue is the jets in blazars that appear to be pointed almost straight at us. If the quasar is ejecting a jet of matter at nearly the velocity of light and that jet is pointed nearly toward Earth, a blob of matter in the jet could appear to move away from the quasar faster than the velocity of light. This is called the **relativistic jet model.**

To see how this can happen, consider ▮ Figure 17-19. A blob of matter in the jet emits radiation that will reach Earth in the year 2005. In the next 35 years, that radiation travels 35 ly toward Earth, but the blob, traveling at 98 percent the velocity of light, travels 34 ly closer to Earth. Radiation that it emits from this second position is only 1 ly behind the radiation it emitted 35 years before. The newly emitted radiation will arrive on Earth in 2006, and we will see the blob move in only 1 year a distance that actually took 35 years. This is a simplified explanation that omits certain relativistic effects, but it clearly shows that a blob of gas in a relativistic jet can appear to travel faster than light if the jet points nearly toward Earth.

In fact, this interesting bit of physics has been observed closer to home. The jet in M87 has been observed to have an apparent velocity along a portion of its length of 2.5 times the speed of light. Evidently, the jet twists and turns as it leaves the nucleus of the galaxy, and one of those turns points the jet almost directly at Earth. Although the gas is moving at slightly over 0.9 times the speed of light, along that segment pointing toward us the gas appears to move 2.5 times faster than light. Even closer to home, within our own galaxy, a binary star containing a neutron star or a black hole has been observed ejecting material at speeds of about 1.25 times the speed of light. The gas is actually traveling at only 0.92 times the speed of light, but the material was ejected within 19° of our line of sight, and that produces the apparent superluminal motion.

Once the nature of superluminal expansion was understood, it was no longer an objection to quasars being located in very distant galaxies. Today the evidence seems very strong, and astronomers are able to combine evidence and theory in a logical argument that describes how a quasar works.

A MODEL QUASAR

The most satisfying part of science is combining evidence with theory to understand some aspect of nature. We have learned to interpret quasar red shifts, and we have reviewed the basic characteristics of quasars. Can we now assemble the evidence and theories in a logical argument that describes quasars? That is, can we create a model of a quasar?

Our model quasar is based on the geometry of a rotating accretion disk ejecting jets along its axis of rotation (page 363). Disks are very common in the universe. We have discovered rotating disks around protostars, neutron stars, and black holes. Our own galaxy has a disk shape, and in a later chapter we will see that our solar system has a disk shape. We have also seen that some disks can produce powerful jets of matter and radiation directed along the axis of rotation. Bipolar flows from protostars are low-energy examples, while the jets expelled by the X-ray binary SS433 are much more powerful. Jets expelled by active galactic nuclei are even more powerful. To create our model quasar, we imagine a galaxy containing this disk-jet structure at its core. Our model is, in fact, just an extension of the model of active galactic nuclei discussed earlier in this chapter.

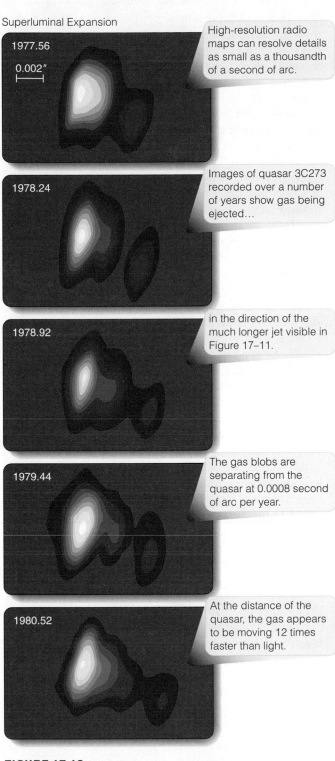

Superluminal Expansion

1977.56

0.002"

High-resolution radio maps can resolve details as small as a thousandth of a second of arc.

1978.24

Images of quasar 3C273 recorded over a number of years show gas being ejected…

1978.92

in the direction of the much longer jet visible in Figure 17–11.

1979.44

The gas blobs are separating from the quasar at 0.0008 second of arc per year.

1980.52

At the distance of the quasar, the gas appears to be moving 12 times faster than light.

FIGURE 17-18
Very-high-resolution radio images of a few quasars show gas that seems to be moving faster than the speed of light.

To make the energy typical of a quasar we need a black hole of 10 million to a billion solar masses surrounded by a large, opaque disk. The innermost part of the accretion disk, much less than a light-year in diameter, would be intensely hot. If matter amounting to

about 1 solar mass per year flows through the accretion disk into the black hole, the energy released could power a quasar. Exactly how some of this matter and energy are focused into two oppositely directed high-speed jets isn't well understood, but magnetic fields are probably involved. In any case, as matter flows into the black hole, some is ejected in jets and a great deal of energy is radiated away. One astronomer described the situation by commenting, "Black holes are messy eaters."

This model could explain the different kinds of radiation we receive from quasars (see Figure 17-6). A small percentage of quasars are strong radio sources, and the radio radiation may come from synchrotron radiation produced in the high-energy gas and magnetic fields in the jets. Much of the light we see from a quasar is spread in a continuous spectrum, and that is the light that fluctuates quickly. Thus, it must come from a small region. Our model could produce this light in the innermost region surrounding the black hole. Because this is a very small region, roughly the size of our solar system, rapid fluctuations can occur, probably because of random fluctuations in the flow of matter into the

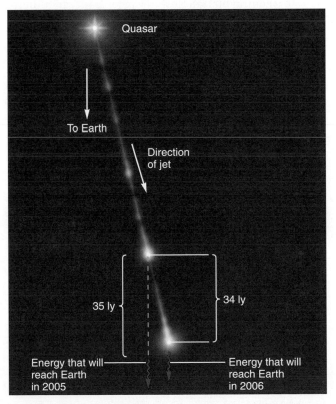

FIGURE 17-19
The relativistic jet model explains superluminal expansion by supposing that the quasar emits a jet that points nearly at Earth. If a blob of gas in the jet, traveling at nearly the velocity of light, emits radio energy toward Earth at two different times separated by 35 years, the first bundle of energy will have only a 1 ly lead. As the energy arrives at Earth, the blob will appear to move in 1 year a distance that actually took 35 years.

accretion disk. The emission lines in quasar spectra don't fluctuate rapidly, and that suggests they are produced in a larger region. In our model, emission lines are emitted by clouds of gas surrounding the core in a region many light-years in diameter. These gas clouds are excited by the intense synchrotron radiation streaming out of the central cavity.

Just as in the unified model of active galactic nuclei, the orientation of the accretion disk is important (see Figure 17-6). If the disk faces us so that one of its jets points directly toward us, we may see one kind of quasar. But if the disk is tipped slightly so the jet is not pointing straight at us, we may see a slightly different kind of quasar. Astronomers are now using this unified model to sort out the different kinds of quasars and active galaxies so we can understand how they are related. For example, using infrared radiation to penetrate dust, astronomers observed the core of the double-lobed radio galaxy Cygnus A (see page 362) and found an object much like a quasar. Astronomers have begun to refer to such objects as buried quasars.

As in the case of active galaxies, many quasars are found in distorted galaxies that are interacting with nearby companions. Such a collision could throw matter into a central black hole and trigger a quasar outburst.

QUASARS THROUGH TIME

When we look at a distant quasar, the look-back time is large, and we see the universe as it was long ago. That means we might be able to tell the story of how quasars formed, evolved, and died out. Perhaps we can trace the history of quasars through time.

In the next chapter, we will see evidence that the universe began about 14 billion years ago. The first clouds of gas began forming stars and falling together to form galaxies, and we suspect that some of that matter formed supermassive black holes at the centers of star clouds that became the nuclear bulges of galaxies. The abundance of matter flooding into these black holes could have triggered outbursts that we see as quasars.

We should also note that galaxies were closer together when the universe was young and had not expanded very much. Because they were closer together, the forming galaxies collided more often, and we have seen how collisions between galaxies could throw matter into supermassive black holes and trigger eruptions. We observe that quasars are often located in host galaxies that are distorted as if they were interacting with other galaxies.

The evidence suggests that when the universe was young, forming galaxies created supermassive black holes and, aided by interactions with other galaxies, large amounts of matter fell into those black holes, triggering quasar eruptions.

Quasars are most common with red shifts of about 2 and less common with red shifts above 2.7. The largest quasar red shifts are over 6, and such high-red-shift quasars are quite rare. Evidently, when we look at quasars with red shifts of 2 or so, we are looking back to an age when galaxies were actively forming, colliding, and merging. In that age of quasars, they were about 1000 times more common than they are now. When we look back to higher red shifts, we see fewer quasars because we are looking back to an age when the universe was so young it had not yet begun to form many galaxies. Nevertheless, even during the age of quasars, quasar eruptions must have been unusual. At any one time, only a few galaxies had quasars erupting in their cores.

Then where are all the dead quasars? An astronomer commented recently, "There is no way to get rid of supermassive black holes, so all of the galaxies that had short-lived quasars still have those black holes." Why don't we see those dead quasars today? Actually, astronomers have found quite a few. The quasar 3C273 is surprisingly near us in space. Evidently, the age of quasars is not quite over.

Many dead quasars are not truly dead—only sleeping. We have discovered that most galaxies contain supermassive black holes, and those black holes may have suffered quasar eruptions when the universe was younger, when galaxies were closer together, and when dust and gas were more plentiful. Quasar eruptions became less common as galaxies became more stable and as the abundance of gas and dust in the centers of galaxies was exhausted. The dormant black holes sleeping at the centers of the galaxies today can be triggered to become active galactic nuclei by collisions. Such AGN are much less energetic than quasars, evidently because there is now much less gas and dust available to feed the supermassive black holes.

Most galaxies contain sleeping supermassive black holes into which matter slowly trickles. Our own Milky Way Galaxy is a good example. It could have been a quasar a long time ago, but today its supermassive black hole is resting.

REVIEW CRITICAL INQUIRY

Why are most quasars so far away?

To analyze this question we must combine two factors. First, quasars are the active cores of galaxies, and only a small percentage of galaxies contain active cores. In order to sample a large number of galaxies in our search, we must extend our search to great distances. Most of the galaxies lie far away from us because the amount of space we search increases rapidly with distance. Just as most seats in a baseball stadium are far from home plate, most galaxies are far from our Milky Way Galaxy. Thus, most of the galaxies that might contain quasars lie at great distances.

But a second factor is much more important. The farther we look into space, the farther back in time we look. It seems that there was a time in the distant past when quasars were more common, and thus we see most of those quasars at large look-back times, meaning at large distances. For these two reasons, most quasars lie at great distances.

If our model of quasars is correct, then quasars must be triggered into eruption. What observational evidence supports that argument?

Our study of active galaxies has led us far out in space and back in time to quasars. The light now arriving from the most distant quasars left them when the universe was only a fraction of its present age. The quasars stand at the very threshold of the study of the history of the universe itself, the subject of the next chapter.

SUMMARY

Some galaxies have peculiar properties. Seyfert galaxies have small, luminous nuclei whose spectra contain emission lines broadened by violently moving gas. Double-lobed radio galaxies emit radio energy from the lobes on either side of the galaxies. Some have jets extending from the core of the galaxy into the radio lobes. Some giant elliptical galaxies have small, energetic cores, some of which have jets of matter rushing outward.

The proposed explanation for double-lobed radio galaxies is called the double-exhaust model. This model assumes that the galaxy ejects jets of relativistic particles in two beams that push into the intergalactic medium and blow up the radio lobes like balloons. The impact of the beams on the intergalactic medium produces the hot spots detected on the leading edges of many radio lobes. We can see instances where galaxies moving through the intergalactic medium leave behind trails of hot gas from their jets. Spiral jets suggest that the source of the jets is precessing.

The source of energy in the active galaxies is believed to be supermassive black holes in their centers. M87, a giant elliptical galaxy with a small, active core and a high-velocity jet, contains very high orbital velocities close to its center. This implies that its small core contains billions of solar masses, probably in the form of a supermassive black hole.

Active galactic nuclei seem to occur in galaxies involved in collisions or mergers. Seyfert galaxies, for example, are three times more common in interacting galaxies than in isolated galaxies. This suggests that an encounter between galaxies can throw matter into the central regions of the galaxies, where it can feed a black hole and release energy. Galaxies not recently involved in collisions will not have matter flowing into their central black hole and will not have active galactic nuclei.

The unified model of active galaxies supposes that what we see depends on the orientation of the black hole and its accretion disk. If we can see into the core, we see broad spectral lines and rapid fluctuations. If we see the disk edge-on, we see narrower spectral lines. If the jet points directly at us, we see a BL Lac object, also known as a blazar.

The quasars appear to be related objects. Their spectra show emission lines with very large red shifts. Their large red shifts imply that they are very distant, and the fact that they are visible at all implies that they are superluminous. Because they fluctuate rapidly, we conclude that they must be very small. A typical quasar can be 100 times more luminous than the entire Milky Way Galaxy but only a few times larger than our solar system.

Because quasars lie at great distances, we see them as they were when the universe was young. The highest-resolution images show that many quasars are the active cores of very distant galaxies. Furthermore, these galaxies are often distorted and have close companions. That suggests that the quasars contain supermassive black holes and are triggered into activity by interactions between galaxies. Quasars may also be linked to the formation of supermassive black holes as material fell together to form the first galaxies. In any case, heavy mass flow into a supermassive black hole could heat the gas and produce the properties of quasars we observe.

Superluminal expansion appears as blobs of material that appear to be rushing away from some quasars at speeds as high as 12 times the speed of light. Of course, it is impossible for the matter to exceed the speed of light, but we can understand it as matter traveling at slightly less than the speed of light in a jet directed almost exactly toward Earth. Then we would see the illusion of superluminal expansion. Thus, superluminal expansion does not contradict our understanding of quasars.

NEW TERMS

radio galaxy	unified model
active galaxy	BL Lac object
active galactic nucleus (AGN)	blazar
	quasar
Seyfert galaxy	relativistic Doppler formula
double-lobed radio source	gravitational lens
double-exhaust model	superluminal expansion
hot spot	relativistic jet model

REVIEW QUESTIONS

Ace✪Astronomy™ Assess your understanding of this chapter's topics with additional quizzing and animations at **http://astronomy.brookscole.com/seeds8e**

1. What is the difference between a type 1 Seyfert galaxy and a type 2 Seyfert galaxy? How might we explain the differences?

2. What evidence do we have that radio lobes are inflated by jets from active galactic nuclei?

3. What is the significance of spiral jets from galaxies?

4. What evidence do we have that AGNs contain black holes?

5. What properties of SS433 resemble the energy source in an AGN?

6. If you located a galaxy that had not interacted with another galaxy recently, would you expect it to be active or inactive? Why?

7. Seyfert galaxies are three times more common in close pairs of galaxies than in isolated galaxies. What does that suggest?

8. What evidence do we have that quasars are small?

9. What evidence do we have that quasars are very far away?

10. How do quasars resemble the AGN in Seyfert galaxies?

11. How does our model quasar account for the different components of a quasar's spectrum?

12. What does it mean that quasars are most common at a red shift of about 2?

13. Where are all the dead quasars?

5. If the active core of a galaxy contains a black hole of 10^6 solar masses, what will the orbital period be for matter orbiting the black hole at a distance of 0.33 AU? (*Hint:* See circular velocity, Chapter 5.)

6. If a quasar is 1000 times more luminous than an entire galaxy, what is the absolute magnitude of such a quasar? (*Hint:* The absolute magnitude of a bright galaxy is about −21.)

7. If a quasar in Problem 6 were located at the center of our galaxy, what would its apparent magnitude be? (*Hint:* Use the magnitude–distance formula.)

8. What is the apparent velocity of recession of 3C48 if its red shift is 0.37?

9. If the Hubble constant is 70 km/s/Mpc, how far away is the quasar in Problem 8? (*Hint:* Use the Hubble law.)

10. The hydrogen Balmer line H β has a wavelength of 486.1 nm. It is shifted to 563.9 nm in the spectrum of 3C273. What is the red shift of this quasar? (*Hint:* What is $\Delta\lambda$?)

DISCUSSION QUESTIONS

1. Do you think that our galaxy has ever been an active galaxy? Could it have hosted a quasar when it was young?

2. If a quasar is triggered in a galaxy's core, what would it look like to people living in the outer disk of the galaxy? Could life continue in that galaxy? (Begin by deciding how bright a quasar would look seen from the outer disk, considering both distance and dust.)

PROBLEMS

1. The total energy stored in a radio lobe is about 10^{53} J. How many solar masses would have to be converted into energy to produce this energy? (*Hints:* Use $E = m_0c^2$. One solar mass equals 2×10^{30} kg.)

2. If the jet in NGC5128 is traveling 5000 km/s and is 40 kpc long, how long will it take for gas to flow from the core of the galaxy out to the end of the jet?

3. Cygnus A is 225 Mpc away, and its jet is about 50 seconds of arc long. What is the length of the jet in parsecs? (*Hint:* Use the small-angle formula.)

4. Use the small-angle formula to find the linear diameter of a radio source with an angular diameter of 0.0015 second of arc and a distance of 3.25 Mpc.

CRITICAL INQUIRIES
FOR THE WEB

1. What object currently holds the distinction as the farthest known galaxy? Search the Web for information on this distant object and find out its red shift and distance. What is the look-back time for this object?

2. The text discusses the superluminal expansion of material from 3C273, but how common is this faster-than-light effect? Search the Internet for information on superluminal motion related to quasars and blazars and list several examples—with the expansion speeds implied by their motion. What is the highest superluminal expansion rate you can find?

3. Gravitational lenses were first predicted by Einstein in 1936 but were not observed until recently. Search the Web for instances of gravitational lensing of galaxies and quasars. For a particular case, discuss how the lens effect allows astronomers to determine information about the lensing and/or lensed objects that might not have been available without the alignment.

 Visit the Seeds *Foundations of Astronomy* companion Web site for critical thinking exercises, articles, and additional readings from InfoTrac College Edition, Brooks/Cole's online student library.

COSMOLOGY IN THE 21ST CENTURY

The Universe, as has been observed before, is an unsettlingly big place,
a fact which for the sake of a quiet life most people tend to ignore.

Douglas Adams, The Restaurant at the End of the Universe

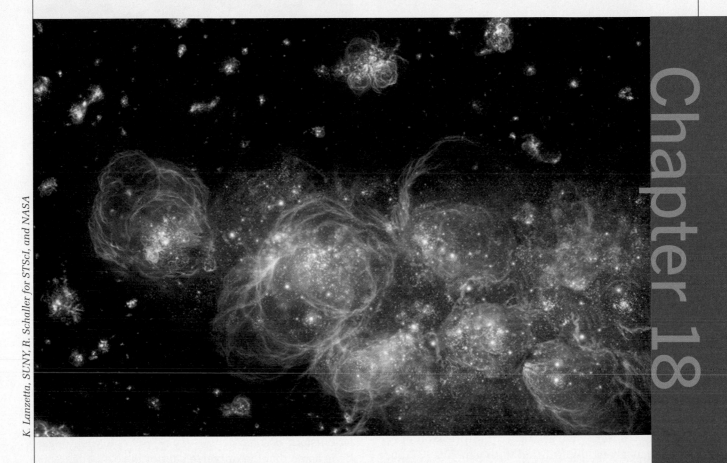

K. Lanzetta, SUNY, R. Schaller for STScI, and NASA

Chapter 18

GUIDEPOST

This chapter marks a watershed in our study of astronomy. Since Chapter 1, our discussion has focused on learning to understand the universe. Our outward journey has discussed the appearance of the night sky, the birth and death of stars, and the interactions of the galaxies. Now we reach the limit of our journey in space and time, the origin and evolution of the universe as a whole.

The ideas in this chapter are the biggest and the most difficult in all of science. Our imaginations can

hardly grasp such ideas as the edge of the universe and the first instant of time. Perhaps it is fitting that the biggest questions are the most challenging.

But this chapter is not an end to our story. Once we complete it, we will have a grasp of the nature of the universe, and we will be ready to focus on our place in that universe—the subject of the rest of this book.

VIRTUAL LABORATORIES

EVIDENCE FOR DARK MATTER

FATE OF THE UNIVERSE

What is the biggest number? A billion? How about a billion billion? That is only 10^{18}. How about 10^{100}? That very large number is called a googol, a number that is at least a billion billion times larger than the total number of atoms in all of the galaxies we can observe. Can you name a number bigger than a googol? Try a billion googols, or, better yet, a googol of googols. You can go even bigger than that—10 raised to a googol. (That's a googolplex.) No matter how large a number you name, we can immediately name a bigger number. That is what infinity means—big without limit.

If the universe is infinite in size, then you can name any distance you want, and the universe is bigger than that. Try a googol to the googol light-years. If the universe is infinite, there is more universe beyond that distance.

Is the universe infinite or finite? In this chapter, we try to answer that question and others, not by playing games with big numbers, but through **cosmology**, the study of the universe as a whole.

Studying the entire universe is difficult because we have expectations about the universe that are in some cases wrong. This chapter is carefully organized to begin with our common expectations, test them against observations, compare them with theories, and build a modern understanding of cosmology. Like wilderness explorers, we must proceed carefully step by step to reach our goal.

Visual + infrared image

FIGURE 18-1

The entire sky is filled with galaxies. Some lie in clusters of thousands, and others are isolated in nearly empty voids between the clusters. In this image of a typical spot on the sky, bright objects with spikes caused by diffraction in the telescope are nearby stars. All other objects are galaxies ranging from the nearby face-on spiral at upper right to the most distant galaxies visible only in the infrared, shown as red in this composite image. (*R. Williams, STScI HDF-South Team, NASA*)

18-1 INTRODUCTION TO THE UNIVERSE

Everyone knows what a gold mine looks like, but few people have actually visited a gold mine. If you explored a real gold mine, you might be surprised at what you find. Similarly, we all have an impression of the universe as a vast depth filled with galaxies (Figure 18-1), but as we begin exploring the universe, we must sort out our expectations so they do not mislead us.

THE EDGE–CENTER PROBLEM

In our daily lives, we are accustomed to boundaries. Rooms have walls, athletic fields have boundary lines, countries have borders, oceans have shores. It is natural for us to think of the universe as having an edge, but that idea leads us astray.

If the universe had an edge, imagine going to that edge. What would you find there: a wall of cardboard? A great empty space? Nothing? We are not talking about

an edge to the distribution of matter, but an edge to space itself, so presumably we could not reach beyond the edge and feel around.

An edge to the universe seems to violate common sense, and modern cosmologists assume that the universe cannot have an edge. That is, the universe must be unbounded.

If the universe has no edge, then it cannot have a center. We find the centers of things—galaxies, globular clusters, oceans, pizzas—by referring to their edges. If we agree that the universe cannot have an edge, then it cannot have a center.

We must think carefully about the universe and avoid using the ideas of an edge or a center. Those ideas are misleading. Of course, if the universe is infinite, then it has no edge or center. We have no edge–center problem if we think of the universe as infinite. Could the universe be finite—could it contain a limited volume? We will see later in this chapter how the universe could be finite but have no edge or center. Before we go further with that idea, we must deal with another of our expectations, the beginning of the universe.

THE NECESSITY OF A BEGINNING

We have all noticed that the night sky is dark. However, reasonable assumptions about the geometry of the universe can lead us to the conclusion that the night sky should glow as brightly as a star's surface. This conflict between observation and theory is called **Olbers's para-**

dox after Heinrich Olbers, a Viennese physician and astronomer, who discussed the problem in 1826.

However, Olbers's paradox is not Olbers's, and it isn't a paradox. The problem of the dark night sky was first discussed by Thomas Digges in 1576 and was further analyzed by astronomers such as Johannes Kepler in 1610 and Edmund Halley in 1721. Olbers gets the credit through an accident of scholarship on the part of modern cosmologists who did not know of previous discussions. What's more, Olbers's paradox is not a paradox. We will be able to understand why the night sky is dark by revising our assumptions about the nature of the universe.

To begin, let's state the so-called paradox. Suppose we assume that the universe is infinite and uniformly filled with stars. (The aggregation of stars into galaxies makes no difference to our argument.) If we look in any direction, our line of sight must eventually reach the surface of a star. Look at ▌Figure 18-2, where we use the analogy of lines of sight in a forest (Window on Science 18-1). When we are deep in a forest, every line of sight

a

FIGURE 18-2

(a) Every direction we look in a forest eventually reaches a tree trunk, and we cannot see out of the forest. *(Photo courtesy Janet Seeds)* (b) If the universe is infinite and uniformly filled with stars, then any line from Earth should eventually reach the surface of a star. This assumption predicts that the night sky should glow as brightly as the surface of the average star, a puzzle commonly referred to as Olbers's paradox.

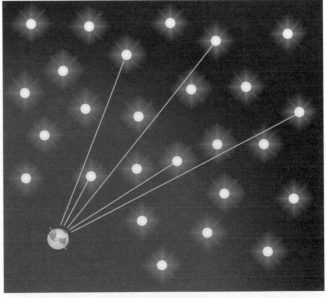

b

terminates on a tree trunk. In our example, every line of sight out into space should terminate on the surface of a star, and the entire sky should be as bright as the surface of an average star. It should not get dark at night.

Of course, the more distant stars would be fainter than nearby stars because of the inverse square law. However, the farther we look into space the larger the volume we see, and the two effects cancel out. Then given our assumptions, every spot on the sky must be occupied by the surface of a star, and the night sky should not be dark.

Imagine the entire sky glowing with the brightness of the surface of the sun. The glare would be overpowering. In fact, the radiation would rapidly heat Earth and all other celestial objects to the average temperature of the surface of the stars, a few thousand degrees. Thus, we can pose Olbers's paradox in another way: Why is the universe so cold?

Olbers assumed that the sky was dark because clouds of matter in space absorb the radiation from distant stars. But this interstellar medium would gradually heat up to the average surface temperature of the stars, and the gas and dust clouds would glow as brightly as the stars.

It is interesting to note that Copernicus and other astronomers of his time saw no paradox in the dark night sky because they believed in a finite universe. Once they looked beyond the sphere of stars, there was nothing to see except, perhaps, the dark floor of heaven. Soon after the time of Galileo and Newton, western astronomers began to think of an infinite universe, and that led to the paradox.

Today, cosmologists believe they understand why the sky is dark, and the universe is cold. Olbers's paradox makes the incorrect prediction that the night sky should be bright because it is based on an incorrect assumption. The universe is not eternal. That is, it is not infinitely old.

This solution to Olbers's paradox was first suggested by Edgar Allan Poe in 1848. He proposed that the night sky was dark because the universe was not infinitely old but had been created at some time in the past. The more distant stars are so far away that light from them has not reached us yet. That is, if we look far enough, the look-back time is greater than the age of the universe, and we look back to a time before stars began to shine. Thus, the night sky is dark because the universe is not infinitely old.

This is a powerful idea because it clearly illustrates the difference between the universe and the observable universe. The universe is everything that exists, and it could be infinite. But the **observable universe** is the part that we can see. We will learn later that the universe is about 14 billion years old. In that case, the observable universe has a radius of about 14 billion light-years. Do not confuse the observable universe, which is finite, with the universe as a whole, which could be infinite.

When we described Olbers's paradox, we assumed without noticing that the universe was infinitely old. This illustrates the importance of assumptions in cosmology and serves as a warning that our commonsense expectations are not dependable. All of astronomy is reasonably unreasonable—that is, reasonable assumptions often lead to unreasonable results. That is especially true in cosmology, so we must proceed with care.

Our expectation that the universe be eternal and unchanging is clearly wrong. Olbers's paradox makes a beginning a necessity. In the next section we will discover that we should not expect the universe to be static; observations show that it is expanding.

COSMIC EXPANSION

In the early 20th century, astronomers began photographing the spectra of galaxies, and they noticed that the spectral lines in galaxy spectra had red shifts. In 1929, Edwin P. Hubble published his discovery that the size of the red shift was proportional to the distance to the galaxy. (We discussed this as the Hubble law in Chapter 16, where we used it to estimate the distances to galaxies.) When interpreted using the Doppler effect, these red shifts imply that the galaxies are receding from each other and that the universe is expanding.

The red shifts of the galaxies are dramatic. Nearby galaxies have small red shifts, but when we look at more distant galaxies, the red shifts are quite large. ▮ Figure 18-3 shows spectra of galaxies in galaxy clusters at various distances. The Virgo cluster is relatively nearby, and its red shift is small. The Hydra cluster is very distant, and its red shift is so large that the two dark lines formed by once-ionized calcium are shifted from the near ultraviolet well into the visible part of the spectrum.

The important point about Hubble's discovery is that the red shifts are proportional to distance. That shows that the galaxies are receding from each other and the universe is expanding uniformly. But notice that the expansion does not imply that we are at the center of the universe. To see why, look at ▮ Figure 18-4, which shows an analogy in which we imagine baking raisin bread. As the dough rises, the raisins are pushed away from each other uniformly at velocities that are proportional to distance. Two raisins that were originally close to each other are pushed apart slowly, but two raisins that were far apart, having more dough between them, are pushed apart faster. If bacterial astronomers lived on a raisin in our raisin bread, they could observe the red shifts of the other raisins and derive a bacterial Hubble law. They could conclude that their universe was expanding uniformly. But it does not matter which raisin the bacterial astronomers lived on. There is no center to the expansion, and they would get the same Hubble law no matter what raisin they lived on. Similarly, there is no center to the expansion

Unshifted position of calcium lines

Name of cluster containing the galaxy

17 Mpc
Virgo
1,200 km/s

215 Mpc
Ursa Major
15,000 km/s

Red shifts shown as red arrows

310 Mpc
Corona Borealis
22,000 km/s

550 Mpc
Boötes
39,000 km/s

860 Mpc
Hydra
61,000 km/s

ACTIVE FIGURE 18-3

These galaxy spectra extend from the near ultraviolet at left to the blue part of the visible spectrum at right. The two dark absorption lines of once-ionized calcium are prominent in the near ultraviolet. The red shifts in galaxy spectra are expressed here as velocities of recession. Note that the apparent velocity of recession is proportional to distance. *(Caltech)*

Ace⊚Astronomy™ Go to AceAstronomy and click Active Figures to see "Cosmic Red Shift." Graph your own observations of red shift and distance.

of the universe and astronomers in any galaxy will see the same law of expansion.

When we look at Figure 18-4, we see the edge of the loaf of raisin bread and we can identify a center. This happens because our raisin bread analogy breaks down when we look at the crust of the bread. Remember that our universe cannot have an edge or a center, so there can be no center to the expansion. We must think and speak very carefully as we discuss cosmology (Window on Science 18-2).

THE NECESSITY OF A BIG BANG

The discovery of the expansion of the universe led astronomers to an astonishing conclusion. The universe must have begun with an event of cosmic fury.

Imagine that we have a videotape of the expanding universe, and we run it backward. We see the galaxies moving toward each other. There is no center to the expansion of the universe, so we do not see galaxies approaching a single spot; but, rather, we see the distances between all galaxies decreasing, and eventually the galaxies begin to merge. If we run our videotape far enough back, we see the matter and energy of the universe compressed into a high-density, high-temperature state. The expanding universe must have begun from this moment of extreme conditions, which modern astronomers call the **big bang.**

We must not allow ourselves to think of the big bang as an explosion. We have concluded that the universe cannot have a center, so the big bang cannot have occurred at a single point. The entire universe had been filled with the high-density, high-temperature state we call the big bang. Our challenge in this chapter is to understand how that could have happened without a center or an edge. We will resolve that problem when we refine our understanding of space and time.

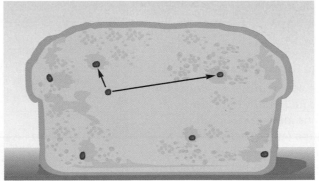

ACTIVE FIGURE 18-4

The raisin bread analogy for the expansion of the universe. As the dough rises, raisins are pushed apart with velocities proportional to distance. The expansion has no center. A colony of bacteria living on any raisin will find that the red shifts of the other raisins are proportional to their distances.

Ace⊚Astronomy™ Go to AceAstronomy and click Active Figures to see "Raisin Bread." You can watch raisin bread rise.

Why Scientists Speak Carefully: Words Lead Thoughts

There are certain words we should never say, not even as a joke. We should never call our friend "fool," for example, because we might begin to think of our friend as a fool. Words lead thoughts. It works in advertising and politics, and it can work in science, too. Using words carelessly can lead us to think carelessly.

In science, there are certain ways to say things, not because scientists are sticklers for good grammar, but because if we say things wrongly we begin to think things wrongly. For example, a biologist would never let us say a beehive knows it must store food. "No, don't say it that way," the biologist would object. "The hive is just a collection of individual creatures and it can't 'know' anything. Even the individual bees don't really 'know' things. The instinctive behavior of individual bees causes food to accumulate in the hive. That's the way to say it."

All scientists are careful of language because careless words can mislead us.

We would never refer to the center of the universe, for example, but we must also be careful not to say things like "galaxies flying away from the big bang." Those words imply a center to the expansion of the universe, and we know the expansion must be centerless. Rather, we should say "galaxies flying away from each other."

Whatever science you study, notice the customary ways of using words. It is not just a matter of convention. It is a matter of careful thought.

For now, we can solve an easier problem. How long ago did the universe begin? We can estimate the age of the universe with a simple calculation. If we must drive to a city 100 miles away and we can travel 50 miles per hour, we divide distance by rate of travel and learn the travel time—in this example, 2 hours. To find the age of the universe, we can divide the distance between two galaxies by the velocity with which they are separating and find out how long they have taken to reach their present separation.

Dividing distance by velocity tells us the age of the universe, and the Hubble constant (Chapter 16) simplifies our task further. The Hubble constant H has the units km/s per Mpc, which is a velocity divided by a distance. If we calculate $\frac{1}{H}$ we have a distance divided by velocity. To make the division give us an age, we must convert megaparsecs to kilometers, and then the distances will cancel out and we will have an age in seconds. To get years, we must divide by the number of seconds in a year. If we make these simple changes in units, the age of the universe in years is approximately 10^{12} divided by H in its normal units, km/s/Mpc.

$$T \approx \frac{1}{H} \times 10^{12} \text{ years}$$

This is known as the **Hubble time.** For example, if H is 70 km/s/Mpc, then the age of the universe can be no older than $10^{12}/70$, which equals 14 billion years.

The Hubble time gives us an estimate of the age of the universe. We will fine tune our estimate later in this chapter, but for the moment we can conclude that basic observations of the recession of the galaxies require that the universe began with a big bang about 14 billion years ago.

Our instinct is to think of the big bang as a historical event, like the Gettysburg Address—something that happened long ago and happened in a particular place. But the big bang isn't really over. The look-back time makes it possible for us to observe the big bang directly. The look-back time to nearby galaxies is only a few million years, and the look-back to more distant galaxies is a large fraction of the age of the universe (■ Figure 18-5). Suppose we look between the distant galaxies, back to the time of the big bang. We should be able to detect the hot gas that filled the universe long ago.

Although our imaginations try to visualize the big bang as a localized event, we must keep firmly in mind that the big bang did not occur at a single place but filled the entire volume of the universe. The matter of which we are made was part of that big bang, so we are inside the remains of that event, and the universe continues to expand around us. We cannot point to any

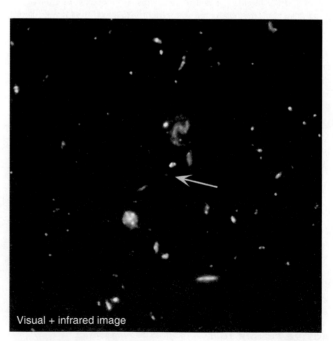

Visual + infrared image

FIGURE 18-5

This faint galaxy (arrow) is believed to be so distant that its light has been traveling for most of the history of the universe. We see it as it was about 1 billion years after the big bang. The galaxy is bright in the infrared but it is not detectable at visual wavelengths. *(Ken Lanzetta and Amos Yahil, Stony Brook and NASA)*

particular place and say, "The big bang occurred over there." The big bang occurred everywhere, and, in whatever direction we look, at great distance we look back to the age when the universe was filled with hot gas (Figure 18-6).

The radiation that comes to us from this great distance has a tremendous red shift. The most distant visible objects are faint galaxies and quasars, with red shifts a bit over 6. In contrast, the radiation from the hot gas of the big bang has a red shift of about 1100. Thus the light emitted by the big bang gases arrives at Earth as infrared radiation and short radio waves. We can't see it with our eyes, but we should be able to detect it with infrared and radio telescopes. Unlike the Gettysburg Address, the big bang isn't over, and we should be able to see it happening if we can detect the radiation it emitted. That amazing discovery is the subject of the next section.

THE COSMIC BACKGROUND RADIATION

If radiation is now arriving from the big bang, then it should be detectable. The story of that discovery begins in the mid-1960s when two Bell Laboratories physicists, Arno Penzias and Robert Wilson, were using a horn antenna to measure the radio brightness of the sky (Figure 18-7a). Their measurements showed a peculiar noise in the system, which they at first attributed to droppings from pigeons living inside the antenna. Perhaps they would have enjoyed cleaning the antenna more if they had known they would win the 1978 Nobel Prize for physics for the discovery they were about to make.

When the antenna was cleaned, they measured the brightness of the sky at radio wavelengths and again found the low-level radio noise. The pigeons were innocent, but what was causing the signal?

The explanation for the noise goes back to 1948, when George Gamow predicted that the early big bang would be very hot and would emit copious black body radiation. A year later, physicists Ralph Alpher and Robert Herman pointed out that the large red shift of the big bang would lengthen the wavelengths of the radiation into the far infrared and radio part of the spectrum. There was no way to detect this radiation until the mid-1960s, when Robert Dicke at Princeton concluded the radiation should be just strong enough to detect with newly developed techniques. Dicke and his team began building a receiver.

When Penzias and Wilson heard of Dicke's work, they recognized the noise they had detected as radiation from the big bang, the **cosmic microwave background radiation.**

The background radiation was measured at many wavelengths as astronomers tried to confirm that it did indeed follow a black body curve. Some measurements could be made from the ground, but balloon and rocket observations were critical and difficult. A few disturbing

a   A region of the universe during the big bang

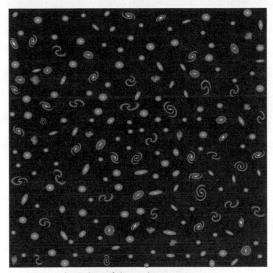

b   A region of the universe now

c
The present universe as it appears from our galaxy

FIGURE 18-6

Three views of a small region of the universe centered on our galaxy. (a) During the big bang, the region is filled with hot gas and radiation. (b) Later, the gas has formed galaxies, but we can't see the universe this way because the look-back time distorts what we see. (c) Near us we see galaxies, but farther way we see young galaxies (dots), and at a great distance we see radiation (arrows) coming from the hot gas of the big bang.

observations suggested that the radiation departed inexplicably from a black body curve and thus might not come from the big bang.

In November 1989, the orbiting Cosmic Background Explorer (COBE) began taking the most accurate measurements of the background radiation ever made. Within two months, the verdict was in (▌ Figure 18-7b). To an accuracy of better than 1 percent, the radiation follows a black body curve with a temperature of 2.725 ± 0.002 K—in good agreement with accepted theory.

It may seem strange that the hot gas of the big bang seems to have a temperature of only 2.7 K, but recall the tremendous red shift. We see light that has a red shift of about 1100—that is, the wavelengths of the photons are about 1100 times longer than when they were emitted. The gas clouds that emitted the photons had a temperature of about 3000 K, and they emitted black body radiation with a λ_{max} of about 1000 nm (Chapter 7). Although this is in the near infrared, the gas would also have emitted enough visible light to glow orange-red. But the red shift has made the wavelengths about 1100 times longer, so λ_{max} is about 1 million nm (equivalent to 1 mm). Thus, the hot gas of the big bang seems to be 1100 times cooler, about 2.7 K.

The cosmic microwave background radiation is evidence that a big bang really did occur. In fact, the evidence is so strong that an alternative theory of the universe was abandoned. The **steady state theory,** popular with some astronomers during the 1950s and 1960s, held that the universe was eternal and unchanging. If the universe had no beginning, then it could never have had a big bang. According to the theory, new matter continuously appeared from nothing to maintain the density of the expanding, steady-state universe. The discovery of the cosmic microwave background radiation in 1965 led astronomers to abandon the steady-state theory within a few years. We know there was a big bang because, using radio and infrared telescopes, we can see it happening.

THE STORY OF THE BIG BANG

Simple observations of the darkness of the night sky and the red shifts of the galaxies tell us that the universe must have begun with a big bang. In fact, we have direct observational evidence of the big bang in the form of the cosmic microwave background radiation. Theorists can combine these observations with modern physics to tell the story of how the big bang occurred. As we review this story, remember that the big bang did not occur at a specific place. It filled the entire volume of the universe from that first moment.

We cannot begin our history at time zero, because we do not understand the physics of matter and energy under such extreme conditions, but we can come close. If we could visit the universe when it was very young, only 10 millionths of a second old, for instance, we

a

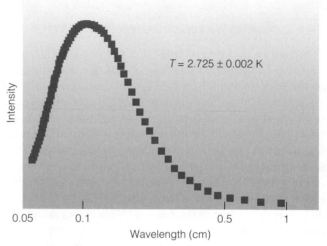

T = 2.725 ± 0.002 K

Intensity

0.05   0.1   0.5   1

Wavelength (cm)

b

FIGURE 18-7

(a) Robert Wilson (left) and Arno Penzias pose before the horn antenna with which they discovered the cosmic microwave background radiation. *(AT&T Archives)* (b) The background radiation from the big bang is observed in the infrared and radio part of the spectrum. At first, the critical measurements near the peak of the curve had to be made from balloons and rockets, but the COBE satellite observed from orbit. These COBE data show that the radiation fits a black body curve with a temperature of 2.725 K.

would find it filled with high-energy photons having a temperature well over 1 trillion (10^{12}) K and a density greater than 5×10^{13} g/cm³, nearly the density of an atomic nucleus. When we say the photons had a given temperature, we mean that the photons were the same as black body radiation emitted by an object of that temperature. Thus, the photons in the early universe were gamma rays of very short wavelength and therefore very high energy. When we say that the radiation had a certain density, we refer to Einstein's equation $E = m_0 c^2$. We can express a given amount of energy in the form of radiation as if it were matter of a given density.

If photons have enough energy, two photons can combine and convert their energy into a pair of particles—a particle of normal matter and a particle of **antimatter.** When an antimatter particle meets its matching particle of normal matter—when an antiproton meets a normal proton, for example—the two particles annihilate each other and convert their mass into energy in the form of two gamma rays. In the early universe, the photons had enough energy to produce proton–antiproton pairs; but when the particles collided, they converted their mass back into photons. Thus, the early universe was a dynamic soup of energy flickering from photons into particles and back again.

While all this went on, the expansion of the universe cooled the gamma rays to lower energy. By the time the universe was 0.0001 second old, its temperature had fallen to 10^{12} K. By this time, the average energy of the gamma rays had fallen below the energy equivalent to the mass of a proton or a neutron, so the gamma rays could no longer produce such heavy particles. The particles combined with their antiparticles and quickly converted most of the mass into photons.

It would seem from this that all of the protons and neutrons should have been annihilated with their antiparticles, but for reasons that are poorly understood a small excess of normal particles existed. For every billion protons annihilated by antiprotons, one survived with no antiparticles to destroy it. Consequently, we live in a world of normal matter, and antimatter is very rare.

Although the gamma rays did not have enough energy to produce protons and neutrons, they could produce electron–positron pairs, which are about 1800 times less massive than protons and neutrons. This continued until the universe was about 4 seconds old, at which time the expansion had stretched and cooled the gamma rays to the point where they could no longer create electron–positron pairs. The protons, neutrons, and electrons of which our universe is made were produced during the first 4 seconds of its history.

This soup of hot gas and radiation continued to cool and eventually began to form atomic nuclei. High-energy gamma rays can break up a nucleus, so the formation of such nuclei could not occur until the universe had cooled somewhat. By the time the universe was about 2 minutes old, protons and neutrons could link to form deuterium, the nucleus of a heavy hydrogen atom; by the end of the next minute, further reactions began converting deuterium into helium. But no heavier atoms could be built because no stable nuclei exist with atomic weights of 5 or 8 (in units of the hydrogen atom). Cosmic element building during the big bang had to proceed rapidly, step by step, like someone hopping up a flight of stairs (❚ Figure 18-8). The lack of stable nuclei at atomic weights of 5 and 8 meant there were missing steps in the stairway, and the step-by-step reactions had great difficulty jumping over

these gaps. A tiny amount of lithium was produced but nothing heavier.

By the time the universe was 30 minutes old, it had cooled sufficiently that nuclear reactions had stopped. About 25 percent of the mass was helium nuclei, and the rest was in the form of protons—hydrogen nuclei. This is the cosmic abundance we see today in the oldest stars. (The heavier elements, remember, were built by nucleosynthesis inside many generations of massive stars.) The cosmic abundance of helium was fixed during the first minutes of the universe.

At first, the universe was dominated by radiation. The gamma rays interacted continuously with the matter, and they cooled together as the universe expanded. The gas was ionized because it was too hot for the nuclei to capture electrons to form neutral atoms, and the free electrons made the gas very opaque. A photon could not travel very far before it collided with an electron and was deflected. Thus, radiation and matter were locked together.

As the young universe expanded, it went through three important changes. First, when the universe reached an age of roughly 50,000 years, the density of the energy present as photons became less than the density of the gas. Before this, matter could not clump together because the intense sea of photons smoothed the gas out. Once the density of the radiation fell below that of matter, the matter could begin to draw together and form the clouds that eventually became stars and galaxies.

As the universe continued to expand, the ionized gas became less dense, which meant that the free electrons were spread farther apart. As the universe reached the age of a few hundred thousand years, the second important change began. The free electrons were spread so far apart, the photons could travel for thousands of parsecs before getting scattered off an electron. That is, the universe began to grow more transparent. At about the same time, the third change happened. As the

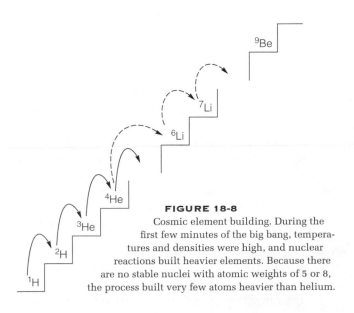

FIGURE 18-8
Cosmic element building. During the first few minutes of the big bang, temperatures and densities were high, and nuclear reactions built heavier elements. Because there are no stable nuclei with atomic weights of 5 or 8, the process built very few atoms heavier than helium.

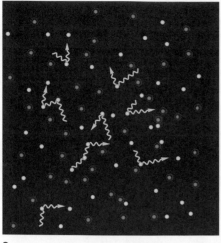
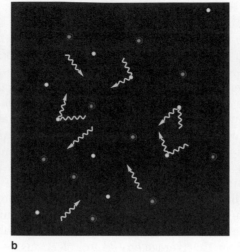
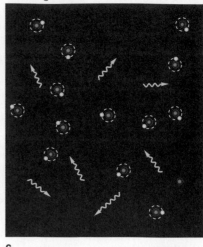

a   b   c

FIGURE 18-9

Photons scatter from electrons (blue) easily but hardly at all from the much more massive protons (red). (a) When the universe was very dense and ionized, photons could not travel very far before they scattered off an electron. This made the gas opaque. (b) As the universe expanded, the electrons were spread further apart, and the photons could travel farther; this made the gas more transparent. (c) After recombination, most electrons were locked to protons to form neutral atoms, and the gas was highly transparent.

falling temperature of the universe reached 3000 K, protons were able to capture and hold free electrons to form neutral hydrogen, a process called **recombination.** As the free electrons were gobbled up, the gas finally became transparent and the photons could travel through the gas without being absorbed (▌ Figure 18-9). Although the gas continued to cool, the photons no longer interacted with the gas, and consequently the photons retained the black body temperature that the gas had at recombination. Those photons, with a black body temperature of 3000 K, are what we observe today as the cosmic microwave background radiation. Remember that the large red shift makes that gas appear to have a temperature of about 2.7 K.

Recombination left the gas of the big bang neutral and transparent. At first the universe was filled with the glow of the hot gas; but, as the universe expanded and cooled, the glow faded into the infrared, and the universe entered what cosmologists call the **dark age,** a period lasting hundreds of millions of years during which the universe expanded in darkness.

The dark age ended as the first stars began to form. The gas from which these first stars formed contained almost no metals and was consequently highly transparent. Models show that, because the first stars formed from this metal-poor gas, they were very massive, very luminous, and very short lived. That first violent burst of star formation produced enough ultraviolet light to begin ionizing the gas, and today's astronomers, looking back to the most distant visible quasars and galaxies, can see traces of that **reionization** of the universe. (▌ Figure 18-10)

▌ Figure 18-11 summarizes our story of the big bang, from the formation of helium in the first three minutes, through energy–matter equality, recombination, and finally reionization. Reionization marks the end of the dark ages and the beginning of the age of stars and galaxies in which we live today.

REVIEW CRITICAL INQUIRY

How do we know there was a big bang?

The cosmic microwave background radiation consists of photons emitted by the hot gas of the big bang, so when we detect those photons, we are "seeing" the big bang. Of course, all scientific evidence must be interpreted, so we must understand how the big bang could produce radiation all around us before we can accept the background radiation as evidence. First, we must recognize that the big bang event filled all of the universe with hot, dense gas. The big bang didn't happen in a single place; it happened everywhere. At recombination, the expansion of the universe reached the stage where the matter became transparent and the radiation was free to travel through space. Today we see that radiation from the age of recombination arriving from all over the sky. It is all around us because we are part of the big bang event, and as we look out into space to great distance, we look back in time and see the hot gas in any direction we look. We can't see the radiation as light because of the large red shift that has lengthened the wavelengths by a factor of 1100 or so, but we can detect the radiation as photons with infrared and short radio wavelengths.

With this interpretation, the cosmic microwave background radiation is powerful evidence that there was a big bang. That tells us how the universe began. Why do we think the universe cannot have a center?

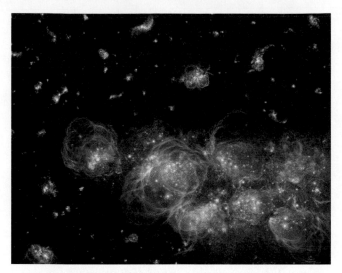

FIGURE 18-10

In this artist's conception of reionization, the first stars produce floods of ultraviolet photons that ionize the gas in expanding bubbles. Such a storm of star formation ended an age when the universe had expanded in darkness. Spectra of the most distant quasars reveal that those first galaxies were surrounded by neutral gas that had not yet been fully ionized. Thus the look-back time allows modern astronomers to observe the age of reionization. *(K. Lanzetta, SUNY, A. Schaller for STScI, and NASA)*

So far in this chapter we have compared our simple expectations with basic observations, and we have discovered some fundamental properties of the universe. Now it is time to stretch our imaginations and refine our expectations. The universe is much more interesting than we have imagined to this point.

18-2 THE SHAPE OF SPACE AND TIME

Our simple expectations have led us to a puzzle. How can the universe expand if it does not have an edge or a center? Solving that problem is the key to understanding modern cosmology, but our everyday expectations about how the world works are not a help. To solve the puzzle, we must put our reasonable expectations on hold and look carefully at how space and time behave on cosmic scales.

LOOKING AT THE UNIVERSE

Let's begin by carefully examining the general appearance of the universe. Not only do we see no edge to the universe and no center, there is no preferred direction.

The gas and dust in our own galaxy block our view of the distant universe along the Milky Way, but once we correct for that, the universe looks the same over the entire sky. Astronomers refer to this as **isotropy,** the property of being the same in all directions. No matter what direction we look we see the same kinds of galaxies and galaxy clusters. The universe seems isotropic.

Furthermore, the cosmic microwave background radiation is isotropic. Astronomers must correct for Doppler shifts caused by the motion of the Earth around the sun, the sun around our galaxy, and our galaxy around the center of mass of the local group, but those are only local variations. Once those local variations are taken into account, the background radiation is almost perfectly uniform over the entire sky.

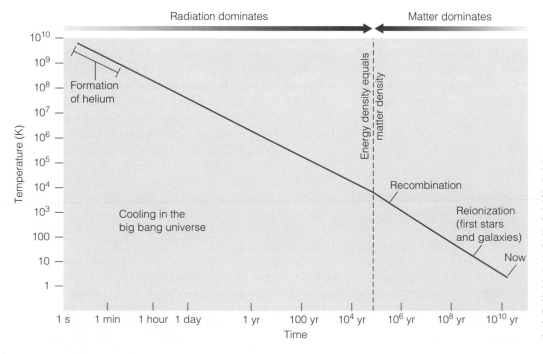

FIGURE 18-11

During the first few minutes of the big bang, some hydrogen was fused to produce helium, but the universe quickly became too cool for such fusion reactions to occur. The rate of cooling increased as matter began to dominate over radiation. Recombination freed the radiation from the influence of the gas, and reionization was caused by the birth of the first stars.

It certainly appears that the universe is the same everywhere. Of course, there are local variations; some points in space are inside galaxies, and some are isolated far from any galaxy, but overall, in its general properties, one place in the universe is the same as any other place. That is, there are no special places in the universe. We cannot point at any particular location in the universe and say, "That location is different from any other place in the universe." Astronomers refer to this as **homogeneity,** the property of being uniform. On the largest scales, the universe appears to be homogeneous.

Astronomers observe that the universe is isotropic and homogeneous on the largest scales, and that leads to an assumption that is critically important in cosmology. The **cosmological principle** says that any observer in any galaxy sees the same general features of the universe. Of course some observers live in spiral galaxies and some live in elliptical galaxies, but those are only local variations. According to the cosmological principle, when we look at the general features of the universe, we should all see the same kind of universe no matter where our planets are located.

Evolutionary changes are not included in the cosmological principle. If the universe is expanding and the galaxies are evolving, observers living at different times may see galaxies at different stages. The cosmological principle says that once observers correct for evolutionary changes, they should see the same general features.

The cosmological principle is actually an extension of the Copernican principle. Copernicus said that Earth is not in a special place; it is just one of a number of planets orbiting the sun. The cosmological principle says that there are no special places in the universe. Local irregularities aside, one place is just like another. Our location in the universe is typical of all other locations.

If we accept the cosmological principle, then we may not imagine that the universe has an edge or a center. Such locations would not be the same as all other locations. Observations confirm this conclusion. No matter what direction we look in the sky, no matter where we look in space, the universe is generally the same. We see no edge and no center.

THE SHAPE OF THE EXPANDING UNIVERSE

How can the universe expand if it does not have an edge or a center? We are now ready to resolve that central puzzle of cosmology. To do so we must understand the nature of space and time. Remember that we are small creatures, and we live on a small world. The distances and times we experience are quite small. It should not surprise us that space and time are actually more complicated than our expectations suggest.

Study ▌ "The Nature of Space-Time" on pages 392 and 393 and notice three important ideas. First, the red shifts of the galaxies are not produced by the Doppler effect. The galaxies do not move at high velocities through space. Space is expanding and carrying the galaxies away from each other and stretching protons to longer and redder wavelengths. Second, notice that the curvature of space-time allows the universe to have no edge or center even if it is finite. Finally, notice that we could make measurements in our three-dimensional universe to determine the curvature of space-time.

It seems to violate our common sense that the universe can expand without having a center or an edge, but we must not depend on our limited imagination as three-dimensional creatures. We must speak carefully and thoughtfully when we discuss cosmology so as not to mislead ourselves (Window on Science 18-2). We must not try to visualize the expansion of the universe as an outer edge moving into previously unoccupied space. Open, flat, or closed, the universe has no edge, so it does not need additional room to expand. The universe contains all of the volume that exists, and the expansion is a gradual change in the nature of space-time that causes that volume to increase.

MODEL UNIVERSES

Modern cosmologists have been able to use general relativity to construct mathematical models of the universe under different assumptions For example, they can describe the nature of the universe if it is open, closed, or flat. These models have had a strong influence on the development of cosmology.

The general curvature of the entire universe is determined by its density. If the average density of the universe is equal to the **critical density,** or 9×10^{-30} g/cm^3, space-time will be flat. If the average density of the universe is less than the critical density, the universe is negatively curved and open. If the average density is greater than the critical density, the universe is positively curved and closed.

We decided earlier that an open universe and a flat universe are both infinite. If their behavior is ruled entirely by their matter density, both kinds of universes will expand forever. The gravitation of the matter in the universe, present as a curvature of space-time, will cause the expansion to slow, but, if the universe is open, it will never come to a stop (▌ Figure 18-12). If the universe is flat, it will barely slow to a stop after an infinite time. Thus the models predict that an open or flat universe will expand forever, and the galaxies will eventually become black, cold, solitary islands in a universe of darkness.

These models predict a different fate for a closed universe. In a closed universe, the gravitational field, present as curved space-time, is sufficient to slow the expansion to a stop and make the universe contract. Eventually, the contraction will compress all matter and energy back into the high-energy, high-density

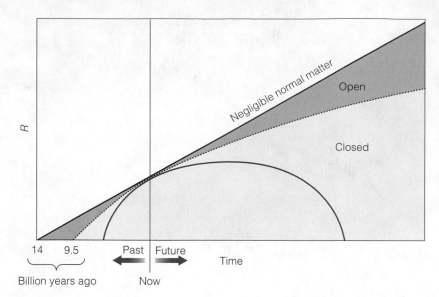

FIGURE 18-12
If the behavior of the universe is determined by the density of matter and energy, then its fate is linked to its geometry. In these models, R is a measure of the extent to which the universe has expanded. Open-universe models expand without end, and the corresponding curves fall in the region shaded orange. Closed models expand and then contract back to a high-density state (red curve). Curves representing closed models fall in the region shaded blue. The dotted line represents a flat universe, the dividing line between open and closed models. Note that the estimated age of the universe depends on the rate at which the expansion is slowing down. (This figure assumes $H = 70$ km/s/Mpc.)

state from which the universe began. This end to the universe, a big bang in reverse, has been called the big crunch. Nothing in the universe could avoid being destroyed by this "crunch."

Some theorists have suggested that the big crunch will spring back to produce a new big bang and a new expanding universe. This theory, called the **oscillating universe theory,** predicts that the universe undergoes alternate stages of expansion and contraction. Theoretical work, however, suggests that successive bounces of an oscillating universe would be smaller until the oscillation ran down, and the theory is no longer taken seriously.

Notice in Figure 18-12 that we must know the geometry of the universe before we can determine the age. Earlier we calculated the Hubble time $\frac{1}{H}$ as an estimate of the age of the universe. Figure 18-12 shows that the Hubble time is the age of a model universe that contains so little matter its expansion is hardly slowed at all. If the density of a model universe is higher, the expansion will be slowed, and that means the true age will be less that $\frac{1}{H}$. If the universe is flat, its true age will be $\frac{2}{3}$ of $\frac{1}{H}$. If H equals 70 km/s/Mpc, then $\frac{1}{H}$ equals 14 billion years. But if the universe is flat and its evolution is controlled by its curvature, then its true age is $\frac{2}{3}$ of 14 billion years, which equals a bit over 9 billion years. This is less than the age of the globular clusters, so the value of H combined with theoretical models of the universe suggest that the density of normal matter should be quite low.

Whether the universe is open, closed, or flat depends on its density, but it is quite difficult to measure the density of the universe. We could count galaxies in a given volume, multiply by the average mass of a galaxy, and divide by the volume; but we are not sure of the average mass of a galaxy, and many small galaxies are too faint to see even if they are nearby. Studies of light coming from distant galaxies suggest that large amounts of normal matter lie in great, nonluminous clouds between the galaxies, so our estimates must include that matter. Also, we must include the mass equivalent to all the energy in the universe because mass and energy are related. When all of the mass in the universe is added up, it provides a density of less than 5 percent of the critical density needed to close the universe.

Many astronomers will admit that, for theoretical reasons we will discuss later, they prefer to believe the universe is flat. Science, however, is not ruled by preferences or theory, but rather by evidence. Observations of normal matter do not reveal enough density to make the universe flat, but we have not included an important factor. We have clear evidence that there is a large amount of dark matter in the universe. Could there be enough dark matter to make the universe flat?

DARK MATTER IN COSMOLOGY

There is more to the universe than meets the eye. In Chapter 16, we discovered that galaxies have much stronger gravitational fields than we would expect based on the amount of matter we see. Even when we add in the nonluminous gas and dust that we expect to find in galaxies, their gravitational fields are much stronger than we would expect. We concluded that galaxies and clusters of galaxies must contain dark matter.

This is not a small issue. Judging by their gravitational fields, galaxies and clusters of galaxies contain as much as 10 times more matter than we expect from what we see. To correct for this dark matter, we would not just add a small percentage. We have to multiply by a *factor* as large as 10.

If we have any doubt how insubstantial the visible universe is, we have only to look at gravitational lensing of distant galaxies. In Chapter 17 we saw how light from distant quasars could be bent by the gravitational

The Nature of Space-Time

In 1929, Edwin Hubble discovered that the red shifts of the galaxies are proportional to distance — a relationship now known as the Hubble law. It was taken as dramatic evidence that the universe is expanding — that the galaxies are moving away from each other. That leads us to consider not only the nature of the expansion but the nature of space and time.

Distance is the separation between two points in space, and time is the separation between two events. Albert Einstein showed how to relate space to time and treat the whole as space-time. We can think of space-time as a canvas on which the universe is painted, a canvas that can stretch.

Astronomers often express the red shifts of the galaxies as velocities, but these red shifts aren't really Doppler shifts. They are caused by the expansion of space-time.

That explains why the relativistic Doppler formula in Chapter 16 isn't quite correct. It describes the motion of an object through space and not the behavior of space itself.

Notice that our description of the expansion of the universe means the galaxies are not really moving any more than the raisins are swimming through the bread dough. Except for orbital motion among neighboring galaxies, the galaxies are motionless in space-time, and it is space-time that is stretching like a rubber sheet and carrying the galaxies away from each other.

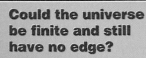

Could the universe be finite and still have no edge?

How can it expand if it has no edge?

How can the universe expand if there is no space outside for it to expand into?

To answer these questions we must use Einstein's theory of general relativity, which predicts that the presence of mass can curve space-time, a curvature we experience as gravity. Furthermore, the theory predicts that even empty space-time can have a curvature, so what we see in the universe depends on the overall curvature of space-time. Finally, the theory shows that space-time could have the property of expansion. That is, the expansion of space-time is not caused by external influences, but is part of space-time itself.

What are the red shifts?

A distant galaxy emits a short-wavelength photon toward our galaxy.

Grid shows expansion of space-time.

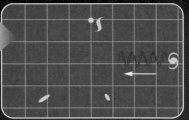

The expansion of space-time stretches the photon to longer wavelength as it travels.

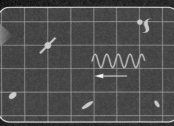

The farther the photon has to travel, the more it is stretched.

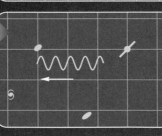

When the photon arrives at our galaxy, we see it with a longer wavelength — a red shift that is proportional to distance.

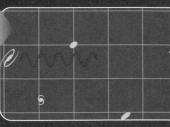

The stretching of space-time not only moves the galaxies away from each other, but it lengthens the wavelength of photons traveling through space-time.

To think about space-time curvature, we can use an analogy of a two-dimensional ant on an orange. If he is truly two-dimensional, he will not be able to understand up and down. He can only travel forward and back, right and left. Then as he walks over the surface of his spherical universe, he will eventually realize he has been everywhere because his universe is covered by his footprints. He will conclude that he lives in a finite universe that has no edge and no center.

Similarly, our universe could be curved back on itself so that it could be finite but we would never find an edge. To find out how our universe is really curved, astronomers must study very distant objects.

Notice that the center of the orange cannot be the center of the ant's two-dimensional universe because the center of the orange does not lie on the surface. That is, the center of the orange is not part of the ant's universe.

Analogy Ahead

Don't forget: These analogies are only two-dimensional. When we think about our universe, we must think of a three-dimensional universe. It is as difficult for us to imagine curvature in our three-dimensional universe as it is for ants to imagine curvature in their two-dimensional universes.

Aha!

There are three possible ways our universe might be curved. It could have positive curvature, analogous to the surface of a sphere. Models of such universes are called **closed universes** because they are finite. Of course, ours could be a **flat universe** with zero curvature. Another possibility is that our universe has negative curvature, analogous to a two-dimensional universe shaped like a saddle or a potato chip. Models of these universes are called **open universes.**

A positively curved universe, also called a closed universe, is finite and has no edge. In our two-dimensional analogy, ants on an orange would notice that the areas of large circles were less than πr^2.

Hey! C'mere!

A flat universe must be infinite or it will have an edge. In a two-dimensional analogy, ants on a flat sheet of paper would find that all circles have areas of πr^2.

Although our ant might be unable to sense a third dimension, it could still measure the curvature of its universe by drawing circles. On a flat universe, the area of a circle would always be πr^2 no matter how big the circle was. But on a spherical surface, large circles would contain less than πr^2. On a saddle-shaped surface, large circles would contain more than πr^2. In the same way, we can detect the curvature of our three-dimensional universe, but we must make measurements over very large distances.

Whether our universe is closed, flat, or open, it cannot have an edge or a center.

What?

A negatively curved universe, also called an open universe, must be infinite or it will have an edge. In a two-dimensional analogy, ants on a saddle shape will discover that large circles have areas greater than πr^2.

FIGURE 18-13

The mass of galaxy cluster 0024+1654 bends the light of a much more distant galaxy to produce arcs that are actually distorted images of the distant galaxy. This gravitational lens effect reveals that many galaxy clusters contain much more mass than is visible as galaxies. Thus galaxy clusters must contain large amounts of dark matter. *(W. N. Colley and E. Turner, Princeton, and J. A. Tyson, Bell Labs, and NASA)*

field of an intervening galaxy to create false images of the quasar. Using the largest telescopes in the world, astronomers are finding that light from the most distant galaxies can be bent by the mass of a nearby cluster of galaxies. The distant galaxies are extremely faint and blue, perhaps because of rapid star formation. As the light from these faint galaxies passes through a nearby cluster of galaxies, the images of the distant galaxies can be distorted into short arcs like rainbows (▮ Figure 18-13). The amount of gravitational lensing can tell astronomers the mass of the nearby galaxy cluster, and, in some cases, the mass is over 10 times greater than the mass visible as galaxies. Evidently such clusters contain over 90 percent dark matter.

Looking at the universe of visible galaxies is like looking at a ham sandwich and seeing only the mayonnaise. Most of the universe is invisible, and most astronomers now believe that the invisible matter is not the normal matter of which you and the stars are made. To follow that line of evidence, we must look for the atoms made during the big bang.

During the first few minutes of the big bang, nuclear reactions converted some protons into helium and a small amount into other elements. How much of these elements was created depends critically on the density of the material. Deuterium, for example, is an isotope of hydrogen in which the nucleus contains a proton and a neutron. The amount of deuterium produced in the big bang depends strongly on the density of normal matter. If there were a lot of normal particles such as protons and neutrons, then they would have collided with the deuterium nuclei and converted them into helium. If the density of normal particles was less, more deuterium would survive.

Lithium is another nucleus that could have been made in small amounts during the big bang. Figure 18-8 shows that there is a gap between helium and lithium; there is no stable nucleus with atomic mass 5, so regular nuclear reactions during the big bang could not convert helium into lithium. If, however, the density of normal matter such as protons and neutrons was high enough, a few nuclear reactions could have leaped the gap and produced a few lithium atoms.

Deuterium is so easily converted into helium that none can be made in stars. In fact, stars destroy what deuterium they have by converting it into helium. Lithium too is destroyed in stars. Using the largest telescopes, astronomers have been able to measure the chemical abundance of gas clouds near quasars. The look-back times to these gas clouds is so great that we see them as they were before stars could have altered the abundance of the elements. The observed abundances of deuterium and lithium tell us that the normal matter in the big bang, the protons and neutrons, could not have amounted to more than 5 percent of the mass needed to make the universe flat (▌ Figure 18-14). Yet observations show that galaxies and galaxy clusters contain large amounts of dark matter. The protons and neutrons that make up normal matter belong to a family of subatomic particles called *baryons,* so cosmologists believe that the dark matter must be **nonbaryonic matter**. No more than about 5 percent of the mass in the universe can be baryonic; the dark matter must be nonbaryonic.

This nonbaryonic matter must be made up of particles that don't often interact with normal matter. If the nonbaryonic particles don't interact very often, we would not notice them except for their gravitational influences. Particles called axions and neutralinos, for example, are predicted by some theories but have never been detected in the laboratory. If they exist, their mass could exert gravitational forces, and they could be part of the dark matter. Data from a large particle detector show that related particles called WIMPs, weakly interacting massive particles, may indeed exist, but the result is still being tested. To what extent WIMPs contribute to the dark matter is unknown.

For some years, astronomers thought that neutrinos could be an important part of the dark matter. We saw in Chapter 8 how observations made by underground detectors suggested that neutrinos oscillate, and that is important because, according to quantum mechanics, if neutrinos oscillate, they cannot have zero mass. Although the mass of neutrinos has not been measured, it is clearly very small. Even a small mass might be important because there are so many neutrinos in the universe—10^8 neutrinos for every normal particle.

Theoretical models of galaxy formation in the early universe tell us that the dark matter can't be mostly neutrinos. Dark matter composed of neutrinos and similar particles that travel at or near the speed of light is called **hot dark matter.** Such fast-moving particles do not clump together easily and could not have stimulated the formation of objects as small as galaxies and clusters of galaxies. The most successful models of galaxy formation require that the dark matter be made up of **cold dark matter,** meaning the particles move slowly and can clump into smaller structures. WIMPs, for example, are massive and slow moving and could be part of the cold dark matter.

Nonbaryonic dark matter does not interact significantly with normal matter or with photons, which is why we can't detect it easily. But that means that dark matter was not affected by the intense radiation that prevented normal matter from contracting into denser clouds right after the big bang. The gravitation of dark

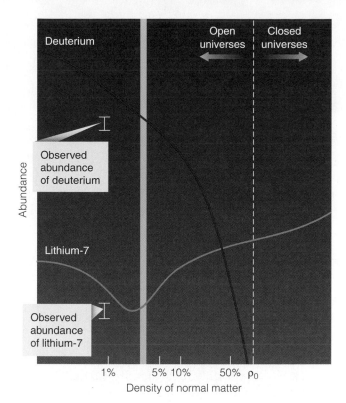

FIGURE 18-14

This graph plots the abundance of deuterium and lithium-7 versus the present density of the universe. Observations of the abundance of deuterium and lithium-7 are uncertain, but they limit the density of normal matter in the universe to a narrow range (green bar). The density of normal matter cannot be more than 5 percent of the critical density ρ_0.

matter could pull it into clouds; and, once the density of radiation fell low enough, normal matter could begin falling into the clouds of dark matter to form the first galaxies. Thus, cold, nonbaryonic dark matter may have given the universe a head start in forming galaxies.

Now that we understand more about dark matter we can evaluate the possibility that the halos of galaxies contain large numbers of very-low-mass stars, brown dwarfs, or planets too faint to see. These **MACHOs,** for Massive Compact Halo Objects, can be detected if they pass between our telescopes on Earth and a distant star. The gravitational field of the MACHO acts as a gravitational lens, focusing the light of the star and making it grow brighter for a period of a few weeks as the MACHO passes between the star and Earth. Extensive searches have detected such events (▌ Figure 18-15) but not in large numbers. Furthermore, searches have turned up large numbers of white dwarfs in the halo of our galaxy. However common these halo objects are, they are made of baryonic matter, and the abundance of deuterium and lithium tells us that the dark matter is nonbaryonic. It does not appear that MACHOs and white dwarfs can make up a significant part of the dark matter.

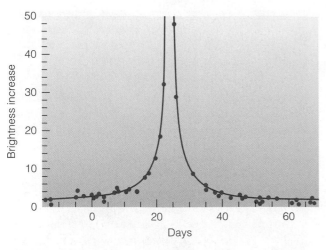

FIGURE 18-15
Measurements of the brightness of a distant star (dots) increased and then decreased exactly as predicted by general relativity (curve) for a star's passage behind a gravitational lens caused by a MACHO. In this example, the star brightened by a factor of over 40.

Although there is good evidence that dark matter exists, it has proven difficult to identify. No form of dark matter has been found that is abundant enough to provide the critical density needed to make the universe flat. In fact, all forms of dark matter appear to add up to less than 30 percent of the critical density. As we will see later in this chapter, there is more to the universe than meets the eye, and more even than the dark matter.

REVIEW CRITICAL INQUIRY

Why do astronomers think that dark matter can't be baryonic?

Isotopes like deuterium and lithium-7 were produced in the first minutes of the big bang, and the abundance of those elements depends strongly on the density of protons and neutrons. Because these particles belong to the family of particles called baryons, astronomers refer to normal matter as baryonic. Measurements of the abundance of deuterium and lithium-7 show that the universe cannot contain more baryons than 5 percent of the critical density. Yet observations of galaxies and galaxy clusters show that dark matter must make up almost 30 percent of the critical density. Consequently, astronomers conclude that the dark matter must be made up of nonbaryonic particles.

Finding the dark matter is important because the density of matter in the universe determines its curvature. How does curvature allow us to avoid an edge or center in a finite model of the universe?

By adding the curvature of space-time and effects of dark matter to our cosmology, we have made our theories much more sophisticated. We are now ready to explore the most recent and most exciting advances in cosmology.

18-3 21ST-CENTURY COSMOLOGY

Cosmology made dramatic advances around the beginning of the 21st century. It may be that astronomers are finally beginning to understand the overall nature of the universe.

INFLATION

In 1980, astronomers faced a problem. The big bang models of the universe could not explain two important features of the universe. To solve those two problems, a new theory of the big bang was created, and that theory has been startlingly successful. To introduce the new theory, we begin with the two problems.

One of the problems is called the **flatness problem.** The universe seems to be balanced near the boundary between an open and a closed universe. That is, it seems nearly flat. Given the vast range of possibilities, from zero to infinite, it seems peculiar that the density of the universe is within a factor of 10 of the critical density that would make it flat. If dark matter is as common as it seems, the density may be even closer than a factor of three to being perfectly flat.

Even a small departure from critical density when the universe was young would be magnified by subsequent expansion. To be so near critical density now, the density of the universe during its first moments must have been within 1 part in 10^{49} of the critical density. So the flatness problem is: Why is the universe so nearly flat?

The second problem with the big bang theory is the isotropy of the primordial microwave background radiation. When we correct for the motion of our galaxy, we see the same background radiation in all directions to at least 1 part in 1000. Yet when we look at background radiation coming from two points in the sky separated by more than a degree, we look at two parts of the big bang that were not causally connected when the radiation was emitted. That is, when recombination occurred and the gas of the big bang became transparent to the radiation, the universe was not old enough for any signal to have traveled from one of these regions to the other. Thus, the two spots we look at did not have time to exchange heat and even out their temperatures. Then how did every part of the entire big bang universe get to be so nearly the same temperature by the time of recombination? This is called the **horizon problem** because the two spots are said to lie beyond their respective light-travel horizons.

The key to these two problems and to others involving subatomic physics may lie with the theory called the **inflationary universe** because it predicts a sudden expansion when the universe was very young, an expansion even more extreme than that predicted by the big bang theory.

To understand the inflationary universe, we must recall that physicists know of only four forces—gravity, the electromagnetic force, the strong force, and the weak force (Chapter 8). We are familiar with gravity, and the electromagnetic force is responsible for making magnets stick to refrigerator doors and cat hair stick to wool sweaters charged with static electricity. The strong force holds atomic nuclei together, and the weak force is involved in certain kinds of radioactive decay.

For many years, theorists have tried to unify these forces; that is, they have tried to describe the forces with a single mathematical law. A century ago, James Clerk Maxwell showed that the electric force and the magnetic force were really the same effect, and we now count them as a single electromagnetic force. In the 1960s, theorists succeeded in unifying the electromagnetic force and the weak force in what they called the electroweak force, effective only for processes at very high energy. At lower energies, the electromagnetic force and the weak force behave differently. Now theorists have found ways of unifying the electroweak force and the strong force at even higher energies. These new theories are called **grand unified theories, or GUTs**.

According to the inflationary universe, the universe expanded and cooled until about 10^{-35} second after the big bang, when it became so cool that the forces of nature began to disconnect from each other and behave in different ways. This released tremendous energy, which suddenly inflated the universe by a factor between 10^{20} and 10^{30} (▌ Figure 18-16). At that time the part of the universe that we can see now, the entire observable universe, was no larger than the volume of an atom, but it suddenly inflated to the volume of a cherry pit and then continued its slower expansion to its present extent.

That sudden inflation can solve the flatness problem and the horizon problem. The sudden inflation of the universe would have forced whatever curvature it had toward zero, just as inflating a balloon makes a small spot on its surface flatter. Thus, we now see the universe nearly flat because of that sudden inflation long ago. In addition, because the part of the universe we can see was once no larger in volume than an atom, it had plenty of time to equalize its temperature before the inflation occurred. Now we see the same temperature for the background radiation in all directions.

The inflationary universe is based, in part, on quantum mechanics, and a slightly different aspect of quantum mechanics may explain why there was a big bang at all. Theorists believe that a universe totally empty of matter could be unstable and decay spontaneously by creating pairs of particles until it was filled with the hot, dense state we call the big bang. This theoretical discovery has led some cosmologists to believe that the universe could have been created by a chance fluctuation in space-time. In the words of physicist Frank Wilczyk, "The reason there is something instead of nothing, is that 'nothing' is unstable."

The inflationary theory predicts that the universe is almost perfectly flat. That is, the true density must equal the critical density. A theory can never be used as evidence, but the beauty of the inflationary theory has given many cosmologists confidence that the universe must be flat. Observations, however, seem to tell us the universe does not contain enough matter (baryonic plus dark) to be flat, so we must search further to understand the nature of the universe.

THE ACCELERATION OF THE UNIVERSE

Ever since Hubble discovered the expansion of the universe, astronomers have known what to expect. They have also known that expectations are not reliable; and, as the 21st century approached, astronomers made a mind-boggling discovery that contradicted everyone's expectations.

Both common sense and mathematical models suggest that as the galaxies recede from each other, the expansion should be slowed by gravity trying to pull the galaxies toward each other. How much the expansion

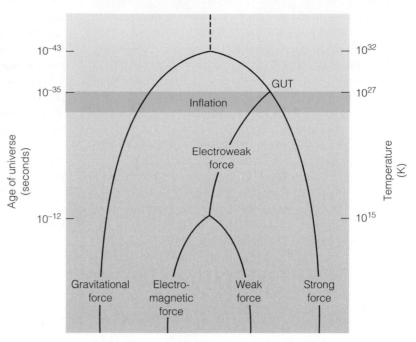

FIGURE 18-16

When the universe was very young and hot (top), the four forces of nature were indistinguishable. As the universe began to expand and cool, the forces separated and triggered a sudden inflation in the size of the universe.

is slowed should depend on the amount of matter in the universe. If the density of matter is less than the critical density, the expansion should be slowed only slightly, and the universe should expand forever. If the density of matter in the universe is greater than the critical density, the expansion should be slowing down dramatically, and the universe should eventually stop expanding and begin contracting. Notice that this is the same as saying a low-density universe should be open and a high enough density universe should be closed.

For decades, astronomers struggled to measure the Hubble constant and detect the slowing of the expansion. A direct measurement of the rate of slowing would reveal the true curvature of the universe. This was one of the key projects for the Hubble Space Telescope, and two teams of astronomers spent years making the measurements.

Detecting a change in the rate of expansion is a difficult project because it requires accurate measurements of the distances to very remote galaxies, and both teams used the same technique. They calibrated type Ia supernovae as distance indicators. A type Ia supernova occurs when a white dwarf gains matter from a companion star, exceeds the Chandrasekhar limit, and collapses in a supernova explosion. Because all such white dwarfs should collapse at the same mass limit, they should all produce explosions of the same luminosity, and that makes them good distance indicators. The teams calibrated type Ia supernovae by locating such supernovae occurring in nearby galaxies whose distance was known from Cepheid variables and other reliable distance indicators. Once the peak luminosity of type Ia supernovae had been determined, they could be used to find the distance to much more distant galaxies.

Both teams announced their results in 1998. The expansion of the universe is not slowing down. It is speeding up! That is, the expansion of the universe is accelerating (❙ Figure 18-17).

The announcement that the expansion of the universe is accelerating was exciting, but astronomers immediately began testing it. It depends critically on the calibration of type Ia supernovae as distance indicators, and some astronomers suggested that the calibration might be wrong (Window on Science 15-1). Some unknown kind of dust might make very distant supernovae too faint without making them too red. No reddening had been detected, so normal dust seemed to be ruled out. Or perhaps the very distant supernovae were too faint because we see them at great look-back times, and some unknown process made those early supernovae a bit less luminous than more recent supernovae. Those problems with the calibration seem to be ruled out by a supernova that occurred in 1997. Known as SN1997ff, it was recognized in 2001 as a type Ia supernova with the very large red shift of 1.7, implying a distance of 10 billion light-years (❙ Figure 18-18). The ob-

servations of the maximum brightness are uncertain, but they seem to rule out problems with the calibration of type Ia supernovae. The universe really does seem to be expanding faster and faster.

If the expansion of the universe is accelerating, then there must be a force of repulsion in the universe, and astronomers are struggling to understand what it could be. One possibility leads us back to 1916.

When Albert Einstein published his theory of general relativity in 1916, he recognized that his equations describing space and time implied that space had to contract or expand. The galaxies could not float unmoving in space because their gravity would pull them together. The only solutions seemed to be a universe that was contracting under the influence of gravity or a universe in which the galaxies were rushing away from each other so rapidly that gravity could not pull them together. In 1916, astronomers did not yet know that the universe was expanding, so Einstein made what he later said was a mistake.

To balance the attractive force of gravity, Einstein added a constant to his equations called the **cosmological constant,** represented by an upper-case lambda (Λ). The constant plays the mathematical role of a force of repulsion that balances the gravitation in the universe so it does not have to contract or expand. Thirteen years later, in 1929, Edwin Hubble announced that the universe was expanding, and Einstein said introducing the cosmological constant was his biggest blunder. Modern astronomers aren't so sure.

One explanation for the acceleration of the universe is that the cosmological constant is not zero but rather represents a force that drives a continuing acceleration in the expansion of the universe. Because the cosmological constant remains constant with time, the universe would have to have experienced this acceleration throughout its history.

Another solution is to suppose that totally empty space, the vacuum, contains energy, which drives the acceleration. This is an interesting possibility because for years theoretical physicists have discussed an energy inherent in empty space. Astronomers have begun referring to this energy of the vacuum as **quintessence.** Unlike the cosmological constant, quintessence need not remain constant over time, so it may be more successful in explaining the observed distribution of galaxies and galaxy clusters.

Whichever explanation is right, the cosmological constant or quintessence, the acceleration alerts us to a form of energy spread throughout space. From Einstein's most famous quotation, $E = mc^2$, we know that energy and mass are equivalent, so the tremendous energy that drives the repulsion behaves like a mass spread through space, and cosmologists think that the invisible mass–energy can provide the density needed to make the universe flat. Astronomers refer to this en-

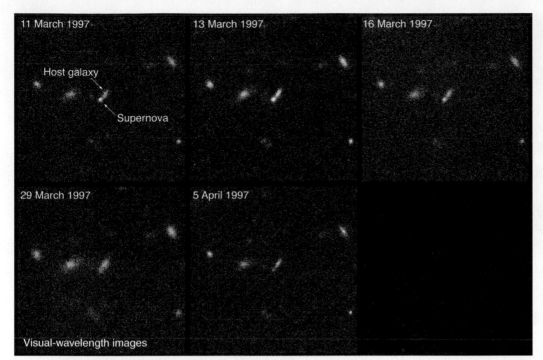

FIGURE 18-17

(a) A type Ia supernova erupts in a very distant galaxy and begins to fade. By calibrating these supernovae, astronomers were able to find the distances to some of the farthest visible galaxies. *(ESO)* (b) Type Ia supernovae in distant galaxies are about 25 percent too faint, which must mean the galaxies are farther away from us than they would be in a universe expanding at a constant rate. This diagram shows observations of supernovae compared with a decelerating, flat universe dominated by dark matter (blue line). The red line shows the relationship for an accelerating, flat universe.

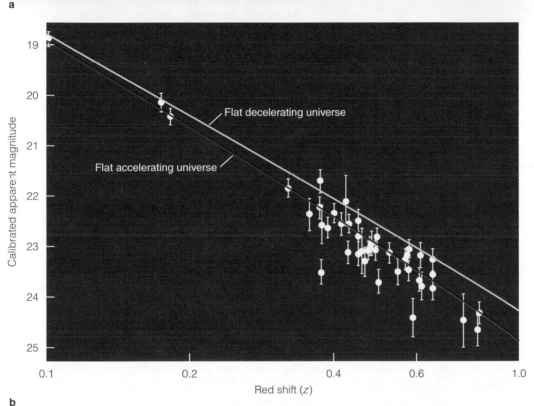

ergy as **dark energy,** the energy inherent in empty space that drives the acceleration of the universe but does not contribute to the formation of starlight or the cosmic microwave background radiation.

The inflationary theory of the universe was proposed to solve two problems, the horizon problem and the flatness problem, but it makes the specific prediction that the universe is flat. The discovery of dark energy seems to confirm that discovery. Baryonic matter plus dark matter makes up about a third of the critical density, and dark energy appears to make up two-thirds. That is, the total density of the universe seems to equal the critical density, which means that the geometry of the universe is flat.

FIGURE 18-18

In a 1997 follow-up image of the Hubble Deep Field made in 1995, astronomers noticed a faint galaxy that was brighter in the second image (lower left). Subtracting the earlier image from the later image revealed a supernova cataloged as SN1997ff (lower right). The unexpected faintness of the supernova confirms that the universe is accelerating. *(Adam Riess, STScI/NASA)*

Acceleration helps with another problem. If the Hubble constant equals 70 km/ s/Mpc, then the Hubble time is about 14 billion years. If the universe is flat, then the age of the universe is two-thirds of the Hubble time, which equals about 9 billion years. That is younger than the globular clusters, and that is a problem that has been hard to explain. However, if the expansion of the universe has been accelerating, then it must have been expanding slower in the past, and that means it can be older than two-thirds of the Hubble time. The latest estimates suggest that the true age of the universe is about 14 billion years, and that solves the age problem.

Acceleration opens many possibilities that astronomers had not considered before. For example, the fate of the universe may not be determined by its geometry. A closed universe that is accelerating, for example, does not have to fall back. It could expand forever, depending on the way the acceleration changes with time.

The discovery of an acceleration of the universe is one of the most important discoveries in modern astronomy, but it leaves us wanting confirmation for this amazing result. Is there some other way to show that the universe is flat? One way to measure the curvature of space-time is to draw circles on it and measure the areas of the circles. But that is only practical in a simple cartoon analogy. In reality, astronomers must measure the size of things at great look-back times, and that measurement is yielding yet more exciting results, which are described in the next section.

THE ORIGIN OF STRUCTURE AND THE CURVATURE OF THE UNIVERSE

For a number of years, astronomers have studied the distribution of galaxies in space, what they refer to as structure. Those studies have led to three astonishing discoveries that are changing our understanding of the universe.

When we look at galaxies in the sky we see them in clusters ranging from a few to thousands, and those clusters appear to be grouped into **superclusters** (❚ Figure 18-19). The Local Supercluster, in which we live, is a roughly disk-shaped swarm of galaxy clusters 50 to 75 Mpc in diameter. By measuring the red shifts and positions of over 100,000 galaxies in great slices across the sky, astronomers have been able to create maps revealing that the superclusters are distributed in long, narrow filaments and thin walls that outline great voids nearly empty of galaxies (❚ Figure 18-20). These filaments and walls of superclusters stretching half a billion light-years are the largest structures in the universe. No larger objects exist.

These vast structures are a problem for cosmologists because the cosmic microwave background radiation is very uniform, and that means the gas of the big bang must have been extremely uniform at the time of recombination. Yet the look-back time to the furthest galaxies is about 93 percent of the way back to the big bang. How did the uniform gas at the time of recombination coagulate so quickly to form galaxies? Worse, how did that highly uniform gas form such large structures as the filaments, voids, and walls?

Baryonic matter is so rare in the universe it does not have enough gravity to pull itself together rapidly. Astronomers believe the nonbaryonic dark matter must provide the gravity, and models have attempted to describe how dark matter can form structure. As we have seen earlier in this chapter, hot dark matter does not clump into small enough structures, so the dark matter must be cold dark matter.

But what started the clumping of matter? Theorists say that subatomic quantum fluctuations in space would have been stretched at the moment of inflation to very large but very subtle variations in gravitational fields that could have stimulated the formation of clusters, filaments, and walls. The structure we see in Figure 18-20 may be the ghostly traces of quantum fluctuations in the infant universe.

The inflationary theory of the universe makes very specific predictions about the sizes of the fluctuations we should see in the cosmic microwave background radiation. Observations made by the COBE satellite in 1992 detected the largest variations, but detecting the smaller variations is critical for testing the theory. A half-dozen teams of astronomers have built specialized telescopes to make these observations. Some have flown under balloons high in the atmosphere, and others have observed from the ice of Antarctica. The most ex-

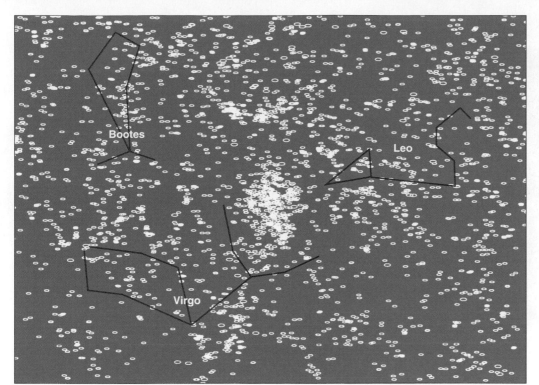

FIGURE 18-19
The distribution of brighter galaxies in the sky reveals the great Virgo cluster (center), containing over 1000 galaxies only about 17 Mpc away. Other clusters fill the sky, such as the more distant Coma cluster just above the Virgo cluster in this diagram. The Virgo cluster is linked with others to form the Local Supercluster.

tensive observations have been made by the Wilkinson Microwave Anisotropy Probe (WMAP), a robotic infrared telescope that observed from space.

The background radiation is very isotropic—it looks almost exactly the same in all directions. However, when the average intensity is subtracted from each spot on the sky, small irregularities are evident. That is, some spots on the sky look a tiny bit hotter and brighter than other spots. (▌Figure 18-21). Those irregularities contain lots of information.

If the irregularities in the cosmic microwave background radiation were produced by inflation, then their size must depend on the velocity of the pressure waves (sound) that could travel through the gas at the time of recombination. Theory predicts most of the irregularities should be about one degree in diameter if the universe is flat. If the universe were open, the most common irregularities would be smaller. Careful measurements of the size of the irregularities in the cosmic background radiation show that the observations fit the

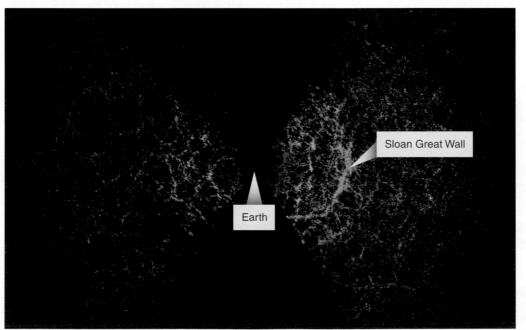

FIGURE 18-20
Nearly 70,000 galaxies are plotted in this double slice of the universe extending outward in the plane of Earth's equator. The nearest galaxies are shown in red and the more distant in green and blue. The galaxies form filaments and walls enclosing empty voids. The Sloan Great Wall is almost 1.4 billion light-years long and is the largest known structure in the universe. The most distant galaxies in this diagram are roughly 3 billion light-years from Earth. *(Sloan Digital Sky Survey)*

FIGURE 18-21
Data from the WMAP spacecraft were used to make this far-infrared map of the entire sky. The infrared glow from dust in our solar system and in the Milky Way Galaxy has been removed to reveal tiny irregularities in the cosmic microwave background radiation. Red and yellow spots are slightly warmer than the green and blue spots. *(NASA/WMAP Science Team)*

theory very well for a flat universe (▌Figure 18-22). Not only is the theory of inflation confirmed, an exciting result itself, but these data tell us that the universe is flat, which indirectly confirms the existence of dark energy and the acceleration of the universe.

Cosmologists can analyze the irregularities using sophisticated mathematics to find out how commonly the spots of different sizes recur. The mathematics con-

firm that spots about 1 degree in diameter are the most common, but spots of other sizes recur as well, and it is possible to plot a graph such as ▌Figure 18-23 to show how common different size irregularities are.

The results from the WMAP observations tell us a great deal about the universe. The universe is flat, accelerating, and will expand forever. The age of the universe derived from the data is 13.7 billion years. Fur-

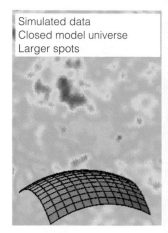
Simulated data
Closed model universe
Larger spots

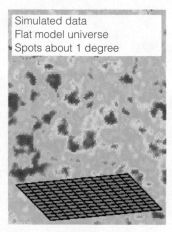
Simulated data
Flat model universe
Spots about 1 degree

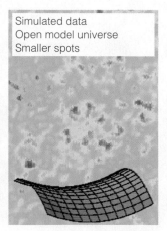
Simulated data
Open model universe
Smaller spots

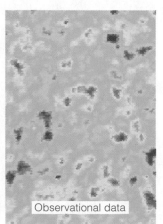
Observational data

FIGURE 18-22
The size of the irregularities in the background radiation can be seen in this diagram. Compare the observations at right with the three simulations at left. The observed size of the irregularities fits best with cosmological models having flat geometry. Detailed mathematical analysis confirms our visual impressions. The universe is flat. *(Courtesy of the BOOMERANG Collaboration)*

thermore, the smaller peaks in the curve reveal that the universe contains 4 percent baryonic (normal) matter, 23 percent dark matter, and 73 percent dark energy. The Hubble constant is confirmed to be 71 km/s/Mpc. The inflationary theory is confirmed, and the data support the cosmological constant version of dark energy, although quintessence is not ruled out. Hot dark matter is ruled out. The dark matter needs to be cold dark matter to clump together so rapidly after the big bang. In fact, the first stars began to produce light when the universe was only about 200 million years old. This is much earlier than most astronomers had expected.

WMAP and other studies of the cosmic microwave background radiation have revolutionized cosmology. At last, astronomers have accurate observations against which to test theories. The basic constants are known to a percent or so. On reviewing these results, one cosmologist announced that "Cosmology is solved!" but that may be premature. We don't understand dark matter or the dark energy that drives the acceleration, so over 90 percent of the universe is not understood. Hearing this, another astronomer suggested a better phrase was "cosmology in crisis." Certainly there are further mysteries to be explored, but cosmologists are growing more confident that they can describe the overall properties of the universe.

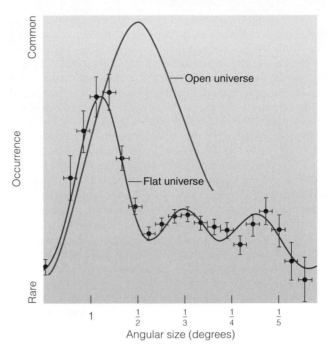

FIGURE 18-23

This graph shows how commonly irregularities of different sizes occur in the cosmic microwave background radiation. Irregularities of about 1 degree in diameter are most common. Models of the universe that are open or closed are ruled out. The data fit a flat model of the universe very well. Crosses on data points show the uncertainty in the measurements

REVIEW CRITICAL INQUIRY

How does inflation theory solve the flatness problem?

The flatness problem can be stated as a question: Why is the universe so nearly flat? After all, the density of matter in the universe could be anything from zero to infinite, but the observed density of baryonic and dark matter adds up to nearly 30 percent of the critical density that would produce a flat universe. Furthermore, the density must have been astonishingly close to the critical density when the universe was very young. Otherwise it would not be so close now. The inflationary theory solves this problem by proposing that the universe underwent a moment of rapid inflation when it was a tiny fraction of a second old. That inflation drove the universe toward flatness just as the inflation of a balloon makes a spot on the balloon nearly flat.

Understanding theory in cosmology is critically important, but science depends on evidence. What evidence do we have that the expansion of the universe is accelerating?

Although we have traced the origin of the universe, the origin of the elements, and the birth and death of stars, we have left out one important class of objects—planets. How do Earth and other planets fit into this grand scheme of origins? That is the subject of the following chapters.

SUMMARY

We conclude that it is impossible for the universe to have an edge, and because we find the centers of things by reference to their edges, we also conclude that the universe cannot have a center.

The fact that the night sky is dark leads us to conclude that the universe is not infinitely old. If it were infinite in extent and age, then every spot on the sky would glow as brightly as the surface of a star. This problem, known incorrectly as Olbers's paradox, forces us to conclude that the universe had a beginning.

Edwin Hubble's 1929 discovery that the red shift of a galaxy is proportional to its distance is known as the Hubble law. Tracing this expansion backward in time, we come to an initial high-density, high-temperature state commonly called the big bang. Although the expanding universe began from this big bang, it has no center. The galaxies recede from each other.

The cosmic microwave background radiation is black body radiation with a temperature of 2.73 K uniformly spread over the entire sky. It is the light from the big bang freed from the gas at the moment of recombination and red shifted by a factor of 1100. The background radiation is strong evidence that the universe began with a big bang.

During the first three minutes of the big bang, nuclear fusion converted some of the hydrogen into helium but was unable to make many other heavy atoms because no stable nuclei exist with weights of 5 or 8.

The universe appears to be isotropic and homogeneous, leading to the cosmological principle, the assumption that the universe looks the same from every location.

The geometry of space-time explains the red shift as a stretching of the photons as they travel through expanding space-time. The universe could be open if its density is less than the critical density and closed if its density is more. If its density equals the critical density, it is flat. The dark matter in the universe is important because it contributes to the density and helps determine the overall curvature. Observations of the abundance of deuterium and lithium-7 tell us that baryonic matter can make up roughly 4 percent of the critical density. Cold dark matter appears to make up less than 30 percent.

The inflationary theory of the universe solves the flatness problem and the horizon problem by proposing that the universe expanded dramatically when it was a tiny fraction of a second old. Inflation predicts that the universe is flat.

Observations of type Ia supernovae reveal that the expansion of the universe is accelerating because of the cosmological constant or because of quintessence. The dark energy of empty space makes up about 73 percent of the critical density and makes the universe flat.

Observation of structure in the universe—irregularities in the cosmic microwave background radiation and the clusters of galaxies—provide strong evidence that the universe is flat and accelerating. These irregularities appear to be subatomic quantum fluctuations in space-time magnified when inflation occurred. The observations show that the age of the universe is 13.7 billion years.

NEW TERMS

cosmology

Olbers's paradox

observable universe

big bang

Hubble time

cosmic microwave background radiation

steady-state theory

antimatter

recombination

dark age

reionization

isotropy

homogeneity

cosmological principle

closed universe

flat universe

open universe

critical density

oscillating universe theory

nonbaryonic matter

hot dark matter

cold dark matter

MACHOs

flatness problem

horizon problem

inflationary universe

grand unified theories (GUTs)

cosmological constant (Λ)

quintessence

dark energy

supercluster

REVIEW QUESTIONS

Ace⊙Astronomy™ Assess your understanding of this chapter's topics with additional quizzing and animations at http://astronomy.brookscole.com/seeds8e

1. How does the darkness of the night sky tell us something important about the universe?

2. How can we be located at the center of the observable universe if we accept the Copernican principle?

3. Why can't an open universe have a center? Why can't a closed universe have a center?

4. What evidence do we have that the universe is expanding? that it began with a big bang?

5. Why couldn't atomic nuclei exist when the universe was younger than 2 minutes?

6. Why is it difficult to determine the present density of the universe?

7. How does the inflationary universe theory resolve the flatness problem? the horizon problem?

8. If the Hubble constant were really 100 km/s/Mpc, much of what we understand about the evolution of stars and star clusters must be wrong. Explain why. (*Hint:* What would the age of the universe be?)

9. Why do we conclude that the universe must have been very uniform during its first million years?

10. What is the difference between hot dark matter and cold dark matter? What difference does it make to cosmology?

11. What evidence do we have that the expansion of the universe is accelerating?

12. What evidence do we have that the universe is flat?

DISCUSSION QUESTIONS

1. Do you think Copernicus would have accepted the cosmological principle? Why or why not?

2. If we reject any model of the universe that has an edge in space because we can't comprehend such a thing, shouldn't we also reject any model of the universe that has a beginning or an ending? Are those just edges in time, or is there a difference?

PROBLEMS

1. Use the data in Figure 18-3 to plot a velocity–distance diagram, find *H,* and estimate the Hubble time.

2. If a galaxy is 8 Mpc away from us and recedes at 456 km/s, what is *H*? What is the Hubble time? How old would the universe be if it were flat and there were no acceleration?

3. At what wavelength of maximum intensity did the gas of the big bang radiate at the time of recombination? By what factor is that different from the wavelength of maximum of the cosmic microwave background radiation?

4. If the average distance between galaxies is 2 Mpc, and the average mass of a galaxy is 10^{11} solar masses, what is the average density of matter in the universe? (*Hints:* The volume of a sphere is $\frac{4}{3}\pi r^3$, and the mass of the sun is 2×10^{33} g.)

5. Figure 18-12 is based on an assumed Hubble constant of 70 km/s/Mpc. How would you change the diagram to fit a Hubble constant of 50 km/s/Mpc?

6. Hubble's first estimate of the Hubble constant was 530 km/s/Mpc. If his distances were too small by a factor of 7, what answer should he have obtained?

7. What was the maximum age of the universe predicted by Hubble's first estimate of the Hubble constant?

8. If the value of the Hubble constant were found to be 60 km/s/Mpc, what would the Hubble time be? How old would the universe be if it were flat and there were no acceleration? How would acceleration change your answer?

CRITICAL INQUIRIES FOR THE WEB

1. What is the latest news concerning acceleration and the flatness of the universe? Search the Web for the most recent observations.

2. The steady-state theory was once a rival cosmology of the big bang. Search for Web sites that provide information on steady-state cosmology. (Be careful to locate legitimate sites that discuss the theory, rather than sites where individuals use steady-state ideas as part of nonscientific arguments on cosmology.) What were the key predictions of steady-state cosmology? How has recent evidence led to its decline?

3. Search for the Web pages of large surveys of galaxies, such as the 2dF Deep Field Survey and the Sloan Digital Sky Survey. What are they discovering about the largest and most distant objects in the universe?

4. What is the Microwave Anistropy Probe? Where is it? What are the latest results?

EXPLORING *THESKY*

1. Search for galaxy clusters. (*Hint:* Use the **View** menu to turn off everything but galaxies, deep sky objects, constellations, and labels. Zoom in until the field of view is 40° or smaller and the fainter galaxies appear. Search for clusters of galaxies in Ursa Major, Canes Venatici, Lynx, Virgo, and Coma Bereneces.)

2. Locate small clusters of galaxies and compare the brightness of the galaxies with those in large clusters. Can you tell that the smaller clusters tend to be further away, or is the range of galaxy luminosities too great? (*Hint:* Click on a galaxy to find its magnitude.)

3. Can you find galaxies and clusters of galaxies along the Milky Way? Turn on the Milky Way and search for galaxy clusters within its outline.

 Visit the Seeds *Foundations of Astronomy* companion Web site for critical thinking exercises, articles, and additional readings from InfoTrac College Edition, Brooks/Cole's online student library.

LIFE ON OTHER WORLDS

Did I solicit thee from darkness to promote me?

John Milton, Paradise Lost

Peter Sawyer

Chapter 19

GUIDEPOST

This chapter is either unnecessary or critical, depending on our point of view. If we believe that astronomy is the study of the physical universe above the clouds, then this chapter does not belong here. But if we believe that astronomy is the study of our position in the universe, not only our physical position but also our role as living beings in the origin and evolution of the universe, then everything else in this book is just preparation for this chapter.

Astronomy is the only science that truly acts as a mirror. In studying the universe up there, we learn what we are down here. Astronomy is not really about stars, galaxies, and planets; it is about us.

VIRTUAL LABORATORY

EXTRASOLAR PLANETS

406

The Nature of Scientific Explanation

Science is a way of understanding the world around us, and the heart of that understanding is the explanations that science gives us for natural phenomena. Whether we call these explanations stories, histories, theories, or hypotheses, they are attempts to describe how nature works based on fundamental rules of evidence and intellectual honesty. While we may take these explanations as factual truth, we should understand that they are not the only explanations that satisfy the rules of logic.

A separate class of explanations involves religion, and those explanations can be quite logical. The Old Testament description of the creation of the world, for instance, does not fit scientific observations, but if we accept the existence of an omnipotent being, then the biblical explanation is internally logical and acceptable. Of course, it is not a scientific explanation, but religion is a matter of faith and not subject to the rules of evidence. Religious explanations follow their own logic, and we would be wrong to demand that they follow the rules of evidence that govern scientific explanations.

If scientific explanations are not the only logical explanations, then why do we give them such weight? First, we must notice the tremendous success of scientific explanations in producing technological advances in our daily lives. Diseases such as chickenpox are more childhood irritations than life-threatening illnesses thanks to the application of scientific explanations to modern medicine. The power of science to shape our world can lead us to think that its explanations are unique. Second, the process we call science depends on the use of evidence to test and perfect our explanations, and the logical rigor of this process gives us great confidence in our conclusions.

Scientific explanations have given us tremendous insight into the workings of nature, and consequently both scientists and nonscientists tend to forget that there can be other logical explanations. The so-called conflict between science and religion has been symbolized for centuries by the trial of Galileo. That conflict is easier to understand when we consider the nature of scientific explanations and the role of evidence in testing scientific understanding.

As living things, we have been promoted from darkness. We are made of heavy atoms that could not have formed at the beginning of the universe. Successive generations of stars fusing light elements into heavier elements have built the atoms so important to our existence. When a dark cloud of interstellar gas enriched in these heavy atoms fell together to form our sun, a small part of the cloud gave birth to the planet we inhabit.

Are there intelligent beings living on other planets? That is the last and perhaps the most challenging question in our study of astronomy. We will try to answer it in three steps, each dealing with a different aspect of life.

First, we must decide what we mean by *life.* Life is not so much a form of matter as it is a behavior. Living matter extracts energy from its environment to modify itself and its surroundings so as to preserve itself and to create offspring. Thus, life is based on information, the recipe for the process of survival and reproduction.

Second, we must study the origin of life on Earth. If we can understand how life began on our planet, then we can try to estimate the likelihood that it has originated on other worlds as well.

Third, we must try to understand how the first primitive living organism on Earth could give rise to the tremendously diverse life forms that now inhabit our world. By understanding how living things evolve to fit their environment, we will see evidence that intelligence is a natural development in the evolution of life.

If life can originate on other worlds, and if intelligence is a natural result of the evolution of life forms, then we might expect that other intelligent races inhabit other worlds. Visits between worlds seem impossible, but perhaps we can detect their existence by radio.

Alien life forms could be quite different from us, but if they are alive, then they must share with us certain characteristics. Our goal in this chapter is to use our knowledge of science to explain the greatest of mysteries—the origin and evolution of life on Earth and on other worlds (Window on Science 19-1).

19-1 THE NATURE OF LIFE

What is life? Philosophers have struggled with that question for thousands of years, so it is unlikely that we will answer it here. But we must agree on a working model of life before we can speculate on its occurrence on other worlds. To that end, we will identify in living things two important aspects: a physical basis and a unit of controlling information.

THE PHYSICAL BASIS OF LIFE

On Earth, the physical basis of life is the carbon atom (❙ Figure 19-1). Because of the way this atom bonds to other atoms, it can form long, complex, stable chains that are capable of extracting, storing, and utilizing energy. Other chemical bases of life may exist. Science fiction stories and movies abound with silicon creatures, living things whose body chemistry is based on silicon rather than carbon. However, silicon forms weaker bonds than carbon does, and it cannot form double bonds as easily. Consequently, it cannot form the long, complex, stable chains that carbon can. Silicon is 135 times more common on Earth than carbon is, yet there are no silicon creatures among us. All Earth life is carbon based. Thus, the likelihood that distant planets are inhabited by silicon people seems small, but we cannot rule out life based on noncarbon chemistry.

In fact, nonchemical life might be possible. What is required is some mechanism capable of supporting the extraction and utilization of energy that we have identified as life. One could at least imagine life based on electromagnetic fields and ionized gas. No one has ever met such a creature, but science fiction writers conjure up all sorts.

Clearly, we could range far in space and time, theorizing about different bases for alien life, but to make progress we must discuss what we know best—carbon-based life on Earth. How can a lump of carbon-rich matter live? The answer lies in the information that guides its life processes.

INFORMATION STORAGE AND DUPLICATION

Living cells are tiny chemical factories. They must store all of the recipes for those chemicals in a safe place, use them to fulfill the cell's task, and hand down duplicates of the recipes to offspring. That information is encoded on long molecules.

Study ▌ "DNA: The Code of Life" on pages 410 and 411 and notice three important points. First, the chemical recipes of life are stored as templates on DNA molecules. The templates automatically guide specific chemical reactions within the cell. Second, the instruc-

a

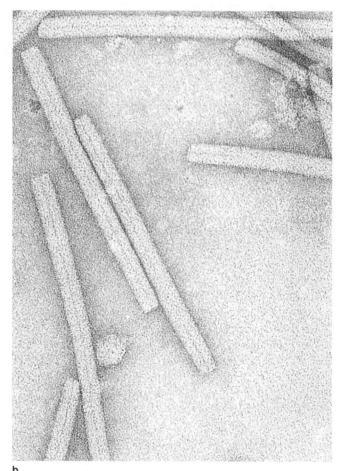

b

FIGURE 19-1

All living things on Earth are based on carbon chemistry. Even the long molecules that carry genetic information, DNA and RNA, have a framework defined by chains of carbon atoms. (a) Katie, a complex mammal, contains about 30 AU of DNA. *(Michael Seeds)* (b) Each rod of the tobacco mosaic virus contains a single spiral strand of RNA about 0.01 mm long. *(L. D. Simon)* All life on Earth stores its genetic information in such carbon-chain molecules.

tions stored in DNA are the genetic information handed down to offspring. When people say "You have your mother's eyes," they are talking about DNA codes. Finally, notice how the DNA molecule reproduces itself when a cell divides so that each new cell contains a copy of the original information.

Although the DNA molecule must preserve its coded information from damage and make accurate copies, it must also be capable of making mistakes now and then. To see why, we must consider how new DNA recipes are created.

MODIFYING THE INFORMATION

If living things are to survive for many generations, then the information stored in their DNA must change as the environment changes. Without change in DNA, a slight warming of the climate, for example, might kill a species of plant, in turn starving the rabbits, deer, and other plant eaters, and leaving the hawks, wolves, and mountain lions with no prey. If the information stored in DNA could never change, then environmental changes would quickly drive life forms to extinction. If life is to survive in a changing world, then the information in DNA must be changeable. Living things must evolve.

Species evolve by **natural selection**. Each time an organism reproduces, its offspring receive the data stored in the DNA, but some variation is possible. For example, most of the rabbits in a litter may be normal, but it is possible for one to get a DNA recipe that gives it stronger teeth. If it has stronger teeth, it may be able to eat something other than the plant the others depend on, and if that plant is becoming scarce, the rabbit with stronger teeth has a survival advantage. It can eat other plants and so will be healthier than its littermates and have more offspring. Some of these offspring will also have stronger teeth, as the altered DNA recipes are handed down to the new generation. Thus nature selects and preserves those attributes that contribute to the survival of the species. Those creatures that are unfit die. Natural selection is merciless to the individual, but it gives the species the best possible chance to survive in a changing environment.

The only way nature can obtain new DNA patterns from which to select the best is from DNA molecules that have changed. This can happen through chance mismatching of base pairs—errors—in the reproduction of the DNA molecule. Another way this can occur is through damage to reproductive cells from exposure to radioactivity such as cosmic rays or natural radioactivity in the soil. In any case, an offspring born with altered DNA is called a **mutant.** Most mutations make no difference at all because they change segments of DNA that are not being used. Many mutations are fatal, and the individual dies long before it can have offspring of its own. But in rare cases, a mutation may give a species a new survival advantage. Then natural selec-

tion makes it likely that the new DNA message will survive and be handed down, making the species more capable of surviving.

Evolution is not random. Of course, the errors that occur in the DNA code are indeed random, but natural selection is not random. Those changes in the DNA that help a species survive are selected and preserved in future generations. With each passing generation, the species becomes more fit to survive in its environment.

REVIEW CRITICAL INQUIRY

Why can't the information in DNA be permanent?

The information stored in a creature's DNA provides the recipes that make the creature what it is. For example, the DNA in a starfish must contain all the recipes for making the various kinds of proteins needed to consume and digest food. That information must be passed on to offspring starfish, or they will be unable to survive. But the information must be changeable because our environment is changeable. Ice ages come and go, mountains rise, lakes dry up, and ocean currents shift. If the environment changes in some way, one or more of the recipes may no longer work. In our example, a change in the temperature of the ocean water may kill off the specific shellfish the starfish eat. If they can't digest other shellfish, the entire species will become extinct. Natural variation in DNA means that among all the infant starfish in any generation, some of the recipes are different; if the environment changes, all of the old-style starfish may die, but a few—those with the different DNA—can carry on.

The survival of life depends on this delicate balance between reliable reproduction and the introduction of small variations in DNA information. What are some of the ways these small changes in DNA can arise?

Life is based not only on information, but also on the duplication of information. Today, that process seems so complex that it is hard to imagine how it could have begun.

19-2 THE ORIGIN OF LIFE

If life on Earth is based on the storage of information in these long, complex, carbon-chain molecules, how could it have ever gotten started? Obviously, 4.5 billion chemical bases didn't just happen to drift together to form the DNA formula for a human being. The key is evolution. Once a life form begins to reproduce itself, natural selection preserves the most advantageous traits. Over long periods of time spanning thousands, perhaps millions, of generations, the life form becomes more fit to survive. This often means the life form becomes more complex. Thus life could have begun as a

DNA: The Code of Life

The key to understanding life is information — the information that guides all of the processes in an organism. In most living things on Earth, that information is stored on a long spiral molecule called DNA (deoxyribonucleic acid).

The DNA molecule looks like a spiral ladder with rails made of phosphates and sugars. The rungs of the ladder are made of four chemical bases arranged in pairs. The bases always pair the same way. That is, base A always pairs with base T, and base G always pairs with base C.

Information is coded on the DNA molecule by the order in which the base pairs occur. To read that code, molecular biologists have to "sequence the DNA." That is, they must determine the order in which the base pairs occur along the DNA ladder.

DNA automatically combines raw materials to form important chemical compounds. The building blocks of these compounds are relatively simple **amino acids**. Segments of DNA act as templates that guide the amino acids to join together in the correct order to build specific **proteins,** chemical compounds important to the structure and function of organisms. Some proteins called **enzymes** regulate other processes. In this way, DNA recipes regulate the production of the compounds of life.

Mike Seeds

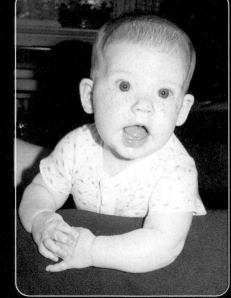

The traits we inherit from our parents, the chemical processes that animate us, and the structure of our bodies are all encoded in our DNA.

The Four Bases

A — Adenine

C — Cytosine

G — Guanine

T — Thymine

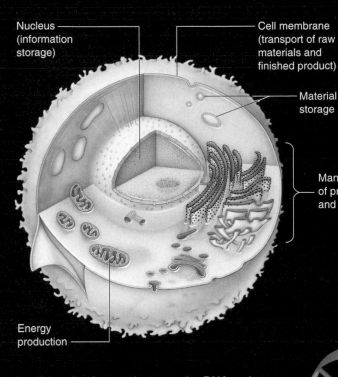

Nucleus
(information
storage)

Cell membrane
(transport of raw
materials and
finished product)

Material
storage

Manufacture
of proteins
and enzymes

Energy
production

A single cell from a human being contains about 1.5 meters of DNA containing about 4.5 billion base pairs — enough to record the entire works of Shakespeare 200 times. A typical human contains a total of about 600 AU of DNA. Yet the DNA in each cell, only 1.5 meters in length, contains all of the information to create a new human. A clone is a new creature created from the DNA code found in a single cell.

Original DNA

A cell is a tiny factory that uses the DNA code to manufacture chemicals. Most of the DNA remains safe in the nucleus of a cell, and the code is copied to create a molecule of **RNA (ribonucleic acid)**. Like a messenger carrying blueprints, the RNA carries the code out of the nucleus to the work site where the proteins and enzymes are made.

DNA, coiled into a tight spiral, makes up the **chromosomes** that are the genetic material in a cell. A **gene** is a segment of a chromosome that controls a certain function. When a cell divides, each of the new cells receives a copy of the chromosomes, as genetic information is handed down to new generations.

Copy DNA

To divide, a cell must duplicate its DNA. The DNA ladder splits, and new bases match to the exposed bases of the ladder to build two copies of the original DNA code. Because the base pairs almost always match correctly, errors in copying are rare. One set of the DNA code goes to each of the two new cells.

Copy DNA

Cell Reproduction by Division

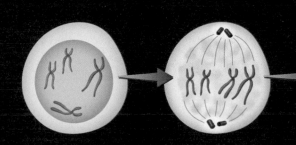

As a cell begins to divide, its DNA duplicates itself.

The duplicated chromosomes move to the middle.

The two sets of chromosomes separate, and . . .

the cell divides to produce . . .

two cells, each containing a full set of the DNA code.

very simple process that gradually became more sophisticated as it was modified by evolution.

We begin our search on Earth, where fossils and an intimate familiarity with carbon-based life give us a glimpse of the first living matter. Once we discover how Earthly life could have begun, we can look for signs that life began on other planets in our solar system. Finally, we can speculate on the chances that other planets, orbiting other stars, have conditions that give rise to life.

THE ORIGIN OF LIFE ON EARTH

The oldest fossils hint that life began in the sea. The first living things on Earth left behind a very poor fossil record. They were at first single-celled creatures, and even when they became more complex, multicellular creatures, they contained no hard parts such as bones or shells that form good fossils. What fossils can be found are clearly the remains of ocean creatures.

Although the oldest rocks contain no obvious fossils, microscopes reveal traces of ancient life. The oldest widely accepted evidence is 2.5-billion-year-old fossils of bacteria. Rock from western Australia that is nearly 3.5 billion years old contains microscopic features that may be fossils of living things (▮ Figure 19-2), but that conclusion is highly controversial. While the experts debate the details, we can conclude that life was present on Earth as simple organisms at least 2.5 billion years ago and possibly a billion years before that.

A little over a half-billion years ago something happened, perhaps a change in Earth's climate, and life exploded into a wide diversity of complex forms. This marks the beginning of the **Cambrian period,** and is sometimes called the *Cambrian explosion.* The best known of the Cambrian creatures may be the trilobites (▮ Figure 19-3). Although the Cambrian creatures were diverse, there are no Cambrian fossils of land plants or animals. Evidently, land surfaces were totally devoid of life until only 400 million years ago.

The fossil record shows that life began as simple organisms in the sea soon after Earth formed. Only recently in geological terms has life become complex.

The key to the origin of this life may lie in an experiment performed by Stanley Miller and Harold Urey in 1952. This **Miller experiment** sought to reproduce the conditions on Earth under which life began. In a closed glass container, the experimenters placed water (to represent the oceans); the gases hydrogen, ammonia, and methane (to represent the primitive atmosphere); and an electric arc (to represent lightning bolts). The apparatus was sterilized, sealed, and set in operation (▮ Figure 19-4).

After a week, Miller and Urey stopped the experiment and analyzed the material in the flask. Among the many compounds the experiment produced, they found four amino acids (building blocks of protein), various

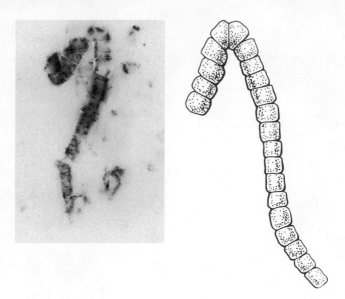

FIGURE 19-2

The microphotograph at left shows a feature found in 3.5-billion-year-old rock from northwestern Australia. The artist's reconstruction at right interprets the feature as a chain of bacteria-like organisms. Experts do not at present agree on the age of the rocks, the way the rocks formed, or the nature of the tiny features. Consequently the search for the oldest fossils on Earth is a challenging quest. *(Courtesy J. William Schopf)*

FIGURE 19-3

Trilobites made their first appearance in the Cambrian oceans. The smallest were almost microscopic, and the largest were bigger than dinner plates. This example, about the size of a human hand, lived 400 million years ago in an ocean floor that is now a limestone deposit in Pennsylvania. *(Grundy Observatory photograph)*

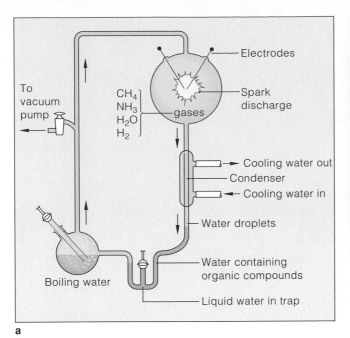

FIGURE 19-4

(a) The Miller experiment circulated gases through water in the presence of an electric arc. This simulation of primitive conditions on Earth produced amino acids, the building blocks of proteins. (b) Stanley Miller with a Miller apparatus. *(Courtesy Stanley Miller)*

fatty acids, and urea, a molecule common to many life processes. Evidently, the energy from the electric arc had molded the atmospheric gases into some of the basic components of living matter. Other energy sources, such as hot silica (to simulate hot lava spilling into the sea) and ultraviolet radiation (to simulate sunlight), give similar results.

Recent studies of the composition of meteorites and models of planet formation suggest that Earth's first atmosphere did not resemble the gases used in the Miller experiment. Earth's first atmosphere was probably composed of carbon dioxide, nitrogen, and water vapor. This finding, however, does not invalidate the Miller experiment. When such gases are processed in a Miller apparatus, a tarry gunk rich in organic molecules soon coats the inside of the chamber.

The Miller experiment did not create life, nor did it necessarily imitate the exact conditions on the young Earth. Rather, it is important because it shows that complex organic molecules form naturally in a wide variety of circumstances. The chemical deck is stacked to deal nature a hand of complex molecules. If we could travel back in time, we would probably find Earth's first oceans filled with a rich mixture of organic compounds in what some have called the **primordial soup.**

The next step on the journey toward life is for the compounds dissolved in the oceans to link up and form larger molecules. Amino acids, for example, can link together to form proteins. This linkage occurs when amino acids join together end to end and release a water molecule (❙ Figure 19-5). For many years, experts have assumed that this process must have happened in sun-warmed tidal pools where evaporation concentrated the broth. But recent studies suggest that the young Earth was subject to extensive volcanism and large meteorite impacts that periodically modified the climate enough to destroy any life forms exposed on the surface. Thus, the early growth of complex molecules likely took place among the hot springs along the midocean ridges. These complex molecules would not have been photosynthetic—taking energy from sunlight. Rather they would have taken energy from the heat and chemicals emerging from the hot springs of the midocean ridge. That energy could have powered the growth of long protein chains. Deep in the oceans, they would have been safe from climate changes.

Although these proteins might have contained hundreds of amino acids, they would not have been alive. Not yet. Such molecules would not have reproduced but would have merely linked together and broken apart at random. Because some molecules are more stable than others, however, and because some molecules bond together more readily than others, this blind **chemical evolution** would have led to the concentration of the varied smaller molecules into the most stable larger forms. Eventually, somewhere in the oceans, a molecule took shape that could reproduce itself. At that point the chemical evolution of molecules became the biological evolution of living things.

An alternative theory proposes that primitive living things such as reproducing molecules did not originate on Earth but came here in meteorites or comets.

FIGURE 19-5
Amino acids can link together through the release of a water molecule to form long carbon-chain molecules. The amino acid in this hypothetical example is alanine, one of the simplest.

Water

Growing carbon-chain molecule · Amino acid · Amino acid · Amino acid

Radio astronomers have found a wide variety of organic molecules in the interstellar medium, and some studies have found similar compounds inside meteorites (Figure 19-6). Such molecules form so readily that we would be surprised if they were not present in space. A few investigators, however, have speculated that living, reproducing molecules originated in space and came to Earth as a cosmic contamination. If this is true, every planet in the universe is contaminated with the seeds of life. However entertaining this theory may be, it is presently untestable, and an untestable theory is of little use in science.

Whether life originated in the oceans or in space, we still face the problem of how life began. Experts studying the origin of life proceed on the assumption that life began as reproducing molecules in Earth's oceans.

Which came first, reproducing molecules or the cell? Because we think of the cell as the basic unit of life, this question seems to make no sense, but in fact the cell may have originated during chemical evolution. If a dry mixture of amino acids is heated, the acids form long, proteinlike molecules that, when poured into water, collect to form microscopic spheres that function in ways similar to cells (Figure 19-7). They have a thin membrane surface, they can absorb material from their surroundings, they grow in size, and they can divide and bud just as cells do. They contain no large molecule that copies itself, however. Thus the structure of the cell may have originated first and the reproducing molecules later.

FIGURE 19-6
A sample of the Murchison meteorite, a carbonaceous chondrite that fell in 1969 near Murchison, Australia. Analysis of the interior of the meteorite revealed evidence of amino acids. Whether the first building blocks of life originated in space is unknown, but the amino acids found in meteorites illustrate how commonly amino acids and other complex molecules occur even in the absence of living things. (Courtesy Chip Clark, National Museum of Natural History)

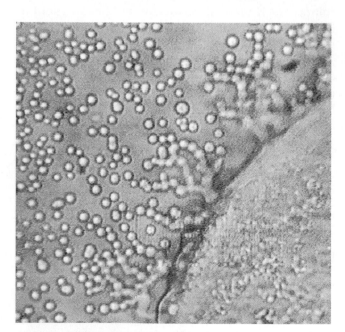

FIGURE 19-7
Single amino acids can be assembled into long proteinlike molecules. When such material cools in water, it can form microspheres, microscopic spheres with double-layered boundaries similar to cell membranes. Microspheres may have been an intermediate stage in the evolution of life between complex molecules and cells holding molecules reproducing genetic information. (Courtesy Sidney Fox and Randall Grubbs)

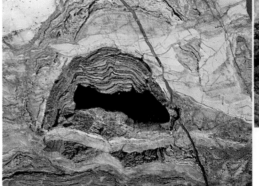

An alternative theory proposes that the replicating molecule developed first. Such a molecule would have been exposed to damage if it had been bare, so the first to manufacture or attract a protective coating of protein would have had a significant survival advantage. If this was the case, the protective cell membrane was a later development of biological evolution.

The first living things must have been single-celled organisms much like modern bacteria. Some of the oldest fossils known are **stromatolites,** structures produced by communities of photosynthesizing bacteria that grew in mats and, year by year, deposited layers of minerals that were later fossilized. One of the oldest such fossils known is believed to be 3.5 billion years old (❚ Figure 19-8). If such bacteria were common when Earth was young, the early atmosphere may have contained a small amount of oxygen produced by the photosynthesis. Recent studies suggest that an oxygen abundance of only 0.1 percent would have been sufficient to provide an ozone screen that would protect organisms from the sun's ultraviolet radiation.

How evolution shaped creatures to live in the ancient oceans, to photosynthesize and respire, to become multicellular, and to reproduce sexually is a fascinating story, but we cannot explore it in detail here. We can see that life could have begun through simple chemical re-

actions building complex molecules, and that once some DNA-like molecule formed, it protected its own survival with selfish determination. Over billions of years, the genetic information stored in living things kept those qualities that favored survival and discarded the rest. As Samuel Butler said, "The chicken is the egg's way of making another egg." In that sense, all living matter on Earth is merely the physical expression of DNA's mindless determination to continue its existence.

Perhaps this seems harsh. Human experience goes far beyond mere reproduction. *Homo sapiens* has art, poetry, music, philosophy, religion, science. Perhaps all of the great accomplishments of our intelligence represent more than mere reproduction of DNA. Nevertheless, intelligence, the ability to analyze complex situations and respond with appropriate action, must have begun as a survival mechanism. For example, a fixed escape strategy stored in the DNA is a disadvantage for a creature that frequently moves from one environment to another. A rodent that always escapes from predators by automatically climbing the nearest tree would be in serious jeopardy if it met a hungry fox in a treeless clearing. Even a faint glimmer of intelligence might allow the rodent to analyze the situation and, finding no trees, to choose running over climbing. Thus, intelligence, of which *Homo sapiens* is so proud,

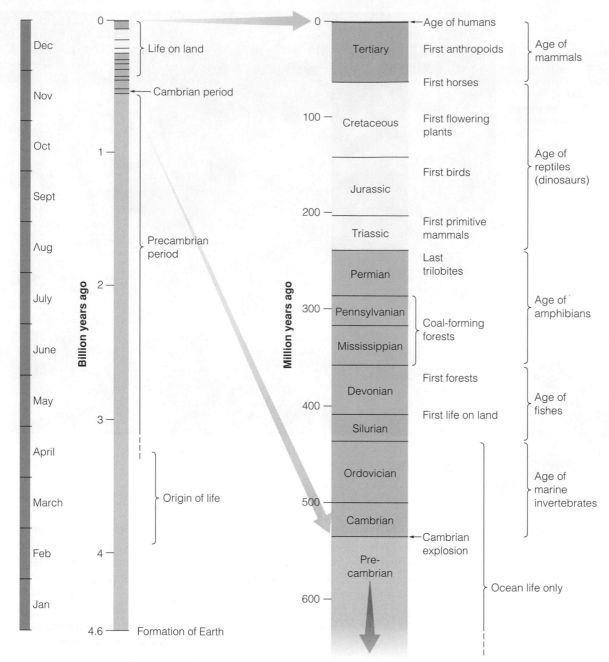

ACTIVE FIGURE 19-9
Complex life has developed on Earth only recently. If the entire history of Earth were represented
in a time line (left), we would have to magnify the end of the line to see details such as life leaving the
oceans and dinosaurs appearing. The age of humans would still be only a thin line at the top of our
diagram. If the history of Earth were a year-long videotape, humans would not appear until the last
hours of December 31.

Ace ☯Astronomy™ Go to AceAstronomy and click Active Figures to see "Earth Calendar." Click on the
calendar to explore through time.

may have developed in ancient creatures as a way of
making them more versatile.

Any discussion of the evolution of life seems to involve highly improbable coincidences until we consider how many years have passed in the history of
Earth. We read the words—4.6 billion years—so easily,
but it is in truth hard to grasp the meaning of such a
long period of time.

GEOLOGIC TIME

Humanity is a very new experiment on planet Earth.
We can fit all of the evolution that leads from the primitive life forms in the oceans of the Cambrian period
600 million years ago to fishes, amphibians, reptiles,
and mammals into a single chart such as ▌ Figure 19-9.
We can take comfort in thinking that creatures like us
have walked on Earth for roughly 3 million years, but

when we add our history to the chart, we discover that the entire history of humanity makes up no more than a thin line at the top. In fact, if we tried to represent the entire 4.6-billion-year history of Earth on the chart, the portion describing the rise of life on the land would be an unreadably small segment.

One way to represent the evolution of life is to compress the 4.6-billion-year history of Earth into a 1-year-long videotape. In such a program, Earth forms as the video begins on January 1, and through all of January and February it cools and is cratered, and the first oceans form. Search as we might, we would find no trace of life in these oceans until sometime in March or early April, when the first living things develop. The slow development of these simplest of living forms grinds on slowly through the spring and summer of our videotape. The entire 4-billion-year history of Precambrian evolution lasts until the video reaches mid-November, when the primitive ocean life explodes into the more complex Cambrian organisms such as trilobites.

While our year-long videotape plays on and on, we might amuse ourselves by looking at the land instead of the oceans, but we would be disappointed. The land is a lifeless waste with no plants or animals of any kind. Not until November 28 in our video does life appear on the land; but, once it does, it evolves rapidly into a wide range of plants and animals. Dinosaurs, for example, appear about December 12 and vanish by Christmas evening as mammals and birds flourish.

Throughout the 1-year run of our video there have been no humans; and even during the last days of the year, as the mammals rise to dominate the landscape, there are no people. In the early evening of December 31, vaguely human forms move through the grasslands, and by late evening they begin making stone tools. The Stone Age lasts till about 11:45 PM, and the first signs of civilization, towns and cities, do not appear until 11:54 PM. The Christian era begins only 14 seconds before the New Year, and the Declaration of Independence is signed with but 1 second to spare.

By converting the history of Earth into a year-long videotape, we have placed the rise of life in perspective. Tremendous amounts of time were needed for the first simple living things to evolve in the oceans, and even more time was needed for the evolution of complex creatures that could colonize the land. As life became more complex, it evolved and diversified faster and faster, as if evolution were drawing on a growing library of solutions that had been previously invented with great effort to solve earlier problems. The burst of diversity on land led eventually to the rise of intelligent creatures like us, a process that has taken 4.6 billion years.

If life could originate on Earth and develop into intelligent creatures, perhaps the same thing could have happened on other planets. This raises three questions.

First, could life originate on another world if conditions were suitable? We can't be certain, but we have found natural, chemical processes that could lead to living things, so our answer to this question should probably be yes. The second question is, will life always evolve toward intelligence? Perhaps intelligence arose on Earth under unusual conditions, but we must recall that intelligence appears to be a way to make a species more versatile and thus more likely to survive. If that is true, then intelligence may develop under a wide range of conditions if enough time is available. But what of the third question: Are suitable conditions so rare that life almost never gets started? The only way to answer that is to search for life on other planets. We begin, in the next section, with the other planets in our solar system.

LIFE IN OUR SOLAR SYSTEM

Although we can imagine life based on something other than carbon chemistry, we know of no examples to tell us how such life might arise and survive. We must limit our discussion to life as we know it and the conditions it requires. The most important requirement is the presence of liquid water, not only as part of the chemical reactions of life, but also as a medium to transport nutrients and wastes within the organism. Also, it seems that life on Earth began in the oceans and developed there for nearly 4 billion years before it was able to emerge onto the land. Certainly, any world where we hope to find life must have liquid water, and that means it must have moderate temperatures.

The water requirement automatically eliminates many worlds in our solar system. The moon is airless, and although some data suggest ice frozen in the soil at its poles, it has never had liquid water on its surface. In the vacuum of the lunar surface, liquid water would boil away rapidly. Mercury too is airless and cannot have had liquid water on its surface for long periods of time. Venus has some traces of water vapor in its atmosphere, but it is much too hot for liquid water to survive. If there were any lakes or oceans of water on its surface when it was young, they must have evaporated quickly. Even if life began there, no traces would be left now.

The inner solar system seems too hot, and the outer solar system seems too cold. The Jovian planets have deep atmospheres, and at a certain level, they have moderate temperatures where water might condense into liquid droplets. But it seems unlikely that life could begin there. The Jovian planets have no surfaces where oceans could nurture the beginning of life, and currents in the atmosphere seem destined to circulate gas and water droplets from regions of moderate temperature to other levels that are much too hot or too cold for life to survive.

A few of the satellites of the Jovian planets might have suitable conditions for life. Jupiter's moon Europa seems to have a liquid-water ocean below its icy crust, and minerals dissolved in that water would provide a rich broth of possibilities for chemical evolution. Nevertheless, Europa is not a promising site to search for life because conditions may not have remained stable for the billions of years needed for life to evolve beyond the microscopic stage. The subsurface ocean is kept from freezing by tidal heating. If Jupiter's moons interact gravitationally and modify their orbits, Europa may have been frozen solid at some points in history. Such periods of freezing would probably prevent life from developing. Drilling through the icy crust of Europa to search for life in its ocean will be a wonderful adventure for future generations, but Europa does not seem to be a good bet to harbor any form of complex life.

Saturn's moon Titan has an atmosphere of nitrogen, argon, and methane and may have oceans of liquid methane and ethane on its surface. Sunlight can convert the methane in the atmosphere to organic smog particles that settle to the surface. The chemistry of life that might crawl or swim on such a world is unknown, but life there may be unlikely because of the temperature. The surface of Titan is a deadly $-179°C$ ($-290°F$). Chemical reactions occur slowly or not at all at such low temperatures, so the chemical evolution needed to begin life may never have occurred on Titan.

Mars is the most likely place for life in our solar system. The evidence, however, is not encouraging. In 1976, two robotic spacecraft, Viking 1 and Viking 2, landed at two different places on the Martian surface. The spacecraft scooped up soil samples and subjected them to tests for the presence of living organisms. For example, they gave some soil samples a dose of nutrient-rich water and watched for signs the nutrients were taken up by biological processes. The results seem negative. Although some peculiar chemical processes were detected, no evidence was found that clearly indicated the presence of any living things in the soil.

The chemical experiments performed by the Viking landers were sophisticated, but they could only search for traces of currently living things. They could not search for signs of past life, such as fossils. For that, we need to send a geologist to Mars or bring rocks from Mars back to Earth. Nature has performed the latter task for us.

Meteorite ALH84001 (❚ Figure 19-10a) was found on the Antarctic ice in 1984. Years later an analysis of its chemical composition showed that it had originated on Mars. It was probably part of debris ejected into space by a large impact on Mars. Some of that debris fell to Earth, and a small number of these meteorites have been found. ALH84001 is important because a team of scientists studied it and announced in 1996 that it contained chemical and physical traces of ancient life on Mars (Figure 19-10b).

This discovery was received with great excitement by the press and the public. It would, indeed, be dramatic evidence if true, because it would show that life could begin on another world.

a

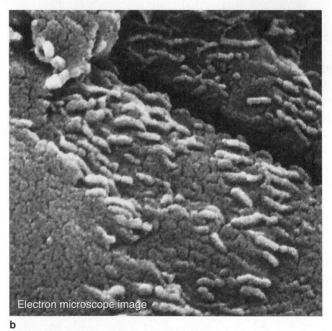

Electron microscope image

b

FIGURE 19-10

(a) Meteorite ALH84001 is one of a dozen meteorites known to have originated on Mars. It was claimed that the meteorite contained chemical and physical traces of ancient life on Mars, including what appear to be fossils of microscopic organisms (b). The evidence has not been confirmed, and the validity of the claim is highly questionable. *(NASA)*

Scientists were excited too, but being professionally skeptical, they began testing the results immediately. In many cases, the results did not confirm the conclusion that life once existed on Mars. Some chemical contamination from water on Earth has occurred, and some chemicals in the meteorite may have originated without the presence of life. The physical features that look like fossil bacteria may be mineral formations in the rock. Although studies of ALH84001 continue, it does not give us unchallenged evidence that life once existed on Mars.

Spacecraft now visiting Mars may help us understand the past history of water there and paint a more detailed picture of present conditions. Landers are planned that will eventually land on Mars, collect rocks, and return them to Earth. Nevertheless, conclusive evidence may have to wait until a geologist in a space suit can wander the dry streambeds of Mars cracking open rocks and searching for fossils.

We are left to conclude that, so far as we know, our solar system is bare of life except for Earth. Consequently our search for life in the universe takes us to other planetary systems.

LIFE IN OTHER PLANETARY SYSTEMS

Might life exist in other solar systems? To consider this question, let us try to decide how common planets are and what conditions a planet must fulfill for life to originate and evolve to intelligence. The first question is astronomical; the second is biological. Our ability to discuss the problem of life outside our solar system is severely limited by our lack of experience.

Planets form as a natural by-product of star formation, and a number of extrasolar planets have been found circling nearby stars. From this we can conclude that planetary systems are very common.

If a planet is to become a suitable home for life, it must have a stable orbit around its sun. This is simple in a solar system like our own, but in a binary system most planetary orbits are unstable. Most planets in such systems would not last long before they were swallowed up by one of the stars or ejected from the system.

Thus, single stars are the most likely to have planets suitable for life. Because our galaxy contains at least 10^{11} stars, half of which are single, there could be roughly 5×10^{10} planetary systems in which we should look for life.

A few million years of suitable conditions does not seem to be enough time to originate life. On our planet it took at least 0.5 to 1 billion years for the first cells to evolve and 4.6 billion years for intelligence to evolve. Clearly, conditions on a planet must remain acceptable over a long time. This eliminates massive stars that remain stable on the main sequence for only a few million years. If it takes a few billion years for life to originate and evolve to intelligence, no star hotter than about F5 will do. This is not really a serious restriction, because upper-main-sequence stars are rare anyway.

In previous sections, we decided that life, at least as we know it, requires liquid water. That requirement defines a **life zone** (or ecosphere) around each star, a region within which a planet has temperatures that permit the existence of liquid water.

The size of the life zone depends on the temperature of the star (▌ Figure 19-11). Hot stars have larger life zones because the planets must be more distant to remain cool. But the short main-sequence lives of these stars make them unacceptable. M stars have small life zones because they are extremely cool—only planets very near the star receive sufficient warmth. However, planets that are close to a star would probably become tidally coupled, keeping the same side toward the star. This might cause the water and atmosphere to freeze in the perpetual darkness of the planet's night side and end all chance of life. Also, M stars are subject to sudden flares that might destroy life on a planet close to the star. Thus, the life zone restricts our search for life to main-sequence G and K stars.

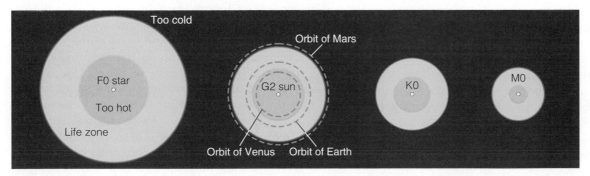

FIGURE 19-11

The life zone around a star (green) is the region where a planet would have a moderate temperature. M stars are not good candidates because they are cool and their life zone is small. F stars have a large life zone but evolve too quickly to allow the rise of complex life. K and G stars like our sun seem the best candidates in a search for intelligent life.

Even a star on the main sequence is not perfectly stable. Main-sequence stars gradually grow more luminous as they convert their hydrogen to helium, and thus the life zone around a star gradually moves outward. A planet might form in the life zone, and life might begin and evolve for billions of years, only to be destroyed as the slowly increasing luminosity of its star moves the life zone outward, evaporates the planet's oceans, and drives off its atmosphere. If a planet is to remain in the life zone for 4 to 5 billion years, it must form on the outer edge of the zone. This may be the most serious restriction we have yet discussed.

We should pause here to note that our concept of a life zone may be very self-centered. We admit that life might exist on worlds such as Europa and Titan, but they lie outside our conventionally defined life zone. Furthermore, scientists on Earth are discovering life in highly unlikely places such as in the Antarctic ice, in hot springs on the ocean floor, and even in solid rock deep underground. As we learn more about living things, we may have to extend our definition of the life zone to include a wider range of conditions.

The fundamental question in our discussion is simple. If conditions are right, will life begin? Early in this chapter we decided that life could begin through simple chemical reactions, so perhaps we should change our question and ask: What could prevent life from beginning? Given what we know about life, it should arise whenever conditions permit, and our galaxy should be filled with planets that are inhabited with living creatures.

REVIEW CRITICAL INQUIRY

What evidence do we have that life is at least possible on other worlds?

Our evidence is limited almost entirely to Earth, but it is promising. The fossils we find show that life originated in Earth's oceans almost 4 billion years ago, and biologists have proposed relatively simple chemical processes that could have created these first reproducing molecules. The fossils show that life developed from a very slow beginning into more and more complex creatures that filled the oceans. The pace of evolution quickened dramatically about half a billion years ago, the beginning of the Cambrian period, when life began taking on complex forms; later, when life emerged onto the surface of the land, it evolved rapidly to produce the tremendous diversity we see around us. Human intelligence has been a very recent development; it is only a few million years old.

If this process occurred on Earth, then it seems reasonable that it could have occurred on other worlds as well. Thus, the evidence suggests we can expect that life might begin and evolve to intelligent forms on any world where conditions are right. What are the conditions we should expect of other worlds that host life?

It is both easy and fun to speculate about life on other worlds, but it leads to a simple question: Is there really life beyond Earth? If we can't leave Earth and visit other worlds, then the only life we can detect will be beings intelligent enough to communicate with us. Why haven't we heard from them?

19-3 COMMUNICATION WITH DISTANT CIVILIZATIONS

If other civilizations exist, perhaps we can communicate with them in some way. Sadly, travel between the stars is more difficult in real life than in science fiction—and may in fact be impossible. If we can't physically visit, perhaps we can communicate by radio. Again, nature places restrictions on such conversations, but the restrictions are not too severe. As we will see, the real problem lies with the life expectancy of civilizations.

TRAVEL BETWEEN THE STARS

Practically speaking, roaming among the stars is tremendously difficult because of three limitations: distance, speed, and fuel. The distances between stars are almost beyond comprehension. It does little good to explain that if we use a golf ball in New York City to represent the sun, the nearest star would be another golf ball in Chicago. It is only slightly better to note that the fastest commercial jet would take about 4 million years to reach the nearest star.

The second limitation is a speed limit—we cannot travel faster than the speed of light. Though science fiction writers invent hyperspace drives so their heroes can zip from star to star, the speed of light is a natural and unavoidable limit that we cannot exceed. This, combined with the large distances between stars, makes interstellar travel very time consuming.

The third limitation is that we can't even approach the speed of light without using a fantastic amount of fuel. Even if we ignore the problem of escaping from Earth's gravity, we must still use energy stored in fuel to accelerate to high speed and to decelerate to a stop when we reach our destination. To return to Earth, assuming we wish to, we have to repeat the process. These changes in velocity require a tremendous amount of fuel. If we flew a spaceship as big as a large yacht to a star 5 light-years (1.5 pc) away and wanted to get there in only 10 years, we would use 40,000 times as much energy as the United States consumes in a year.

Travel for a few individuals might be possible if we accept very long travel times. That would require some form of suspended animation (currently unknown) or colony ships that carry a complete, though small, society in which people are born, live, and die generation after

UFOs and Space Aliens

When we discuss life on other worlds, we might be tempted to use UFO sightings and supposed visits by aliens from outer space as evidence to test our hypotheses. We don't do so for two reasons, both related to the reliability of these observations.

First, the reputation of the sources of UFO sightings and alien encounters does not give us confidence that these data are reliable. Most people hear of such events via grocery store tabloids, daytime talk shows, or sensational "specials" on viewer-hungry cable networks. We must take note of the low reputation of the media that report UFOs and space aliens. Most of these reports are simply made up for the sake of sensation, and we cannot use them as reliable evidence.

Second, the remaining UFO sightings, those not simply made up, do not survive careful examination. Most are mistakes or unconscious misinterpretations of natural events made by honest people. A number of unbiased studies have found no grounds for believing in UFOs.

In short, there is no conclusive evidence that Earth has been visited by aliens from space.

That's too bad. A confirmed visit by intelligent creatures from beyond our solar system would answer many of our questions. It would be exciting, enlightening, and, like any real adventure, a bit scary. But none of the UFO sightings is dependable, and we are left with no direct evidence of intelligent life on other worlds.

---

generation. Whether the occupants of such a ship would retain the social characteristics of humans over a long voyage is questionable.

These three limitations not only make it difficult for us to leave our solar system, but also would make it difficult for aliens to visit Earth. Reputable scientists have studied "unidentified flying objects" (UFOs) and related phenomena and have never found any evidence that Earth is being visited or has ever been visited by aliens from other worlds (Window on Science 19-2). Thus, humans are unlikely ever to meet an alien face to face. The only way we can communicate with other civilizations is via radio.

RADIO COMMUNICATION

Nature places two restrictions on our ability to communicate with distant societies by radio. One has to do with simple physics, is well understood, and merely makes the communication difficult. The second has to do with the fate of technological civilizations, is still unresolved, and may severely limit the number of societies we can detect by radio.

Radio signals are electromagnetic waves that travel at the speed of light. Because even the nearest civilizations must be a few light-years away, this limits our ability to carry on a conversation with distant beings. If we ask a question of a creature 10 light-years away, we will have to wait 20 years for a reply. Clearly, the give-and-take of normal conversation will be impossible.

Ace⟲Astronomy™ Go to AceAstronomy and click Active Figures to see "Interstellar Communications." Click on the sun and on another star to see how long communication could take.

Instead, we could simply broadcast a radio beacon of friendship to announce our presence. In fact, we are already broadcasting a recognizable beacon. Short-wavelength radio signals, such as TV and FM, have been leaking into space for the last 50 years or so. Any civilization within 50 light-years might already have detected us.

If we intentionally broadcast a signal, we can anticode it. That is, we can arrange it to make it easy to decode. We could transmit pulses to represent 1s and gaps to represent 0s. A message counting through the first few prime numbers would distinguish our signal from natural sources of radio noise. We could even transmit a picture by sending a string of 1s and 0s that can be arranged in only two ways. One way produces nonsense, but the other way produces a meaningful picture (▌Figure 19-12).

In 1974, at the dedication of the 1000-ft radio telescope at Arecibo, radio astronomers transmitted such a signal toward the globular cluster M13, which is located 26,000 ly from Earth. When the signal finally arrives, any aliens who detect it will be able to arrange its 1679 pulses in only two ways, as 23 rows of 73 or 73 rows of 23. The second arrangement will form a picture that describes life on Earth (▌Figure 19-13).

It took only minutes to transmit the Arecibo message. If more time were taken, a more detailed picture could be sent, and if we were sure our radio telescope was pointed at a listening civilization, we could send a long series of pictures. With pictures we could teach aliens our language and tell them all about our life, our difficulties, and our accomplishments.

If we can think of sending such signals, aliens can think of it too. If we point our radio telescopes in the right direction and listen at the right wavelength, we might hear other intelligent races calling out to one another. This raises two questions: Which stars are the best candidates, and what wavelengths are most likely? We have already answered the first question. Main-sequence G and K stars have the most favorable characteristics. But the second question is more complex.

An anticoded message

1	0	1	0	0	1	1
1	1	1	1	1		
0	0	1	0	1	0	0
0	1	0	0			
0	1	0	1	0	0	1
0	1	0				
0	1	0	1	0		

5 rows of 7

7 rows of 5

FIGURE 19-12

An anticoded message is designed for easy decoding. Here a string of 35 radio pulses, represented as 1s and 0s, can be arranged in only two ways, as 5 rows of 7 or 7 rows of 5. The second way produces a friendly message. Any number of pulses can be used so long as it is the product of two prime numbers. Then the pulses can be arranged in only two ways.

Only certain wavelengths are useful for communication. We cannot use wavelengths longer than about 30 cm because the signal would be lost in the background radio noise from our galaxy. Nor can we go to wavelengths much shorter than 1 cm because of absorption within our atmosphere. Thus only a certain range of wavelengths, a radio window, is open for communication.

This communications window is very wide, so a radio telescope would take a long time to tune over all the wavelengths searching for intelligent signals. Nature may have given us a way to narrow the search, however. Within the communications window lie the 21-cm line of neutral hydrogen and the 18-cm line of OH. The interval between these two lines has been dubbed the **water hole** because the combination of H and OH yields water (H_2O). Water is the fundamental solvent in our life form, so it might seem natural for similar water creatures to call out to each other at wavelengths in the water hole (❚ Figure 19-14). But even silicon creatures would be familiar with the 21-cm line of hydrogen.

This is not idle speculation. A number of searches for extraterrestrial radio signals have been made, and some major searches are now under way. The field has become known as **SETI**, Search for Extra-Terrestrial Intelligence, and it has generated heated debate. Some scientists and philosophers argue that life on other worlds can't possibly exist, so it is a waste of money to search. Others argue that life on other worlds is common and could be detected with present technology. Congress funded a NASA search for a short time but then ended support in the early 1990s because political leaders feared public reaction. In fact, the annual cost of a major search is only about as much as a single Air Force attack helicopter. The controversy may spring in part from the theological and philosophical controversy that would result from the discovery of intelligent life on another world.

In spite of the controversy, searches are under way. The NASA SETI project canceled by Congress is now supported by private funds as Project Phoenix. It is located at the SETI Institute and has been using radio telescopes at major radio astronomy observatories to listen to the radio emissions of a list of candidate stars. About two billion radio-frequency bands must be examined for each star, so the project depends heavily on computers to search for

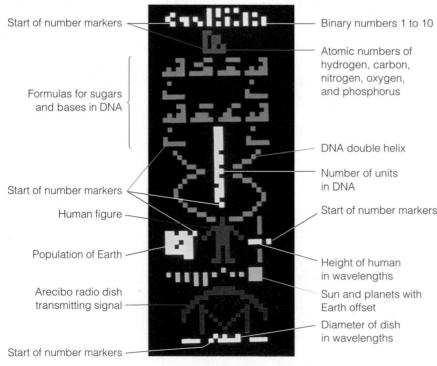

Start of number markers — Binary numbers 1 to 10

Atomic numbers of hydrogen, carbon, nitrogen, oxygen, and phosphorus

Formulas for sugars and bases in DNA

DNA double helix

Number of units in DNA

Start of number markers

Human figure

Start of number markers

Population of Earth

Height of human in wavelengths

Arecibo radio dish transmitting signal

Sun and planets with Earth offset

Diameter of dish in wavelengths

Start of number markers

FIGURE 19-13

The Arecibo message to M13 begins by counting from 1 to 10 and goes on to describe our solar system and the biochemistry of life on Earth (color added for clarity). Binary numbers give the height of the human figure (1110) and the diameter of the telescope dish (100101111110) in units of the wavelength of the signal, 12.3 cm. *(NASA)*

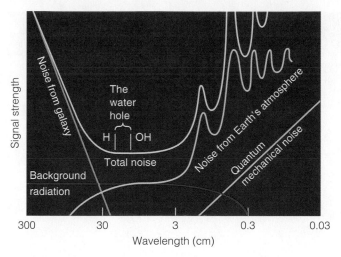

FIGURE 19-14

Radio noise from various sources makes it difficult to detect distant signals at wavelengths longer than 30 cm or shorter than 1 cm. In this range, radio emission from H atoms and from OH marks a small wavelength range dubbed the water hole, which may be a likely place for communication.

signs of artificial radio transmissions in the recorded radio emissions of each star.

Astonishing amounts of computer power are needed to search for weak or unusual signals. Consequently, the Berkeley SETI team, with the support of the Planetary Society, has recruited owners of personal computers linked to the Internet to participate in a project called seti@home. Participants download a screen saver that searches data files from the Arecibo radio telescope for meaningful signals whenever the owner is not using the computer. About 4 million people have signed up and are providing 1000 years of computer time per day for the project. For information, locate the seti@home project at http://setiathome.ssl.berkeley.edu/.

Since 1985, META, Megachannel Extra-Terrestrial Assay, has been searching the entire sky at millions of frequency bands in the water hole. Funded by the Planetary Society, the search uses radio antennas at Harvard and near Buenos Aires, Argentina. The project has instituted a second, even more efficient search called BETA. Since it began, META has detected dozens of candidate signals. That is, it has found dozens of radio signals that satisfy its most basic criteria. Unfortunately, none of these signals has proved to be continuous, and no signal has been detected when the radio telescopes were redirected at any of the candidate stars. Many of these signals are probably unusual noise from sources on Earth, but the candidate signals are still under study.

One ingenious search is called SERENDIP, Search for Extraterrestrial Radio Emission from Nearby Developed Intelligent Populations. Rather than monopolize an entire radio telescope, SERENDIP rides piggyback on the 305-m Arecibo Telescope. Wherever the radio astronomers point the telescope, the receiver samples the signal, looking for intelligent signals over millions

of frequency bands. If candidate signals are found, the receivers note the position of the radio telescope for future investigation.

Radio noise pollution is a serious problem for these searches. Year by year our society generates more radio signals from sources on Earth and satellites in orbit. Poorly designed radio transmitters generate noise at unexpected frequencies, and other electronic devices, such as computers, emit radio frequency signals. Furthermore, society, industry, and government press to use wider and wider sections of the electromagnetic spectrum. Radio astronomers struggle to hear through the radio babble, and SETI searches suffer because many of the noise signals mimic the patterns expected from other civilizations. It would be ironic if we fail to detect signals from another world because our own world has become too noisy.

What should we expect if signals are found? By international agreement, those searching for signals have agreed to share candidate signals and obtain reliable confirmation before making public announcements. False alarms would be embarrassing. The chances of success depend on the number of inhabited worlds in our galaxy, and that number is difficult to estimate.

HOW MANY INHABITED WORLDS?

The technology exists, and given enough time the searches will find other inhabited worlds, assuming there are at least a few out there. If intelligence is common, then we should find the signals soon—in the next few decades—but if intelligence is rare in the universe, it may be a very long time before we confirm that we are not alone.

Simple arithmetic can give us an estimate of the number of technological civilizations with which we might communicate, N_c. The first proposed formula for N_c is now known as the **Drake equation,** named after radio astronomer Frank Drake, a pioneer in the search for extraterrestrial intelligence. We will use a version of the Drake equation modified slightly to make it a bit easier to understand. The formula gives N_c, the number of communicative civilizations in a galaxy, as

$$N_c = N^* \cdot f_P \cdot n_{LZ} \cdot f_L \cdot f_I \cdot F_S$$

N^* is the number of stars in a galaxy, and f_P represents the fraction of all stars that have planets. If all single stars have planets, f_P is about 0.5. The factor n_{LZ} is the average number of planets in a solar system suitably placed in the life zone, f_L is the fraction of suitable planets on which life begins, and f_I is the fraction of life forms that evolve to intelligence. These factors can be roughly estimated, but the remaining factor is much more uncertain.

F_S is the fraction of a star's life during which the life form is communicative. Here we assume that a star lives about 10 billion years. If a society survives at a technological level for only 100 years, our chances of

TABLE 19-1
The Number of Technological Civilizations per Galaxy

	Variables	Estimates Pessimistic	Estimates Optimistic
N^*	Number of stars per galaxy	2×10^{11}	2×10^{11}
f_P	Fraction of stars with planets	0.01	0.5
n_{LZ}	Number of planets per star that lie in life zone for longer than 4 billion years	0.01	1
f_L	Fraction of suitable planets on which life begins	0.01	1
f_I	Fraction of life forms that evolve to intelligence	0.01	1
F_S	Fraction of star's life during which a technological society survives	10^{-8}	10^{-4}
N_C	Number of communicative civilizations per galaxy	2×10^{-5}	10×10^6

communicating with it are small. But a society that stabilizes and remains technological for a long time is much more likely to be in the communicative phase at the proper time to signal to us. If we assume that technological societies destroy themselves in about 100 years, F_S is 100 divided by 10 billion, or 10^{-8}. But if societies can remain technological for a million years, then F_S is 10^{-4}. The influence of the factors in the formula is shown in ▌Table 19-1.

Ace◉Astronomy™ Go to AceAstronomy and click Active Figures to see "Drake Equation." Make your own estimates and see how common life is on other worlds.

If the optimistic estimates are true, there may be a communicative civilization within a few dozen light-years of us, and we could locate it by searching through only a few thousand stars. On the other hand, if the pessimistic estimates are correct, we may be the only planet in our galaxy capable of communication. We may never know until we understand how technological societies function.

REVIEW CRITICAL INQUIRY

Why does the number of inhabited worlds we might hear from depend on how long civilizations survive at a technological level?

When we turn radio telescopes toward the sky and scan millions of frequency bands, we take a snapshot of the universe at a particular time when we are living and able to build radio telescopes. To detect other civilizations, they must be in a similar technological stage so they will be broadcasting either intentionally or accidentally. If we search for decades and detect no other signal, it may mean not that life is rare but rather that civilizations do not survive at a technological level for very long. Only a few decades ago, our civilization was threatened by nuclear war, and now we are threatened by the pollution of our environment. If nearly all civilizations in our galaxy are either on the long road up from primitive life forms in oceans or on the long road down

from nuclear war or environmental collapse, there may be no one transmitting during the short interval when we are capable of building radio telescopes with which to listen.

Radio communication between inhabited worlds is limited because it requires very fast computers to search many frequency intervals. Why must we search so many frequencies when we suspect that the water hole would be a good place to listen?

Are we the only thinking race? If we are, we bear the sole responsibility to understand and admire the universe. Then we are the sole representatives of that state of matter called intelligence. The mere detection of signals from another civilization would demonstrate that we share the universe with others. Although we might never leave our solar system, such communication would end the self-centered isolation of humanity and stimulate a reevaluation of the meaning of our existence. We may never realize our full potential as humans until we communicate with nonhuman intelligent life.

SUMMARY

To discuss life on other worlds, we must first understand something about life in general, life on Earth, and the origin of life. In general, we can identify two properties of living things: a physical basis and a controlling unit of information. The process of life must extract energy from the surroundings, maintain the organism, and modify the surroundings to promote the organism's survival. The physical basis is the arrangement of matter and energy that implements the life process. On Earth, all life is based on carbon chemistry. The controlling information is the data necessary to maintain the organism's function. Data for Earth life are stored in long carbon-chain molecules called DNA.

The DNA molecule stores information in the form of chemical bases linked together like the rungs of a ladder. When these patterns are copied by RNA molecules, they can direct the manufacture of proteins and enzymes. Thus the DNA information is the chemical formulas the cell needs to function. When a cell divides, the DNA molecule splits lengthwise and duplicates itself so that each of the new cells has a copy of the information. Errors in the duplication or damage to the DNA molecule can produce mutants, organisms that contain new DNA information and have new properties. Natural selection determines which of these new organisms are best suited to survive, and the species evolves to fit its environment.

The Miller experiment suggests that energy sources such as lightning could have caused amino acids and other complex molecules to form. Chemical evolution would have connected these together in larger and more complex, but not yet living, molecules. When a molecule acquired the ability to produce copies of itself, natural selection began improving the organism through biological evolution. Though this may have happened in the first billion years, life did not become diverse and complex until the Cambrian period, about 0.5 billion years ago. Life emerged from the oceans about 0.4 billion years ago, and humanity developed only a few million years ago.

It seems unlikely that there is now life on other planets in our solar system. Most of the planets are too hot or too cold. Mars may have been suitable in the past, but it is now a cold, dry desert, and any liquid water is trapped below the crust. Jupiter's moon Europa has a liquid-water ocean below its icy crust, but conditions there may not have remained stable long enough for life to develop. Saturn's moon Titan may have organic materials on its surface, but it may be too cold for chemical reactions to lead to life.

To find life, we must look beyond our solar system. Because we suspect that planets form from the leftover debris of star formation, we suspect that most stars have planets. The rise of intelligence may take billions of years, however, so short-lived massive stars and binary stars with unstable planetary orbits must be discarded. The best candidates are G and K main-sequence stars.

The distances between stars are too large to permit travel, but communication by radio could be possible. A certain wavelength range called a radio window is suitable, and a small range between the radio signals of H and OH, the so-called water hole, is especially likely.

We now have the technology to search for other intelligent life in the universe. Although such searches are controversial, a number of radio astronomers are now searching for radio signals from extraterrestrial civilizations. If life is common in the universe, success seems inevitable.

NEW TERMS

DNA (deoxyribonucleic acid)

amino acid

protein

enzyme

RNA (ribonucleic acid)

chromosome

gene

natural selection

mutant

Cambrian period

Miller experiment

primordial soup

chemical evolution

stromatolite

life zone

water hole

SETI

Drake equation

REVIEW QUESTIONS

Ace◐Astronomy™ Assess your understanding of this chapter's topics with additional quizzing and animations at http://astronomy.brookscole.com/seeds8e

1. If life is based on information, what is that information?

2. What would happen to a life form if the information handed down to offspring was always the same? How would that endanger the future of the life form?

3. How does the DNA molecule produce a copy of itself?

4. Give an example of natural selection acting on new DNA patterns to select the most advantageous characteristics.

5. Why do we believe that life on Earth began in the sea?

6. Why do we think that liquid water is necessary for the origin of life?

7. What is the difference between chemical evolution and biological evolution?

8. What was the significance of the Miller experiment?

9. How does intelligence make a creature more likely to survive?

10. Why are upper-main-sequence stars unlikely sites for intelligent civilizations?

11. Why do we suspect that travel between stars is nearly impossible?

12. How does the stability of technological civilizations affect the probability that we can communicate with them?

13. What is the water hole, and why would it be a good place to look for other civilizations?

DISCUSSION QUESTIONS

1. What would you change in the Arecibo message if humanity lived on Mars instead of Earth?

2. What do you think it would mean if decades of careful searches for radio signals for extraterrestrial intelligence turned up nothing?

PROBLEMS

1. A single human cell encloses about 1.5 m of DNA containing 4.5 billion base pairs. What is the spacing between these base pairs in nanometers? That is, how far apart are the rungs on the DNA ladder?

2. If we represent the history of the Earth by a line 1 m long, how long a segment would represent the 400 million years since life moved onto the land? How long a segment would represent the 3-million-year history of human life?

3. If a human generation, the time from birth to childbearing, is 20 years, how many generations have passed in the last million years?

4. If a star must remain on the main sequence for at least 5 billion years for life to evolve to intelligence, how massive could a star be and still harbor intelligent life on one of its planets? (*Hint:* See Chapter 12.)

5. If there are about 1.4×10^{-4} stars like the sun per cubic light-year, how many lie within 100 light-years of Earth? (*Hint:* The volume of a sphere is $\frac{4}{3}\pi r^3$.)

6. Mathematician Karl Gauss suggested planting forests and fields in a gigantic geometric proof to signal to possible Martians that intelligent life exists on Earth. If Martians had telescopes that could resolve details no smaller than 1 second of arc, how large would the smallest element of Gauss's proof have to be? (*Hint:* Use the small-angle formula.)

7. If we detected radio signals with an average wavelength of 20 cm and suspected that they came from a civilization on a distant planet, roughly how much of a change in wavelength should we expect to see because of the orbital motion of the distant planet? (*Hint:* See the Doppler shift in Chapter 7.)

8. Calculate the number of communicative civilizations per galaxy from your own estimates of the factors in Table 19-1.

CRITICAL INQUIRIES FOR THE WEB

1. The popular movie *Contact* focused interest on the SETI program by profiling the work of a radio astronomer dedicated to the search for extraterrestrial intelligence. Visit Web sites that give information about the movie, SETI programs, and radio astronomy, and discuss how realistic the movie was in capturing how such research is done.

2. Where outside the solar system would you look for habitable planets? NASA has increasingly focused its interest on this question. Look for information online about programs dedicated to detecting which planets might support life as we know it. What criteria are used to choose targets for the planned searches? What methods will be used to carry out the searches?

 Visit the Seeds *Foundations of Astronomy* companion Web site for critical thinking exercises, articles, and additional readings from InfoTrac College Edition, Brooks/Cole's online student library.

AFTERWORD

The aggregate of all our joys and sufferings, thousands of confident religions, ideologies and economic doctrines, every hunter and forager, every hero and coward, every creator and destroyer of civilizations, every king and peasant, every young couple in love, every hopeful child, every mother and father, every inventor and explorer, every teacher of morals, every corrupt politician, every superstar, every supreme leader, every saint and sinner in the history of our species, lived there on a mote of dust, suspended in a sunbeam.

Carl Sagan (1934–1996)

Earth photographed by Voyager 1
from the edge of the solar system.

Our journey is over, but before we part company, there is one last thing to discuss—the place of humanity in the universe. Astronomy gives us some comprehension of the workings of stars, galaxies, and planets, but its greatest value lies in what it teaches us about ourselves. Now that we have surveyed astronomical knowledge, we can better understand our own position in nature.

To some, the word *nature* conjures up visions of furry rabbits hopping about in a forest glade dotted with pastel wildflowers. To others, nature is the blue-green ocean depths filled with creatures swirling in a mad struggle for survival. Still others think of nature as windswept mountaintops of gray stone and glittering ice. As diverse as these images are, they are all Earthbound. Having studied astronomy, we can view nature as a beautiful mechanism composed of matter and energy interacting according to simple rules to form galaxies, stars, planets, mountaintops, ocean depths, and forest glades.

Perhaps the most important astronomical lesson is that we are a small but important part of the universe. Most of the universe is lifeless. The vast reaches between the galaxies appear to be empty of all but the thinnest gas, and the stars are much too hot to preserve the chemical bonds that seem necessary for life to survive and develop. Only on the surfaces of a few planets, where temperatures are moderate, could atoms link together to form living matter.

If life is special, then intelligence is precious. The universe must contain many planets devoid of life, planets where the wind has blown unfelt for billions of years. There may also exist planets where life has developed but has not become complex, planets on which the wind stirs wide plains of grass and rustles dark forests. On some planets, insects, fish, birds, and animals may watch the passing days unaware of their own existence. It is intelligence, human or alien, that gives meaning to the landscape.

Science is the process by which intelligence tries to understand the universe. Science is not the invention of new devices or processes. It does not create home computers, cure the mumps, or manufacture plastic spoons—that is engineering and technology, the adaptation of scientific understanding for practical purposes. Science is understanding nature, and astronomy is understanding on the grandest scale. Astronomy is the science by which the universe, through its intelligent lumps of matter, tries to understand its own existence.

As the primary intelligent species on this planet, we are the custodians of a priceless gift—a planet filled with living things. This is especially true if life is rare in the universe. In fact, if Earth is the only inhabited planet, our responsibility is overwhelming. In any case, we are the only creatures who can take action to preserve the existence of life on Earth, and, ironically, our own actions are the most serious hazards.

The future of humanity is not secure. We are trapped on a tiny planet with limited resources and a population growing faster than our ability to produce food. In our efforts to survive, we have already driven some creatures to extinction and now threaten others. If our civilization collapses because of starvation, or if our race destroys itself somehow, the only bright spot is that the rest of the creatures on Earth will be better off for our absence.

But even if we control our population and conserve and recycle our resources, life on Earth is doomed. In 5 billion years, the sun will leave the main sequence and swell into a red giant, incinerating Earth. However, Earth will be lifeless long before that. Within the next few billion years, the growing luminosity of the sun will first alter Earth's climate and then boil its atmosphere and oceans. Earth, like everything else in the universe, is only temporary.

Ace⊘Astronomy™ Go to AceAstronomy and click Active Figures to see "Future of the Sun." See how the sun's evolution limits the future of life on Earth.

To survive, humanity must eventually leave Earth and search for other planets. Colonizing the moon and other planets of our solar system will not save us, because they will face the same fate as Earth when the sun dies. But travel to other stars is tremendously difficult and may be impossible with the limited resources we have in our small solar system. We and all of the living things on Earth may be trapped.

This is a depressing prospect, but a few factors are comforting. First, everything in the universe is temporary. Stars die, galaxies die, perhaps the entire universe will someday end. That our distant future is limited only assures us that we are a part of a much larger whole. Second, we have a few billion years to prepare, and a billion years is a very long time. Only a few million years ago, our ancestors were learning to walk erect and communicate. A billion years ago, our ancestors were microscopic organisms living in the primeval oceans. To suppose that a billion years hence we humans will still be human, that we will still be the dominant species on Earth, or that we will still be the sole intelligence on Earth is the ultimate conceit.

Our responsibility is not to save our race for all eternity but to behave as dependable custodians of our planet, preserving it, admiring it, and trying to understand it. That calls for drastic changes in our behavior toward other living things and a revolution in our attitude toward our planet's resources. Whether we can change our ways is debatable—humanity is far from perfect in its understanding, abilities, or intentions. However, we must not imagine that we and our civilization are less than precious. We have the gift of intelligence, and that is the finest thing this planet has ever produced.

> We shall not cease from exploration
> And the end of all our exploring
> Will be to arrive where we started
> And know the place for the first time.
> —*T. S. Eliot, "Little Gidding"*

APPENDIX A

UNITS AND ASTRONOMICAL DATA

INTRODUCTION

The metric system is used worldwide as the system of units, not only in science but also in engineering, business, sports, and daily life. Developed in 18th-century France, the metric system has gained acceptance in almost every country in the world because it simplifies computations.

A system of units is based on the three fundamental units for length, mass, and time. Other quantities such as density and force are derived from these fundamental units. In the English (or British) system of units (commonly used only in the United States, Tonga, and Southern Yemen, but not in Great Britain) the fundamental unit of length is the foot, composed of 12 inches. The metric system is based on the decimal system of numbers, and the fundamental unit of length is the meter, composed of 100 centimeters.

To see the advantage of having a decimal-based system, try computing the volume of a bathtub that is 5′9″ long, 1′2″ deep, and 2′10″ wide. In the metric system the length of the tub is 1 m and 75 cm, or 1.75 m. The other dimensions are 0.35 m and 0.86 m, and the volume is just $1.75 \times 0.35 \times 0.86$ m^3. To make the computation in English units, we must first convert inches to feet by dividing by 12. We can convert centimeters to meters by the simpler process of moving the decimal point. Thus, the computation is much easier if we measure the tub in meters and centimeters instead of feet and inches.

Because the metric system is a decimal system, it is easy to express quantities in larger or smaller units as is convenient. We can express distances in centimeters, meters, kilometers, and so on. The prefixes specify the relation of the unit to the meter. Just as a cent is $\frac{1}{100}$ of a dollar, so a centimeter is $\frac{1}{100}$ of a meter. A kilometer is 1000 m, and a kilogram is 1000 g. The meanings of the commonly used prefixes are given in ▮ Table A-1.

THE SI UNITS

Any system of units based on the decimal system would be easy to use, but by international agreement,

TABLE A-1
Metric Prefixes

Prefix	Symbol	Factor
Mega	M	10^6
Kilo	k	10^3
Centi	c	10^{-2}
Milli	m	10^{-3}
Micro	μ	10^{-6}
Nano	n	10^{-9}

the preferred set of units, known as the *Système International d'Unités* (SI units) is based on the meter, kilogram, and second. These three fundamental units define the rest of the units, as given in ▮ Table A-2.

The SI unit of force is the newton (N), named after Isaac Newton. It is the force needed to accelerate a 1 kg mass by 1 m/s^2, or the force roughly equivalent to the weight of an apple at Earth's surface. The SI unit of energy is the joule (J), the energy produced by a force of 1 N acting through a distance of 1 m. A joule is roughly the energy in the impact of an apple falling off a table.

EXCEPTIONS

Units help us in two ways. They make it possible to make calculations, and they help us to conceive of cer-

TABLE A-2
SI Metric Units

Quantity	SI Unit	English Unit
Length	Meter (m)	Foot
Mass	Kilogram (kg)	Slug (sl)
Time	Second (s)	Second (s)
Force	Newton (N)	Pound (lb)
Energy	Joule (J)	Foot-pound (fp)

tain quantities. For calculations the metric system is far superior, and we use it for our calculations throughout this book.

But Americans commonly use the English system of units, so for conceptual purposes we can express quantities in English units. Instead of saying the average person would weigh 133 N on the moon, we could express the weight as 30 lb. Thus, the text commonly gives quantities in metric form followed by the English form in parentheses: the radius of the moon is 1738 km (1080 miles).

In SI units, density should be expressed as kilograms per cubic meter, but no human can enclose a cubic meter in his or her hand, so this unit does not help us grasp the significance of a given density. This book refers to density in grams per cubic centimeter. A gram is roughly the mass of a paperclip, and a cubic centimeter is the size of a small sugar cube, so we can conceive of a density of 1 g/cm³, roughly the density of water. This is not a bothersome departure from SI units because we will not make complex calculations using density.

CONVERSIONS

To convert from one metric unit to another (from meters to kilometers, for example), we have only to look at the prefix. However, converting from metric to English or English to metric is more complicated. The conversion factors are given in ▮ Table A-3.

Example: The radius of the moon is 1738 km. What is this in miles? Table A-3 indicates that 1 mile equals 1.609 km, so

$$1738 \text{ km} \times \frac{1 \text{ mile}}{1.609 \text{ km}} = 1080 \text{ miles}$$

TEMPERATURE SCALES

In astronomy, as in most other sciences, temperatures are expressed on the Kelvin scale, although the centigrade (or Celsius) scale is also used. The Fahrenheit scale commonly used in the United States is not used in scientific work.

Temperatures on the Kelvin scale are measured from absolute zero, the temperature of an object that contains no extractable heat. In practice, no object can be as cold as absolute zero, although laboratory apparatuses have reached temperatures less than 10^{-6} K. The scale is named after the Scottish mathematical physicist William Thomson, Lord Kelvin (1824–1907).

The centigrade scale refers temperatures to the freezing point of water (0°C) and to the boiling point of

TABLE A-3 Conversion Factors	
1 inch = 2.54 centimeters	1 centimeter = 0.394 inch
1 foot = 0.3048 meter	1 meter = 39.36 inches = 3.28 feet
1 mile = 1.6093 kilometers	1 kilometer = 0.6214 mile
1 slug = 14.594 kilograms	1 kilogram = 0.0685 slug
1 pound = 4.4482 newtons	1 newton = 0.2248 pound
1 foot-pound = 1.35582 joules	1 joule = 0.7376 foot-pound
1 horsepower = 745.7 joules/s	1 joule/s = 1 watt

water (100°C). One degree centigrade is 1/100th the temperature difference between the freezing and boiling points of water. Thus the prefix *centi*. The centigrade scale is also called the Celsius scale after its inventor, the Swedish astronomer Anders Celsius (1701–1744).

The Fahrenheit scale fixes the freezing point of water at 32°F and the boiling point at 212°F. Named after the German physicist Gabriel Daniel Fahrenheit (1686–1736), who made the first successful mercury thermometer in 1720, the Fahrenheit scale is used only in the United States.

It is easy to convert temperatures from one scale to another using the information given in ▮ Table A-4.

POWERS OF 10 NOTATION

Powers of 10 make writing very large numbers much simpler. For example, the nearest star is about 43,000,000,000,000 km from the sun. Writing this number as 4.3×10^{13} km is much easier.

Very small numbers can also be written with powers of 10. For example, the wavelength of visible light

TABLE A-4 Temperature Scales			
	Kelvin (K)	Centigrade (°C)	Fahrenheit (°F)
Absolute zero	0 K	−273°C	−459°F
Freezing point of water	273 K	0°C	32°F
Boiling point of water	373 K	100°C	212°F

Conversions:
$$K = °C + 273$$
$$°C = \frac{5}{9} (°F - 32)$$
$$°F = \frac{9}{5} °C + 32$$

is about 0.0000005 m. In powers of 10 this becomes 5×10^{-7} m.

The powers of 10 used in this notation appear below. The exponent tells us how to move the decimal point. If the exponent is positive, we move the decimal point to the right. If the exponent is negative, we move the decimal point to the left. Thus, 2×10^3 equals 2000.0, and 2×10^{-3} equals 0.002.

$$\vdots$$

$$10^5 = 100,000$$
$$10^4 = 10,000$$
$$10^3 = 1,000$$
$$10^2 = 100$$
$$10^1 = 10$$
$$10^0 = 1$$
$$10^{-1} = 0.1$$
$$10^{-2} = 0.01$$
$$10^{-3} = 0.001$$
$$10^{-4} = 0.0001$$

$$\vdots$$

If you use scientific notation in calculations, be sure you correctly enter numbers into your calculator. Not all calculators accept scientific notation, but those that can have a key labeled EXP, EEX, or perhaps EE that allows you to enter the exponent of ten. To enter a number such as 3×10^8, press the keys 3 EXP 8. To enter a number with a negative exponent, we must use the change-sign key, usually labeled $+/-$ or CHS. To enter the number 5.2×10^{-3}, we press the keys 5.2 EXP $+/-$ 3. Try a few examples.

To read a number in scientific notation from a calculator you must read the exponent separately. The number 3.1×10^{25} may appear in a calculator display as 3.1 25 or on some calculators as 3.1 10^{25}. Examine your calculator to determine how such numbers are displayed.

***

Astronomy, and science in general, is a way of learning about nature and understanding the universe we live in. To test hypotheses about how nature works, scientists use observations of nature. The tables that follow contain some of the basic observations that support our best understanding of the astronomical universe. Of course, these data are expressed in the form of numbers, not because science reduces all understanding to mere numbers, but because the struggle to understand nature is so demanding we must use every tool available. Quantitative thinking, reasoning mathematically, is one of the most powerful tools ever invented by the human brain. Thus these tables are not nature reduced to mere numbers, but numbers supporting humanity's growing understanding of the natural world around us.

TABLE A-5
Constants

Astronomical unit (AU)	$= 1.495979 \times 10^{11}$ m
Parsec (pc)	$= 206,265$ AU
	$= 3.085678 \times 10^{16}$ m
	$= 3.261633$ ly
Light-year (ly)	$= 9.46053 \times 10^{15}$ m
Velocity of light (c)	$= 2.997925 \times 10^8$ m/s
Gravitational constant (G)	$= 6.67 \times 10^{-11}$ m³/s²kg
Mass of Earth ($M_\oplus$)	$= 5.976 \times 10^{24}$ kg
Earth equatorial radius ($R_\oplus$)	$= 6378.164$ km
Mass of sun ($M_\odot$)	$= 1.989 \times 10^{30}$ kg
Radius of sun ($R_\odot$)	$= 6.9599 \times 10^8$ m
Solar luminosity ($L_\odot$)	$= 3.826 \times 10^{26}$ J/s
Mass of moon	$= 7.350 \times 10^{22}$ kg
Radius of moon	$= 1738$ km
Mass of H atom	$= 1.67352 \times 10^{-27}$ kg

TABLE A-6
Units Used in Astronomy

1 angstrom (Å)	$= 10^{-8}$ cm
	$= 10^{-10}$ m
1 astronomical unit (AU)	$= 1.495979 \times 10^{11}$ m
	$= 92.95582 \times 10^6$ miles
1 light-year (ly)	$= 6.3240 \times 10^4$ AU
	$= 9.46053 \times 10^{15}$ m
	$= 5.9 \times 10^{12}$ miles
1 parsec (pc)	$= 206,265$ AU
	$= 3.085678 \times 10^{16}$ m
	$= 3.261633$ ly
1 kiloparsec (kpc)	$= 1000$ pc
1 megaparsec (Mpc)	$= 1,000,000$ pc

Properties of Main-Sequence Stars

Spectral Type	Absolute Visual Magnitude (M_v)	Luminosity*	Temp. (K)	λ_{max} (nm)	Mass*	Radius*	Average Density (g/cm³)
O5	−5.8	501,000	40,000	72.4	40	17.8	0.01
B0	−4.1	20,000	28,000	100	18	7.4	0.1
B5	−1.1	790	15,000	190	6.4	3.8	0.2
A0	+0.7	79	9900	290	3.2	2.5	0.3
A5	+2.0	20	8500	340	2.1	1.7	0.6
F0	+2.6	6.3	7400	390	1.7	1.4	1.0
F5	+3.4	2.5	6600	440	1.3	1.2	1.1
G0	+4.4	1.3	6000	480	1.1	1.0	1.4
G5	+5.1	0.8	5500	520	0.9	0.9	1.6
K0	+5.9	0.4	4900	590	0.8	0.8	1.8
K5	+7.3	0.2	4100	700	0.7	0.7	2.4
M0	+9.0	0.1	3500	830	0.5	0.6	2.5
M5	+11.8	0.01	2800	1000	0.2	0.3	10.0
M8	+16	0.001	2400	1200	0.1	0.1	63

*Luminosity, mass, and radius are given in terms of the sun's luminosity, mass, and radius.

TABLE A-8
The Brightest Stars

Star	Name	Apparent Visual Magnitude (m_v)	Spectral Type	Absolute Visual Magnitude (M_v)	Distance (ly)
α CMa A	Sirius	−1.47	A1	1.4	8.7
α Car	Canopus	−0.72	F0	−3.1	98
α Cen	Rigil Kentaurus	−0.01	G2	4.4	4.3
α Boo	Arcturus	−0.06	K2	−0.3	36
α Lyr	Vega	0.04	A0	0.5	26.5
α Aur	Capella	0.05	G8	−0.6	45
β Ori A	Rigel	0.14	B8	−7.1	900
α CMi A	Procyon	0.37	F5	2.7	11.3
α Ori	Betelgeuse	0.41	M2	−5.6	520
α Eri	Achernar	0.51	B3	−2.3	118
β Cen AB	Hadar	0.63	B1	−5.2	490
α Aql	Altair	0.77	A7	2.2	16.5
α Tau A	Aldebaran	0.86	K5	−0.7	68
α Cru	Acrux	0.90	B2	−3.5	260
α Vir	Spica	0.91	B1	−3.3	220
α Sco A	Antares	0.92	M1	−5.1	520
α PsA	Fomalhaut	1.15	A3	2.0	22.6
β Gem	Pollux	1.16	K0	1.0	35
α Cyg	Deneb	1.26	A2	−7.1	1600
β Cru	Beta Crucis	1.28	B0.5	−4.6	490

Name	Absolute Magnitude (M_v)	Distance (ly)	Spectral Type	Apparent Visual Magnitude (m_v)
Sun	4.83		G2	−26.8
Proxima Cen	15.45	4.28	M5	11.05
α Cen A	4.38	4.3	G2	0.1
B	5.76	4.3	K5	1.5
Barnard's Star	13.21	5.9	M5	9.5
Wolf 359	16.80	7.6	M6	13.5
Lalande 21185	10.42	8.1	M2	7.5
Sirius A	1.41	8.6	A1	−1.5
B	11.54	8.6	white dwarf	7.2
Luyten 726-8A	15.27	8.9	M5	12.5
B (UV Cet)	15.8	8.9	M6	13.0
Ross 154	13.3	9.4	M5	10.6
Ross 248	14.8	10.3	M6	12.2
ε Eri	6.13	10.7	K2	3.7
Luyten 789-6	14.6	10.8	M7	12.2
Ross 128	13.5	10.8	M5	11.1
61 CYG A	7.58	11.2	K5	5.2
B	8.39	11.2	K7	6.0
ε Ind	7.0	11.2	K5	4.7
Procyon A	2.64	11.4	F5	0.3
B	13.1	11.4	white dwarf	10.8
Σ 2398 A	11.15	11.5	M4	8.9
B	11.94	11.5	M5	9.7
Groombridge 34 A	10.32	11.6	M1	8.1
B	13.29	11.6	M6	11.0
Lacaille 9352	9.59	11.7	M2	7.4
τ Ceti	5.72	11.9	G8	3.5
BD + 5° 1668	11.98	12.2	M5	9.8
L 725-32	15.27	12.4	M5	11.5
Lacaille 8760	8.75	12.5	M0	6.7
Kapteyn's Star	10.85	12.7	M0	8.8
Kruger 60 A	11.87	12.8	M3	9.7
B	13.3	12.8	M4	11.2

TABLE A-10
Properties of the Planets

PHYSICAL PROPERTIES (EARTH = ⊕)

Planet	Equatorial Radius (km)	Equatorial Radius (⊕ = 1)	Mass (⊕ = 1)	Average Density (g/cm³)	Surface Gravity (⊕ = 1)	Escape Velocity (km/s)	Sidereal Period of Rotation	Inclination of Equator to Orbit
Mercury	2439	0.382	0.0558	5.44	0.378	4.3	58.646^d	0°
Venus	6052	0.95	0.815	5.24	0.903	10.3	244.3^d	177°
Earth	6378	1.00	1.00	5.497	1.00	11.2	23^{h}56^{m}04.1^s	23°27′
Mars	3396	0.53	0.1075	3.94	0.379	5.0	24^{h}37^{m}22.6^s	25°19′
Jupiter	71,494	11.20	317.83	1.34	2.54	61	9^{h}55^{m}30^s	3°5′
Saturn	60,330	9.42	95.147	0.69	1.16	35.6	10^{h}13^{m}59^s	26°24′
Uranus	25,559	4.01	14.54	1.19	0.919	22	17^{h}14^m	97°55′
Neptune	24,750	3.93	17.23	1.66	1.19	25	16^{h}3^m	28°48′
Pluto	1185	0.18	0.0022	2.0	0.06	1.2	6^{d}9^{h}21^m	122°

ORBITAL PROPERTIES

Planet	Semimajor Axis (a) (AU)	Semimajor Axis (a) (10⁶ km)	Orbital Period (P) (y)	Orbital Period (P) (days)	Average Orbital Velocity (km/s)	Orbital Eccentricity	Inclination to Ecliptic
Mercury	0.3871	57.9	0.24084	87.969	47.89	0.2056	7°0′16″
Venus	0.7233	108.2	0.61515	224.68	35.03	0.0068	3°23′40″
Earth	1	149.6	1	365.26	29.79	0.0167	0°
Mars	1.5237	227.9	1.8808	686.95	24.13	0.0934	1°51′09″
Jupiter	5.2028	778.3	11.867	4334.3	13.06	0.0484	1°18′29″
Saturn	9.5388	1427.0	29.461	10,760	9.64	0.0560	2°29′17″
Uranus	19.18	2869.0	84.013	30,685	6.81	0.0461	0°46′23″
Neptune	30.0611	4497.1	164.793	60,189	5.43	0.0100	1°46′27″
Pluto	39.44	5900	247.7	90,465	4.74	0.2484	17°9′3″

TABLE A-11
Principal Satellites of the Solar System

Planet	Satellite	Radius (km)	Distance from Planet (10^3 km)	Orbital Period (days)	Orbital Eccentricity	Orbital Inclination
Earth	Moon	1738	384.4	27.322	0.055	5°8′43″
Mars	Phobos	14 × 12 × 10	9.38	0.3189	0.018	1°.0
	Deimos	8 × 6 × 5	23.5	1.262	0.002	2°.8
Jupiter	Metis	20	126	0.29	0.0	0°.0
	Adrastea	12 × 8 × 10	128	0.294	0.0	0°.0
	Amalthea	135 × 100 × 78	182	0.4982	0.003	0°.45
	Thebe	50	223	0.674	0.0	1°.3
	Io	1820	422	1.769	0.000	0°.3
	Europa	1565	671	3.551	0.000	0°.46
	Ganymede	2640	1071	7.155	0.002	0°.18
	Callisto	2420	1884	16.689	0.008	0°.25
	Leda	~8	11,110	240	0.146	26°.7
	Himalia	~85	11,470	250.6	0.158	27°.6
	Lysithea	~20	11,710	260	0.12	29°
	Elara	~30	11,740	260.1	0.207	24°.8
	Ananke	15	21,200	631	0.169	147°
	Carme	22	22,350	692	0.207	163°
	Pasiphae	35	23,300	735	0.40	147°
	Sinope	20	23,700	758	0.275	156°
Saturn	Pan	10	133.570	0.574	0.000	0°
	Atlas	20 × 15 × 15	137.7	0.601	0.002	0°.3
	Prometheus	70 × 40 × 50	139.4	0.613	0.003	0°.0
	Pandora	55 × 35 × 50	141.7	0.629	0.004	0°.05
	Epimetheus	70 × 50 × 50	151.42	0.694	0.009	0°.34
	Janus	110 × 80 × 100	151.47	0.695	0.007	0°.14
	Mimas	196	185.54	0.942	0.020	1°.5
	Enceladus	250	238.04	1.370	0.004	0°.0
	Tethys	530	294.67	1.888	0.000	1°.1
	Calypso	17 × 11 × 12	294.67	1.888	0.0	~1°.?
	Telesto	12	294.67	1.888	0.0	~1°?
	Dione	560	377	2.737	0.002	0°.0
	Helene	20 × 15 × 15	377	2.74	0.005	0°.15
	Rhea	765	527	4.518	0.001	0°.4
	Titan	2575	1222	15.94	0.029	0°.3
	Hyperion	205 × 130 × 110	1484	21.28	0.104	~0°.5
	Iapetus	720	3562	79.33	0.028	14°.72
	Phoebe	110	12,930	550.4	0.163	150°
Uranus	Cordelia	20	49.8	0.3333	~0	~0°
	Ophelia	15	53.8	0.375	~0	~0°
	Bianca	25	59.1	0.433	~0	~0°
	Cressida	30	61.8	0.462	~0	~0°
	Desdemona	30	62.7	0.475	~0	~0°
	Juliet	40	64.4	0.492	~0	~0°
	Portia	55	66.1	0.512	~0	~0°

(continued)

TABLE A-11
(continued)

Planet	Satellite	Radius (km)	Distance from Planet (10³ km)	Orbital Period (days)	Orbital Eccentricity	Orbital Inclination
Uranus	Rosalind	30	69.9	0.558	~0	~0°
(continued)	Belinda	30	75.2	0.621	~0	~0°
	Puck	85 ± 5	85.9	0.762	~0	~0°
	Miranda	242 ± 5	129.9	1.414	0.017	3°.4
	Ariel	580 ± 5	190.9	2.520	0.003	0°
	Umbriel	595 ± 10	266.0	4.144	0.003	0°
	Titania	805 ± 5	436.3	8.706	0.002	0°
	Oberon	775 ± 10	583.4	13.463	0.001	0°
	Caliban	40	7164	579	0.082	139.2°
	Stephano	~20	7900	676		
	Sycorax	80	12,174	1284	0.509	152.7°
	Prospero	~20	16,100	1950		
	Setebos	~20	17,600	2240	0.539	
Neptune	Naiad	30	48.2	0.296	~0	~0°
	Thalassa	40	50.0	0.312	~0	~0°
	Despina	90	52.5	0.333	~0	~0°
	Galatea	75	62.0	0.396	~0	~0°
	Larissa	95	73.6	0.554	~0	~0°
	Proteus	205	117.6	1.121	~0	~0°
	Triton	1352	354.59	5.875	0.00	160°
	Nereid	170	5588.6	360.125	0.76	27.7°
Pluto	Charon	626	19.7	6.38718	~0	122°

TABLE A-12
Meteor Showers

Shower	Dates	Hourly Rate	Radiant R.A.	Radiant Dec.	Associated Comet
Quadrantids	Jan. 2–4	30	15ʰ24ᵐ	50°	
Lyrids	April 20–22	8	18ʰ4ᵐ	33°	1861 I
η Aquarids	May 2–7	10	22ʰ24ᵐ	0°	Halley?
δ Aquarids	July 26–31	15	22ʰ36ᵐ	−10°	
Perseids	Aug. 10–14	40	3ʰ4ᵐ	58°	1982 III
Orionids	Oct. 18–23	15	6ʰ20ᵐ	15°	Halley?
Taurids	Nov. 1–7	8	3ʰ40ᵐ	17°	Encke
Leonids	Nov. 14–19	6	10ʰ12ᵐ	22°	1866 I Temp
Geminids	Dec. 10–13	50	7ʰ28ᵐ	32°	

TABLE A-13
Greatest Elongations of Mercury

Evening Sky	Morning Sky
March 29, 2004*	May 14, 2004
July 27, 2004	Sept. 9, 2004
Nov. 21, 2004	Dec. 29, 2004
March 12, 2005*	April 26, 2005
July 9, 2005	Aug. 23, 2005
Nov. 3, 2005	Dec. 12, 2005
Feb. 24, 2006*	April 8, 2006
June 20, 2006	Aug. 7, 2006
Oct. 16, 2006	Nov. 25, 2006*
Feb. 7, 2007	March 22, 2007
June 2, 2007	July 20, 2007
Sept. 29, 2007	Nov. 8, 2007*
Jan. 22, 2008	March 3, 2008
May 14, 2008	July 1, 2008
Sept. 11, 2008	Oct. 22, 2008*

Elongation is the angular distance from the sun to a planet.
*Most favorable elongations.

TABLE A-14
Greatest Elongations of Venus

Evening Sky	Morning Sky
March 29, 2004	Aug. 17, 2004
Nov. 3, 2005	March 25, 2006
June 9, 2007	Oct. 28, 2007
Jan. 14, 2009	June 5, 2009
Aug. 20, 2010	Jan. 8, 2011
March 27, 2012	Aug. 15, 2012
Nov. 1, 2013	March 22, 2014
June 6, 2015	Oct. 26, 2015

TABLE A-15
The Greek Alphabet

A, α	alpha	H, η	eta	N, ν	nu	T, τ	tau
B, β	beta	Θ, θ	theta	Ξ, ξ	xi	Y, υ	upsilon
Γ, γ	gamma	I, ι	iota	O, o	omicron	Φ, ϕ	phi
Δ, δ	delta	K, κ	kappa	Π, π	pi	X, χ	chi
E, ε	epsilon	Λ, λ	lambda	P, ρ	rho	Ψ, ψ	psi
Z, ζ	zeta	M, μ	mu	Σ, σ	sigma	Ω, ω	omega

TABLE A-16
Periodic Table of the Elements

Atomic masses are based on carbon-12. Numbers in parentheses are mass numbers of most stable or best-known isotopes of radioactive elements.

Key:
- Atomic number → 11
- Symbol → Na
- Atomic mass → 22.99

Group IA(1)	IIA(2)	IIIB(3)	IVB(4)	VB(5)	VIB(6)	VIIB(7)	VIII (8)	(9)	(10)	IB(11)	IIB(12)	IIIA(13)	IVA(14)	VA(15)	VIA(16)	VIIA(17)	Noble Gases (18)
1 H 1.008																	2 He 4.003
3 Li 6.941	4 Be 9.012											5 B 10.81	6 C 12.01	7 N 14.01	8 O 16.00	9 F 19.00	10 Ne 20.18
11 Na 22.99	12 Mg 24.31											13 Al 26.98	14 Si 28.09	15 P 30.97	16 S 32.06	17 Cl 35.45	18 Ar 39.95
19 K 39.10	20 Ca 40.08	21 Sc 44.96	22 Ti 47.90	23 V 50.94	24 Cr 52.00	25 Mn 54.94	26 Fe 55.85	27 Co 58.93	28 Ni 58.7	29 Cu 63.55	30 Zn 65.38	31 Ga 69.72	32 Ge 72.59	33 As 74.92	34 Se 78.96	35 Br 79.90	36 Kr 83.80
37 Rb 85.47	38 Sr 87.62	39 Y 88.91	40 Zr 91.22	41 Nb 92.91	42 Mo 95.94	43 Tc 98.91	44 Ru 101.1	45 Rh 102.9	46 Pd 106.4	47 Ag 107.9	48 Cd 112.4	49 In 114.8	50 Sn 118.7	51 Sb 121.8	52 Te 127.6	53 I 126.9	54 Xe 131.3
55 Cs 132.9	56 Ba 137.3	57* La 138.9	72 Hf 178.5	73 Ta 180.9	74 W 183.9	75 Re 186.2	76 Os 190.2	77 Ir 192.2	78 Pt 195.1	79 Au 197.0	80 Hg 200.6	81 Tl 204.4	82 Pb 207.2	83 Bi 209.0	84 Po (210)	85 At (210)	86 Rn (222)
87 Fr (223)	88 Ra 226.0	89** Ac (227)	104 Rf (261)	105 Db (262)	106 Sg (263)	107 Bh (262)	108 Hs (265)	109 Mt (266)	110 Uun (269)	111 Uuu (272)	112 Uub (277)	114 Uuq (285)			116 Uuh (289)		

Period is indicated at left: 1, 2, 3, 4, 5, 6, 7. "Transition Elements" spans groups IIIB–IIB.

Inner Transition Elements

Lanthanide Series 6 *

58 Ce 140.1	59 Pr 140.9	60 Nd 144.2	61 Pm (145)	62 Sm 150.4	63 Eu 152.0	64 Gd 157.3	65 Tb 158.9	66 Dy 162.5	67 Ho 164.9	68 Er 167.3	69 Tm 168.9	70 Yb 173.0	71 Lu 175.0

Actinide Series 7 **

90 Th 232.0	91 Pa 231.0	92 U 238.0	93 Np 237.0	94 Pu (244)	95 Am (243)	96 Cm (247)	97 Bk (247)	98 Cf (251)	99 Es (252)	100 Fm (257)	101 Md (258)	102 No (259)	103 Lr (260)

OBSERVING THE SKY

Observing the sky with the naked eye is of no more importance to modern astronomy than picking up pretty pebbles is to modern geology. But the sky is a natural wonder unimaginably bigger than the Grand Canyon, the Rocky Mountains, or any other natural wonder that tourists visit every year. To neglect the beauty of the sky is equivalent to geologists neglecting the beauty of the minerals they study. This supplement is meant to act as a tourist's guide to the sky. We analyzed the universe in the regular chapters, but here we will admire it.

The brighter stars in the sky are visible even from the centers of cities with their air and light pollution. But in the countryside only a few miles beyond the cities, the night sky is a velvety blackness strewn with thousands of glittering stars. From a wilderness location, far from the city's glare, and especially from high mountains, the night sky is spectacular.

USING STAR CHARTS

The constellations are a fascinating cultural heritage of our planet, but they are sometimes a bit difficult to learn because of Earth's motion. The constellations above the horizon change with the time of night and the seasons.

Because Earth rotates eastward, the sky appears to rotate around us westward. A constellation visible in the southern sky soon after sunset will appear to move westward, and in a few hours it will disappear below the horizon. Other constellations will rise in the east, so the sky changes gradually through the night.

In addition, Earth's orbital motion makes the sun appear to move eastward among the stars. Each day the sun moves about twice its own diameter eastward along the ecliptic, and each night at sunset, the constellations are about 1° farther toward the west.

Orion, for instance, is visible in the evening sky in January, but as the days pass, the sun moves closer to Orion. By March, Orion is difficult to see in the western sky soon after sunset. By June, the sun is so close to Orion it sets with the sun and is invisible. Not until late July is the sun far enough past Orion for the con-

stellation to become visible rising in the eastern sky just before dawn.

Because of the rotation and orbital motion of Earth, we must use more than one star chart to map the sky. Which chart we select depends on the month and the time of night. The charts given in this appendix show the evening sky for each month.

Two sets of charts are included for two typical locations on Earth. The northern hemisphere charts show the sky as seen from a northern latitude typical of the United States and central Europe. The southern hemisphere star charts are appropriate for readers in Earth's southern hemisphere, including Australia, southern South America, and southern Africa.

To use the charts, select the appropriate chart and hold it overhead as shown in ▌Figure B-1. If you face south, turn the chart until the words *southern horizon* are at the bottom of the chart. If you face other directions, turn the chart appropriately.

FIGURE B-1
To use the star charts in this book, select the appropriate chart for the date and time. Hold it overhead, and turn it until the direction at the bottom of the chart is the same as the direction you are facing.

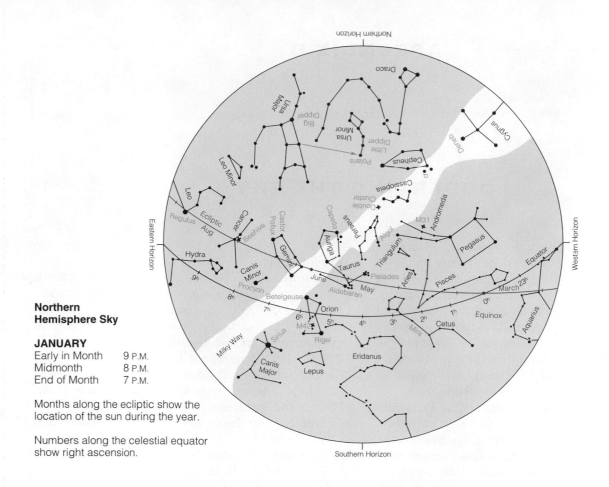

**Northern
Hemisphere Sky**

JANUARY

Early in Month	9 P.M.
Midmonth	8 P.M.
End of Month	7 P.M.

Months along the ecliptic show the location of the sun during the year.

Numbers along the celestial equator show right ascension.

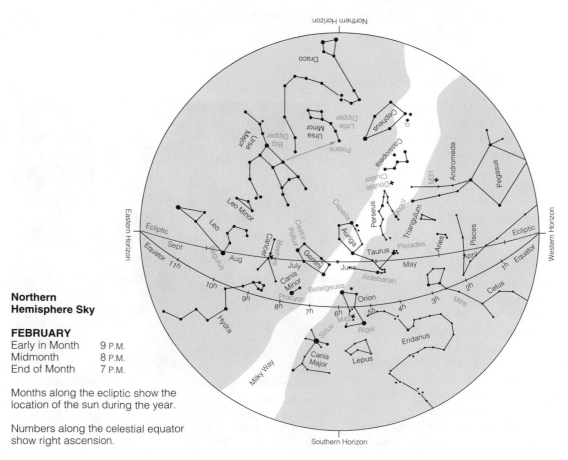

**Northern
Hemisphere Sky**

FEBRUARY

Early in Month	9 P.M.
Midmonth	8 P.M.
End of Month	7 P.M.

Months along the ecliptic show the location of the sun during the year.

Numbers along the celestial equator show right ascension.

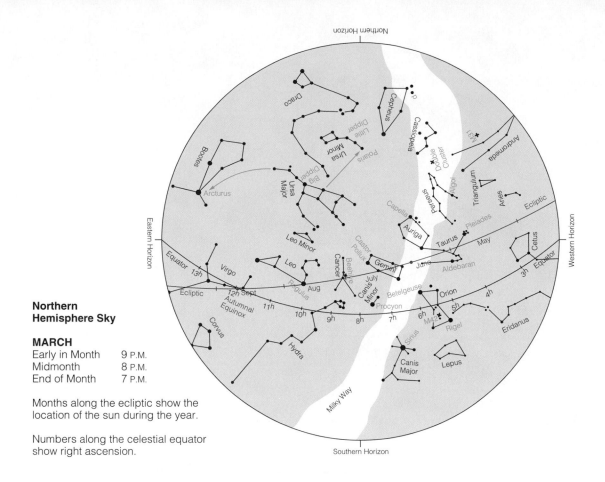

Northern Hemisphere Sky

MARCH

Early in Month	9 P.M.
Midmonth	8 P.M.
End of Month	7 P.M.

Months along the ecliptic show the location of the sun during the year.

Numbers along the celestial equator show right ascension.

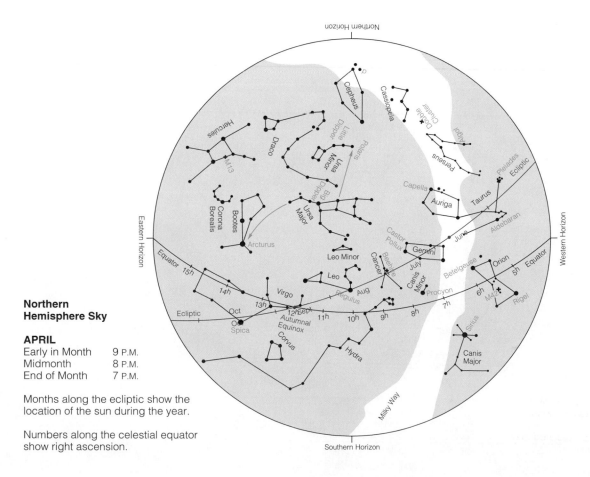

Northern Hemisphere Sky

APRIL

Early in Month	9 P.M.
Midmonth	8 P.M.
End of Month	7 P.M.

Months along the ecliptic show the location of the sun during the year.

Numbers along the celestial equator show right ascension.

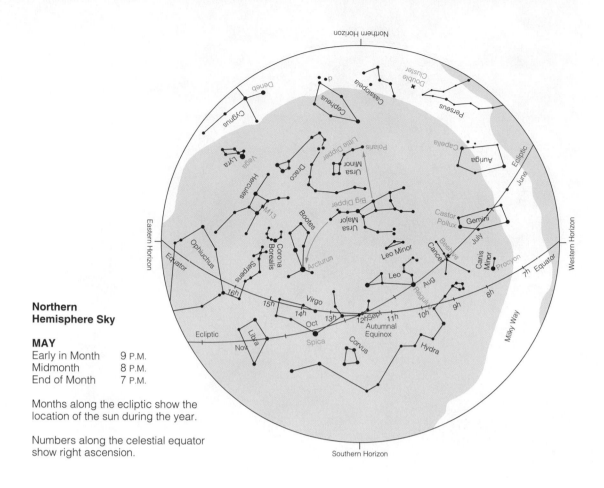

Northern Hemisphere Sky

MAY

Early in Month	9 P.M.
Midmonth	8 P.M.
End of Month	7 P.M.

Months along the ecliptic show the location of the sun during the year.

Numbers along the celestial equator show right ascension.

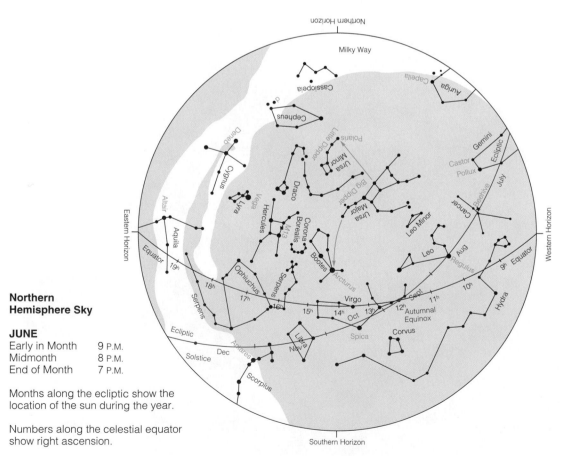

Northern Hemisphere Sky

JUNE

Early in Month	9 P.M.
Midmonth	8 P.M.
End of Month	7 P.M.

Months along the ecliptic show the location of the sun during the year.

Numbers along the celestial equator show right ascension.

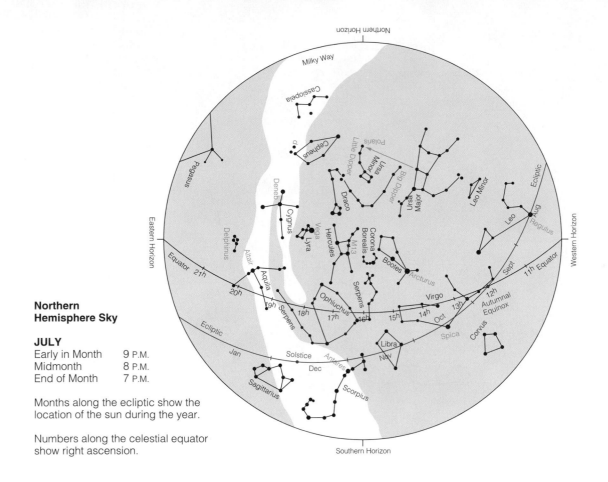

Northern Hemisphere Sky

JULY

Early in Month	9 P.M.
Midmonth	8 P.M.
End of Month	7 P.M.

Months along the ecliptic show the location of the sun during the year.

Numbers along the celestial equator show right ascension.

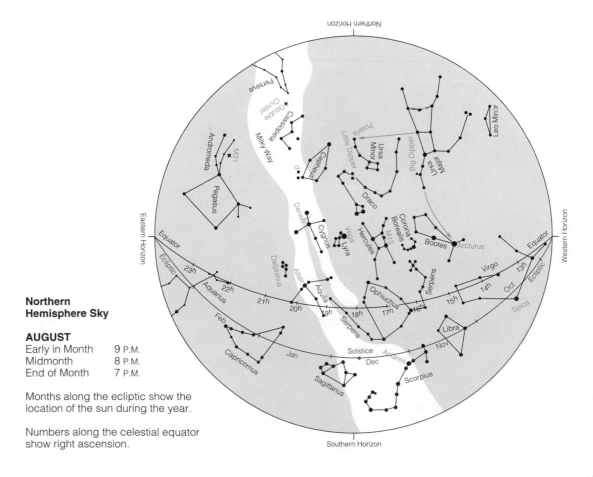

Northern Hemisphere Sky

AUGUST

Early in Month	9 P.M.
Midmonth	8 P.M.
End of Month	7 P.M.

Months along the ecliptic show the location of the sun during the year.

Numbers along the celestial equator show right ascension.

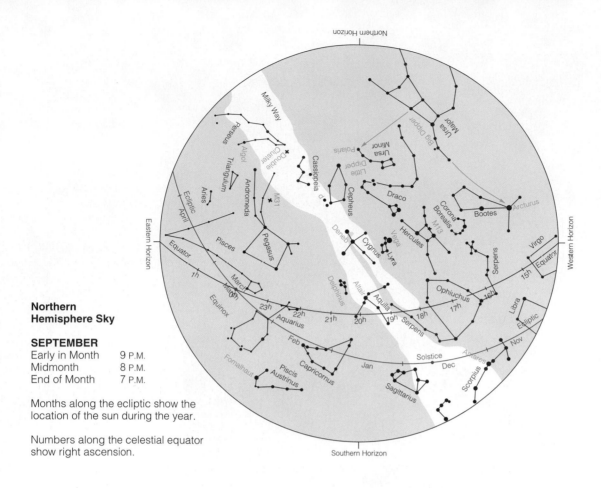

**Northern
Hemisphere Sky**

SEPTEMBER

Early in Month	9 P.M.
Midmonth	8 P.M.
End of Month	7 P.M.

Months along the ecliptic show the
location of the sun during the year.

Numbers along the celestial equator
show right ascension.

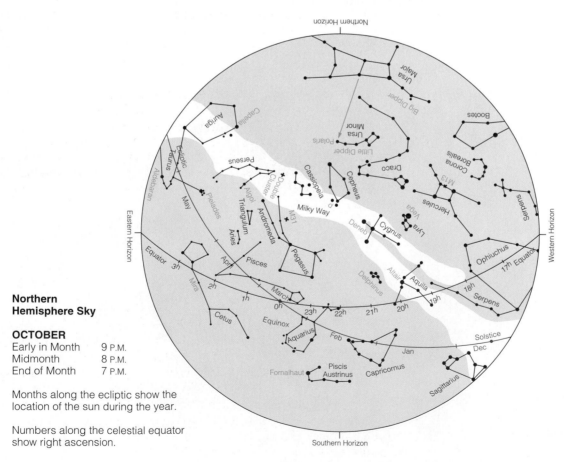

**Northern
Hemisphere Sky**

OCTOBER

Early in Month	9 P.M.
Midmonth	8 P.M.
End of Month	7 P.M.

Months along the ecliptic show the
location of the sun during the year.

Numbers along the celestial equator
show right ascension.

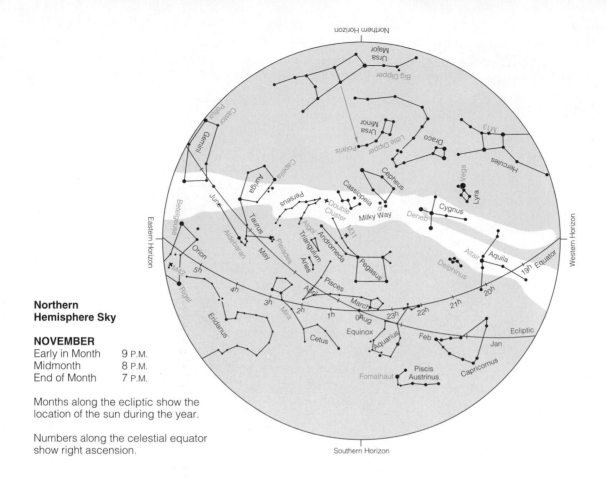

Northern Hemisphere Sky

NOVEMBER

Early in Month	9 P.M.
Midmonth	8 P.M.
End of Month	7 P.M.

Months along the ecliptic show the location of the sun during the year.

Numbers along the celestial equator show right ascension.

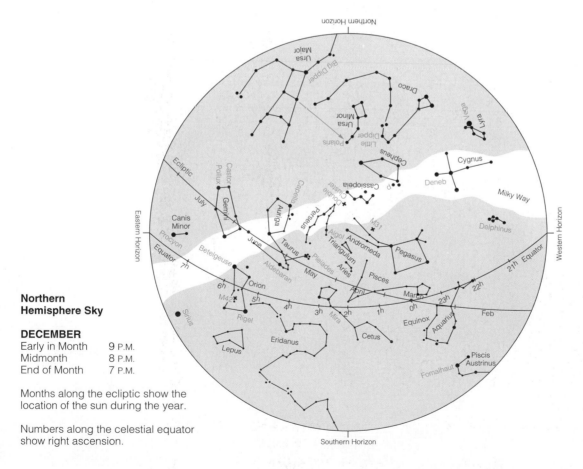

Northern Hemisphere Sky

DECEMBER

Early in Month	9 P.M.
Midmonth	8 P.M.
End of Month	7 P.M.

Months along the ecliptic show the location of the sun during the year.

Numbers along the celestial equator show right ascension.

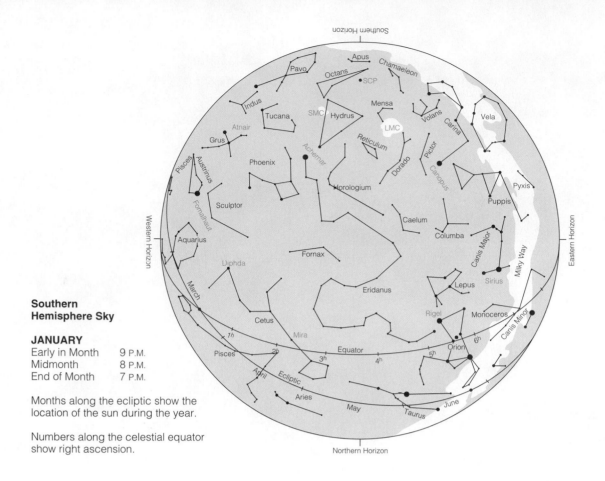

**Southern
Hemisphere Sky**

JANUARY

Early in Month	9 P.M.
Midmonth	8 P.M.
End of Month	7 P.M.

Months along the ecliptic show the
location of the sun during the year.

Numbers along the celestial equator
show right ascension.

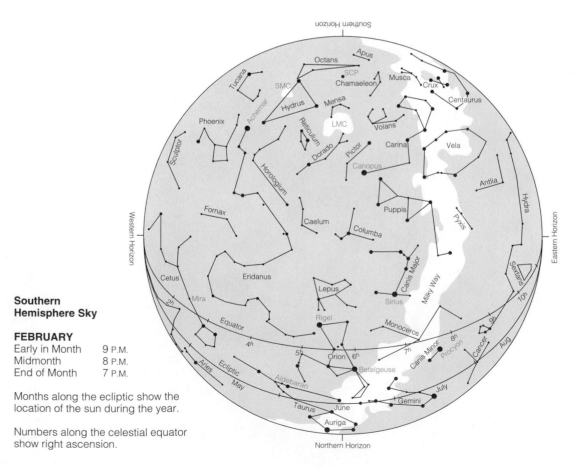

**Southern
Hemisphere Sky**

FEBRUARY

Early in Month	9 P.M.
Midmonth	8 P.M.
End of Month	7 P.M.

Months along the ecliptic show the
location of the sun during the year.

Numbers along the celestial equator
show right ascension.

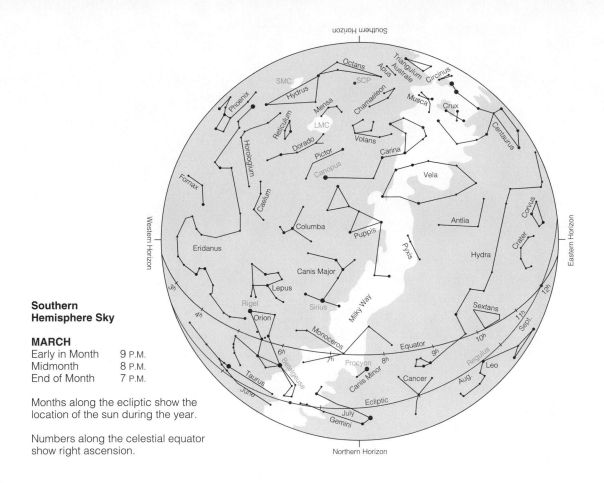

**Southern
Hemisphere Sky**

MARCH

Early in Month	9 P.M.
Midmonth	8 P.M.
End of Month	7 P.M.

Months along the ecliptic show the
location of the sun during the year.

Numbers along the celestial equator
show right ascension.

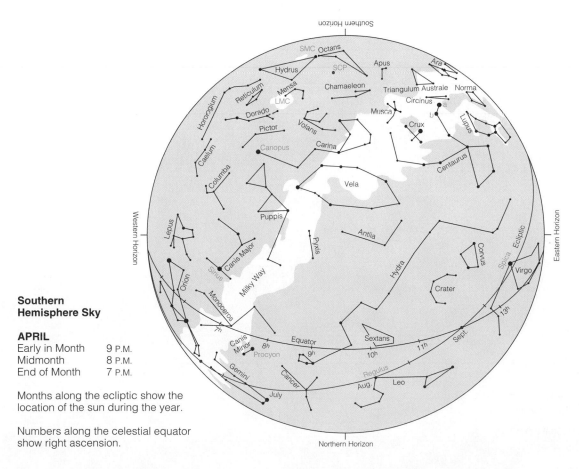

**Southern
Hemisphere Sky**

APRIL

Early in Month	9 P.M.
Midmonth	8 P.M.
End of Month	7 P.M.

Months along the ecliptic show the
location of the sun during the year.

Numbers along the celestial equator
show right ascension.

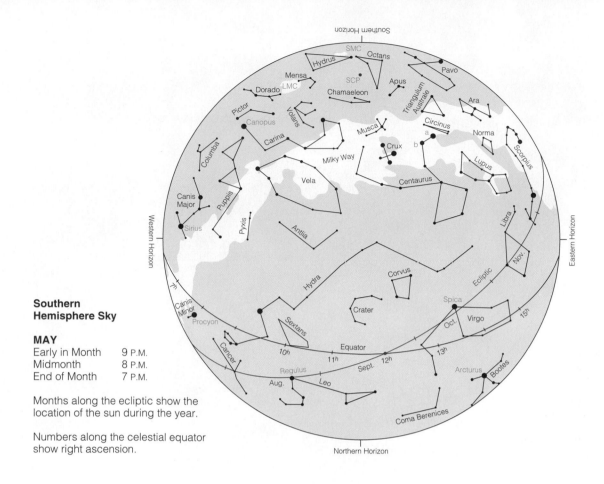

**Southern
Hemisphere Sky**

MAY

Early in Month	9 P.M.
Midmonth	8 P.M.
End of Month	7 P.M.

Months along the ecliptic show the location of the sun during the year.

Numbers along the celestial equator show right ascension.

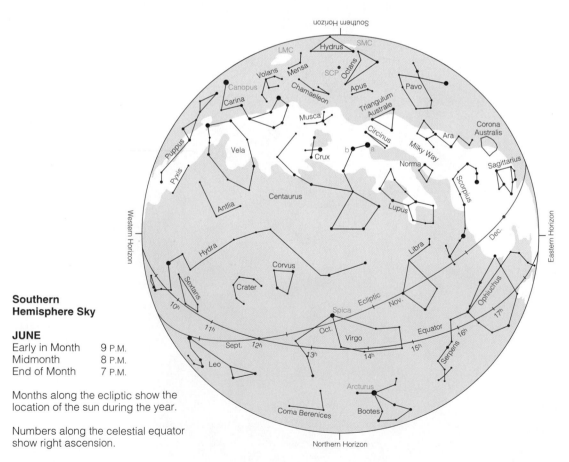

**Southern
Hemisphere Sky**

JUNE

Early in Month	9 P.M.
Midmonth	8 P.M.
End of Month	7 P.M.

Months along the ecliptic show the location of the sun during the year.

Numbers along the celestial equator show right ascension.

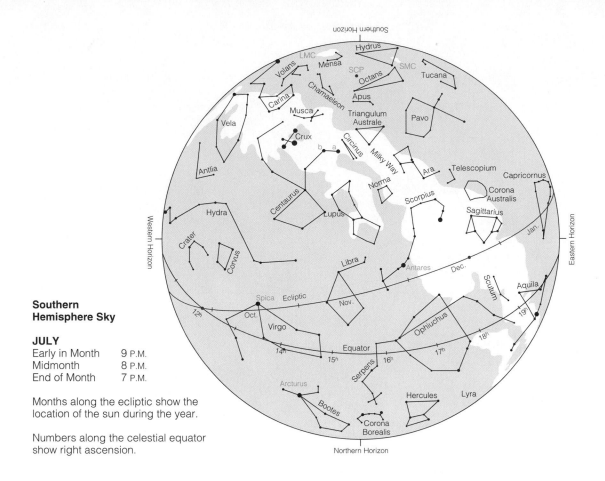

Southern Hemisphere Sky

JULY

Early in Month   9 P.M.
Midmonth   8 P.M.
End of Month   7 P.M.

Months along the ecliptic show the location of the sun during the year.

Numbers along the celestial equator show right ascension.

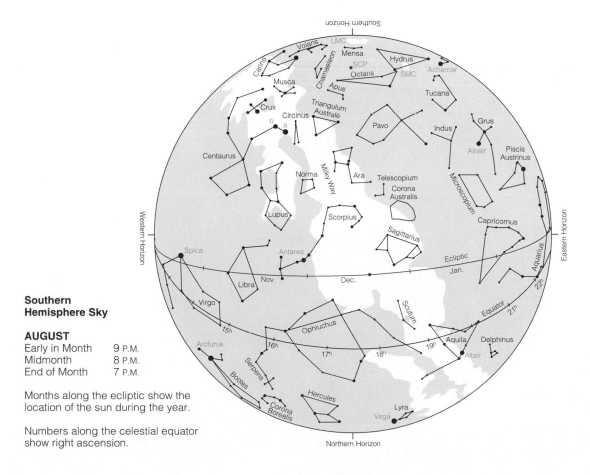

Southern Hemisphere Sky

AUGUST

Early in Month   9 P.M.
Midmonth   8 P.M.
End of Month   7 P.M.

Months along the ecliptic show the location of the sun during the year.

Numbers along the celestial equator show right ascension.

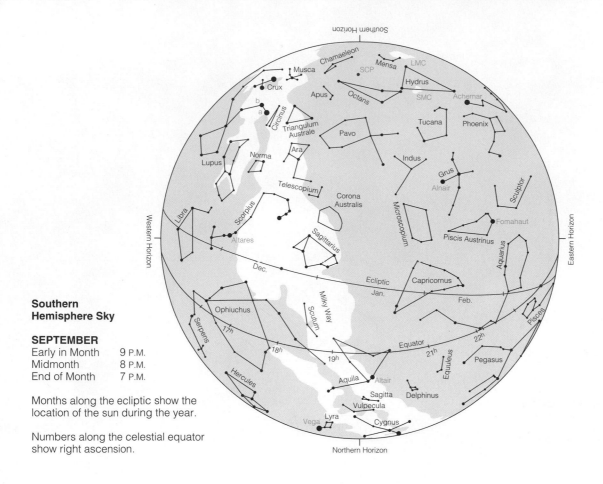

Southern Hemisphere Sky

SEPTEMBER
Early in Month	9 P.M.
Midmonth	8 P.M.
End of Month	7 P.M.

Months along the ecliptic show the location of the sun during the year.

Numbers along the celestial equator show right ascension.

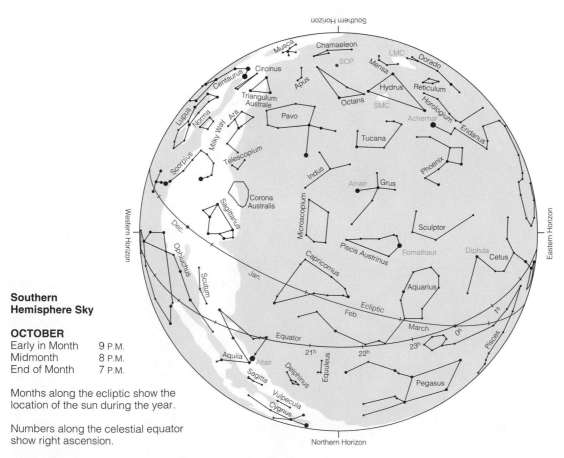

Southern Hemisphere Sky

OCTOBER
Early in Month	9 P.M.
Midmonth	8 P.M.
End of Month	7 P.M.

Months along the ecliptic show the location of the sun during the year.

Numbers along the celestial equator show right ascension.

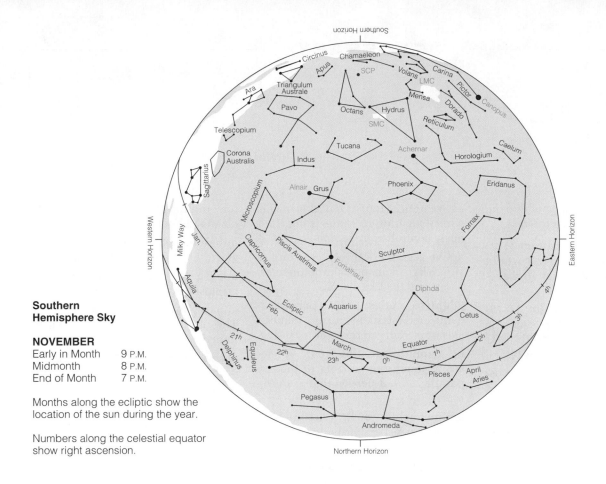

Southern Hemisphere Sky

NOVEMBER

Early in Month   9 P.M.
Midmonth   8 P.M.
End of Month   7 P.M.

Months along the ecliptic show the location of the sun during the year.

Numbers along the celestial equator show right ascension.

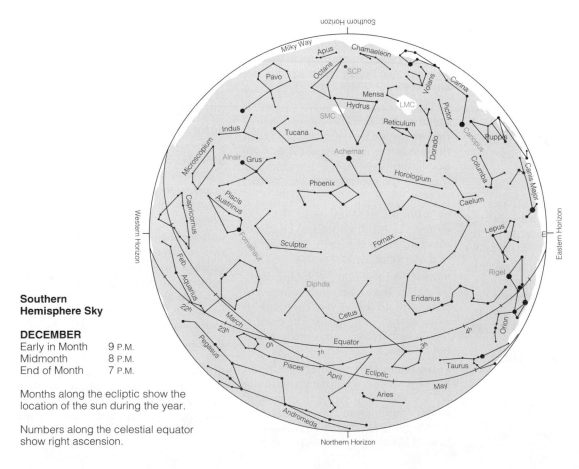

Southern Hemisphere Sky

DECEMBER

Early in Month   9 P.M.
Midmonth   8 P.M.
End of Month   7 P.M.

Months along the ecliptic show the location of the sun during the year.

Numbers along the celestial equator show right ascension.

GLOSSARY

Numbers refer to the page where the term is first discussed in the text.

absolute bolometric magnitude (176) The absolute magnitude we would observe if we could detect all wavelengths.

absolute visual magnitude (M_v) (175) Intrinsic brightness of a star; the apparent visual magnitude the star would have if it were 10 pc away.

absolute zero (124) The lowest possible temperature; the temperature at which the particles in a material, atoms or molecules, contain no energy of motion that can be extracted from the body.

absorption line (132) A dark line in a spectrum; produced by the absence of photons absorbed by atoms or molecules.

absorption spectrum (dark-line spectrum) (132) A spectrum that contains absorption lines.

acceleration (81) A change in a velocity; a change in either speed or direction. (See **velocity**.)

acceleration of gravity (78) A measure of the strength of gravity at a planet's surface.

accretion disk (264) The whirling disk of gas that forms around a compact object such as a white dwarf, neutron star, or black hole as matter is drawn in.

achromatic lens (101) A telescope lens composed of two lenses ground from different kinds of glass and designed to bring two selected colors to the same focus and correct for chromatic aberration.

active galactic nucleus (AGN) (358) The central energy source of an active galaxy.

active galaxy (358) A galaxy that is a source of excess radiation, usually radio waves, X rays, gamma rays, or some combination.

active optics (109) Optical elements whose position or shape is continuously controlled by computers.

active region (153) An area on the sun where sunspots, prominences, flares, and the like, occur.

adaptive optics (109) Computer-controlled telescope mirrors that can at least partially compensate for seeing.

alt-azimuth mounting (109) A telescope mounting capable of motion parallel to and perpendicular to the horizon.

amino acids (409) Carbon-chain molecules that are the building blocks of protein.

Angstrom (Å) (99) A unit of distance; $1 \text{ Å} = 10^{-10}$ m; often used to measure the wavelength of light.

angular diameter (17) A measure of the size of an object in the sky; numerically equal to the angle in degrees between two lines extending from the observer's eye to opposite edges of the object.

angular distance (17) A measure of the separation between two objects in the sky; numerically equal to the angle in degrees between two lines extending from the observer's eye to the two objects.

angular momentum (86) The tendency of a rotating body to continue rotating; mathematically, the product of mass, velocity, and radius.

annular eclipse (43) A solar eclipse in which the solar photosphere appears around the edge of the moon in a bright ring, or annulus. The corona, chromosphere, and prominences cannot be seen.

antimatter (387) Matter composed of antiparticles, which upon colliding with a matching particle of normal matter annihilate and convert the mass of both particles into energy. The antiproton is the antiparticle of the proton, and the positron is the antiparticle of the electron.

aphelion (23) The orbital point of greatest distance from the sun.

apogee (43) The orbital point of greatest distance from Earth.

apparent visual magnitude (m_v) (14) The brightness of a star as seen by human eyes on Earth.

archaeoastronomy (52) The study of the astronomy of ancient cultures.

association (219) Group of widely scattered stars (10 to 1000) moving together through space; not gravitationally bound into a cluster.

asterism (12) A named group of stars not identified as a constellation, e.g., the Big Dipper.

astronomical unit (AU) (4) Average distance from Earth to the sun; 1.5×10^8 km, or 93×10^6 miles.

atmospheric window (101) Wavelength regions in which our atmosphere is transparent—at visual wavelengths, infrared, and at radio wavelengths.

aurora (163) The glowing light display that results when a planet's magnetic field guides charged particles toward the north and south magnetic poles, where they strike the upper atmosphere and excite atoms to emit photons.

autumnal equinox (22) The point on the celestial sphere where the sun crosses the celestial equator going southward. Also, the time when the sun reaches this point and autumn begins in the northern hemisphere—about September 22.

Babcock model (158) A model of the sun's magnetic cycle in which the differential rotation of the sun winds up and tangles the solar magnetic field in a 22-year cycle. This is thought to be responsible for the 11-year sunspot cycle.

Balmer series (133)   Spectral lines in the visible and near-ultraviolet spectrum of hydrogen produced by transitions whose lowest orbit is the second.

barred spiral galaxy (338)   A spiral galaxy with an elongated nucleus resembling a bar from which the arms originate.

big bang (383)   The theory that the universe began with a violent explosion from which the expanding universe of galaxies eventually formed.

binary stars (185)   Pairs of stars that orbit around their common center of mass.

binding energy (129)   The energy needed to pull an electron away from its atom.

bipolar flow (221)   Oppositely directed jets of gas ejected by some protostellar objects.

birth line (218)   In the H–R diagram, the line above the main sequence where protostars first become visible.

black body radiation (125)   Radiation emitted by a hypothetical perfect radiator; the spectrum is continuous, and the wavelength of maximum emission depends only on the body's temperature.

black dwarf (259)   The end state of a white dwarf that has cooled to low temperature.

black hole (292)   A mass that has collapsed to such a small volume that its gravity prevents the escape of all radiation; also, the volume of space from which radiation may not escape.

blazar (365)   See **BL Lac objects.**

BL Lac objects (365)   Objects that resemble quasars; thought to be highly luminous cores of distant galaxies.

blue shift (139)   The shortening of the wavelengths of light observed when the source and observer are approaching each other.

Bok globule (220)   Small, dark cloud only about 1 ly in diameter that contains 10 to 1000 $M_{\odot}$ of gas and dust; believed related to star formation.

bright-line spectrum (132)   See **emission spectrum.**

brown dwarf (238)   A very cool, low-luminosity star whose mass is not sufficient to ignite nuclear fusion.

butterfly diagram   See **Maunder butterfly diagram.**

Cambrian period (412)   A geological period 0.6 to 0.5 billion years ago during which life on Earth became diverse and complex. Cambrian rocks contain the oldest easily identifiable fossils.

carbon deflagration (270)   The process in which the carbon in a white dwarf is completely consumed by nuclear fusion, producing a type Ia supernova explosion.

carbon–nitrogen–oxygen (CNO) cycle (223)   A series of nuclear reactions that use carbon as a catalyst to combine four hydrogen atoms to make one helium atom plus energy; effective in stars more massive than the sun.

Cassegrain focus (108)   The optical design of a reflecting telescope in which the secondary mirror reflects light back down the tube through a hole in the center of the objective mirror.

celestial equator (16)   The imaginary line around the sky directly above Earth's equator.

celestial sphere (16)   An imaginary sphere of very large radius surrounding Earth and to which the planets, stars, sun, and moon seem to be attached.

center of mass (85)   The balance point of a body or system of bodies.

Cepheid variable star (247)   Variable star with a period of 1 to 60 days; the period of variation is related to luminosity.

Chandrasekhar limit (259)   The maximum mass of a white dwarf, about 1.4 $M_{\odot}$; a white dwarf of greater mass cannot support itself and will collapse.

charge-coupled device (CCD) (111)   An electronic device consisting of a large array of light-sensitive elements used to record very faint images.

chemical evolution (413)   The chemical process that led to the growth of complex molecules on the primitive Earth. This did not involve the reproduction of molecules.

chromatic aberration (101)   A distortion found in refracting telescopes because lenses focus different colors at slightly different distances. Images are consequently surrounded by color fringes.

chromosome (410)   One of the bodies in a cell that contains the DNA carrying genetic information.

chromosphere (43)   Bright gases just above the photosphere of the sun.

circular velocity (84)   The velocity required to remain in a circular orbit about a body.

circumpolar constellation (17)   Any of the constellations so close to the celestial pole that they never set (or never rise) as seen from a given latitude.

closed orbit (85)   An orbit that returns to its starting point; a circular or elliptical orbit. (See **open orbit.**)

closed universe (393)   A model universe in which the average density is great enough to stop the expansion and make the universe contract.

cluster method (344)   The method of determining the masses of galaxies based on the motions of galaxies in a cluster.

CNO cycle (223)   See **carbon–nitrogen–oxygen cycle.**

cocoon (217)   The cloud of gas and dust around a contracting protostar that conceals it at visible wavelengths.

cold dark matter (395)   Invisible matter in the universe composed by heavy, slow-moving particles such as WIMPs.

collapsar (302)   See **hypernova.**

collisional broadening (141)   The smearing out of a spectral line because of collisions among the atoms of the gas.

color index (127)   A numerical measure of the color of a star.

compact object (259)   A star that has collapsed to form a white dwarf, neutron star, or black hole.

comparison spectrum (112)   A spectrum of known spectral lines used to identify unknown wavelengths in an object's spectrum.

conservation of energy law (234)   One of the basic laws of stellar structure. The amount of energy flowing out of the top of a shell must equal the amount coming in at the bottom plus whatever energy is generated within the shell.

conservation of mass law (234)   One of the basic laws of stellar structure. The total mass of the star must equal the sum of the masses of the shells, and the mass must be distributed smoothly throughout the star.

constellation (11)   One of the stellar patterns identified by name, usually of mythological gods, people, animals, or objects; also, the region of the sky containing that star pattern.

continuous spectrum (132)   A spectrum in which there are no absorption or emission lines.

convection (148)   Circulation in a fluid driven by heat; hot material rises, and cool material sinks.

corona (42)   The faint outer atmosphere of the sun; composed of low-density, very hot, ionized gas.

coronagraph (150)   A telescope designed to photograph the inner corona of the sun.

coronal gas (207)   Extremely high-temperature, low-density gas in the interstellar medium.

coronal hole (163)   An area of the solar surface that is dark at X-ray wavelengths; thought to be associated with divergent magnetic fields and the source of the solar wind.

coronal mass ejection (CME) (163) Gas trapped in the sun's magnetic field and blown outward through the sun's atmosphere.

cosmic microwave background radiation (385)   Radiation from the hot clouds of the big bang explosion. Because of its large red shift, it appears to come from a body whose temperature is only 2.7 K.

cosmic ray (118)   A subatomic particle traveling at tremendous velocity that strikes Earth's atmosphere from space.

cosmological constant (Λ) (398)   Einstein's constant that represents a repulsion in space to oppose gravity.

cosmological principle (390)   The assumption that any observer in any galaxy sees the same general features of the universe.

cosmology (380)   The study of the nature, origin, and evolution of the universe.

Coulomb barrier (164)   The electrostatic force of repulsion between bodies of like charge; commonly applied to atomic nuclei.

Coulomb force (129)   The repulsive force between particles with like electrostatic charge.

critical density (390)   The average density of the universe needed to make its curvature flat.

dark age (388)   The period of a few hundred million years during which the universe expanded in darkness. Extends from soon after the big bang glow faded into the infrared to the formation of the first stars.

dark energy (398)   The energy of empty space that drives the acceleration of the expanding universe.

dark-line spectrum   See **absorption spectrum.**

dark matter (314)   Nonluminous material that is detected only by its gravitational influence.

dark nebula (199)   A nonluminous cloud of gas and dust visible because it blocks light from more distant stars and nebulae.

deferent (59)   In the Ptolemaic theory, the large circle around Earth along which the center of the epicycle moved.

degenerate matter (242)   Extremely high-density matter in which pressure no longer depends on temperature, due to quantum mechanical effects.

density (142)   The amount of matter per unit volume in a material; measured in grams per cubic centimeter, for example.

density wave theory (324)   Theory proposed to account for spiral arms as compressions of the interstellar medium in the disk of the galaxy.

deuterium (165)   An isotope of hydrogen in which the nucleus contains a proton and a neutron.

diamond ring effect (43)   A momentary phenomenon seen during some total solar eclipses when the ring of the corona and a bright spot of photosphere resemble a large diamond set in a silvery ring.

differential rotation (156)   The rotation of a body in which different parts of the body have different periods of rotation; this is true of the sun, the Jovian planets, and the disk of the galaxy.

diffraction fringe (104)   Blurred fringe surrounding any image caused by the wave properties of light. Because of this, no image detail smaller than the fringe can be seen.

disk component (311)   All material confined to the plane of the galaxy.

distance indicator (337)   Object whose luminosity or diameter is known; used to find the distance to a star cluster or galaxy.

distance modulus ($m_v - M_v$) (175) The difference between the apparent and absolute magnitude of a star; a measure of how far away the star is.

distance scale (341)   The combined calibration of distance indicators used by astronomers to find the distances to remote galaxies.

DNA (deoxyribonucleic acid) (409) The long carbon-chain molecule that records information to govern the biological activity of the organism. DNA carries the genetic data passed to offspring.

Doppler broadening (141)   The smearing of spectral lines because of the motion of the atoms in the gas.

Doppler effect (139)   The change in the wavelength of radiation due to relative radial motion of source and observer.

double-exhaust model (360)   The theory that double radio lobes are produced by pairs of jets emitted in opposite directions from the centers of active galaxies.

double-lobed radio source (360)   A galaxy that emits radio energy from two regions (lobes) located on opposite sides of the galaxy.

Drake equation (423)   A formula for the number of communicative civilizations in our galaxy.

dynamo effect (157)   The process by which a rotating, convecting body of conducting matter, such as Earth's core, can generate a magnetic field.

east point (16)   The point on the eastern horizon exactly halfway between the north point and the south point; exactly east.

eccentric (56)   In astronomy, an off-center circular path.

eccentricity (e) (70)   A measure of the flattening of an ellipse. An ellipse of $e = 0$ is circular. The closer to 1 e becomes, the more flattened the ellipse.

eclipse season (45)   That period when the sun is near a node of the moon's orbit and eclipses are possible.

eclipse year (46)   The time the sun takes to circle the sky and return to a node of the moon's orbit; 346.62 days.

eclipsing binary system (185) A binary star system in which the stars eclipse each other.

ecliptic (20) The apparent path of the sun around the sky.

electromagnetic radiation (98) Changing electric and magnetic fields that travel through space and transfer energy from one place to another—for example, light, radio waves, etc.

electron (125) Low-mass atomic particle carrying a negative charge.

ellipse (70) A closed curve enclosing two points (foci) such that the total distance from one focus to any point on the curve back to the other focus equals a constant.

elliptical galaxy (338) A galaxy that is round or elliptical in outline; it contains little gas and dust, no disk or spiral arms, and few hot, bright stars.

emission line (132) A bright line in a spectrum caused by the emission of photons from atoms.

emission nebula (198) A cloud of glowing gas excited by ultraviolet radiation from hot stars.

emission spectrum (bright-line spectrum) (132) A spectrum containing emission lines.

energy (87) The capacity of a natural system to perform work—for example, thermal energy.

energy level (130) One of a number of states an electron may occupy in an atom, depending on its binding energy.

enzyme (409) Special protein that controls processes in an organism.

epicycle (59) The small circle followed by a planet in the Ptolemaic theory. The center of the epicycle follows a larger circle (deferent) around Earth.

equant (59) The point off-center in the deferent from which the center of the epicycle appears to move uniformly.

equatorial mounting (109) A telescope mounting that allows motion parallel to and perpendicular to the celestial equator.

ergosphere (293) The region surrounding a rotating black hole within which one could not resist being dragged around the black hole. It is possible for a particle to escape from the ergosphere and extract energy from the black hole.

escape velocity (85) The initial velocity an object needs to escape from the surface of a celestial body.

evening star (21) Any planet visible in the sky just after sunset.

event horizon (292) The boundary of the region of a black hole from which no radiation may escape. No event that occurs within the event horizon is visible to a distant observer.

excited atom (130) An atom in which an electron has moved from a lower to a higher orbit.

eyepiece (101) A short-focal-length lens used to enlarge the image in a telescope; the lens nearest the eye.

false-color image (111) A representation of graphical data in which the colors are altered or added to reveal details.

field (83) A way of explaining action at a distance; a particle produces a field of influence (gravitational, electric, or magnetic) to which another particle in the field responds.

filament (149, 162) On the sun, a prominence seen silhouetted against the solar surface.

filtergram (149) An image (usually of the sun) taken in the light of a specific region of the spectrum—e.g., an H-alpha filtergram.

flare (163) A violent eruption on the sun's surface.

flatness problem (396) In cosmology, the circumstance that the early universe must have contained almost exactly the right amount of matter to close space-time (to make space-time flat).

flat universe (393) A model of the universe in which space-time is not curved.

flocculent (325) Woolly, fluffy; used to refer to certain galaxies that have a woolly appearance.

flux (175) A measure of the flow of energy onto or through a surface. Usually applied to light.

focal length (101) The distance from a lens to the point where it focuses parallel rays of light.

forbidden line (197) A spectral line that does not occur in the laboratory because it depends on an atomic transition that is highly unlikely.

free-fall contraction (217) The early contraction of a gas cloud to form a star during which internal pressure is too low to resist contraction.

frequency (99) The number of times a given event occurs in a given time; for a wave, the number of cycles that pass the observer in 1 second.

galactic cannibalism (350) The theory that large galaxies absorb smaller galaxies.

galactic corona (314) The low-density extension of the halo of a galaxy; now suspected to extend many times the visible diameter of the galaxy.

galactic fountain (317) A region of the galaxy's disk in which gas heated by supernova explosions throws gas out of the disk where it can fall back and spread metals through the disk.

galaxy (6) A very large collection of gas, dust, and stars orbiting a common center of mass. We live in the Milky Way Galaxy.

gamma-ray burster (300) An object that produces a sudden burst of gamma rays; thought to be associated with neutron stars and black holes.

gene (410) A unit of DNA containing genetic information that influences a particular inherited trait.

general theory of relativity (91) Einstein's more sophisticated theory of space and time, which describes gravity as a curvature of space-time.

geocentric universe (58) A model universe with Earth at the center, such as the Ptolemaic universe.

geosynchronous satellite (84) A satellite in an eastward orbit whose period is 24 hours. A satellite in such an orbit remains above the same spot on Earth's surface.

giant molecular clouds (205) Very large, cool clouds of dense gas in which stars form.

giants (178) Large, cool, highly luminous stars in the upper right of the H–R diagram; typically 10 to 100 times the diameter of the sun.

glitch (282) A sudden change in the period of a pulsar.

globular cluster (248) A star cluster containing 50,000 to 1 million stars in a sphere about 75 ly in diameter; generally old, metal-poor, and found in the spherical component of the galaxy.

grand unified theories (GUTs) (397) Theories that attempt to unify (describe in a similar way) the electromagnetic, weak, and strong forces of nature.

granulation (148) The fine structure visible on the solar surface caused by rising currents of hot gas and sinking currents of cool gas below the surface.

grating (112) A piece of material in which numerous microscopic parallel lines are scribed; light encountering a grating is dispersed to form a spectrum.

gravitational lens (370) The effect of focusing of light from a distant galaxy or quasar by an intervening galaxy to produce multiple images of the distant body.

gravitational radiation (287) As predicted by general relativity, expanding waves in a gravitational field that transport energy through space.

gravitational red shift (294) The lengthening of the wavelength of a photon due to its escape from a gravitational field.

ground state (130) The lowest permitted electron orbit in an atom.

halo (307) The spherical region of a spiral galaxy containing a thin scattering of stars, star clusters, and small amounts of gas.

heliocentric universe (60) A model of the universe with the sun at the center, such as the Copernican universe.

helioseismology (151) The study of the interior of the sun by the analysis of its modes of vibration.

helium flash (243) The explosive ignition of helium burning that takes place in some giant stars.

Herbig–Haro object (221) Small nebula that varies irregularly in brightness; believed to be associated with star formation.

Hertzsprung–Russell diagram (177) A plot of the intrinsic brightness versus the surface temperature of stars; it separates the effects of temperature and surface area on stellar luminosity; commonly absolute magnitude versus spectral type, but also luminosity versus surface temperature or color.

HI cloud (201) An interstellar cloud of neutral hydrogen.

HII region (198) A region of ionized hydrogen around a hot star.

high-velocity star (313) A star with a large space velocity. Such stars are halo stars passing through the disk of the galaxy at steep angles.

homogeneity (390) The assumption that, on the large scale, matter is uniformly spread through the universe.

horizon (16) The line that marks the apparent intersection of Earth and the sky.

horizon problem (396) In cosmology, the circumstance that the primordial background radiation seems much more isotropic than could be explained by the standard big bang theory.

horizontal branch (249) In the H–R diagram of a globular cluster, the sequence of stars extending from the red giants toward the blue side of the diagram; includes RR Lyrae stars.

horoscope (24) A chart showing the positions of the sun, moon, planets, and constellations at the time of a person's birth; used in astrology to attempt to read character or foretell the future.

hot dark matter (395) Invisible matter in the universe composed of low-mass, high-velocity particles such as neutrinos.

hot spot (362) In radio astronomy, a bright spot in a radio lobe.

H–R diagram (180) (See **Hertzsprung–Russell diagram.**)

Hubble constant (*H*) (342) A measure of the rate of expansion of the universe; the average value of velocity of recession divided by distance; presently believed to be about 70 km/s/megaparsec.

Hubble law (342) The linear relation between the distance to a galaxy and its radial velocity.

Hubble time (384) An upper limit on the age of the universe derived from the Hubble constant.

hydrostatic equilibrium (225) The balance between the weight of the material pressing downward on a layer in a star and the pressure in that layer.

hypernova (302) The explosion produced as a very massive star collapses into a black hole; thought to be responsible for at least some gamma-ray bursts.

hypothesis (71) A conjecture, subject to further tests, that accounts for a set of facts.

inflationary universe (396) A version of the big bang theory that includes a rapid expansion when the universe was very young.

infrared cirrus (206) A fine network of filaments covering the sky detected in the far infrared by the IRAS satellite; believed to be associated with dust in the interstellar medium.

infrared radiation (100) Electromagnetic radiation with wavelengths intermediate between visible light and radio waves.

instability strip (251) The region of the H–R diagram in which stars are unstable to pulsation; a star passing through this strip becomes a variable star.

intercloud medium (201) The hot, low-density gas between cooler clouds in the interstellar medium.

interferometry (110) The observing technique in which separated telescopes combine to produce a virtual telescope with the resolution of a much-larger-diameter telescope.

interstellar absorption lines (200) Dark lines in some stellar spectra that are formed by interstellar gas.

interstellar dust (200) Microscopic solid grains in the interstellar medium.

interstellar extinction (200) The dimming of starlight by gas and dust in the interstellar medium.

interstellar medium (197) The gas and dust distributed between the stars.

interstellar reddening (200) The process in which dust scatters blue light out of starlight and makes the stars look redder.

intrinsic variable (247) A variable driven to pulsate by processes in its interior.

inverse square law (82) The rule that the strength of an effect (such as gravity) decreases in proportion as the distance squared increases.

ion (128) An atom that has lost or gained one or more electrons.

ionization (128) The process in which atoms lose or gain electrons.

irregular galaxy (339) A galaxy with a chaotic appearance, large clouds of gas and dust, and both population I and population II stars, but without spiral arms.

island universe (336) An older term for a galaxy.

isotopes (128) Atoms that have the same number of protons but a different number of neutrons.

isotropy (389) The assumption that in its general properties the universe looks the same in every direction.

joule (J) (87) A unit of energy equivalent to a force of 1 newton acting over a distance of 1 meter; 1 joule per second equals 1 watt of power.

Kelvin temperature scale (124)   The temperature, in Celsius (centigrade) degrees, measured above absolute zero.

Keplerian motion (314)   Orbital motion in accord with Kepler's laws of planetary motion.

Kerr black hole (293)   A solution to the equations of general relativity that describes the properties of a rotating black hole.

kiloparsec (kpc) (307)   A unit of distance equal to 1000 pc, or 3260 ly.

Kirchhoff's laws (132)   A set of laws that describes the absorption and emission of light by matter.

Lagrangian point (263)   Point of stability in the orbital plane of a binary star system, planet, or moon. One is located 60° ahead and one 60° behind the orbiting bodies; another is located between the orbiting bodies.

L dwarf (136)   A type of star that is even cooler than the M stars.

life zone (419)   A region around a star within which a planet can have temperatures that permit the existence of liquid water.

light curve (186)   A graph of brightness versus time commonly used in analyzing variable stars and eclipsing binaries.

light-gathering power (102)   The ability of a telescope to collect light; proportional to the area of the telescope objective lens or mirror.

lighthouse model (282)   The explanation of a pulsar as a spinning neutron star sweeping beams of radio radiation around the sky.

light pollution (105)   The illumination of the night sky by waste light from cities and outdoor lighting, which prevents the observation of faint objects.

light-year (ly) (5)   The distance light travels in 1 year.

limb (148)   The edge of the apparent disk of a body, as in "the limb of the moon."

limb darkening (148)   The decrease in brightness of the sun or other body from its center to its limb.

line of nodes (46)   The line across an orbit connecting the nodes; commonly applied to the orbit of the moon.

line profile (140)   A graph of light intensity versus wavelength showing the shape of an absorption line.

local bubble or void (207)   A region of high-temperature, low-density gas in the interstellar medium in which the sun happens to be located.

look-back time (341)   The amount by which we look into the past when we look at a distant galaxy; a time equal to the distance to the galaxy in light-years.

luminosity (L) (176)   The total amount of energy a star radiates in 1 second.

luminosity class (180)   A category of stars of similar luminosity; determined by the widths of lines in their spectra.

lunar eclipse (37)   The darkening of the moon when it moves through Earth's shadow.

Lyman series (133)   Spectral lines in the ultraviolet spectrum of hydrogen produced by transitions whose lowest orbit is the ground state.

MACHOs (395)   Massive Compact Halo Objects; low-luminosity objects such as planets and brown dwarfs that contribute to the mass of the halo.

Magellanic Clouds (306)   Small, irregular galaxies that are companions to the Milky Way; visible in the southern sky.

magnetar (286)   A class of neutron stars that have exceedingly strong magnetic fields; thought to be responsible for soft gamma-ray repeaters.

magnetic carpet (151)   The widely distributed, low-level magnetic field extending up through the sun's visible surface.

magnifying power (104)   The ability of a telescope to make an image larger.

magnitude–distance formula (175)   The mathematical formula that relates the apparent magnitude and absolute magnitude of a star to its distance.

magnitude scale (12)   The astronomical brightness scale; the larger the number, the fainter the star.

main sequence (177)   The region of the H–R diagram running from upper left to lower right, which includes roughly 90 percent of all stars.

mass (81)   A measure of the amount of matter making up an object.

mass–luminosity relation (190)   The more massive a star is, the more luminous it is.

Maunder butterfly diagram (156)   A graph showing the latitude of sunspots versus time; first plotted by W. W. Maunder in 1904.

Maunder minimum (160)   A period of less numerous sunspots and other solar activity from 1645 to 1715.

megaparsec (Mpc) (337)   A unit of distance equal to 1 million pc.

metals (315)   In astronomical usage, all atoms heavier than helium.

metastable level (197)   An atomic energy level from which an electron takes a long time to decay; responsible for producing forbidden lines.

Milankovitch hypothesis (25)   The hypothesis that small changes in Earth's orbital and rotational motions cause the ice ages.

Milky Way (6)   The hazy band of light that circles the sky, produced by the combined light of billions of stars in our Milky Way Galaxy.

Milky Way Galaxy (6)   The spiral galaxy containing our sun; visible at night as the Milky Way.

Miller experiment (412)   An experiment that reproduced the conditions under which life began on Earth and manufactured amino acids and other organic compounds.

millisecond pulsar (289)   A pulsar with a period of approximately a millisecond, a thousandth of a second.

minute of arc (17)   An angular measure; each degree is divided into 60 minutes of arc.

molecular cloud (205)   An interstellar gas cloud that is dense enough for the formation of molecules; discovered and studied through the radio emissions of such molecules.

molecule (128)   Two or more atoms bonded together.

momentum (80)   The tendency of a moving object to continue moving; mathematically, the product of mass and velocity.

morning star (21)   Any planet visible in the sky just before sunrise.

mutant (411)   Offspring born with altered DNA.

nadir (16)   The point on the bottom of the sky directly under our feet.

nanometer (nm) (99)   A unit of length equal to 10^{-9} m.

natural law (71)   A conjecture about how nature works in which scientists have overwhelming confidence.

natural motion (78)   In Aristotelian physics, the motion of objects toward their natural places—fire and air upward and earth and water downward.

natural selection (411)   The process by which the best traits are passed on, allowing the most able to survive.

neap tide (36)   Ocean tide of low amplitude occurring at first- and third-quarter moon.

nebula (197)   A cloud of gas and dust in space.

neutrino (165)   A neutral, massless atomic particle that travels at or nearly at the speed of light.

neutron (127)   An atomic particle with no charge and about the same mass as a proton.

neutron star (279)   A small, highly dense star composed almost entirely of tightly packed neutrons; radius about 10 km.

Newtonian focus (108)   The focal arrangement of a reflecting telescope in which a diagonal mirror reflects light out the side of the telescope tube for easier access.

node (45)   A point where an object's orbit passes through the plane of Earth's orbit.

nonbaryonic matter (395)   In cosmology, a suspected component of the dark matter composed of matter that does not contain protons and neutrons.

north celestial pole (16)   The point on the celestial sphere directly above Earth's North Pole.

north point (16)   The point on the horizon directly below the north celestial pole; exactly north.

nova (256)   From the Latin "new," a sudden brightening of a star, making it appear as a "new" star in the sky; believed associated with eruptions on white dwarfs in binary systems.

nuclear bulge (307)   The spherical cloud of stars that lies at the center of spiral galaxies.

nuclear fission (161)   Reaction that splits nuclei into less massive fragments.

nuclear fusion (164)   Reaction that joins the nuclei of atoms to form more massive nuclei.

nucleosynthesis (315)   The production of elements heavier than helium by the fusion of atomic nuclei in stars and during supernovae explosions.

nucleus (of an atom) (127)   The central core of an atom containing protons and neutrons; carries a net positive charge.

O association (219)   A large, loosely bound cluster of very young stars.

objective lens or mirror (101)   The main optical element in an astronomical telescope. The large lens at the top of the telescope or large mirror at the bottom.

observable universe (382)   The part of the universe that we can see from our location in space and time.

Olbers's paradox (380)   The conflict between observation and theory as to why the night sky should or should not be dark.

opacity (224)   The resistance of a gas to the passage of radiation.

open cluster (248)   A cluster of 10 to 10,000 stars with an open, transparent appearance and stars not tightly grouped; usually relatively young and located in the disk of the galaxy.

open orbit (85)   An orbit that does not return to its starting point; an escape orbit. (See **closed orbit**.)

open universe (393)   A model universe in which the average density is less than the critical density needed to halt the expansion.

oscillating universe theory (391)   The theory that the universe begins with a big bang, expands, is slowed by its own gravity, and then falls back to create another big bang.

paradigm (63)   A commonly accepted set of scientific ideas and assumptions.

parallax (58)   The apparent change in the position of an object due to a change in the location of the observer. Astronomical parallax is measured in seconds of arc.

parsec (pc) (172)   The distance to a hypothetical star whose parallax is one second of arc; 1 pc = 206,265 AU = 3.26 ly.

partial eclipse (38, 40)   A lunar eclipse in which the moon does not completely enter Earth's shadow; a solar eclipse in which the moon does not completely cover the sun.

Paschen series (133)   Spectral lines in the infrared spectrum of hydrogen produced by transitions whose lowest orbit is the third.

path of totality (41)   The track of the moon's umbral shadow over Earth's surface. The sun is totally eclipsed as seen from within this path.

penumbra (37)   The portion of a shadow that is only partially shaded.

penumbral eclipse (39)   A lunar eclipse in which the moon enters the penumbra of Earth's shadow but does not reach the umbra.

perigee (43)   The orbital point of closest approach to Earth.

perihelion (23)   The orbital point of closest approach to the sun.

period–luminosity relation (247)   The relation between period of pulsation and intrinsic brightness among Cepheid variable stars.

permitted orbit (129)   One of the energy levels in an atom that an electron may occupy.

photon (99)   A quantum of electromagnetic energy; carries an amount of energy that depends inversely on its wavelength.

photosphere (42)   The bright visible surface of the sun.

planet (4)   A nonluminous object, larger than a comet or asteroid, that orbits a star.

planetary nebula (258)   An expanding shell of gas ejected from a star during the latter stages of its evolution.

polar axis (109)   The axis around which a celestial body rotates.

poor cluster (347)   An irregularly shaped cluster that contains fewer than 1000 galaxies, many spiral, and no giant ellipticals.

population I star (315)   Star rich in atoms heavier than helium; nearly always a relatively young star found in the disk of the galaxy.

population II star (315)   Star poor in atoms heavier than helium; nearly always a relatively old star found in the halo, globular clusters, or the nuclear bulge.

precession (18)   The slow change in the direction of Earth's axis of rotation; one cycle takes nearly 26,000 years.

pressure (203)   A force exerted over a surface. Expressed as force per unit area.

primary lens or mirror (101)   The main optical element in an astronomical telescope. The large lens at the top of the telescope tube or the large mirror at the bottom.

prime focus (108)   The point at which the objective mirror forms an image in a reflecting telescope.

primordial soup (413)   The rich solution of organic molecules in Earth's first oceans.

prominence (43, 162)   Eruption on the solar surface; visible during total solar eclipses.

proper motion (173)   The rate at which a star moves across the sky; measured in seconds of arc per year.

protein (409)   Complex molecule composed of amino acid units.

proton (127)   A positively charged atomic particle contained in the nucleus of atoms; the nucleus of a hydrogen atom.

proton–proton chain (165)   A series of three nuclear reactions that build a helium atom by adding together protons; the main energy source in the sun.

protostar (217)   A collapsing cloud of gas and dust destined to become a star.

protostellar disk (218)   A gas cloud around a forming star flattened by its rotation.

pulsar (281)   A source of short, precisely timed radio bursts; believed to be a spinning neutron star.

pulsar wind (282)   The flow of high-energy particles that carries most of the energy away from a spinning neutron star.

quantum mechanics (129)   The study of the behavior of atoms and atomic particles.

quasar (quasi-stellar object, or QSO) (367)   Small, powerful source of energy believed to be the active core of very distant galaxies.

quasi-periodic oscillation (QPO) (297)   A high-speed flickering in the radiation from an accretion disk evidently caused by material spiraling inward.

quintessence (398)   The proposed energy of empty space that causes the acceleration of the expanding universe.

radial velocity (V_r) (139)   That component of an object's velocity directed away from or toward Earth.

radio galaxy (358)   A galaxy that is a strong source of radio signals.

radio interferometer (114)   Two or more radio telescopes that combine their signals to achieve the resolving power of a larger telescope.

recombination (388)   The stage within 10^6 years of the big bang when the gas became transparent to radiation.

reconnection (163)   The process in the sun's atmosphere by which opposing magnetic fields combine and release energy to power solar flares.

red dwarf (178)   Cool, low-mass star on the lower main sequence.

red shift (139)   The lengthening of the wavelengths of light seen when the source and observer are receding from each other.

reflecting telescope (101)   A telescope that uses a concave mirror to focus light into an image.

reflection nebula (198)   A nebula produced by starlight reflecting off of dust particles in the interstellar medium.

refracting telescope (101)   A telescope that forms images by bending (refracting) light with a lens.

reionization (388)   The stage in the early history of the universe when ultraviolet photons from the first stars ionized the gas filling space.

relativistic Doppler formula (369)   The formula for the Doppler shift produced by an object moving at a significant fraction of the speed of light.

relativistic jet model (374)   An explanation of superluminal expansion based on a high-velocity jet from a quasar directed approximately toward Earth.

resolving power (104)   The ability of a telescope to reveal fine detail; depends on the diameter of the telescope objective.

retrograde motion (58)   The apparent backward (westward) motion of planets as seen against the background of stars.

revolution (19)   The motion of an object in a closed path about a point outside its volume; Earth revolves around the sun.

rich cluster (347)   A cluster containing over 1000 galaxies, mostly elliptical, scattered over a volume about 3 Mpc in diameter.

ring galaxy (349)   A galaxy that resembles a ring around a bright nucleus; believed to be the result of a head-on collision of two galaxies.

RNA (ribonucleic acid) (410)   Long carbon-chain molecules that use the information stored in DNA to manufacture complex molecules necessary to the organism.

Roche lobe (263)   In a system with two bodies orbiting each other, the volume of space dominated by the gravitation of one of the bodies.

Roche surface (263)   In a system with two bodies orbiting each other, the outer boundary of the volume of space dominated by the gravitation of one of the bodies.

rotation (19)   The turning of a body about an axis that passes through its volume; Earth rotates on its axis.

rotation curve (314, 343)   A graph of orbital velocity versus radius in the disk of a galaxy.

rotation curve method (343)   The procedure for finding the mass of a galaxy from its rotation curve.

RR Lyrae variable star (247)   Variable star with a period of 12 to 24 hours; common in some globular clusters.

Sagittarius A* (328)   The powerful radio source located at the core of the Milky Way Galaxy.

saros cycle (47)   An 18-year, $11\frac{1}{3}$-day period after which the pattern of lunar and solar eclipses repeats.

Schmidt-Cassegrain focus (108)   The optical design of a reflecting telescope in which a thin correcting lens is placed at the top of a Cassegrain telescope.

Schwarzschild radius (R_s) (292)   The radius of the event horizon around a black hole.

scientific model (15)   An intellectual concept designed to help us think about a natural process without necessarily being a conjecture of truth.

scientific notation (4)   The system of recording very large or very small numbers by using powers of 10.

secondary mirror (108)   In a reflecting telescope, the mirror that reflects the light to a point of easy observation.

second of arc (17)   An angular measure; each minute of arc is divided into 60 seconds of arc.

seeing (104)   Atmospheric conditions on a given night. When the atmosphere is unsteady, producing blurred images, the seeing is said to be poor.

self-sustaining star formation (325)   The process by which the birth of stars compresses the surrounding gas clouds and triggers the formation of more stars; proposed to explain spiral arms.

semimajor axis (*a*) (70)   Half of the longest axis of an ellipse.

SETI (605)   Search for Extra-Terrestrial Intelligence.

Seyfert galaxy (358)   An otherwise normal spiral galaxy with an unusually bright, small core that fluctuates in brightness; believed to indicate the core is erupting.

Shapley–Curtis Debate (336)   The 1920 debate between Harlow Shapley and Heber Curtis over the nature of the spiral nebulae.

shock wave (216)   A sudden change in pressure that travels as an intense sound wave.

sidereal drive (109)   The motor and gears on a telescope that turn it westward to keep it pointed at a star.

sidereal period (35)   The period of rotation or revolution of an astronomical body referred to the stars.

singularity (291)   The object of zero radius into which the matter in a black hole is believed to fall.

small-angle formula (40)   The mathematical formula that relates an object's linear diameter and distance to its angular diameter.

soft gamma-ray repeater (SGR) (301)   An object that produces repeated bursts of lower-energy gamma rays; thought to be produced by magnetars.

solar constant (160)   A measure of the energy output of the sun; the total solar energy striking 1 m² just above Earth's atmosphere in 1 second.

solar eclipse (40)   The event that occurs when the moon passes directly between Earth and the sun, blocking our view of the sun.

solar system (4)   The sun and the nonluminous objects that orbit it, including the planets, comets, and asteroids.

solar wind (151)   Rapidly moving atoms and ions that escape from the solar corona and blow outward through the solar system.

south celestial pole (16)   The point of the celestial sphere directly above Earth's South Pole.

south point (16)   The point on the horizon directly above the south celestial pole; exactly south.

special relativity (90)   The first of Einstein's theories of relativity, which dealt with uniform motion.

spectral class or type (135)   A star's position in the temperature classification system O, B, A F, G, K, and M. Based on the appearance of the star's spectrum.

spectral sequence (135)   The arrangement of spectral classes (O, B, A, F, G, K, M) ranging from hot to cool.

spectrograph (112)   A device that separates light by wavelength to produce a spectrum.

spectroscopic binary system (184)   A star system in which the stars are too close together to be visible separately. We see a single point of light, and only by taking a spectrum can we determine that there are two stars.

spectroscopic parallax (181)   The method of determining a star's distance by comparing its apparent magnitude with its absolute magnitude as estimated from its spectrum.

spherical component (311)   The part of the galaxy including all matter in a spherical distribution around the center (the halo and nuclear bulge).

spicule (149)   Small, flamelike projection in the chromosphere of the sun.

spiral arms (6)   Long, spiral patterns of bright stars, star clusters, gas, and dust that extend from the center to the edge of the disk of spiral galaxies.

spiral galaxy (338)   A galaxy with an obvious disk component containing gas; dust; hot, bright stars; and spiral arms.

spiral nebula (335)   Nebulous object with a spiral appearance observed in early telescopes; later recognized as a spiral galaxy.

spiral tracers (321)   Objects used to map the spiral arms (e.g., O and B associations, open clusters, clouds of ionized hydrogen, and some types of variable stars).

spring tide (36)   Ocean tide of high amplitude that occurs at full and new moon.

standard candle (337)   Object of known brightness which astronomers use to find distance—for example, Cepheid variable stars and supernovae.

star (4)   A celestial object composed of gas held together by its own gravity and supported by nuclear fusion occurring in its interior.

starburst galaxy (352)   A bright blue galaxy in which many new stars are forming, believed caused by collisions between galaxies.

steady state theory (386)   The theory (now generally abandoned) that the universe does not evolve.

stellar model (234)   A table of numbers representing the conditions in various layers within a star.

stellar parallax (*p*) (172)   A measure of stellar distance. (See **parallax.**)

stromatolite (415)   A layered fossil formation caused by ancient mats of algae or bacteria that build up mineral deposits season after season.

strong force (161)   One of the four forces of nature; the strong force binds protons and neutrons together in atomic nuclei.

summer solstice (22)   The point on the celestial sphere where the sun is at its most northerly point; also, the time when the sun passes this point, about June 22, and summer begins in the northern hemisphere.

sunspot (147)   Relatively dark spot on the sun that contains intense magnetic fields.

supercluster (400)   A cluster of galaxy clusters.

supergiant (178)   Exceptionally luminous star 10 to 1000 times the sun's diameter.

supergranule (148)   A large granule on the sun's surface including many smaller granules.

superluminal expansion (373)   The apparent expansion of parts of a quasar at speeds greater than the speed of light.

supernova (256)   A new star appearing in Earth's sky and lasting for a year or so before fading. Caused by the violent explosion of a star.

supernova remnant (271)   The expanding shell of gas marking the site of a supernova explosion.

supernova (type I) (269)   The explosion of a star, believed caused by the transfer of matter to a white dwarf.

supernova (type II) (269)   The explosion of a star, believed caused by the collapse of a massive star.

synchrotron radiation (271)   Radiation emitted when high-speed electrons move through a magnetic field.

synodic period (35)   The period of rotation or revolution of a celestial body with respect to the sun.

T association (219)   A large, loosely bound group of T Tauri stars.

T dwarf (136)   A very-low-mass star at the bottom end of the main sequence with a cool surface and a low luminosity.

temperature (124)   A measure of the velocity of random motions among the atoms or molecules in a material.

theory (71)   A system of assumptions and principles applicable to a wide range of phenomena that have been repeatedly verified.

thermal energy (125)   The energy stored in an object as agitation among its atoms and molecules.

thermal pulse (258)   Periodic eruptions in the helium fusion shell in an aging giant star; thought to aid in ejecting the surface layers of the stars to form planetary nebulae.

time dilation (293)   The slowing of moving clocks or clocks in strong gravitational fields.

total eclipse (37, 40)   A solar eclipse in which the moon completely covers the bright surface of the sun; a lunar eclipse in which the moon completely enters Earth's dark shadow.

transition (133)   The movement of an electron from one atomic orbit to another.

transition region (149)   The layer in the solar atmosphere between the chromosphere and the corona.

transverse velocity (139)   The velocity of a star perpendicular to the line of sight.

triple alpha process (243)   The nuclear fusion process that combines three helium nuclei (alpha particles) to make one carbon nucleus.

T Tauri star (220)   Young star surrounded by gas and dust, believed to be contracting toward the main sequence.

turnoff point (248)   The point in an H–R diagram where a cluster's stars turn off of the main sequence and move toward the red giant region, revealing the approximate age of the cluster.

21-cm radiation (202)   Radio emission produced by cold, low-density hydrogen in interstellar space.

ultraluminous infrared galaxy (350)   A highly luminous galaxy so filled with dust that most of its energy escapes as infrared photons emitted by warmed dust.

ultraviolet radiation (101)   Electromagnetic radiation with wavelengths shorter than visible light but longer than X rays.

umbra (37)   The region of a shadow that is totally shaded.

unified model (365)   The attempt to explain the different kinds of active galaxies and quasars by a single model.

uniform circular motion (58)   The classical belief that the perfect heavens could move only by the combination of uniform motion along circular orbits.

variable star (247)   A star whose brightness changes periodically.

velocity (81)   A rate of travel that specifies both speed and direction.

velocity dispersion method (344)   A method of finding a galaxy's mass by observing the range of velocities within the galaxy.

vernal equinox (22)   The place on the celestial sphere where the sun crosses the celestial equator moving northward; also, the time of year when the sun crosses this point, about March 21, and spring begins in the northern hemisphere.

violent motion (78)   In Aristotelian physics, motion other than natural motion. (See **natural motion.**)

visual binary system (183)   A binary star system in which the two stars are separately visible in the telescope.

water hole (422)   The interval of the radio spectrum between the 21-cm hydrogen radiation and the 18-cm OH radiation; likely wavelengths to use in the search for extraterrestrial life.

wavelength (99)   The distance between successive peaks or troughs of a wave; usually represented by λ.

wavelength of maximum intensity (λ_{max}) (126)   The wavelength at which a perfect radiator emits the maximum amount of energy; depends only on the object's temperature.

weak force (161)   One of the four forces of nature; the weak force is responsible for some forms of radioactive decay.

west point (16)   The point on the western horizon exactly halfway between the north point and the south point; exactly west.

white dwarf (178)   Dying star that has collapsed to the size of Earth and is slowly cooling off; at the lower left of the H–R diagram.

winter solstice (22)   The point on the celestial sphere where the sun is farthest south; also, the time of year when the sun passes this point, about December 22, and winter begins in the northern hemisphere.

X-ray burster (297)   An object that produces repeated bursts of X rays.

Zeeman effect (155)   The splitting of spectral lines into multiple components when the atoms are in a magnetic field.

zenith (16)   The point on the sky directly overhead.

zero-age main sequence (ZAMS) (239)   The locus in the H–R diagram where stars first reach stability as hydrogen-burning stars.

zodiac (21)   The band around the sky centered on the ecliptic within which the planets move.

ANSWERS TO EVEN-NUMBERED PROBLEMS

Chapter 1: **2.** 3475 km; **4.** 1.05×10^8 km; **6.** about 1.2 seconds; **8.** 75,000 years; **10.** about 27

Chapter 2: **2.** 4; **4.** 2800; **6.** A is brighter than B by a factor of 170; **8.** 66.5°; 113.5°

Chapter 3: **2.** a) full; b) first quarter; c) waxing gibbous; d) waxing crescent; **4.** 12; 12.4. The moon orbits Earth and moves eastward about 13° per day; **6.** 6850 arc seconds or about 1.9°; **8.** a) The moon won't be full until Oct. 17; b) The moon will no longer be near the node of its orbit; **10.** August 12, 2026 [July 10, 1972 + 3 × (6585$\frac{1}{3}$ days)]. In order to get Aug. 12 instead of Aug. 11, you must take into account the number of leap days in the interval.

Chapter 4: **2.** Retrograde motion: Jupiter, Saturn, Uranus, Neptune, and Pluto; Never seen as crescents: Jupiter, Saturn, Uranus, Neptune, and Pluto; **4.** Mars, about 18 seconds of arc; the maximum angular diameter of Jupiter is 50 seconds of arc; **6.** $\sqrt{27} = 5.2$ years

Chapter 5: **2.** The force of gravity on the moon is about $\frac{1}{6}$ the force of gravity on Earth; **4.** 7350 m/s; **6.** 5070 s (1 hr and 25 min); **8.** The cannonball would move in an elliptical orbit with Earth's center at one focus of the ellipse; **10.** 6320 s (1 hr and 45 min)

Chapter 6: **2.** 3m; **4.** Either Gemini telescope has a light-gathering power that is 1.02 million times greater than the human eye; **6.** No, his resolving power should have been about 5.8 seconds of arc at best; **8.** 0.013 m (1.3 cm or about 0.5 inches); **10.** about 50 cm (From 400 km above, a human is about 0.25 seconds of arc from shoulder to shoulder.)

Chapter 7: **2.** 150 nm; **4.** by a factor of 16; **6.** 250 nm; **8.** a) B; b) F; c) M; d) K; **10.** about 0.58 nm

Chapter 8: **2.** 730 km; **4.** about 3.5 times; **6.** 400,000 years; **8.** 9×10^{16} J; **10.** 0.222 kg

Chapter 9: **2.** 63 pc; absolute magnitude is 2; **4.** about B7; **6.** about 1580 solar luminosities; **8.** 160 pc; **10.** a, c, c, d (use Figure 9-12 to determine the absolute magnitudes);

12. 3.69 days or 0.010 years; about 1.2 solar masses; about 0.67 and 0.53 solar masses; **14.** 1.38×10^6 km, about the size of the sun

Chapter 10: **2.** 60,000 nm; **4.** 0.0001; **6.** 1.5×10^6 years; **8.** 4.2×10^{35} kg or 210,000 solar masses (*Note:* Each hydrogen molecule contains two H atoms.)

Chapter 11: **2.** 24.5 km/s; **4.** 2.98 km/s; **6.** 9.46×10^{10} s (about 3000 years); **8.** There are four ^{1}H nuclei in the figure. They are used in a series of reactions to build up nuclei from ^{12}C to ^{15}N. In the last reaction (with the ^{15}N) a ^{12}C and a ^{4}He are produced. Because the ^{12}C can be used in the next cycle, the net is four ^{1}H nuclei in with one ^{4}He nucleus out; **10.** 7.9×10^{33} kg

Chapter 12: **2.** about 9.8×10^6 years for a 16-solar-mass star; about 5.7×10^5 years for a 50-solar-mass star; **4.** about 1×10^6 times less than present or about 1.4×10^{-6} g/cm^3; **6.** 2.4×10^{-9} or 1/420,000,000; **8.** about 3 pc; **10.** 3.04 minutes early after 1 year; 30.4 minutes early after 10 years

Chapter 13: **2.** about 1.7 ly; **4.** about 16,000 years old; **6.** about 940 years ago (approximately 1060 AD); **8.** 2200 km/s; **10.** about 300 years ago

Chapter 14: **2.** 7.1×10^{25} J/s or about 0.19 solar luminosity; **4.** 820 km/s (assuming mass is one solar mass); **6.** about 11 seconds of arc; **8.** about 490 seconds

Chapter 15: **2.** about 16 percent; **4.** 3.8×10^6 years; **6.** overestimate by a factor of 1.58; **8.** about 21 kpc; **10.** 7.8×10^{10} solar masses; **12.** 1500 K

Chapter 16: **2.** 2.58 Mpc; **4.** 131 km/s; **6.** 28.6 Mpc; **8.** 1.64×10^8 yr; **10.** 4.49×10^{41} kg

Chapter 17: **2.** 7.8×10^6 years; **4.** 0.024 pc; **6.** −28.5; **8.** about 91,000 km/s; **10.** 0.16

Chapter 18: **2.** 57; 17.5 billion years; 11.7 billion years; **4.** 1.6×10^{-30} gm/cm^3; **6.** 76 km/Mpc; **8.** 16.6 billion years; 11 billion years; the universe could be older

Chapter 19: **2.** 8.9 cm; 0.67 mm; **4.** about 1.3 solar masses; **6.** 380 km; **8.** pessimistic, 2×10^{-5}; optimistic, 10^7

CREDITS

Chapter 2, The Sky Around Us, page 16, Star trails: AURA/NOAO/NSF.

Chapter 4, The Ancient Universe, page 58, Ancient universe: from *Cosmographics* by Peter Apian, 1539.

Chapter 5, Orbiting Earth, page 85, Astronaut: NASA.

Chapter 6, Modern Astronomical Telescopes, page 108, 4-meter telescope: AURA/NOAO/NSF; **page 109,** Keck I Mirror: © Russ Underwood/ Keck Observatory; Adaptive optics: Paul Kalas.

Chapter 7, Atomic Spectra, page 132, Stamp: Courtesy Ron Reese and Michelangelo Fazio; **page 133,** Lagoon Nebula: AURA/NOAO/NSF.

Chapter 8, Magnetic Solar Phenomena, page 162, UV image of sun: Trace/NASA; Quiescent prominence: SOHO/EIT, ESA and NASA; Loop prominence: Sacramento Peak Observatory; **page 163,** Solar flare: NASA; Waves from solar flare: SOHO/MDI, ESA, and NASA; Coronal holes: NASA and ISAS; Coronal mass ejection: SOHO/LASCO, ESA, and NASA; Aurora: Copyright Jan Curtis.

Chapter 9, The Family of Stars, page 193, Star field: Hubble Heritage Team.

Chapter 10, Three Kinds of Nebulae, page 198, Star-forming region in Ara: ESO; Reflection nebulae in Orion: AATB; **page 199,** Pleiades: Caltech; Lagoon and Trifid Nebulae: Daniel Good; Bernard 86: AATB; Horsehead Nebulae: Hubble Heritage Team; Merope Nebula: NASA.

Chapter 11, Observational Evidence of Star Formation, page 220, S Mon Nebula: AATB; Bok globules: NASA; **page 221,** HH34: Richard Mundt, Calar Alto 3.5 m telescope; Jet inset: NASA; Eagle Nebula: AURA/NOAO/NSF; EGGs:NASA.
Star Formation in the Orion Nebula, page 230, Orion Nebula: Daniel Good; Low-Mass stars: NASA and K. L. Luhman (Harvard-Smithsonian Center for Astrophysics, Cambridge, Mass.); and G. Schneider, E. Young, G. Rieke, A

Cotera, H. Chen, M. Rieke, R. Thompson (Steward Observatory, University of Arizona, Tucson, Ariz.); X-ray image: NASA/CXC/SAO; **page 231,** Orion image: European Southern Observatory; Jets of gas: AATB; Infrared sources: Dan Gezari, Dana Backman, and Mike Werner; Trapezium: NASA; Disks around stars: John Bally, Dave Devine, and Ralph Sutherland and NASA.

Chapter 12, Star Cluster H–R Diagrams, page 248, The Jewel Box: AURA/NOAO/NSF; 47 Tucanae: AATB; **page 249,** NGC2264: AATB; Pleiades: CalTech; M67: Nigel Sharp, Mark Hanna/AURA/NOAO/NSF.

Chapter 13, The Formation of Planetary Nebulae, page 260, IC3568: H. Bond, B. Ballick, and NASA; Ring Nebula: Hubble Heritage Team, AURA/STScI/NASA; **page 261,** Planetary nebulae: H. Bond, B. Ballick, and NASA; Hour Glass: Raghvendra Sahai and John Trauger, JPL; WFPC2 Science Team NASA; Cats Eye: NASA/CXC/SAO, Soltan G. Levay, Space Telescope Science Institute, J. P. Harrington and K. J. Borkowski, University of Maryland and NASA; Egg Nebula: Rodger Thompson, Marcia Rieke, Glenn Schneider, Dean Hines. University of Arizona; Raghvendra Sahai, JPL; NICMOS Team; and NASA.

Chapter 14, The Lighthouse Model of a Pulsar, page 285, HST image of Crab Pulsar: Jeff Hester and Paul Scowen, Arizona State University, NASA; X-ray, visual and radio image of Crab Pulsar: Jeff Hester et al., NASA/CXC/SAO, HST and VLA/NRAO; Puppis A: Steve Snowden, ROSAT/NASA; HST neutron star: Fred Walter, State University of New York at Stony Brook, NASA.

Chapter 15, Sagittarius A*, page 328, Sagittarius in IR: 2MASS; Center of Galaxy in Radio: NRL; VLA Image of Sgr A: NRAO; **page 329,** High res Sgr A*: N. Killeen and Kwok-Yung Lo; S2 field: ESO; Sgr A* X-ray: NASA/CXC/SAO.

Chapter 16, Galaxy Classification, page 338, M87: AURA/NOAO/NSF; Leo I dwarf elliptical: Anglo-Australian Telescope Board; NGC1365: Anglo-Australian Telescope Board; NGC3623, NGC3627, NGC2997: Anglo-Australian Telescope Board; **page 339,** NGC4013: The Hubble Heritage Team and NASA; NGC2207 and IC2163: NASA Hubble Heritage Team, STScI; S0 galaxy: Rudolph E. Schild; Small Magellanic Cloud: AURA/NOAO/NSF; Large Magellanic Cloud: Copyright R. J. Dufour, Rice University; IC4182: G. J. Jacoby, M. J. Pierce/AURA/NOAO, NSF.

Chapter 17, Cosmic Jets and Radio Lobes, page 362, Cygnus A radio image: NRAO; Cygnus A photo: Used with permission, Fosbury R. A. E., Vernet, J., Villar-Martin, M., Cohen, M. H., Ogle, P. M., Tran, H. D. & Hook, R. N. 1998, Optical continuum structure of Cygnus A. "KNAW colloquium on: The most distant radio galaxies," Amsterdam, 15–17 October 1997, Roettgering H, Best P and Lehnert M eds, Reidel, astro-ph/ 9803310; NGC5128: X-ray: NASA/ CXC/M. Karovska et al., Radio: NRAO/ VLA/Schiminovich, et al., Radio: NRAO/ VLA/J. Condon et al., Optical: Digitized Sky Survey U.K. Schmidt Image/STScI; NGC5128 Infrared: E. Schreier, STScI, NASA; NGC128 radio: NRAO, AUI; NGC128 X-ray; NASA/CXC/SAO; **page 363,** 3C449: Richard A. Perley, 3C129: Laurence Rudnick, University of Minnesota, and Jack Burns, University of New Mexico; 3C129 model: Vincent Icke, University of Minnesota and *The Astrophysical Journal,* published by the University of Chicago Press, © 1981 American Astronomical Society; 3C75: NRAO/AUI, F. N. Owen, C. P. O'Dea, and M. Inoue; Disk-jet artwork: adapted from a diagram by Ann Field, NASA, STScI.

Chapter 19, DNA: The Code of Life, page 409, Baby: Mike Seeds.

INDEX

giant molecular clouds and, 215–217

interstellar medium and, 208–210, 215–222

Orion Nebula and, 228, 230–231

protostars and, **217–218**

self-sustaining, **325,** 327

spiral arms and, 325–326, 327

stellar structure and, 224–228

starquakes, 282

stars, **4**

asterisms, **12**

binary, **182–189,** 263–266

brightness of, 5, 12–14, 174–176, 432

color index of, 127

composition of, 136–139

constellations, **11**–12, 439

death of, 255–277

density of, 190

diameters of, 176–182

distances to, 171–174

dwarf, 178–179

energy transport in, 224–225

evolution of, 233–254

family of, 192–193

formation of, 208–210, 214–232

giant, **178,** 240–246

high-velocity, **314**

life expectancies of, 240

life zone around, **419–420**

light emitted from, 124–127

loss of mass from, 258, 259, 262

luminosity of, **176,** 177, 180–181, 190, 192–193

magnetic cycles on, 159–160

main-sequence, 177–178, 226, 234–240

masses of, 182–189, 190, 239–240, 258, 259, 262, 280

mass-luminosity relation, **190**

motion and velocity of, 139–140

names of, 12

nearest, 433

neutron, **279**–290

nuclear fusion in, 223, 227, 243–245, 267–268

observing, 439–451

proper motion of, **173**–174

protostars and, **217**–218

pulsating, 250–252

rotation of, 159

sizes of, 178–179, 181

spectral classification of, 135–136, 192–193

spots on, 155–156, 159

stability of, 227

structure of, 224–228, 234–235

supergiant, **178**

surveying, 191

temperature of, 124, 126, 134–135, 177

travel between, 420–421

variable, **247,** 250–253

young, 228

star trails, 16

statistical evidence, 360

steady-state theory, **386**

Stefan-Boltzmann law, 126, 127

Stella Nova, De (The New Star) (Tycho Brahe), 67

stellar death, 255–277

binary stars and, 263–265, 266

massive stars and, 267–275

mass loss and, 258, 259, 262

nova explosions and, 265

planetary nebulae and, **258,** 260–261

red dwarfs and, 256

sunlike stars and, 256–258

supernova explosions and, 268–269, 274–275

white dwarfs and, 258–259, 262–263

stellar evolution, 233–254

binary stars and, 263–266

life expectancies and, 240

main-sequence stars and, 234–240

mass and, 280

post-main-sequence evolution, 240–246

star clusters and, 246–247, 248–249

uncertainties related to, 246

variable stars and, 247, 250–253

stellar models, **234**–235

stellar parallax, **172**–173

stellar populations, 316

stellar structure, 224–228

energy transport and, 224–225

four laws of, 234–235

hydrostatic equilibrium and, **225,** 235–236

pressure-temperature thermostat and, 227

Stonehenge, 52, 53

Stonehenge Decoded (Hawkins), 52

Stratospheric Observatory for Infrared Astronomy (SOFIA), 116

stromatolites, **415**

strong force, **161**

Sudbury Neutrino Observatory, 166

summer solstice, **22,** 23

sun, 145–169

activity of, 153–161

age of, 421

angular diameter of, 40, 43, 44

annual motion of, 19–20

atmosphere of, 146–153

chromosphere of, **43,** 147, 148–150

composition of, 138, 142

corona of, **42**–43, 147, 150–151

death of, 262, 266

distance from Earth to, **4**

eclipses of, 40–44

evolution of, 239

galactic orbit of, 326

heat flow in, 147

helioseismology of, **151**–152

magnetic phenomena in, 154–155, 156–159, 160, 162–163

mass loss from, 258

nearest star to, 5

neutrino mystery and, 165–167

nuclear fusion in, 161, 164–167

observation of, 152, 153

photosphere of, **42,** 125, 147–148

rotation of, 156, 157, 161

size of, 4, 146

solar constant and, **160**–161

spectral analysis of, 142, 148–151

statistical data about, 146

structure of, 225–226

temperature of, 147, 149

sunspots, **147,** 153–159

active regions and, 153–156

cycles of, 156, 157

discovery of, 64

Earth's climate and, 160–161

magnetic behavior of, 154–155, 156–159

observing, 153

superclusters, 6–7, **400,** 401

supergiants, **178**

supergranules, **148**

superluminal expansion, **373**–374, 375

supermassive black holes, 330, 345, 361, 364, 365, 366–367

supernova (supernovae), **256,** 268–275

distance calculations and, 341

expansion of the universe and, 398, 399

explosions of, 268–269, 274–275

life on Earth and, 274–275

neutrinos released by, 273–274

neutron stars and, 286–287

observations of, 270–271

remnants of, 271, 272

SN1987A, 272–274, 286

SN1997ff, 398, 400

star formation and, 219, 222

synchrotron radiation and, 271, 273

types of, 269–270

supernova remnants, **271,** 272, 286, 317–318, 328

surveyor's method, 171–172

synchrotron radiation, **271,** 273

synodic period, **35**

Ace ◐ Astronomy™

What do you need to learn NOW?

Take charge of your learning with **AceAstronomy,**™ a powerful new learning tool for astronomy! **AceAstronomy** is Web-based and FREE with every new copy of this text. This powerful and interactive resource helps you gauge your unique study needs, and then gives you a personalized learning plan that focuses your study time on the concepts you most need to master. By providing you with a better understanding of exactly what you need to focus on, **AceAstronomy** helps you maximize your study time

. . . bringing you a step closer to success!

What does an integrated learning system do?

This unmatched resource and the new edition of this text were developed in concert, to enhance one another and provide you with a seamless integrated learning system. As you work through the text, you will see notes that direct you to the dynamic media-enhanced activities and helpful tutorials in **AceAstronomy**. This precise page-by-page integration means you'll spend less time flipping through pages or navigating Web sites and more time preparing for exams and reviewing the most critical terms and concepts.

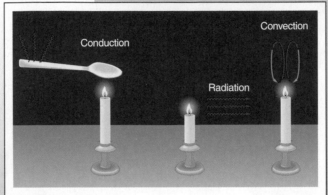

ACTIVE FIGURE 12-11
The three modes by which energy may be transported from the flame of a candle, as shown here, are the three modes of energ[y] within a star.

Ace ◐ Astronomy™ Go to AceAstronomy and click Active [Figure to] see "Conduction, Convection, and Radiation." Experiment [with a] candle flame.

How does it work?

You simply log on to **AceAstronomy** from the book's Companion Web Site at **http://astronomy.brookscole.com/seeds4e** by using the free passcode packaged with this text. You'll immediately notice the system's simple browser-based format—as easy to use as surfing the Web. Just a click of the mouse gives you the freedom to enter and explore the system at any point. You can build a complete personalized learning plan for yourself by taking advantage of all three of **AceAstronomy's** powerful components: *What Do I Know?, What Do I Need to Learn?,* and *What Have I Learned?*

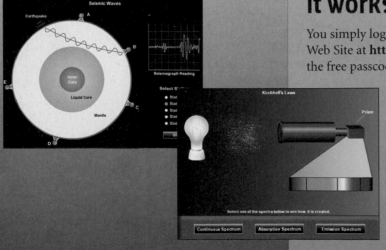